Very
Large
Data
Bases

PROCEEDINGS

PROCEEDINGS OF THE FIFTEENTH INTERNATIONAL CONFERENCE ON VERY LARGE DATA BASES

August 22-25 / 1989
Amsterdam
The Netherlands

Editors: Peter M.G. Apers
Gio Wiederhold

Major Contributors

VLDB Endowment, USA
Elsevier Science Publishers, The Netherlands
Royal Netherlands Academy for Sciences, The Netherlands
Dutch Computer Society (NGI), The Netherlands
Vrije Universiteit, The Netherlands
Universiteit Twente, The Netherlands
Centre for Mathematics and Computer Science (CWI), The Netherlands

Contributors

INRIA, France
Science and Engineering Research Council, United Kingdom
Air Co. Ltd., Japan
Osaka University, Japan
University of Stockholm, Sweden
SION (Dutch Foundation for Computer Science Research)
Neotech AB, Sweden

Industry Supporters

COMPUTER ASSOCIATES®
Software superior by design.

DCE
INFORMATION MANAGEMENT CONSULTANCY

IBM

UNISYS

sun microsystems

datex

WANG

Additional Copies

Additional copies of this proceedings may be obtained from:
Morgan Kaufmann Publishers, Inc.
P.O. Box 50490
Palo Alto, Ca 94303-9953
USA

ISBN 1-55860-101-5

ORGANIZATION

General Conference Chairman
Reind P. van de Riet
Vrije Universiteit
Amsterdam, The Netherlands

European Program Committee Chairman
Peter M.G. Apers
University of Twente
Enschede, The Netherlands

American Program Committee Chairman
Gio Wiederhold
Stanford University
Stanford, USA

Tutorials and Publicity Coordinator
Wiebren de Jonge
Vrije Universiteit
Amsterdam, The Netherlands

North American Coordinator
S. Bing Yao
College of Business and Management
University of Maryland, USA

North American Publicity Coordinator
R. Ahad
College of Business and Management
University of Maryland, USA

Organizing Committee Chairman
Martin L. Kersten
Centre for Mathematics and Computer Science
Amsterdam, The Netherlands

Treasurer
Marien J. Krijger
Postbank
Amsterdam, The Netherlands

Sponsoring Coordinator
Gerard A.M. Otten
DCE Information Management Consultancy
Nieuwegein, The Netherlands

Chairman Exhibition Committee
Warwick Longman
Wang European Headquarters
Brussels, Belgium

PROGRAM COMMITTEE MEMBERS

Serge Abiteboul, France
Rakesh Agrawal, USA
Malcolm Atkinson, United Kingdom
Tom Atwood, USA
Francois Bancilhon, France
Don Batory, USA
David Bell, Ireland
Dina Bitton, USA
Henk Blanken, The Netherlands
Alex Borgida, USA
Giamio Bracchi, Italy
James Bradley, USA
Janis Bubenko Jr, Sweden
Alex Buchmann, USA
Michael Carey, USA
Stefano Ceri, Italy
Qiming Chen, China
Peter Dadam, FR Germany
Umesh Dayal, USA
Robert Demolombe, France
Klaus Dittrich, FR Germany
Eckhard Falkenberg, The Netherlands
Hector Garcia-Molina, USA
George Gardarin, France
Georg Gottlob, Austria
Peter Gray, United Kingdom
Laura Haas, USA
Kees van Hee, The Netherlands
Tomasz Imielinski, USA
Yannis Ioannidis, USA
Matthias Jarke, FR Germany
Hannu Kangassalo, Finland
Arthur Keller, USA
Roger King, USA
Walter Kohler, USA
Ravi Krishnamurthy, USA
Per-Ake Larson, Canada
Withold Litwin, France
Peter Lockemann, FR Germany
Guy Lohman, USA
David Lomet, USA
Peter Lyngbaek, USA
Dennis McLeod, USA
Robert Meersman, The Netherlands
Michele Missikoff, Italy
Matthew Morgenstern, USA
Eliot Moss, USA
J. Motiwalla, Singapore
John Mylopoulos, Canada
Sham Navathe, USA
Shojiro Nishio, Japan
Shir Nijssen, Australia
Z. Meral Ozsoyoglu, USA
P. Paolini, Italy
Jan Paredaens, Belgium
Alain Pirotte, Belgium
Hans Schek, FR Germany
Günther Schlageter, FR Germany
Joachim Schmidt, FR Germany
Timos Sellis, USA
Amilcar Sernadas, Portugal
Stefano Spaccapietra, France
Peter Stocker, United Kingdom
Stanley Su, USA
K. Subieta, Poland
Constantino Thanos, Italy
Dennis Tsichritzis, Switzerland
Patrick Valduriez, USA
Yannis Vassiliou, Greece
Gerry Wilson, USA
Mark Wallace, FR Germany
Carlo Zaniolo, USA

LIST OF REFEREES

Aerts, A.T.M.
Ahlsen, M.
Ahmed, R.
Annevelink, J.
Atzeni, P.
Balsters, H.
Barrera, R.
Berkel, T.
Boman, M.
Boettcher, S.
Brinkkemper, J.N.
Bruza, P.D.
Bry, F.
Bueltzingsloewen von, G.
Byeon, K.J.
Caloini, A.
Carzotto, F.
Chakravarthy, U.
Chen, Y.-H.
Chiang Lee
Chimenti, D.
Cholvy, L.
Coburn, N.
Cuppens, F.
Dalal, M.
Danforth, S.
Davis, J.
Desai, D.
Dubey, R.
Effelsberg, W.
Eiben, A.
Fang, D.
Fiadeiro, J.L.
Finkelstein, S.
Flach, P.
Franklin, M.
Freeston, M.
Frenkel, G.
Freytag, J.C.
Gagliardi, R.
Gala, S.
Gamboa, R.
Goldhirsch, D.
Guo, M.
Hannatani, Y.
Hasan, W.
Heiler, S.
Herrmann, U.
Houben, G.J.P.M.
Hulin, G.
I-Min Amy Chen
Jenq, P.
Johannesson, P.
Johannsen, W.
Kandzia, P.
Karlapalem, K.
Katoh, N.
Kemp, R.
Kent, W.
Kifer, M.
Kirn, S.
Klahold, P.
Knudsen, E.
Kotz, A.
Kuespert, K.
Lausen, G.
Lee, B.
Lee, C.
Lee, Y.
Ling, L.
Linnemann, V.
Lundh, J.
Maindreville de, C.
Mainetti, S.
Mainetto, G.
Manthey, R.
Masuyama, S.
Matthes, F.
Ming-Chien Shan
Mingsen Guo
Missikoff, M.
Moerkotte, G.
Motro, A.
Muthukumar, K.
Naqvi, S.
Nilsson, B.
Norrie, M.
Orenstein, J.
Ozrow, L.
Pels, H.J.
Pernici, B.
Perry, D.
Pieper, F.T.A.M.
Pissinou, N.
Pistor, P.
Pucheral, P.
Ra, M.
Remmen, F.
Risch, T.
Roelants, D.
Rothermel, K.
Royer, V.
Rozenshtein, D.
Saake, G.
Sebastiani, F.
Sernadas, C.
Simon, E.
Smith, M
Soederlund, L.
Spyratos, N.
Sullivan, T.
Tanca, L.
Thakore, A.
Thevenin, J.M.
Troyer de, O.
Unland, R.
Vardi, M.
Vauclair, M.
Verharen, E.
Vieille, L.
Voorhoeve, M.
Wangler, B.
Weddell, G.
Weide van der, Th.P.
Wilkes, W.
Xinhua Wu
Yaseen, R.
Yaw-Huei Chen
Yokota, K.
Yun-Ping Hsu
Zimanyi, E.

VLDB Endowment Board of Trustees

STEERING COMMITTEE

REGIONAL COORDINATORS

CONTENTS

Session 7B: Allocation and Optimization

Session 8A: Panel Session

Session 8B: Statistics and Statistical Databases

Session 9A: Complex Objects

Session 9B: Recovery and Concurrency Control

Session 10A: Object Management

Session 10B: Priority Scheduling

Session 11A: Languages for OODB

Session 11B: Panel Session

Session 12A: Active databases

Session 12B: Panel Session

PREFACE

It's been a pleasure to have the 15-th Conference on Very Large Data Bases in Amsterdam. An old Dutch saying is that All Good Things Come in Three. We as organizers of the conference were very keen on the following three:

× The scientific quality of the conference. The two program committees, chaired by Peter Apers and Gio Wiederhold have selected from almost 200 submitted papers a very interesting set of 45 which were presented at the conference and reproduced in this Proceedings. Our thanks go to the 71 members of the two program committees, who did a tremendous amount of work, each one reading more than ten papers, the two chairmen, for organizing the distribution of the papers, chairing the meetings and coordinating the work of the two committees. Our thanks are also due to the authors of the accepted papers for their presentation and preparation of the manuscripts. Finally, we thank the many authors of the papers which were not chosen, for preparing their manuscripts. We hope that the remarks of the referees have been helpful and clear.

  Also we wish to thank our key-note speaker: Hervé Gallaire for his speech on database research carried out in beautiful Laguna's and Open Waters. For someone working in Munich quite an experience, indeed.

  We also thank the panel members for their contributions.

× The tutorial program as given by Michael Brodie, John Mylopoulos, Chris Date, Shir Nijssen, Krzysztof Apt and Stan Zdonik. Their contributions with state-of-the-art exposé's attracted many people, which also could participate in the scientific program.

× The cultural events, of which we organized many in order to highlight Amsterdams cultural place in the world. Whether it was the Dutch evening, the mini-concerts or the pictures in the Van Gogh Museum, it was all worth while to do.

The three crosses above have also been used in our logo and were borrowed from the coat of arms of Amsterdam. The artist Vincent Bus created this logo as his impression of joining our field of research (data on diks) and the location of the conference.
A conference like the VLDB can only become a success when so many people put their skills together and do lots of work. I am, in particular thankful to Martin Kersten as the chairman of the organizing committee, to Wiebren de Jonge for coordinating the mailings and organizing the tutorial program, to Gerard Otten for a very successful sponsorship program, to Marien Krijger for handling so carefully the finances and to Warwick Longman for organizing the very interesting exhibition. It goes without saying that I am very grateful to the Organization Bureau of the RAI, in particular mrs. Isabella Oosterhout and Marlette Vercauteren, for doing the ground work.
Finally, I would like to express my gratitude towards the VLDB Endowment, in particular its president Alfonso Cardenas, for giving us the opportunity to organize the fifteenth VLDB conference. Looking back at a fruitful cooperation during four years of preparation, I wish the Endowment and its local organizers much success in future VLDB conferences.

Reind van de Riet,

General Conference Chairman VLDB 89

Key-note Speech
From a laguna to open waters:
Another view on the next generations of databases

H. Gallaire
ECRC

The Laguna Beach report [1] identiefies major trends for the evolution of databases. While interesting issues are analysed and debated in that report, the views expressed are conservative and short-sighted. While [1] emphasizes new application domains such as CASE and CIM, I believe that decision support systems (DSS) will also become important for pushing the limits of database systems. DSS combine requirements for better modeling techniques that they share with CASE and CIM—of which the object-oriented models of [1] are but a special case—to the needs of dynamic views and queries in ways which cannot be limited to simple rule firing and require the power of a declarative programming language to be available in the context of a database system. Logic-based systems remain the first candidates in this area. The talk will present an example of a DSS, review its functionality, discuss a suitable architecture. Further points to be discussed will involve the availability of workstations and very large memories and its influence on the architecture of a database system.

[1] *Future Directions in DBMS research* - ICS (eds. E. Neuhold & M. Stonebraker)

PANEL SESSION
Knowledge to Mediate from User's Workstations to Databases

Panel chairman:
Gio Wiederhold
Stanford University, USA

Information systems that are now becoming available are reaching the capabilities envisaged by Vannevar Bush in 1945 when he specified the 'memex'. We can sit at our workstations and view data from many different sources. But any integration of the information still takes place in our minds, and we now have a common complaint of information overload. New hardware and software technologies address aspects of this problem.
In this panel we want to develop a vision which integrates new directions to carry us beyond 'Memex'. We use the term 'Mediation' for the software which fits between the users' agent – the workstation, and the diverse data sources.
Specifically, the ability of computers to communicate effectively opens up possibilities for information processing which greatly exceed those on our immediate horizon. User's on their workstations should be able to get information, not just data, from the private and public databases. To achive such services mediation modules are needed, which combine expert knowledge about accessing and processing data, so that only relevant information is fed to the users' stations. The mediators must be maintainable, and hence limited in scope, just like the experts which they represent. Mediation encompasses concepts as views, computations involving closure, ranking of results, generalization, navigational search, etc.
The user, mediator, and server modules can be mapped onto the powerful distributed hardware that is becoming available, so that it forms an architecture for information processing into the foreseeable future. We will argue that the modularity is not only a goal, but also enables the goal to be reached, since these systems will need autonomous modules to permit growth and enable them to survive in a rapidly changing world.
The discussion should not be oriented towards any specific design or implementation, but focus on the architecture and critical sub-tasks needed to devlop a new conceptual framework.

On the design and implementation of information systems from deductive conceptual models

Antoni Olivé
Facultat d'Informàtica. Universitat Politècnica de Catalunya
Pau Gargallo, 5. 08028 Barcelona. Catalonia

ABSTRACT

Deductive conceptual models (DCMs) aim at providing a complete specification of information systems, expressing only its logic component. It has been shown that DCMs have some advantages with respect to traditional, operational conceptual models, but they are more difficult to implement. We present a new approach to the design and implementation from a DCM. It consists in deriving from the DCM a new model, which we call the internal events model. This model describes the actions the information system must perform in terms of the inputs. The use of this model for data base and transactions design is discussed.

1. INTRODUCTION

During the last years we have seen the emergence and rapid growth of the logic programming field. This is due to a number of reasons, perhaps one of the most important being the clear distinction found in logic programming between the logic and control components of a program. As Kowalski says, "Logic programs express only the logic component of a program. The control component is exercised by the program executor, either following its own autonomously determined control decisions or else following control instructions provided by the programmer" [7].

There is a parallel trend at the information systems level in the conceptual modelling field [1]. Conceptual models of information systems express more and more the logic component of the system, but unfortunately they still somehow define (part of) the control component. The field is still dominated, both in theory and practice, by the "operational" approach to conceptual modelling. By this we mean models in which:

a) Changes in the Information Base (IB), corresponding to changes in the Universe of Discourse (UoD), are defined by means of **operations**. Occurrence of a real-world, external event triggers the execution of an operation (transaction), which reflects the effects of the event on the IB. These effects are usually expressed by some kind of insertions, updates or deletions to the IB.

b) Operations, as well as queries and integrity constraints, can only access to the current state of the IB.

Deductive Conceptual Models (DCMs) aim at providing a complete specification of an information system expressing only its logic component. The specification of the control component is entirely left to the subsequent phases of information systems development. A DCM relates the contents of the IB to the external events in terms of **deduction rules**. On the other hand, they allow to define integrity constraints and queries as if a complete history of the IB were available.

A detailed comparison of the operational and deductive approaches can be found in [3,13]. The main conclusions are that DCMs provide more local definitions, favour the understanding of informations, are easier to change and to accomodate new requirements, and provide more design freedom. However, DCMs are much more difficult to implement than operational models. The reason is that an operational model already embeddes some architectural design decisions, which are not made in DCMs.

It is clear that progress in deductive conceptual modelling is hampered by the lack of efficient implementation methods. We can expect that progress in the fields of knowledge bases and deductive data bases will make implementation of DCMs easier. However, an ideal solution would be to have methods for the design and implementation of DCMs in conventional hardware/software environments.

This paper is a step in that direction. We present a formal method to derive from a DCM a new model, called the internal events model, which is much easier to implement. The paper is structured as follows. In Section 2 we present the DCMs, including an example used throughout the paper. In Section 3 we review some approaches to design from DCMs. Section 4 presents a method for the derivation of the internal events model, which is the main part of the paper. The use of this model for the design and implementation is briefly discussed in Section 5. Finally, Section 6 gives the conclusions and points out future work.

Proceedings of the Fifteenth International Conference on Very Large Data Bases
Amsterdam, 1989

2. DEDUCTIVE CONCEPTUAL MODELS

There are several languages for deductive conceptual modelling of information systems. Among them are CIAM [5], DADES [12] and IPL [16]. These languages differ in many aspects, but all share a common approach. Incidentally, this approach is very similar to the one taken by the recent Event Calculus [8] for representing and reasoning about time and events within a logic programming framework. In this paper, we will abstract from the details of specific languages and try to characterize the deductive conceptual modelling approach in a first order logic framework.

Time plays a major role in this approach. Every possible information **i** about the UoD is associated with a time point T(**i**), which states when the information holds in the UoD. It can be called the occurrence or observation time. This time point is a component of the information proper, so informations are self-describing in this respect. We will assume that occurrence times are always expressed in a unique time unit (such as second, day, etc.) small enough to avoid ambiguities.

By life span T of an information system we mean the time interval in which the system operates. It is defined as an ordered set of consecutive time points $T = \{t_0,...,t_f\}$, where t_0 and t_f are the initial and final times, respectively, and where each $t \varepsilon T$ is expressed in the given time unit. We can then say that, for any information **i**, T(**i**)εT.

A DCM consists of a set B of base predicates, a set D of derived predicates, a set DR of deduction rules, a set IC of integrity constraints, and a set OR of output requirements. Each of them is described in the following, and illustrated with references to the DCM example of figure 1.

2.1 Base predicates

Base predicates model the external event types of the UoD. They are the inputs to the information system. Each fact of a base predicate, called base fact, corresponds to an occurrence of an external event. We assume, by convention, that the last term of a base fact gives the time when the external event occurred. If $p(a_1,...,a_n,t_i)$ is a fact we say that $p(a_1,...,a_n)$ is true or holds at t_i.

In the example of figure 1 we have three base predicates: start, end and assign. A base fact start(p1,e1,d1,t1) means that project p1 has started at time t1, is planned for termination at time e1, and corresponds to department d1. A base fact end(p1,t1) reports that project p1 effectively ends at time t1. A base fact assign(p1,pg1,t1) means that at time t1, programmer pg1 is assigned to project p1.

<u>Base predicates</u>

start(project,end,department,time)
end(project,time)
assign(project,programmer,time)

<u>Derived predicates</u>

DR.1 active(p,t) $\leftarrow$ start(p,e,d,t1) $\wedge$ t1 $\leq$ t $\wedge$ $\neg$ completed(p,t)
DR.2 p-end(p,e,t) $\leftarrow$ start(p,e,d,t1) $\wedge$ t1 $\leq$ t $\wedge$ active(p,t)
DR.3 dpt(p,d,t) $\leftarrow$ start(p,e,d,t1) $\wedge$ t1 $\leq$ t $\wedge$ active(p,t)
DR.4 completed(p,t) $\leftarrow$ end(p,t1) $\wedge$ t1 $\leq$ t
DR.5 assigned(p,pg,t) $\leftarrow$ assign(p,pg,t1) $\wedge$ t1 $\leq$ t $\wedge$ active(p,t)
DR.6 has-worked(d,pg,t) $\leftarrow$ starts-working(d,pg,t1) $\wedge$ t1 $\leq$ t
DR.7 starts-working(d,pg,t) $\leftarrow$ assign(p,pg,t) $\wedge$ dpt(p,d,t)
DR.8 new-in-dept(d,pg,t) $\leftarrow$ starts-working(d,pg,t) $\wedge$ $\neg$has-worked(d,pg,t-1)
DR.9 critical(p,t) $\leftarrow$ p-end(p,e,t) $\wedge$ t > e - 8 $\wedge$ t $\leq$ e

<u>Integrity constraints</u>

IC.1 $\leftarrow$ start(p,e,d,t) $\wedge$ e $\leq$ t
IC.2 $\leftarrow$ start(p,e,d,t) $\wedge$ active(p,t1) $\wedge$ t1 < t
IC.3 $\leftarrow$ end(p,t) $\wedge$ $\neg$active(p,t-1)
IC.4 $\leftarrow$ assign(p,pg,t) $\wedge$ assigned(p,pg,t-1)
IC.5 $\leftarrow$ assign(p,pg,t) $\wedge$ $\neg$active(p,t-1)

<u>Output requirements</u>

OR.1 O1(t) $\leftarrow$ Q1(t)
O1(t) = {<p,e> | critical(p,t) $\wedge$ p-end(p,e,t)}
OR.2 O2(t1) $\leftarrow$ Q2(t1,t) $\wedge$ t1 $\leq$ t
O2(t1) = {<p,d> | dpt(p,d,t1)}
OR.3 O3(pg,t) $\leftarrow$ new-in-dept(d,pg,t)
O3(pg,t) = {<pg,p> | assigned(p,pg,t)}

Figure 1. Example of Deductive Conceptual Model

2.2 Derived predicates

Derived predicates model the relevant types of knowledge about the UoD. Each fact of a derived predicate, called derived fact, represents an information about the state of the UoD, at a particular time point. We will also assume that the last term of a derived fact gives the time when the information holds. In the example there are nine derived predicates: active, p-end, dpt, completed, assigned, has-worked, starts-working, new-in-dept and critical. The semantics of a derived predicate is given by its deduction rules.

2.3 Deduction rules

There are one or more deduction rules for each derived predicate. Let p(x1,...,xn,t) be a derived predicate, with n+1 terms. A deduction rule for p has the form p(x1,...,xn,t) $\leftarrow \phi$ where ϕ is a literal or a conjunction of (positive or negative) literals, and all variables are assumed to be universally quantified.

The terms in the rule head must be distinct variables. The terms in ϕ (rule body) must be constants or variables. We assume every rule to be range-restricted, i.e. every variable occurring in the head, or in a negative literal in ϕ, occurs in a positive literal in ϕ as well [4]. We also assume every rule to be time-restricted. This means that, for every base or derived predicate q occurring in the body as a positive literal q(...,t1), the condition $\phi \rightarrow t1 \leq t$ must hold. This condition ensures that p(x1,...,xn,t) is defined in terms of q-facts holding at time t or before.

In the example, the nine deduction rules are (hopefully) self-explanatory. Note the use of variable t-1 as a syntactic convention instead of t´ and t´ = t - 1. Thus, in DR.8, new-in-dept(d,pg,t) holds if programmer pg starts working for department d at time t, and pg had not worked for d at time t - 1.

2.4 Integrity constraints

Integrity constraints are closed formulae that base and/or derived facts must satisfy to be consistent. An integrity constraint can only be falsified by the presence (or absence) of new base facts, and it is assumed that some mechanism of integrity constraints enforcement will reject (or require) those facts to maintain the informations consistent. We deal here (as in [9]) with constraints that have the form of a denial $\leftarrow L_1 \wedge \ldots \wedge L_n$ where the L_i are literals, and variables are assumed to be quantified over the whole formula.

In the example, there are five integrity constraints. If a project starts at t, its planned end time must be greater than t (IC.1), and it can not have been active before (IC.2). IC. 3 requires a project to be active at time t-1 if it ends at t. Programmer pg can be assigned to a project p at time t if pg was not assigned to p at t-1 (IC.4) and project p was active at t-1 (IC.5).

2.5 Output requirements

An output requirement defines an output that must be produced by the information system. It has the form:

Output $\leftarrow$ Condition
Output = Contents

meaning that whenever the condition is satisfied, the system must produce an output with the defined contents.

In its simplest form, the condition is a query Qi with some parameters. In the example, OR.1 and OR.2 are of this form. The meaning is:
- When query Q1 is issued at time t, the system must provide the set of projects which are critical at t, and their planned end times.
- When query Q2 is issued at time t, with parameter t1, the system must provide the projects that were active at time t1, and their corresponding departments. Note that this query requires the system to know the complete history of predicate dpt.

In its most general form, the condition is any base or derived predicate (in fact, we regard queries as base predicates although, for convenience, we do not require to define them explicitly as such). An example is predicate new-in-dept(d,pg,t) in OR.3. The meaning is that as soon as a new fact of new-in-dept is known, the output must be produced. Thus, every time that a programmer pg is new in a department, the system must give the set of projects to which pg is assigned at t.

2.6 DCM vs. deductive data bases

As can be observed, there is a strong similarity in form between a DCM and a deductive data base. However, there are some fundamental differences between both, and it may be worthwhile to comment them briefly before closing this section.

A deductive data base consists of a set of base facts, a set of deduction rules and a set of integrity constraints [4]. The base facts are explictly stored in the data base, the deduction rules are used to augment the knowledge of the data base, and the integrity constraints are used to control its consistency.

A DCM does not include any fact. It is a model of a future information system, independent of any particular fact. On the other hand, a DCM does not assume any particular data base contents. It is left to the designer to decide which facts will be explictly stored and which ones will be deduced, and when and how will this deduction take place. Furthermore, a DCM includes a list of output requirements that the system must satisfy. This list (completed with quantitative data such as frequence, volume and response times) is the basis from which the designer must develop an efficient implementation. Finally, a DCM adopts a temporal view of informations, instead of the "single state" view of most current deductive data bases.

3. DESIGNING FROM A DCM

In this Section we first present the main decisions a designer must make when designing an information system from a DCM, then review some relevant work in this context, and finally outline our contribution.

3.1 Design decisions

The two main design decisions are: Data base design and Transactions design. In turn, data base design consists of data base contents design and data base schema design. In data base contents design, the designer must decide which predicates of the Information Base will be explicitly stored in the data base and, for each of them, the time interval for which the informations will be stored. There are, of course, many valid alternatives in this decision, and they must be evaluated with respect to the design objectives (cost, time, etc.).

In data base schema design, the designer must decide how to represent the data base contents in the data model used. Several options can also be available. If, for example, we use the relational data model, we can group two or more predicates in a single relation scheme. Full time history of a predicate can be represented by means of "events" data bases [6] or by associating a "time start" and "time end" components to each tuple [10]. When only the current state is stored we can eliminate the time component of the facts, making it implicit.

In transactions design, the designer must decide which transactions will exist and, for each of them, when will be executed, its pre-conditions and the actions to be performed, including data base updates and output production.

3.2 Previous work

The CIAM methodology [5] discusses in detail the design decisions required once the DCM is defined, but it does not provide a formal procedure for making them. CIAM suggests two phases: Conceptual information processing system (CIPS) design and DBMS adaptation. During CIPS design, the designer defines two models: A Conceptual data base model (CDBM), and a Conceptual processing model (CPM). The former describes the information which will be stored in the data base, while the latter decribes the processes for the maintanance of the data described in the CDBM. The CIPS level is independent of the DBMS that will be used. Adaptation to the chosen DBMS is done in the second phase.

The DADES methodology [12] also discusses the design decisions, and emphasizes the verification aspect, but again no precise method is given. It suggests an initial transaction structure, which is then refined to improve efficiency.

The DADES/GP prototype generator system [15] generates prototypes of information systems from DCMs written in the DADES language, with a minimum of designer intervention. DADES/GP stores in the data base the full history of all base and derived predicates. When a new fact is received all derived facts implied by it are generated and stored. Outputs are derived directly from the data base. The prototypes are thus a convenient means for validation purposes, but inefficient for the final system.

The same applies to the implementation discussed by Weigand [17]. He uses DCMs written directly in PROLOG, which are then interpreted. New base facts are added to the facts base. Derived facts are produced only when requested in some output, and never stored. Output response times can thus be unacceptable for a final system.

Some work from the deductive data bases field is also relevant here. Nicolas and Yazdanian [11] discuss the use of deduction rules for the generation of derived facts when a new fact is inserted in the data base. Their strategy is not directly applicable to a DCM, mainly because a "single state" view of the data base is taken. If that strategy were to be used in our example, this would imply that facts of predicates like active and has-worked would be generated for each instant of the life span. They cannot recognize that a project remains active until terminated and, thus, it is not necessary to generate and store the state of a project in all time points. We will see, however, that their strategy can be used once the DCM is transformed.

3.3 Our work

Our work is based on the idea of internal events. We define an internal event as a change in the extension of a predicate, and thus it corresponds to an update of the Information Base. An internal event rule is a rule that defines when an internal event happens. Internal events rules can be derived formally from the DCM. The set of internal events rules corresponding to a DCM is the internal events model. The internal events model allows us to know which internal events are implied by a given occurrence of an external event (base fact). We can then use the model as a basis for data base and transaction design. Both uses are discussed in section 5.

4. INTERNAL EVENTS MODEL

4.1 Classification of derived predicates

In trying to develop a method for the design and implementation of an information system from a DCM, we have observed that not all derived predicates should be treated in the same way. There are fundamental differences among several kinds of derived predicates, and an effective method should be able to recognize them. For this reason, we have built a classification of derived predicates that captures those differences. It consists of two dimensions, that we call the Positive and Negative dimensions. In the former, we distinguish among five types, and among two types in the latter. Each derived predicate has a type in the Positive dimension (P-type) and a type in the Negative one (N-type).

We may infer from the deduction rules to which P- and N-type a given predicate belongs to. However, we believe that our classification is so natural, and so intrinsically related to the intended meaning of a predicate, that we would be in favour of requiring the designer to explicitly define these types. Once the model is complete, we can then formally deduce them. In case of disagreement, either the deduction rules are incorrect or the perception the designer has of the predicate meaning is wrong.

Positive dimension

Let p be a derived predicate, **x** a vector of variables, **c** a vector of constants, and p(**c**) any fact that is true at time t-1. What can we say about the truth value of p(**c**) at time t?. Three cases are possible:

a) p(**c**) will be true at t. We say that p is P-steady.
b) p(**c**) will be false at t. We say that p is P-momentary.
c) p(**c**) may be true or false at t.

In this third case, assume now that no external events happen at time t. We find then three subcases:

c1) p(**c**) will be true at t. We say that p is P-state.
c2) p(**c**) will be false at t. We say that p is P-transient.
c3) p(**c**) will be true or false, depending on the truth value of a predicate $\phi(\mathbf{c},t)$. We say that p is P-spontaneous.

In what follows we define each case formally, and give some examples.

P-steady A predicate p is P-steady iff:

(4.1) $\forall \mathbf{x},t\ [p(\mathbf{x},t-1) \rightarrow p(\mathbf{x},t)]$

That is, if p(**x**) is true at time t-1, then it is also true at time t, independently of the external events that have occurred at t, and of any internal condition. In the example of figure 1, predicates completed and has-worked are of this type. If a project is completed at a given time, then it remains completed at any later time.

P-momentary A predicate p is P-momentary iff:

(4.2) $\forall \mathbf{x},t\ [p(\mathbf{x},t-1) \rightarrow \neg p(\mathbf{x},t)]$

In the example, new-in-dept is of this type. In a programmer is new in a department at time t-1, he cannot be new again in the same department at t.

P-state A predicate p is P-state iff:

(4.3) $\forall \mathbf{x},t\ [p(\mathbf{x},t-1) \wedge \text{noextev}(t) \rightarrow p(\mathbf{x},t)]$

where noextev(t) means that no external events have occurred at time t. Formally, if $p_1(\mathbf{y}_1,t),\ldots,p_n(\mathbf{y}_n,t)$ are the base predicates:

$\forall t\ [\text{noextev}(t) \leftrightarrow \neg(\exists \mathbf{y}_1 p_1(\mathbf{y}_1,t) \vee \ldots\ldots \vee \exists \mathbf{y}_n p_n(\mathbf{y}_n,t))]$

In the example, predicates active, p-end, dpt and assigned are of this type.

P-transient A predicate p is P-transient iff:

(4.4) $\forall \mathbf{x},t\ [p(\mathbf{x},t-1) \wedge \text{noextev}(t) \rightarrow \neg p(\mathbf{x},t)]$

In the example, predicate starts-working is of this type. If a programmer starts working for a department at time t-1, and is not assigned to any project at t, he will not "start working for" any department at t.

P-spontaneous A predicate p is P-spontaneous iff:

(4.5) $\forall \mathbf{x},t\ [p(\mathbf{x},t-1) \wedge \text{noextev}(t) \wedge \phi(\mathbf{x},t) \rightarrow p(\mathbf{x},t)]$
and $\forall \mathbf{x},t\ [p(\mathbf{x},t-1) \wedge \text{noextev}(t) \wedge \neg\phi(\mathbf{x},t) \rightarrow \neg p(\mathbf{x},t)]$

where $\phi(\mathbf{x},t)$ is distinct from the true predicate. We call $\phi(\mathbf{x},t)$ the persistence condition. The meaning is: If p(**x**) holds at t-1 and there are not external events at t, and the persistence condition is true, then p(**x**) holds at t (the fact persists). If the persistence condition is false, then p(**x**) does not hold at t. In the example, predicate critical is of this type.

Negative dimension

Let p be a derived predicate and let p(**c**) be any fact that is false at time t-1. Assume that no external events happen at t. What can we say about the truth value of p(**c**) at t?. Two cases are possible:

a) p(**c**) will remain false at t. We say that p is N-state.
b) p(**c**) will be true or false at t, depending on the truth value of a predicate $\phi'(\mathbf{c},t)$. We say that p is N-spontaneous.

In the Negative dimension we do not consider the types equivalent to P-steady, P-momentary and P-transient, because they do not seem to be of any practical interest. In what follows, we define both cases formally, and give some examples.

N-state A predicate p is N-state iff:

(4.6) $\forall \mathbf{x},t\ [\neg p(\mathbf{x},t-1) \wedge \text{noextev}(t) \rightarrow \neg p(\mathbf{x},t)]$

In the example, all derived predicates, except critical, are of this type.

N-spontaneous A predicate p is N-spontaneous iff:

(4.7) $\forall \mathbf{x},t\ [\neg p(\mathbf{x},t-1) \wedge \text{noextev}(t) \wedge \phi'(\mathbf{x},t) \rightarrow p(\mathbf{x},t)]$
and $\forall \mathbf{x},t\ [\neg p(\mathbf{x},t-1) \wedge \text{noextev}(t) \wedge \neg\phi'(\mathbf{x},t) \rightarrow \neg p(\mathbf{x},t)]$

where $\phi'(\mathbf{x},t)$ is distinct from true. We call $\phi'(\mathbf{x},t)$ the creation condition. In the example, predicate critical is of this type.

4.2 Internal events

Internal events play a critical role in our approach. As we will see in Section 5, we use internal events as triggers for data base updates, output production and process execution. There are two classes of internal events: insertion and deletion. To each base or derived predicate p corresponds an insertion internal event predicate ιp, with the same number of terms, and to each P-state or P-spontaneous predicate q corresponds a deletion internal event δq, with the same number of terms.

Insertion events are defined as follows. If p is a P-steady, P-state or P-spontaneous predicate then:

(4.8) $\forall x,t\ [\iota p(x,t) \leftrightarrow p(x,t) \wedge \neg p(x,t\text{-}1)]$

That is, event $\iota p(c)$ occurs at t if p(c) is true at t and it was false at time t-1. If p is a P-momentary, P-transient or base predicate:

(4.9) $\forall x,t\ [\iota p(x,t) \leftrightarrow p(x,t)]$

That is, event $\iota p(c)$ occurs at t if p(c) is true at t. Note that we also associate an internal event to each base fact (external event).

Similarly, deletion events are defined as follows. If p is a P-state or P-spontaneous predicate:

(4.10) $\forall x,t\ [\delta p(x,t) \leftrightarrow p(x,t\text{-}1) \wedge \neg p(x,t)]$

Note that we do not define deletion events for P-steady, P-momentary, P-transient and base predicates. Deletion events for P-steady predicates would never happen, because if p(c) holds at t-1 it also holds at any t' > t. We are not interested in deletion events of P-momentary, P-transient and base predicates, because we require an insertion event for every fact.

Rules 4.8, 4.9 and 4.10 above are called internal events deduction rules. They allow us to deduce at any time which internal events occur at that time. Thus, for example, applying rule 4.8 to predicate active we have:

(4.11) $\forall p,t\ [\iota active(p,t) \leftrightarrow active(p,t) \wedge \neg active(p,t\text{-}1)]$

Evaluating this rule at time t (and using DR.1) we get the internal events ιactive(c) that occur at t. However, it is obvious that this evaluation would be greatly inefficient, and that there are alternative rules much more efficient. In this case, for example, the rule:

(4.12) $\forall p,t\ [\iota active(p,t) \leftrightarrow \iota start(p,e,d,t)]$

is equivalent to 4.11 and quite easy to evaluate. Rule 4.12 says that an occurrence of event ιstart at time t produces an occurrence of ιactive at t. The rule can be obtained from 4.11 and DR.1 by taking into account that active is P-state and N-state, and the deduction rules and integrity constraints of the DCM. This is explained in the next sections.

4.3 Transformation of deduction rules

We first transform the deduction rules of the DCM into a set of equivalent ones. Let $p(x,t) \leftarrow B(x,y,t)$ be a deduction rule, where B(x,y,t) is a conjunction of one or more literals. The idea is to replace each literal in B, corresponding to a base or derived predicate, and whose time variable may range in the rule over a set including t, by an equivalent expression containing internal event predicates, whose time variables range only over {t}, and base or derived predicates, whose time variables range over a set not including t.

Let L1,...,Ln be the literals in B corresponding to base or derived predicates and whose time variable may range in the rule over a set including t. The transformation we apply to each Li (i = 1,...,n) is:

<u>If Li = q(z,t) or Li = ¬q(z,t)</u>

<u>If q is P-momentary, P-transient or base predicate</u>

$$q(z,t) \equiv \iota q(z,t)$$
$$\neg q(z,t) \equiv \neg \iota q(z,t)$$

<u>If q is P-steady</u>

$$q(z,t) \equiv q(z,t\text{-}1) \vee \iota q(z,t)$$
$$\neg q(z,t) \equiv \neg q(z,t\text{-}1) \wedge \neg\, \iota q(z,t)$$

<u>If q is P-state or P-spontaneous</u>

$$q(z,t) \equiv (q(z,t\text{-}1) \wedge \neg \delta q(z,t)) \vee \iota q(z,t)$$
$$\neg q(z,t) \equiv (\neg q(z,t\text{-}1) \wedge \neg \iota q(z,t)) \vee \delta q(z,t))$$

<u>If Li = q(z,t1) or Li = ¬q(z,t1)</u>
with t1 distinct from t and ranging over a set including t:

<u>If q is P-momentary, P-transient or base predicate</u>

$$q(z,t1) \equiv [t1 < t \wedge q(z,t1)] \vee [t1 = t \wedge \iota q(z,t)]$$
$$\neg q(z,t1) \equiv [t1 < t \wedge \neg q(z,t1)] \vee [t1 = t \wedge \neg \iota q(z,t)]$$

<u>If q is P-steady</u>

$$q(z,t1) \equiv [t1 < t \wedge q(z,t1)] \vee [t1 = t \wedge \iota q(z,t)]$$
$$\neg q(z,t1) \equiv [t1 < t \wedge \neg q(z,t1)] \vee [t1 = t \wedge \neg q(z,t\text{-}1) \wedge \neg \iota q(z,t)]$$

<u>If q is P-state or P-spontaneous</u>

$$q(z,t1) \equiv [t1 < t \wedge q(z,t1)] \vee [t1 = t \wedge q(z,t\text{-}1) \wedge \neg \delta q(z,t)] \vee [t1 = t \wedge \iota q(z,t)]$$
$$\neg q(z,t1) \equiv [t1 < t \wedge \neg q(z,t1)] \vee [t1 = t \wedge \neg q(z,t\text{-}1) \wedge \neg \iota q(z,t)] \vee [t1 = t \wedge \delta q(z,t)]$$

As an example, consider the application of the above rules to DR.1 (where the time variable of start ranges over the set $\{t_0,...,t\}$ and the time variable of completed ranges over {t}):

$$active(p,t) \leftarrow ([start(p,e,d,t1) \wedge t1 < t] \vee [\iota start(p,e,d,t) \wedge t1 = t]) \wedge t1 \leq t \wedge \neg completed(p,t\text{-}1) \wedge \neg \iota completed(p,t)$$

which is equivalent to:

$$active(p,t) \leftarrow start(p,e,d,t1) \wedge t1 < t \wedge \neg completed(p,t\text{-}1) \wedge \neg \iota completed(p,t)$$
$$active(p,t) \leftarrow \iota start(p,e,d,t) \wedge \neg completed(p,t\text{-}1) \wedge \neg \iota completed(p,t)$$

Similarly, the application to DR.4 gives:

$$completed(p,t) \leftarrow end(p,t1) \wedge t1 < t$$
$$completed(p,t) \leftarrow \iota end(p,t)$$

4.4 Normalized internal events deduction rules

Once we have transformed the deduction rules, we easily derive the normalized internal events deduction rules, which form the internal events model. These rules, we believe, are a useful basis from which the designer can develop an efficient implementation. The normalized internal events deduction rules are the same as rules 4.8, 4.9 and 4.10 (in Section 4.2), but replacing literal p(x,t) by the transformed body of the corresponding deduction rule. Thus, if for a predicate p we have n transformed deduction rules of the form:

$$p(x,t) \leftrightarrow \phi_i(x,y_i,t) \qquad i = 1,...,n$$

and we use rule 4.8, replacing p(x,t) yields:

$$\iota p(x,t) \leftrightarrow [\phi_1(x,y_1,t) \ \vee ... \vee \ \phi_n(x,y_n,t)] \ \wedge \neg p(x,t\text{-}1)$$

which is equivalent to the rules:

$$\iota p(x,t) \leftarrow \phi_i(x,y_i,t) \ \wedge \ \neg p(x,t\text{-}1) \qquad i = 1,...,n$$

As an example, using rules 4.8 and 4.10 for predicate active, we have:

(4.13) $\iota active(p,t) \leftarrow start(p,e,d,t1) \wedge t1 < t \wedge \neg completed(p,t\text{-}1) \wedge \neg \iota completed(p,t) \wedge \neg active(p,t\text{-}1)$

(4.14) $\iota active(p,t) \leftarrow \iota start(p,e,d,t) \wedge \neg completed(p,t\text{-}1) \wedge \neg \iota completed(p,t) \wedge \neg active(p,t\text{-}1)$

(4.15) $\delta active(p,t) \leftarrow active(p,t\text{-}1) \wedge \neg[start(p,e,d,t1) \wedge t1 < t \wedge \neg completed(p,t\text{-}1) \wedge \neg \iota completed(p,t)] \wedge \neg[\iota start(p,e,d,t) \wedge \neg completed(p,t\text{-}1) \wedge \neg \iota completed(p,t)]$

Similarly, using rule 4.8 for predicate completed, we get:

(4.16) $\iota completed(p,t) \leftarrow end(p,t1) \wedge t1 < t \wedge \neg completed(p,t\text{-}1)$

(4.17) $\iota completed(p,t) \leftarrow \iota end(p,t) \wedge \neg completed(p,t\text{-}1)$

4.5 Simplification of rules

These normalized rules can now be subjected to a set of simplifications, by combining them with the properties of the predicate types, (4.1) to (4.7), the deduction rules and the integrity constraints of the DCM.

We can only show here the potential of these simplifications, by means of an example. Let us take rules 4.16 and 4.17. Being completed a N-state predicate, we have, by 4.6:

$$noextev(t) \rightarrow \neg \iota completed(p,t)$$

and, if ιend is false no ιcompleted events can happen, which means that no p, t1, t can satisfy the body of rule 4.16. Therefore, we eliminate rule 4.16. On the other hand, we have by IC.3:

(4.18) $\iota end(p,t) \rightarrow active(p,t\text{-}1)$

and by DR.1: $active(p,t\text{-}1) \rightarrow \neg completed(p,t\text{-}1)$

and, thus, we can remove literal $\neg completed(p,t\text{-}1)$ from rule 4.17. So, rules 4.16 and 4.17 are equivalent to:

(4.17') $\iota completed(p,t) \leftarrow \iota end(p,t)$

We give in figure 2 the resulting set of rules after simplification.

IDR.1 $\iota active(p,t) \leftarrow \iota start(p,e,d,t)$

IDR.2 $\delta active(p,t) \leftarrow \iota completed(p,t)$

IDR.3 $\iota p\text{-}end(p,e,t) \leftarrow \iota start(p,e,d,t)$

IDR.4 $\delta p\text{-}end(p,e,t) \leftarrow p\text{-}end(p,e,t\text{-}1) \wedge \delta active(p,t)$

IDR.5 $\iota dpt(p,d,t) \leftarrow \iota start(p,e,d,t)$

IDR.6 $\delta dpt(p,d,t) \leftarrow dpt(p,d,t\text{-}1) \wedge \delta active(p,t)$

IDR.7 $\iota completed(p,t) \leftarrow \iota end(p,t)$

IDR.8 $\iota assigned(p,pg,t) \leftarrow \iota assign(p,pg,t) \wedge \neg \delta active(p,t)$

IDR.9 $\delta assigned(p,pg,t) \leftarrow assigned(p,pg,t\text{-}1) \wedge \delta active(p,t)$

IDR.10 $\iota has\text{-}worked(d,pg,t) \leftarrow \iota starts\text{-}working(d,pg,t) \wedge \neg has\text{-}worked(d,pg,t\text{-}1)$

IDR.11 $\iota starts\text{-}working(d,pg,t) \leftarrow \iota assign(p,pg,t) \wedge dpt(p,d,t\text{-}1) \wedge \neg \ \delta dpt(p,d,t)$

IDR.12 $\iota new\text{-}in\text{-}dept(d,pg,t) \leftarrow \iota starts\text{-}working(d,pg,t) \wedge \neg has\text{-}worked(d,pg,t\text{-}1)$

IDR.13 $\iota critical(p,t) \leftarrow p\text{-}end(p,e,t\text{-}1) \wedge \neg \delta p\text{-}end(p,e,t) \wedge t\text{-}1 = e\text{-}8$

IDR.14 $\iota critical(p,t) \leftarrow \iota p\text{-}end(p,e,t) \wedge t > e\text{-}8$

IDR.15 $\delta critical(p,t) \leftarrow critical(p,t\text{-}1) \wedge \delta p\text{-}end(p,e,t)$

IDR.16 $\delta critical(p,t) \leftarrow critical(p,t\text{-}1) \wedge p\text{-}end(p,e,t\text{-}1) \wedge t = e+1$

Figure 2. Internal events model example

5. USE OF THE INTERNAL EVENTS MODEL

In this Section we outline a possible use of the internal events model for the design and implementation of information systems, and show its application to the example.

5.1 Data base design

An alternative of data base contents DB(t) is characterized by:

- A set of base or derived predicates of the DCM to be explicitly stored in the data base, and
- For each of them, a time interval for which its facts will be stored.

We will consider here only two extreme cases of time intervals: either current state or full history. If p(x,t) is a stored predicate, in the first case we store in the data base the set $\{<x,t> \mid p(x,t)\}$, and in the second case the set $\{<x,t1> \mid p(x,t1) \ \wedge t1 \leq t\}$. We usually do not store current states of P-momentary, P-transient and base predicates, since this would imply to store their facts when produced (at time t), and to delete them immediately (also at time t).

A valid alternative of data base contents DB(t) is one that satisfies [12]:

1) Internal events that happen at t can be derived from DB(t-1) and base facts received at t. DB(t_0-1) is assumed to be empty.
2) Contents of the outputs to be produced at t can also be derived from DB(t-1) and base facts received at t.
Note that the alternative of storing the full history of only base predicates is always a valid one.

In our example, base facts are ιstart, ιend and ιassign. A valid alternative would then be:

$$DB(t) = \{<p,d,t1> \mid dpt(p,d,t1) \wedge t1 \leq t\}, \{<p,e,t> \mid p\text{-}end(p,e,t)\}, \{<p,pg,t> \mid assigned(p,pg,t)\}, \{<d,pg,t> \mid has\text{-}worked(d,pg,t)\}$$

The database schema corresponding to a given DB(t) can also be designed systematically, although we do not have developed the method yet. If we decide to represent each stored predicate as an independent relation then the following rules could be used:

1) For a current state predicate p(**x**,t) define a relation scheme p(**x**).

2) For a full history predicate p(**x**,t):

If p is P-momentary, P-transient or base, define a relation scheme p(**x**,t).

If p is a P-steady predicate, define a relation scheme p(**x**,t_s), where attribute t_s will represent the first time p(**x**) is true.

If p is a P-state or P-spontaneous predicate, define a relation scheme p(**x**,t_S,t_E), where attributes t_S , t_E represent an interval in which p(**x**) holds.

In our example, a possible data base schema could be:

DEPARTMENT(PROJECT,DPT,T-START,T-END)
P-END(PROJECT,END)
ASSIGNED(PROJECT,PROGRAMMER)
HAS-WORKED(DEPT,PROGRAMMER)

5.2 Transaction design

Once the data base has been designed we can proceed to transaction design. Again, many alternatives exist, and several structuring mechanisms could be used. We only want to outline here a simple approach, that can work well in some applications, as an illustration of the use of the internal events model.

We first define **relevant internal events** as follows:

1) Insertion and deletion events corresponding to stored predicates are relevant internal events.
2) Insertion events corresponding to output requirements conditions are relevant internal events.
In the example, relevant internal events are: ιdpt, δdpt, ιp-end, δp-end, ιassigned, δassigned, ιhas-worked and ιnew-in-dept.

We then build a transaction type for each base fact type. The purpose of each transaction will be:

1) Read the corresponding base fact.
2) Verify the applicable integrity constraints.
3) Derive the relevant internal events produced by the base fact. This can be easily determined bottom-up from the internal events model. The generation strategy described in [11] can be used for this purpose.
4) For each internal event thus produced, if it corresponds to a stored predicate perform the appropiate data base operation; and if it corresponds to an output condition produce the defined output contents.

If some relevant internal event corresponds to a P- or N-spontaneous predicate, then we would also need a special transaction, to be executed at each time point of the life span.

Finally, we also build a transaction type for each query type. The purpose of these transactions will be:

1) Read the query parameters, if any, and
2) Produce the defined output contents.

As an example, consider transaction end(p,t). The actions to be performed are:

1) Read project identifier p (time t is implicit).
2) Verify integrity constraints (not considered here).
3) Delete tuple with PROJECT = p from P-END relation. (This is deduced from IDR.7, 2 and 4).
4) Read DEPARTMENT with PROJECT = p from the data base, and mark it as inactive: T-END = t (This is deduced from IDR.6).
5) Remove all tuples with PROJECT = p from ASSIGNED relation. (This is deduced from IDR.9).

6. CONCLUSIONS

We have first characterized deductive conceptual models (DCMs) in a first order logic framework. Then we have discussed the main design decisions involved in the design and implementation of an information system from a DCM. A new approach has been presented, which consists in deriving an internal events model from a DCM. This model is a useful basis from which several design alternatives can be systematically developed and evaluated. We have outlined a possible use of the internal events model in data base and transaction design, but other uses can be imagined as well. We also expect that the internal events model, and the classification of predicates we have proposed, can be useful in the field of deductive data bases, although this has not been elaborated in the paper.

Several research directions can be explored from the work reported here. First, the problem of integrity constraints enforcement should be taken into account. This has not been considered in the paper, but it is felt that work from deductive data bases can be easily integrated here [2]. Second, specific procedures for the simplification of normalized internal events deduction rules should be developed, perhaps integrated in an interactive man-machine system. Finally, one or more approaches to design and implementation from internal events models should be completely developed, and integrated in existing or new methodologies.

ACKNOWLEDGMENTS

I would like to thank D. Costal, J.A. Pastor, M.R. Sancho and J. Sistac for many useful comments on earlier drafts of this paper.

7. REFERENCES

[1] Brodie,M.;Mylopoulos,J.;Schmidt,J. "On conceptual modelling: Perspectives from Artificial Intelligence, Databases and Programming Languages", Springer-Verlag, 1983.

[2] Bry,F.; Decker,H.; Manthey,R. "A uniform approach to constraint satisfaction and constraint satisfiability in deductive databases". Proc. 1st. Int. Conf. Extending Data Base Technology (EDBT), March 1988.

[3] Bubenko,J.; Olivé,A. "Dynamic or temporal modelling?. An illustrative comparison". SYSLAB Working Paper No. 117, University of Stockholm, 1986.

[4] Gallaire,H.; Minker,J.; Nicolas,J-M. "Logic and databases: A deductive approach". Computing Surveys, Vol. 16, No. 2, June 1984, pp. 153-185.

[5] Gustafsson,M.; Karlsson,T.; Bubenko, J. "A declarative approach to conceptual information modelling", In [14], pp. 93-142.

[6] Kobayashi,I. "Temporal aspects of databases: Interaction between state and event relations". IFIP WG 2.6 Working Conference DS-1, Belgium, Jan. 1985.

[7] Kowalski,R. A. "Logic for problem solving", North-Holland, 1979.

[8] Kowalski,R.A.; Sergot,M.J. "A logic based Calculus of Events", New Generation Computing, 4, No. 1, 1986, pp. 67-95.

[9] Kowalski,R.;Sadri,F.;Soper,P. "Integrity checking in deductive databases". Proc. 13th. VLDB, Sept. 1987, pp. 61-69.

[10] Navathe,S.B.; Ahmed,R. "TSQL- A language interface for history databases", in Proc. WG 8.1 Conf. on Temporal aspects in information systems, Sophia-Antipolis, France, May 1987, pp. 113-128.

[11] Nicolas, J-M.; Yazdanian, K. "An outline of BDGEN: A deductive DBMS", IFIP 1983, North-Holland, pp. 711-717.

[12] Olivé,A. "DADES: A methodology for specification and design of information systems", In [14], pp. 285-334.

[13] Olivé,A. "A comparison of the operational and deductive approaches to conceptual information systems modelling", Proc. IFIP-86, Dublin, pp. 91-96.

[14] Olle,T.W.; Sol,H.G.; Verrijn-Stuart,A.A. (Eds.) "Information systems design methodologies: A comparative review", North-Holland, 1982.

[15] Sistac,J.; Olivé,A. "DADES/GP: A prototype generator from deductive conceptual models of data base systems", Convention Informatique 1986, Paris.

[16] van Griethuysen, J.J. (Ed.)."Concepts and terminology for the conceptual schema and the information base", ISO/TC97/SC5/WG3, March 1982.

[17] Weigand,H. "Conceptual models in PROLOG", IFIP WG 2.6 Working Conference DS-1, Belgium, Jan. 1985.

A Deductive Method for Entity-Relationship Modeling
(Extended Abstract)

Giuseppe Di Battista and Maurizio Lenzerini

Dipartimento di Informatica e Sistemistica, Università degli Studi di Roma "La Sapienza"
Via Buonarroti 12, 00185 Roma, Italia

Abstract

We present a simple entity-relationship-oriented model, which essentially includes the notion of class, together with different types of relations among classes, such as is-a, part-of, and disjointness. We define the semantics of the model in terms of first order logic, and present a sound, complete, and efficient inference algorithm for such a model. We argue that our model and the associated inference capabilities provide a suitable formal basis for designing an effective environment supporting conceptual modeling.

1 Introduction

Recent works on data base modeling show a growing interest in the object-oriented paradigm (see, for instance, [Ul87]). One of the basic motivations for such an interest comes from the need of embedding several types of abstraction capabilities into the existing data models. Object-oriented data models provide many concepts which seem particularly suited for such a requirement. Perhaps, the most important one is inheritance, which represents a powerful mechanism for explicitely stating that the properties of a class propagate over other related classes.

It is interesting to note that the above objectives were already present in the research on semantic data models (see [HK87]), which has addressed many problems related to the semantic aspects of data description, with the specific goals of defining the basic modeling primitives needed in a data base formalism, and studying their characteristics. We believe that one novelty of the investigation on object-orientation is the concern on the deductive capabilities of the data model, especially those related to inheritance.

Recent works (see [AM86], [MZ86], [AS87], [Le87a], [Le87b]) have carried out an investigation on the basic modeling primitives of object-oriented data models.

Proceedings of the Fifteenth International Conference on Very Large Data Bases

All of these works deal with simple representational mechanisms, which essentially include the notion of class (an abstraction for a set of objects) and various forms of containment assertions on classes, with the main goals of investigating on the inherent complexity of reasoning about these assertions, and providing algorithms for the inference problem. For example, Arisawa and Miura [AM86] consider assertions of the form: (A and B and C is-a D and E), which states that the intersection of the (set of instances of the) classes A,B and C is contained in the intersection of D and E, and propose a polynomial time algorithm for computing inferences on these assertions. Aggregation relations on classes, allowing for representing a class as a property of another class, as well as other features of object-oriented data models, related to the definition of operations on classes, are not taken into account in these works.

The goal of this paper is to go one step further in the above direction, by considering aggregation relations among classes. In particular, we present a simple object-oriented data model, called ERL, which allows one to deal with entities (simple classes) and relationships (aggregated classes), and provides several assertional mechanisms for describing how such classes relate to each other. Both positive assertions (stating that a certain relation holds among classes) and negative assertions (stating that a relation does not hold) are expressible in this language.

Four types of basic semantic relations are taken into account, namely subsetting, disjointness, typing, and mandatory participation of an entity in a given relationship. For example, we can assert in ERL that the set of instances of a given class A is included in the set of instances of another class B (subsetting), or that classes A and B cannot have common instances (disjointness). As an example of negative assertion, we can state that A and B are not disjoint, i.e. there is at least one object that is both an instance of A and an instance of B. Typing allows for asserting that objects participating in a given relationship B are instances of an entity A. Conversely, we can state that every instance of A participates in at least one instance of B (mandatory participation).

Our modeling language is formally defined in terms of first order logic. Besides providing the formal tool for expressing the semantics of the language, logic allows us to devise sound and complete inference procedures for ERL. In fact, we present a sound, complete, and polynomial time

Amsterdam, 1989

algorithm for computing inferences on a set of ERL assertions.

It is our opinion that the resulting inference technique may constitute the formal basis for building a sound and effective environment for conceptual modeling. During the past years, many efforts have been devoted to provide the data base designer with useful methodologies and tools supporting her/his activities. Many of these works are based on the entity-relationship data model (see, for example, [Ce83], [BLN86], [TYF86]). These proposals are often based on pragmatical criteria for achieving "good" qualities of the conceptual schema, such as consistency, clarity, and minimality. In our approach, a formal definition of such concepts is provided, based on the precise definition of the semantics of the language. Moreover, the inference techniques associated with the language can be directly used to build a system which is able to automatically check the representation against several correctness and minimality criteria, and to support the designer in all the activities requiring reasoning about the description, and exploring different modeling choices.

The paper is organised as follows. In Section 2 we informally define the language, and briefly discuss its expressive power. In Section 3, we present the formal definition of the language. In Section 4, we describe our method for performing inference on a set of assertions expressed in the language. In Section 5, we briefly discuss the use of the inference technique in the definition of an environment supporting entity-relationship modeling. Finally, in Section 6 we present the conclusions and outline future developments of our research.

In the remainder of the paper, we assume that the reader is familiar with the basic notions of first order logic (at the level of [Me64]). In this extended abstract, the proofs of the theorems are omitted (the interested reader is referred to [DL89] for the full paper).

2 An Entity-Relationship-based modeling language

In this section we informally discuss the basic characteristics of a simple ER-based modeling language, called ERL (Entity Relationship Language), which can be used to define entities and relationships and to state several types of assertions on how they relate to each other. The formal definition of ERL appears in section 3.

In ERL, the universe of discourse is partitioned into two levels, called extensional and intensional.

In the extensional level, both individual objects, and tuples of individual objects are represented. An individual object is an atomic object, identifiable through a unique name, whereas a tuple is an aggregation of individual objects. The number of objects which are components of a tuple is called the arity of the tuple. Each component is referenced by its position in the tuple: for example, the individual objects a,b and c are said to be the 1-component, 2-component and 3-component of the tuple <a,b,c>, respectively.

The objects belonging to the intensional level represent classes of either individual objects, or tuples. Classes are distinguished into entities and relationships.

An entity class (simply entity in the following) is a class of individual objects, whereas a relationship class (relationship) is a class of tuples of the same arity, which is called the arity of the relationship. The objects belonging to a given class constitute the set of instances of that class.

Note that, for the sake of simplicity, we do not explicitely deal with attributes, which are usually considered in the ER model. However, it is easy to see that what follows applies with minor changes to a language incorporating attributes.

In order to represent meaningful properties of entities and relationships, ERL provides the modeler with several types of assertions, which are discussed in the following.

Typing: Individual objects which are components of the tuples belonging to a given relationship, can be asserted to belong to a certain entity. For example, if Tutoring is a relationship of arity 2, we can assert that for each instance t of Tutoring, the 1-component of t is an instance of the entity Person, and the 2-component of t is an instance of the entity Course. In the classical ER model, typing corresponds to the fact that relationships are defined on a fixed collection of entities. Conversely, in ERL there is no limitation on the number of typing assertions for a given relationship: for example, the i-component of the tuples belonging to the relationship R, can be asserted to be both of type A and of type B.

Subsetting: One class can be asserted to be a subset of another class: in this case, every instance of the former, is also an instance of the latter. For each assertion of this type, we require the two classes to be either both entities, or both relationships of the same arity. For example, if Person and Student are entities, Tutoring and Summer_Tutoring are relationships of arity 2, we can assert Professor is subset of Person and Summer_Tutoring is subset of Tutoring. Notice that, in the literature, the subset relationship is often referred to as is-a relationship.

Disjointness: Two types of disjointness assertions can be expressed in ERL. Assertions of the first type are used to state that the extensions of two classes are disjoint. In this case, we require the two classes to be either both entities or both relationships of the same arity. For example we can state that the two entities Graduate_Student and Undergraduate_Student are disjoint. The second type of disjointness assertion can be used to state that the set of objects that are i-components of the tuples belonging to a given relationship is disjoint from the set of instances of a certain entity. Assertions of this type are used to represent the fact that an entity cannot participate in a given relationship in a specified role. For example, we can state that the instances of Undergraduate_Student cannot participate in the relationship Tutoring as 1-components.

Existence: The instances of a given entity can be

asserted to mandatorily participate in a certain relationship for a specified role. In other words, every instance of the entity must be the i-component of at least one instance of the relationship. For example, we can assert that the instances of the entity Teacher mandatorily participate in the relationship Course.

Negation: All the above assertions can appear in a negative form, in order to represent that a certain property does not hold. For example, we can assert that entity A is not related to B by means of the subset relationship, i.e. that there is at least one object which is included in the extension of A, but not in the extension of B. Another example of negative assertion is the one stating that A and B are not disjoint, i.e. there always exists one object that is in the extension of both classes. Notice that negative assertions are not usually considered in semantic data models.

In Section 3, we provide the formal definition of ERL syntax and semantics, whereas in Section 4 we address the problem of computing inference on ERL assertions. Such a problem consists of checking whether a set Σ of ERL assertions implies a single assertion σ, i.e. whether the fact that all the assertions in Σ hold, implies that σ holds as well. In Section 4, a polynomial time algorithm for this problem is presented.

Assertions of the type considered in ERL, have been taken into account in several recent works both on data bases, and knowledge representation. We already mentioned in Section 1 a number of works dealing with the problem of performing inference on containment assertions on classes. As we said before, such works do not consider aggregation relationships among classes.

The interaction of containment assertions with aggregation relationships has been considered in the context of the relational data base theory. We are referring, in particular, to those papers investigating the so-called inclusion dependencies. An inclusion dependency is a statement of the form: $R(A_1, ..., A_m) \supseteq S(B_1,...,B_m)$, where R and S are two relation schemes, and $A_1, ..., A_m, B_1, ..., B_m$ are attributes. Such a dependency holds in a relational database D, if each of the tuples in the projection of s over $B_1, ..., B_m$, is also in the projection of r over $A_1,...,A_m$, where r and s are the relations in D corresponding to the relation schemes R and S, respectively.

In [CFP84], it is shown that the inference problem for inclusion dependencies is P-space complete in the general case. However, the problem can be solved in linear time if the dependencies are unary (m=1). This fact is exploited in [CaVi83], where the inference problem is studied for a language including a restricted form of inclusion and exclusion dependencies.

Notice that ERL do not limit its expressive power to unary dependencies. Indeed, it allows both the subset, and the disjointness relationship to be established both between entities, and between relationships. Moreover, ERL allows typing, existence, and negative assertions, to be expressed without any limitation. Nevertheless, the results reported in this paper show that the inference problem for ERL is tractable.

Inclusion dependencies have also been studied in their interaction with functional dependencies. Chandra and Vardi [CV85] have shown that the inference problem for a language including both dependencies, is undecidable; Kannellakis, Cosmadakis, and Vardi [KCV83], have shown that a polynomial time solution exists, if inclusion dependencies are restricted to be unary. Note that no construct is currently provided by ERL for expressing functional dependencies.

3 Logical formulation

In this section, we give a formal account of ERL in terms of first order logic. We rely on a first order language that includes constants, variable and predicate symbols. Constants and constant tuples correspond to individual and aggregated objects, respectively; 1-place and n-places (n>1) predicate symbols correspond to entities and relationships, respectively. Finally, the assertions discussed in Section 2 are expressed by means of sentences in the first order language.

Before delving into details, some remarks on the notation are in order. We shall use several metasymbols (possibly with subscripts), whose meaning is as follows: greek upper case letters for predicate symbols; x, y, z for variable symbols, and **x**, **y**, **z** for variable tuples; c, d for constant symbols, and **c**,**d** for constant tuples; w for constant or variable symbols, and **w** for tuples whose each component is either a constant or a variable; $\mathbf{w}^i$ for the i-th component of the tuple **w**. Finally, we write $\exists\mathbf{x}$ instead of $\exists x_1,...,\exists x_n$ and $\forall\mathbf{x}$ instead of $\forall x_1,...,\forall x_n$ (where $\mathbf{x}=<x_1,...,x_n>$).

Formally, a set of ERL assertions is simply a first order theory (i.e., a first order language, plus a set of sentences in that language), as specified by the following definition.

Definition 1 An ERL-theory is a first order theory <L,A>, where L is a first order language consisting of a set of constant, variable and predicate symbols, and A, the axioms of the theory, is a set of formulas of L, with the constraint that each element of A has one of the following forms:

$\forall\mathbf{x}\ (\neg\ \Phi_1(\mathbf{x})\ \vee\ \Phi_2(\mathbf{y}))$ where **y** is a subtuple of **x**;

$\forall\mathbf{z}\ (\neg\ \Phi_1(\mathbf{x})\ \vee\ \neg\ \Phi_2(\mathbf{y}))$ where **x** and **y** are subtuples of **z**;

$\forall\mathbf{x}\ \exists y\ (\neg\ \Phi_1(\mathbf{x})\ \vee\ \Phi_2(\mathbf{z}))$ where **z** contains **x** and all the variables in y;

$\exists\mathbf{x}\ (\Phi_1(\mathbf{x})\ \wedge \neg\ \Phi_2(\mathbf{y}))$ where **y** is a subtuple of **x**;

$\exists\mathbf{z}\ (\Phi_1(\mathbf{x})\ \wedge\ \Phi_2(\mathbf{y}))$ where **x** and **y** are subtuples of z;

$\exists x \, \forall y \; (\Phi_1(x) \wedge \neg \Phi_2(z)) \;)$ where **z** contains x and all the variables in y;

It is well known that the notion of interpretation is used for assigning semantics to any logical formula (see [Me79]). As far as ERL-theories are concerned, the notion of interpretation specialises as follows.

Definition 2 An interpretation for an ERL-theory T is a pair <DOM,EXT>, where DOM is a finite set, called the domain of the interpretation, and EXT is a mapping assigning an element of DOM to each constant symbol of T, and a subset of DOM^n to each predicate symbol P of T (n is the arity of P).

Taking into account the semantics of the axioms expressible in ERL-theories, we now show how they are used to express typing, subsetting, disjointness, existence, and negative assertions.

Axioms of the type:

$$\forall \mathbf{x} \; (\neg \Phi_1(x) \vee \Phi_2(y)),$$

can be used to express both typing and subsetting. In particular, if Φ_2 is a 1-place predicate symbol representing an entity ($|y| = 1$), Φ_1 is a n-place predicate symbol representing a relationship, and $\mathbf{x}^i = y$, then the above axiom specifies that every i-component of Φ_1 instances, is an instance of Φ_2, i.e. that the i-components of the relationship Φ_1 are typed with entity Φ_2. For example, the relationship Tutoring can be typed with entities Person and Course as follows:

$$\forall x_1, x_2 \, (\neg \, \text{Tutoring}(x_1, x_2) \vee \text{Person}(x_1))$$

$$\forall x_1, x_2 \, (\neg \, \text{Tutoring}(x_1, x_2) \vee \text{Course}(x_2)).$$

Observe that the above axiom allows a more general form of typing than the one discussed in Section 2. Indeed, the cardinality of y needs not to be 1 and, therefore, the axiom may express a dependency between two relationships.

If $\mathbf{x} = \mathbf{y}$, then the axiom expresses subsetting, on either entities ($|x| = 1$), or relationships ($|x| > 1$). For example, in order to express the subset relationship between entities Professor and Person, and between relationships Summer_Tutoring and Tutoring, we can use the following axioms:

$$\forall x_1 \, (\neg \, \text{Professor}(x_1) \vee \text{Person}(x_1)),$$

$$\forall x_1, x_2 \, (\neg \, \text{Summer\_Tutoring}(x_1, x_2) \vee \text{Tutoring}(x_1, x_2)).$$

Axioms of the type:

$$\forall \mathbf{z} \, (\neg \Phi_1(\mathbf{x}) \vee \neg \Phi_2(\mathbf{y}))$$

are disjointness axioms. If $\mathbf{z} = \mathbf{x} = \mathbf{y}$, then the axiom expresses the disjointness relationship between either two entities ($|x| = 1$), or two relationships ($|x| > 1$). For example, disjointness between the entities Graduate_Student and Undergraduate_Student, and between the relationships Tutoring and Attends, is expressed as:

$$\forall x_1 \, (\neg \, \text{Graduate\_Student}(x_1) \vee \neg \, \text{Undergraduate\_Student}(x_1))$$

$$\forall x_1, x_2 \, (\neg \, \text{Tutoring}(x_1, x_2) \vee \neg \, \text{Attends}(x_1, x_2)).$$

Disjointness relationships between entities and relationships can be expressed by imposing $|x| = 1$, and $\mathbf{z} = \mathbf{y}$. For example the disjointness relationship between Undergraduate_Student and Tutoring in the 1-component, is expressed by means of the following axiom:

$$\forall x_1, x_2 \, (\neg \, \text{Undergraduate\_Student}(x_1) \vee \neg \, \text{Tutoring}(x_1, x_2)).$$

Axioms of the type:

$$\forall \mathbf{x} \, \exists \mathbf{y} \; (\neg \Phi_1(x) \vee \Phi_2(z)),$$

express existence assertions, where Φ_1 represents an entity ($|x| = 1$) and Φ_2 represents a relationship ($|z| > 1$). For example, the mandatory participation of the entity Course in the relationship Teaching as 1-component, is expressed by:

$$\forall x \, \exists y \; (\neg \, \text{Course}(x) \vee \text{Teaching}\,(x,y)).$$

Up to now, we have discussed the first three forms of allowable axioms in ERL-theory. With ragard to negative assertions, it is easy to see that they correspond to the last three forms reported in definition 1. As an example, the fact that it is not true that Tutoring is typed with the entity Professor as 1-component, can be expressed by the following axiom:

$$\exists x \exists y \; (\text{Tutoring}(x,y) \wedge \neg \text{Professor}(y)).$$

After the formal definition of ERL, we are now ready to address the main point of this paper, which is the problem of computing inference on ERL-theories. Such a problem is dealt with in the next section, in which we present a sound and complete inference algorithm for ERL-theories.

4 Checking ERL-theories for satisfiability

Our method for computing inference on ERL-theories exploits the notion of satisfiability in logic. Recall that a theory is said to be satisfiable, if it admits at least one model (i.e. an interpretation which satisfies all the axioms of the theory). A theory which is not satisfiable is said to be unsatisfiable. Logical implication is the problem of determining whether a given sentence τ is true in every model of a theory T. Such a problem can be reduced to the one of checking a theory for unsatisfiability: T logically implies τ, just in case there cannot be models of T which satisfy $\neg\tau$, i.e. $T \cup \{\neg\tau\}$ is unsatisfiable. Notice that if T is an ERL-theory, and τ is any formula which has the form of an admissible axiom for T, then $T \cup \{\neg\tau\}$ is an ERL-theory as well.

The goal of this section is to present an efficient

algorithm for checking an ERL-theory for satisfiability. The above considerations show that such an algorithm can be used to perform inference on ERL-theories.

4.1 Clausal form for ERL-theories

The method we present requires an ERL-theory to be transformed into a set of clauses, which are in turn represented by means of a graph. A clause is simply a disjunction of literals, each literal being either an atomic formula (i.e. a predicate symbol with the corresponding arguments, called positive literal), or the negation of an atomic formula (negative literal). A negative (positive) clause is a clause constituted by negative (positive) literals. All the variables appearing in a clause are implicitely universally quantified. Well known results in first order logic ensure us that any set of formulas can be transformed into a set of clauses which is satisfiable if and only if the original set is satisfiable.

Table 1 shows the rules for obtaining a set of clauses from the axioms of an ERL-theory (axioms are on the left and the corresponding clauses on the right).

$\forall x\,(\neg\,\Phi_1(x) \vee \Phi_2(y))$	$\neg\,\Phi_1(x) \vee \Phi_2(y)$
$\forall z\,(\neg\,\Phi_1(x) \vee \neg\,\Phi_2(y))$	$\neg\,\Phi_1(x) \vee \neg\,\Phi_2(y)$
$\forall x\,\exists y\,(\neg\,\Phi_1(x) \vee \Phi_2(z))$	$\neg\,\Phi_1(x) \vee \Phi_2(v)$
$\exists x\,(\Phi_1(x) \wedge \neg\,\Phi_2(y))$	$\Phi_1(c),\ \neg\,\Phi_2(d)$
$\exists z\,(\Phi_1(x) \wedge \Phi_2(y))$	$\Phi_1(c),\ \Phi_2(d)$
$\exists x\,\forall y\,(\Phi_1(x) \wedge \neg\,\Phi_2(z))$	$\Phi_1(c),\ \neg\,\Phi_2(w)$

Table 1

The first two axiom forms are easily translated into clauses. Due to the presence of existential quantifiers, the remaining four forms require the introduction of Skolem symbols (constants and functions). In the third clause form, **v** indicates a tuple that is obtained from **z** by leaving unchanged the **x** variables, and substituting each $\mathbf{y}^i$ with $f_i^j(x)$, where f_i^j is a Skolem function and j is a number identifying the original axiom (we are assuming that the axioms of the ERL-theory are numbered). Concerning the fourth and fifth axiom forms, each of them is transformed into two clauses forms, that are simply obtained by substituting **x** and **y** with tuples of Skolem constants, **c** and **d**, of the same arity. Finally, in the sixth form, **c** is a tuple of Skolem constants of the same arity of **x**, and **w** is obtained from **z** by substituting the **x** variables with the coresponding constants of **c**.

Example 1 We show how to transform a simple ERL-theory into the corresponding set of clauses:

$\forall x$ (¬Professor(x) ∨ Person(x))
$\forall x$ (¬Student(x) ∨ Person(x))
$\forall x \forall y$ (¬Undergraduate_Student(x) ∨ ¬Tutoring(x,y))
$\exists x \exists y$ (Tutoring(x,y) ∧ ¬Professor(y))
$\forall x \exists y$ (¬Course(x) ∨ Tutoring(x,y))

¬Professor(x) ∨ Person(x)
¬Student(x) ∨ Person(x)
¬Undergraduate_Student(x) ∨ ¬Tutoring(x,y)
Tutoring(d_1,d_2)
¬Professor(d_2)
¬Course(x) ∨ Tutoring(x,f_2^5(x)).

In what follows, we call ERL-clause any clause obtained from an axiom of an ERL-theory by means of the above transformation. Notice that ERL-clauses are Horn clauses, i.e. they contain at most one positive literal. Moreover, due to the presence of function symbols, the Herbrand universe of a set of ERL-clauses is infinite.

The following lemma allows us to transform a set of ERL-clauses into another set, in which every Skolem function is replaced by a constant symbol. Obviously, the Herbrand universe of the resulting set is finite.

Lemma 1 Let C be a set of ERL-clauses, and C' the set of ERL-clauses obtained from C by substituting each $f_i^j(x)$ with a new constant symbol z_i^j. Then C is satisfiable if and only if C' is satisfiable.

In the following, we take advantage of Lemma 1, and consider only ERL-clauses in which Skolem functions have been eliminated.

4.2 The algorithm for satisfiability checking

Our algorithm for satisfiability checking is based on a variant of the resolution. Resolution is an inference rule that has been shown to be sound and complete for the clausal form of logic (see [CL73]): a resolution refutation (i.e. a derivation of the empty clause obtained by repeated applications of resolution steps) exists for a set of clauses if and only if such a set is unsatisfiable. Positive unit resolution (see [CL73]) is a special form of resolution, requiring every step to be applied to at least one positive unit clause, i.e. a clause containing just one positive literal. For example, a positive unit resolution step can be applied to A(c) and (¬A(x) ∨ B(x)), yielding the positive clause B(c).

Definition 3 Let C be a set of ERL-clauses. A positive derivation chain (or simply a chain) of C from P(c) to Q(d) is a sequence $<\sigma_1,...,\sigma_n>$ of literals, with $\sigma_i \neq \sigma_j$ for $i \neq j$, where σ_1 is P(c), σ_n is Q(d), and for each i (i=2,...,n), σ_i is obtained by positive unit resolution from a clause of C of the form $(\neg\Phi_1(w_1) \vee \Phi_2(w_2))$ and from

σ_{i-1}; *n* is called the length of the chain $<\sigma_1,...,\sigma_n>$.

The following theorem establishes an important property of ERL-clauses, namely that we can check any set C of ERL-clauses for unsatisfiability, by looking for particular positive derivation chains of C.

Theorem 2 A set C of ERL-clauses is unsatisfiable if and only if it includes a negative clause $\Gamma = (\neg\Phi_1(\mathbf{w}_1) \vee \neg\Phi_2(\mathbf{w}_2))$ of C such that there is a chain of C from $\psi_1(\mathbf{c}_1)$ to $\Phi_1(\mathbf{d}_1)$, and a chain of C from $\psi_2(\mathbf{c}_2)$ to $\Phi_2(\mathbf{d}_2)$, where $\psi_1(\mathbf{c}_1)$ and $\psi_2(\mathbf{c}_2)$ are positive unit clauses of C, and $(\neg\Phi_1(\mathbf{d}_1) \vee \neg\Phi_2(\mathbf{d}_2))$ is a ground instance of Γ.

We are now in a position to define the concept of graph associated with a set of ERL-clauses, which is used by our algorithm for satisfiability checking.

Definition 4 Given a set of ERL-clauses C, we associate with it a directed graph G(P,D,I,A), called ERL-graph, labeled on nodes and arcs. The set of nodes is partitioned into three subsets, called P, D, and I; A is the set of arcs. P-nodes, I-nodes, and D-nodes are called predicate nodes, implication nodes, and disjointness nodes, respectively. The ERL-graph G(P,D,I,A) is defined according to the following rules:

1. for each predicate Q of C define a P-node, labeled with Q;
2. for each clause of the form $(\neg\Phi(\mathbf{w}))$, define a D-node d and an arc $e=<\Phi,d>$, labeled with $\mathbf{u}=<1,...,n>$, where $n=|\mathbf{w}|$;
3. for each clause of the form $(\neg\Phi_1(\mathbf{x}) \vee \neg\Phi_2(\mathbf{y}))$, define a D-node d and arcs $e_1=<\Phi_1,d>$, and $e_2=<\Phi_2,d>$; the label $\mathbf{u}$ of e_1, and the label $\mathbf{v}$ of e_2 are defined as follows: $\mathbf{u}$ is constituted by all the variables of $\mathbf{x}$ that appear in $\mathbf{y}$, whereas $\mathbf{v}$ is such that $|\mathbf{v}| = |\mathbf{u}|$, and for each i, $\mathbf{v}^i = j$, where j is the integer such that $\mathbf{y}^i = \mathbf{x}^k$, where $k=\mathbf{u}^i$.
4. for each clause of the form $(\neg\Phi_1(\mathbf{x}) \vee \Phi_2(\mathbf{y}))$, define an I-node m and arcs $<\Phi_1,m>$, labeled with $\mathbf{u}$, and $<m,\Phi_2>$, labeled with $\mathbf{v}$; $\mathbf{u}$ and $\mathbf{v}$ are defined as follows:

$\mathbf{u} = <1,...,n>$, where $n = |\mathbf{x}|$;

$\mathbf{v} = <v^1,...,v^p>$, where $p = |\mathbf{y}|$, and, for each i:

$$v^i = \begin{cases} j & \text{if } \mathbf{y}^i = \mathbf{x}^j; \\ z_h{}^k & \text{if } \mathbf{y}^i = z_h{}^k. \end{cases}$$

Example 2 The ERL-graph associated with the set of clauses of example 1 is shown in figure 1.

For an arc e, head(e) denotes the reaching node of e; for example, if e=<n,m>, then head(e)=m.

We now present an algorithm whose purpose is to store the information about the derivation chains of C (where C is a set of ERL-clauses) from a given literal P(c).

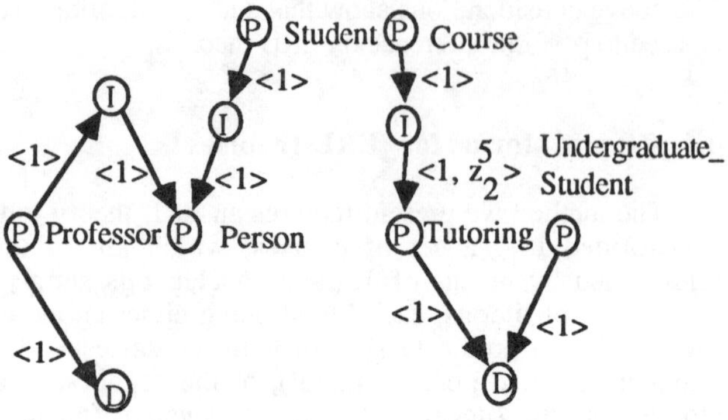

Figure 1

We assume that associated with each P-node Q of an ERL-graph, there is a set of tuples, denoted by TERM_SET(Q). The goal of the algorithm is to insert the tuple **d** into TERM_SET(Q) whenever there is a chain of C from P(**c**) to Q(**d**). The algorithm makes use of a function *select* which, given a constant tuple **c**, and an arc e, with label(e) = **u**, returns a new constant tuple **r** of the same arity as **u**, according to the following rule; select(**c**,e) = **r** where:

$$\mathbf{r}^i = \begin{cases} c^j & \text{if } \mathbf{u}^i = j \text{ (j integer)} \\ \mathbf{u}^i & \text{if } \mathbf{u}^i \text{ is a constant} \end{cases}$$

Algorithm chain(G,**c**,P)
Input an ERL-graph G corresponding to a set C of ERL-clauses, a function TERM_SET, a constant tuple **c**, and a node P of G
Output the algorithm inserts **d** in TERM_SET(Q), for each **d** and Q such that there is a chain of C from P(**c**) to Q(**d**)

begin
  if P is a P-node and $\mathbf{d} \notin$ TERM_SET(P)
  then *begin*
    insert **c** into TERM_SET(P)
    for each outgoing arc e of P
    do chain(G, select(**c**,e), head(e))
    end
  else *if* P is an I-node with outgoing arc e
    then chain(G, select(**c**,e), head(e))
end

Theorem 3 shows that algorithm *chain* is correct.

Theorem 3 Let C be a set of ERL-clauses, G the associated graph, and P,Q two of its P-nodes. During the execution of chain(G,**c**,P), **d** is inserted into TERM_SET(Q), if and only if there is a chain of C from P(**c**) to Q(**d**).

The algorithm *chain* is the basis of our method for checking a set C of ERL-clauses for unsatisfiability. Given the ERL-graph G, we set TERM_SET(P) to empty, for each P-node of G. Then we execute the algorithm *chain* for each **c** and P such that P(**c**) is a clause of C. Finally, we check a special condition on the D-nodes of G, as specified by the following algorithm.

Algorithm Unsatisfiable?(C)
Input a set C of ERL-clauses
Output true, if C is unsatisfiable, false otherwise
begin
construct the ERL-graph G associated with C
for each P-node P *do* set TERM_SET(P) to { }
for each positive clause (P(**c**)) in C *do* chain(G,**c**,P)
if there is a D-node D of G, with ingoing arcs
$e_1=<Q_1,D>$ and $e_2=<Q_2,D>$ such that there is a
h_1 in TERM_SET(Q_1) and a h_2 in
TERM_SET(Q_2) such that
select(h_1,e_1) = select(h_2,e_2)
or
there is a D-node D of G, associated with ¬Q(**w**),
with one ingoing arc $e=<Q,D>$ such that there is
a h in TERM_SET(Q) such that select(h,e)=**w**',
where **w**' is a ground instance of **w**
then return (true)
else return (false)
end

Theorem 4 Let C be a set of ERL-clauses. Then *Unsatisfiable?*(C) returns true if and only if C is unsatisfiable.

As far as the computational complexity of our method is concerned, the next theorem shows that algorithm *Unsatisfiable?* is polynomial in the size of the graph.

Theorem 5 Let C be a set of ERL-clauses. Algorithm *Unsatisfiable?* takes $O(E \cdot K \cdot N \cdot \log(K \cdot N))$ time, where E is the number of clauses of C, K is the number of constants of C, and N is the number of predicates of C.

5. An environment for conceptual modeling

The purpose of this section is to demonstrate that, using the language presented in Section 3 and the algorithm presented in Section 4, one can define an effective environment supporting conceptual modeling. The basic observation is that our method for satisfiability checking can be directly used to perform inferences on ERL-theories. Therefore, we can easily devise a deductive environment which allows the designer to check wether a given sentence is logically implied by an ERL-theory.

In what follows, we shall consider several crucial problems in conceptual modeling, and analyse the features provided by our approach in order to deal with these problems.

5.1 Consistency

All the methodologies and tools for conceptual modeling insist on the need for checking the consistency of the conceptual schema. However, such a concept is often defined using informal arguments, and the methodologies provide only some intuitive guidelines for achieving the consistency of the representation. In our approach, consistency can be naturally defined in terms of satisfiability. In particular, if we allow the ERL-theory to be built incrementally, the consistency of the schema can be automatically checked as follows: every time a new axiom α is added to an ERL-theory T, the system automatically checks $T \cup \{\alpha\}$ for satisfiability, and the axiom is rejected if $T \cup \{\alpha\}$ is unsatisfiable. Notice that in this case the system is also able to provide the designer with useful information about the contradictory axioms, in particular using the positive chains which lead to the contradiction (see [DL89]). In this case, the algorithm *Unsatisfiable?* is used for ensuring the global consistency of the representation.

Besides the global consistency, other consistency checks can be performed by the system. For example, an entity-relationship schema which is satisfiable may suffer from another anomaly, called class unsatisfiability ([LN87]). A class S (either an entity or a relationship) is said to be unsatisfiable in an ERL-theory T, if S has no instances in all the models of T. For example, in the following ERL-theory:

$\forall x \, \exists y \, (\neg A(x) \vee R(x,y))$
$\forall x, y \, (\neg B(x) \vee \neg R(x, y))$
$\forall x \, (\neg S(x) \vee A(x))$
$\forall x \, (\neg S(x) \vee B(x))$

the object class S is unsatisfiable, although the theory is satisfiable. In general, we can check if a class S is usatisfiable in T as follows: if C is the set of clauses corresponding to T, and z is a constant symbol not appearing in C, then S is unsatisfiable in T if and only if $C \cup \{S(z)\}$ is unsatisfiable. Notice that $C \cup \{S(z)\}$ is a set of ERL-clauses.

Finally, our environment can support prototyping facilities, which are very important for allowing the designer to directly test and verify the schema. To this end, it is easy to extend ERL with axioms of the type: $\Phi(\mathbf{c})$ and $\neg \Phi(\mathbf{c})$, where **c** is a constant tuple. Such axioms assert that **c** (does not) belongs to an entity (relationship) Φ. Clearly the resulting set of clauses is still a set of ERL-clauses, and the method presented in section 4 can be used to perform inference on such a set.

In summary, the designer can incrementally build an ERL-theory, insert instances into classes, assert that an object is not an instance of a class, and pose queries to the system anytime she/he needs to know whether a certain sentence logically follows from the defined ERL-theory.

5.2 Minimality

Another important quality of object descriptions is minimality. In our approach minimality can be formally defined in terms of derivability of axioms. An ERL-theory Σ is minimal if no axiom α of Σ can be derived from $\Sigma - \{\alpha\}$. Thus, it is possible to test if an ERL-theory Σ is minimal by simply testing if for all axioms $\alpha \in \Sigma$, $\Sigma - \{\alpha\} \cup \{\neg(\alpha)\}$ is satisfiable.

Example 3 Consider the fragment of entity relationship schema shown in figure 2, and suppose that the following properties hold for such a schema:

- the Scientific_Faculty_Enrollment relationship is a subset of the Enrollment relationship (represented by an arrow in the figure);
- Mandatory_Tutoring is a subset of Tutoring;
- relationship Tutoring is typed with Scientific_Faculty_Student in 1-component;
- Suspended_Student cannot participate in the relationship Enrollment;
- Scientific_Faculty_Student must participate in the relationship Scientific_Faculty_Enrollment (minimum cardinality equal to 1);
- Scientific_Faculty_PhD_Student must participate in the relationship Mandatory_Tutoring.

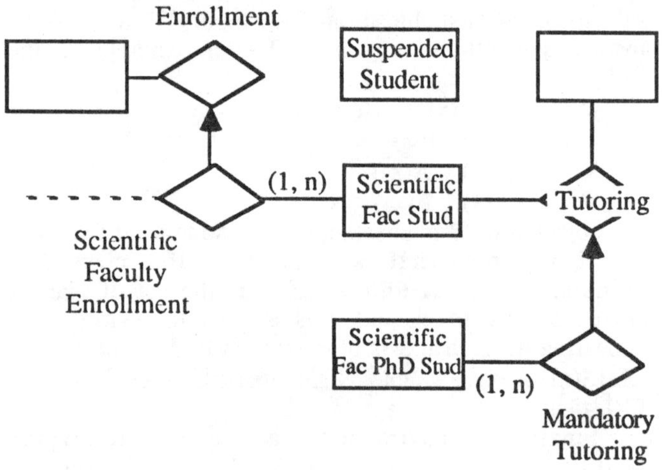

Figure 2

The schema and the above properties can be represented by the following ERL-theory T:

$\forall x_1, x_2 (\neg$ Scientific_Faculty_Enrollment$(x_1, x_2) \vee$ Enrollment$(x_1, x_2))$

$\forall x_1, x_2 (\neg$ Mandatory_Tutoring$(x_1, x_2) \vee$ Tutoring$(x_1, x_2))$

$\forall x_1, x_2 (\neg$ Tutoring$(x_1, x_2) \vee$ Scientific_Faculty_Student$(x_1))$

$\forall x_1, x_2 (\neg$ Suspended_Student$(x_2) \vee \neg$Enrollment$(x_1, x_2))$

$\forall x_1 \exists x_2 (\neg$ Scientific_Faculty_Student$(x_2) \vee$ Scientific_Faculty_Enrollment$(x_1,x_2))$

$\forall x_1 \exists x_2 (\neg$ Scientific_Faculty_PhD_Student$(x_1) \vee$ Mandatory Tutoring$(x_1,x_2))$.

Now, suppose we want to assert that the two entities Suspended_Student and Scientific_Faculty_PhD_Student are disjoint, i.e. we want to add to the theory the axiom :

$\forall x_1 (\neg$ Scientific_Faculty_PhD_Student $(x_1) \vee \neg$Suspended_Student$(x_1))$.

In order to maintain the theory minimal we have to check if such an assertion is already implied by the theory T, that is done by adding its negation to T, and checking the resulting theory T' for unsatisfiability. The negation of the above axiom is:

$\exists x_1$ (Scientific_Faculty_PhD_Student$(x_1) \wedge$ Suspended_Student(x_1)),

which in clausal form yields:

Scientific_Faculty_PhD_Student(c_1), and

Suspended_Student(c_1).

The ERL-graph corresponding to T' is shown in figure 3. Algorithm *chain* propagates the Skolem constant c_1 to the nodes Suspended_Student and Enrollment, which are connected to the disjointness node D. Therefore, the algorithm *Unsatisfiable?* returns true as result. Hence, we can conclude that the above disjointness axiom is impled by T.

Other types of minimality of the description can be considered by using ERL. For example, one may want to obtain the minimal set of subset relations which is implied by an ERL-theory T. In this case, we can first compute the set of all the subset relations implied by T, and then compute the transitive reduction of such a set. Again, this can be done by using the deductive capabilities of the system.

A further example of how minimality can be dealt with in ERL, concerns the existence of equivalent entities or relationships in the schema. Two classes A and B are said to be equivalent in T if they have the same set of instances in all the models of T. This condition can be checked by testing if both $\forall x (\neg B(x) \vee A(x))$ and $\forall x (\neg A(x) \vee B(x))$ are derivable from T.

6 Conclusions and further research

We have presented ERL, a simple object modeling language that allows one to express positive and negative assertions concerning several semantic relations among object classes.

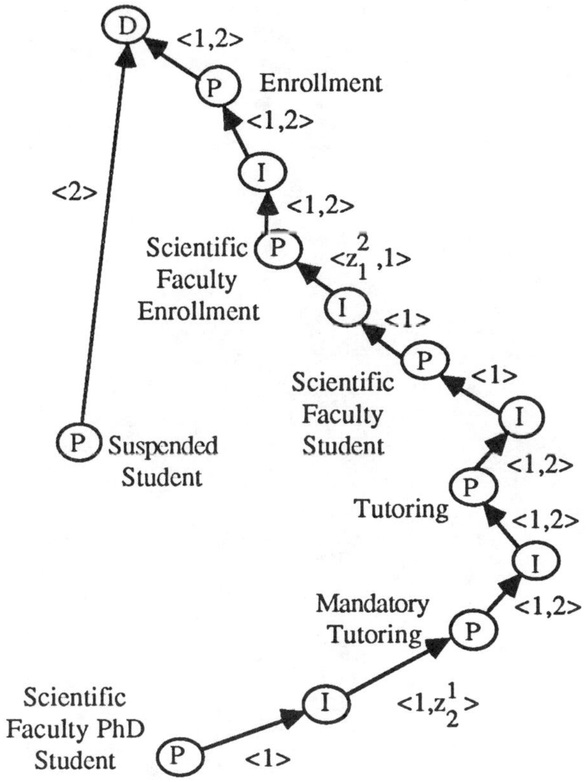

Figure 3

The semantics of such a language has been defined in terms of first order logic. A polynomial time algorithm has been provided for computing inference in ERL, which has been proven to be sound and complete with respect to the semantics. Our work extends previous results on containment inference in object-based models, by considering aggregation relations among classes. Moreover, we have shown how ERL and the associated inference capabilities can be used as a kernel for a well-founded environment for conceptual modeling.

In the future, we aim at extending the expressive power of ERL by allowing different levels of aggregation among classes, and by considering other meaningful semantic relations, with the constraint that such extensions do not affect the computational tractability of the inference problem.

References

[AM86] Arisawa H., and Miura T., "On the Properties of Extended Inclusion Dependencies", *Proc. of the 12th Very Large Data Bases (VLDB) Conference*, Kyoto, 1986.

[AS87] Atzeni P., and Stott Parker D., "Set Containment Inference", *Proc. of the International Conference on Database Theory, LNCS*, N.243, Springer-Verlag New York Inc., 1987.

[BLN86] Batini C., Lenzerini M., and Navathe S.B., "A Comparative Analysis of Methodologies for Database Schema Integration", *ACM Computing Surveys*, Vol. 18, N. 4, 1986.

[CFP84] Casanova M.A., Fagin R., and Papadimitriou C.H., "Inclusion Dependencies and Their Interaction with Functional Dependencies", *Journal of Computer and System Sciences*, 28, 1984.

[CaVi83] Casanova M.A., and Vidal V.M.P., "Towards a Sound View Integration Methodology", *Proc. of the 2nd ACM Symposium on Principles of Database Systems*, 1983.

[Ce83] Ceri S. (ed.), *Methodology and Tools for Database Design*, North Holland, 1983.

[CV85] Chandra A.K., and Vardi M.Y., "The Implication Problem for Functional and Inclusion Dependencies is Undecidable", *SIAM Journal on Computing*, Vol. 14, 1985.

[CL73] Chang C.L., and Lee R.C.T., *Symbolic Logic and Mechanical Theorem Proving*, Academic Press, New York, 1973.

[DL89] Di Battista G., and Lenzerini M., "A Deductive Method for Entity-Relationship Modeling", Tech. Rep. Dipartimento di Informatica e Sistemistica, Universita' di Roma "La Sapienza", 1989 .

[HK87] Hull R., and King R., "Semantic Database Modeling: Survey, Applications, and Research Issues", *ACM Computing Surveys*, Vol. 19, 1987.

[KCV83] Kannellakis P.C., Cosmadakis S.S., Vardi M.Y., "Unary Inclusion Dependencies Have Polynomial Time Inference Problems", *Proc. of the 15th ACM Symposium on Theory of Computer Science* , 1983.

[Le87a] Lenzerini M., "Covering and Disjointness Constraints in Type Networks", *Proc. of the 3rd IEEE International Conference on Data Engineering*, Los Angeles, Ca, February 1987.

[Le87b] Lenzerini M., "Class Hierarchies and their Complexity", *Proc. of the Workshop on Database Programming Languages* , Roscoff, 1987.

[LN87] Lenzerini M., and Nobili P., "On the Satisfiability of Dependency Constraints in Entity-Relationship Schemata", *Proc. of the 13th Conference on Very Large Data Bases* (VLDB), Brighton, UK, 1987.

[MZ86] McAllester D., and Zabih R., "Boolean Classes", *Proc. of ACM Object Oriented Prog. Systems, Languages and Applicatons (OOPSLA) Conference*, 1986.

[Me64] Mendelson E., *Introduction to Mathematical Logic*, Van Nostrand Reinhold, New York, 1964.

[TYF86] Teorey T.J., Yang D., and Fry J.P., "A Logical Design Methodology for Relational Databases Using the Extended Entity-Relationship Model", *ACM Computing Surveys*, Vol. 18, 2, 1986.

[Ul87] Ullman J.D., "Database Theory: Past and Future", *Proc. of ACM Conf. on Principles of Database Systems*, 1987.

A Family of Incomplete Relational Database Models†

Adegbemiga Ola# and Gultekin Ozsoyoglu

Department of Computer Engineering and Science, Case Western Reserve University

ABSTRACT

In this paper, we utilize intervals for unknown values in incomplete relational databases. We use tables to represent unknown relations. First, we define three partial tuple types in a table to specify incompleteness relationships among tuples of the same table. For tuples of different tables, we distinguish between the cases where incompleteness are introduced at the *relation level, tuple level* or *attribute-value level*. And, based on these relationships among tuples in different tables, we present a family of incomplete relational database models.

For each of the models, the query evaluation is *sound* (i.e., no incorrect results are derivable). None of the models is *complete* (i.e., all valid conclusions are derivable). We briefly compare two of the models in the family with other approaches.

Considering each table tuple as a set of d-dimensional cubes, each model in the family of models presented in this paper can be considered as a geometric database model. We are presently implementing a version of one of the models. We briefly summarize the geometric operations and the primitive update semantics being utilized in the implementation.

1. Introduction

Null values and incomplete information in databases have received much attention in recent years, for instance see [Codd 79, Vass 79, Lips 79, Gran 79, Gran 80, Bisk 81, Bisk 83, ImiL 84, Zani 84, AbKG 87]. Partial information in databases in the form of possible values is allowed in [Lips 79] and [Gran 80]. Imilienski and Lipski [ImiL 84] give conditions that ensure *soundness* and *completeness* of query evaluation in a certain sense, as opposed to the correctness of individual operators.

In this paper, we introduce a family of incomplete information models for the relational model. The main characteristic of all the models in this family is that the partial knowledge about an unknown value is specified as possible values in a set of intervals rather than an arbitrary set. For example, the unknown tuple t=(5, 9) is represented by the tuple, say, ([1,10], 9), in which the first component is the interval [1,10] containing the unknown value 5 in tuple t. Figure 1 illustrates tuples p_1=([3,5], [8,10], [3,5]) and p_2=([10,12], {[13,15],[20,22]}, [8,10]) of table U with scheme U(A_1, A_2, A_3). The points or tuples in the set of cubes (also called d-rectangles) are called *candidate tuples* for the unknown tuple t, and exactly one of the points is the unknown tuple. This approach allows database operations to be transformed into operations in Computational Geometry, leading to efficient operator evaluations [OlaO 88a].

Following [ImiL 84] we use the terminology that a table in the incomplete database environment *represents* a relation some tuples of which are unknown. Tables contain *partial tuples* (i.e., tuples with incomplete components) as well as total tuples (i.e., tuples whose components are all known).

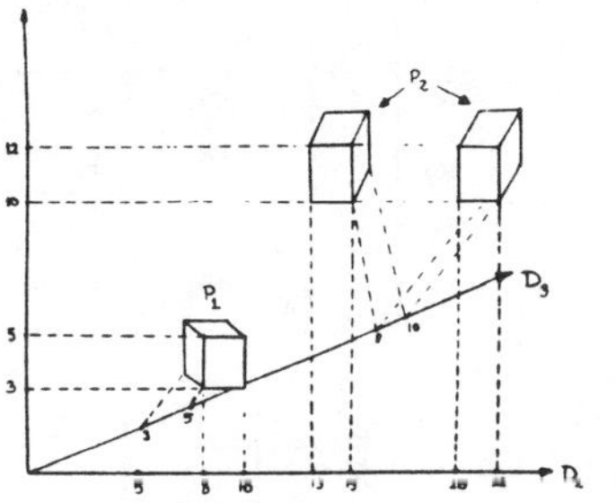

Figure 1. Geometric View of Partial Tuples

Table Tuple Type	Semantics of the Represented Relation tuple
type 0	Total tuple
type 1	Exists and Is-uniquely-represented
type 2	Exists and May-be-represented-already
type 3	May-not-exist

Figure 2. Semantics of Tuple Types

We now define new tuple types. A partial tuple consists of two or more tuples, exactly one of which is the unknown tuple. Let a table *U represent* a relation r, then the following tuple types are allowed in U.

(a) a *type 0 total* tuple is the usual relation tuple without unknown values.

(b) a *type 1 partial* tuple in *U* represents an unknown tuple t which *exists,* and no other tuple in U represents t.

(c) a *type 2* partial tuple p in *U* represents an unknown tuple t of r which is known to *exist.* However, another tuple p' of U, physically distinct than p, may also represent t. That is, p may represent a tuple in r which is already represented by some other tuple in U.

(d) a *type 3* partial tuple p in U represents an unknown tuple in r that may exist (i.e., a *maybe* tuple); the set of possible tuples for p includes the *null tuple* (a special tuple Ø denoting

† This research is supported by the National Science Foundation under Grants DCR-8605554 and IRI-8811057.

The present address of A. Ola is Department of Computer Science, University of Dayton, Dayton, OH 45469.

the nonexistent tuple).
Figure 2 summarizes the different tuple types. We give an example.

Example 1: Let us consider relation r in figure 3, consisting of four tuples t_1, t_2, t_3 and t_4. If the last digits of the second components in t_1, t_2 and t_4 are unknown then r is represented by table U. The partial tuples in U are type 1 because they represent tuples which are known to be distinct. Table V is the projection on attribute A_2 of U. The tuples p_1' and p_2' are type 1 in the projection because the unknown tuples represented by p_1 and p_2 of U must have distinct A_2 values. Tuple p_3' is of type 2 because the A_2 value of the unknown tuple represented by p_3 may be the same as that of tuple t_1. Table W is the result of applying a selection formula $A_2 > 55$ to table V; both tuples p_1' and p_2' have a chance of satisfying the formula, and are therefore included in W as type 3.

r

	A_1	A_2
t_1	4	53
t_2	4	56
t_3	5	41
t_4	3	41

U

	A_1	A_2	TYPE
p_1	4	[50,59]	1
p_2	4	[50,59]	1
t_1	5	41	0
p_3	3	[40,49]	1

V

	A_2	TYPE
p_1'	[50,59]	1
p_2'	[50,59]	1
t_1'	41	0
p_3'	[40,49]	2

W

A_2	TYPE
[50,59]	3
[50,59]	3

Figure 3. Relation r, Table U, and Operations on Tables.

Clearly, as the type of a table tuple increases from 0 to 3, it becomes less informative. For example, a type 1 tuple is more informative than a type 3 tuple. In a practical environment, users may perhaps be interested in only type 0 and type 1 tuples. However, even if the base tables in a relational database contains only type 0 and type 1 tuples, the relational algebra operators introduce type 2 and type 3 tuples.

The semantic difference between type 2 and type 3 tuples is as follows. Consider a type 1 tuple p of a table U. When a set union of U with another table is performed, p in the resulting table may represent the same unknown relation tuple (i.e., p becomes type 2); but p will not become a type 3 (i.e., a maybe) tuple. On the other hand, as illustrated in example 1, when a selection on U is performed, the unknown tuple represented by p may or may not be selected in which case p becomes a type 3 (i.e., a maybe) tuple. Also, n number of type 2 tuples in a table represent at least one tuple in the corresponding unknown relation whereas n number of type 3 tuples may not represent any tuples in the corresponding relation.

By taking into consideration the known relationships among tuples from different tables of the database one can derive more information in query responses. Such relationships vary from one environment to the other. In this paper, we identify various assumptions about relationships among tuples in different tables. Then, a family of extended models (*M-1, M-2, M-3*, and *M-4*) based on these assumptions is introduced. This approach, we believe, is practical because we are able to put at the users' disposal the variety of models, one of which possibly meets the users' assumptions. We now describe the family of models.

In model *M-1*, we model an environment where errors or incompleteness are introduced at the **relation level** meaning that the same unknown tuple may be represented by different tuples in different tables. Thus, within the database, different errors may occur in different occurrences of a given tuple in different relations.

Basic Assumption in Model M-1 (*Relation Level*):
Consider two tuples z_1 and z_2 occurring in different tables. If the candidate tuples of z_1 and z_2 do not intersect then z_1 and z_2 represent different (relation) tuples, otherwise they may represent the same (relation) tuple.

Example 2: In figure 4, relation r_2 with tuple t_1 contains no errors, and is represented "as-is" as a table (with the addition of the TYPE column). On the other hand, the first digit of the second component of tuple t_1 of r_1 is unknown, and r_1 is represented by table U. Tuple t_1 of r_1 is represented by the partial tuple p_1 in U. But the occurrence of t_1 in r_2 is total. The inference we can make is that two tuples from different tables do **not** represent the same relation tuple if at least one of their tuple components do not intersect.

r_1

	A_1	A_2
t_1	4	53
t_2	4	56

U

	A_1	A_2	TYPE
p_1	4	[13,93]	1
t_2	4	56	0

r_2

	A_1	A_2
t_1	4	53
t_3	4	55

Figure 4. Table Interpretations in Model *M-1*

M-2 models an environment where errors or incompleteness are introduced at the level of attribute values at a source, before the tuples are inserted into different tables in the database. Hence the same identifier can be attached to the same unknown value wherever it occurs in the database.

Basic Assumption in Model M-2 (*Attribute Value Level*):
An unknown value has a unique identifier. Different occurrences of the same unknown value in the database have the same identifier and the same range value. If two identifiers for two unknown values τ_1 and τ_2 are the same then $\tau_1 = \tau_2$; otherwise τ_1 may or may not be equal to τ_2.

In *M-2*, we have the advantage that duplicate unknown tuples can be identified in base tables using identifiers, resulting in inexpensive Union and Difference operations. Clearly, in such an environment, the base tables of the database contain only total and type 1 tuples.

In *M-3*, errors are introduced at the attribute-value level as in *M-2*, but any two different identifiers are known to represent different values.

Basic Assumption in Model M-3 (*Attribute Value Level*):
Two identifiers for two unknown values τ_1 and τ_2 are the same iff $\tau_1 = \tau_2$.

M-4 models an environment where errors are introduced at the **tuple level** before tuple insertions are made into the various tables of the database. A given tuple with unknown values is represented across the database by the same partial tuple. However, identifiers are not attached to unknown attribute values appearing in relations with different schemes. Rather, identifiers are attached to tuples.

Basic Assumption in Model M-4 (*Tuple Level*):
Two table tuples in different tables represent the same unknown (relation) tuple iff they have the same identifiers.

In each of the models, we extend the five basic relational algebra (RA) operators. As a rule, we retain the semantics of the regular RA operators for total tuples, and extend them for partial tuples. All of the extended operators in our models are *faithful* [Maie 83] in the sense that they reduce to the usual RA operators when the tables consist of only total tuples. We show for each of the models that our extension is *sound* (i.e., no incorrect results are derivable when an RA expression is evaluated) in the Imilienski-Lipski sense [ImiL 84]. However, query evaluation is not always complete; that is, some valid conclusions may not be derivable. In [OlaO 88b], we present a version of the model M-2 in which RA operators are restricted to the cases where the query evaluation is sound and complete.

The rest of the paper is arranged as follows. Section 2 gives the terminology and definitions. Section 3 discusses the correctness notions for the extended models. In section 4, models *M-1* and *M-2* are presented. In section 5, we compare models *M-1* and *M-2* in our family of models with other approaches. Section 6 discusses the common geometric operations needed in RA operator implementations of all the models in the family. We are presently implementing a version of model M-1 in a main-memory database system [OlaO 88a, She 88]. Section 7 briefly discusses the parameters being measured in the implementation.

2. Terminology and Definitions

Let *tuples*(p) denote the set of total tuples contained in a partial tuple p, and t_p be the unknown tuple represented by p. t_{p_i} represented by a type 1 tuple p_i is different from any total tuple or t_{p_j}, i≠j, for p_j in the same table. Relational algebra operators for tables (as opposed to relations) are superscripted by *.

Tables are denoted as U_j's, and relations are denoted as r_j's. The letters t and p, with or without subscripts, denote total and partial tuples, respectively. The letter z is used to refer to either total or partial tuples. And, the predicate that the tuple z is in table U is written as T^U(z), T_1^U(z), T_2^U(z) or T_3^U(z) when z is total, type 1, type 2 or type 3, respectively. U_T, U_{T_1}, U_{T_2}, U_{T_3}, respectively, denote the collection of total, type 1, type 2 or type 3 tuples in table U. Different combinations of the numbers 0, 1, 2 and 3 as subscripts of T refer to a combination of tuple types; for instance, $U_{T_{023}}$ denotes total, or type 2 or type 3 tuples in U.

U_{T_1}, U_{T_2}, U_{T_3} are *multisets* (i.e., sets with duplicates). Therefore, we use the predicate that two tuples z_1 and z_2 in a table U with possibly identical tuple components are both "physically" kept in U, written as $z_1 \neq_p z_2$. The usual equality predicate, =, holds (does not hold) between any two distinct tuples with identical (nonidentical) tuple components.

z[Y], the projection of a type 1 or type 2 tuple z on a set of attributes Y, is total, written as T(z[Y]), if there are no unknown values in the Y-values of z. However, z[Y] for a type 3 tuple z is always type 3.

The *intersection of two tuples* z_1 and z_2, written as $z_1 \cap z_2$, consists of those tuples occurring in both *tuples*(z_1) and *tuples*(z_2). z_1 and z_2 are said to *intersect* if $z_1 \cap z_2 \neq \emptyset$. Given a type 1 tuple p_1 in U, the *candidate tuples* for t_{p_1} consists of tuples in *tuples*(p_1) that are not in U_T.

Given tables U and V, the *effective intersection*, written as $p_1 \tilde{\cap} p_2$, between a type 1 tuple p_1 in U
(a) and a type 1 tuple p_2 in V is the intersection of the candidate tuples of p_1 and p_2,
(b) and a type 2 or type 3 tuple p_2 in V consists of the tuples in *tuples*$(p_1) \cap$ *tuples*(p_2) that are *not* in U_T,
(c) and a total tuple t in V is t if t ∈ *tuples*(p_1) and t is not in U_T, otherwise the effective intersection is empty.
The effective intersection of tuples z_1 and z_2 has the effect of eliminating from the intersection $(z_1 \cap z_2)$ those tuples that can not possibly be t_{z_1} or t_{z_2}.

The *information content* of a table is defined [ImiL 84] by the mapping *rep* which maps a table U to *rep*(U), the set of possible relations for the unknown relation represented by U. An example of the *rep* mapping for the model *M-1* is given in example 4. Notions similar to *rep* are also used in [Gran 80] and [Bisk 83]. Let r be a relation in *rep*(U) of table U; a tuple z in U has a corresponding tuple, denoted as (t_z)r, in r. We call (t_z)r the *representative tuple* of z in r, or z is said to be mapped in r to t_z.

3. Correctness Notion

For each of the models in section 4, the soundness and completeness of a query is examined using revised versions of concepts from [ImiL 84].

Definition 1 : Let U_1 and U_2 be two tables. Let α and θ be unary and binary operators, respectively, i.e., $\alpha \in \{\pi, \sigma\}$ and $\theta \in \{-, \cup, \bowtie\}$. Let α* and θ* be the extended versions of α and θ, respectively.
(a) An extended model is *sound* if, for every relation r in $(rep(U_1)\ \theta\ rep(U_2))$ or $\alpha\ (rep(U_1))$, there is a relation s in $rep(U_1\ \theta^*\ U_2)$ or $rep(\ \alpha^*\ (U_1))$, respectively, such that $s \subseteq r$.
(b) An extended model is *complete* if, for every relation s in $rep(U_1\ \theta^*\ U_2)$ or $rep(\ \alpha^*\ (U_1))$, there is a relation r in $(rep(U_1)\ \theta\ rep(U_2))$ or $\alpha\ (rep(U_1))$, respectively, such that r

$\subseteq$ s.
$rep(U_1)\ \theta\ rep(U_2)$ is understood to be $\{r_1\ \theta\ r_2 \mid r_1 \in rep(U_1)$ and $r_2 \in rep(U_2)\}$, while $\alpha(rep(U)) = \{\alpha(r) \mid r \in rep(U)\}$.

Clearly, in definition 1, $rep(U_1\ \theta^*\ U_2)$ is the "imperfect" representation that attempts to capture the "ideal" representation $rep(U_1)\ \theta\ rep(U_2)$. One may also think of alternative soundness criteria. Let us look at one.

Definition 2 : An extended model is sound* if, for every relation s in $rep(U_1\ \theta^*\ U_2)$ or $rep\ (\alpha^*(U_1))$, there is a relation r in $(rep(U_1)\ \theta\ rep(U_2))$ or $\alpha\ (rep(U_1))$, respectively, such that $s \subseteq r$.

Definition 2, however, is not acceptable. We give an example.
Example 3: Let a given α and α^* be such that $\alpha(rep(U))$ and $rep(\alpha^*(U))$ consist of the relation sets $\{r_1, r_2\}$ and $\{s_1, s_2, s_3\}$ in figure 5 and figure 6, respectively.

r_1

A_1	A_2
1	2
3	4
5	6
6	7

r_2

A_1	A_2
1	2
4	5

Figure 5. An Example $\alpha(rep(U))$

s_1

A_1	A_2
1	2
3	4

s_2

A_1	A_2
1	2
5	6
3	4

s_3

A_1	A_2
1	2
3	4
6	7

Figure 6. An Example $rep(\alpha^*(U))$

The $\alpha\ (rep(U))$ and $rep(\ \alpha^*\ (U))$ definitions in figures 5 and 6 do satisfy definition 2. One can also observe that when the unknown relation represented by U is r_1, all three "imperfect" relations s_1, s_2, and s_3 are contained in the unknown relation r_1, which is desirable. However, when the unknown relation represented by U is r_2, then none of s_1, s_2, and s_3 is contained in r_2, and definition 2 does not detect this problem. In contrast, the soundness criteria of definition 1 is violated in this example.

The condition (a) in definition 1 when considered separately from (b) ensures the soundness of query evaluation. However, consider the case in which $rep(U_1\ \theta^*\ U_2)$ or $rep(\ \alpha^*\ (U))$ contain an empty relation (This may occur, for example, in a set difference operation). In such a case, the soundness criterion of definition 1 will be trivially satisfied. Thus, whenever condition (a) is satisfied for a given extended operator, we have to see to it that the operator definition ensures a *nontrivial soundness*. Therefore, in the rest of the paper, we use the following definition of soundness.

Definition 3: An extended model is *sound* if, for every relation r in $(rep(U_1)\ \theta\ rep(U_2))$ or $\alpha\ (rep(U_1))$, there is a relation s in $rep(U_1\ \theta^*\ U_2)$ or $rep(\ \alpha^*\ (U_1))$, respectively, such that s = r. That is, $rep(\alpha^*(U)) \supseteq \alpha(rep(U))$ and $rep(U_1\ \theta^*\ U_2) \supseteq rep(U_1)\ \theta\ rep(U_2)$.

An operation α^* that satisfies definition 3 is said to be *adequate* for α [Maie 83]. Therefore, if each of the extended operator of a given model is adequate, we have a nontrivial soundness (for the query). Thus, we would like every extended operator in our models to be adequate. We can also observe that definition 3 implies condition (a) of definition 1. Let us denote the condition (b) of definition 1 by $rep(U_1)\ \theta\ rep(U_2) \subseteq^* rep(U_1\ \theta^*\ U_2)$.

In the next section, for each binary operator θ , we examine the relationship that holds between $rep(U_1\ \theta^*\ U_1)$ and $rep(U_1)\ \theta\ rep(U_2)$ with respect to definition 3 and condition (b) of definition 1. Similarly, for each unary operator α, the relationship between $rep(\alpha^*(U))$ and $\alpha(rep(U))$ is examined.

4. Family of Extended Models

Based on the basic assumptions, we now define the extended model *M-1*. Due to space considerations, we briefly describe *M-2*, and omit the discussions of models *M-3* and *M-4*. The basic assumptions define the *rep* mapping; the mappings in models *M-1* and *M-2* are denoted as rep_1 and rep_2, respectively. In defining a *rep*, we use the following version of Reiter's [Reit 78] *Closed World Assumption* in [Bisk 83]: If a total tuple t is not in U_T and there is no partial tuple p in U and relation r in $rep(U)$ such that $(t_p)r$ = t, then t is not in the unknown relation represented by U.

4.1. The Model M-1

In model *M-1* we can not always recognize different occurrences of the same tuple in different tables. The relationship between any two tuples is defined solely by the intersection of their candidate tuples.

Definition (*rep* in *M-1*):

$rep_1(U) = \{r \mid (\forall t)(T^U(t) \rightarrow t \in r)$

$\wedge\ (\forall p)(T_1{}^U(p) \rightarrow (\exists t_1)(t_1 \in tuples(p) \wedge t_1 \in r\ \wedge$

$(\forall z)((z \in U \wedge z \neq_p p) \rightarrow t_1 \neq (t_z)_r)))$

$\wedge\ (\forall p)(T_2{}^U(p) \rightarrow (\exists t_1)(t_1 \in tuples(p) \wedge t_1 \in r))$

$\wedge\ (\forall p)(T_3{}^U(p) \rightarrow (\exists t_1)((t_1 \in tuples(p) \wedge t_1 \in r)$

$\vee\ (t_1 = \varnothing \wedge t_1 \notin r)))\}$

$rep_1(U)$ for a table U is defined independent of other tables in the database. Also, $rep_1(U)$ consists of a finite number of relations because (a) the set of candidate tuples for an unknown tuple is finite, and (b) the Closed World Assumption is used.

Example 4: In figure 7, $rep_1(U) = \{r_1, r_2, r_3\}$. In both relations r_1 and r_3, the type 3 tuple in U is mapped either to the null tuple or to tuple (1 , 2).

In order to define the *Union* and *Difference* operators of the extended relational algebra, we have to answer the question of when a tuple in one table is a duplicate in another one. The following *type-based membership rules* are used for that

U

A_1	A_2	TYPE
1	2	0
1	[2,4]	1
1	[2,3]	3

r_1

A_1	A_2
1	2
1	3

r_2

A_1	A_2
1	2
1	4
1	3

r_3

A_1	A_2
1	2
1	4

Figure 7. Illustration of rep_1(U)

purpose.

Rule 1: Given a table U, a total tuple t in table V is
(a) a *duplicate* in U if there is a t′ in U_T such that t=t′,
(b) *not a duplicate* in U if there is no tuple z in U such that z∩t=t,
(c) *may be a duplicate* in U if (a) and (b) do not hold.
Rule 2: Given a table U and a type 1 tuple p in table V, the unknown tuple t_p represented by p is
(a) *a duplicate* in U if all the candidates for t_p are total tuples in U,
(b) *not a duplicate* in U if for all tuples z in U, p $\hat{\cap}$ z = ∅,
(c) *may be a duplicate* in U if (a) and (b) do not hold.
Rule 3: Given a table U and a (type 2 or type 3) tuple p in V, the unknown tuple t_p represented by p is
(a) *a duplicate* in U if all the tuples in *tuples*(p) are in either U_T or V_T,
(b) *may be a duplicate* in U if there is a tuple in *tuples*(p) that is not a total tuple in either U or V.

We now discuss the extended versions of the relational algebra operators in the model *M-1*. Formal definitions are given in [OlaO 88b].

A. Union

The union of two tables U and V, denoted as U ∪* V, consists of
(a) the set of total tuples in either V or U,
(b) type 1 tuples p such that
(i) T_1^U(p) holds, and t_p is *not a duplicate* in V, or
(ii) T_1^V(p) holds, and t_p is *not a duplicate* in U,
(c) type 2 tuples p such that
(i) T_1^U(p) or T_2^U(p) hold, and t_p *may be a duplicate* in V,
(ii) T_1^V(p) or T_2^V(p) hold, and t_p *may be a duplicate* in U,
(d) type 3 tuples p such that T_3^U(p) or T_3^V(p) hold, and t_p *may be a duplicate* in V or U, respectively.

Please note that the union operator in *M-1* does not migrate tuples into the type 3 class. That is, if a tuple of U or V is not type 3 then it will not be converted into a type 3 tuple in the union.

Example 5: Figure 8 illustrates the union operation. In the union, each of the three partial tuples in U and V is changed to a type 2 tuple because we can not express the fact that t_{p_3} may be equal to either t_{p_1} or t_{p_2} but not both. This information loss is one source of incompleteness in query evaluation in model *M-1*.

U

	A_1	A_2	TYPE
p_1	2	[1,5]	1
p_2	2	[2,6]	1
	2	4	0

V

	A_1	A_2	TYPE
	2	4	0
p_3	2	[3,6]	1

U ∪* V

A_1	A_2	TYPE
2	4	0
2	[1,5]	2
2	[2,6]	2
2	[3,6]	2

s

A_1	A_2
2	4
2	3

Figure 8. The Extended Union

Remark: There exist tables U and V such that rep_1(U) ∪ rep_1(V) ⊄* rep_1(U ∪* V).
Proof: Using the tables in figure 8, relation s is in rep_1(U ∪* V), but there are no relations $r_1 \in rep_1$(U) and $r_2 \in rep_2$(V) such that s ⊇ $r_1 \cup r_2$ because there are at least three tuples in any r ∈ rep_1(U). *Q.E.D.*

Union fails to satisfy the completeness because we can not identify those type 1 tuples that originate from the same table and are changed to type 2 in the union.

B. Difference

The difference of tables U and V, denoted as U −* V, consists of
(a) the set of total tuples t in U such that t is *not a duplicate* in V,
(b) type 1 tuples p such that T_1^U(p) holds, and t_p is *not a duplicate* in V,
(c) type 2 tuples p such that T_2^U(p) holds, and, for all tuples z in V, z ∩ p = ∅ .
(d) type 3 tuples z such that $T^U{}_{0123}(z)$ holds, and t_z *may be a duplicate* in V.
Please note that some tuples of U that are not type 3 may become type 3 in the difference. That is, difference operation may create a migration of tuples into the type 3 class of tuples.

Example 6: Figure 9 illustrates the difference. Again, we do not have any way of specifying that at most one of the two type 1 tuples in U can represent the same unknown tuple as the type 1 tuple in V. Hence both type 1 tuples in U are changed to type 3 in the difference. This example also illustrates the semantic difference between type 2 and type 3 tuples: the type 3 tuples in the difference may not exist; thus, they cannot possibly be type 2 tuples in the difference.

Remark: There exist tables U and V such that rep_1(U) − rep_1(V) ⊄*rep_1(U −* V).

U	A1	A2	TYPE
	1	2	0
	2	[2,4]	1
	2	[2,6]	1
	1	[2,6]	2

V	A1	A2	TYPE
	2	[2,4]	1

U −* V	A1	A2	TYPE
	1	2	0
	2	[2,4]	3
	2	[2,6]	3
	1	[2,6]	2

Figure 9. The Extended Difference

C. Projection

The tables in figure 10 illustrate the projection operation.

U	A1	A2	TYPE
	[2,3]	[2,3]	1
	[2,3]	[2,3]	1
	[2,3]	[2,3]	1
	5	[4,7]	1
	5	[5,8]	3

$\pi^*_{A_2}(U)$	A2	TYPE
	[2,3]	2
	[2,3]	2
	[2,3]	2
	[4,7]	1
	[5,8]	3

Figure 10. The Extended Projection

$p[A_2]$, the projection of a type 1 tuple p in table U remains as type 1 in $\pi^*_{A_2}(U)$ if $p[A_2]\cap z[A_2] = \emptyset$ for all the tuples z in U or whenever $p[A_2]\cap z[A_2] \neq \emptyset$, p and z have equal nonnull A_1 values. In figure 10, only the tuple (5 , [4,7]) satisfies this condition. The intervals [4,7] and [5,8] intersect, but the unknown tuples represented by tuples (5 , [4,7]) and (5 , [5,8]) -if they exist- can only have A_1 value of 5, and thus different A_2 values. Projection on type 2 and type 3 tuples remain as the same type. We now formalize these observations.

Let R(Z) be the scheme for table U such that $X \subset Z$. We say that two tuples z_1 and z_2 are *X-indistinguishable* if $tuples(z_1[X]) = tuples(z_2[X])$ and $tuples(z_1[X])$ has a cardinality of 1.

Please note that a tuple z_1 and a type 3 tuple z_2 can be X-indistinguishable even though $z_2[X]$ is a type 3 tuple (not a total tuple). Also please note that z[Y], the projection of tuple z on the Y attributes, is total (written as T(z[Y])) if z is either total, type 1 or type 2, and z[Y] does not contain unknown values. However, a projection on a type 3 tuple cannot be total.

Let X = (Attributes of U − Y), then the extended projection of U on Y, denoted as $\pi^*_Y(U)$, consists of
(a) total tuples z[Y] such that T(z[Y]) holds - z can be total, type 1 or type 2,
(b) type 1 tuples z[Y] such that $T_1{}^U(z)$ holds and whenever $z_1[Y]\cap z[Y] \neq \emptyset$, $z_1 \in U$, z and z_1 are X-indistinguishable,
(c) type 2 types z[Y] such that
(i) $T_1{}^U(z)$ holds, $z[Y]\cap z_1[Y] \neq \emptyset$, and z and z_1 are not X-indistinguishable, for a tuple z_1 in U, or
(ii) $T_2{}^U(z)$ holds, and there is a tuple in z[Y] which is not in $(\pi^*_Y(U))_T$.
(d) type 3 tuples z[Y] such that $T_3{}^U(z)$ holds, and there is a tuple in z[Y] which is not in $(\pi^*_Y(U))_T$.

Remark: There exists a table U such that $rep_1(U)^{\pi_Y} \not\subseteq^* rep_1(\pi^*_Y(U))$, where $rep_1(U)^{\pi_Y} = \{r \mid r_1$ is a relation in $rep_1(U)$ and $r = \pi_Y(r_1)\}$.

The information loss in the projection operation of *M-1* occurs because the combinatorial implications arising from having finite candidate values for an unknown value are too expensive to be incorporated into the definition of projection operation.

D. Equi-join

The equi-join is illustrated in figure 11.

U	A1	A2	TYPE
	3	[2,5]	1
	2	2	0
	[6,9]	6	2

V	A2	A3	TYPE
	2	[2,4]	1
	6	[5,6]	2
	2	6	0

U ⋈* V	A1	A2	A3	TYPE
	3	2	[2,4]	3
	3	2	6	3
	2	2	[2,4]	1
	2	2	6	0
	[6,9]	6	[5,6]	2

s	A1	A2	A3
	3	2	2
	2	2	3
	6	6	5
	2	2	6

Figure 11. The Extended Equi-Join

In general, let $R_1(XY)$ and $R_2(XZ)$, $X\cap Y=\emptyset$ and $X\cap Z=\emptyset$, be the schemes for tables U and V, respectively. For any two tuples z_1 in U and z_2 in V such that $T(z_1[X])$ and $z_1[X]=z_2[X]$ hold, the equi-join U ⋈* V contains z_1 concatenated by z_2. And whenever $z_1[X]\cap z_2[X]\neq\emptyset$ and at least one of $z_1[X]$ and $z_2[X]$ is partial, there is a possibility that t_{z_1} and t_{z_2} are "joinable".

Let z_1 be in U and z_2 be in V. The equi-join of U and V, denoted as U ⋈* V, is defined using z_1 and z_2 as follows:
(a) If $T(z_1[X])$ holds and $z_1[X]=z_2[X]$, a concatenation of z_1 and z_2 is made; the resulting tuple of the equi-join is
(i) total if both z_1 and z_2 are total,
(ii) type 1 if one of z_1 and z_2 is type 1, and the other is total or type 1,
(iii) type 2 if either of z_1 or z_2 is type 2.
(b) If at least one of $z_1[X]$ and $z_2[X]$ is partial (i.e., contains an unknown value) and $z_1[X]\cap z_2[X]\neq\emptyset$; the resulting tuple of the equi-join is type 3.

Remark: There exist tables U and V such that $rep_1(U) \bowtie rep_1(V) \not\subseteq^* rep_1(U \bowtie^* V)$.
Proof: In figure 11, relation s is in $rep_1(U \bowtie^* V)$, but there are no relations $r_1 \in rep_1(U)$ and $r_2 \in rep_1(V)$ such that $r_1 \bowtie r_2 \subseteq$ s. *Q.E.D.*

E. Selection

The resulting table when a selection formula F is applied to a table U, denoted as $\sigma^*_F(U)$, consists of

(a) total tuples t such that T^U(t)holds, and F(t) holds,

(b) type 1 tuples p such that $T_1{}^U$(p) holds, and all the candidate tuples of p satisfy F,

(c) type 2 tuples p such that $T_2{}^U$(p) holds, and all the tuples in *tuples*(p) satisfy F,

(d) type 3 tuples p such that $T_{123}{}^U$(p)holds, and at least one t_1, $t_1 \in$ *tuples*(p), satisfies F, and there is a $t_2 \in$ *tuples*(p) that does not satisfy F.

Figure 12 illustrates the extended selection operation.

U

A_1	A_2	TYPE
1	2	0
1	[3,4]	1
2	[2,4]	1
2	[2,5]	1
[2,5]	3	2

$\sigma^*_{A_2>2}(U)$

A_1	A_2	TYPE
1	[3,4]	1
2	[2,4]	3
2	[2,5]	3
[2,5]	3	2

s

A_1	A_2
1	3
2	3

Figure 12. The Extended Selection

Remark: There exists a table U such that $\sigma_F(rep_1(U)) \not\subseteq^* rep_1(\sigma^*_F(U))$.

Proof: In figure 12, relation s is in $rep_1(\sigma^*_F(U))$, and there is no relation r in $rep_1(U)$ such that $\sigma_F(r) \subseteq s$. *Q.E.D.*

• Correctness of Query Evaluation in M-1

Lemma 1: Let U and V be two tables, and α and θ be any unary and binary RA operators, respectively, i.e., $\alpha \in \{\pi, \sigma\}$ and $\theta \in \{-, \cup, \bowtie\}$. Let α^* and θ^* be the extended versions of α and θ in $M-1$, respectively.

(a) $rep_1(U\ \theta^*\ V) \supseteq rep_1(U)\ \theta\ rep_1(V)$, and

(b)$rep_1(\ \alpha^*\ (U)) \supseteq \alpha\ (rep_1(U))$

Theorem 1: Query evaluation in the extended model *M-1* is sound.

Proof: Follows from lemma 1, and a straightforward induction on the number of operators of an RA expression.

4.2 The Model M-2

In *M-2*, the base tables of the database contain only total and type 1 tuples. However, the projection operation introduces type 2 tuples in model *M-2*. In other words, tuples can not be uniquely identified in intermediate tables resulting from a projection operation (on a base or intermediate table). Selection is defined as in *M-1*; hence type 3 tuples are also introduced in model *M-2*.

The *identifier of a tuple* z in a table with scheme $U(A_1, A_2, ..., A_d)$, denoted by id(z), is the tuple $(i_1, ..., i_d)$ where i_j is $z[A_j]$ if $z[A_j]$ is a known value, otherwise i_j is the identifier for the unknown value $z[A_j]$.

rep_2, the *rep* mapping for *M-2*, is similar to rep_1 except that the same unknown-value identifier, wherever it occurs in the database, must be mapped to the same value. Hence $rep_2(U)$ is defined relative to the other tables in the database D of tables. Please note that a partial tuple can appear in intermediate tables as a different type tuple; when this is the case, a type 3 tuple in an intermediate relation obtained from a type 1 tuple, for instance, can be mapped to the null tuple. D represents the database of tables.

Definition (*rep* in *M-2*):

$$rep_2(U)/D = \{r \mid (r \in rep_1(U) \wedge$$
$$(\forall U_1)(\forall r_1)(\forall z_1)(\forall z)((U_1 \in D \wedge r_1 \in rep_1(U_1)$$
$$\wedge\ z_1 \in U_1 \wedge z \in U \wedge id(z_1[A_i]) = id(z[A_j]))$$
$$\rightarrow (\ ((t_{z_1})r_1)[A_i] = ((t_z)r)[A_j] \vee (t_{z_1})r_1 = \varnothing \vee (t_z)r = \varnothing)))$$
$$\}$$

The RA operators of *M-2* are similar to the RA operators of *M-1*, except that we now recognize tuples of any type with the same identifier, and keep only one copy of such tuples in the resulting table. Definitions of the RA operators of M-2 are in [OlaO 88b].

After substituting "M-2" for "M-1", lemma 1 and theorem 1 still hold. That is, similar to *M-1*, *query evaluation in M-2* is sound, but not complete. A special case of *M-2* for which query evaluation is sound and complete is given in [OlaO 88b].

5. Comparison of Models M-1 and M-2 with Other Models

M-1 is unique in the sense that it models incomplete data environments that can not be handled by models based on Codd tables [Codd 79 and Bisk 83] where unknown values are denoted by a special null symbol, or the models based on V-tables [ImiL 84] where unknown values are represented by variables.

Let us consider relation r in figure 13. In *M-1*, r is represented as table U if the last digit of the A_2 value of the tuples in r are unknown.

r

A_1	A_2
4	53
4	56
5	41
3	41

U

A_1	A_2	TYPE
4	[50,59]	1
4	[50,59]	1
5	[40,49]	1
3	[40,49]	1

V

A_2
[50,59]
[50,59]
[40,49]
[40,49]

Figure 13. Duplicate Partial Tuples in [Gran 80]

The incomplete relation would be represented in [Codd 79 and Bisk 83] by table R in figure 14, where "STATUS=d" denotes definite tuples, and ω is the null symbol. Only one occurrence of tuple (4 , ω) is stored because one stored tuple *represents* several, but unspecified, number of model (unknown) tuples. In R, we are not able to represent the fact that there are exactly four unknown tuples. This has significant implications, especially when the *COUNT* of the tuples satisfying a given query is important. In *M-1*, using the tuple types, we are able to provide a tighter range for *COUNT*.

R

A_1	A_2	STATUS
4	ω	d
5	ω	d
3	ω	d

U

A_1	A_2
4	x_1
4	x_2
5	x_3
3	x_3

s

A_1	A_2
4	50
5	42
3	42

Figure 14. V-Tables and Codd Tables

[Gran 80] allows duplicate partial tuples as in *M-1*. Thus, the incomplete relation r in figure 13 is represented by table U as in *M-1*. However, in [Gran 80], the relationship among the tuples of an intermediate table can not be specified. For example, in figure 13, table V represents the projection of U on attribute A_2 in [Gran 80]. There is no way of stating the fact that the two occurrences of tuple ([50,59]) represent different unknown values, while the two occurrences of tuple ([40,49]) may denote the same value.

The model *M-2* is similar to the model based on V-tables in [ImiL 84], where unknown values are represented by distinct variables which take values from infinite domains. In comparison, in *M-2*, the range of an unknown value is finite and the tuple types (in a projection, for instance) distinguish variables that are known to take different values (type 1's) and those that may take the same value (type 2's).

Both Codd tables and V-tables can be used to model the *M-2* environment. Using Codd tables, every unknown value is denoted by a null symbol; but, as stated above, duplicate partial tuples are not allowed. In V-tables, the unknown values are represented by variables. An unknown value takes the same variable name wherever it occurs in the database. For instance, the incomplete relation in figure 13 is represented by the V-table U in figure 14. However, a V-table is not quite the same as a table in *M-2*. The interpretation of a V-table is such that a variable may take any value from an appropriate domain; different variables may take the same value. For example, relation s in figure 14, consisting of only three tuples, is in the set of relations represented by the V-table U.

6. Geometric Operations on d-rectangles

We now briefly discuss the geometric operations needed to implement the RA operators of model M-1 (and, also, the models M-2, M-3 and M-4).

A d-dimensional rectilinearly-oriented rectangle (d-rectangle) is the cartesian product of d closed intervals, one on each coordinate axis. A *partial tuple* p in a table of degree d is a set of disjoint d-rectangles in d-space. The *candidate tuples* of p is the set of total tuples contained in p.

Efficient manipulation of d-rectangles in the geometric model is central to the query evaluation process. Geometric operations that have to be performed include

(a) Finding the intersection of d-rectangles (used in all the five basic RA operations).

(b) Finding the complement of a set of d-rectangles (used in the projection operation).

(c) Testing for the containment of a d-rectangle in a union of d-rectangles (used in the selection operation).

Below we discuss these operations.

6.1. Intersection of d-rectangles

The intersection of two partial tuples is obtained by intersecting the corresponding d ordered sets (one per tuple component), each of which may have up to f intervals. Since the intersection of two ordered sets of intervals can be determined by comparing the two sets sequentially in $O(f)$ time, the intersection of two partial tuples is performed in $O(df)$ time.

Given a partial tuple p, finding all partial tuples of a table U which intersect with p is also needed in all five RA operators. The most straightforward approach is to intersect p sequentially with each partial tuple of U, which takes $O(ndf)$ time. Using the following result from [EdeM 81, Edel 83], one can give a different time complexity for this task as $O(f \log^d nf + kf)$ where k is the number of partial tuples of U that intersect with p.

Let S be an arbitrary collection of n d-rectangles. Then, there exists a data structure which reports in $O(\log^d n + k)$ time the k d-rectangles intersecting a given d-rectangle. The data structure is a combination of *segment trees* and *range trees*(i.e, a *tree-of-trees* structure). The construction of the data structure requires $O(n \log^{d-1} n)$ space and $O(n \log^d n)$ time. An insertion and deletion of a d-rectangle from S can be accomplished in $O(\log^d n)$.

The disadvantage of the approach in [EdeM 81, Edel 83] is that a complex tree-of-trees structure has to be constructed and maintained separately, which is expensive.

Another question that arises in union, difference and projection operations is to find whether any of the total tuples of a relation (or the candidate tuples of a partial tuple) intersects with (i.e. are contained in) a given partial tuple. The following lemma gives the complexity of such a task.

Lemma 2: Whether there is a total tuple in a d-degree relation of size n, sorted on an attribute A_i, that intersects with k_i of a partial tuple p = $k_1 x k_2 x \ldots x k_d$ can be determined in O(f logn) comparisons, where f is the degree of fragmentation.

The RA operator evaluations also utilize the counting problem of finding the number of candidate tuples of a partial tuple that intersect with the total tuples of a table. This operation can be implemented using the intersection operation of lemma 2.

6.2. Finding the Complement of a d-rectangle

We define the **complement** C' of a given rectangle C as the set of rectangles that contains all the points in $(D_{A_1}, D_{A_2}, \ldots, D_{A_d})$ except those in C. The complement $C_p{}'$ of the region defined by a partial tuple p can in general result in a set with disjoint $(2f+1)^d - f^d$ d-rectangles. However, lemma 3 states that $C_p{}'$ can be represented by at most (f+1)d d-rectangles.

Lemma 3: Given a partial tuple represented by a set of d-rectangles C_p, the *complement* $C_p{}'$ of C_p can be represented by the union of at most (f+1)d d-rectangles which are not necessarily disjoint.

6.3. Testing for Containment

In order to determine if a partial tuple p satisfies a selection formula Q, we test for the containment of C_p, the region defined by p, in C_Q, the query region defined by Q. For this test, one can use the concept of *d-measure* [Guy 77]. The d-measure of a single rectangle is a numerical value of the product of its intervals. For d=2 and d=3, this corresponds to area and volume, respectively. In finding the measure of a union of rectangles, points appearing in two or more rectangles are "counted" once. Thus, $C_p = C_p \cap C_Q$, and hence $C_p \subseteq C_Q$ holds only if the measure of C_p is the same as the measure of $C_p \cap C_Q$. However, the query region of Q, and consequently $C_p \cap C_Q$, may consist of a union of arbitrary d-rectangles. Finding the measure of n d-rectangles may take up to $0(n^{d-1})$ time [LeeW 81]; this is quite expensive for large d.

For tables with large degrees, Q may be transformed into the Conjunctive Normal Form, and p may be tested iteratively against every conjunct. The details of this approach are in [OlaO 88].

7. Implementation Effort

There are two parameters that directly effect the performance of any of the gemetric incomplete database models given in this paper. The first is the **degree of fragmentation f**, i.e., the maximum number of intervals allowed in a tuple component. Clearly, there is a tradeoff between the number of intervals for an unknown value and the cost of operator evaluation. The second parameter is the **percentage of partial tuples in tables.**

In [OlaO88a], for each of the extended operators, an evaluation algorithm is given. The worst-case time costs of the algorithms are analyzed in contrast with the costs of evaluating the usual RA operators in a particular DBMS. For more accurate time cost estimates, the distribution of nulls in partial tuples and the degree of fragmentation must be taken into consideration. We have come to the conclusion that a prototype development is needed to provide a better understanding of the feasibility of the models proposed in this paper.

As a first step in evaluating the performance of the models in this paper, we are implementing a version of model M-1 within a main-memory-only DBMS [She 88]. In our implementation, we are using the basic assumption of model $M-1$, and its type 0 and 1 tuple types. Type 2 and type 3 tuples of $M-1$ are merged into a single tuple type of type 2'. This change simplifies the RA operator definitions of M-1 significantly, and allows for a more efficient implementation. The goals of the implementation are
(a) to obtain empirical information about the effects of the parameters
(i) the degree of fragmentation, and
(ii) the proportion of partial tuples in tables, on the performance of the system, and
(b) to evaluate the feasibility of the models proposed in this paper.
We directly use in the implementation the geometric operations of section 6.

8. References

[AbKG 87] Abiteboul, S., Kanellakis, P. and Grahne, G., "On the Representation and Querying of Sets of Possible Worlds", *proc. ACM-SIGMOD conf.*, 1987.

[Bisk 81] Biskup, J., "A Formal Approach to Null Values in Database Relations", in *Advances in Database Theory* (H. Gallaire, J. Minker and J.M. Nicolas, Eds), vol 1, 1981, pp. 299-341.

[Bisk 83] Biskup, J., "Foundations of Codd's Relational Maybe Operations", *ACM TODS* vol 8(4) 1983, pp. 608-636.

[Codd 79] Codd, E.F., "Extending the Database Relational Model to Capture More Meaning", *ACM TODS*, vol 4(4) 1979, pp. 397-434.

[EdeM 81] Edelsbrunner, H. and Maurer, H.A., "On the Intersection of Orthogonal Objects", *Inf. Proc. Letters*, 13, 1981, pp 177-181.

[Edel 83] Edelsbrunner, H., "A New Approach to Rectangle Intersections", *Int. J. of Computer Math.*, vol 13, 1983, pp 209-219.

[Gran 79] Grant, J., "Partial Values in Tabular Database Model", *Inf. Proc. letters*, 9, 3, 1979, pp. 97-99.

[Gran 80] Grant, J., "Incomplete Information in a Relational Database", *Fundamenta Informaticae III.3* (1980), pp. 363-378.

[Guy 77] Guy, R., "Research Problems", in *American Math. Monthly*, 84(4), 1977, pp. 284-285.

[ImiL 84] Imielinski, T. and Lipski, W., "Incomplete Information in Relational Databases", *JACM* vol 31(4) 1984, pp. 761-791.

[LeeW 81] Leeuwen, J and Wood, D., "The measure problem for rectangular ranges in d-space", *Journal of Algorithms*, 2, 1981, pp.282-300.

[Lips 79] Lipski, W., "On semantic Issues Connected with Incomplete Information", *ACM TODS* Sept. 1979, pp. 262-296.

[Maie 83] Maier, D., *The Theory of Relational Databases*, Computer Science Press, 1983.

[OlaO 88a] Ola, A. and Ozsoyoglu, G., "Geometric Modeling of Incomplete Relational Databases", Tech. Report, CWRU, May 1988.

[OlaO 88b] Ola, A. and Ozsoyoglu, G., "A Family of Incomplete Relational Database Models", Tech. Report, CWRU, Aug. 1988.

[Reit 78] Reiter, R., "On Closed World Databases" in *Logic and Databases* (Gallaire and Minker, Eds), Plenum Press, 1978, pp.55-76.

[She 88] Shen, Jun, "A Main-Memory Incomplete Information DBMS and its Performance", MS thesis, CWRU, in progress, 1988.

[SixW 82] Six, H.W. and Wood, D., "Counting and Reporting Intersections of d_Ranges", *IEEE TC*, vol C-31, 3, 1982, pp 181-187.

[Vass 79] Vassiliou, Y., "Null Values in Database Management: A Denotational Semantics Approach", *ACM-SIGMOD* 1979, pp. 162-169.

[Zani 84] Zaniolo, C., "Database Systems With Null Values", *JCSS*, 1984.

Gral: An Extensible Relational Database System for Geometric Applications

Ralf Hartmut Güting

Fachbereich Informatik, Universität Dortmund
D-4600 Dortmund 50, West Germany

Abstract: We describe the architecture of a relational database system that is extensible by user-defined data types and operations, including relation operations. The central concept is to use languages based on many-sorted algebra to represent queries as well as query execution plans. This leads to a simple and clean extensible system architecture, eases the task of an application developer by providing a uniform framework, and also simplifies rule-based optimization. As a case study the extensions needed for a geometric database system are considered.

1. Introduction

Much of the database research of recent years was aimed at providing a better support for non-standard applications such as office information systems, geographic information systems, CAD databases, etc. A common need of these applications is the representation and manipulation of more complex objects than those representable by a tuple of a relation in the traditional relational model, for example, an office form, a complete map or a river, say, in a geographic information system, or the design of a VLSI circuit.

A fundamental choice for the representation of a complex object is whether its structure should be visible or hidden at the level of the data model. Speaking in terms of the relational model, for representation the question is whether the object should be described by a collection of tuples from various relations, or by a single attribute value from a specific domain for this kind of objects. For manipulation, the distinction is whether the internal structure is accessible to the general facilities of the query language (selection of subobjects, for example) or only to domain-specific operations. The two ways of handling complex objects have been called structural and behavioural object orientation, respectively [Di86].

Both approaches are obviously needed and are appropriate for certain applications. For example, one should be able to define the internal structure of an office document within the data model and to access it through the query language. On the other hand, a river can well be represented as an atomic value of an abstract data type LINE with suitable operations, for instance, a function returning the length.

A lot of work has been done to support the modeling of visible object structures. Enhancements to the relational model have been proposed by linking together tuples to represent an object either explicitly [Co79, HaL82] or implicitly, through the use of nested relations [ScS86]. Most of the more recent data models and system proposals do also support structural object orientation, for example [MaD86, CaDV88, PiT86].

The idea of allowing application-specific abstract data types as base types, or attribute domains, of a database system was perhaps first put forward in [StRG83]. Since base types need to be implemented in a programming language and because they are application-specific, a user must be able to implement such a type and to add it to a database system. This observation has led to efforts by several groups to construct *extensible* database systems. Two directions can be distinguished. One is to select a data model and to implement for this data model a system with well-defined interfaces for user extensions. This is the approach chosen by the POSTGRES [StR86] and Starburst [Schw86] projects, based on the relational model, and within the PROBE project [Daya87] for an extended functional data model. A different view is taken in the EXODUS [Care86] and GENESIS [Bato86] projects where a collection of powerful tools for *building* a database system is provided. In this case an application developer ("users" providing extensions have also been called "database implementor" [Care86] or "database customizer" [Lo88]) implements the whole database system, aided by the tools. With this approach there is more flexibility for the application even to select a specific data model.

In this paper we describe an (implemented and running) relational database system, called Gral[1], extensible by user-defined data types and operations. This means that an application developer (AD) can

- add data types and operations to the query language,
- add the representation of a type to the system,
- add new relation representations or index structures to support type-specific storage or searching,
- add the implementation of type-specific operations, and
- add rules to the system to describe the implementation of a new query language operation and to support optimization for this operation.

The key concept embodied in Gral, which distiguishes it from other work, is *many-sorted algebra*. Gral uses many-sorted algebra as a query language and also as an executable language to describe query plans (access plans). The whole system is centered around this formal concept and we show that this leads to a very simple and clean extensible system

[1] Gral stands for "**G**eo-**R**elational **Al**gebra" which is the query language of the Gral system.

architecture. For example, there is a layer in the system corresponding to the sorts of the executable algebra and another layer corresponding to the operators.

Algebraic query languages have been used before [Gr84] and are perhaps now viewed as old-fashioned. However, they have clear advantages for the use in an extensible system, because they are inherently modular and extensible. SQL- or QUEL-like languages accomodate easily new functions or predicates for new atomic types (e.g. in where-clauses). Yet it is awkward to add new relation operators to such languages. As we shall see, non-standard applications do require special relation operations (see also [OsH86]). Many-sorted algebra and the Gral system allow to add any kind of user-defined relation mapping. The particular kind of algebra that we use includes very powerful retrieval operations such as selection or join with an arbitrary algebra expression as condition and the embedding of derived values (described by algebra expressions) into relations.

The optimizer translates query language expressions to query execution plans. In Gral both are algebraic languages with the same structure. This eases considerably the application developer's task who needs to be familiar with both language levels. In particular it makes it relatively easy to formulate the rules driving the query optimizer.

The Gral project started with the goal of building an efficient database system for geometric applications. While this is still our goal, the need for extensibility was soon recognized and the focus shifted initially to provide a clean and simple extensible architecture. We find that having in mind a specific application is quite helpful to see clearly what kind of extensibility is required.

We consider an extensible system to have a model and a system part and, orthogonally, an application-independent and an application-specific part. We use the following terminology. The application-independent parts are called the *Gral model frame* and the *Gral kernel system.* Speaking of application extensions in general we use the terms *Gral application model* and *Gral application (system).* Finally, the running Gral prototype is one specific application; its data model is *geo-relational algebra* and we call it the *Gral system.* The following three sections describe Gral's query language, its executable language, and the system architecture. Related work is mentioned in Section 5.

2. An Extensible Data Model and Query Language

Basically Gral uses an extended relational algebra as a query language. More precisely, Gral's data model and query language are based on the formal concept of a *many-sorted algebra.* A many-sorted algebra is a collection of sets and functions between these sets; it is described by an *S-sorted signature* Σ where S is a set of *sorts* (names for the sets) and Σ a family of sets $\Sigma_{w,s}$ of operator symbols (names for the functions), where $w \in S^*$ and $s \in S$ describe the functionality of operators in $\Sigma_{w,s}$ [GoTW78]. For Gral, this algebra has one sort for relations and further sorts for atomic data types; the operations include relational operators such as selection or join and operations on atomic data types such as integer multiplication or a test whether two polygons intersect. More precisely, again, the structured data objects in Gral are not relations (sets of tuples) but *sequences of tuples*, or ordered relations. This allows, for example, to include sequence and sorting operators in the algebra.

Since Gral is an extensible database system together with one specific extension we can distinguish between the application-independent *model frame* and the application-specific data model called *geo-relational algebra.* The model frame describes any application as a many-sorted algebra; it defines the application-independent sorts and operators and is abstract with respect to the application-specific sorts and operators. The set of sorts is $S = \{s_1, \dots, s_n\}$ with s_1 = REL and s_2 = BOOL. The collection of operators (signature) of the model frame is $\Sigma = \Sigma_G \cup \Sigma_A$ (G for "general", A for "application-specific") of which Σ_G is defined within the model frame. Some operators of Σ_G are shown in the following table.

REL × REL	→ REL	**union, product, join, concat**	_ _ #
REL	→ REL	**select, project, extend, ord, head**	_ #
REL	→ ATOM	**extract**	(_ #)
BOOL × BOOL	→ BOOL	**and, or**	(_ # _)
BOOL	→ BOOL	**not**	# (_)

The model frame contains sorts for relations (sequences of tuples) and Boolean values and their associated operations. The rightmost column shows the syntax for using an operator within an expression (a query); "_" denotes an operand and "#" the operator, parentheses are to be used as indicated. For example "rel1 rel2 **union**" and "(x **and** (y **or** z))" are well-formed expressions. Some operators take parameters which appear in square brackets behind the operator symbol, as, for example, in "cities **select** [cpop > 500000] **project** [cname, cpop]", for a "cities" relation with attributes for city name and population.

The set of sorts S has a subset ATOM = $\{s_2, \dots, s_n\}$ which contains the atomic data types. The model frame provides also a set ORD $\subseteq$ ATOM containing the atomic data types that are one-dimensional ordered domains so that a relation can be sorted by attributes of such a type.

Any application-specific database system must specify its atomic data types and may provide within Σ_A any operations on these atomic types, but also operations involving relations. Standard types such as integer, real, and string with suitable operations are likely to occur in most applications. Sorts INT, REAL, and STR are also part of the geo-relational algebra together with the following operations:

$STD_i \times STD_i$	→ BOOL	=, ≠, <, ≤, ≥, >	(_ # _)
INT × INT	→ INT	+, -, *, **div, mod**	(_ # _)
INT × INT	→ REAL	/	(_ # _)
REAL × NUM	→ REAL	+, -, *, /	(_ # _)
NUM × REAL	→ REAL		
REL	→ INT	**count**	(_ #)
NUM*	→ REAL	**avg**	(_ #)
NUM_i*	→ NUM_i	**sum, min, max**	(_ #)

In this listing some abbreviations are used. STD and NUM denote sets of sorts, namely, STD = {INT, REAL, STR} and NUM = {INT, REAL}. The notation X_i with such a set identifier means that X_i is to be consistently replaced by the same element of set X. So the first line allows comparisons of any two operands of equal type. The notation X* denotes a relation operand where an attribute of type X is specified, or, intuitively, a set of X values. The attribute is given as a parameter of the operator, for example, "(cities **avg** [cpop])" returns the average population of cities (within the cities relation) as a real number. The last line is also an abbreviation for two lines with the functionalities INT* → INT and REAL* → REAL.

We can now illustrate the operators of the model frame by a few examples.

(1) Determine the five cities with the highest population numbers.

cities **ord** [cpop -] **head** [5]

The **ord** operator sorts cities descending by population; from the resulting sequence **head** returns the first five tuples.

(2) Add to each city tuple an attribute giving its population in percent of the maximal population.

cities
extend [(cpop/(cities **max** [cpop]))*100 {percent}]

The **extend** operator takes a list of algebra expressions as parameters, each expression is followed by a new attribute name. For each expression, the relation will have a new attribute whose value is determined for each tuple by evaluating the expression.

(3) Find pairs of cities whose population number differs by less than 50000.

cities cities {*->*2} **join** [(cpop - cpop2) < 50000]

For the second operand relation, the attributes are first renamed (appending a suffix "2") before the **join** operator is applied. Note that the **select** as well as the **join** operator take arbitrary algebra expressions with Boolean result as parameters.

(4) How many people live in Dortmund, in thousands?

(cities **extract** [cname="Dortmund"; cpop]) **div** 1000

The **extract** operator allows one to extract from a relation a single atomic value which is then available for further operations. Parameters of **extract** are an algebra expression (as a selection condition which must return exactly one tuple) and an attribute name.

An application may decide to choose the collection of standard types and operators shown here "from the shelf". However, since all atomic types, except for BOOL, are considered application-specific, it may instead also use very long integers, text fields containing formatting instructions, etc.

An application will then introduce its own more specific data types. Gral's data model, the geo-relational algebra, is designed to support geometric applications in the plane, perhaps with a bias towards supporting geographic information systems. Gral has therefore types for points, lines, and regions in the plane. A line is conceptually a finite sequence of straight line segments. A region is a hole-free, non-selfintersecting polygon. Actually there are two types of regions called PGON and AREA; for type AREA the restriction holds that no two polygons that are attribute values of the same relation may intersect. Type AREA is used to model subdivisions of the plane as they often occur in maps. So the complete set of sorts of the geo-relational algebra is S = {REL, BOOL, INT, REAL, STR, POINT, LINE, PGON, AREA}. To simplify the signature there is also a hierarchy within the geometric types: REG = {PGON, AREA} (regions), EXT = {LINE} ∪ REG (extended objects), and GEO = {POINT} ∪ EXT (any geometric object). A selection of the operators is shown below:

GEO × REG	→ BOOL	**inside**	(_ # _)
EXT × EXT	→ BOOL	**intersects**	(_ # _)
LINE* × LINE*	→ POINT*	**intersection**	_ _ #
LINE* × REG*	→ LINE*		
PGON* × REG*	→ PGON*		
POINT* × POINT	→ REL	**closest**	_ _ #
POINT × POINT	→ REAL	**dist**	# (_ , _)
LINE	→ REAL	**length**	# (_)
REG	→ REAL	**perimeter**, **area**	# (_)

The **inside** operator tests whether any kind of geometric object is wholly contained within a region. The **intersects** operator checks extended objects for intersection. The **intersection** operator (with functionality LINE* × LINE* → POINT*) is applied to two relation operands; for each operand an attribute of type LINE is specified. It returns a relation which "looks like" the cartesian product of the operand relations extended by a POINT attribute. To be precise, the result relation contains one tuple for each intersection point of a line of the first operand with a line of the second operand. This tuple consists of all attributes of the (tuple with the) first line, all attributes of the second line, and the intersection point. Intuitively, the functionality should be read "The intersection of two sets of lines is a set of points". Forming the intersection of lines is embedded in this way into a relation operation because even the intersection of two single lines is generally a set of points, so one cannot define an atomic operator with functionality LINE × LINE → POINT. The **intersection** operator can be applied similarly to a set of lines and a set of regions, or two sets of regions. The **closest** operator is given a relation with a point attribute specified, and a (query) point; it returns the tuple (or those tuples) of the operand relation whose point attribute has minimal distance to the query point. Finally, our selection shows operators to compute the distance between two points, the length of a line, and perimeter and area of a region. Some example queries, based on relations representing cities, states, rivers, and highways of West Germany, illustrate the use of these operators.

cities	cname STR	center POINT	cpop INT
states	sname STR	region AREA	spop INT
rivers	rname STR	route LINE	
highways	hname STR	way LINE	

(5) Associate with each city the state containing it.
cities states **join** [center **inside** region] **project** [cname, sname]

(6) Find rivers passing through the state of Bayern (Bavaria).
states **select** [sname="Bayern"] rivers **join** [route **intersects** region] **project** [rname]

(7) Which city with more than 100000 inhabitants is closest to the point where highway A1 passes the Rhine?
(rivers **select** [rname="Rhein"]
highways **select** [hname="A1"]
intersection [route, way, {crossing}]
extract [true; crossing]) {crosspoint};
cities **select** [cpop > 100000] crosspoint
closest [center]

The query consists of two steps. In the first step the intersection point of river Rhine and highway A1 is computed and made available as a POINT object named "crosspoint". In the second step the **closest** operator is applied to the cities with more than 100000 inhabitants, and the point "crosspoint".

(8) Sort cities by distance from München (Munich).
(cities **extract** [sname="Muenchen"; center])
{Munich};
cities **extend** [**dist** (center, Munich) {MunDist}]
ord [MunDist +]

By a *data model* we mean a mathematical model for data and operations on data. In this sense, the Gral model frame is a data model; its structured objects (sequences of tuples) and operations have been formally defined in [Gü88b]. The essential part of the semantics definition is the treatment of nested algebra expressions (since an operator, for example **select, join, extend** may have an algebra expression as a parameter); this and most of the operator definitions can also be found in [GüZC88]. For the application-specific extension *geo-relational algebra* semantics have also been formally defined in [Gü88b], that is, a model for LINE and PGON objects is given and the operations shown above (as well as the remaining operators) have been defined in terms of these object models. Geo-relational algebra as such is discussed in more detail and illustrated by examples in [Gü88a].

The Gral model frame is extensible in that new sorts for atomic data objects as well as new operators involving old or new atomic data types and relations can be added. To perform such a data model extension an application developer should provide a mathematical model for a new data type. For operators the functionality and allowed parameters must be specified and the mapping associated with the operator be defined. This corresponds to a formal specification of a program to be implemented. In many cases application developers will omit the specification of extensions at the level of the formal data model and just implement new types and operators. However, in case of more complex data types, and to be able to check the correctness of the implementation, it is advisable to also extend the formal model.

We have seen that the algebra is used rather directly as a query language. To extend the query language, besides introducing new types and operators, one needs to specify syntactical aspects such as a pattern for using the new operator, e.g. "(_ # _)", and the structure of the parameter list, if the operator has parameters. To make the query language somewhat homogeneous and to allow the Gral system parser to analyze expressions, certain rules have to be followed:

(A) Use postfix notation whenever an operator takes at least one relation as an operand.

(B) Put parentheses if and only if the resulting object is atomic.

There is a lot of flexibility in the allowed syntactical patterns:

(1) Prefix outside parentheses # (_), # (_ , _), ...
(2) Prefix or infix in parentheses (# _), (# _ _), ...
(_ # _)
(_ # _ _), (_ _ # _), ...
(3) Postfix in parentheses (_ #), (_ _ #), ...
(4) Postfix without parentheses _ #, _ _ #, ...

Parameter lists may also have a fairly general structure, basically a parameter list may consist of several sublists, separated by semicolons, where each list may have a fixed or variable number of items, separated by commas. How syntactical patterns and the structure of parameter lists are made known to the parser is discussed in Section 4.3.

3. Descriptive and Executable Algebra

Before describing the Gral system architecture it is important to introduce the distinction between *descriptive* and *executable* algebra. In any Gral application a query is formulated as an expression of the application-defined many-sorted algebra. It should be clear that such a query is to be understood as the user's specification of some data object (relation or atomic object) that the system should return; this data object is described by a *hypothetical* series of applications of operators to the stored objects. It is the task of the optimizer to translate this query to an executable query plan and, in fact, to select among many available query plans the most efficient one.

In an extensible database system it is necessary that the application developer understands this "language of query plans" since obviously it must also be extensible. A central idea of the Gral system is that the language of query plans should also be an algebra, with the same syntactical structure, although with different sorts and operators, as the query language algebra. This simplifies the task of the application developer since he/she needs to understand only a single formalism; it is surely preferable to introducing a quite different way of describing query plans. The optimization process can then be understood (and partially be formulated) by the application developer as a translation from one algebra expression to another one. To distinguish, we call the query language algebra the *descriptive* and the "language of query plans" the *executable* algebra.

Whereas for the descriptive algebra the foremost design goals are expressive power and simplicity, and efficiency plays no role at all, the major issue in the design of an executable algebra is efficiency. On the other hand, because of the requirement of extensibility the executable algebra should still be as simple as possible, so a reasonable balance between efficiency and simplicity must be found.

Some important ideas to achieve efficiency at the level of the executable algebra are the following:

- Use index structures (access paths).
- Support processing in "tuple mode", that is, as far as possible perform a series of operations on a single tuple instead of copying the relation for each operation.

- Use complex operators. That means, whenever an operator needs to process a relation, it should do as much as possible during this access. For example, a sorting operator should include a selection condition to eliminate tuples before sorting, and allow to process each tuple further after sorting.

These are well-known strategies for the implementation of relational database systems. We show how they can be incorporated into the design of an executable algebra.

In the descriptive algebra a sort is associated with a set of objects and an operator with a function; neither is a representation specified for the objects nor an algorithm for computing the function. In contrast, in the executable algebra a sort corresponds to a specific object representation within the system and each operator has associated a fixed algorithm or procedure in the system. Generally for a Gral application system the set of sorts will consist of the following subsets and specific sorts:

- REL - each sort in REL describes one kind of physical representation (primary index) of a relation.
- INDEX - each sort describes one specific secondary index structure.
- TUPLE - a sort describing a tuple in memory.
- TIDSEQ - a sort describing a collection of tuple identifiers (through which tuples represented in one of the REL representations can be accessed).
- ATOM - these are the sorts for atomic objects that are also present in the descriptive algebra.

In the currently running Gral prototype there is only a single sort SREL in REL corresponding to a simple sequential representation of a relation. Primary and secondary index structures will be added soon. To illustrate the general approach we describe an executable algebra already containing index structures. In this design we have

REL = {SREL, STIDREL, BTREE}

SREL is a sequential representation used, for example, for temporary relations during the processing of a query. STIDREL is a permanent representation providing tuple identifiers; a relation represented in this way can be accessed sequentially or through a secondary index. BTREE is a B*-tree representation ordered by one of the attributes. It can be accessed through the primary index, through a secondary index, or sequentially. Let TIDREL = {STIDREL, BTREE} be the structures accessible through a secondary index.

INDEX = {SBTREE, SGRID, SMLGRID}

All of these structures allow to obtain a collection of tuple identifiers. SBTREE is a B*-tree used as a secondary index, SGRID a grid file [NiHS84], supporting geometric searches on a set of points in the plane, SMLGRID a multi-layer grid file [SiW88] supporting searches on a set of rectangles in the plane. The stored rectangles in an SMLGRID index are the bounding boxes of extended objects such as lines or polygons.

Sorts TUPLE and TIDSEQ are also present and we have

ATOM = {BOOL, INT, REAL, STR, POINT, LINE, PGON, RECT}

The executable algebra has an additional sort for rectangles which are needed as bounding boxes of LINE and PGON objects.

The set of atomic types with ordered domains is ORD = {BOOL, INT, REAL, STR}.

A subset of the operators of the current Gral executable algebra extended by some operators for index structures is shown in Figure 3-1.

REL	→ SREL	***scan, sort***	_ #
REL × REL	→ SREL	***product, smjoin, inside_join***	_ _ #
REL	→ ATOM	***extract***	(_ #)
TUPLE	→ TUPLE	***sel, proj, extend***	_ #
BTREE(ORD_i) × ORD_i	→ SREL	***exactmatch***	_ _ #
BTREE(ORD_i) × ORD_i × ORD_i	→ SREL	***rangescan***	_ _ _ #
SGRID(POINT) × RECT	→ TIDSEQ	***geo_range_search***	_ _ #
SMLGRID(RECT) × RECT	→ TIDSEQ	***geo_intersection_search***	_ _ #
TIDSEQ × TIDREL	→ SREL	***tidscan***	_ _ #
EXT	→ RECT	***bbox***	# (_)
POINT × PGON	→ BOOL	***pr_inside***	(_ # _)

Figure 3-1: Operators of the Executable Algebra

The executable algebra contains also atomic operations on standard types; these are as in the descriptive algebra. Also operators for SBTREEs have been omitted; they are similar to those on primary BTREEs but return sets of tuple identifiers.

The ***scan*** operator has a parameter list of the form [E_1; E_2] where E_1 is an algebra expression of result type BOOL and E_2 an expression of result type TUPLE. This operator scans a relation given in any of the REL representations. E_1 is a *filter* expression which is evaluated for each tuple of the operand relation; those tuples passing the filter are then made available to the second expression E_2 by the keyword *tuple* to which tuple operators can be applied. So E_2 allows processing in "tuple mode". For example, the query "List the names of big cities" looks in descriptive (D) and executable (E) algebra as follows:

D cities **select** [cpop > 500000] **project** [cname]
E cities ***scan*** [cpop > 500000; tuple ***proj*** [cname]]

The ***sort*** operator realizes the **ord** operator of the descriptive algebra but has also an input filter expression and a tuple expression applicable to tuples after sorting.

Similarly the binary relation operators have two input filter expressions and a tuple expression to be applied to the concatenated tuples resulting from the operation. For example, the ***product*** operator computes the Cartesian product of the operand relations and has a parameter list of the form [E_1, E_2; E_3] where E_1 and E_2 are the filter expressions and E_3 is the tuple expression. It can be used to answer query (5) from Section 2 "Associate with each city the state containing it":

D cities states **join** [center **inside** region]
  project [cname, sname]

E cities states ***product*** [true, true; tuple ***sel*** [center ***pr_inside*** region] ***proj*** [cname, sname]]

The selection operator on tuples ***sel*** either returns the operand tuple or a special object called **notuple**; any tuple operator receiving **notuple** as an operand also returns **notuple**. The same query could also have been realized using the ***inside_join*** operator. This operator can be applied to (a relation with) a set of points and (a relation with) a set of regions; it performs a plane-sweep [PrS85] to join tuples of the two operand relations where the point is contained in the region. We consider a slightly more complicated example query with this operator "List for big cities their population number as a percentage of their state's population":

D cities **select** [cpop > 500000] states
  join [center **inside** region]
  extend [(cpop/spop)*100 {percent}]
  project [cname, percent]

E cities states ***inside_join*** [cpop > 500000, true;
  center, region;
  tuple ***extend*** [(cpop/spop)*100 {percent}]
  proj [cname, percent]]

The ***smjoin*** operator performs a sort/merge join. Its parameter list has the form [E_1, E_2; E_3 cop E_4; E_5]. E_1, E_2 and E_5 are filter and tuple expressions, as before. E_3 and E_4 are expressions involving only attributes of the first and second operand relation, respectively, and *cop* is one of the comparison operators {<, ≤, =, >, ≥}. The two operand relations are first sorted by the value of their respective expression E_3 or E_4 and then merged according to the comparison operator. - The ***extract*** operator of the executable algebra has no parameters; it can be applied to a relation with a single tuple and a single attribute and returns the attribute value as an atomic object of the attribute's type (see the example below).

Access to primary or secondary index structures is also realized by executable operators. A BTREE is organized by attribute values of one of the ordered data types and can be accessed with a single value of the same type or a range of such values through the ***exactmatch*** and ***rangescan*** operators. These operators also have filter and tuple expressions like the ***scan*** operator to impose additional restrictions on the returned tuples and to process these tuples further. For the following examples we assume that relations and index structures are organized as follows:

cities:	BTREE(STR) ordered by the cname attribute
cpop_index:	SBTREE(INT) on the cpop attribute
center_index:	SGRID(POINT) on the center attribute
states:	BTREE(STR) ordered by the sname attribute
region_index:	SMLGRID(RECT) on the region attribute

The query "Retrieve the tuple representing the city Essen" can then be formulated as:

D cities **select** [cname = "Essen"]

E cities "Essen" ***exactmatch*** [true; tuple]

An SGRID index stores a set of points; the ***geo_range_search*** operator returns the tuple identifiers belonging to points inside a search rectangle given as the second operand. Similarly an SMLGRID index stores a set of rectangles; the ***geo_intersection_search*** operator returns the tuple identifiers belonging to rectangles intersecting a search rectangle. The returned tuple identifiers can be processed further through the ***tidscan*** operator which accesses the original relation and allows to filter and process tuples. As an example we consider the query "Find cities in Bayern!":

D (states **extract** [sname = "Bayern"; region])
  {Bayern};
 cities **select** [center **inside** Bayern]

E (states "Bayern" ***exactmatch*** [true;
  tuple ***proj*** [region]] ***extract***) {Bayern};
 center_index ***bbox*** (Bayern) ***geo_range_search***
  states ***tidscan*** [center ***pr_inside*** Bayern; tuple]

Let us reconsider the executable algebra from the point of view of extensibility. The Gral kernel system will contain generally useful relation representations such as those given above in REL and, say, SBTREEs as a secondary index structure, as well as sorts TUPLE, TIDSEQ, and BOOL. It will have executable operators needed to implement the descriptive operators of the model frame, for example ***scan***, ***sort***, ***smjoin***, ***product***, ***extract***, ***exactmatch***, ***tidscan***, etc. Standard types such as INT, REAL, STR together with their operations will be available as a module. A Gral application system may then add its own special types such as POINT, LINE, PGON, and provide operators for these types as well as special index structures such as SGRID, SMLGRID with suitable operators for searching these structures.

This distinction between Gral kernel system and Gral application system arises from a practical point of view. Note, however, that the Gral architecture to be described next is *absolutely uniform* with respect to *all* sorts and operators. For example, ***scan*** and ***smjoin*** are brought into the system through the same extension mechanism as any other operator and they might as well be application-defined. Parser, query evaluator, and optimizer handle these operators in the same way as any application-defined operator. We believe that only in this way a clean architecture can be achieved.

4. An Extensible System Architecture

4.1. Survey

The Gral system prototype has been implemented during the last two years in a project at the University of Dortmund. The implementation language is Modula-2; the system runs on Sun workstations under UNIX. The interactive interface uses SunWindows and SunCGI for window management and graphic representation of geometric objects; this interface is mostly written in C. The system currently offers the following functionality:

- Database operations: create, destroy, open, close, save a database, that is, a collection of relations.
- Relation operations: create, destroy a relation; insert a tuple into a relation.
- Load/unload: move a relation out of the system into a UNIX text file and load it from such a text file. This is currently the only way to change the contents of a relation (through a standard text editor).
- Show: display the contents of a relation. Attribute types that are textually representable are shown in a text window; types that are graphically representable (such as geometric types) are shown in a graphics window. The text window shows tuples and their attributes sequentially; one can scroll forward or backward or access directly any position within this textual representation of the relation. The graphics window shows a section of the real cartesian plane which can be selected by the user; geometric objects occurring as attribute values of the displayed relation are visible as far as they intersect this section. It is possible to overlay the graphical representations of several relations.
- Query: enter a query either in descriptive or in executable algebra. The resulting relation or atomic object is then displayed. The descriptive algebra implemented so far is a subset of the geo-relational algebra described in [Gü88a].

Note that the interactive interface offering this functionality is entirely *data type independent*. For example, the program that displays a relation does not know how to display any attribute value. Instead, it calls through the *type manager* a type-specific procedure to display a value of this type which was provided by the implementor of this type. So this program is prepared to handle any attribute type that will be added to the system in the future. According to the philosophy of the Gral model frame the only object type built into the system is that of relations.

In this section we focus on the *layered architecture for query evaluation* and emphasize and explain extensibility at the various layers. This architecture is shown in Figure 4-1.

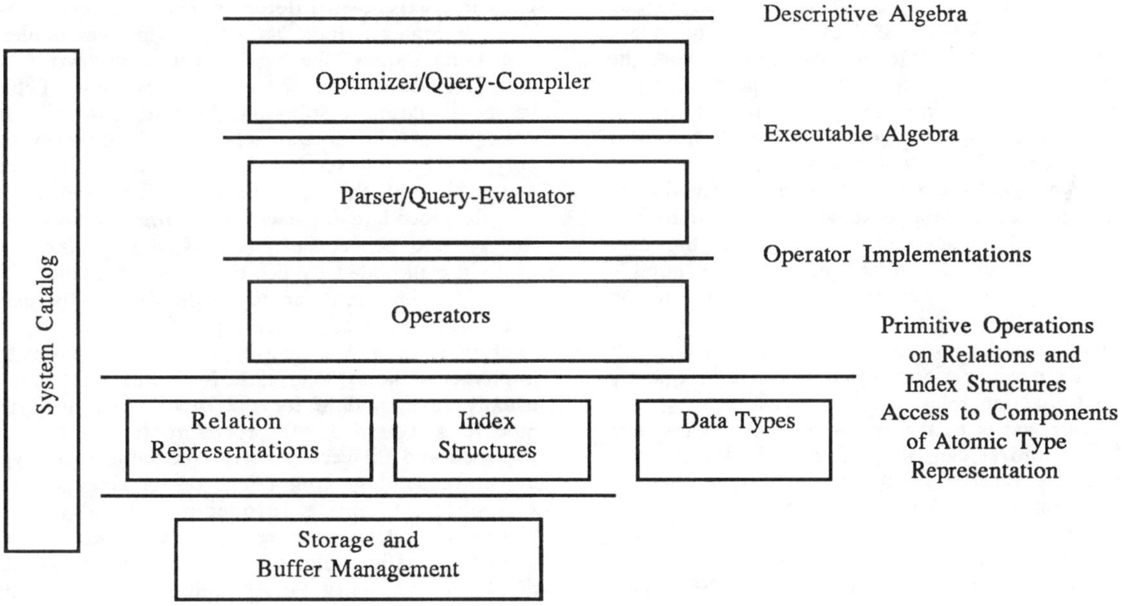

Figure 4-1: Architecture for Query Evaluation

Although query processing proceeds top-down, the architecture of the system is best explained bottom-up. The bottom layer provides rather traditional *storage and buffer management* except that page sequences are available to hold data objects occurring as attribute values (it is generally assumed that "atomic" data objects may have arbitrary size) and facilities for fixing page sequences in the buffer. Currently we see no need for this layer to be extensible. The second layer provides *representations* for the sorts of the executable algebra; the implementation of relation representations and index structures uses the functions offered by the storage management layer. The second layer provides also a higher level representation of a tuple in memory with suitable operations and a scan component offering uniform scanning operations for all relation representations. This layer is extensible by atomic data types, index structures, and relation representations.

The third layer consists of *implementations of executable operators*. These are based on the primitive access operations offered by relations or index structures and the representation of atomic types realized by the layer below. For example, the ***scan*** operator of the executable algebra uses the scan component of the representation layer. A "point inside polygon" test operator (***pr_inside***) accesses the representations of point and polygon objects implemented in that layer. Each executable operator is implemented by two procedures. One of these describes the *schema transformation* (ST) performed by the operator (for example, the ***product*** operator creates a new relation schema by concatenating the two operand schemas). The second procedure implements the *occurrence transformation* (OT), that is, the mapping of the relation instances or atomic data objects. There are two separate procedures because the ST procedure is already called during the parsing of an executable algebra expression whereas the OT procedure is called during query evaluation, after parsing is completed. This layer is, of course, extensible by operator implementations.

The next layer contains the *parser* and *query evaluation* components. Given an executable algebra expression, the parser produces an operator tree representation in memory. For this purpose it accesses information in the system catalog, in particular the "Operators" table which contains for each operator its name (or symbol), the functionality, and a description of the parameter list structure, if there is one. As mentioned before, the parser also calls an operator's ST procedure to apply the operator to an operand schema (which may in turn be the result of parsing a subexpression). The parser has no built-in knowledge of any operator. The same is true for the query evaluator. This component "understands", however, the structure of the operator trees which the parser builds. The query evaluator traverses the operator tree, computes bindings for the leaves, representing objects, and calls for each operator represented by an internal node the corresponding OT procedure. The query evaluator takes (references to) the operand objects from the tree and passes them to the procedure as parameters; it receives the resulting object from the procedure and links it into the tree. This layer is extensible by operator descriptions being entered into the system catalog.

Up to this layer we have a complete system capable to evaluate queries in executable algebra. As mentioned before, in the Gral system this capability is present at the user interface, that is, a user can enter queries in executable algebra. It is necessary to expose this interface to an application developer so that he/she can test new operator implementations in queries for correctness and performance. The knowledge gained in performance tests can subsequently be built into optimization rules.

The top layer consists of the *optimizer/query compiler* component. The optimizer calls the parser of the layer below to analyze a given descriptive algebra expression and to produce an internal tree representation; the parser again receives the necessary information from the "Operators" table in the system catalog. The same parser can be used because descriptive and executable algebra have the same syntactical structure. The information needed for optimization and query translation itself comes from a text file "OptimizationRules". This file contains various classes of optimization/translation rules in a specific syntax and order. When the Gral system is started the contents of the file are analyzed and represented by internal data structures so that patterns occurring in rules can be matched efficiently against parts of the tree representing the query. How the optimizer works is described in some more detail in Section 4.4. The result is an expression in executable algebra which can be shown in text form to the user, if desired. For further processing of the query, however, it is passed to the query evaluator directly in internal form, so it need not be parsed again. The top layer is extensible by descriptive operators and optimization/translation rules.

In the following three subsections extensibility by data types, executable operators, and descriptive operators is discussed in a little more detail.

4.2. Extensibility by Data Types

The implementation of a data type in Gral consists of:

(1) A Modula-2 type, usually opaque, together with operations for building an object accossiated with a variable of this type, and operations to access the components of an object. This part constitutes the Modula-2 implementation of an abstract data type.

(2) Three classes of procedures:

I Procedures for storing and loading an object. These convert the object representation in a variable to a bytestring and store the object as an attribute value in a relation and vice-versa. Whenever possible the bytestring representation produced by the Modula-2 compiler is used so that no conversion is necessary. Optionally there may be procedures to load an object only partially.

II Procedures for interactive input/output. For example, a procedure to read a string value from the keyboard or a procedure to display a polygon on the graphics screen.

III Procedures to convert from/to textual form.

There exist two different kinds of programs using a data type: programs written *before* the type was implemented and that do not "know" the type, and programs written *afterwards* with knowledge about the type. An example of the first class is the program displaying a relation at the screen. An example of the second class is a program testing two polygons for intersection. Programs of the first class access a type through the *type manager*. This module provides a generic procedure for each task within classes I - III above; the generic procedure just switches to the type-specific procedure provided by the type implementation. The generic procedure is implemented in Modula-2 by an **array of procedure** indexed by the type.

Hence, to add a data type to the Gral system, an AD implements the type as described under (1) and (2) above, usually in a module for the type. Then the type manager module is extended: the procedures in the type module are imported and for each generic procedure in the type manager the corresponding component of its array of procedure is assigned the type's procedure. Finally a procedure "InsertType" is called to enter the name of the type and some other information (e.g. whether the type is sortable, can be displayed textually, graphically, etc.) into the system catalog and the system relation about types.

Similar ideas for providing type extensions have been reported in [StRG83, Ong84, OsH86] although the details differ. For example, in Gral types are not defined within the system's data definition language to avoid inconsistencies with the source code which must be present in the system anyway. What is perhaps new about our treatment of types is the possibility of *partial loading*. Since we generally assume that objects may be arbitrarily large it is useful to allow to load only some part of an object into memory. For example, a polygon is represented by a record with fields for the bounding box, the area, the perimeter, and the vertices (given as an array). An operator testing two polygons for intersection may initially only load the bounding boxes and only load the vertices later if the bounding boxes intersect.

4.3. Extensibility by Executable Operators

Adding an executable operator to the system consists of three steps. First, one implements its schema and occurrence transformation procedures, using the primitives of the representation layer. (That layer contains also a module "SchemaTransformations" providing a data structure to represent a schema and operations on schemas such as concatenating schemas, adding an attrribute, etc.). The new operator then needs to be linked into the "OpManager" module. The OpManager has separate arrays of procedure for ST and OT procedures, one array for each arity of an operator. So the second step is to assign the new procedures to appropriate array components. Through the OpManager these procedures can then be called by parser and query evaluator.

The third step in adding an operator is to make its name, functionality, syntax, and parameter list structure known in the system catalog where this information is then available for the parser. Similar as for types this is achieved by calling a procedure "InsertOp" with text parameters containing corresponding descriptions. The system catalog (and the system relation about operators) has four entries for each operator:

(1) Operator name or symbol.

(2) A description of functionality and syntax in the form *pattern* → *result type*. For example, for the ***scan*** and ***pr_inside*** operators the descriptions are:

```
REL scan -> REL
(POINT pr_inside PGON) -> BOOL
```

(3) A description of the parameter list if there is one, otherwise the word "None". Parameter lists may be either in *standard form* or *irregular*. To be in standard form means that a parameter list may be structured into a fixed number of *sublists*, separated by ";", each sublist may have a fixed or variable number of *items*, separated by ",". An item may have a fixed number of parts, separated by blanks or user-defined delimiters. Parts may be recognizable for the system parser or not. Recognizable parts are constants of some atomic type, attribute names, (re)namings, and algebra expressions of specified result type. In Gral these rules are built into a grammar which defines the allowed parameter list descriptions and thus the allowed parameter lists. Parameter list descriptions for ***scan*** and ***proj***, for example, are as follows:

scan: `[Expr(BOOL); Expr(TUPLE)]`

proj: `[Attrname *]`

If a parameter list is irregular, a procedure for parsing the parameter list must be provided by the AD with the operator implementation. In the current Gral system this was never necessary.

(4) The last entry is either "Executable" or "Descriptive" since descriptive operators are described in exactly the same way and appear also in the operator table.

The parser transforms a parameter list encountered in an algebra expression according to the description in (3) to a nested list data structure and connects this representation to the operator's node in the operator tree. When a parameter list contains an algebra expression, the parser recursively parses this expression and builds an operator tree which it links into the nested list structure. In this way nested algebra expressions are represented in the operator tree. During query evaluation, the nested list representation of a parameter list is passed to the operator's OT procedure.The implementor of this procedure (the AD) relies on the parser having built the right structure; thus the description (3) serves as an interface definition between parser and OT procedure. When an OT procedure has to process an algebra expression occurring as a parameter it simply calls the query evaluator on the tree representation of this expression. This is an important point: A user-defined OT procedure calls the system's query evaluator to handle nested algebra expressions, as they occur, for example, in ***scan*** or ***extend*** operators.

4.4. Extensibility by Descriptive Operators and Optimization/Translation Rules

In this section we briefly sketch how the optimizer/query compiler works and how rules are described and arranged in the file "OptimizationRules"; a detailed report is to follow [BeG89]. The translation of a given descriptive algebra expression proceeds in the following stages:

---*Descriptive multistep query*---

1. *Normalize:* In a multistep query names of objects resulting from expressions can be replaced by these expressions without changing the semantics. This is done to obtain a single expression which can be optimized further as a whole.

---*Descriptive expression*---

2. *Decompose:* Certain operators, such as selection, join, extend, are decomposable. For example, a selection with several and-connected conditions can be transformed into several selections with simple conditions. Decomposable operators are taken apart, so that their parts (e.g. simple selections) can move freely.
3. *Improve:* Change the order of operators according to heuristics (classical algebraic optimization).
4. *Compose:* Put subsequent operators together (for example, a series of selections).
5. *Eliminate* constant and common subexpressions.

---*Final descriptive multistep query*---

For each expression in the multistep query:

---*Descriptive expression*---

6. *Translate* descriptive subexpressions to executable subexpressions. (During this stage, the optimizer works on a mixed descriptive and executable expression.)

---*Executable expression*---

7. *Combine* sequences of executable operators into a single operator.

---*Final executable expression (and multistep query)*---

All stages except 1 and 4 are supported by rules; these are given in the file OptimizationRules which is structured roughly according to the stages into sections. In the first stage a query formulated by a user in several steps is transformed (by embedding expressions for names) into a single expression (*normal form*) so that optimization of the whole query is possible. The steps selected by the user might otherwise prevent certain optimizations. Stage 2 (Decompose) is supported by a corresponding section in OptimizationRules:

```
DECOMPOSING RULES
...
RULE
  REL_1 select [BOOL_1 and BOOL_2]
  -> REL_1 select [BOOL_1] select [BOOL_2].

RULE
  REL_1 REL_2 join [BOOL_1]
  -> REL_1 select [BOOL_1] REL_2 product
     if BOOL_1.attr c REL_1.attrnames,
  -> REL_1 REL_2 select [BOOL_1] product
     if BOOL_1.attr c REL_2.attrnames.
...
END
```

A rule consists in general of a pattern followed by one or more replacement patterns each of which may have an associated condition. The condition may involve properties of the expression, such as the attribute names occurring, and properties of the involved objects, such as the number of tuples in an operand relation or the physical sizes of attributes or atomic objects, etc. A rule may be preceded by a specification part to simplify notation within the rule. Each rule described in this way can be viewed as a collection of rules of a production system [Ni80]. For most stages an irrevocable control strategy is used and rules are applied until no further rule is applicable. In these stages the first replacement pattern whose condition is fulfilled is selected to replace the original pattern in the expression. Different control strategies which explore alternatives are used in stages 3 and 6.

The goal of stage 3 is to arrange operators in an order which is "good" according to heuristics. For this purpose it is possible to specify a partial order on descriptive operators to indicate a *desired* order of operators in the expression - whether that order can be achieved is another matter. This is a general facility to formulate such familiar heuristics as "selection and projection should be done as early as possible". The partial order is given by a section "OPERATOR ORDER". How operators may actually be exchanged to achieve a good order is described in the section "IMPROVING RULES", one example is shown:

```
SPEC
  op_1 in {select, project, ord};
RULE
  REL_1 op_1 [Param_1] select [BOOL_1]
  -> REL_1 select [BOOL_1] op_1 [Param_1].
```

The control strategy for stage 3 attempts to arrange operators according to the partial order in such a way that the number of "runs" (ordered subsequences) is minimized. By a suitable definition of operator order and rules, one can let a run correspond to a single operator of the executable algebra and so minimize the number of executable operators needed. The section "COMPOSING RULES" contains rules for stage 4 (no example is shown). Together, stages 2 - 4 realize the classical algebraic optimization.

No rules are necessary to support stage 5 (Eliminate). A constant subexpression is one that is evaluated once for each tuple of a relation but does not depend on the tuple's attribute values. In query (2) of Section 2 "(cities **max** [cpop])" is a constant subexpression. Stage 4 translates this to a two-step query where the constant subexpression is evaluated only once initially:

(cities **max** [cpop]) {Constant_1};
cities **extend** [cpop/Constant_1 {percent}]

The next section contains rules for the translation of descriptive to executable subexpressions. This is the point where descriptive operators are associated with their executable counterparts.

```
TRANSLATING RULES
...
RULE
  REL_1 select [Attrname_1 = Const_1]
  -> REL_1 Const_1 exactmatch [true; tuple]
     if ReprIs (REL_1, Attrname_1, BTREE),
  -> Index (REL_1, Attrname_1) Const_1
     indexexactmatch REL_1 tidscan [true; tuple]
     if ExistsIndex (REL_1, Attrname_1, SBTREE),
  -> REL_1 scan [Attrname_1 = Const_1; tuple].

RULE
  REL_1 REL_2 join[Attrname_1 inside Attrname_2]
  -> REL_1 REL_2 inside_join [true, true;
        Attrname_1, Attrname_2; tuple]
     if Attrname_1 in REL_1.attrnames and
        Attrname_2 in REL_2.attrnames and
        REL_1.size > 100,
  -> REL_1 REL_2 product [true, true; tuple sel
        [Attrname_1 pr_inside Attrname_2]].
...
END
```

The first rule describes how a simple selection can be implemented either by access through primary or secondary index structures, if present, or otherwise by a relation scan. The second rule shows how the **inside** operator of the descriptive algebra, defined on atomic objects, can actually be implemented by a plane-sweep algorithm which is (under certain conditions) the most efficient way to realize this kind of geometric join. This example shows how the system architecture does indeed support the efficient processing of geometric queries, which was one of the initial goals of the Gral development. The rule is also a simple example of how the optimizer can switch between various implementations: for small sets of points (here up to 100 points) forming the cartesian product with subsequent selection is preferable.

The last section "COMBINING RULES" of the file OptimizationRules contains rules for combining executable operators. The previous translation stage which replaces descriptive operators independently of each other by executable operators may, for example, have produced series of ***scan*** operators, or ***scan*** followed by ***product***, etc., which can now be combined into one:

```
RULE
  REL_1 REL_2 product [BOOL_1, BOOL_2; TUPLE_1]
    scan [BOOL_3; tuple]
  -> REL_1 REL_2 product [BOOL_1, BOOL_2;
       TUPLE_1 sel [BOOL_3]].
```

We have given here only a rough survey of the Gral extensible optimizer; most details and the more sophisticated techniques have been left out. For example, the optimizer can be advised by rules to search systematically all permutations of n and-connected simple conditions, within a selection or join, say, and for the translation stage (6) each replacement pattern in a rule can be assigned a numeric expression. The replacement pattern can then be selected according to the value of that expression. These facilities allow one to search the possibilities for implementing a single operator, such as a complex selection, exhaustively, and to make a good choice among these. For example, a simple condition supported by an index structure will be evaluated first. These techniques are described in more detail in [BeG89].

It is now obvious how a descriptive operator is added to the system: Its name, functionality and syntax, and parameter list structure are entered into the system catalog as for executable operators. The descriptive operator is then "implemented" by a collection of rules in the file OptimizationRules. Of course, rules can be added to or removed from the file at any time independently from adding operators.

5. Related Work

We have described the query processing architecture of an extensible relational database system based on many-sorted algebra and shown some of the specific extension mechanisms used in the Gral system. Obviously much of the previous work on extensible database systems is related in some way; some of it can be associated with specific layers of our architecture. Extensibility by relation representations and index structures and in particular the interaction between these two in update processing have been studied in [LiMP87]; their work fits nicely into our representation layer. Methods for adding atomic types, functions, and predicates have been studied in [StRG83, OsH86]. The RAD system [OsH86] uses also a query language based on relational algebra and in principle allows relation mappings (called "transformations"). However, these have not been implemented and no extension mechanisms are described. Techniques for adding operator implementations have also been studied in [St86, GrD87]. In contrast to our approach, in the INGRES [St86] or POSTGRES [StR86] project the architecture of the database system is not affected as a whole by extensibility. That is, there is a collection of built-in types which are treated differently from user-defined types; also all the standard parts of the query language are implemented in a traditional way and user-defined operations are handled separately. These are essentially restricted to simple functions on atomic types; relation mappings are not foreseen. The approach of adding interfaces to a traditional architecture shows in the way how user-defined operations need to specify whether certain methods of the fixed part of the system are applicable, e.g. whether hash-join is feasible for this operator. In Gral, hash-join could in principle as well be a user-defined operation.

Rule-based optimization has also been studied by Freytag [Fr87] and in the EXODUS project [GrD87]. Our techniques have some similarity to Freytag's proposal. Differences are that he describes only the translation stage, so there is no algebraic optimization before or after translation, and that his rules describe how to generate all possible query execution plans but no strategy for selecting a good one. Gral contains a complete extensible optimizer which actually produces a good query plan. Generally one could say that we have filled in many details to his proposal and implemented a relatively complete query language including aggregate functions, sorting, and geometric operators. Freytag considers essentially only selection and join. Graefe and DeWitt [GrD87] focus on the aspect of heuristically searching the space of query plans to find a good one quickly. In their relational test implementation they also consider only selection and join. Another approach to extensible query optimization using the rules of a grammar to construct query plans is described in [Lo88]. - We shall compare the methods for extensible optimization in more detail in [BeG89].

Acknowledgment

Past and present members of the Gral development team are Ludger Becker, Thaddäus Behnke, Anja Huesmann, Dirk Markert, Rainer Matenia, Stefan Pölt, Markus Schneider, Norbert Suchhart, Jacqueline Teschner, Reinhilde Uphaus, Bernd Vielhauer and Leonhard Wawrzinek. Their contribution to the design and implementation of the Gral prototype is gratefully acknowledged.

References

[Bato86] Batory, D., J. Barnett, J. Garza, K. Smith, K. Tsukuda, C. Twichell, and T. Wise, GENESIS: A Reconfigurable Database Management System. University of Texas at Austin, Dept. of Computer Science, Report TR-86-07, 1986.

[BeG89] Becker, L., and R.H. Güting, An Optimizer for an Extensible Geometric Database System. Universität Dortmund, Fachbereich Informatik, Manuscript, 1989.

[Care86] Carey, J.M., D.J. DeWitt, D. Frank, G. Graefe, M. Muralikrishna, J.E. Richardson, and E.J. Shekita, The Architecture of the EXODUS Extensible DBMS. In [Di86], 52-65.

[CaDV88] Carey, J.M., D.J. DeWitt, and S.L. Vandenberg, A Data Model and Query Language for EXODUS. Proc. ACM SIGMOD 1988, 413-423.

[Co79] Codd, E.F., Extending the Database Relational Model to Capture More Meaning. *ACM Transactions on Database Systems 4 (1979)*, 397-434.

[Daya87] Dayal, U., F. Manola, A. Buchman, U. Chakravarthy, D. Goldhirsch, S. Heiler, J. Orenstein, and A. Rosenthal, Simplifying Complex Objects: The PROBE Approach to Modelling and Querying Them. In: H.J. Schek and G. Schlageter (eds.), Proc. BTW 87, Springer 1987, 17-37.

[Di86] Dittrich, K.R., Object-Oriented Database Systems: the Notion and the Issues. Proc. of the IEEE/ACM International Workshop on Object-Oriented Database Systems, Pacific Grove, California (September 1986), 2-4.

[Fr87] Freytag, J.C., A Rule-Based View of Query Optimization. Proc. ACM SIGMOD 1987, 173-180.

[GoTW78] Goguen, J.A., J.W. Thatcher, and E.G. Wagner, An Initial Algebra Approach to the Specification, Correctness, and Implementation of Abstract Data Types. In: R. Yeh (Ed.), Current Trends in Programming Methodology, Vol. IV, Prentice-Hall 1978, 80-149.

[GrD87] Graefe, G., and D.J. DeWitt, The EXODUS Optimizer Generator. Proc. ACM SIGMOD 1987, 160-172.

[Gr84] Gray, P.M.D., Logic, Algebra, and Databases. Ellis Horwood Ltd., Chichester, 1984.

[Gü88a] Güting, R.H., Geo-Relational Algebra: A Model and Query Language for Geometric Database Systems. In: J.W. Schmidt, S. Ceri, M. Missikoff (eds.), Proc. EDBT 1988, 506-527.

[Gü88b] Güting, R.H., Modeling Non-Standard Database Systems by Many-Sorted Algebras. Universität Dortmund, Fachbereich Informatik, Report 255, 1988.

[GüZC88] Güting, R.H., R. Zicari, and D.M. Choy, An Algebra for Structured Office Documents. Universität Dortmund, Fachbereich Informatik, Report 254, 1988, to appear in *ACM Transactions on Office Information Systems (Jan. 1989).*

[HaL82] Haskin, R.L., and R.A. Lorie, On Extending the Functions of a Relational Database System. Proc. ACM SIGMOD 1982, 207-212.

[LiMP87] Lindsay, B., J. McPherson, and H. Pirahesh, A Data Management Extension Architecture. Proc. ACM SIGMOD 1987, 220-226.

[Lo88] Lohman, G.M., Grammar-like Functional Rules for Representing Query Optimization Alternatives. Proc. ACM SIGMOD 1988, 18-27.

[MaD86] Manola, F., and U. Dayal, PDM: An Object-Oriented Data Model. In [Di86], 18-25.

[NiHS84] Nievergelt, J., H. Hinterberger, and K.C. Sevcik, The Grid File: An Adaptable, Symmetric Multikey File Structure. *ACM Transactions on Database Systems 9 (1984)*, 38-71.

[Ni80] Nilsson, N.J., Principles of Artificial Intelligence. Tioga Publ. Company, Palo Alto, CA, 1980.

[Ong84] Ong, J., D. Fogg, and M. Stonebraker, Implementation of Data Abstraction in the Relational System INGRES. ACM SIGMOD Record 14, No. 1 (March 1984), 1-14.

[OsH86] Osborn, S.L., and T.E. Heaven, The Design of a Relational Database System with Abstract Data Types for Domains. *ACM Transactions on Database Systems 11 (1986)*, 357-373.

[PiT86] Pistor, P., and R. Traunmüller, A Database Language for Sets, Lists, and Tables. *Information Systems 11 (1986)*, 323-336.

[PrS85] Preparata, F.P., and M.I. Shamos, Computational Geometry: An Introduction. Springer 1985.

[ScS86] Schek, H.J., and M.H. Scholl, The Relational Model with Relation-Valued Attributes. *Information Systems 11 (1986)*, 137-147.

[SiW88] Six, H.W., and P. Widmayer, Spatial Searching in Geometric Databases. Proc. IEEE Data Engineering Conf. 1988, 496-503.

[St86] Stonebraker, M., Inclusion of New Types in Relational Data Base Systems. Proc. 2nd Intl. Conf. on Data Engineering (Los Angeles, CA, February 1986), 262-269.

[StR86] Stonebraker, M., and L.A. Rowe, The Design of POSTGRES. Proc. of the 1986 SIGMOD Conf. (Washington, DC, May 1986), 340-355.

[StRG83] Stonebraker, M., B. Rubenstein, and A. Guttmann, Application of Abstract Data Types and Abstract Indices to CAD Databases. Proc. 1983 ACM Engineering Design Applications, 107-114.

[Schw86] Schwarz, P., W. Chang, J.C. Freytag, G. Lohman, J. McPherson, C. Mohan, and H. Pirahesh, Extensibility in the Starburst Database System. In [Di86], 85-92.

The LSD tree: spatial access to multidimensional point and non point objects *

Andreas Henrich
FernUniversität Hagen
5800 Hagen
West Germany

Hans-Werner Six
FernUniversität Hagen
5800 Hagen
West Germany

Peter Widmayer
Universität Freiburg
7800 Freiburg
West Germany

Abstract

We propose the Local Split Decision tree (LSD tree, for short), a data structure supporting efficient spatial access to geometric objects. Its main advantages over other structures are that it performs well for all reasonable data distributions, cover quotients (which measure the overlapping of the data objects), and bucket capacities, and that it maintains multidimensional points as well as arbitrary geometric objects. These properties demonstrated by an extensive performance evaluation make the LSD tree extremely suitable for the implementation of spatial access paths in geometric databases. The paging algorithm for the binary tree directory is interesting in its own right because a practical solution for the problem of how to page a (multidimensional) binary tree without access path degeneration is presented.

1. Introduction

In non-standard applications such as cartography, CAD and robotics, Database Management Systems have to organize large sets of multidimensional geometric objects on secondary storage such that these objects can quickly be retrieved according to their spatial locations. Typical spatial queries are the retrieval of an object by its coordinates (exact match) and range queries where all objects geometrically intersecting the query region are selected for further processing or presentation on a screen. Since the set of objects varies over time, insertions and deletions have to be performed as well.

Data structures, which efficiently support spatial access to geometric objects, usually divide the data space into cells and store all objects located in a cell in an associated data bucket. As far as multidimensional points are concerned, various efficient data structures have been proposed (see e.g. [Free87], [HSW88a], [HSW88b], [KS86], [KS88], [KW85], [NHS84], [Otoo86], [Rob81]). In typical applications, however, most objects are arbitrary geometric, i.e. non-point, objects. In many situations, it has proven to be useful to represent non-point objects by their (minimal) bounding boxes, serving as simple geometric keys. We therefore concentrate on multidimensional intervals, as far as non-point objects are concerned.

An obvious approach for storing intervals uses data structures for points. Here, intervals need not be entirely contained in a cell, but may instead intersect several cells. Hence, in order to perform range queries efficiently, information about an interval must be stored in each bucket, whose corresponding cell intersects the interval. Using this so-called *clipping technique*, the space requirements increase substantially due to the redundant information.

The clipping approach is based on data structures which partition the data space into pairwise disjoint cells. To avoid clipping problems, in R-Trees ([Gut84], [FSR87]), respectively multilayer grid files [SW88], the data space is divided into *overlapping cells*, such that all, respectively most, intervals are entirely contained in a cell, i.e. need not be clipped. Unfortunately, the R-Tree suffers from a poor exact match performance and often from inefficient range queries because cells may overlap considerably in a dynamic setting. In the multilayer grid file with each layer a grid file is associated inducing a directory overhead which deteriorates the efficiency of operations concerning few objects.*

In the so-called *transformation technique* ([Hin85], [SK88]), k-dimensional intervals are interpreted as points in a 2k-dimensional space, in order to use point data structures in a standard way. For instance, a 1-dimensional interval [a, b] may be interpreted as point (a, b). Since $a \leq b$ holds, the image space is a triangle. The main drawbacks of this approach are that the point distribution in the triangular image space is extremely skew and that a (bounded) range query on intervals becomes a partly unbounded range query on image points.

From a wide spectrum of performance tests we have got the experience that the efficiency of spatial data structures depends at least on the object distribution, the cover quotient defined as the sum of all object areas divided by the area of the data space, and the bucket capacity, i.e. the maximal number of objects in a bucket. For an increasing cover quotient as well as for small bucket capacities, clipping and overlapping cell techniques deteriorate substantially. On the other hand, all non-tree structures (see e.g. [HSW88a], [KS86], [KS88], [NHS84]) degenerate for skew object distributions.

In this paper, we propose a data structure supporting spatial access to k-dimensional points as well as k-dimensional intervals. The access to intervals is based on the transformation technique. A sophisticated directory tree together with a refined splitting technique eliminates the pre-

* This work has been supported by DFG grants Si 374/1 and Wi 810/2.

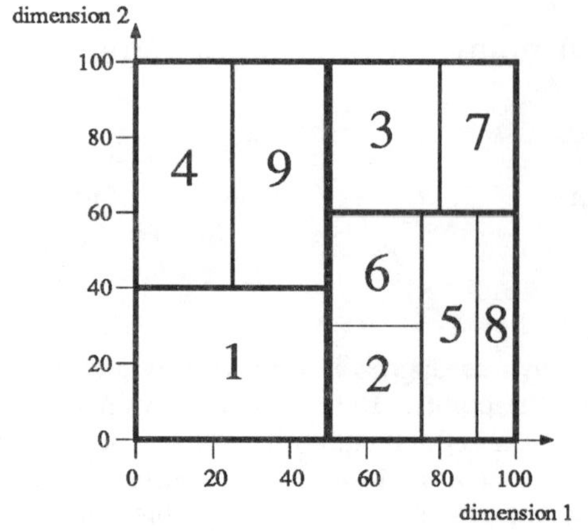

Figure 2.1: Possible partition of the data space for an LSD tree

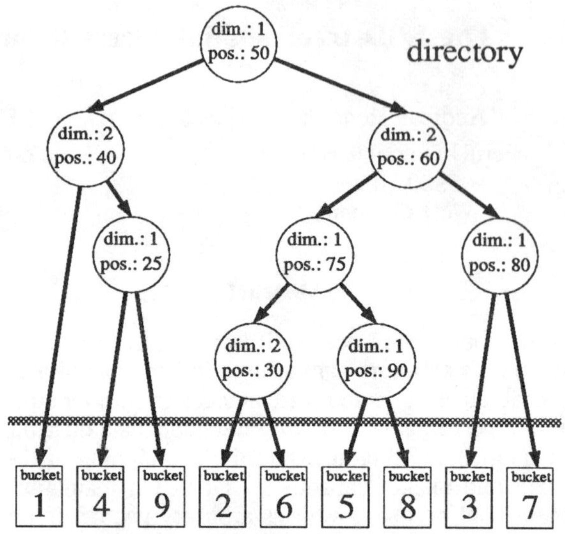

Figure 2.2: LSD tree associated with the data space partition of Figure 2.1

vious drawbacks of this technique. The performance evaluation convincingly demonstrates that the new structure is well qualified for maintaining large sets of geometric objects. However, the main advantage of the new structure is not only the efficient spatial access but also its robustness. By robustness we mean that the new structure behaves well for all reasonable data distributions, cover quotients and bucket capacities.

Section 2 explains the new data structure for point objects while in section 3 the generalization to non-point objects is provided. In section 4 a performance evaluation of the new structure and a comparison with the multilayer grid file is presented. Section 5 concludes the paper.

2. The LSD tree for points

2.1 Basic ideas and properties

Like most structures, the new structure partitions the data space into pairwise disjoint cells with associated buckets of fixed size. In contrast to the grid file [NHS84], however, it is not grid oriented, i.e. all cell boundaries may occur at arbitrary positions. The free choice of split positions is the basis of the graceful adaptation to arbitrary skew object distributions. Since a new split position can be chosen locally optimal, i.e. optimal with respect only to the cell to be split and independent from other existing cell boundaries, we call the new structure **Local Split Decision tree** (LSD tree, for short). Figure 2.1 shows a possible partition of a 2–dimensional data space for an LSD tree.

The LSD directory maintaining the flexible data space partition is a binary tree similar to a k-d tree [Ben75]. Each node of this tree represents one split decision by storing the split dimension and the split position. Figure 2.2 illustrates the LSD tree associated with the data space partition of Figure 2.1.

It should be obvious that the directory provides the freedom for using the split strategy best suitable for the actual application. This is an important advantage over other structures (see e.g. [Free87], [HSW88a], [KS86], [KS88], [NHS84]) where split decisions are more or less influenced by previous split decisions. Furthermore, the size of the directory is directly related to the number of buckets, i.e. for n buckets the directory contains n-1 nodes. This is in contrast to the grid file where several entries in the directory may point to the same bucket.

Besides the advantages of the LSD tree directory there are some drawbacks which are typical for multidimensional binary tree structures:

1. A multidimensional binary tree may become unbalanced, i.e. may contain long paths with almost no branches, and
2. no suitable method for paging a multidimensional binary tree is known. (The interesting paging technique presented in [LZL88] is suitable only for the one-dimensional case.)

We overcome these problems by introducing a paging algorithm which preserves the following **external balancing property**:

The number of external directory pages which are traversed on any two paths from the directory root to a bucket differs by at most 1.

When geometric objects are inserted into an initially empty LSD tree the directory grows up to a size when it cannot be kept in the dedicated part of the main memory any longer. Then the paging algorithm determines a subtree to be paged on secondary storage such that the external balancing property is preserved. If the subtree consists of n_s nodes, the main memory is then able to receive additional n_s nodes until a further invocation of the paging algorithm must take place.

Figure 2.3 shows the overall structure of the LSD tree.

2.2 A closer look

In this section, we discuss the LSD tree in more detail by explaining the insertion of a new geometric object into the structure.

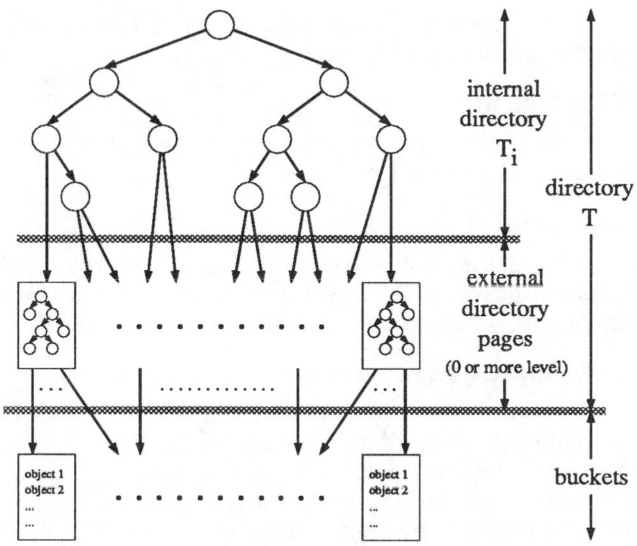

Figure 2.3: Overall structure of the LSD tree

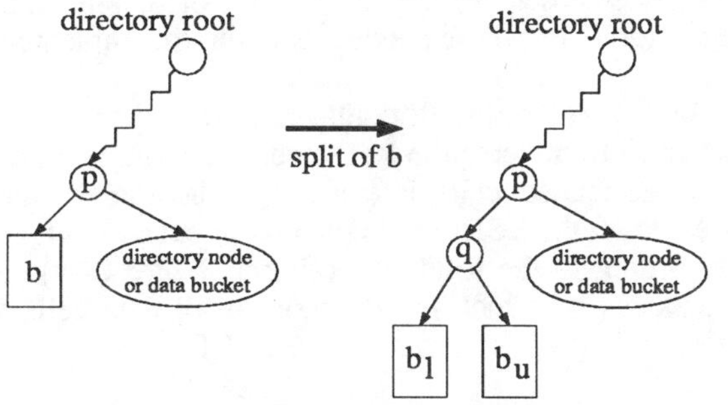

Figure 2.4: Effect of a bucket split

The search for the bucket b which will receive the new object is guided by the directory as in k-d trees. If b does not overflow, the insertion is finished, otherwise the **bucket split algorithm** creates two new buckets b_l and b_u from b according to a *split strategy* we describe afterwards. In case of a bucket split the pointer in the directory referencing b is changed to a pointer referencing a new directory node q representing the split decision concerning b, i.e. the new node q is inserted into the directory by calling the *directory insertion algorithm* explained later. The new node q points to the new buckets b_l and b_u. Figure 2.4 depicts the effect of a bucket split.

We distinguish between two inherently distinct types of **split strategies**:

1. *Data dependent split strategies*
   These strategies depend only on the objects stored in the bucket to be split. A typical example for such a strategy is to choose for the split position the average of all object coordinates with respect to a certain dimension.
2. *Distribution dependent split strategies*
   These strategies choose the split dimension and the split position independently of the actual objects stored in the bucket to be split. A typical example for such a strategy is to split a cell into two cells of equal areas. Note that this "halving split strategy" relies on the assumption of a uniform distribution of the objects.

Since the LSD directory is a binary tree, any type of split strategy can be implemented in an easy and efficient manner. Note that data dependent split strategies cannot be realized by data structures based on hashing (see e.g. [HSW88a], [KS86], [KS88], [NHS84]).

We now turn our attention to the directory of the LSD tree. As already mentioned in the previous section, if the number of nodes in the directory T exceeds the maximal possible number of internal nodes, a subtree of T is written onto secondary storage, i.e. stored in a directory page. In such a directory page a subtree is organized as a sequential heap of fixed height h_p. Hence, whenever the height of the associated subtree exceeds h_p after an additional insertion, a directory page split has to be performed. The **directory page split algorithm** is simple: the left and right subtree of the root are stored in two distinct directory pages and the root is inserted into the directory T by calling the *directory insertion algorithm.*

We are now in a position to describe how a new node q, resulting from a bucket or a directory page split, is inserted into the directory T by the **directory insertion algorithm.** The heart of this algorithm is the paging algorithm we explain afterwards. In the following, we assume that the main memory capacity reserved for the directory T is n_i. The directory insertion algorithm assures that the internal prefix tree T_i of the directory T contains at most n_i-1 nodes. Two cases may occur:

case 1: The father p of the new node q is an internal node.

- **case 1.1:** The number of internal nodes is less than n_i-1. Then insert q into T_i and finish.
- **case 1.2:** The number of internal nodes is equal to n_i-1. Then insert q into T_i, call the *paging algorithm* for T, and finish. Note that after the execution of the paging algorithm the number of internal nodes is at most n_i-1.

case 2: The father p of the new node q is a node in a subtree T_p of T stored in an external directory page.

- **case 2.1:** After the insertion of q the height of T_p is at most h_p; then finish.
- **case 2.2:** After the insertion of q the height of T_p is greater than h_p. Then call the *directory page split algorithm* for T_p and finish.

The **paging algorithm** is called when after an insertion of an additional node the size of the internal prefix tree T_i reaches the maximal possible number n_i of internal nodes. The algorithm searches for a subtree T_s in T_i such that paging T_s preserves the external balancing property defined in section 2.1. This property is preserved if T_s is a *paging candidate*, i.e. T_s fulfills the following properties:

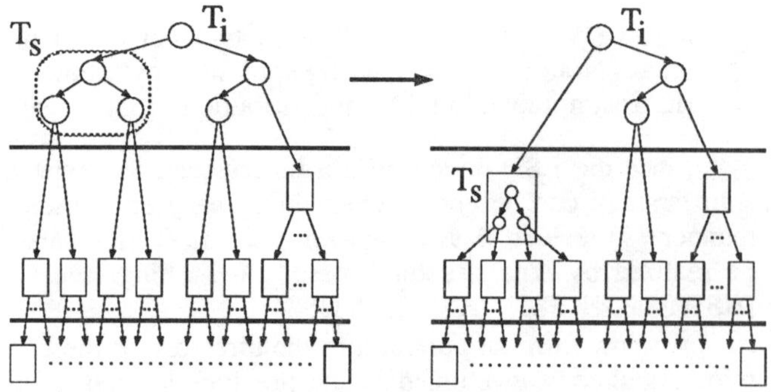

Figure 2.5: Directory before and after paging subtree T_s

1. Any path from the root of T_s down to a bucket contains the minimal number of external directory pages (of all paths in the directory T).
2. The height of T_s is at most h_p.

Figure 2.5 shows a directory before and after paging the subtree T_s.

If more than one paging candidate occurs in T_i, the paging algorithm chooses a candidate with the maximal possible number of nodes.

In order to direct the search for a paging candidate in T_i the following numbers are attached to each node v in T_i:

$nep_{min}(v)$, resp. $nep_{max}(v)$,: the minimal, resp. maximal, number of external directory pages occurring on any path in T containing v.

s(v): the number of nodes of the biggest paging candidate which can be reached from v.

h(v): The height of the subtree with root v in T_i.

The paging algorithm moves down the internal directory T_i branching at each node w on the search path according to the following criteria:

1. If nep_{min}(left son of w) $\neq$ nep_{min}(right son of w), continue with the son with lower nep_{min}.
2. If nep_{min}(left son of w) $=$ nep_{min}(right son of w), continue with the son with greater s.

The root r of a paging candidate T_s is determined if

1. $h(r) \leq h_p$ and
2. $nep_{min}(r) = nep_{max}(r)$.

The second condition assures that after paging T_s the external balancing property is preserved for T.

It should be clear that after the insertion of a new node q into the internal directory T_i, resp. the paging of a subtree T_s of T_i, the numbers nep_{min}, nep_{max}, s and h must be updated for each node w on the path P from the root of T_i to the father of q, resp. to the father of the root of T_s (now stored in an external directory page). These numbers can easily be recomputed from the existing numbers of the nodes (and their direct sons) on the path P.

2.3 The operations

2.3.1 Exact match

In an exact match operation the directory is traversed until the corresponding bucket is determined. The bucket is scanned until the object searched for is located or the search ends unsuccessfully.

2.3.2 Insertion

The insertion algorithm has been explained in detail in the previous section.

2.3.3 Deletion

The deletion of a node in the directory T basically works inversely to the insertion of a node. Due to space limitations we cannot discuss this topic in more detail.

2.3.4 Range query

In a range query all points located in the query region are reported. According to Fredman [Fred80], a query region may be a rectangle (*orthogonal range query*), a circle (*circular range query*) or a polygon (*polygonal range query*). In the following, we restrict the discussion to orthogonal range queries, because the algorithm is the same for all query types, except for the procedures evaluating whether a data region is enclosed by, intersected by, or disjoint from the query region. But these are details left to the implementation level.

In order to report all points located in the query region Q we have to traverse the LSD tree to determine all buckets whose associated cells intersect Q. The query algorithm moves down the LSD tree branching at each directory node w according to the following criteria: (Here D(w) denotes the data region which is the union of all data cells whose corresponding buckets can be reached from w.)

1. If $Q \cap D(\text{right son of w}) = \emptyset$, continue with the left son of w.
2. If $Q \cap D(\text{left son of w}) = \emptyset$, continue with the right son of w.
3. Otherwise continue with both sons of w.

Note that $Q \cap D(w) \neq \emptyset$ is the invariant condition of the loop of the query algorithm. Hence, in 1., resp. 2., $Q \cap D$(left son of w) $\neq \emptyset$, resp. $Q \cap D$(right son of w) $\neq \emptyset$, holds.

3. The LSD tree for non-point objects

We explain the non-point situation for k-dimensional intervals which serve as bounding boxes for arbitrary geometric objects in many applications. We restrict the discussion to the 2-dimensional situation, i.e. to rectangles in the plane, because a generalization to higher dimensions is straightforward.

To store a set of rectangles in the LSD tree we use the *transformation technique* ([Hin85], [SK88]), i.e. 2-dimensional rectangles are stored as 4-dimensional points. We choose the simple corner representation [SK88] which considers for each of the two dimensions the lower and upper bounds of the rectangles to be distinct dimensions.

The idea is simple but several severe problems arise from this approach. First, there is a strong correlation between upper and lower bounds, because for each dimension the upper bound of a rectangle is always greater than (or equal to) the lower bound. Because of the correlation all points are located in a triangular shaped subspace of the image space. Furthermore, since in almost all applications all rectangles are small compared to the data space, the points are located in a small strip above the diagonal.

Data structures which rely on a rectangular shaped data space and partition the data space into rectangular cells tend to degenerate for such applications, especially if they are based on hashing techniques. However, the LSD tree overcomes the drawbacks of the transformation technique if a refined bucket split strategy is used. Since the split strategy is crucial to the efficiency of the LSD tree for non-point objects, we devote the next section to this topic.

3.1 The split strategies

In this section, we discuss split strategies suitable for the skew data distributions induced by the transformation technique. First of all, a suitable split strategy must take into account the correlation between lower and upper bounds of rectangles in the original dimension 1, resp. 2, stored in dimensions 1 and 2, resp. 3 and 4.

We will explain two different split strategies, a data dependent and a distribution dependent one. The *data dependent split strategy* is simple: For the split dimension under concern the average over all coordinates of objects stored in the bucket to be split including the object to be inserted is chosen as the split position.

The *distribution dependent split strategy* is a combination of two basic (distribution dependent) split strategies each of them designed for an extreme situation. The first split strategy relies on the (fictitious) assumption that all rectangles are degenerated to points, i.e. the upper and lower bounds coincide for each dimension. Here, all image points are located on the diagonal w.r.t the dimensions 1 and 2, resp. 3 and 4. A suitable split strategy for this case is to split the data cell into two cells containing equally long parts of the diagonal. The split position achieved by this split strategy is denoted bp SP_1 in Figure 3.1.

The second basic split strategy relies on the assumption that all image points are uniformly distributed over the triangular subspace of the image space built from dimensions 1 and 2, resp. 3 and 4. Here, a suitable split strategy halves the data cell into two cells of equal areas. The split position achieved by this split strategy is denoted bp SP_2 in Figure 3.1.

The split position SP calculated by the combined split strategy is the weighted sum of SP_1 and SP_2:

$$SP = \alpha SP_1 + (1 - \alpha) SP_2 , \qquad \text{where}$$

$$\alpha = \sqrt[10]{\frac{\textit{data cell area above diagonal}}{(U_d - L_d)^2}} .$$

(Here L_d denotes the lower and U_d the upper bound of the data space w.r.t the split dimension d.)

The effect of the choice of α is that for large data cells SP approaches SP_1 while for small cells SP approaches

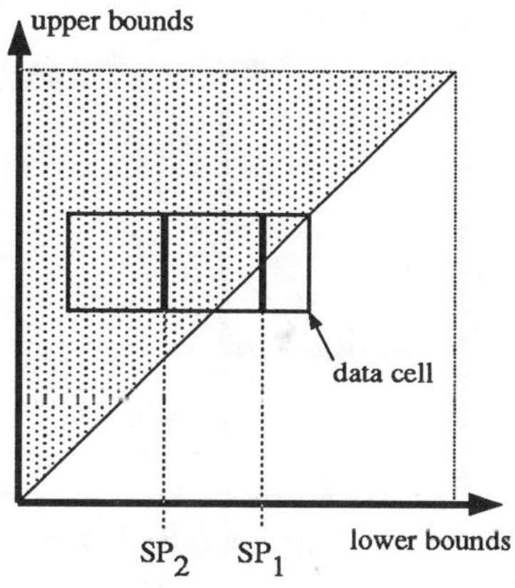

Figure 3.1: Split positions achieved by two basic split strategies

SP_2. This effect is desirable for the usual situation where rectangles tend to be small compared to the data space and hence the image points tend to be located in a strip above the diagonal. We performed simulations with other roots but the 10th root behaved well in all cases.

3.2 The operations

In this section, we discuss the LSD tree operations for non-point objects. The operations *exact match, insertion* and *deletion* are identical to the corresponding operations for 4-dimensional points. In the case of a *range query* the situation is different, because the original 2-dimensional query region and the 4-dimensional image query region differ substantially because of the different representations of the objects.

In a range query, the query region can either be an orthogonal rectangle, a circle or a polygon. Independent of the three kinds of query regions we distinguish between two query types for a set of rectangles $\Re$ (see [SK88]):

1. *Rectangle intersection*:
   Given a query region Q find all $R \in \Re$ s.t. $Q \cap R \neq \emptyset$.
2. *Rectangle enclosure*:
   Given a query region Q find all $R \in \Re$ s.t. $R \subseteq Q$.

In contrast to the point situation, the algorithm for circular and polygonal range queries is different from the algorithm for orthogonal range queries. However, due to space limitations of the paper we discuss only orthogonal range queries.

We begin explaining the *rectangle intersection problem for orthogonal query regions.* In this case, the original 2-dimensional query region for rectangles $[l_1, u_1] \times [l_2, u_2]$ is transformed into a 4-dimensional query region for points.

For each original dimension $d \in \{1, 2\}$ we define

$$\varphi([l_d, u_d]) \stackrel{\text{def}}{=} [L_d, u_d] \times [l_d, U_d] .$$

Then for the original query region $[l_1, u_1] \times [l_2, u_2]$ the image region is given by $\varphi([l_1, u_1]) \times \varphi([l_2, u_2])$. Figure 3.2 illustrates the transformation of a query interval $[l_d, u_d]$.

The area of the image region can be reduced by using the transformation φ' instead of φ. Let E_d denote the

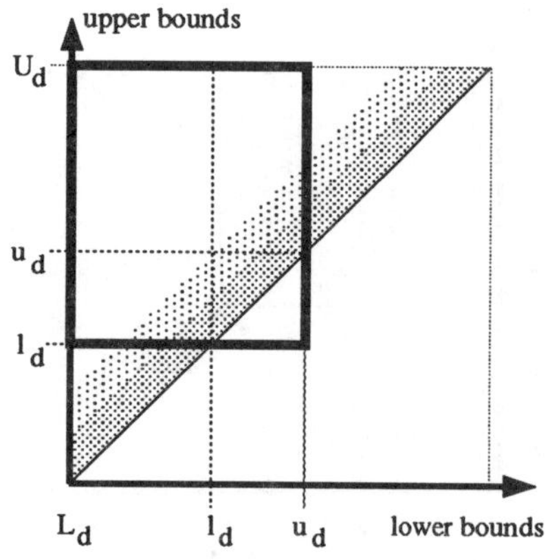

Figure 3.2: Transformation of query interval $[l_d, u_d]$

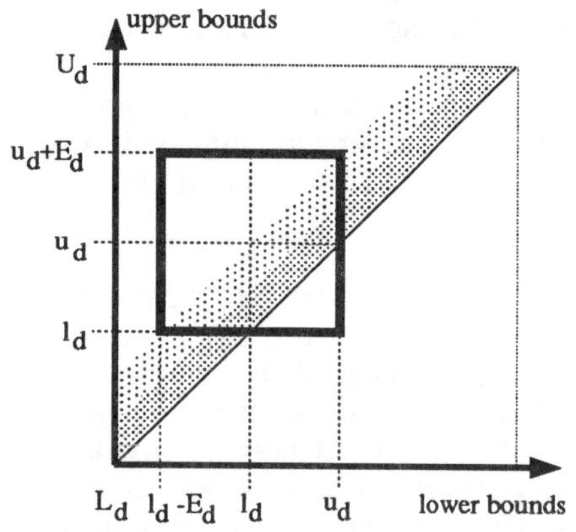

Figure 3.3: Improved transformation of query interval $[l_d, u_d]$

greatest extension of an inserted rectangle for dimension d, then

$$\varphi'([l_d, u_d]) \stackrel{\text{def}}{=} [l_d - E_d, u_d] \times [l_d, u_d + E_d] .$$

Figure 3.3 illustrates the improved transformation of a query interval $[l_d, u_d]$.

For the image region the range query algorithm for points can directly be used.

We continue the discussion with the *rectangle enclosure problem for orthogonal query regions*. Since in this case for each dimension d both, the lower and the upper bound of a rectangle, must be enclosed in the query interval $[l_d, u_d]$, we use the simple transformation

$$\vartheta([l_d, u_d]) \stackrel{\text{def}}{=} [l_d, u_d] \times [l_d, u_d] .$$

The image query region for the rectangle enclosure problem is smaller than for the rectangle intersection problem. Hence, an enclosure query can be performed more efficiently than an intersection query. Note that this holds only for transformation techniques and not for clipping or overlapping cell techniques.

Figure 4.1: "Uniformly distributed" rectangles

4. Performance evaluation

To assess the merits of the LSD tree we have evaluated the performance for rectangles in the plane. We do not discuss the efficiency of the LSD tree for points, because the performance for rectangles is an upper bound of the performance for points. We have implemented an LSD tree on a SUN workstation in Modula-2.

The *internal directory* is stored in an array storing 1000 nodes. An *external directory page* of size 512 bytes contains subtrees up to a height of 6 organized as sequential heaps. We choose bucket capacities of 5 and 50 rectangles.

The simulations are based on a sophisticated random rectangle generator creating sets of 10,000 and 100,000 rectangles according to two different distributions. These distributions are illustrated for 1,000 rectangles in Figures 4.1 and 4.2. Since the cover quotient remains constant at 2.5, in the case of 10,000, resp. 100,000, rectangles the average area of a rectangle is 10, resp. 100, times smaller than for 1,000 rectangles.

Bucket splits are performed according to the split strategies described in section 3.1. In the case of the data dependent split strategy we used a refined insertion procedure: If a new rectangle causes the split of a bucket b_1 which has a brother bucket b_2, i.e. both buckets stem from the same bucket split with split line S, and the capacity of b_2 is not exhausted, the object which is closest to S in b_1 is moved to b_2 and S is updated in the directory. Then the new rectangle can be inserted without a bucket split. Otherwise, b_1 is split.

First, we focus on the *directory evaluation*. We have randomly inserted 10,000, resp. 100,000, uniformly distributed rectangles and 10,000, resp. 100,000, skew distributed rectangles into an initially empty LSD tree. To simulate a "worst case" situation, 100,000 uniformly distributed rectangles have been inserted in "sorted" order. The "sorting" has been carried out by random insertions into an LSD

test data				size of the directory						storage utilization		
number of rect-angles	rectangle distribu-tion	split strategy	bucket capac-ity	nodes (internal and external)	internal nodes	height of the internal direc-tory	number of external directory levels	number of directory pages	average utilization of directory pages	buckets	bucket utilia-tion	overall storage utiliza-tion (100 byte per objekt)
100,000	uniform	data	50	2,864	976	13	1	41	73.5 %	2,865	69.8 %	68.2 %
10,000	"	"	"	281	281	9	0	0	-	282	70.9 %	68.1 %
100,000	skew	"	"	2,837	962	13	1	40	74.8 %	2,838	70.5 %	68.8 %
10,000	"	"	"	289	289	9	0	0	-	290	69.0 %	66.3 %
100,000	uniform	distrib.	"	3,577	986	15	1	150	28.6 %	3,302	60.6 %	59.0 %
100,000	skew	"	"	3,526	985	16	1	156	27.0 %	3,266	61.2 %	59.6 %
100,000	skew	data	5	25,882	980	14	2	1,172	34.7 %	25,883	77.3 %	71.8 %
10,000	"	"	"	2,592	979	17	1	67	39.2 %	2,593	77.2 %	71.7 %
100,000	"	distrib.	"	34,688	997	17	2	4,031	14.6 %	30,678	65.2 %	56.0 %
10,000	"	"	"	3,419	994	21	1	221	18.7 %	3,058	65.4 %	58.4 %
100,000	sorted	data	5	23,085	997	204	2	4,836	8.7 %	23,086	86.6 %	69.6 %
100,000	"	distrib.	"	33,787	985	16	2	3,501	16.2 %	29,967	66.7 %	58.1 %

Table 4.1: Size of the directory and storage utilization (directory page size = 512 bytes; max. number of internal nodes = 1000)

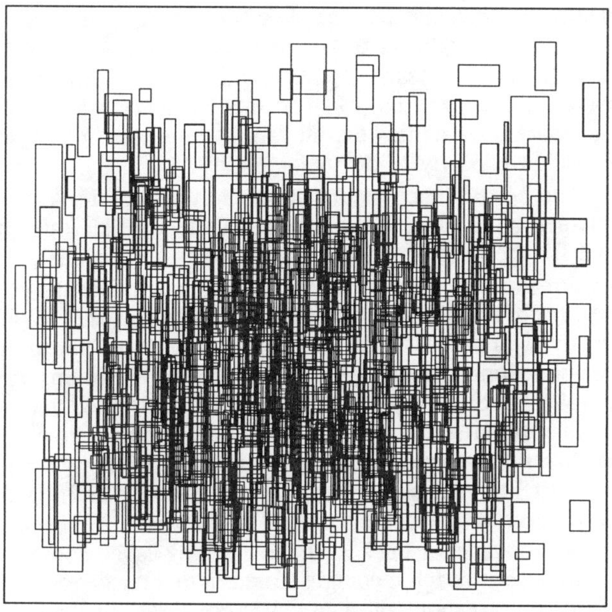

Figure 4.2: "Skew distributed" rectangles

tree with bucket capacity 5 followed by a left to right scan through the LSD leaves. The results are shown in Table 4.1.

It comes out very clearly that the size of the directory does not depend on the data distribution but on the split strategy (and of course on the size of the data set and the bucket capacity). For unsorted situations, the data dependent split strategy performs significantly better than the distribution dependent variant, while, as expected, in the sorted case the distribution dependent split strategy is the winner. For the sorted case and the data dependent split strategy the unbalance of the directory is reflected mainly by the height of the internal directory. Because of the external balancing property the number of external directory levels is 2 as for the distribution dependent split strategy. The utilization of the directory pages is mainly influenced by the split strategy and the bucket capacity (and, for the data dependent split strategy, of course by the order of insertion).

For the same test set we have also measured the *bucket utilization* which is defined as

$$\frac{\text{number of stored objects}}{\text{number of buckets} \times \text{bucket capacity}}$$

and the *overall storage utilization* defined as

$$\frac{\text{number of stored objects} \times 100\,\text{bytes}}{\text{bytes needed for the LSD tree}}$$

which includes the storage space needed for the directory and some administrative informations. Empty buckets are not allocated but represented by nil-pointers in the directory. The results are shown in Table 4.1.

For the data dependent split strategy the bucket utilization is independent of the object distribution and slightly above the theoretical value of 69.3% (ln 2). Due to the refinement of the insertion procedure which prefers small bucket capacities the utilization is even higher for bucket capacity 5. The bucket utilization of 86.6% for the sorted situation is a consequence of the same effect. For the distribution dependent split strategy the bucket utilization is

number of rectangles	rectangle distribution	split strategy	bucket capacity	range query type 1 (0.5% of the data space)			range query type 2 (5% of the data space)		
				objects found	bucket accesses	directory page accesses	objects found	bucket accesses	directory page accesses
100,000	uniform	data	50	579.7	35.6	1.9	5272.8	199.3	6.3
10,000	"	"	"	76.4	9.7	-	585.1	30.2	-
100,000	skew	"	"	818.9	49.9	1.8	6625.9	244.6	6.3
10,000	"	"	"	110.7	13.5	-	744.4	38.2	-
100,000	uniform	distrib.	"	579.7	39.5	3.8	5272.8	229.4	14.2
100,000	skew	"	"	818.9	52.0	4.2	6625.9	276.1	17.7
100,000	skew	data	5	818.9	332.3	26.0	6625.9	1995.3	109.9
10,000	"	"	"	110.7	70.3	4.1	744.4	277.3	11.0
100,000	"	distrib.	"	818.9	347.8	66.9	6625.9	2282.3	350.0
10,000	"	"	"	110.7	73.5	8.7	744.4	304.9	27.5
100,000	sorted	data	5	579.7	537.6	139.6	5272.8	2059.5	633.9
100,000	"	distrib.	"	579.7	251.8	44.5	5272.8	1790.1	249.9

Table 4.2: Range query performance (directory page size = 512 bytes; max. number of internal nodes = 1000)

below 69% but still above 60% and hence not bad at all. A comparison of the overall storage utilization and the bucket utilization convincingly demonstrates that the storage space needed to accommodate the directory is rather small compared to the data storage space.

We now turn our attention to the performance of the LSD tree operations. Clearly, in an *exact match* at most i directory pages and one bucket must be read if the directory contains i external levels. The performance of the *insertion procedure* is easy to estimate, too. For bucket capacity 5, resp. 50, we have between 4 and 5, resp. less than 3, external accesses (directory page and bucket I/O) per inserted object, if 100,000 objects are inserted into an initially empty LSD tree irrespective of the split strategy used.

Hence, we focus on *range queries*. We concentrate on the intersection query because its performance is an upper bound of the enclosure query performance. (Experiments show that the enclosure queries can be carried out 10% faster than intersection queries on the average.) Table 4.2 shows the average number of external accesses for two types of range queries. For square regions of sizes 0.5% and 5% of the size of the data space, we have performed 20 range queries each, at random positions.

As with all other data structures larger query regions lead to fewer disk accesses per found object, because the number of buckets completely contained in the query region grows faster than the number of buckets intersected by the region boundary.

Another important characteristic number is the *hit ratio*, defined as

$$\frac{\text{number of objects found}}{\text{bucket capacity} \times \text{disk accesses}}\ .$$

The hit ratio is higher for smaller bucket capacities, because of the higher selectivity. For the data dependent split strategy, bucket capacity 5, skew distribution, and query type 2, the hit ratio is 66.4% if only bucket accesses are counted. This is nearly optimal with respect to a normal bucket utilization of 69.3%. For smaller query regions and larger bucket capacities the hit ratio deteriorates: Changing the bucket capacity to 50 yields 54.2%.

The performance results in the sorted situation can be explained by the fact that the data cells tend to be long and small for the data dependent split strategy in this case.

For the remainder of this section we compare the *performance of the LSD tree* and the *multilayer grid file* [SW88] using 5 layers (5L-GF for short). According to the multilayer philosophy layer 5 is implemented as a clipping grid file. We have inserted 50,000 uniform distributed rectangles in random order into both initially empty structures. The cover quotient is 2.5 and the bucket capacity varies from 5 to 30 in steps of 5. It turns out that the 5L-GF is not able to work with bucket capacity 5. After the insertion of 20,974 rectangles a bucket of the clipping layer could not be split because each rectangle stored in this bucket covered the whole corresponding data cell. (For the skew data distribution the 5L-GF runs into a similar error situation even for bucket capacity 10.)

Figure 4.3 shows the bucket utilization for the LSD tree with the data dependent split strategy (LSD_{data}), the LSD tree with the distribution dependent split strategy (LSD_{dis}), and the 5L-GF.

The range query performance is illustrated in Figures 4.4 and 4.5. We have used the same query types as before. For the first, resp. second type, 308, resp. 2712, objects are selected on the average.

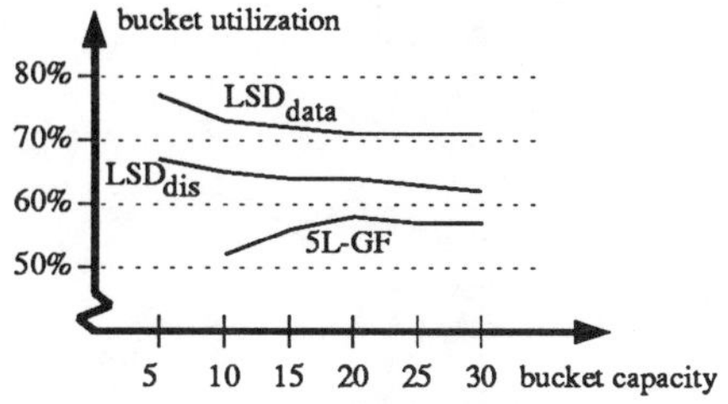

Figure 4.3: Bucket utilization

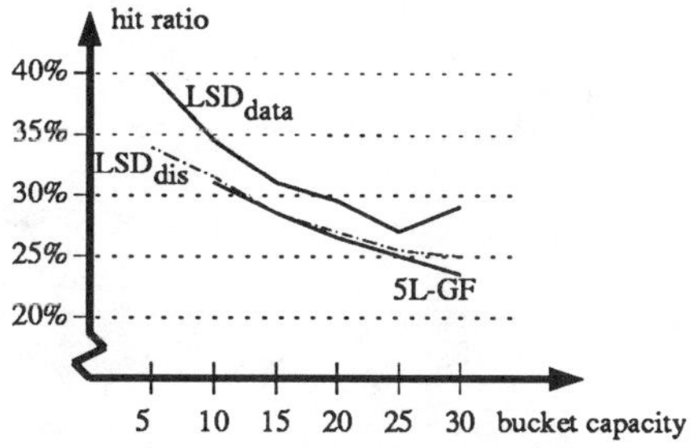

Figure 4.4: Range query performance (0.5% of the data space)

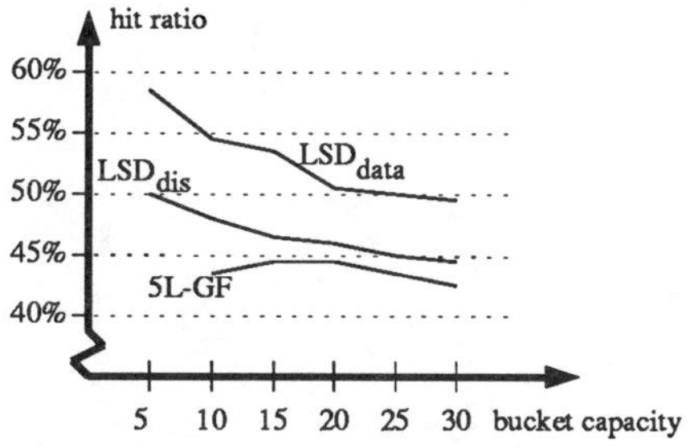

Figure 4.5: Range query performance (5% of the data space)

It turns out that the LSD tree with the data dependent split strategy clearly outperforms the 5L-GF while the LSD tree with the distribution dependent split strategy is at least as efficient as the 5L-GF. We have not compared the exact match and insertion performance because it is obvious that the 5 layers of the 5L-GF do not allow a competitive performance.

It should be noted that besides its better overall performance the LSD tree is much easier to implement than the 5L-GF and does not need an additional (completely different) "overflow" data structure for storing objects which do not fit into the main structure.

5. Conclusion

We have proposed the LSD tree, a data structure supporting efficient spatial access to geometric objects. Its main advantages over other structures are that it performs well for all reasonable data distributions, cover quotients, and bucket capacities, and that it maintains multidimensional points as well as arbitrary geometric objects. These properties make the LSD tree extremely suitable for the implementation of spatial access paths in geometric databases.

In addition to the performance evaluation an analysis of the expected storage utilization and the expected external height proves the efficiency of the LSD tree [HSW89]. At the moment, we are implementing more general (spatial) operations, like non-orthogonal range queries, point queries [SK88] and queries where geometric as well as standard attributes are qualified. Furthermore, we are embedding the LSD tree as spatial access path into the geometric database system Gral [Güt89]. Hence, an empirical study about the benefits of the LSD tree in such an environment can be carried out in the near future.

References

[Ben75] Bentley, J.L.: 'Multidimensional Binary Search Trees Used in Database Applications', Communications of the ACM, Vol. 18, 9, 509-517, 1975

[FSR87] Faloutsos, C., Sellis, T., Roussopoulos, N.: 'Analysis of Object Oriented Spatial Access Methods', Proc. ACM SIGMOD Int. Conf. on Management of Data, 426-439, 1987

[Fred80] Fredman, M.L.: 'The Inherent Complexity of Dynamic Data Structures Which Accommodate Range Queries', IEEE, CH1498–5/80, 1980

[Free87] Freeston, M.: 'The BANG file: a new kind of grid file', Proc. ACM SIGMOD Int. Conf. on Management of Data, 260-269, 1987

[Gut84] Guttman, A.: 'R-Trees: A Dynamic Index Structure for Spatial Searching', Proc. ACM SIGMOD Int. Conf. on Management of Data, 47-57, 1984

[Güt89] Güting, R.H.: 'Gral An Extensible Relational Database System for Geometric Applications', Proc. 15th Int. Conf. on VLDB (1989), to appear

[Hin85] Hinrichs, K.: 'The Grid File System: Implementation and Case Studies of Applications', Doctoral Thesis No. 7734, ETH Zürich, 1985

[HSW88a] Hutflesz, A., Six, H.-W., Widmayer, P.: 'Globally Order Preserving Multidimensional Linear Hashing', Proc. IEEE 4th Int. Conf. on Data Engineering, 572-579, 1988

[HSW88b] Hutflesz, A., Six, H.-W., Widmayer, P.: 'Twin Grid Files: Space Optimizing Access Schemes', Proc. ACM SIGMOD Int. Conf. on Management of Data, 183-190, 1988

[HSW89] Henrich, A., Six, H.-W., Widmayer, P.: 'Paging binary trees with external balancing', Proc. Int. Workshop on Graphtheoretic Concepts in Computer Science (WG '89), Springer Lecture Notes in Comp. Science, to appear

[KS86] Kriegel, H.-P., Seeger, B.: 'Multidimensional Order Preserving Linear Hashing with Partial Expansions', Proc. Int. Conf. on Database Theory, 203-220, 1986

[KS88] Kriegel, H.-P., Seeger, B.: 'PLOP-Hashing: A Grid File without Directory', Proc. IEEE 4th Int. Conf. on Data Engineering, 369-376, 1988

[KW85] Krishnamurthy, R., Whang, K.-Y.: 'Multilevel Grid Files', IBM Research Report, Yorktown Heights, 1985

[LZL88] Litwin, W., Zegour, D., Levy, G.: 'Multilevel Trie Hashing', Proc. Int. Conference Extending Database Technology (EDBT '88), Springer Lecture Notes in Comp. Science, 309–335, 1988

[NHS84] Nievergelt, J., Hinterberger, H., Sevcik, K.C.: 'The Grid File: An Adaptable Symmetric Multikey File Structure', ACM Transactions on Database Systems, Vol. 9, 1, 38-71, 1984

[Otoo86] Otoo, E.J.: 'Balanced Multidimensional Extendible Hash Tree', Proc. 5th ACM SIGACT / SIGMOD Symposium on Principles of Database Systems, 100-113, 1986

[Rob81] Robinson, J.T.: 'The K-D-B-Tree: A Search Structure for Large Multidimensional Dynamic Indexes', Proc. ACM SIGMOD Int. Conf. on Management of Data, 10-18, 1981

[SK88] Seeger, B., Kriegel, H.-P.: 'Techniques for Design and Implementation of Efficient Spatial Access Methods', Proc. 14th Int. Conf. on VLDB, 360–371, 1988

[SW88] Six, H.-W., Widmayer, P.: 'Spatial Searching in Geometric Databases', Proc. IEEE 4th Int. Conf. on Data Engineering, 496–503, 1988

Managing Complex Objects in an Extensible Relational DBMS

Georges GARDARIN(i), J-Pierre CHEINEY(ii), Gerald KIERNAN(i),
Dominique PASTRE(iii), Hervé STORA(iv)

Abstract

This paper presents the management of complex objects in an extensible relational database management system. Complex objects are built using the list constructor. Complex object types are defined as relation domains encapsulated within methods written in LISP or C. User-defined domains may be described using a "IS-A" hierarchy, which makes possible the inheritance of methods among domains. The domain methods are directly used in the external data manipulation language, which is an extended version of SQL with object oriented features (i.e., structured complex objects, methods and inheritance). When executing a query, the system selects the correct methods to apply on complex objects according to all parameter types given in method calls. With the specific LISP interpreter embedded in the DBMS, errors in functions on complex objects are detected at run time. Thus, the DBMS is protected from errors arising in user programs. Methods can also be programmed in C code and dynamically loaded when C programs are referenced in queries. Relations can be clustered according to the result of user-defined methods applied to basic or user-defined domains. Graphical interface procedures are available to input and display non-standard data. A fully running version of the system is commercially available for SUN UNIX workstations.

(i) Institut National de Recherche en Informatique et Automatique
(ii) Ecole Nationale Supérieure des Télécommunications
(iii) Eurosoft
(iv) Infosys

This work has been partly supported by the ESPRIT project ISIDE (1133) and by CNET contract n° 857.B077 and 867.B078.

Proceedings of the Fifteenth International Conference on Very Large Data Bases

Amsterdam, 1989

1. Introduction

The appeal of relational database systems for standard applications is due to the simplicity of the relational model. However, relational systems fail to cater to new applications such as office automation, CAD/CAM, CASE or geographical applications, which are characterized by complex data types and operations. For such applications, relational database systems may serve as storage systems on which are built the application software. A first approach to extend relational systems consists in adding an interface layer over an existing relational database system. The interface simulates the extended model by converting schemas and queries into their relational counterpart. This is the approach used in GEM [Tsur84], [Zaniolo83,85] which offers an entity-relationship database interface. The attractiveness of such an approach lies in its inexpensive implementation using reliable existing technology. However, the shortfall is performances. The greater the difference between the end-user model and the database model, the more complex is the translation process leading to eventual inefficiency.

Recent publications point out the need to include user-defined data types within the relational systems [Gardarin89]. User-defined data types allow the users of relational systems to tailor the database system to the needs of their specific applications [Ong84], [Osborn86], [Stonbraker83,86]. User-defined domains are operationaly defined, that is, the semantics of new data-types are the operations which can be performed over them. The programs which manipulate these new data types are registered within the DBMS system and dynamically linked at DBMS run-time. Other attempts to extend relational systems have been done using non first normal form data models. Complex objects retain more of their semantics by using richer data models which permit to capture for example the hierarchical quality of data [Bancilhon86], [Schek86], [Verso86], [Zaniolo85].

The approach proposed in this paper extends the notion of domain to support complex objects. In most implementations of relational systems, available domains are limited to integers, real number and character strings. Certain extensions have added date and money. In Codd's relational model [Codd70,79], the notion of domain is defined by a set of values. No restriction is made on the types of values which can be represented as domains. Extending relational domains to include user-defined domains has been initially presented in [Stonebraker83,86], [Ong84], for the INGRES database system and in [Osborn86] for the RAD database system. In both these systems, the relational model is extended to include user-defined domains with their associated methods. The external query languages of these systems are extended to include user-defined methods within relational expressions. User-defined methods can appear in any clause standard operators appear (projection, restriction, ...). The user programs methods in a programming language which can be compiled or interpreted. Other aspects of the database system have also to be reconsidered. Efficiency is maintained by clustering relations with frequently applied methods [Cheiney88], [Wilms88] and also by cost evaluations for query optimization [Schwartz86], [Carey86].

Different implementations of User-defined Data Types (UDT) can vary according to the following points :

(i) The possibility of defining new functions only on user-defined domains or both on UDT and on basic domains ;
(ii) The choice of programming language in which to implement user-defined operations ;
(iii) The possibility to use existing code to implement new operations ;
(iv) The existence of an "IS-A" hierarchy among domains to establish operation inheritance ;
(v) The execution of operations : dynamically linked or interpreted ;
(vi) The presence of clustering methods for UDT.

Based on these criteria, our approach :

(i) allows to define new methods (i.e., functions attached to objects) on basic or complex domains ;
(ii) uses an object oriented version of LISP for UDT programming ;
(iii) allows to reuse any method or code which is registered within the DBMS to build new UDT ;
(iv) defines UDT using an "IS-A" hierarchy to apply method inheritance among domains ;
(v) is based on a special implementation of a LISP interpreter. Although, UDT operations can be programmed in C source code and compiled into object code which is linked dynamically at run-time. LISP code can be interfaced with the C code. Errors arising in C programs are not as easily managed as those in LISP.
(vi) allows a complex object clustering method based on predicate trees.

Using these principles, we extend an existing relational DBMS named SABRINA, developed at INRIA in the beginning of the 80s [Gardarin87]. In the extended system, a UDT is defined as a method which is used to create and validate the occurrences of the domain, and methods that are applicable to this domain. The name of the method which is used to create and validate the occurrences of the domain is also used as the name of the domain. The declaration of a generalization hierarchy for each complex domain allows dynamic inheritance; a method is selected according to its parameters types. This provides greater flexibility and consistency than having each method attached to a unique domain type. When methods are programmed in LISP, the run-time is interpreted; this allows managing programming errors. However, it is also possible to dynamically link and run C object code. This feature provides better performances than interpreting LISP code. However, if no C code is used, no external files need to be managed; the method code and relative information are stored within the database system and hold in one relation. To maintain performances with LISP, an appropriate design of LISP has been implemented. This design differs from standard LISP implementations in error handling and garbage collection. No global environment variables are permitted thus simplifying and optimizing the garbage collection process.

This article presents the manipulation of complex objects in an extended relational DBMS. Complex objects are integrated as user-defined data types (UDT) at the domain level, which makes the system extensible. The extensible system is currently available as a product (SABRINA, version 7). Apart from this introduction, section 2 details the definition of UDTs using an object oriented version of the LISP language based on a subset of the LeLisp syntax [Chailloux86]. Section 3 describes the extensions to the external SQL interface for manipulating UDTs. The overall system architecture is described in section 4 with special emphasis on the alterations brought to the system for UDT manipulation. The next section describes the partitioning of relations on disk for optimal retrieval of relations containing UDT. This is done using a clustering strategy based on predicate trees and partitioning relations on the results of frequently applied methods. Section 6 deals with further strategies for optimizing query processing in the UDT framework. Section 7 discusses special interface methods for managing graphical information. The conclusion terminates the paper.

2. User-defined Data Types : Concepts and Language

2.1. User Data Types and Methods

The basic domain types commonly available in commercial relational systems (integers, real numbers and character strings) are insufficient to represent the types of data manipulated by new and diverse applications (graphical data types, lists, etc.). Each data type should be manipulated with a set of appropriate methods which may or may not be limited in number. Because of the diversity of types required by the new applications, it is insufficient to extend the database system to include a specific and limited set of data types. For this reason, the object oriented approach allows the user to define methods on objects in an incremental way. Our approach is similar in the sense that it lets the user define the data types and methods which are specific to the application area. Geographical applications may require geometric data types such as polygons. These can be built from simpler types such as Cartesian coordinates (points). A polygon can be represented as a list of points. Geographical districts can then be described as a specific form (or specialization) of polygon. Specialized methods for geographical districts can then be defined, for example, the surface method returns in acres the surface area of a district. If districts are described as a specialization of polygons, then all methods applicable to polygons are also applicable to districts.

While the implementation remains altogether relational, this extends the relational system to cater to a wider variety of application areas. It also integrate in the relational system a few concepts of the object oriented approach, namely complex objects, methods, messages inheritance and extensibility. The LISP language was chosen as a basis to build a UDT programming environment for the following reasons : i) LISP is a powerful language for manipulating complex structured objects built with the list constructor; structures such as lists, sets and trees are easily manipulated in LISP; because LISP is a weakly typed language, UDT could be defined operationaly instead of structuraly; ii) the functional approach seemed appropriate for UDT methods; new methods can be built from existing ones; iii) domain hierarchies and inheritance can be handled by LISP; iv) user programming errors could be easily managed by specific controls over basic LISP functions; v) the programming environment could be completely integrated within the database environment having function code retrieved from the database selectively; vi) specialized LISP functions could be easily added to manipulate UDT; for example, for storing and modifying UDT contained in the database. The LISP interpreter is an entry point from which UDT are updated and registered within the database system [Kiernan87].

2.2. Complex Domain Definition

To create a new UDT, the user writes an initial method that creates and validates instances of the new domain. This method will be used automatically by the integrity mechanisms either when instances are inserted or modified. The name of this method is used as the name of the domain. Once a domain name is defined, additional methods can be declared with references to these domains as parameter types. Each method parameter must be typed using a basic domain (integer, real or text) or a complex domain declared in the inheritance hierarchy of complex domains whose root is denoted # (complex object). In this way, method inheritance can be applied. For example, the surface method can be inherited from rectangle to square. The complex domain hierarchy is represented using the notion of package found in the LeLisp language [Chailloux86]. Thus, a domain is a path name from the root (labeled as #) to a terminal name in the hierarchy. For example, the square domain is described as a specialization of rectangles as follows : #:rectangle:square.

Hence, registering a new domain requires defining the domain integrity in the form of a validation method and specifying the place of the domain in the generalization hierarchy. A UDT is described as a LISP function and comprises the domain name, the generic domain names, and the validation function body definition. All methods applicable to a generic domain are also applicable to the domain itself unless it has been redefined at that level. The dd function is used to define new UDT. The syntax of the dd function is as follows :

```
(dd <domain name> (<parameter>) <function body>)
```

For example, the following function defines the RECTANGLE domain :

```
(dd #:RECTANGLE (x)
  (and (listp x)
    (numberp (car x))
    (numberp (car (cdr x)))
    (null (cdr (cdr x)))))
```

According to this function, for an object x to be a rectangle, it has to be a list of two elements where both elements are numbers. The first element is the height of the rectangle, the second is the width.

2.3. Method Definition

A UDT method is defined also as a LISP function. The de and the df functions are used to define UDT methods. These two functions are standard in current implementations of LISP. The general syntax of these functions is :

```
(de<method name> (<parameter list>)<method body>)
```

The method name is the name of the UDT method being defined. The list of parameters are the arguments to which method values will be bound to at method run-time. The method body implements the method. Other UDT can be referenced in the method body. The method name is composed of a name which is preceded by a list of domain types, one per parameter in the parameter list; thus identifying each parameter's type. For example, the LENGTH method which is applicable to RECTANGLES will be named #:(#:RECTANGLE):LENGTH. This notation insures that all parameter types are taken into consideration when selecting a method. Hence, methods may be overloaded according to all parameter types.

The following methods are defined for RECTANGLES :

```
(de #:(#:RECTANGLE):HEIGHT (x) (car x))

(de #:(#:RECTANGLE):WIDTH (x) (car (cdr x)))

(de #:(#:RECTANGLE):SURFACE (x) (* (:width x)(:height x)))

(de #:(#:RECTANGLE #:RECTANGLE):HIGHEST (x y)
  (if (> (:height x)(:height y)) x y))
```

The HEIGHT method extracts the first number in the two element list as the height of the rectangle. The WIDTH method does the same for the second element. The SURFACE method multiplies the height by the width to obtain the surface. The HIGHEST method returns the rectangle with the greatest height. Methods can also be programmed in C source code, compiled and dynamically linked with the DBMS program at run-time. For example, the surface method may defined as follows :

```
(dc #:(#:RECTANGLE):SURFACE (x)
"/usr/mydir/surface.o")
```

The dc function tells the interpreter that the SURFACE method is programmed in C. LISP and C function can be mixed. Programming errors occurring in C functions may cause the DBMS program to terminate abnormally while error occurring in LISP functions are controlled by the interpreter. Although C code runs faster than LISP code, there is a non negligible amount of time required by the loader to load and link object code to the DBMS run-time.

To define new UDTs, the user obtains the general UDT programming environment from within the external SQL language by typing the LISP command. An example of a UDT programming session follows. In this session, a new UDT method named BIGGEST is defined for RECTANGLES. The BIGGEST method accepts two arguments of type RECTANGLE and returns as result, the RECTANGLE with the greatest surface value. Once the method is defined, it is tested on trial data and then registered within the DBMS system. The SAVE function serves this purpose. The SAVE function takes two arguments: the name of the UDT and the type of result returned by the UDT. The result type is needed by the DBMS to manage coherency in relational expressions. END terminates the session and control is returned to SQL. At which point the user can include the BIGGEST method in appropriate relational expressions.

```
> LISP ;

? (de #:(#:RECTANGLE #:RECTANGLE):BIGGEST
(x y)
?  (if (> (:SURFACE x) (:SURFACE y)) x y))

= #:(#:RECTANGLE #:RECTANGLE):BIGGEST

? (:BIGGEST '(4 5) '(2 3))
= (4 5)

? (save #:(#:RECTANGLE
#:RECTANGLE):BIGGEST #:RECTANGLE)
= #:(#:RECTANGLE #:RECTANGLE):BIGGEST

? END
[]
```

3. OBJECT-SQL : The External Language Interface

3.1. Complex Object Creation

The relational DBMS implements an SQL interface which is based on the SQL norm. The language has been extended to manipulate UDT. OBJECT-SQL is the name of this extended language interface. The database administrator extends the DBMS by defining new UDT which are then made available for use in the external language. In this section, the various extensions brought to SQL to include UDT are considered.

Relations are created using the CREATE TABLE command. For example, the RECTANGLES relation is created in the following:

```
CREATE TABLE RECTANGLES  (
    R#          integer,
    COLOR       text,
    SIDES       rectangle) ;
```

This relation contains three attributes where the first two are of standard domains and the last one is a complex domain. The SIDES attribute takes its values from the RECTANGLE domain which has been defined as a list of two numbers. Once the rectangles relation has been created, values may be inserted into the relation using the INSERT command. For example,

```
INSERT INTO RECTANGLES VALUES (1, BLEU,
(4 5)) ;
```

When new values are inserted into relations, the UDT method which implements domain integrity constraints validation are run over the new values to determine if the values qualify as occurrences of the domain. The same check applies when UDT values are updated. Here, the value (4 5) qualifies as an occurrence of the rectangle domain. Note that the system does not implement object identity. However, it could be possible to identify complex objects within the validation function; that would make possible referential sharing among complex objects (i.e., complex domain values). This is left for further research.

3.2. Complex Object Selection

A UDT method can be used in any clause of a relational expression (projection, restriction, aggregation, sort) and is applicable to one or more attributes. A UDT method F applied to a number n of arguments is written as F(P1..Pn). The parameters Pi can be constants, attributes or UDT methods applied to other parameters. The F function will be selected according to all the parameter types. Methods may also appear according to their complete name (as in their declaration) and thereby bypassing the inheritance based selection mechanism. The following examples demonstrate the various possibilities:

Example 1 : A UDT method appears in the projection clause. The query selects all attributes in the relation in addition to the surface value of the sides attribute.

```
SELECT *, SURFACE (SIDES)
FROM RECTANGLES ;
```

Example 2 : UDT methods are used in a restriction clause. This query selects those rectangles with a height greater than their width.

```
SELECT *
FROM RECTANGLES
WHERE  HEIGHT (SIDES) > WIDTH (SIDES) ;
```

Example 3 : A UDT method is used in a join expression. This query selects rectangles with different surface values and displays the greatest of the two values.

```
SELECT *, BIG (R1.SIDES, R2.SIDES)
FROM  RECTANGLES AS R1, RECTANGLES
AS R2
WHERE SURFACE (R1.SIDES) <> SURFACE
(R2.SIDES) ;
```

4. The System Architecture

The system which implements the ideas described in this paper is version 7 of the SABRINA relational DBMS [Gardarin87]. In this section, the general system architecture is detailed. Then, the modifications which have been introduced to manipulate UDTs will be highlighted. The architecture is a three layered architecture which spans from the external user interfaces to the disk storage system :

(1) The interface machine comprises the outer layer of the system. It is responsible for allowing different end-users to interact with the system. These processes transform queries into an internal representation called Data Manipulation Protocol (DMP). Different types of interfaces are available in the system.

(2) The assertional machine is the intermediary layer of the system. This layer is responsible for transforming the operations of relational calculus in an optimized extended relational algebra tree, for managing data security and integrity, and for managing views.

(3) The algebraic machine is the internal layer of the system. This layer performs the operations of relational algebra. To maintain performances, access methods are used, cache memory is managed and efficient join and selection algorithms are implemented. Moreover, the algebraic machine manages concurrency and reliability.

Each of these machines comprises a number of functional processors. The global architecture was not modified by the integration of UDT. However, a new processor was added to manage UDT at the algebraic machine level. The meta-base which describes the relations, attributes and integrity constraints was extended to describe complex objects and manage domain types which are not base types. Figure 1 describes the overall system architecture. Note that several external languages are available on top of the system, including PROLOG and a specific rule language called RDL1 [Maindreville88]. Rules are interpreted at the level of the assertional machine.

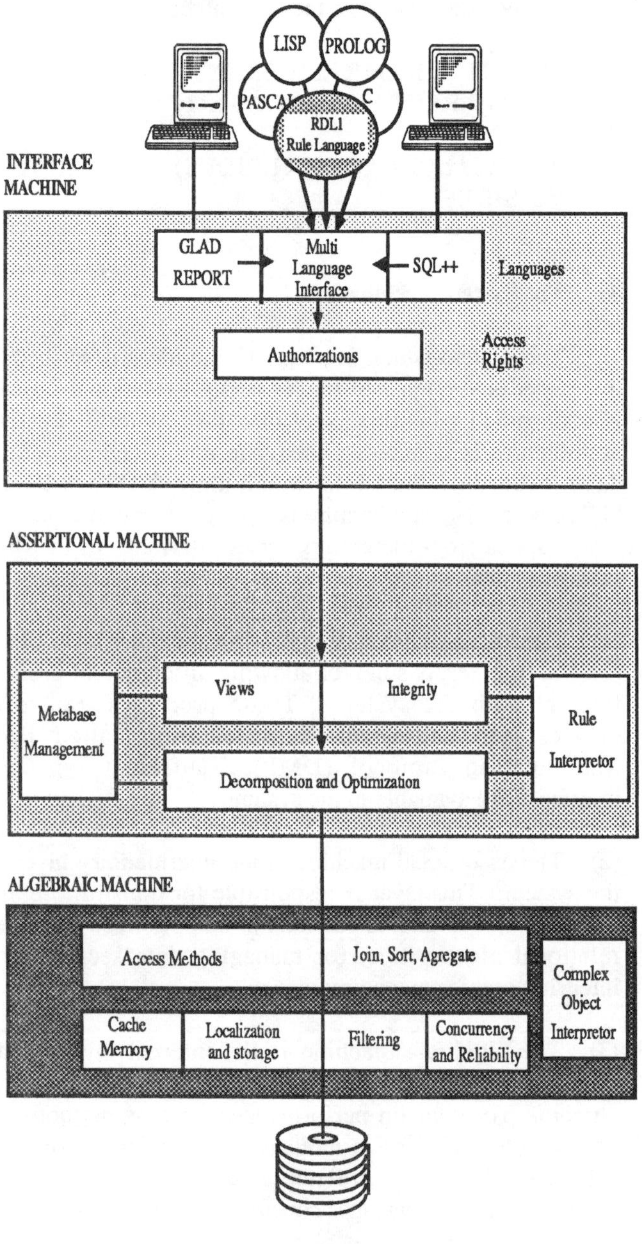

Figure 1 : Functional System Architecture

A LISP language processor is available at the interface level to allow the definition of new UDT. All information pertaining to UDT is stored in a meta-base relation. No external files need be used with only one exception: source and object files for C code. The relation which is used to manage UDT has four attributes : i) the method name defines the name of the UDT, ii) the result type defines the domain type of result returned by the function, iii) the function type describes whether the function implements a domain or a method, iv) the function text is the source code which implements the method. A sample of this relation is given in Figure 2. A UDT processor is associated to the algebraic machine to evaluate UDT in relational expressions. This processor is called by the filter when method evaluation against relations is required. The processor is also called when clustering relations on UDT.

FUNCTION_NAME	RESULT_TYPE	FUNCTION_TYPE	FUNCTION_TEXT
#:POINT	LIST	DOMAIN	(DD #:POINT(x) (and (listp x) (numberp (car x)) (numberp (car (cdr x))) (null (cdr (cdr x)))))
#:POINT:ABCISSE	INTEGER	OPERATOR	(DE #:POINT:ABCISSE (x) (car x))

Figure 2 : A sample of the UDT relation

5. Clustering Using UDT

5.1. <u>Principles and state of the art</u>

A crucial problem in introducing UDT for managing complex objects in a relational system is maintaining performances at a reasonable level. While main memory storage is increasing in size, disk I/O remains the bottleneck for such systems. New applications need to manage a large number of objects where individual objects can be big in size. Applications such as geographic applications employ graphics and consume important amounts of memory and processing time. Considering the important cost of disk I/O, appropriate clustering methods are essential.

INGRES [Stonebraker86] implements several clustering methods (B-trees, hashing), and allows adding appropriate clustering methods for complex objects. Thus, it is possible to add clustering methods designed for spatial objects such as R-trees [Gutman84] and k-d-trees [Bentley75]. This approach allows the user to integrate complex and appropriate clustering for UDT; however a programming error is hard to avoid and can cause the DBMS program to terminate abnormally.

In [Valduriez86], two techniques for clustering hierarchical objects are presented. An initial approach for storing objects consists in physically clustering object and sub-object. The method favors access to entire objects over access to sub-objects. The storage model is referred to as the Direct Storage Model (DSM). A second approach called Normalized Storage Model (NSM), stores atomic objects in flat relations. The method favors access to sub-objects which can be clustered according to one or more attribute values. Accessing entire objects requires rebuilding the object with nested join operations. However, rebuilding objects can be accelerated by maintaining join indices. These join indices join tuples on predetermined attribute values. This approach is appealing in that it does not require any special clustering method and in that it is easy to implement.

These procedures do not allow to define clustering of tuples in terms of general selection criteria. They allow clustering directly on attribute values but not on the result of methods frequently applied to complex attributes. Graphical applications seldom require accessing entire objects but more often require the result of a method applied to an object. Therefore, access to such objects is functional. For example, in the relation Rectangles (R#, Color, Sides), the sides attribute is of the rectangle domain which is a UDT. Users can require accessing rectangles based on the surface value of attribute Sides. This query will result in applying the surface method against the values of the Sides attribute. It is therefore important that clustering be also functional. That is, that tuples be grouped according to the surface value of Sides. Thus, tuples must be clustered according to the most frequently used methods in queries.

5.2. Clustering using UDT methods

The SABRINA relational DBMS implements a meta-method for clustering tuples that allows describing clusters using a Predicate Tree (PT) [Gardarin84], [Valduriez84]. Tuples qualifying a same criteria are clustered in the same branch of the PT. Each level of the PT divides the relation into a set of disjoint relations. The depth traversal of the PT allows partitioning the relation into finer and finer clusters dividing the sub-relation itself into a set of disjoint relations. A leaf of the PT corresponds to the set of tuples that qualify the conjunction of criteria from the root of the PT to that leaf. A catalogue manages the relationship between a leaf of the PT and physical disk blocks. The catalogue allows logical and physical independence between the PT and corresponding disk blocks.

When dealing with UDT, clustering has to be done on tuples qualifying a method result. Selecting rectangles according to Side ='(4 5)' can be optimized by clustering directly on the attribute's values. However, if rectangles are selected with surface value equal to 12, then clustering must be done based on the result of the surface method applied to Sides.

The extended version of SABRINA allows the user to cluster tuples according to results obtained from frequently applied methods defined as LISP or C functions. An appropriate partitioning of tuples using method results reduces the number of disk blocks needed to be scanned and thus the time required to process queries. This approach doesn't require implementing specific access methods for UDT. It allows multi-dimensional clustering on simple values and on UDT method results. The same methods used in queries can be used for clustering. Furthermore, in the case of LISP written methods, errors are managed by the interpreter thereby protecting the DBMS from abnormal termination.

Each level of the PT partitions the relation according to a specific value. The clustering predicates are of the form *f (Attri) op value* where f is a function, Attri is an attribute of the relation, op is a comparison operator among {=, <>, <,<=, >, >=}. The f function must be known to the system, that is, it has been previously coded and registered with the DBMS as a complex object method.

Different possibilities to partition a relation follow. The example will be based on the Rectangles (R#, Color, Sides) relation. Figure 3 illustrates a four level partitioning of the Rectangles relation. The first level partitions the relation by applying a hashing function on the rectangle number. The second level partitions the relation by again applying a hashing function but to the result of the surface method applied to the SIDES attribute. The third level uses a ranking function to partition the relation according to a list of predicates of the form Attri = value. The forth level again uses the ranking function but at this level, the list of predicates includes UDT methods.

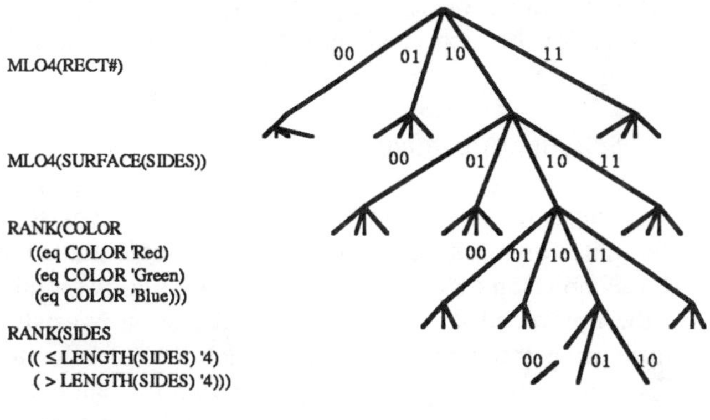

Figure 3 : Example of a Predicate Tree manipulating UDT methods

Branch 0 at level 3 constitutes the 'other' branch reserved for tuples qualifying none of the predicates in the list of predicates. The 'other' branch of level 4 is never used because the union of both predicates at this level coincide with all values included in this domain. The external language command which is used to define the clustering of a relation is the following :

```
CLUSTER RECTANGLES ACCORDING TO
(R#               ML04)
(surface(Sides)   ML04)
(Color            =RED =GREEN =BLUE)
(Length(Sides)    <=4  >4)
```

The PT is stored in a meta-base relation called CLUSTER.

5.3. Determining the location of a tuple

When a tuple is inserted, it is necessary to determine to which cluster belongs the tuple. For each level of the PT, the branch to which to the tuple belongs is established. At a certain level, the tuple can only belong to a single branch since predicates create disjoint sets of tuples. The branch to which a tuple belongs can be obtained with the following procedure :

i) a standard hashing function (modulo, folding, etc.) is applied to the result of a UDT method.

ii) the rank of the predicate to which the tuple qualified indicates the branch to which the tuple belongs.

The ranking procedure makes it possible to directly enumerate values or intervals. However, an 'other' branch is imperative because a tuple may not qualify any of the predicates at that level. For one level, clustering is determined according to a single attribute. The branch to which a tuple belongs is determined by successively applying each predicate to the tuple to determine the corresponding branch. When the tuple qualifies the predicate, the process is finished for that level and it resumes for the next level until the full depth of the PT is reached.

5.4. Selection Algorithm

The selection algorithm is divided into two steps. The first is an optimization that qualifies usable predicates. The second step determines branch numbers corresponding to the optimized PT. An initial simplification consists in eliminating those predicates Qij which do not refer to a clustering attribute by replacing them with the value 'true'. Then, the following rules are applied : the predicates Qij and 'true' are replaced by Qij, the predicates Qij or 'true' are replaced by 'true'. This simplification is independent of the functions used in the query or in the PT.

Each selection predicate is then considered successively. For each level of the tree, the predicates which can be qualified are determined along with those which are contradictory with the selection predicate. Determining the non-contradiction of two predicates is a complex problem. Only two predicates of the form *f (Attri) op value* will be considered. The general form of the predicates is the following :

Selection predicate : SP: f1 (Attri) op1 value1 when enumerating values or intervals, the clustering predicate has the following form :

Clustering predicate : BPi : f2 (Attri) op2 value2

If a standard hashing function is used (F is a hashing by division, interpolation or digital), then F(f2 (Attri)) defines the set of tuples at that level. If the clustering predicate is not atomic (as in the case of an interval), then the evaluation will be done on each basic predicate. For example, 10 < R# < 20 will be evaluated by R# > 10 and R# > 20.

To eliminate clusters which do not participate in the result of a query, selection predicates and clustering predicates must be comparable. For example, the query :

```
SELECT  *
FROM    RECTANGLES
WHERE   COLOR = "RED"
AND     SURFACE (SIDES) = 12 ;
```

uses level 2 and 3 of the PT in figure 2 and not level 4. *Surface (Sides) = 12* qualifies branch 00 at the second level with *ML04 (12) = 0*. The selection predicate *Color ="RED"* determines branch 01 at level 3 of the PT. However, no information on *Length(Sides)* can be drawn from *Surface (Sides) = 12*. The Length and Surface methods are independent.

At a given level in the PT, evaluating which branches qualify a selection predicate is done by applying hashing functions or by running the SelectBranch function which returns the list of possible branches where results may be found. When the equality operator is used in the selection predicate, then only one branch contains results. If the selection function is the identity function and clustering at a level is organized around a method applied to that attribute, then clusters can be qualified. The f2 function is applied to elements to which are applied the identity function. Then, clusters can be established using the function result. For example, a level clustering on the result of the method Surface (Sides) is useful in a query selecting tuples based on the Surface method applied to the Sides attribute: f1 and f2 are the Surface method and op is equality. Now consider for example that selection is done on identity as for Sides = (3 4). To use a cluster based on Surface (Sides), the Surface method is first applied to the constant (3 4) to obtain the result. With this result, clusters may be determined. The simplified SelectBranch algorithm is given in Figure 4.

---

function SelectBranch (SP, BPi) : Ri

{SP is the selection predicate}
{BPi is the list of predicates at a level}
{Ri is the result, i.e., a list of possible branches}

begin
 if f1 = identity **then** calculate f2 (Attri) and substitute SP ;
 if f1 <> f2 **then** select all branches
 else for each branch i **do**
  if Sp qualifies BPi **then** select branch i ;
end;

Figure 4 : The branch selection algorithm

The number of calls to the LISP interpreter by the clustering process must be limited as much as possible. The proposed algorithm reduces the number of calls when f2 is a LISP function and f1 is the identity function. All calls to the interpreter are avoided when f1 = f2. Although the clustering method is complex, little processing time is required when accessing data.

6. Optimizing the evaluation of LISP functions

6.1. Function evaluation cycles

Methods on complex objects are interpreted by a specific LISP interpreter embedded in the algebraic machine of the relational DBMS. The processing of expressions (i.e., complex objects structured as lists) by a LISP interpreter must be reconsidered for the database context where performances need to be maintained. In this section, modifications and simplifications brought to Lisp procedures are presented. The purpose for these alterations is to lessen the overhead of expression evaluation.

In this implementation of the interpreter, values contained in tuples are considered to be non symbolic. Atoms which would normally be considered symbolic are treated as plain text. This avoid having to manage the dictionary when expressions in tuples are compiled by the READ procedure. Once the expression has been compiled, the EVAL procedure decomposes the UDT methods into simpler operations which are the basic LISP functions. The cost of evaluating an expression is directly proportional to its complexity.

Once the evaluation cycle has finished, control is transferred to the PRINT procedure which transforms S-expr from their internal representation into an external readable format. The cost of printing a structure is directly proportional to its complexity. However, simple values like real numbers, integers or character strings are returned in their actual representation so no conversion operation takes place. The PRINT procedure is the least costly of the four basic procedures.

6.2. Garbage collection

Once the result has been printed, the garbage collection procedure is run to trace unused memory fragments (in certain implementations of LISP, this process does not necessarily occur after the print procedure). The GC process is also responsible for the important processing overhead of LISP.

A simplification brought to the interpreter is to allow only local variables in functions and no property lists. Therefore, only function code persists between evaluation cycles. Since the interpreter is essentially a UDT method evaluator and no user environment needs to be maintained between evaluation cycles, this restriction is not cumbersome to the UDT programmer. These two restrictions greatly simplify the task of identifying garbage in this implementation. Function code is marked as non garbage and thus valid for the lifetime of the transaction. All memory consumed during a cycle can be reused during the next cycle. The memory allocation procedure has been altered to allocate old memory fragments until those fragments have been depleted. At which point, basic system calls are run to allocate new memory. Indexes provide quick access to memory elements. So garbage collection does not hinder the performances in this context.

7. End-user high-level interface

Standard relational language interfaces offer end-users the responses to their queries in tabular form. This view of data is well adapted to the relational model and to the needs of most applications. However, the flat representation of tabular forms do not capture graphical data in a significant way.

The approach used to extend the SABRINA relational DBMS does not alter the relational view of data. The results are thus naturally expresses in the form of tables. Moreover, complex data are stored in character strings where the hierarchical quality of data is displayed with parenthesis, thus directly appreciable by the end-user. Nevertheless, users should be able to enter and display data in a manner which is appropriate with their view of it. A domain is composed of instances and methods (which can be inherited from other domains) which are user-defined. Among the set of methods applicable to an object, a user can defined methods to display and enter data in a way appropriate to human understanding.

Graphical results are represented in tabular form. These are portrayed as icons. To display the contents of such a result, the particular icon to display is pointed to with the help of a mouse. Clicking the icon displays the contents of the data in graphical form. More than one object may be displayed and superimposed on the same window. Figure 5 illustrates this possibility.

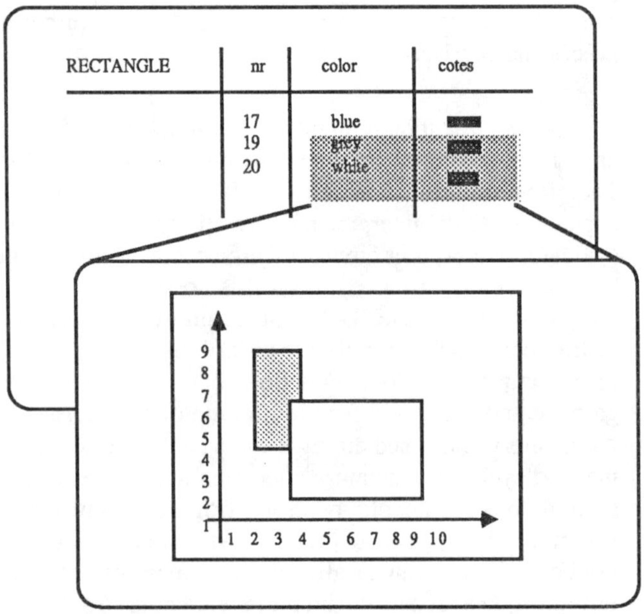

Figure 5 : Displaying graphical data

8. Conclusion

This article has described an extension to the SABRINA relational DBMS to manage User defined Data Types (UDT), which encapsulate a complex object within a set of methods. The current implementation has been described and along with the main strategies used to maintain performances. UDT are represented and programmed as LISP structures. The integrated LISP environment allows simple manipulation and implementation of UDTs. Queries can be optimized by a clustering method which allows clustering tuples according to frequently applied methods. Special methods can be defined to enter and display data in a manner appropriate to each domain type. While the interpreted environment insures protection against run-time errors in user programs, it is possible to implement UDT in C language code and dynamically link C object code to the DBMS system at run-time. C procedures are indeed called by the LISP interpreter.

To further optimize the interpretation cycle, work is being done to speed up the READ and also the PRINT phase of the evaluation cycle. At present, memory management has been simplified and the complex problem of garbage collection during method evaluation is avoided. Moreover, expressions coding complex objects in the tuples of relations are considered to be non symbolic which reduces the overhead of the system by having otherwise to manage a symbol dictionary.

In summary, UDT extends the possibilities of a relational DBMS beyond standard applications. In particular, new applications that manipulate graphical data are sought. A fully commercial version of the system is operational on SUN UNIX machines.

9. References

[AIDA87] "Aïda Reference Manual", Version 1.1, ILOG, July 1987.

[Bancilhon86] BANCILHON F., KHOSHAFIAN S., *"ACalculus for Complex Objects"*. Proc. of ACM PODS, Boston, March 1986.

[Barbedette86] BARBEDETTE G., et RICHARD P., *"VOOD : The VERSO Oriented Object Data Model"*, INRIA, Research Report No. 580, Nov. 86

[Bentley75] BENTLEY J.L. , *"Multidimentional Binary Trees Used for Associative Searching"* Communications of the ACM, N°9, Vol 18, 1975.

[Carey86] CAREY M. et al., *"The Architecture of the EXODUS Extensible DBMS"*, Proc. of the International Workshop on Object-Oriented Database Systems, Sept 1986, Pacific Grove, California, pp.52-65.

[Chailloux86] CHAILLOUX J., *"Le_Lisp Version 15.2 Reference Manual"*, Ed. INRIA, Nov 86, 3rd edition.

[Cheiney88] CHEINEY J.P., KIERNAN G., *"A Functional Clustering Method For Optimal Access To Complex Domaims In A Relational DBMS"*, Proc. 4th Int. Conf. on Data Engineering, Los Angeles, feb. 1988.

[Codd70] CODD E.F., "*A Relational Model for Large Shared Data Banks*", CACM, Vol 13, N°6, June 1970.

[Codd79] CODD E.F., "*Extending the Relational Model to Capture More Meaning*", ACM Transactions on Database System, Vol 4, N°4, Dec 1979.

[Gardarin84] GARDARIN G., VALDURIEZ P., VIEMONT Y. , "*Predicates Trees: A Way For Optimizing Relational Queries*", Proceedings of the IEEE Computer Engineering Conference, Los Angeles, 1984.

[Gardarin87] GARDARIN G., JEAN-NOEL M., KERHERVE B., PASQUER F., PASTRE D., SIMON E., VALDURIEZ P., VIEMONT Y., VERLAINE L., "*Sabrina, a relational database system developed in a research environment*", Technologiy and Sciences of Informatics, AFCET-Gauthier Villard-John Willey & Sons Ltd, 1987.

[Gardarin89] GARDARIN G., VALDURIEZ P., "*Relational Databases and Knowledge Bases*", Book, Addison Wesley, Reading, Mass., January 1989, 448 pages.

[Gutman84] GUTMAN A., "*R-Trees: A Dynamic Index Structure for Spatial Searching*" Proc. of ACM SIGMOD Conf. on Management of Data, Boston, Juin 1984.

[Kiernan87] KIERNAN G., LE MAOULT R., PASQUER F., "*Support of Complex Domains in SABRINADBMS : An Approach Using A Lisp Interpretor* ", 3ème Journées Bases de Données Avancées, Port-Barcarès, France, May 1987.

[Maindreville88] de MAINDREVILLE C., SIMON E. : "*Modelling Queries and Updates in Deductive Databases.*". Proc of 14th VLDB, Los Angeles, Sept. 1988.

[Ong84] ONG J. et al., "*Implementation of Data Abstraction in the Relational Data Base System INGRES*", SIGMOD, rec. 14, PP1-14, 1984

[Osborn86] OSBORN S. et HEAVEN T., "*The Design of a Relational Database System with Abstract Data Types for Domains*", ACM Transactions of Database Systems, Vol. 11, No. 3, Sept. 86, pp. 357-373.

[Schwarz86] SCHWARZ P., et al., "*Extensibility in the Starburst Database System*", Proc. of the International Workshop on Object-Oriented Database Systems, Sept. 1986, Pacific Grove, California, pp. 85-93

[Schek86] SCHEK H.J., SCHOLL M.H., "*The Relational Model with Relation-Valued Attributes*", Information Systems, V11, N2, 1986.

[Stonebraker83] STONEBRAKER M et. al., "*Application of Abstract Data Types and Abstact Indices to CAD Data Bases*", Proc. Engineering Design Applications of ACM IEEE Database Week, San Jose, 1983.

[Stonebraker86] STONEBRAKER M., "*Inclusion of New Types in Relational Data Base Systems*", Proc. of the 2nd Conf. On Data Engineering, Los Angeles, 1986.

[Tsur84] TSUR S. et ZANIOLO C., "*An Implementation of GEM: supporting a semanticdata model on a relational back-end*", Proc. ACM-SIGMOD Conference on Management of DATA, Boston, 1984.

[Valduriez84] VALDURIEZ P., VIEMONT Y., "*A Multikey Hashing Schema Using Predicate Trees*", ACM SIGMOD Conference on Management of DATA, Boston, 1984.

[Valduriez86] VALDURIEZ P., KHOSHAFIAN S., COPELAND G., "*Implementation Techniques of Complex Objects*", Proc of the 12th Int. Conf. On VLDB, Kyoto, 1986

[Verso86] VERSO, "*A Database Machine based on Non 1NF relations*" reseach Report INRIA n° 523, May 1986.

[Wilms88] WILMS P.F., SCHWARZ P.M., SCHEK H.J., HAAS L.M., "*Incorporating Data Types in an Extensible Database Architecture*", IBM Research Report N° RJ6405, Aug. 1988.

[Zaniolo83] ZANIOLO C., "*The Database Language GEM*"", Proc. of ACM SIGMOD, Conf. on Management of Data, San Jose, 1983.

[Zaniolo85] ZANIOLO C ., "*The Representation and Deductive Retrieval of Complex Objects*", Proc. of the 11th Int. Conf. on VLDB, Stockholm, 1985.

Effective Resource Utilization for Multiprocessor Join Execution*

Marguerite C. Murphy‡
Doron Rotem
Computer Science Research Dept.
Lawrence Berkeley Laboratory
Berkeley, CA 94720

Abstract

Conventional approaches to execution of database queries on general purpose multiprocessors attempt to maximize system throughput using inter-query parallelism with a fixed number of processors. Standard uniprocessor optimization techniques are used to minimize execution time of individual queries. Our approach is to increase performance by utilizing intra-query parallelism as well as minimizing overall resource requirements. Specifically, *processor* and *i/o bandwidth* requirements are minimized by coordinating the order in which data pages are read into memory and page joins assigned to available processors. We present a scheduling strategy based on join indices and prove lower and upper bounds on its resource requirements. We then describe a heuristic for estimating the number of processors required to complete join execution in minimal time. Our simulation results indicate that these techniques are effective with respect to processor utilization and buffer requirements.

1. Introduction

Multiprocessors have recently entered the marketplace as a cost effective high performance alternative to high-end mainframe uniprocessors. Multiprocessor architectures exploit current microprocessor technology by integrating a variable number of processors into a single system with all processors sharing a single main memory and input/output subsystem. High performance is achieved primarily via parallelism and to a lesser extent by sharing of information in main memory. With coarse grained parallelism (i.e. at the process level) existing applications can be executed concurrently without modification by simply assigning each process to a separate processor (assuming the use of system-call compatible operating systems). For example, any of the major UNIX based relational DBMS (Ingres, Oracle, Informix, Unify) can be executed with minimal modifications on a SEQUENT multiprocessor [SEQU88]. Inter-query parallelism is achieved by running multiple (independent) queries in parallel on multiple processors, using standard uniprocessor algorithms and optimization techniques.

Although impressive increases in system throughput can be achieved with coarse grained parallelism, it does not take advantage of the potential concurrency within the individual processes themselves. There are two types of parallelism that might be realized: parallel execution of CPU operations on multiple processors and overlap of CPU and I/O operations. Since the degree of CPU parallelism is potentially unbounded, system performance rapidly becomes limited by the i/o bandwidth--the rate at which data can be transferred to and from the stable storage devices into main memory for computation. In this paper we continue our investigation [MURP89] into new algorithms for parallel execution of relational join operations within the limits imposed by the i/o subsystem bandwidth.

Our approach is based on a decomposition of relational join processing into a collection of page reads and page joins (with dependencies introduced by the requirement that data pages be read into memory before participating in any page joins). We assume that data on secondary storage is organized into fixed size pages with an indexing scheme that allows construction of a "page connectivity graph"-- a bipartite graph with one node corresponding to each page of each relation and one edge connecting each pair of pages which contain at least one matching join attribute value [MERR81]. Page connectivity graphs can be easily constructed from join indices [VALD87], Bc trees [GOYA88] or intermediate results computed during index join processing [BLAS76]. The granularity of processor scheduling is the individual page join and the granularity of i/o scheduling the individual page read.

Given a page connectivity graph describing the page joins to be performed, we first determine how many processors to allocate to the computation, then schedule the order in which data pages are read into memory from disk and finally schedule each join for execution on a processor. Initially we assume that sufficient memory is available to buffer pages until joining pages and processors are

* Issued as tech report LBL-26601. This work is supported by the Applied Mathematical Sciences Research Program of the Office of Energy Research, U.S. Department of Energy under Contract DE-AC03-76SF00098. Authors' electronic mail addresses: rotem@csam.lbl.gov, murphy@lbl-csam.arpa

‡ Also with Computer Science Department, San Francisco State University.

available, implying that a page never needs to be read into memory more than once. In [MURP89] we presented a family of practical read scheduling algorithms to use with FIFO processor scheduling.

In [MURP89] we also present lower bounds on the join execution time as well as bounds on the number of processors required to complete processing in minimum time. These bounds assumed page join processing times were constant and equal to the page read time. In this paper we relax this assumption and present analogous bounds for arbitrary constant page join times as well as a more detailed stochastic performance model which includes both constant and exponentially distributed page join times. In addition, we present a heuristic for estimating the actual number of processors to allocate to a particular join graph and introduce a modified scheduling algorithm to be used in an environment with a bounded buffer pool size.

To summarize, the major contributions of this paper are: (1) a heuristic for estimating the optimal number of processors for a particular join graph, (2) a modified scheduling algorithm to be used in environments with a bounded buffer pool size, (3) bounds on execution time and number of processors assuming constant page join times, (4) extensive simulation results corroborating algorithms, bounds and heuristics.

This paper is organized as follows. The next section summarizes our join processing strategy and the following section our improved bounds on resource requirements. Section four presents a performance model based on queueing theory and the simulation model we implemented. Section five summarizes the results of a series of experiments designed to evaluate the effectiveness of our strategy. In the final section we present our conclusions and recommendations for practical implementation.

2. Join Processing Strategy

In this section we present a condensed review of material presented in [MURP89]. We do this in order to more clearly explain the extensions introduced in this paper as well as to make this paper self-contained. After describing our basic scheduling algorithm, we present a heuristic extension for use in environments with a bounded number of buffers available for join processing.

2.1 Multiprocessor Architecture

We assume a multiprocessor architecture with a flexible number of processors, one main memory and a single shared i/o subsystem. Each processor executes instructions independently of the other processors and can be individually scheduled. Data is stored and transferred in fixed sized units, referred to as blocks on secondary storage, buffers in main memory or simply pages of data. Data is transferred between the i/o subsystem and memory over a channel which has a concurrency of one (that is, at most one i/o operation can be in progress at any time). Once a data page is resident in a buffer, it can be accessed by any number of processors simultaneously. Note that we do not require simultaneous access to individual addressable units of main memory.

The following figure summarizes the basic system architecture, which is typical of that found in existing commercial multiprocessors, i.e. [SEQU88b].

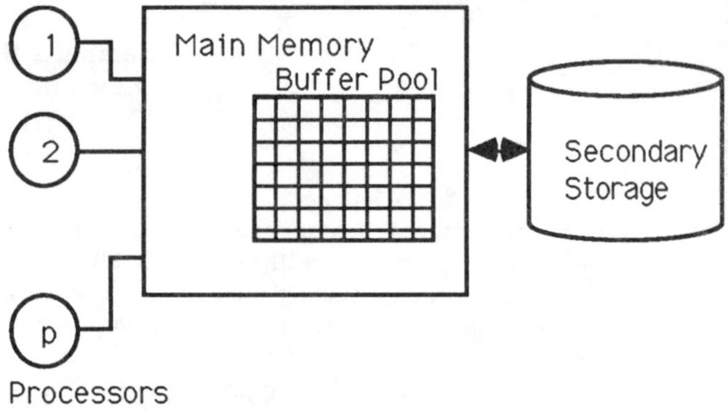

2.2 Preliminaries

The following standard definitions and notations from graph theory are used in the remainder of this paper. A bipartite graph B(V,E) consists of a vertex set $V = V_1 \cup V_2$ and $V_1 \cap V_2 = \emptyset$ and edge set $D \subseteq V_1 \times V_2$, i.e. edges are pairs of the form (x,y) where $x \in V_1$ and $y \in V_2$. A complete bipartite graph is a bipartite graph with the maximum number of edges, i.e., $|E| = |V1| * |V2|$.

For a vertex in V, we denote by d(v) its degree, which is equal to the number of edges incident on it. A connected component C of B is a subgraph of B such that a path exists between every pair of vertices of C and no path exists between a vertex in B-C and a vertex in C. We can find all the connected components of a graph B in $O(|V| + |E|)$ time using a depth-first search algorithm [Aho74].

2.3 Read & Processor Scheduling

We summarize here the scheduling algorithm presented in [MURP89]. The algorithm takes a bipartite graph B(V,E) as its input and produces a schedule S for p processors from it. For expository reasons we divide the process of producing a schedule into four stages. A schedule consists of a read schedule and a join schedule. The first three stages are directed towards producing an efficient read schedule. In the final step we perform the join schedule.

In stage 1, we decompose the graph into its connected components. In stage 2, for each component C_i we derive independently an ordering of its page reads. In stage 3, we compute for each component C_i with m edges and n vertices the function $f(C_i) = m / p - n$. We then concatenate the read schedules of all the components in decreasing order of $f(C_i)$.

At this point we have a read schedule for the whole graph. It now remains to schedule the page joins at each step. This is done in stage 4 in a simple manner. We maintain a queue of unprocessed joins as follows. Initially the queue is empty. Let us assume that vertex v is scheduled to be read at time step i. We remove from the queue the first p edges and schedule their corresponding page joins on processors 1,2,...,p respectively. If the queue contains less than p edges then some (or all) of the processors remain idle at this step. We then insert into the end of the queue all the edges incident on v which have their other endpoint memory resident. By our definitions there are A(v) such edges.

As stages 1, 3 and 4 are relatively simple we only give here a more formal description of stage 2, which is presented as Algorithm 1.

Algorithm 1:

Input: Component C with vertex set V and edge set E.

Output: A read schedule S for C. We build S by concatenating a new vertex to it each time the loop is processed. the variable A(v) keeps track of the actual work for each node. The degree of node v is denoted by d(v).

Step 1: (Initialize) For all $v \in V$, set $A(v) := 0$ and $S := \emptyset$;

Step 2: (First Vertex) Choose a vertex v_x such that $d(v_x) = \max \{ d(v) \text{ where } v \in V\}$, ties may be broken arbitrarily.

Step 3: (Add chosen vertex to S) Set *last* := v_x, append *last* to S.

Step 4: (Update actual): for all $v \in V\text{-}S$, where v is adjacent to *last* set $A(v) := A(v) + 1$;

While $V\text{-}S \neq \emptyset$ do

BEGIN

Step 5: (Choose all vertices with maximum A(v))Let MAX be the set of all vertices not in S with maximum A(v).

$MAX := \{ v \in V\text{-}S \mid A(v) = \max \{A(v) \text{ and } v \in V\text{-}S\}\}$

Step 6: (Maximize potential) Set *last* :=w where $w \in$ MAX with the value of d(w) as large as possible (break ties arbitrarily). Append *last* to S.

Step 7: (Update actual) For all $v \in V\text{-}S$ update actual work as in Step 4.

END

2.4 Restrictions on the Number of Buffers

In case the number of available buffers is restricted to some constant K, we may be forced to replace pages in memory before all their associated joins are completed. Such pages will have to be read again (at least once) to complete the execution of all their joins. We are interested in identifying join graphs which may force such a replacement. For such graphs it is not possible to complete the join execution using only |V| input operations.

In the next theorem we show a connection between the structure of the join graph, the constant K, and page replacements. The following definitions are used in the theorem. A vertex subgraph of a bipartite graph $B(V_1,V_2,E)$, is itself a bipartite graph $B'(V_1',V_2',E')$ where V_i' is a subset of V_i (i=1,2) and E' consists of all edges in E with both endpoints in V_i'. For a graph X, we denote by min(X) the value of the smallest degree of a vertex in X.

Theorem 1: Given a join graph B with |V| vertices, a join execution for B requires at least |V| +1 read operations under a restriction of K buffers, if B contains a vertex subgraph B' such that $\min(B') \geq K$.

Proof: Let us assume that B contains a subgraph B' with $\min(B') \geq K$ and $|V| = r$. We observe that $r \geq 2K$. We will derive a contradiction by assuming that it is possible to complete the join of B with no page replacements.

Let us label the nodes of B' as $v_1,v_2,\ldots,v_r$ such that v_i is the ith node of B' read by the schedule. When v_r, the last vertex of B' is read in, it must join with at least K other nodes from B' by our assumption on min(B'). Since there are at most K-1 available buffers, at least one of these neighbors is not currently present in memory. This means that at least one additional node of B' must be read in contradicting the fact that v_r is the last node of B' read by the schedule. ◻

The above theorem is of theoretical interest only since identifying such a subgraph is an NP-complete problem [GARE79]. In the next section we show how our heuristic algorithm can be modified to operate efficiently under buffer restrictions.

2.5 Modifications to the Heuristic for Buffer Restrictions

In the presence of memory restrictions, Algorithm 1 has to be modified so that nodes can be replaced in memory and then read again. In this section we describe informally how our scheduling can be dynamically adapted to an environment with only K buffers. In this case, some changes must be made to the read schedule when it cannot simply read a new node in because all K buffers contain nodes which must still participate in more joins. At this point we have to make a choice between two alternatives:

(a) In the next step no node will be read in and only joins among memory resident nodes will be performed;

(b) Choose a node in memory (the "victim") and replace it by an input node. The "victim" node will be reread at some later point in order to complete its joins.

Our strategy, as before, is to attempt at maximizing the amount of work in the system by "greedy" decisions. For each node v we keep track of its actual and potential work when it is read in using the counters A(v) and P(v) respectively. We subtract one from each of these counters with each join performed which involves node v. A node for which P(v) becomes zero can be replaced by a new node without any rereads.

As in Algorithm 1, we read in the next node for which A(v) is maximum as long as we have free buffers. Let us assume that the next input node according to our read schedule is x, and there are no free buffers, i.e., the current set of nodes in memory is M where |M| = K and no member v of M has P(v) = 0. We compute the value of the total maximum actual work over all nodes in M and compare it with the total actual work which can be achieved in the system by replacing a node in M with x. If the former value is larger, we simply proceed with alternative (a). Otherwise we choose a node y in memory for which the total actual work in the system (computed after replacement of y with x) is the largest possible. We then read node x into the buffer currently occupied by node y.

The following adjustments must be made:

* All joins involving node y must be removed from the work queue.

* The value of A(v) must be decremented for all neighbors of node y.

* The node y is placed in the queue of unread nodes with its current value of A(y) and P(y) and will be scheduled for reading according to our usual criteria.

* As before, after node x is read in, we need to update the values of A(v) for all nodes in the system which are neighbors of x (increase by 1) and add all joins involving x and a memory resident node into the work queue.

3. Bounds Assuming Constant Page Join Processing Time

In this section we assume that the join graph B(V,E) is a bipartite connected graph where $V = V_1 \cup V_2$. The bounds we derive here are based on knowledge of some simple parameters of the graph such as the cardinalities of the edge and vertex set or the size of the largest degree in the graph. Of course it is possible to derive tighter bounds if we have more information about the graph. Bounds are important since they give us some ideal measures against which we can compare our algorithms. We consider in this section bounds on the number of processors required to complete the join in optimal time. In the next section we deal with bounds on the time required to complete the join with a given fixed number of processors.

We denote by $T_{opt}(B)$ the optimal time to complete a join represented by the graph B, i.e. the minimum number of time steps with an unbounded number of processors where the join time is a constant C.

Lemma 1:

$$T_{opt}(B) = |V| + C.$$

Proof: Let S be an optimal time schedule for the graph B. At each time step we can read exactly one vertex of the graph. Let us denote by v_n the last vertex read by the schedule S. The processing of all edges incident on v_n can be completed only after v_n has been read. Since the graph is connected, there will be at least one edge incident on v_n and therefore at least C additional time units are needed after all the vertices have been read in order to complete the execution.

In the following theorems we compute an upper and a lower bound on $P_{opt}(B)$, the number of processors required to complete the join in optimal time $T_{opt}(B)$. We derive a bound which assumes that all we know about the graph is |V| and |E|.

Lemma 2: The maximum number of page joins a schedule can complete during the first i + C time units (with unbounded number of processors) is

$$\begin{cases} \frac{i^2}{4} & \text{for i even} \\ \frac{(i-1)(i+1)}{4} & \text{for i odd} \end{cases}$$

Proof: Let us denote by $B_i(S)$ the subgraph of B read by a schedule S during its first i steps. This subgraph consists of the set of i vertices read by the schedule and all edges of B with both endpoints in this set. Let V^i_1 and V^i_2 be the vertices of $B_i(S)$ which belong to V_1 and V_2 respectively. At step i+1, the schedule S can perform all joins such that their corresponding edges are in $B_i(S)$. It is easy to see that for any schedule S, the number of edges in $B_i(S)$ is maximized when this graph is a complete bipartite graph with $|V^i_1| = |V^i_2| = i/2$ when i is even and $|V^i_1| = (i-1)/2$ and $|V^i_2| = (i+1)/2$ for i odd. The expression in the statement of the lemma represents the number of edges in the graph corresponding to each of these cases. ¤

Theorem 2: The number of processors required to complete the join in optimal time on the graph B(V,E) satisfies

$$P_{opt}(B) \geq C \times \text{MAX}\left\{\frac{|E|}{t}; \frac{|V| - \sqrt{|V|^2 - 4|E|}}{2}\right\}$$

where $t = |V| + C - 2$.

Proof: We first prove that the number of processors needed is larger than the first term in the curly brackets. Let us assume that schedule S completes the join in optimal execution time with $P_{opt}(B)$ processors. We observe that there are t time units in which processors must complete all |E| joins because no join can take place during the first two time units. Since a single processor can complete at most $\left\lfloor\frac{t}{C}\right\rfloor$ joins during this period we have

$$P_{opt}(B) \geq \frac{|E|\,C}{t}$$

We now prove that the number of processors needed is also larger than the second term in the curly brackets. We denote by c(i) the number of page joins performed by S during its first i+C time units. Then in order for S to complete the join in optimal time, it has to perform the additional |E|-c(i) joins during the remaining time which is |V| - i. Since each processor can complete at most $\left\lfloor\frac{|V| - i}{C}\right\rfloor$ joins during this period, we have the inequality

$$P_{opt}(B) \geq C \times \left\lceil\frac{|E| - c(i)}{|V| - i}\right\rceil$$

By Lemma 2, for $1 \leq i \leq |V|$

$$c(i) \leq \frac{i^2}{4}$$

From which we conclude that for $1 \leq i \leq |V|$

$$P_{opt}(B) \geq C \times \left(\frac{|E| - \frac{i^2}{4}}{|V| - i}\right)$$

In order to make the bound as tight as possible we will find the value of i for which the right hand side achieves its maximum. We use elementary calculus to find that the right hand side achieves its maximum when the value of i is the closest integer to

$$|V| - \sqrt{|V|^2 - 4|E|}$$

By substituting this value of i into the right hand side we obtain

$$P_{opt}(B) \geq \frac{|V| - \sqrt{|V|^2 - 4|E|}}{2}$$

as claimed.□

In the next theorem we exploit more information about the graph to derive an upper bound on $P_{opt}(B)$. We assume that $|E|$, $|V_1|$, $|V_2|$ are given. Without loss of generality let $|V_1| \leq |V_2|$.

Theorem 3: The optimal number of processors satisfies

$$P_{opt}(B) \quad \leq \quad \left\lceil \frac{|E|}{|V_2|} \times M \right\rceil$$

where $M = \lceil C \rceil$ if $C \geq 1$ and $M = \frac{1}{\left\lfloor \frac{1}{C} \right\rfloor}$ if $C < 1$.

Proof: We exhibit a simple schedule S' which completes the join in optimal time using no more than the above number of processors. The schedule S' is characterized by the following rules: Let us call a step in which a vertex of V_1 is read a type I step and all other steps are called type II steps.

3.1 The vertices of V_2 are sorted in non-increasing order of their degrees and relabelled $v_1, v_2, ..., v_n$ such that v_1 is the vertex with the maximum degree and v_n has the minimum degree.
The schedule performs its reads in n+1 rounds each consisting of a type II step followed by zero or more type I steps in the following way:

3.2 In the first time step of round i (i<n+1) v_i is read, this is followed by reading in all the vertices of V_1 connected to it which have not yet been read. Round n+1 consists of the final join.

3.3 All page joins are scheduled to take place as early as possible, i.e. as soon as the two endpoints of an edge have been read in and there is a free processor to perform the join.

The number of new potential joins introduced at the end of round i is at most equal to $d(v_i)$, the degree of v_i. For simplicity, from now on we will assume that page joins are performed only on the first step of each round (from round 2 onwards), i.e. on the first step of round i+1 we will attempt to perform all remaining page joins involving v_i. (this is possible as all endpoints of edges involved have been read). The proof of the theorem has two parts, in Part I we prove that it is sufficient to complete at each round $A = \left\lceil \frac{E}{n} \right\rceil$ joins in order to obtain optimal execution time. We then show in Part II that A x M processors are sufficient to achieve this goal.

Part I:

Intuitively, this result holds since there are $|E|$ joins which must be performed over n rounds and therefore A is roughly the average number of joins per round. More precisely, at the beginning of round i+1 (for i<n) there are at most $d(v_i)$ new joins to be performed. In case only A are performed during this round, there are up to $d(v_i)$-A potential joins which may have to be deferred to some future round.

Note that the value of A is the ceiling of the average degree of vertices in the set V_2, so that there must be a first index j such that $d(v_j) \leq A$. At the beginning of round j, the total number of page joins deferred from all previous rounds is at most

$$\sum_{i=1}^{i=j-1} (d(v_i) - A)$$

On the other hand, by the decreasing order of degrees, there will be a total of at least

$$\sum_{i=j}^{n} (A - d(v_i))$$

available processors to complete these deferred joins during rounds j+1,j+2,...,n+1. Since as we noted before A is equal or larger than the average degree in the set V_2 we have

$$A \quad \geq \quad \frac{1}{n} \sum_{i=1}^{i=n} d(v_i)$$

from which it follows that

$$\sum_{i=j}^{n} (A - d(v_i)) \quad \geq \quad \sum_{i=1}^{i=j-1} (d(v_i) - A)$$

and the number of available processors is sufficient to complete all the page joins.

Part II:

Given a join time of C we will now show how the processors can be scheduled during each round. If $C \leq 1$ each processor can complete $\left\lfloor \frac{1}{C} \right\rfloor$ joins in one time unit. Therefore A x M processors are sufficient to complete A joins in each round. In the case $C > 1$, we can use a group of A processors at the beginning of each round. Since a group of processors is utilized for at most C time units we can reuse it every $\lceil C \rceil$ steps. In this way we will be using at most A x M processors at any given step.□

4. Discrete Event Simulation

The dynamic behavior of our algorithm can be modeled by a simple multi-server queueing model. At each step of the read schedule a data page is read into memory. After the read is complete, one or more joins to other memory resident data pages may be enabled. As soon as a join is enabled, it may be assigned to a processor for execution. If all processors are busy, the enabled joins are queued until they can be processed. The following diagram summarizes this behavior:

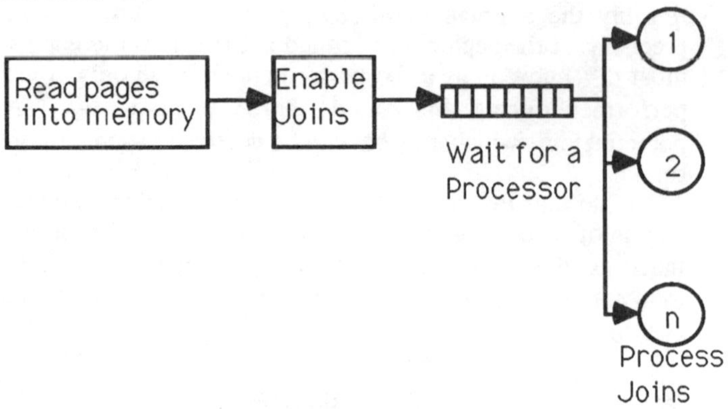

In order to evaluate our algorithms and heuristics, we implemented a stochastic simulation model. Input parameters are summarized in the following table:

Input	Description
NumNodes	Total Number of Nodes in Graph
Relation 1	Number of nodes in first Relation (Opt)
Alpha	Fraction of Edges present in graph
Mean	Mean CPU Processing Time
Distribution	Exponential/Constant Processing Time
Seed	Random Number Seed
Runs	Number of Replications
Lookahead	Depth of Potential Cost Evaluation
NumProcs	Number of Processors (Optional)

The first step of our simulation is to generate a *random graph* by partitioning the nodes into two subsets. If the number of nodes in the first subset is not specified, a random partition is performed. Next the number of edges is determined as the product of alpha and the maximum number of edges possible. We randomly select this number of edges (without replacement) from all possible edges and construct a bipartite graph. This graph represents a page connectivity graph. The page connectivity graph is then used as input to our scheduling algorithm to produce a page read schedule giving the order in which pages (nodes) should be read into memory. This schedule is augmented by a list of joins (edges) ordered by the times at which they become enabled.

This join list is then input into a discrete event simulation [LAWK82] of a multiprocessor with n processors. Each processor has a join processing time (service time) described by a random variable. The random variable has either a constant or exponential distribution with a fixed mean. Throughout the simulation, random numbers are generated using techniques described in [PARK88]. If the number of processors is not specified, the simulation will estimate the number of processors required to complete processing in optimal time (Popt) and use this number during simulation of join execution.

The following table lists the output parameters of our simulation:

Output	Description
Popt	Estimated Optimal Number of Procs
Execution Time	Simulated total execution time
Utilization	Simulated Processor Utilization
NumBufs	Number of Buffers required

4.1 Optimal Number of Processors

The optimal number of processors is estimated by assuming that the page join times are constant and equal to the page read times. In [MURP89] we presented lower bounds on the execution time in terms of the total number of nodes (|V|), the number of edges (|E|) and the number of processors (p):

$$\text{Execution time} \geq \max\left\{|V| + 1,\ 2p + 1 + \left\lceil \frac{|E| - p^2}{p} \right\rceil \right\}$$

In initial experiments we observed that the simulated execution time was almost always equal to the lower bound. Intuitively this should be the case when the processor utilization is sufficiently low that joins rarely need to wait to be processed. We then estimate the optimal number of processors as the value of p for which

$$|V| + 1 = 2p + 1 + \left\lceil \frac{|E| - p^2}{p} \right\rceil$$

In the following experiments we evaluate the use of this estimate when the page join times are constant, but not necessarily equal to the page read time. Intuitively, if the page join times are close to the page read times, the estimated optimal number of processors should be close to the true optimal value.

5. Simulation Experiments and Results

In this section we present the results of an extensive series of simulation experiments designed to verify our algorithms, bounds and heuristics. The charts displayed in this paper are representitive of the results we obtained.

5.1 Comparison with Random Schedule

In this experiment we compared the execution time produced by the read schedule of Algorithm 1 with a random read schedule for three typical graphs. As we can see in Charts 1 and 2, Algorithm 1 is consistently better (by roughly 20%) when the number of processors is less than the optimal. As expected, increasing the number of processors above Popt has little effect on the execution time since no additional improvement is possible.

5.2 Execution Time and Popt

This experiment measured the execution time produced by Algorithm 1 for a wide range of typical join graphs. The parameters of these graphs are given in the two tables below. The trends in Charts 3 and 4 are consistent with our predictions, i.e. the execution time decreases with the number of processors and increases with the number of edges. The charts level off at the optimal execution time slightly before we use Popt processors, i.e. no further improvements in execution time occur after that point.

Graph	Symbol
Split=45-55%, Alpha = .25	◆
Split=45-55%,Alpha = .60	◇
Split=25-75%,Alpha = .25	■
Split=25-75%, Alpha = .60	□

Number of Nodes	Split	Alpha	Popt
75	45-55%	0.25	7
	25-75%	0.25	5
	45-55%	0.6	15
	25-75%	0.6	11
25	45-55%	0.25	3
	28-67%	0.25	3
	45-55%	0.6	6
	28-67%	0.6	5

5.3 Buffers and Popt

Our simulation model assumes that a buffer is freed whenever all joins associated with the node have been completed. In this experiment we measure the maximum number of buffers used by Algorithm 1 under this assumption. We observe a few interesting trends in these Charts (5 and 6):

(1) The relative sizes of the relations and densities of the graphs is significant, i.e. graphs which are more evenly split and/or denser require more buffers.

(2) Adding more processors helps to reduce the number of required buffers as joins have to wait less time for execution. If the number of processors is less than Popt, the number of buffers required increases sharply.

5.4 Join Time and Popt

In the following series of experiments we vary the page join times between 0.1 and 2.0 (the units are fraction of the time required for a page read). In this experiment we measured the execution time (Charts 7 and 9) and processor utilization (Charts 8 and 10) for the optimal number of processors (Popt), the optimal number of processors plus a constant one (Popt +1) and the optimal number of processors less one (Popt -1). We observe that the two metrics are highly correlated with the number of processors. The execution time is constant (and bounded by the optimal time) until the processors "saturate" (i.e. utilization approaches one), at which point the execution time increases roughly linearly with the page join time.

5.5 Exponential vs Constant Join Times

These experiments were designed to examine the effects on processor utilization and total execution time of stochastic variations in the page join times (Charts 11,12,13,14). We compared the performance of two typical join graphs assuming exponentially distributed join times with the performance using constant page join times having the same means. We did not find significant differences in execution time or processor utilization, indicating that our bounds based on an assumption of constant page joins times are reasonable approximations to the situation of exponentially distributed page join times.

6. Conclusions and Future Work

In this paper we studied the problem of optimizing join execution on multiprocessors from both the theoretical and practical point of view. We derived lower bounds on the execution time and optimal number of processors based on the structure of the join graph. We then devised a heuristic scheduling algorithm which produces an order of reading in the pages on the relations and scheduling the joins on the processors. We identified several parameters which might significantly affect the resource utilization requirements of a join plan. These include: size of the graph, density of edges, relative sizes of relations, distribution of join time and number of processors. We conducted a large number of experiments with our heuristic in order to examine the effect of all these variables on the resource requirements and their inter-dependencies.

Our results indicate that the heuristic performs well under a wide range of conditions and the resource utilization achieved by it matches quite closely the theoretical lower bounds.

Future work in this area includes examining more join strategies and also algorithms for additional relational operators. In the near future we plan to integrate buffer restrictions and heuristic strategies into our simulation models. In addition, we plan to implement our algorithms and obtain performance measurements on a commercial multiprocessor such as Sequent.

We believe that in the future query optimizers for databases which run on multiprocessors will have to be enhanced in order to take full advantage of the parallelism offered by such a system. Such optimizers will need to incorporate cost functions based on accurate predictions of resource requirements and execution time in order to correctly evaluate the costs associated with different join plans. The research presented here is a first step in this direction.

7. References

[Aho74] Aho, A. V., J. E. Hopcroft and J. D. Ullman. (1974). The Design and Analysis of Computer Algorithms. Reading, Mass, Addison-Wesley.

[BLAS76] Blasgen, M. W. and K. P. Eswaran. (1976). On the Evaluation of Queries in a Relational Data Base System. IBM Technical Report #RJ1745 (#25553) Computer Science (April 8, 1976).

[GARE79] Garey, M. and D. Johnson. (1979). Computers and Intractability. San Francisco, W.H. Freeman and Company.

[GOYA88] Goyal, P., H. F. Li, E. Regener and F. Sadri. (1988)."Scheduling of Page Fetches in Join Operations Using Bc-Trees." Proceedings of 4th International Conference on Data Engineering,Los Angeles, CA,IEEE.

[LAWK82] Law, A. M. and W. D. Kelton. (1982). Simulation Modeling and Analysis. New York, McGraw Hill.

[MERR81] Merrett, T., Y. Kambayashi and H. Yasuura. (1981)."Scheduling of Page-fetches in Join Operations." Proceedings 7th International Conference on Very Large Data Bases,Cannes, France.

[MURP89] Murphy, M. C. and D. Rotem. (1989)."Processor Scheduling for Multiprocessor Joins." Fifth International Conference on Data Engineering,Los Angeles, CA,IEEE.

[PARK88] Park, S. K. and K. W. Miller. (1988). "Random Number Generators: Good ones are Hard to Find." CACM. **31**(10).

[SEQU88] Sequent Computer and Codd & Date Associates. (1988). Combining the Benefits of Relational Database Technology and Parallel Computing. Technical Seminar, San Francisco, CA, September 28, 1988.

[SEQU88b] Sequent Computer Systems. (1988). Systems Overview. Product Description, Sequent Computer, Beaverton, ORE.

[VALD87] Valduriez, P. (June 1987). "Join Indices." ACM Transactions on Database Systems. **12**(2): 218-246.

Chart 1:

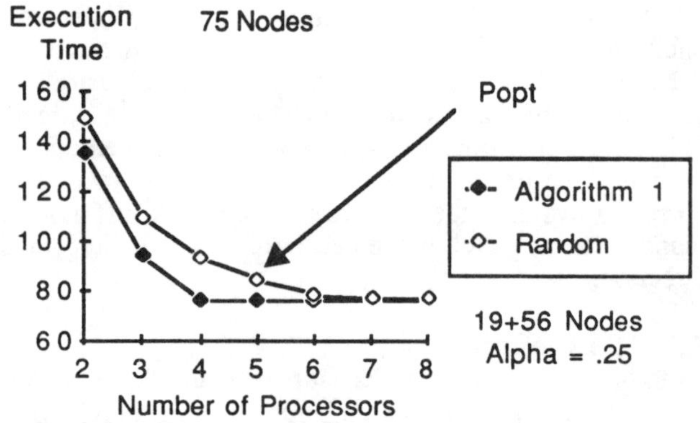

Chart 2:

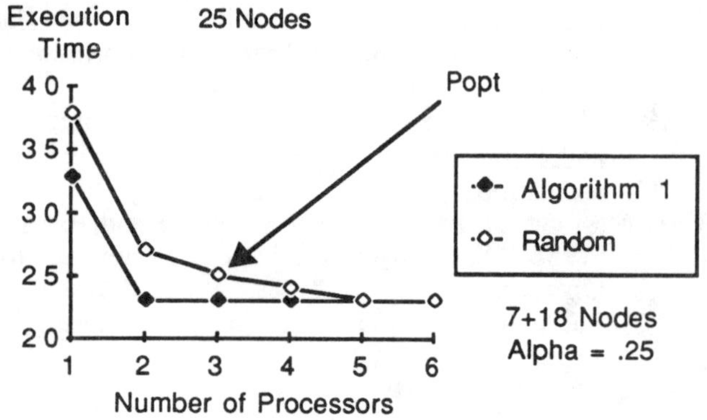

Chart 3:

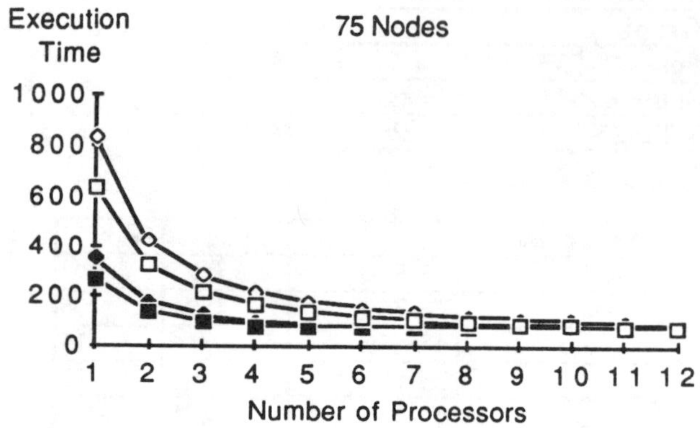

Chart 4:

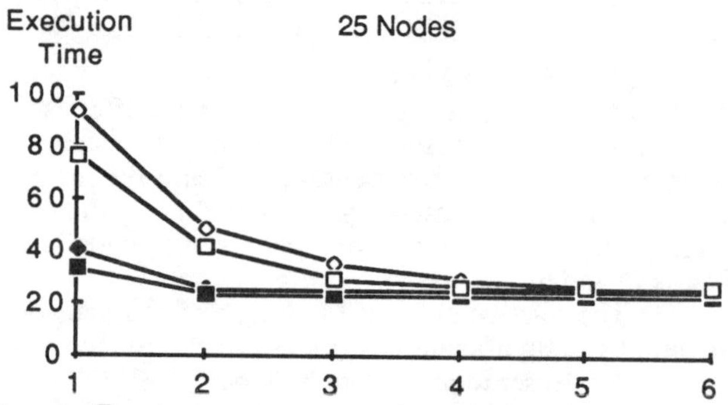

Chart 5:

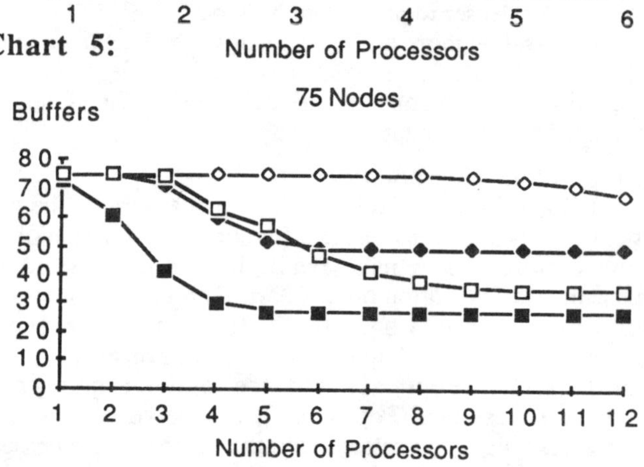

Chart 6:

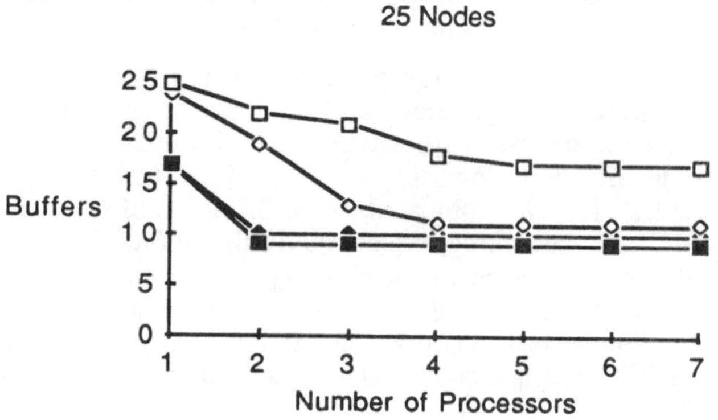

Chart 7:

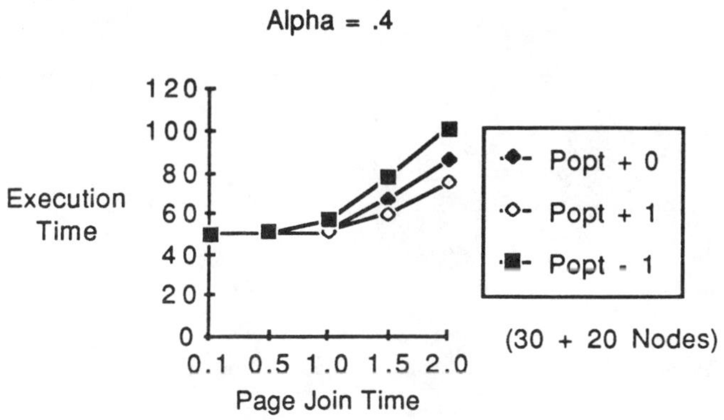

Chart 8:

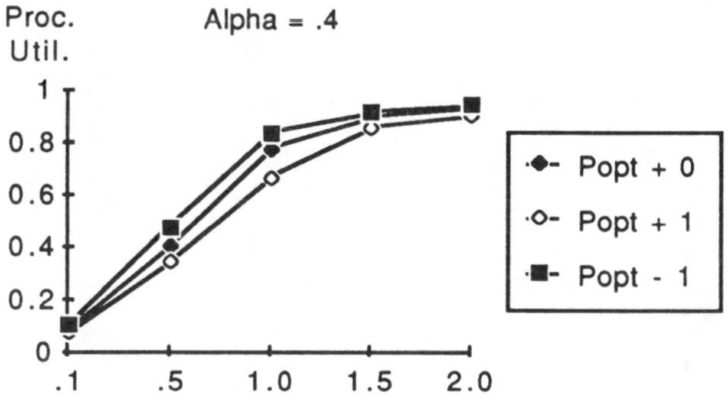

Chart 9:

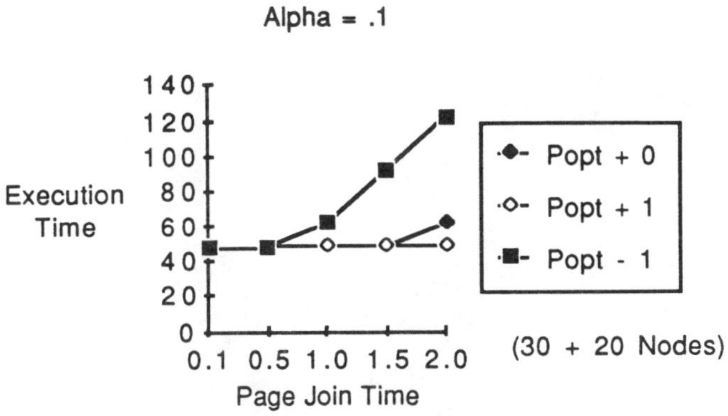

Chart 10:

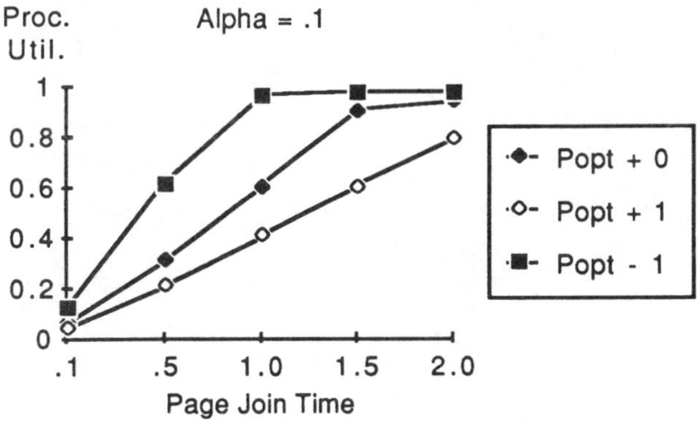

Chart 11:

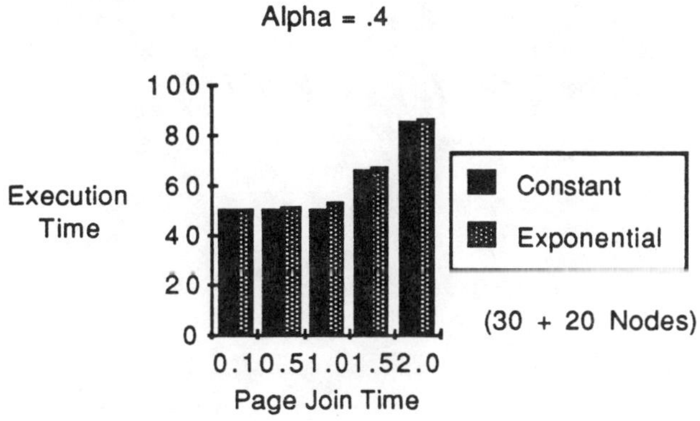

Chart 12:

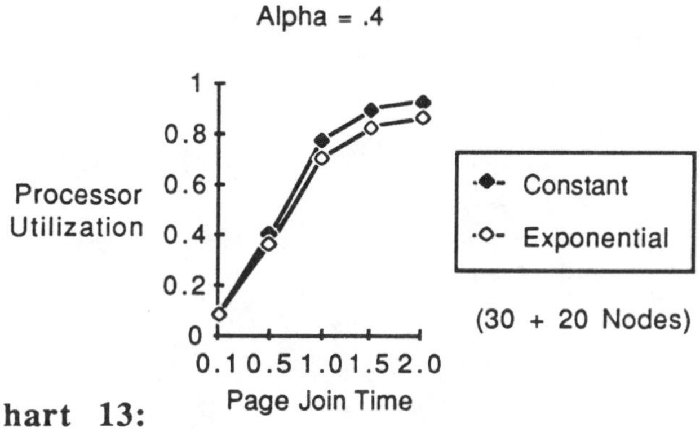

Chart 13:

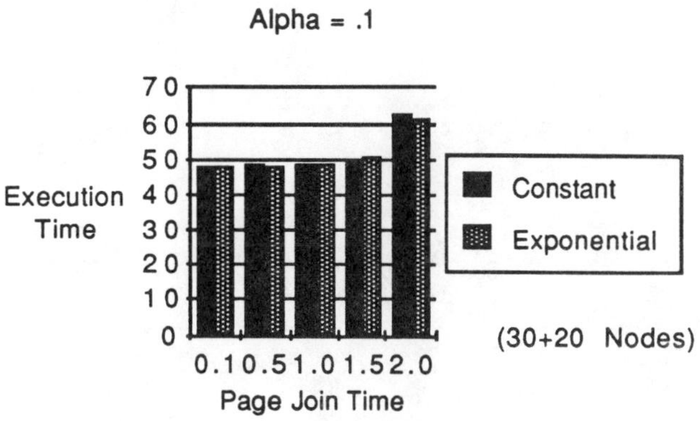

Chart 14:

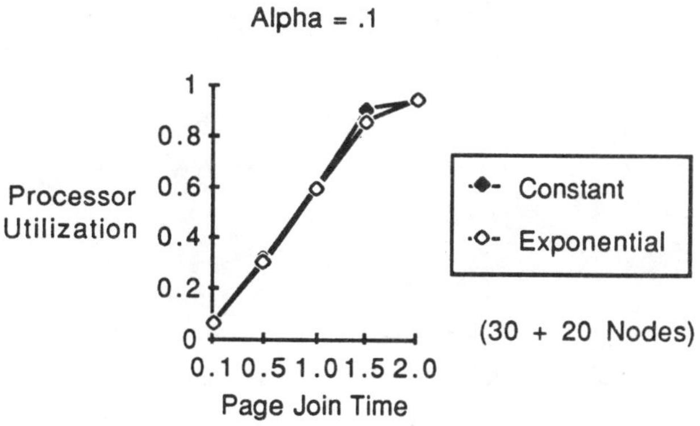

Optimization and Dataflow Algorithms for Nested Tree Queries

M. Muralikrishna
murali@cookie.dec.com
Digital Equipment Corporation
1175 Chapel Hills Drive
Colorado Springs, CO 80920

Keywords: unnesting, optimization, SQL, dataflow, the COUNT bug, outer join, anti-join, correlation predicate.

1. Abstract

The SQL language allows users to express queries that have nested subqueries in them. Optimization of nested queries has received considerable attention over the last few years. Most of the previous optimization work has assumed that at most one block is nested within any given block. The solutions presented in the literature for the general case (where an arbitrary number of blocks are nested within a block) have either been incorrect or have dealt with a restricted subset of queries. The two main contributions of this paper are: (1) optimization strategies for queries that have an arbitrary number of blocks nested within any given block, and (2) a new dataflow algorithm for the execution of nested queries, involving one or more outer joins, in a multi-processor environment such as the one found in GAMMA. The new algorithm cuts down on message and CPU costs over conventional dataflow algorithms.

2. Introduction

Traditionally, database systems have executed nested SQL [Astrahan75] queries using Tuple Iteration Semantics (TIS). It was analytically shown in [Kim82] that executing queries by TIS can be very inefficient. It was first pointed out in [Kim82] that nested queries can be evaluated very efficiently using relational algebra operators or set-oriented operations. The process of obtaining set-oriented operations to evaluate nested queries is known as **unnesting**.

It was later pointed out in [Kiessling84] and [Ganski87] that the unnesting techniques presented in [Kim82] do not always yield the correct results for nested queries that have non equi-join correlation predicates or for queries that have the COUNT function between nested blocks. Unnesting solutions for these types of queries were provided in [Ganski87]. These solutions were further refined and extended in [Dayal87].

In this paper, we will focus our attention on unnesting Join-Aggregate (JA) [Kim82] type queries. These queries have correlation join predicates and an aggregate function (AVG, SUM, MIN, MAX, or COUNT) between the nested blocks. The reason for focusing on JA type queries is that many other nesting predicates (such as EXISTS, NOT EXISTS, ALL, ANY) can be reduced to JA type queries [Ganski87, Dayal87]. An example of a JA type query is:

SELECT R_1.a
FROM R_1
WHERE R_1.b =
    (SELECT COUNT (R_2.b)
    FROM R_2
    WHERE R_1.c > R_2.c)

The predicate ($R_1.c > R_2.c$) is the correlation join predicate. We will explain the meaning of these types of queries in the next section.

We introduce a couple of definitions here:

Definition 1: A **(Nested) Linear Query** is a JA type query in which at most one block is nested within any block.

Definition 2: A **(Nested) Tree Query** is a JA type query in which there is at least one block which has two or more blocks nested within it at the same level.

Proceedings of the Fifteenth International Conference on Very Large Data Bases

Amsterdam, 1989

It is worth pointing out that the unnesting solution presented in [Ganski87] for a linear query with more than two blocks is incorrect (see Section 4). [Dayal87] does not discuss tree queries.

The rest of the paper is organized as follows: In Section 3, we introduce the notation that we will use for JA type queries and explain the meaning of these queries using TIS. In Section 4 we will briefly summarize the results presented in [Dayal87] that enable us to unnest nested linear queries. We will present our solution for tree queries in Section 5. Section 6 discusses a new dataflow algorithm for nested queries in a multi-processor environment (such as the one found in GAMMA [Dewitt86], [Gerber86]). The new dataflow algorithm reduces processing and message costs. Finally, we will present an algorithm that takes a tree query as input and generates an execution tree that can be routed using our new dataflow algorithm.

3. Interpreting JA Type Queries

The easiest way to understand a general nested query is by means of Tuple Iteration Semantics (TIS). TIS provide an algorithm, albeit inefficient, for obtaining the result of a nested query. It is instructive, however, to interpret a nested query using TIS. In order to explain the meaning of JA type queries, we first define a notation which we will use in this paper to represent such queries.

3.1. Notation

A JA type query may be represented as a tree. Each node in the tree corresponds to a SQL query block. Query blocks that are nested within a parent block are represented as child nodes of the node corresponding to the parent block. For ease of explanation, we shall assume that each block has one relation in its FROM clause. By definition, a node is also its own ancestor. A predicate clause in a given block may reference a relation associated with any ancestor block. Predicate clauses may either be selection or join predicates.

The relation associated with block (or node) i is represented by R_i ($i > 0$). Lower case letters (a, b, etc.) represent attribute names. A '*' is used to denote all the attributes of a relation. R_i.# is some unique key of R_i. r_i, r_i', r_i'' are each used to denote a tuple of relation R_i. OP_n ($n > 0$) is any one of the following operators ($=, \neq, <, \leq, >, \geq$). $F_i(R_j)$ represents a selection predicate in the ith block on R_j. To simplify the notation, we will assume that all join predicates are binary[1]. A join predicate in the ith block is then represented as $F_i(R_j, R_k)$, where $j, k > 0$ and $j \neq k$. If a predicate in the ith block does not reference R_i, then it is called an outer predicate. In this paper we will assume that there are no outer predicates in our queries. Outer predicates can be handled as shown in [Dayal87].

3.2. Interpretation using TIS

Consider the following linear JA type Query.

Example 1: A Two Block Linear Query

SELECT R_1.a
FROM R_1
WHERE $F_1(R_1)$
AND R_1.b OP_1
  (SELECT COUNT (R_2.*)
  FROM R_2
  WHERE $F_2(R_2)$ AND $F_2(R_2, R_1)$)

$F_1(R_1)$ and $F_2(R_2)$ are selection predicates on R_1 and R_2 respectively, while $F_2(R_2, R_1)$ is a correlation join predicate between R_1 and R_2.

A run time system that would execute the above query using TIS would proceed as follows: A tuple r_1 from R_1 would be fetched. If $F_1(R_1)$ is false for r_1, tuple r_1 will not be present in the result. Assuming $F_1(R_1)$ is true, the values of the relevant attributes of r_1 would be substituted into predicates at deeper levels ($F_2(R_2, R_1)$). The two block query now becomes a single block query

  SELECT COUNT (R_2.*)
  FROM R_2
  WHERE $F_2'(R_2)$

$F_2'(R_2)$ is a predicate on R_2 and is equivalent to $F_2(R_2)$ AND $F_2(R_2, R_1)$ after values of r_1's attributes have been substituted in $F_2(R_2, R_1)$.

Let the COUNT value returned by this block be C ($C >= 0$). C represents the number of tuples of R_2 that satisfy $F_2'(R_2)$. If (r_1.b OP_1 C) is true, r_1 will be in the result. Notice that each tuple of R_1 can occur in the result at most once. Using TIS, the system executes a query on R_2 (the inner relation) for every tuple of R_1 (the outer relation) leading to a very inefficient execution

[1] n-ary join predicates can be easily incorporated into the solutions presented in this paper.

strategy [Kim82].

One can easily interpret SQL queries with multiple levels of nesting using TIS. For example, consider the three block linear query shown in Example 2.

Example 2: A Three Block Linear Query

SELECT R_1.a
FROM R_1
WHERE $F_1(R_1)$
AND R_1.b OP_1
    (SELECT COUNT (R_2.*)
    FROM R_2
    WHERE $F_2(R_2)$ AND $F_2(R_2, R_1)$
    AND R_2.c OP_2
        (SELECT (COUNT(R_3.*)
        FROM R_3
        WHERE $F_3(R_3)$ AND $F_3(R_3, R_2)$
        AND $F_3(R_3, R_1)$))

After reducing the above query to a two block query, one would substitute values from each tuple of R_1 and R_2 that satisfied $F_1(R_1)$ AND $F_2(R_2)$ AND $F_2(R_2, R_1)$ into the third block and evaluate COUNT(R_3.*). Those tuples of R_2 that satisfy R_2.c OP_2 COUNT(R_3.*) will be returned as part of the two block query. Notice that a COUNT(R_2.*) is associated with every relevant tuple of R_1 (satisfying $F_1(R_1)$) and a COUNT(R_3.*) is associated with every relevant pair of tuples of R_1 and R_2. This implies that the system would have to execute a query on R_3 for every relevant pair of tuples from R_1 and R_2. Pictorially this can be represented as shown in Figure 1.

Figure 1: Pictorial Representation of the Three Block Query of Example 2

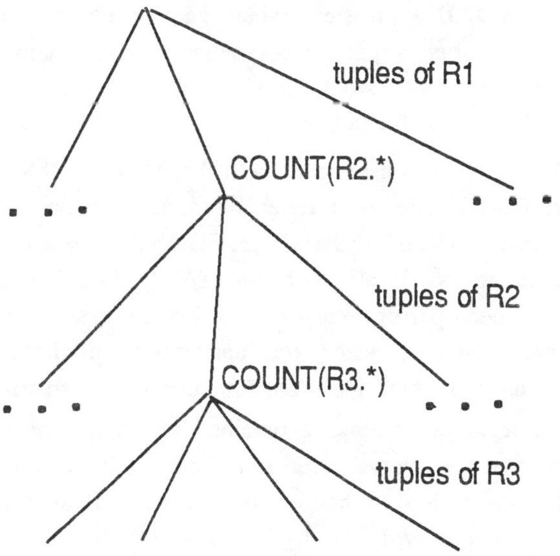

The edges at level 1 in this tree represent all the tuples of R_1 that satisfy $F_1(R_1)$. Only those tuples of R_1 will be present in the result for which (R_1.b OP_1 COUNT(R_2.*)) is also true. Under each tuple of R_1, are those tuples of R_2 for which $F_2(R_2)$ AND $F_2(R_2, R_1)$ is true. Only those tuples of R_2 that also satisfy (R_2.c OP_2 COUNT(R_3.*)) will contribute to COUNT(R_2.*). Similarly, the tuples of R_3 under each relevant pair of tuples of R_1 and R_2 represent tuples that satisfy $F_3(R_3)$ AND $F_3(R_3, R_2)$ AND $F_3(R_3, R_1)$.

4. From TIS to Set-Oriented Semantics

In this section, we briefly summarize the general solution presented in [Dayal87] and show how it can be applied to unnest linear queries. For reasons of space constraints, we do not discuss the more specific solutions that are based on the strategies presented in [Kim82]. Besides, the solutions in [Kim82] are not general and hence can be applied only in special cases (as pointed out in Section 2). However, as pointed out in [Dayal87], the unnesting solutions presented in [Kim82] (when applicable) may yield a more efficient execution strategy than the general solution.

Let us now return to the Query of Example 1 in Section 3.2. A naive unnesting algorithm would join R_1 and R_2 using predicate $F_2(R_2, R_1)$ (after performing the respective selections first). The algorithm would then group the result by R_1.# (some unique key of R_1[2]) and compute COUNT(R_2.*) for each group and select only those groups associated with each tuple of R_1 that satisfy (R_1.b OP_1 COUNT(R_2.*)).

The naive algorithm gives rise to the COUNT bug [Kiessling84, Ganski87]. The COUNT bug arises if there are no joining tuples in R_2 for a particular tuple r_1 in R_1. r_1 would then be lost after the join. However, the COUNT associated with r_1 is 0 and if (r_1.b OP_1 0) is true, tuple r_1 should appear in the result. In order to

[2] A unique key is required in order to avoid the problem with duplicates in R_1 [Ganski87].

preserve tuples in R_1 that have no joining tuples in R_2, an outer join[3] (OJ) is performed when the COUNT function is present between two blocks [Ganski87].

A linear query with multiple blocks will give rise to a 'linear J/OJ expression' where each instance of an operator is either a join or an outer join. A general linear J/OJ expression would look like:

$$R_1 \text{ J/OJ } R_2 \text{ J/OJ } R_3 \text{ J/OJ } \ldots \text{ J/OJ } R_n$$

Relation R_1 is associated with the outermost block, relation R_2 with the next inner block and so on. An outer join is required if there is a COUNT between the respective blocks. In all other cases (AVG, MAX, MIN, SUM), we need perform only a join. The joins and outer joins are evaluated using the appropriate predicates. Since joins and outer joins do not commute with each other[4], a legal order may be obtained by computing all the joins first and then computing the outer joins in a left to right order (top to bottom if you like) [Dayal87]. Thus, the expression R_1 OJ R_2 J R_3 J R_4 OJ R_5 J R_6 can be legally evaluated as ((R_1 OJ (R_2 J R_3 J R_4)) OJ (R_5 J R_6)). Since we can evaluate joins in any order, we can choose the cheapest join order to join R_2, R_3, and R_4.

It is worth pointing out here that the solution presented in Section 9 of [Ganski87] for multiple level queries was incomplete in the sense that it does not discuss legal orderings when joins and outer joins are present in the same expression.

After all the joins and outer joins have been evaluated, the aggregate functions are evaluated in a bottom-up order after grouping the result by the appropriate unique keys. This is best illustrated with an example.

Consider the three block linear query of Example 2 in Section 3.2. The corresponding linear expression is R_1 OJ R_2 OJ R_3 and hence a legal order is (R_1 OJ R_2) OJ R_3. The predicate for R_1 OJ R_2 is $F_2(R_2, R_1)$ and the predicate for the outer join with R_3 is $F_3(R_3, R_2)$ AND $F_3(R_3, R_1)$.

We now show how the query of Example 2 can be evaluated using set-oriented operations. The result is obtained by executing more than one query. The result from one query may be pipelined to the next query. The two queries in this case are (not in strict SQL syntax!):

Query A: SELECT INTO TEMP
R_1.#, R_1.a, R_1.b, R_2.*
FROM R_1, R_2, R_3
WHERE (R_1 OJ R_2) OJ R_3
GROUP BY R_1.#, R_2.#
HAVING R_2.c OP_2 COUNT(R_3.*)

Query B: SELECT R_1.a
FROM TEMP
GROUP BY R_1.#
HAVING R_1.b OP_1 COUNT(R_2.*)

The results from Query A are fed into Query B. Even though the selection predicates ($F_i(R_i)$, i = 1, 2, 3) have not been shown in Query A, they are applied to the respective relations before they participate in the outer joins. The outer join predicates are also implicit in Query A.

4.1. A Few Subtleties

Query A has a few subtleties that were not mentioned in [Dayal87] and deserve to be highlighted. These subtleties will lead us to the development of the new dataflow algorithm (described in Section 6). The outer join between R_1 and R_2 results in two sets of tuples, viz., (R_1- X NULL)[5] and R_1R_2. R_1R_2 denotes the set $\{(r_1, r_2)$: $F_2(R_2)$ AND $F_2(R_2, R_1)$ AND $F_1(R_1)\}$, where the r_1 tuple $\in R_1$ and the r_2 tuple $\in R_2$. Let R_1+ denote the set of tuples of R_1 present in R_1R_2 (tuples of R_1 that participated in the join with R_2). R_1- denotes the set R_1 - (R_1+) (the tuples of R_1 in the anti-join).

Similarly, let $R_1R_2R_3$ denote the set $\{(r_1, r_2, r_3)$: $F_3(R_3)$ AND $F_3(R_3, R_2)$ AND $F_3(R_3, R_1)$ AND $F_2(R_2)$ AND $F_2(R_2, R_1)$ AND $F_1(R_1)\}$. Let the set of (r_1, r_2) tuples in R_1R_2 that joined with at least one tuple of R_3 be denoted by R_1R_2+. The set of (r_1, r_2) tuples that did not join with any tuple of R_3 is denoted by R_1R_2- and is

[3]When we talk about outer joins, we implicitly mean left outer joins.

[4]Dayal proposed the notion of generalized joins (G-Joins) to make joins and outer joins commutable but the equation given in the paper was incorrect. Without repeating the notation used in defining G-Join and the formal definition of G-Join, we simply state that the following equation was given in [Dayal87]: G-Join (R, G-Join(S, T; ∅; J2); R.*; J1) = G-Join(G-Join(R, S; R.*; J1), T; R.*; J2). However, it can be shown that this equation does not hold for the query in Figure 4.1 on page 202 in Dayal's paper.

[5]X represents the cartesian product operation.

equal to R_1R_2 - (R_1R_2+). Thus, the outer join with R_3 may yield up to three distinct sets of tuples, viz., (R_1- X NULL X NULL), (R_1R_2- X NULL), and $R_1R_2R_3$ respectively.

The (GROUP BY ... HAVING) operation in Query A has special semantics associated with it. For a given group of (r_1.#, r_2.#), if (r_2.c OP_2 COUNT(R_3.*)) is true, the (r_1.#, r_2.#) group is passed along to Query B. However, if (r_2.c OP_2 COUNT(R_3.*)) is false, the (r_1.#, r_2.#) group cannot be discarded. If the (r_1.#, r_2.#) is discarded and if this is the only group in which r_1 was present, COUNT(R_2.*) associated with the r_1 tuple is 0 and hence should be preserved. If (r_1.b OP_1 0) is true, r_1 will be part of the result. The (r_1, r_2) tuple that does not satisfy (r_2.c OP_2 COUNT(R_3.*)) should be passed along to Query B as (r_1, NULL). Similarly, for tuples in the set (R_1- X NULL X NULL), the GROUP BY ... HAVING operation passes them as (R_1- X NULL) to Query B because the predicate (r_2.c OP_2 COUNT(R_3.*)) is false as (NULL OP 0) is false.

5. Tree Expressions or Non Linear Expressions

So far we have restricted our discussion to linear queries only. If we permit more than one block to be nested within a given block at the same level (nested tree queries), we can get J/OJ expressions that are arbitrary trees. In this section, we will extend Dayal's solution to tree queries. A simple extension, albeit inefficient, to evaluate a tree expression would be the following: Choose an arbitrary path (perhaps the least expensive one) from the root to a leaf and evaluate the linear expression specified by this path as outlined in the previous section. This will yield a subset R_1' of the tuples of the root relation R_1. Using tuples in R_1', another path is evaluated yielding R_1'', a subset of R_1'. This is repeated until all paths are exhausted and the final set of result tuples are obtained.

The above scheme is inefficient because relations that belong to two or more paths will be accessed more than once. For example, consider the tree query shown in Example 3 whose tree expression is shown in Figure 2. The edges in Figure 2 are labeled either by a J (denoting a join) or by an OJ (denoting an outer join).

Assume that the first path chosen is R_1--R_2--R_3. After evaluating the expression (R_1 OJ R_2) OJ R_3 and computing the respective aggregates bottom up, we will

Example 3: A Tree Query

```
SELECT R1.a
FROM R1
WHERE F1(R1)
AND R1.b OP1
        (SELECT COUNT (R2.*)
        FROM R2
        WHERE F2(R2) AND F2(R2, R1)
        AND R2.c OP2
                (SELECT (COUNT(R3.*)
                FROM R3
                WHERE F3(R3) AND F3(R3, R2)
                AND F3(R3, R1))
        AND R2.d OP3
                (SELECT (AVG(R4.d)
                FROM R4
                WHERE F4(R4) AND F4(R4, R2)
                AND F4(R4, R1)))
```

Figure 2: Tree Expression for the Query in Example 3

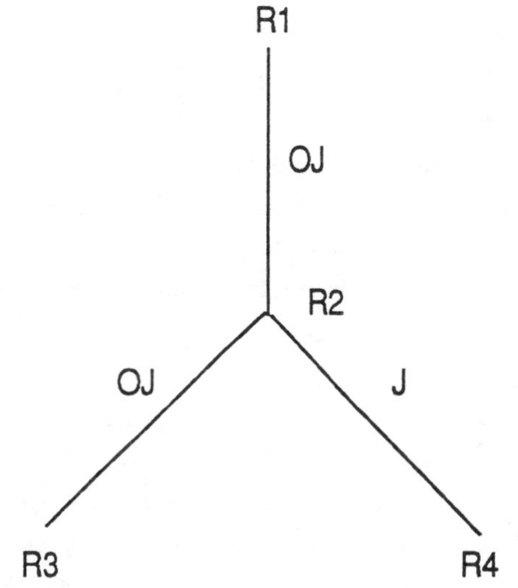

get a subset R_1' of tuples. Using these tuples, we take the other path. The J/OJ expression along this path is R_1' OJ (R_2 J R_4). Thus, the relation R_2 is accessed again. It would be ideal if each relation is accessed only once. After evaluating a J/OJ expression along one path, we would like to compute the aggregates bottom up only up to the point where a new branch begins. The idea is to use the tuples obtained thus far to evaluate the remainder of the J/OJ expression along the new path. However, this cannot be accomplished in a straight forward manner. Two kinds of anomalies may occur. We will illustrate these using the tree expression of Figure 2.

After evaluating (R_1 OJ R_2) OJ R_3 and the aggregate COUNT(R_3.*), there are two distinct sets of tuples. Tuples in the first set are of the form (r_1, NULL) and tuples in the second set are of the form (r_1, r_2) where $r_1 \in R_1$ and $r_2 \in R_2$.

Anomaly 1: Consider a tuple (r_1', NULL) from the first set. If r_1' is not present in the second set, COUNT(R_2.*) associated with r_1' is 0 and will be part of the result if (r_1'.b OP_1 0) is true. However, if (r_1', NULL) is joined with R_4 when the second path is taken, r_1' will be lost (because $F_4(R_4, R_2)$ is false since the R_2 fields are NULL).

Anomaly 2: Consider a tuple (r_1'', r_2') from the second set. If there is no tuple of R_4 such that $F_4(R_4, R_1)$ is true, then the r_1'' tuple will be lost after the join is computed. However, if the second path (R_1--R_2--R_4) was taken first, we would have got (r_1'', NULL, NULL) after evaluating R_1 OJ (R_2 J R_4). In this case, we do not lose r_1''. When the other path (R_1--R_2--R_3) is taken, r_1'' will not be lost as there is already an outer join between the R_2 and R_3 blocks.

The two anomalies demonstrate that tuples will be lost if a join is performed after evaluating an outer join. These tuples can be saved if the join between the R_2 and R_4 blocks is performed as an outer join. This leads to the following lemma.

Lemma: Let R_1--R_2--...--R_i--...--R_p and R_1--R_2--...--R_i-J-R_j--...--R_q be two paths from the root R_1 to the leaves R_p and R_q respectively in a tree expression. Let there be at least one outer join in the shared path R_1--R_2--...--R_i. Assuming we chose the R_1--R_2--...--R_p path first and there is a join between R_i and R_j, we can obtain the correct result by treating the join between R_i and R_j as an outer join.

Note that any joins below R_j are not affected as they will be evaluated before the outer join between the R_i and R_j blocks.

The proof for the lemma can be obtained by following the train of thought of the previous paragraphs and is omitted here. In summary, any join that comes after an outer join in a path must be evaluated as an outer join. Then each relation need be accessed only once.

6. An Improved Dataflow Algorithm

In this section, we will discuss a new dataflow algorithm for nested queries with COUNTs (and hence outer joins) in a multi-processor environment such as the one found in the GAMMA database machine [DeWitt86]. The new dataflow algorithm reduces message and processing costs over the conventional dataflow algorithm. The GAMMA database machine is a dataflow database machine that uses hash-partitioned algorithms for computing joins. A query with multiple joins can be routed very efficiently in GAMMA. The result of one join is pipelined to the next join operation. Details of the dataflow and scheduling algorithms of GAMMA can be found in [DeWitt86] and [Gerber86].

We will illustrate how GAMMA would have routed the three block query presented in Example 2, using the conventional dataflow algorithm. The

Figure 3: The Routing Method in GAMMA

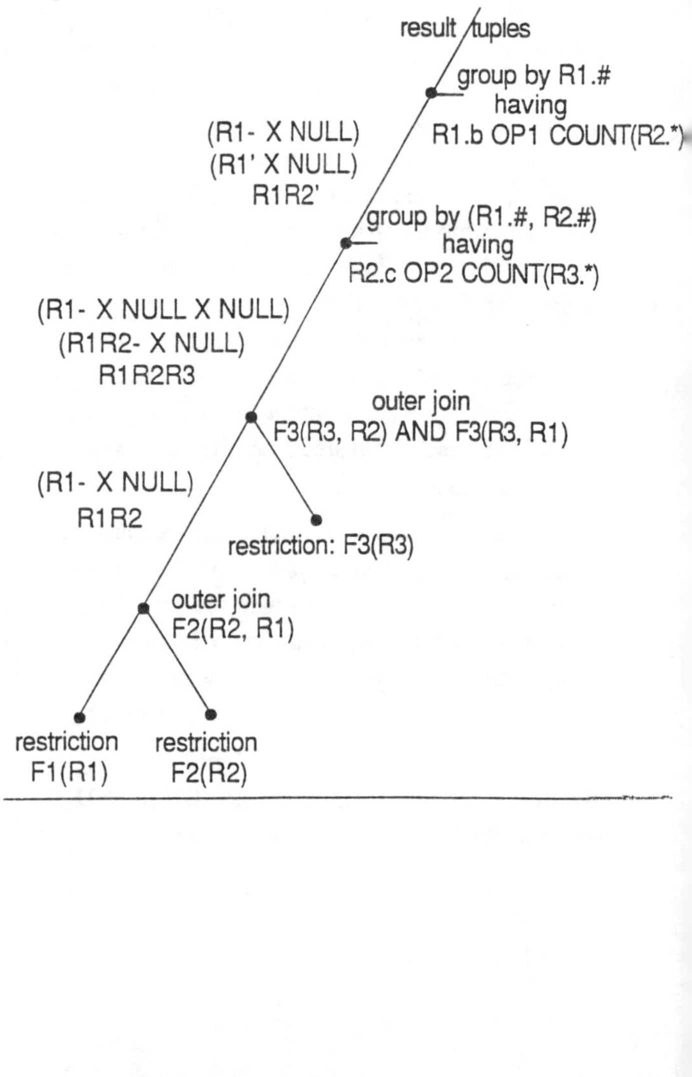

optimizer in GAMMA sends the scheduler an optimized execution tree. The execution tree for the query in Example 2 is shown in Figure 3. The nodes in the execution tree represent operations (restriction/join/group by etc.), while the directed arcs represent information flow. Tuples always flow in an upward direction. In Figure 3, it is important to notice that the conventional dataflow algorithm sends the joining tuples as well as the anti-join tuples to the immediate higher (parent) node. The sets that are propagated between the operators of the execution tree of Figure 3 are shown along the respective edges (using the notation of Section 4.1).

The set R_1R_2' is derived from those tuples in $R_1R_2R_3$ and (R_1R_2- X NULL) that satisfy the predicate (R_2.c OP_2 COUNT(R_3.*)), while the set (R_1' X NULL) is derived from those tuples in $R_1R_2R_3$ and (R_1R_2- X NULL) that don't satisfy the above predicate. Notice that the attributes of R_2 have been replaced by NULL for these tuples.

In the presence of outer joins, a better execution tree may be obtained by sending the anti-join tuples to a node possibly higher than the parent node in the execution tree. This will result in savings in message and processing (CPU) costs. The new execution tree is shown in Figure 4.

In the first execution tree (Figure 3), we are shipping the (R_1- X NULL X NULL) tuples (from the second outer join node to the first group by node) and the (R_1- X NULL) tuples (from the first group by node to the second group by node) unnecessarily. By doing so, we also incur the cost of processing them. In the second execution tree (Figure 4), the (R_1- X NULL) tuples are shipped directly from the first outer join node to the second group by node.

As the depth of nesting increases, the savings in message and processing costs will increase if the anti-join tuples are sent to a (possibly) higher node than the parent node. In the next section we will use the idea presented in Section 4.1 to send tuples that do not satisfy an aggregate predicate to a possibly higher node.

6.1. From J/OJ Tree Expressions to Execution Trees

We now describe how an execution tree is derived from a J/OJ expression tree. There are four different

Figure 4: The Improved Routing Method

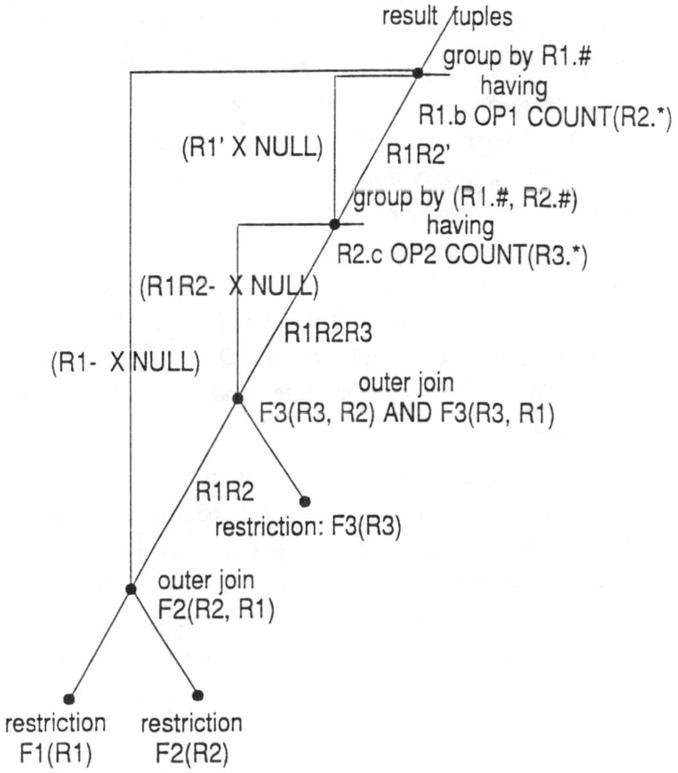

kinds of nodes in an execution tree, each representing a different kind of operation. They are:

1. Restriction nodes (R nodes): These are leaf nodes in the execution tree and represent the restriction operation on the base relations.

2. Group By ... Aggregate nodes (GBA nodes): These nodes group the input tuples by the unique keys of the relevant relations and compute the aggregate for each group. Tuples that satisfy the aggregate predicate are sent to the immediate higher node and, if needed, tuples that do not satisfy the aggregate predicate are sent only to a GBA node that is possibly much higher in the execution tree (after nulling appropriate fields).

3. Join nodes (J nodes): The output of a join is fed to the immediate higher node in the execution tree.

4. Outer join nodes (OJ nodes): The joining tuples are sent to the immediate higher node, but the anti-join tuples (if needed) are sent only to a GBA node that is possibly much higher in the execution tree.

The outdegree of R and J nodes is one while the outdegree of remaining kinds of nodes is at least one and at most two. The indegree of R nodes is 0, while the indegree of J and OJ nodes is two. The indegree of the GBA nodes is at least one. A GBA node can receive input from many internal nodes in the execution tree.

The number of leaf nodes in the execution tree is equal to the number of base relations in the query (or blocks in the query assuming that each block has exactly one relation associated with it). The number of internal nodes in the execution tree is equal to twice the number of edges in the J/OJ expression tree. Half of these nodes are J or OJ nodes while the other half are GBA nodes. This is because each edge in the J/OJ expression tree can be associated with a join/outer join and an aggregate operation.

Having determined the nodes in the execution tree, we need to define how the arcs are constructed. Each arc can be determined by its end nodes. The key is to also classify the nodes in terms of the structure of input tuples they expect and output tuples they produce. A directed arc will then be present between a node that produces a certain kind of output tuples and a node that expects that same kind of tuples as input.

6.2. Input/Output Classification of Execution Tree Nodes

1. R nodes: These leaf nodes only produce output. The output tuples from these nodes are qualifying single relation tuples. Attributes (from output at all nodes) not required for future operations are projected out.

2. J nodes: If relations R_i and R_j are joined at a J node, R_i and R_j tuples form the input. Output tuples have the form (r_i, r_j) where $r_i \in R_i$ and $r_j \in R_j$. R_i and R_j could be base relations or composite relations.

3. OJ nodes: If relations R_i (composed of relations $R_1, R_2, ..., R_n$) and R_j are outer joined at an OJ node, R_i and R_j tuples form the input. Joining output tuples have the form (r_i, r_j) where $r_i \in R_i$ $(r_i = (r_1, r_2, ..., r_n))$ and $r_j \in R_j$. One would expect the anti-join tuples to have the form $(r_i$, NULL). This need not always be the case.

Consider the case when a J node is transformed to an OJ node. The anti-join tuple $(r_1, r_2, ..., r_n$, NULL) can be discarded if there is no outer join in the path from R_1 to R_n. However, if the last outer join in the path from R_1 to R_n occurred between R_k and R_{k+1}, $0 < k < n$, the anti-join tuple $(r_1, r_2, ..., r_k, r_{k+1}, ..., r_n$, NULL) is sent as $(r_1, r_2, ..., r_k$, NULL) to the GBA node computing the COUNT of tuples of R_{k+1}. The COUNT(R_{k+1}.*) associated with the group $(r_1, r_2, ..., r_k)$ may be 0 and hence must be preserved.

We could have sent the tuple $(r_1, r_2, ..., r_n$, NULL) to a much lower GBA node in the execution tree. However, since the aggregate at this GBA node is not a COUNT, the $(r_1, r_2, ...r_k, ..., r_n)$ tuple would not satisfy the aggregate predicate because of the presence of NULL. Hence the tuple will be passed to the next higher node as $(r_1, r_2, ..., r_k, ... r_{n-1}$, NULL). However, we need only to preserve $(r_1, r_2, ..., r_k)$ and hence we can send $(r_1, r_2, ..., r_k$, NULL) tuple directly to the corresponding GBA node.

4. GBA nodes: Assume that this GBA operation occurs between R_{n-1} and R_n in the path from R_1 to R_n. If the aggregate operation at this node was a non COUNT function, then the input to this node is of the form $(r_1, r_2, ..., r_{n-1}, r_n)$. If the aggregate function was a COUNT function tuples of the form $(r_1, r_2, ..., r_{n-1}$, NULL) will also be included in the input. The output formats of a GBA node are exactly similar to those of OJ nodes.

6.3. An Example

We will illustrate the above concepts in deriving an execution tree for the J/OJ expression in Figure 2. The execution tree will have four leaf nodes (R nodes) in the execution tree corresponding to the four base relations. There will be three pairs of J/OJ nodes and GBA nodes corresponding to the three edges. Assuming that we again take the left path (R_1--R_2--R_3) in Figure 2 first, the execution tree will be as shown in Figure 5. The nodes are labeled for ease of reference. Each arc is labeled with the format of tuples flowing along that arc. Notice that the 'anti-join' tuples of node OJ_3 are of the form $(r_1$, NULL) rather than of the form $(r_1, r_2$, NULL). As per the Lemma, the join at this node was converted to an outer join. It would not be useful to send tuples of the form $(r_1, r_2$, NULL) to node GBA_2. Node GBA_2 would simply pass on these tuples to node GBA_3 as $(r_1$, NULL). Instead, one saves processing and message costs by sending $(r_1$, NULL) tuples directly from node OJ_3 to node GBA_3. It turns out that the operation at node OJ_3 is a G-Join [Dayal87]. Table 1 summarizes the savings realized in message costs.

Table 1.

Description	Size (in Bytes)	Conventional Routing Path	New Routing Path	Savings in Message Costs (in Bytes)
anti-join tuples from node OJ_1	M_1	OJ_1-OJ_2-GBA_1-OJ_3-GBA_2-GBA_3	OJ_1-GBA_3	$4*M_1$
tuples not satisfying aggregate comparison at node GBA_1	M_2	GBA_1-OJ_3-GBA_2-GBA_3	GBA_1-GBA_3	$2*M_2$
anti-join tuples from node OJ_3	M_3	OJ_3-GBA_2-GBA_3	OJ_3-GBA_3	M_3

Table 1 shows that for this example, the new dataflow algorithm results in a savings of $(4*M_1 + 2*M_2 + M_3)$ bytes in message costs over the conventional dataflow algorithm. Further, a proportional savings is realized in processing costs.

Figure 5: Execution Tree for J/OJ Expression of Figure 2

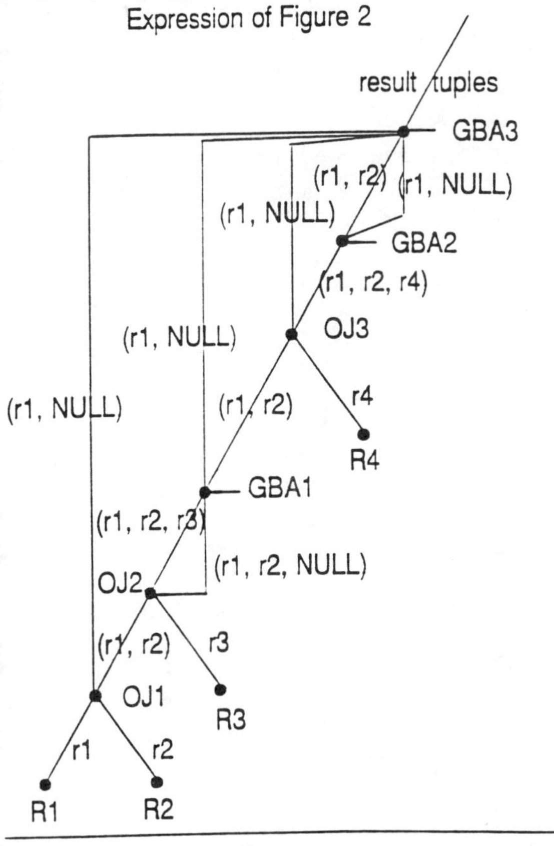

7. Conclusions and Future Work

In this paper, we have presented a scheme for evaluating nested tree queries. We also described a new dataflow algorithm for such queries.

We are in the process of developing an efficient scheduling algorithm to go with our new dataflow algorithm. Our scheduling algorithm will be more complex than that of GAMMA's as we may to have to activate many operators in the execution tree simultaneously.

In this paper, we have assumed that blocks nested within a block at the same level are separated only by AND's. We are investigating optimization techniques for queries that have OR's between blocks that are nested at the same level.

8. Acknowledgments

The author wishes to thank Bob Gerber, Goetz Graefe, Krishna Kulkarni, Shirish Puranik, Jim Reuter, Mateen Siddiqui, Lynn Still, and Wai-Sze Tam for carefully reviewing earlier versions of this paper. Their efforts have considerably improved the quality of the paper.

9. References

[Astrahan75] M. Astrahan, and D. Chamberlin, "Implementation of a structured English query language," Comm. of the ACM, Vol. 18, No. 10, (October 1975).

[Dayal87] U. Dayal, "Of Nests and Trees: A Unified Approach to Processing Queries That Contain Nested Subqueries, Aggregates, and Quantifiers," Proc. VLDB Conf., pp.197-202, (September 1987).

[DeWitt86] D. DeWitt, R. Gerber, G. Graefe, M. Heytens, K. Kumar, and M. Muralikrishna, "GAMMA: A High Performance Dataflow Database Machine," Proc. VLDB Conf., pp.228-237, (August 1986).

[Ganski87] Richard A. Ganski and Harry K. T. Long, "Optimization of nested SQL Queries Revisited", Proc. SIGMOD Conf., pp. 23-33, (May 1987).

[Gerber86] R. Gerber, "Dataflow Query Processing Using Multiprocessor Hash-Partitioned Algorithms," PhD Thesis, Univ. of Wisconsin at Madison, (October 1986).

[Kiessling84] W. Kiessling, "SQL-like and Quel-like correlation queries with aggregates revisited", UCB/ERL Memo 84/75, Univ. of California at Berkeley, (Sept. 1984).

[Kim82] W. Kim, "On Optimizing an SQL-like Nested Query", Trans. on Database Systems, Vol 9, No. 3, (1982).

Parallel Processing of Recursive Queries in Distributed Architectures

Guy Hulin

Philips Research Laboratory
2 avenue Van Becelaere, 1170 Brussels, Belgium
ghulin@prlb2.uucp

Abstract

This paper presents a parallel algorithm for recursive query processing and shows how it can be efficiently implemented in a local computer network. The algorithm relies on an interpretive approach where recursive rule processing and data retrieval are merged in a top-down computation. It employs "sideways information passing" to restrict to relevant facts the information extracted from the relational database. Evaluation is divided into a compilation phase and a dynamic phase. The compilation phase statically constructs a derivation tree that expresses the decomposition of a query into subqueries and the "sideways information passing" strategy. In the dynamic phase, cooperative processes are associated with subsets of "equivalent" nodes of the derivation tree. They communicate by message passing without sharing memory. Some optimizations are discussed for a practical parallel implementation. Gains in efficiency with respect to classical sequential algorithms are also discussed.

1 Introduction

A *distributed database* [5] is a collection of data which are distributed over different computers of a computer network. Relations are partitioned into fragments vertically and/or horizontally. The *vertical fragmentation* of a relation is the subdivision of its attributes into groups; fragments are obtained by projecting the relation on each group. The *horizontal fragmentation* of a relation consists in partitioning its tuples into subsets.

Query evaluation in distributed databases is performed in parallel through the computer network. The existence of several cooperating processors results in increasing performance.

Databases can be given deductive capabilities by the addition of *deduction rules*. Rules are often restricted to be definite Horn clauses. They define new relations and can be recursive.

Proceedings of the Fifteenth International
Conference on Very Large Data Bases

This paper presents a parallel algorithm for evaluating atomic queries in deductive databases and shows how it can be efficiently implemented in a local computer network (locality is necessary because parallelism in the algorithm is relatively fine-grained). It suggests a multi-computer architecture well-fitted for distributed deductive databases where some processors handle fragments of the distributed database while others are devoted to the evaluation of recursive queries. Such an architecture is currently being developed in the database tract of the PRISMA project (PRISMA for PaRallel Inference and Storage MAchine) [13].

From a logical point of view, a *deductive database*[1] is a finite set of function-free definite clauses. It is composed of an *extensional* database and of an *intensional* database. The extensional database is a set of ground atomic formulas called *facts*. The intensional database is a set of *rules*.

Example 1
Consider a directed graph with two types of arcs and two predicates $e_1(X,Y)$ and $e_2(X,Y)$ meaning that there exists an arc, respectively of type 1 or 2, from Node X to Node Y. Then, the following rules

$p(X,Y) \leftarrow e_1(X,Y),$

$p(X,Y) \leftarrow e_2(X,Z), p(Z,T), e_2(T,Y),$

define a virtual predicate p true for couples of nodes (X,Y) such that there exists a path from X to Y all of whose arcs are of type 2 except the central one. •

It can be assumed, without loss of generality, that a fact in the extensional database and an atomic formula in the head of a rule of the intensional database are never instances of the same predicate. The predicates of the deductive database some of whose instances are heads of rules of the intensional database are called *virtual predicates*, the other ones are called *base predicates*.

The semantics of a deductive database B is given by its least Herbrand model $\mathcal{M}_B$ which can be constructed by fixpoint computation. A closed formula F is true in the deductive database B if $\models_{\mathcal{M}_B} F$.

A first approach for evaluating queries in deductive databases consists in adding to the traditional relational algebra a least fixpoint operator *LFP*. If p is a virtual predicate, $LFP(p) = \{a \mid \models_{\mathcal{M}_B} p(a)\}$.

In Example 1, the query $p(a,Y)$? is translated into

[1] For more details about deductive databases, see [12].

Amsterdam, 1989

$\sigma_{1=a}LFP(p)$. Unfortunately, commuting selection and least fixpoint operator is not possible in general and that leads, for Example 1 and the query $p(a, Y)$?, to the necessity of computing the complete extension of p. Yet, in order to answer the query, only part of the extension of p is strictly necessary. The facts that effectively contribute to the answer are said to be *relevant* to the query. A good evaluation strategy should not extract other facts from p. In general, which facts are relevant also depends on an ordering of the predicates in the right-hand sides of the rules [4].

Various proposals have been made to overcome this problem [7,8,10,15,20,27] but they are restrictive in that they are complete only for special (typically linear) types of recursion.

Another approach takes advantage of the decomposition of queries into subqueries and of the possible constant propagation among them. This approach seems more fruitful and has been heavily investigated recently. In [16], an active connection graph is dynamically created, with rule nodes and goal nodes corresponding to queries and subqueries. Solutions flow across the rule nodes to the goal nodes. However, constant propagation is poor and therefore, the algorithm is rather inefficient. Also the APEX algorithm [14] clearly lacks an efficient propagation of constants. The Recursive Query/Subquery (QSQR) and QoSaQ approaches [24,25], the Alexander method [19], and the generalized counting method [4] are algorithms which compute only relevant facts.

A uniform presentation of recursive query evaluation methods can be found in [18]. The algorithms are classified in three classes according to their halting condition. They are also compared on completeness and efficiency criteria.

Another efficiency requirement, which has received little attention until now, is the minimization of the number of database accesses. It can be addressed in two ways. First, when the same query on base predicates is generated at different times, a good evaluation process should access the database only once. Further, the strategies mentioned above repeatedly instantiate subsets $\{x_{i_1}, ..., x_{i_k}\}$ ($k \leq n$) of variables in base predicate $p(x_1, ..., x_n)$, and access the extensional database for all the facts $p(u_1, ..., u_n)$ such that $(u_{i_1}, ..., u_{i_k})$ are the values instantiating $(x_{i1}, ..., x_{ik})$. It is clear that, rather than accessing the extensional database each time an instance is generated, a better strategy should accumulate successive instances and make a single access for solving all of them simultaneously.

The algorithms cited above are sequential and little attention has been paid to parallelism until now. Parallelism is however recognized as a very important optimization feature for recursive query evaluation. A few proposals exist for evaluating transitive closures in distributed database systems [1,9,22]. More generally, several classes of logic programs, e.g., the class of linear single rule programs, can be evaluated in parallel without introducing interprocess communication, or synchronization overhead [26]. A first step towards a parallel algorithm for general recursive queries has been proposed in [23]. A first attempt to implement that algorithm on top of a multi-computer system [21] has not proved satisfactory, due to an excess of duplicated work.

This paper is a continuation of [11] where we presented a new sequential algorithm, in the same vein as the QSQR approach, that accepts general recursion and guarantees termination and completeness. This paper describes a parallel version of the algorithm. The evaluation of a query is decomposed into two parts. The first one is static: it is a compilation of the query into an AND/OR tree. The second one is dynamic: cooperative processes associated with nodes of the tree rule the flow of data traversing the tree. They communicate by message passing and do not share memory. We also discuss possible optimizations for a practical parallel implementation.

The paper is organized as follows. Definitions and notations are introduced in the following section. The compilation phase is described in Section 3 while Section 4 presents the parallel processes of the dynamic phase. Section 5 discusses some possible optimizations: choice of a selection function, minimization of the number of database accesses, and increase of parallelism. Section 6 is a conclusion.

2 Basic definitions

Variables in rules are taken in an alphabet V not containing the symbol $*$. In addition, V^* is a new set of variables where x^* is an element of V^* if and only if x is an element of V.

A *query scheme* associated with a predicate p is any atomic formula $p(x_1, ..., x_n)$ without constants and whose variables are in V^* or in V. Two variables $x \in V$ and $x^* \in V^*$ cannot both be present in the same query scheme. If x^* is present, x is called an *entry variable* of the query scheme. The *exit variables* of a query scheme are those belonging to V.

Two query schemes associated with an n-ary predicate p are said to be *equivalent* if they are identical up to a renaming of their variables respecting their entry or exit character.

An *entry* (resp., *exit*) *value* for a query scheme with n entry (resp., exit) variables is any set of n variable/value pairs where each variable appears once and only once.

A mechanism is assumed, provided for example by a (distributed) DBMS, that, given a query associated with an n-ary base predicate, computes its solution from the extensional database. More precisely, if Q is a query scheme associated with a base predicate, *in-v* is an entry value of Q, and B_E is the database, then,

$$ans(Q, in\text{-}v, B_E) = \{in\text{-}v \cup out\text{-}v \mid out\text{-}v \text{ is an exit value of } Q \;\&\; Q[in\text{-}v \cup out\text{-}v] \in B_E\}$$

where $F[\sigma]$ denotes the result of applying the substitution σ to the formula F.

3 The compilation phase

The result of compiling a query scheme is a *derivation tree* whose nodes are labeled with query schemes.

Derivation trees are special kinds of AND/OR trees [17]. They can be viewed as a kind of hypertree, where, instead of arcs connecting pairs of nodes, hyperarcs connect a parent node to an ordered list of successor nodes. The derivation tree of a query scheme makes explicit its possible decompositions (derived from the deduction rules) into subquery schemes. Like in Prolog, the ordering of subqueries governs the propagation of the instantiations of the entry variables.

The compilation phase constructs the derivation tree by recursively splitting each query scheme in as many sequences of subquery schemes as there are deduction rules defining the query scheme predicate in the intensional database. The resulting derivation tree has the following properties:

- when several equivalent query schemes are present in the derivation tree, only one of them is explicitly decomposed;
- the subquery order in a decomposition is imposed by a selection function. Different selection functions can generate different derivation trees. A good choice of the function is important: the efficiency of the evaluation process is very sensitive to the ordering. Heuristics guiding the choice of a selection function are discussed in Section 5.1;
- a variable in a subquery scheme SQ at the extremity of a hyperarc h is an entry variable if it is an entry variable of the query scheme which is the origin of h or an exit variable of a subquery scheme preceding SQ in the extremity of h.

Example 2

For the query scheme $p(X^*, Y)$ and the database of Example 1, the following tree is a possible derivation tree (a hyperarc is denoted by several binary arcs joined together by a curved line and nodes are numbered for easier reference.):

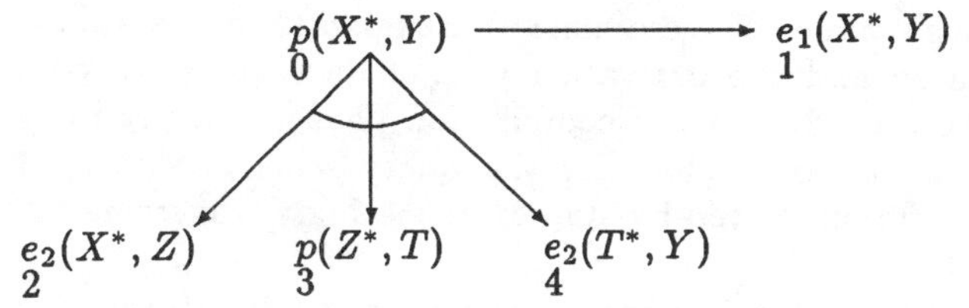

Note, for example, that Z which is an exit variable at Node 2 becomes an entry variable at Node 3. •

Example 3

In the following database, the virtual predicates a_1 and c_1 are mutually recursive:

$a_1(X,Y,Z) \leftarrow b(X,Z_1), a_1(T,Z_1,Z), c_1(Z_1,T,Y).$
$c_1(X,Y,Z) \leftarrow d(X,Z_1), a_1(Z_1,Z,Y).$
$a_1(X,Y,Z) \leftarrow e(X,Y,Z).$
$c_1(X,Y,Z) \leftarrow f(X,Y,Z).$

The predicates b, d, e and f are base predicates. The query scheme $a_1(X^*, Y^*, Z)$ leads to the possible derivation tree:

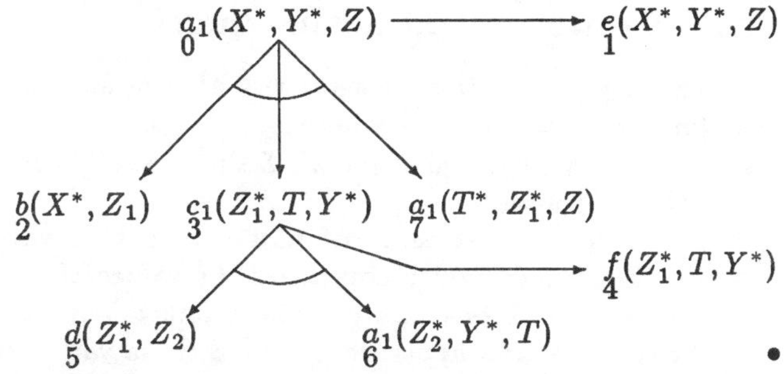

•

Before considering the dynamic phase, some more definitions are needed. A node n_1 of the derivation tree *precedes* a node n_2 (or, equivalently, n_2 *follows* n_1) if they are extremities of the same hyperarc and if n_1 is on the left of n_2. In Example 3, Node 2 precedes Node 3 and Node 7; Node 5 precedes Node 6. A node without a predecessor is called an *initial* node (Nodes 0, 1, 2, 4 and 5) and a node without a successor a *terminal* node (Nodes 0, 1, 4, 6 and 7).

The concept of *father node* has its usual meaning: Node 0 is the father of Nodes 1, 2, 3, and 7.

In the derivation tree, it may happen that several nodes are labeled with equivalent query schemes. The first one encountered during a depth-first traversal of the tree is called *archetype* node. In Example 3, Nodes 0, 1, 2, 3, 4 and 5 are archetype nodes. This concept will be useful for the dynamic phase.

4 The dynamic phase

An *atomic query* is a query scheme together with an entry value for it. It is denoted by the query scheme where entry variables have been instantiated with the entry value. The dynamic phase operates on an atomic query and the derivation tree of its scheme, as constructed during the compilation phase, and produces a set of exit values for the query scheme. This set is computed step-by-step by repeatingly traversing the tree.

The dynamic phase consists of the interleaving of parallel processes. Each process executes in a dynamic *context*. A context remembers the minimal part of the computation state necessary to carry on with the computation. *Active queries* remember the subqueries generated during the evaluation process with their partial solution and the contexts in which they were issued. Contexts and active queries play an essential role in avoiding redundant work as well as in guaranteeing correctness and completeness. They are presented in Section 4.1. Parallel processes are informally described in Section 4.2 and their precise algorithms are given in Section 4.3.

4.1 Contexts and active queries

A context of invocation of a process at a node n has the form $<n, rel\text{-}v, in\text{-}v\text{-}f>$ where

- *in-v-f* is an entry value for n's father node ({} if n is the root node),
- *rel-v* is a *relevant value* of n, i.e., a set of variable/value pairs which comprises the values of
  - entry variables of nodes which follow n in the extremity of a hyperarc provided those variables are also entry variables of the father of n or exit variables of nodes preceding n,
  - exit variables of the father of n which are exit variables of nodes preceding n.

The variables in a relevant value of a node n are called *relevant variables* of n. During evaluation, the relevant value in a context of n remembers the values of variables that are known before evaluating the current query at n and that are still needed after its evaluation.

To illustrate these concepts, consider the following derivation tree:

Example 4

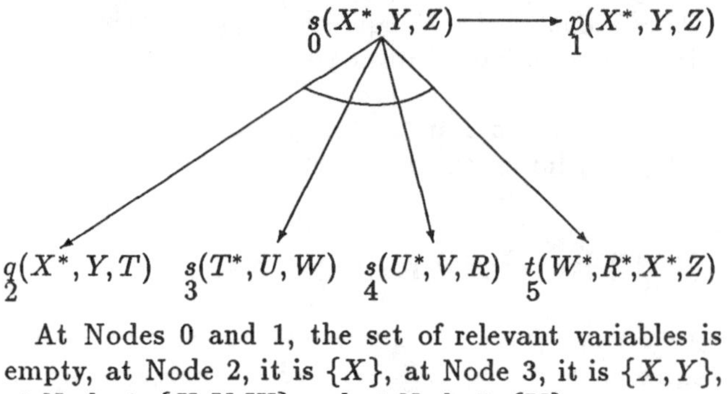

At Nodes 0 and 1, the set of relevant variables is empty, at Node 2, it is $\{X\}$, at Node 3, it is $\{X, Y\}$, at Node 4, $\{X, Y, W\}$ and at Node 5, $\{Y\}$.

At Node 4, a context comprises, besides a reference to Node n itself:

- a relevant value $\{Y/c_1, X/c_2, W/c_3\}$. Y/c_1 is a part of a possible solution of the query at Node 0 with entry value $\{X/c_2\}$ while X/c_2 and W/c_3 are needed at Node 5 to find solutions for Z,
- an entry value for Node 0: $\{X/c_2\}$.

From this context and an exit value from Node 4, a context and an entry value can be constructed for Node 5. A context and an exit value of Node 5 allow to construct an exit value of Node 0. A context of a node remembers the minimal part of the computation state necessary to go on in the computation when an exit value is found at the node. •

During the evaluation process, information units, called *active queries*, are associated with each node to remember queries previously requested at the node, their partial solution, and the contexts in which the requests were issued. As distinct nodes may be labeled with equivalent query schemes, it is enough to associate active queries only with archetype nodes.

An active query at an archetype node n comprises:

- an entry value (it uniquely identifies the active query at the node),
- a set of exit values for that entry value (this set will be used to avoid repeating the same computation and, in particular, to avoid cycling),
- a set of contexts of nodes whose archetype is n (this set will be used to guarantee completeness, that is, to guarantee that exit values are used in all the contexts in which they were requested).

During the evaluation process, the set of exit values of the active queries are not necessarily complete, i.e. do not contain all the solutions of the query. That is why the contexts where queries are generated are remembered: each time a new solution is found for an active query of a node n_i, it is propagated along all contexts remembered in that active query. This guarantees completeness: each exit value of each active query is propagated along each context. At the end of evaluation, each active query has received a complete set of exit values.

Query schemes at equivalent nodes differ by their variables. If n and n' are equivalent nodes and v is an entry or exit value for n, the corresponding value for n' will be denoted $v@n'$. In Example 3, Nodes 0 and 7 are equivalent. An entry value of Node 7 is $<T/a, Z_1/b>$. Then, $<T/a, Z_1/b>@0 = <X/a, Y/b>$ is an entry value of Node 0.

4.2 Parallel processes

There is one process **Evaluate**(n) for each archetype node n. They interactively solve subqueries generated during evaluation.

To each archetype node n_a we associate:

- a process **Evaluate**(n_a) which computes and propagates the answers of active queries at n_a,
- a private memory $Memory(n_a)$, that is, a set of active queries of n_a,
- two buffers, if n_a is a virtual node, to store messages sent by other processes:
  $Request(n_a)$, a set of *request messages* for n_a,
  $Answer(n_a)$, a set of *answer messages* for n_a,
- a buffer $Request(n_a)$, if n_a is a base node.

A *request message* for n_a is a pair made of an entry value and a context for a node whose archetype is n_a. An *answer message* is a set of variable/value pairs for the variables of n_a.

In Example 4, four processes coexist: **Evaluate**(n_i) for $i = 0, 1, 2, 5$. A query such as $s(x, Y, Z)$ (lower case letters denote constants) is solved by executing:

$$\mathbf{Evaluate}(n_0) \parallel \ldots \parallel \mathbf{Evaluate}(n_5)$$

after having sent the message $<\{X/x\},<n_0,\{\},\{\}>>$ to $Request(n_0)$. $P_1 \parallel P_2$ denotes the parallel computation of P_1 and P_2.

The reception of $<\{X/x\},<n_0,\{\},\{\}>>$ in buffer $Request(n_0)$ initiates the computation. The process **Evaluate**(n_0) adds to $Memory(n_0)$ the active query $(\{X/x\},\{\},\{<n_0,\{\},\{\}>\})$. That query is solved by parallel decomposition along each hyperarc issuing from node 0 as follows:

1. A request message $<\{X/x\},< n_1,\{\},\{X/x\}>>$ is sent to $Request(n_1)$. Node n_1 is a base node. The request is solved by **Evaluate**(n_1) which computes $S = ans(p(X^*, Y, Z), \{X/x\}, B_E)$ and sends it to $Answer(n_0)$. That computation is remembered by adding to $Memory(n_1)$ the active query $(\{X/x\}, S, \{<n_1,\{\},\{X/x\}>\})$. If, later, n_1 is queried again with the same entry value X/x, the set of answers is already present in $Memory(n_1)$ and no new access to the extensional database is necessary.
2. A request message $<\{X/x\},<n_2,\{X/x\},\{X/x\}>>$ is sent to $Request(n_2)$. Exit values $\{Y/y, T/t\}$ are computed by **Evaluate**(n_2) by accessing B_E for q (see 1.). For each of them, the request message $<\{T/t\},<n_3,\{X/x,Y/y\},\{X/x\}>>$ is sent to $Request(n_0)$, since n_0 is the archetype of n_3.
   Two situations may then arise, in general, for **Evaluate**(n_0):
   - there exists in $Memory(n_0)$ an active query Q with entry value $\{X/t\}$. In that case, if the context $<n_3,\{X/x,Y/y\},\{X/x\}>$ is not already in the set of contexts of Q, it is added to it. For each $\{Y/y_1, Z/z_1\}$ in the set of exit values of Q, the request message $<\{U/y_1\},<n_4,\{X/x,Y/y,W/z_1\},\{X/x\}>>$ is sent to $Request(n_0)(n_0$ is the archetype of $n_4)$,
   - otherwise, an active query with the entry value $\{X/t\}$, an empty set of exit values and a singleton set of contexts containing $<n_3,\{X/x,Y/y\},\{X/x\}>$ is added to $Memory(n_0)$. Its set of exit values is computed by parallel decomposition as described above for the original query $s(x, Y, Z)$.
3. Each $<\{U/y_1\},<n_4,\{X/x,Y/y,W/z_1\},\{X/x\}>>$ in $Request(n_0)$ is also solved by parallel decomposition. Each answer $\{U/y_1, V/v, R/r\}$ generates a request $<\{W/z_1, R/r, X/x\},<n_5,\{Y/y\},\{X/x\}>>$ in $Request(n_5)$. These requests are solved by **Evaluate**(n_5) by accessing B_E for $t(z_1, r, x, Z)$. Each value z_2 of Z resulting from that evaluation generates an answer message $\{X/x, Y/y, Z/z_2\}$ sent to $Answer(n_0)$.

Answer messages are processed in parallel with request messages. Notice that two active queries in the private memory of a process have different entry values. If a query is encountered more than once at equivalent nodes, there will be, in the private memory of the process associated with their archetype node, an active query with several contexts to remember the different states where the queries appeared. The contexts of a query $(\{X/a\}, S, C)$ at n_0 are of one of the form:

a. $<n_0,\{\},\{\}>$,
b. $<n_3,\{X/x,Y/y\},\{X/x\}>$,
c. $<n_4,\{X/x,Y/y,W/w\},\{X/x\}>$.

During the evaluation process, the set of exit values of the active queries are not necessarily complete, i.e. do not contain all the solutions of the query (except for queries on base predicates). That is why the contexts where queries are generated are remembered: each time a new solution is found for an active query of $Memory(n_i)$, **Evaluate**(n_i) propagates it along all contexts remembered in that active query. This guarantees completeness: each exit value of each active query is propagated along each context. At the end of evaluation, each active query has received a complete set of exit values.

Given a solution $\{X/a, Y/b, Z/c\}$ of $Answer(n_0)$ and the active query $(\{X/a\}, S, C)$ at Node n_0, if $\{Y/b, Z/c\}$ is not already in the current solution S, it is added to S and the solution is propagated in each context in C of the form b or c[2], i.e.,

type b: $<\{U/b\},<n_4,\{X/x,Y/y,W/c\},\{X/x\}>>$ is sent to $Request(n_0)$,

type c: $<\{W/w, R/c, X/x\},<n_5,\{Y/y\},\{X/x\}>>$ is sent to $Request(n_5)$.

[2]There is no propagation in the context of type a which is the context of the initial query.

Termination occurs when all request and answer buffers are empty and when all processes are waiting for new messages.

4.3 Algorithms

Some properties of entry, exit and relevant variables will be useful for writing the algorithms:

1. The entry and relevant variables of an initial node are also entry variables of its father.
2. The entry and relevant variables of a node with a predecessor are among the exit and relevant variables of the predecessor.
3. If n is terminal, then n's relevant and output variables make up the exit variables of n's father.

Let us call $\Theta_n^{ent}(vvp)$ those elements in a list of variable/value pairs *vvp* whose variables are entry variables of n and $\Theta_n^{rel}(vvp)$ those elements whose variables are relevant to n. With these notations, the preceding properties write:

1. if n is an initial node and *in-v-f* is an entry value of n's father, $\Theta_n^{ent}(in\text{-}v\text{-}f)$ is an entry value of n and $\Theta_n^{rel}(in\text{-}v\text{-}f)$ is a relevant value of n,
2. if n has a successor *n-suc*, if *rel-v* is a relevant value of n and *out-v* is an exit value of n, $\Theta_{n\text{-}suc}^{ent}(rel\text{-}v \cup out\text{-}v)$ is an entry value of *n-suc* and $\Theta_{n\text{-}suc}^{rel}(rel\text{-}v \cup out\text{-}v)$ is a relevant value of *n-suc*,
3. if n is terminal and if *rel-v* is a relevant value of n and *out-v* is an exit value of n, $rel\text{-}v \cup out\text{-}v$ is an exit value of n's father.

All the elements necessary to present the processes are now gathered[3].

If n_a is a virtual archetype node:

Evaluate(n_a):
While *true* do
1. if $Request(n_a) \neq \{\}$, then call **action**$_1(n_a)$; *
2. if $Answer(n_a) \neq \{\}$, then call **action**$_2(n_a)$. *

action$_1(n_a)$:
1. take an element $\langle in\text{-}v,c\rangle$ ($c = \langle n,rel\text{-}v,in\text{-}v\text{-}f\rangle$) out of $Request(n_a)$;
2. if there exists an active query $Q = (in\text{-}v@n_a, S, C)$ in $Memory(n_a)$,
   then,
   if c is not in C,
   1. add c to C;
   2. for each *out-v* in S,
      call **propagate**$(out\text{-}v@n,c)$

   else,
   1. create the new active query
      $NQ = (in\text{-}v@n_a, \{\}, \{c\})$;
   2. add NQ to $Memory(n_a)$;
   3. for each hyperarc h issuing from Node n_a, *
      let *n-in* be the origin of h, *
      $n\text{-}in_a$ be the archetype of *n-in*, *
      send $\langle\Theta_{n\text{-}in}^{ent}(in\text{-}v@n_a), \langle n\text{-}in,\Theta_{n\text{-}in}^{rel}(in\text{-}v@n_a), in\text{-}v@n_a\rangle\rangle$ to $Request(n\text{-}in_a)$. *

action$_2(n_a)$:
1. take an element s out of $Answer(n_a)$:
2. let $Q = (in\text{-}v, S, C)$ be the active query of $Memory(n_a)$ such that $in\text{-}v \subset s$
   if $out\text{-}v = s \backslash in\text{-}v$ is not in S,
   then,
   1. add *out-v* to S;
   2. for each $c = \langle n,rel\text{-}v,in\text{-}v\text{-}f\rangle$ in C
      call **propagate**$(out\text{-}v@n,c)$.

propagate$(out\text{-}v,\langle n,rel\text{-}v,in\text{-}v\text{-}f\rangle)$:
If n is a terminal node,
then,
 if n has a father $n\text{-}f_a$,
 then
  send $in\text{-}v\text{-}f \cup rel\text{-}v \cup out\text{-}v$ to $Answer(n\text{-}f_a)$,
else,
 let *n-suc* be the successor of n,
  $n\text{-}suc_a$ be the archetype of *n-suc*,
 send $\langle\Theta_{n\text{-}suc}^{ent}(rel\text{-}v \cup out\text{-}v), \langle n\text{-}suc,\Theta_{n\text{-}suc}^{rel}(rel\text{-}v \cup out\text{-}v), in\text{-}v\text{-}f\rangle\rangle$ to $Request(n\text{-}suc_a)$.

The process associated with a base archetype Node n_a is similar to the one associated with a virtual node except that the lines of **Evaluate**(n_a) and **action**$_2(n_a)$ marked by $*$'s are respectively replaced by:

if $Request(n_a) \neq \{\}$, then call **action**$_1(n_a)$.
and
3. let $\mathcal{Q}$ be the query scheme labeling n_a,
   1. replace the empty set of exit values of NQ by $ans(\mathcal{Q},in\text{-}v@n_a, B_E)$;
   2. for each s in $ans(\mathcal{Q},in\text{-}v@n_a, B_E)$,
      call **propagate**$((s@n)\backslash in\text{-}v,c)$.

As they are presented, the algorithms implement a tuple-oriented evaluation strategy: a message only contains one answer or one request. This is not inherent to our evaluation method. Each process could also accumulate the requests and answers it generates until its own *Answer* and *Request* buffers are empty and then only send sets of requests and answers to the other processes. The evaluation would then be set-oriented.

Any other strategy that lies between those two extremes is also possible. The choice of an optimal strategy depends on the communication costs in the distributed architecture. If they are negligible, a tuple-oriented strategy can be preferable because, as each request and answer are sent as soon as they are generated, the cpu utilization of the processes is optimized.

---

[3] Correctness and completeness of the corresponding sequential algorithm have been proved in [11]. The demonstration could easily be transformed for the parallel version.

Otherwise, a more set-oriented strategy must be chosen.

4.4 Termination

In order to discuss termination, each process is assumed to acknowledge each message it receives as soon as it reads it and to maintain, in a counter, the number of messages that it sent and has not yet been acknowledged of.

Termination occurs as soon as, for every process,

- *Request* and *Answer* buffers are empty,
- there is no message currently being processed,
- the value of counter is equal to zero.

The problem in a distributed architecture is determining asynchronously when all the above conditions are satisfied. Fortunately, that problem has well-known solutions. Their presentations is out of the scope of this paper and the interested reader is invited to refer to [6]. That problem is also shortly discussed in [23].

4.5 Example

We now present the trace of an example. It is short but typical due to the symmetry of the e_2 relation on the couple (a, b). This symmetry is responsible for the formation of an active query with two different contexts. We have made explicit the consequences of the execution of processes. The creations and alterations of the different active queries at the root node are indicated in $M(n_0)$ (for *Memory*(n_0)). For the sake of simplicity and brevity, they are not shown for the base archetype nodes. The *Request* and *Answer* buffers of a node n are denoted $R(n)$ and $A(n)$. They are empty by default. In this simple example, entry and exit values have only one component.

The intensional database is

$$p(X,Y) \leftarrow e_1(X,Y)$$
$$p(X,Y) \leftarrow e_2(X,Z), p(Z,T), e_2(T,Y).$$

The extensional database is

$e_1(b,c)$	$e_2(a,b)$	$e_2(d,e)$
$e_1(d,g)$	$e_2(b,a)$	$e_2(e,f)$
	$e_2(c,d)$	$e_2(g,h)$

The query is $p(a,Y)$. Nodes n_i $(0 \leq i \leq 4)$ are numbered as in Example 2. Nodes n_0, n_1 and n_2 are archetype nodes.

Initially, $R(n_0) = \{<\{X/a\},<n_0,\{\},\{\}>>\}$.

1. **Evaluate**(n_0)
   $M(n_0) = \{(\{X/a\},\{\},\{<n_0,\{\},\{\}>\})\}$
   $R(n_1) = \{<\{X/a\},<n_1,\{\},\{X/a\}>>\}$
   $R(n_2) = \{<\{X/a\},<n_2,\{\},\{X/a\}>>\}$
2. **Evaluate**(n_1) $\|$ **Evaluate**(n_2)
   $R(n_0) = \{<\{Z/b\},<n_3,\{\},\{X/a\}>>\}$
3. **Evaluate**(n_0)
   $M(n_0) = \{(\{X/a\},\{\},\{<n_0,\{\},\{\}>\}),$
   $(\{X/b\},\{\},\{<n_3,\{\},\{X/a\}>\})\}$
   $R(n_1) = \{<\{X/b\},<n_1,\{\},\{X/b\}>>\}$
   $R(n_2) = \{<\{X/b\},<n_2,\{\},\{X/b\}>>\}$
4. **Evaluate**(n_1) $\|$ **Evaluate**(n_2)
   $R(n_0) = \{<\{Z/a\},<n_3,\{\},\{X/b\}>>\}$
   $A(n_0) = \{\{X/b, Y/c\}\}$
5. **Evaluate**(n_0)
   $M(n_0) = \{(\{X/a\},\{\},\{<n_0,\{\},\{\}>,<n_3,\{\},\{X/b\}>\}),$
   $(\{X/b\},\{\{Y/c\}\},\{<n_3,\{\},\{X/a\}>\})\}$
   $R(n_2) = \{<\{T/c\},<n_4,\{\},\{X/a\}>>\}$
6. **Evaluate**(n_2)
   $A(n_0) = \{\{X/a, Y/d\}\}$
7. **Evaluate**(n_0)
   $M(n_0) =$
   $\{(\{X/a\},\{\{Y/d\}\},\{<n_0,\{\},\{\}>,<n_3,\{\},\{X/b\}>\}),$
   $(\{X/b\},\{\{Y/c\}\},\{<n_3,\{\},\{X/a\}>\})\}$
   $R(n_2) = \{<\{T/d\},<n_4,\{\},\{X/b\}>>\}$
8. **Evaluate**(n_2)
   $A(n_0) = \{\{X/b, Y/e\}\}$
9. **Evaluate**(n_0)
   $M(n_0) =$
   $\{(\{X/a\},\{\{Y/d\}\},\{<n_0,\{\},\{\}>,<n_3,\{\},\{X/b\}>\}),$
   $(\{X/b\},\{\{Y/c\},\{Y/e\}\},\{<n_3,\{\},\{X/a\}>\})\}$
   $R(n_2) = \{<\{T/e\},<n_4,\{\},\{X/a\}>>\}$
10. **Evaluate**(n_2)
    $A(n_0) = \{\{X/a, Y/f\}\}$
11. **Evaluate**(n_0)
    $M(n_0) =$
    $\{(\{X/a\},\{\{Y/d\},\{Y/f\}\},\{<n_0,\{\},\{\}>,$
    $<n_3,\{\},\{X/b\}>\}),$
    $(\{X/b\},\{\{Y/c\},\{Y/e\}\},\{<n_3,\{\},\{X/a\}>\})\}$
    $R(n_2) = \{<\{T/f\},<n_4,\{\},\{X/b\}>>\}$
12. **Evaluate**(n_2)

The computation stops here and the set of answers is $\{\{Y/d\},\{Y/f\}\}$.

5 Optimizations

5.1 The selection function

The selection function is of great importance because it orders the subquery schemes generated from a query scheme. The evaluation processes follow this order when they solve queries and propagate solutions.

In the rule $a(X,Y) \leftarrow b(X,Z), a(Z,Y)$, and for solving the query scheme $a(X^*,Y)$, a selection of $a(Z,Y)$ before $b(X^*,Z)$ leads to the computation of the complete relation a, which is clearly unnecessary. A good selection function must order the subquery schemes so that the number of relevant facts extracted from the extensional database becomes minimum.

A simple choice reducing the size of the manipulated relations consists of always selecting the literal with most entry variables, making the assumption (heuristic) that queries with more instantiated variables have smaller set of exit values. Statistics could also be of great help.

Another selection practice is based on *recursion levels*. Base predicates are by convention at level 0. A predicate p is of higher level than a predicate q if an instance of q is in the antecedent of a rule defining p or if there exists an r such that p is of higher level than r and r than q. Of course, p and q are at the same level if p is of higher level than q and conversely.

A *recursion level* is a maximal set of predicates at the same level.

Along this line, a good selection practice is often to prefer subqueries leading to lowest recursion levels. Thus, we propose to select, in order of preference, the base predicates (level 0), then the predicates of strictly lower level than the consequent of the rule, and finally the ones at the same level. The assumption is that before working at a level of recursion, it is better to have computed a maximum of facts from lower levels so that maximum information is available.

A more sophisticated strategy integrates the two selection methods just outlined. It needs the definition of the concept of *basic determined part* of a set of query schemes [10]. If $\{l_1, l_2, \ldots, l_n\}$ is a set of query schemes, l_i $(0 \leq i \leq n)$ is *basic determined* if it is a base literal and if either it has an entry variable as one of its arguments or there exists a j $(0 \leq j \leq n)$ such that l_j is basic determined and such that l_i shares variables with l_j. The basic determined part of a set of query schemes is the set of basic determined query schemes.

The selection practice is then to take the entire basic determined part first. If it is empty, one among the virtual predicates of lowest level with a maximum number of entry variables is selected.

The basic determined part gathers together all the base literals which could have been chosen successively by the first simple selection methods. It is clearly more efficient to treat them as a single conjunctive query for accessing the extensional database instead of making successive elementary accesses. This can dramatically reduce the number of accesses. Moreover, conjunctive queries correspond to join operations, that a good DBMS can handle efficiently.

To accommodate this optimization, the compilation phase must be slightly modified. The nodes of the derivation tree are then either virtual query schemes or conjunctions of base ones (in this case, they are tip nodes).

The selection function is also discussed in [23].

5.2 Minimization of the number of extensional database accesses

The preceding processes access the extensional database whenever a new base query is generated. It is clear that, due to the time needed by the (possibly distributed) DBMS to answer literal queries, there is a risk that requests pile up in the buffers *Request*(n_i) associated with base nodes while other buffers become empty.

A first optimization can be to access the extensional database with queries composed of a disjunction of literal queries. In Example 2, *Request*(n_2) can always be written in the form

$$Request(n_2) = \bigcup_{i=1}^{n}\{<\{X/x_i\}, <n_2, \{\}, \{X/x_i\}>>\} \cup \bigcup_{j=1}^{m}\{<\{Z/z_j\}, <n_4, \{\}, \{X/x'_j\}>>\}.$$

Instead of answering the requests separately, the DBMS can be accessed with the composed query $\bigvee_{i=1}^{n} e_2(x_i, Y) \vee \bigvee_{j=1}^{m} e_2(z_j, Y)$. The answers to the query are then sorted by **Evaluate**(n_2), added to the set of exit values of their corresponding active queries and propagated to *Request*(n_0) or returned to *Answer*(n_0) while another composed query is being solved by the DBMS. Thus, the number of accesses to the DBMS decreases and their efficiency increases.

Further, a special interface process can be in charge of managing all queries coming from processes associated with the base nodes. It could for example give a higher priority to the composed queries which are estimated to have the largest set of answers. The intuition underlying this strategy is that it is wise to extract the relevant information from the extensional database as early as possible in order to make it available for computing the virtual queries. Heuristics must be designed to implement that strategy. They could be based on statistical informations or more simply on the rough assumption that the larger a disjunctive composed query, the larger its set of answers. We will not go further in that kind of optimization, because it is clearly closely dependent on the architecture of the DBMS.

5.3 Increasing parallelism

Parallelism can be increased by associating cooperative processes with *Request* and *Answer* buffers instead of archetype nodes. Indeed, the **Evaluate**(n_a) process associated with a virtual archetype node n_a alternatively answers messages from *Request*(n_a) and *Answer*(n_a). It can clearly be split into two parallel processes separately managing the two buffers. Those two processes would share *Memory*(n_a) and some attention must be paid to their correct design.

If N_b and N_v are the numbers of archetypes nodes labeled with query schemes and respectively associated with base and virtual predicates, then, the number of involved processes becomes $N_b + (2 * N_v)$ instead of $N_b + N_v$.

Parallelism can also be increased by a better use of the or-parallelism of the rules. In Example 2,

- requests in buffer $Request(n_0)$ generate, through **Evaluate**(n_0), requests in buffers $Request(n_1)$ and $Request(n_2)$,
- requests in buffer $Request(n_1)$ generate, through **Evaluate**(n_1), answers in buffer $Answer(n_0)$,
- requests in buffer $Request(n_2)$ generate, through **Evaluate**(n_2), answers in buffer $Answer(n_0)$ and requests in buffer $Request(n_0)$,
- answers in buffer $Answer(n_0)$ generate, through **Evaluate**(n_0), requests in buffer $Request(n_2)$.

Cooperative processes could be associated with those communication channels between buffers rather than with archetype nodes or *Request* and *Answer* buffers of archetype nodes. This would again result in an increase of the number of involved processes.

In Example 2, there are 3 archetype nodes, 4 *Request* and *Answer* buffers, and 6 communication channels between buffers.

6 Conclusion

Performance analysis for sequential recursive query evaluation strategies is not easy as shown in [3] where it has been carried out for several strategies on a set of four queries, over a range of data.

The analysis of parallel evaluation strategies is still more complex. For computing the answer to a query, one must take into account the number of messages, the size of buffers and private memories, the number of concurrent accesses to the extensional (distributed) database and so on. In full generality, this is not straightforward, if not impossible. So we will content ourselves with general considerations.

Our algorithm, in its sequential version [11], is close to the QSQR algorithm of [24]. In its spirit, it is also similar to the Alexander method [19] or to the similar Magic Sets method [2]. In the latter methods, constant propagation is compiled into new Horn rules instead of being interpreted. These algorithms are among the most efficient for general recursive rules and general data [3].

Our main contribution, besides the introduction of parallelism and the attention devoted to minimizing the number of database accesses, is the new concept of contexts which remember the minimal part of the computation state necessary to go on when new exit values are found at nodes of the derivation tree. In QSQR, evaluation of queries and subqueries is repeated from the beginning as long as there are no more new solutions. This leads to duplicated work which is avoided with our contexts. In [2] and [19], contexts are hidden in new predicates of rewritten rules. Albeit very elegant, such a solution increases the number of virtual predicates.

Further introducing parallelism is an undisputable speed-up. Of course, one cannot expect to gain an order of magnitude in the evaluation performance: the chosen architecture is multi-computer, with a finite and rather small number of processors. However, efficiency is significantly enhanced for the following reasons:

- Our algorithm fits quite naturally in the framework of a distributed database. It can take advantage of this framework where access to the extensional database is speeded up.
- Evaluation time heavily depends on the average number of processors in operation at any given moment. The higher is the branching in base relations, the higher the average number of working processors. In the worst case, this number is close to one. This is true for example when computing the ancestors of an individual from a parent relation where every individual has at most one child. Then, the problem is intrinsically sequential and no gain is obtained. In the most favorable case of relations with high branching, most processors are working together and parallel computation reaches its full power with a true improvement in execution time.
- In a sequential architecture, the time spent in accessing the extensional database is crucial. During these often numerous accesses, the sequential process suspends itself in a waiting state. It is a bottleneck that can be responsible for a very bad overall performance, as the total evaluation time is always much larger than the total access time to the database. In a parallel architecture, processors work simultaneously with the accesses to the database, evaluation time often remains close to the total access time. Furthermore, in Section 5.2, we showed that the total access time could be significantly decreased by gathering related queries as single access.

Our work can hardly be compared with [1,9,22]. Those papers are dedicated to the parallel evaluation of transitive closure of binary relations and cannot be generalized to general datalog programs. Moreover, [1] and [22] only can compute the complete transitive closure of a relation and thus do not focus on relevant data.

The approach of [26] is very attractive because it introduces parallelism without the necessity of interprocess communication. However, it is only applicable to a limited class of datalog programs (a superclass of the linear single rule programs).

Our work is closely related to [23] where evaluation

is also divided into a compilation phase that builds a rule/goal graph, and a top-down dynamic phase. We have simplified the compilation phase: our derivation graph is much more simple and less redundant than the rule/goal graph of [23]. The dynamic phase in [23] is very shortly and informally described. On the contrary, we tried to be as precise as possible and to provide clear guidelines and hints for implementation.

In conclusion, we have shown that a multi-computer architecture is a nice framework on top of which to implement distributed databases with deductive capabilities. Recursive evaluation with the help of our algorithm improves the traditional sequential methods.

ACKNOWLEDGEMENTS

This work was supported by the Commission of the European Communities, under Project ESTEAM-316 of the ESPRIT Program. We are grateful to A. Pirotte and D. Roelants, our colleagues of PRLB, for their careful readings of early drafts of this paper.

References

[1] R. Agrawal and H. V. Jagadish. Multiprocessor transitive closure algorithms. In *Proceedings of the International Symposium on Databases in Parallel and Distributed Systems*, Austin, Texas, Dec. 1988.

[2] F. Bancilhon, D. Mayer, Y. Sagiv, and J. Ullman. Magic sets and other strange ways to implement logic programs. In *Proceedings of the fifth ACM Symposium on Principles of Database Systems*, March 1986.

[3] F. Bancilhon and R. Ramakrishnan. Performance evaluation of data intensive logic programs. In J. Minker, editor, *Foundations of Deductive Databases and Logic Programming*, pages 439–518, Morgan Kaufmann, 1988.

[4] C. Beeri and R. Ramakrishnan. On the power of magic. In *Proceedings of Sixth ACM SIGACT-SIGMOD Symposium on Principles of Database Systems*, pages 269–283, San Diego, California, March 1987.

[5] S. Ceri and G. Pelagatti. *Distributed Databases. Computer Science Series*, McGraw-Hill, 1984.

[6] M. Chandy and J. Misra. An example of stepwise refinement of distributed programs: quiescence detection. *ACM Transactions on Programming Language and Systems*, 8(3):326–343, July 1986.

[7] C. L. Chang. On the evaluation of queries containing derived relations in a relational data base. In H. Gallaire and J. Minker, editors, *Advances in Data Base Theory, vol. 1*, pages 235–260, Plenum Press, New York, 1980.

[8] G. Gardarin and C. de Maindreville. Evaluation of database recursive logic programs as recurrent function series. In *Proceedings of the ACM-SIGMOD International Conference on Management of Data*, pages 177–186, Washington, May 1986.

[9] K. Guh and C. T. Yu. Evaluation of transitive closure in distributed database systems. *IEEE Journal on Selected Areas in Communications*, 7(3):399–407, Apr. 1989.

[10] L. Henschen and S. Naqvi. On compiling queries in recursive first order databases. *Journal of the ACM*, 31(1):137–147, 1984.

[11] G. Hulin. *An Efficient Interpretive Algorithm for Recursive Queries.* Technical Report R521, Philips Research Laboratory Brussels, Jan. 1988.

[12] G. Hulin, A. Pirotte, D. Roelants, and M. Vauclair. Logic and databases. In A. Thayse, editor, *From Modal Logic to Deductive Databases*, Wiley, 1989.

[13] M. L. Kersten, P. M. Apers, M. A. W. Houtsma, E. J. A. van Kuijk, and R. L. W. van de Weg. A distributed, main-memory database machine. In *Proceedings of the Fifth International Workshop on Database Machines*, pages 496–512, Karuizawa, 1987.

[14] E. Lozinskii. Evaluating queries in deductive databases by generating. In *IJCAI'85 Proceedings of the Ninth International Joint Conference on Artificial Intelligence, Los Angeles, USA*, pages 173–177, Aug. 1985.

[15] G. Marque-Pucheu, J. Martin-Gallausiaux, and G. Jomier. Interfacing prolog and relational data base management systems. In *ICOD-2 Workshop*, Cambridge, England, Sep. 1983.

[16] D. P. McKay and S. C. Shapiro. Using active connection graphs for reasoning with recursive rules. In *IJCAI'81 Proceedings of the Seventh International Joint Conference on Artificial Intelligence, Vancouver, B. C. Canada*, pages 368–375, 1981.

[17] N. J. Nilsson. *Principles of Artificial Intelligence.* Tioga, Palo Alto, 1980.

[18] D. Roelants. *Recursive Rules in Logic Databases.* Technical Report R513, Philips Research Laboratory Brussels, March 1987.

[19] J. Rohmer and R. Lescoeur. *La Méthode d'Alexandre : Une solution pour traiter les axiomes récursifs dans les bases de données déductives.* Research Report DRAL/IA/45.01, Centre de Recherche Bull, March 1985. (in French).

[20] D. Saccà and C. Zaniolo. The generalized counting method for recursive logic queries. In *Proceedings of the International Conference on Database Theory*, Roma, Sep. 1986.

[21] E. Smagge. *Distributed Evaluation in Pool-T of Horn Clauses.* PRISMA Doc. P0174, Philips Research Laboratories Eindhoven, Sep. 1987.

[22] P. Valduriez. Parallel recursive query processing in a share nothing data server. In *Quatrième Journées Bases de Données Avancées*, pages 213–222, Benodet (France), May 1988.

[23] A. Van Gelder. A message passing framework for logical query evaluation. In *Proceedings of ACM SIGMOD International Conference on Management of Data*, pages 155–165, Washington, May 1986.

[24] L. Vieille. Recursive axioms in deductive databases: the Query/Subquery approach. In *Proceedings of the First International Conference on Expert Database Systems, Columbia, S.C.*, pages 253–267, Columbia, SC, 1986.

[25] L. Vieille. From QSQ towards QoSaQ: global optimization of recursive queries. In *Proceedings of the Second International Conference on Expert Database Systems, Washington, D.C.*, pages 421–436, Washington, DC, 1988.

[26] O. Wolfson. Sharing the load of logic programming evaluation. In *Proceedings of the International Symposium on Databases in Parallel and Distributed Systems*, pages 46–55, Austin, Texas, Dec. 1988.

[27] H. Yokota and S. K. et al. *An enhanced inference mechanism for generating relational algebra queries.* Technical Report TR-026, ICOT, Oct. 1983.

Pasta-3's Graphical Query Language: Direct Manipulation, Cooperative Queries, Full Expressive Power

Michel Kuntz and Rainer Melchert

ECRC, Arabellastrasse 17, 8000 Munich 81, West Germany

Abstract

Graphical Direct Manipulation (GrDM) query languages are characterized and positioned with respect to traditional textual query languages. After a brief survey of the current state of the art, the Pasta-3 GrDM query language is presented, with emphasis on three features that make significant new contributions to this field: **(1)** its GrDM basis (query editing through clicking and dragging of an iconic representation of the query expression), **(2)** its cooperative environment (handy values, automatic path completion, edit-and-reevaluate loop), and **(3)** its expressive power (arbitrarily complex conditions, recursive queries, logical variables and quantification, subqueries, mixing in Prolog code). Several examples taken directly from the nearly complete implementation are discussed in detail.

1. Introduction

Query languages as a data base research topic need no introduction -- nor any justification. Until recently query languages were linear textual languages: of the three usual levels of linguistic analysis -- semantic, syntactic, lexical -- only the first one was considered important. The other two were routine: syntax was relegated to a BNF grammar in an appendix, and the lexical level was merely a matter of character strings, not worth mentioning as such. Besides this, language design was the main research activity, since implementation of such languages posed few problems (optimisation excepted) thanks to the maturity of compilation techniques. Thirdly, using such languages was no matter for discussion: there was little to discuss -- you simply typed in statements.

This one-sided approach was inevitable as long as a small number of professional programmers were the only target user group and the only I/O devices widely available were alpha-numeric terminals and keyboards. Today this is no longer the case, and there is good reason to ask if the triple imbalance just described should not be corrected.

A shift towards a better balance has in fact already begun with the emergence of **graphical** query languages. Work in this specific area has been motivated by two factors:

- the realisation that query languages are not only pure mathematical objects but also part of interfaces intended for actual practical use, and
- the technical and economic availability of hardware (high resolution, bit mapped screens and pointing devices such as mice) and software (window managers and graphics packages) supporting graphical interaction.

The graphical Direct Manipulation (GrDM) interaction paradigm[1] usually involves selection of items from pop-up menus and use of icons that can be clicked on and dragged: visible graphical objects both represent the current state of the interface and can be acted upon by "direct" physical actions: moving the tracking symbol over the screen and pressing mouse buttons. GrDM was first used commercially in the Xerox Star and Apple Lisa systems. It has already contributed to the success of software such as spreadsheets, WYSIWIG text editors, CAD systems, etc.

Graphical querying facilities have become an active research area since the mid 1980s. However the old tendency to privilege semantics and design and to ignore actual use is still alive and well: quite recent work in this area [Campbell 87, Catarci 88, Cruz 87, Dart 88] focusses on abstract definitions of new languages and says very little about how they actually look on the screen or how someone would go about using them. On the other hand, some research [Bryce 86, Goldman 85, Kim 88] has helped to restore the balance and place emphasis on the previously neglected aspects of graphical querying. The work reported in this paper, Pasta-3, is a further step in this direction.

The basic approach adopted here is not to define a new formal language but to provide a sophisticated GrDM editing capability for the original formal linguistic objects (already well-defined semantically in the back-end DBMS: see Section 2.2). Put another way, it is the same semantics associated with a GrDM (rather than a textual) syntax.

The Pasta-3 interface includes GrDM capabilities for **all** types of DBMS manipulation activities.[2] However, the purpose of this paper is only to present certain important features of the Pasta-3 **query language**:

[1] [Shnei 83] and [Hutch 86] are just two of the many publications that provide more complete discussions of GrDM.

[2] See [Kuntz 89a, Kuntz 89b, Kuntz 89c] for the other aspects of Pasta-3.

- its GrDM basis: query editing through clicking and dragging of an iconic representation of the query expression (Section 3),
- its cooperative environment: handy values, automatic path completion, edit-and-reevaluate loop (Section 4),
- its expressive power: arbitrarily complex conditions, recursive queries, logical variables and quantification, subqueries, mixing in Prolog code (Section 5).

All three aspects make significant new contributions to the area of GrDM query facilities.

But before these main topics can be properly explained, three items of background information need to be supplied. Since Pasta-3 is not alone in its field, a short survey of related work to date will help to clarify relative positions (2.1). A query language is difficult to understand without reference to the DBMS it is meant to work with -- the KB2 System in this case (2.2). Finally, Pasta-3's query language is part of a complete, integrated interface and can only be understood in this context (2.3).

A note on the implementation and the conclusion make up the last section (6).

2. Background

2.1. State of the Art

Non application domain specific research[3] has produced at least 20 published results: [Bryce 86], [Campb 87], [Catarci 88], [Cruz 87], [Dart 88], [Du 88], [Elmas 85], [Gimn 88], [Goldm 85], [Kim 88], [King 84], [McDon 75], [Melch 87], [Motro 88], [Teskey 84], [Udaga 82], [Urspr 83], [Wong 82], [Wu 86], [Zhang 83].

Analysing them individually and in detail would be the subject matter for another whole paper.[4] Only a few general observations can be made here.

The term "graphical" is **ambiguous**. One sense is abstract and general: pictorial or diagrammatic (as opposed to textual). Another is concrete and restricted: pertaining to raster operations on bitmaps. A third is mathematical: pertaining to graphs. The literature applies an inclusive OR over all three meanings when speaking of query facilities. All results mentioned above are graphical in the first sense, but in the second and third, some are and some are not.

The **major contribution** of graphical querying has been to reduce the cognitive load on users in several ways: lack of typing skill is no longer a significant handicap, formal syntactical conventions are largely replaced by intuitive GrDM operations, and active memorized knowledge of the DB is rarely required since schema components are already visible on the screen and need only be passively recognized.

All the results cited above are **combinations of graphics and text** -- rather than purely graphical systems. In some cases [Catarci 88, Dart 88, Elmasri 85, Ursprung 83, Wong 82, Wu 86, Zhang 83] graphics are used essentially for schema visualisation, and query formulation relies mainly on form-filling or typed-in expressions.

Some graphical query facilities [Goldman 85, King 84, Teskey 84, Wong 82, Zhang 83] do not rely on any explicit graphical **language** at all. They are defined procedurally, in terms of the operations needed to input a query -- in this case operations other than typing characters. There is no well-defined area of the screen the appearance of which completely expresses the query. One cannot simply "read" some configuration of shapes and text to know what was indeed the intent of the preceding sequence of operations.

On the other hand, some query facilities [Campbell 87, Catarci 88, Cruz 87, Dart 88, Udagawa 82] offer only a naked language, with no accompanying **environment** (in the usual sense: a well-integrated set of facilities that support / enhance actual use of the (query) language). Experience in software engineering has demonstrated the importance of good environments for programming languages -- the same holds true for query languages.

There tends to be a **trade-off** between expressive power and ease of implementation: powerful graphical languages have sometimes not been implemented, and many effectively running graphical query facilities support only very simple queries.

This situation indicates that the three main points where progress needs to be made involve:

1. extending **expressive power** to handle complex queries, while
2. taking better advantage of the **GrDM** style of querying, and
3. combining the benefits of a **language and** an **environment**.

2.2. The KB2 System

Pasta-3 is an interface to KB2 [Walla 86], a knowledge base system embedded in Prolog -- or more exactly -- in EDUCE [Bocca 86], a logic programming data base system that supports both a loose coupling and a tight integration of Prolog and a relational DBMS. KB2 uses the Entity / Relationship data model, extended with inheritance and deductive rules. It was developed at ECRC by Mark Wallace and other members of the Knowledge Base Group.

KB2 provides facilities for structured information storage on disk. Queries and updates expressed in terms of

---

[3] Application domain specific work such as [Herot 80, Chen 87, Egenh 88] is not considered here.

[4] and extending the field of comparison to include graphical (or "visual") programming (cf. [Raeder 85], [Myers 86], [Reiser 88]) and graphical interfaces for other logical constructs such as inference rules (e.g. [Lewis 83], [Poltr 86]) is also regretfully excluded

its extended E-R model are mapped down to their relational equivalents. Both schema and data can be modified at any time. However, all updates are checked for consistency with a set of integrity constraints. The information represented with KB2 includes both basic facts and deduction rules. Enforcing the integrity constraints just mentioned ensures that all information -- what is deducible as well as data stored as such -- is always consistent.

KB2 extends traditional data bases in two main directions:

- it supports **general** information through deduction rules and integrity constraints, and
- it allows applications to have a completely **dynamic** knowledge structure.

This dynamic character means that, throughout the life of an application, new Entities and Relationships can be created, the inheritance lattices extended, and properties of E-R items added or deleted.

KB2 supports **complex logical queries**, including quantification, limited recursion, and mathematical functions, expressed as predicate calculus formulae, which are mapped onto Prolog goals and clauses. The fact that KB2 offers its users a manipulation language of this power has important implications for Pasta-3: if its GrDM query language is not equally expressive, end-users would be reluctant to accept a trade-off of more ease of use against loss of expressiveness.

2.3. Overview of Pasta-3

This section provides general background information on Pasta-3 and gives some brief notes on its top-level architecture.

Figure 1 shows a partial view of one possible screen with several different types of windows.

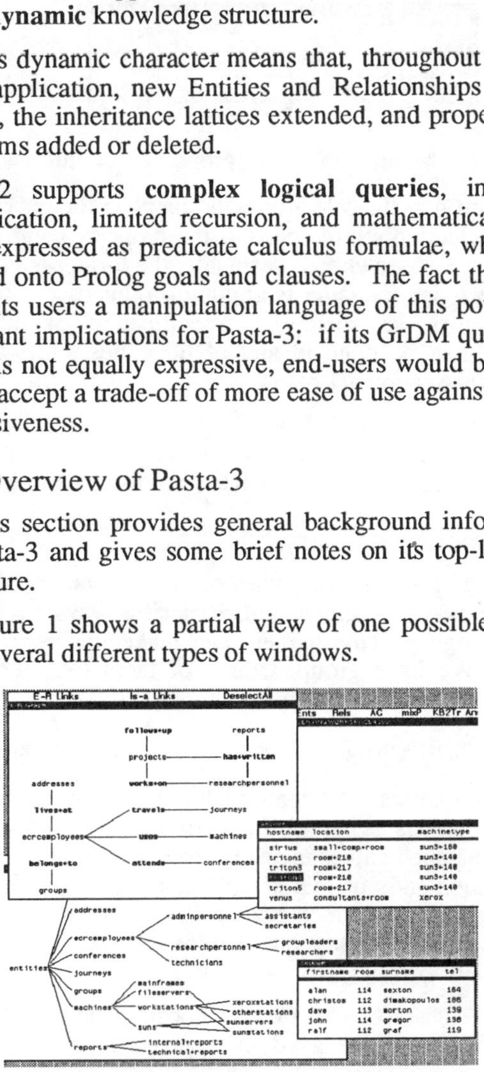

Figure 1: Partial view of screen (one possible layout)

On the left are two graphical schema manipulation windows (E-R schema and Entity inheritance lattice). They provide both schema creation and schema browsing functionality. Two tabular data display windows are visible towards the lower right: they could be the result of browsing or querying. In the upper right quadrant one can see a (still empty) querying workspace.

Figure 2 shows Pasta-3's intermediate position between the workstation and the back-end DBMS, KB2.

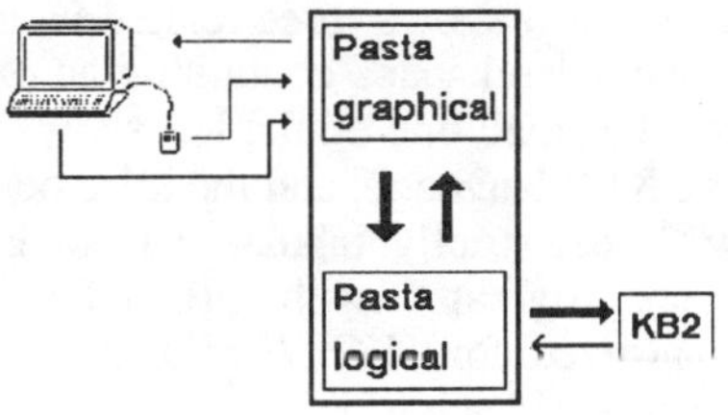

Figure 2: Top-level architecture of the Pasta-3 interface

Pasta-3 is composed of two main parts: **Pasta Logical** and **Pasta Graphical**. The latter is "nearest" to the user and is directly responsible for dealing with all I/O that passes through the interface. In the input direction it translates user activity manifested by means of mouse and keyboard into expressions of the language used as an internal communication protocol connecting to the other main component, Pasta Logical. In the output direction, Pasta Graphical takes character-strings containing information generated at lower levels of the overall system and produces the appropriate visual representation on the screen. The jobs of Pasta Logical are to construct KB2 expressions from the protocol language commands, to invoke KB2 processing, and to collect arbitrary numbers of variable instantiations returned from KB2 into a single result list, suitable for transmission up to Pasta Graphical.

3. Basic GrDM Characteristics

Pasta-3's querying capability has been designed to support all queries that are expressible in KB2's linear language (which is logical rather than navigational). The query specification process takes place in an instance of a dedicated window type (a "workspace") and results in exactly one expression in Pasta-3's two-dimensional language. The steps composing this process can be carried out in whatever order the user prefers.

For presentation purposes, it is helpful to distinguish the core language elements from relatively sophisticated aspects. The basic representations and operations used in querying are illustrated with a rudimentary example having only one Entity in the workspace and a compound selection consisting of one conjunction of selections all with the comparator "=". At this point three of Pasta-3's cooperative features can be explained (Section 4). Next (in 5.1) certain more advanced aspects will be discussed together from the starting point of a query involving several Entities and Relationships in the workspace, possible use of aggregate functions, and a selection consisting of an arbitrarily complex tree of compound selections (AND, OR, NOT) containing selections using various comparators. Finally (5.2) we move on to other powerful options illustrated by four final examples.

3.1. Graphical Representation

The first example query is "name and location of the Xeroxes ?" Fig. 3 below shows three related windows: the large one is the query workspace containing an elementary query expression, the small one on the left shows the query translated into the KB2 language, and the third one is a data display window. This strictly tabular way of expressing elementary queries originated with QBE [Zloof 77] and was previously adopted for HIQUEL [Urspr 83], GLAD [Wu 86], and Pasta-II [Melch 87].

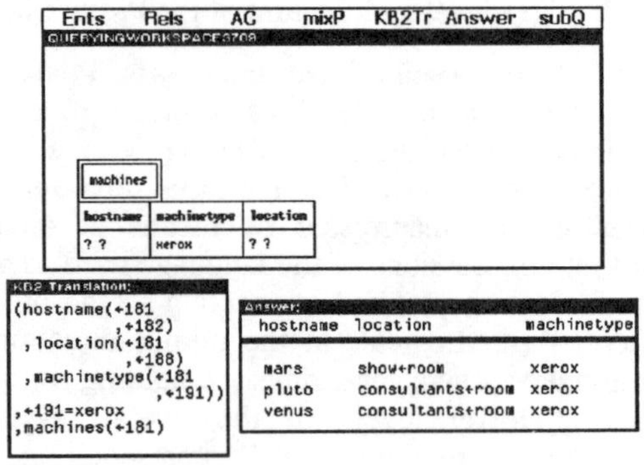

Figure 3: Query workspace, KB2 translation, and answer

The double-bordered rectangle in the lower-left corner of the workspace represents the Entity "machines". Three of its properties ("hostname", "machinetype", and "location") are already available for further work. The "??" projection symbol in the boxes below "hostname" and "location" indicates that results (i.e. data) for these properties are to be displayed. The string "xerox" boxed below "machinetype" represents a selection, the "=" comparator being implicit here by convention (as the conjunction would be if there were several selections).

Display of the linear syntax translation is normally entirely optional: it is necessary only in conjunction with the "Mix in Prolog" capability (see 5.2 below).

3.2. Query Building Operations

How did this very simple query expression come into existence ?

Once a new workspace has been created (via the Pasta-3 main menu), the first step in using it is to indicate which Entities and/or Relationships the query involves. To do this, users have two possibilities: if they have been browsing, they can copy into the workspace E-R items already displayed in other windows (cf. Fig. 1); if they have opted for querying directly, the attached menu at the top of the query workspace window provides the required functionality: two of its items are "Ents" and "Rels". Clicking on these pops up lists of Entities and Relationships from which the ones of interest can be chosen. Either possibility produces the same effects: an icon representing the E-R item in question appears in the workspace and the corresponding icon in the schema graph windows (cf. Fig. 1) is highlighted. In the example, the user has chosen the "machines" Entity.

Clicking the "Properties" item in the pop-up menu for "machines" (leftmost in Fig. 4) displays the list of this Entity's properties from which the user can choose one or more for further work. In Figure 3 above, the user has already chosen three properties: "hostname", "machinetype", and "location".

To indicate that they want results for the "hostname" property to be displayed, users pop up the command menu for properties (second in Fig. 4) by clicking on "hostname" in the workspace in Fig. 3 and choose the "Display" item. This immediately causes the display of the "??" projection symbol in the box below "hostname". They do the same for "location".

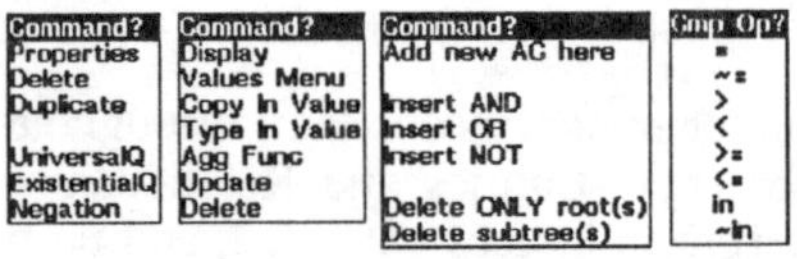

Figure 4: Menus: ER items, properties, conditions, comps

Specifying the selection will complete the query. Two of the selection's three parts -- the property name and the comparator -- are already available: the former is visible on the screen and the latter is implicit (in this case -- for other comparators, see below). Specification of the value "xerox" can be done in any of three possible ways, explained in Section 4.1 below.

Clicking on the "Answer" option in the attached menu of the workspace window invokes query evaluation and displays the answer in a separate data display window.

4. Cooperative Querying

GrDM contributes to ease of use in a non-application-specific way. There are also specific ways in which a query language environment can make life easier for its users. We have grouped three of them together in this section.

4.1. Handy Values

Most queries -- no matter what the language -- contain constant values as part of the specification of the selection. Indicating such values is a very frequent operation in query formulation and one that is particularly error-prone when it involves typing and memory recall. Making it easier would be a big gain for any query interface. Pasta-3 provides three ways to do it.

The easiest one to use is choosing the desired value from the menu displayed in response to a click on the "Values menu" item in the properties pop-up menu (Fig. 4). This item is present in the menu only for those properties that have no more than 15 different possible values.[5] This is shown in Fig. 5.

[5] an empirical and somewhat arbitrary limit: users have no difficulty in interacting with a menu of 15 items in this context, but a larger or smaller number of values could be used.

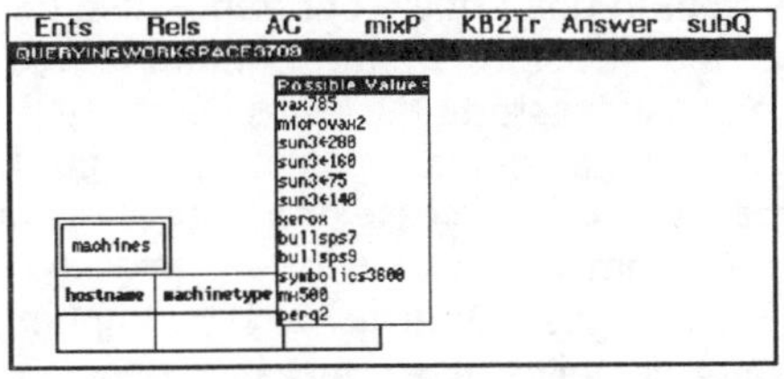

Figure 5: Choosing a value from a menu of possibilities

This method of indicating a value contributes to cooperative querying by anticipating and avoiding some of the problems addressed by work on "cooperative answers" [Kaplan 82, Wahls 85]. Instead of having to compute an explanation of the result "No data found" caused by the use of an unknown value in a restriction, the system encourages (but does not force) users **not to enter** any such values, thus eliminating the empty answer and the need to explain it.[6]

Use of such value menus has its price in interface complexity and computational cost. Cost is mostly incurred once when Pasta-3 is first set-up for a given DB: at that time all the data has to be counted, values collected for properties with less than 15 (or n) different values, and menus built. With the toy DB we have been using for demos this takes about five minutes. With very large DBs the cost might be prohibitive. Interface complexity derives from the fact that whenever updates are performed, Pasta-3 must build, destroy, and/or modify menus dynamically (but this entails no user-perceptible performance overhead).

A more serious problem will arise when multiple users access the same DB: in this case updates occur without the interface knowing of them. This problem is more general than just handy values menus -- it raises fundamental questions about the relationship between interfaces and backends, since the only solution available seems to be modifying the DBMS itself so that: it remembers updates that happened since the last login of user X and informs X's current session with Pasta-3 of these changes, and it does mutual "tipping off" of updates among concurrently running sessions.

The second possibility is to use the "Copy In value" command. This can be invoked when values are already on display elsewhere on the screen in data display tables (as a result of data browsing or as the answer to some previous query) . In this case also, user typing errors are obviously excluded. However, one cannot guarantee that empty answers will be avoided, since there are no constraints on which values can be copied from where to where.

The third possibility is the traditional one: entering a character string from the keyboard.

4.2. Automatic Path Completion

Many complex queries involve several Entities and Relationships. The most frequent case that we have encountered (and the simplest one) concerns two Entities -- that users select correctly -- and one Relationship -- that users forget to select -- even though their intention was that the two Entities should be considered connected. In less simple cases the missing connection can include several Entities and Relationships and be ambiguous (i.e. be a subgraph instead of a path). The difficulties involved here have already been studied [Maier 83, Zhang 83, Pahwa 85]. Traditionally users have a semantic problem to solve in navigating the right connection and syntactic one in building the linear textual expression that corresponds to the chosen connection. Since getting it wrong is easy, helping users to get it right is again an important advantage for any interface.

Pasta-3 is cooperative in this respect thanks to its ability to compute missing connections automatically: if the workspace contains E-R items that are connectable but not yet directly connected, and evaluation is invoked, the system will find the missing items and either offer to add them in (if exactly one connecting path exists) or propose a choice (if ambiguous possibilities exist). Doing these computations is one of the major functions of the logical subsystem of Pasta-3.

4.3. Edit-and-Reevaluate Loop

As soon as a query has been built and evaluated[7], results are displayed. On seeing them, users often realize that the query did not in fact express what they had wanted. In other cases, the query was indeed faithful, but they want to see results for a slightly different query and compare them to the previous ones. Any interface that does not provide convenient means to support these two querying behaviours[8]is not cooperative. Pasta-3 does and is.

The query expression in a workspace can be modified, using the same GrDM techniques as in the building phase, and reevaluated, with the new answer data appearing in another display window, as many times as the end user wishes. Support for such an "edit-and-reuse" way of working constitutes an essential factor contributing to good productivity.

5. Extended Expressiveness

As so far presented, the Pasta-3 query language is very convenient to use but has very little expressive power. One must retain this ease of use (i.e. not revert to typing in statements of a textual language) while greatly extending expressiveness. This section gives some idea of how Pasta-3 solves this problem. However space constraints

[6]except possibly in queries incorporating open (i.e. set) subqueries

[7]KB2 as such supports lazy evaluation, but for the time being Pasta-3 only uses immediate evaluation.

[8]also termed "retrieval by reformulation" in other interfaces (e.g. RABBIT [Willi 84] and KARMA [Bose 86])

make it impossible to include in the discussion of examples explanations of the processing done by the logical subsystem of Pasta-3 -- it must nonetheless be pointed out that much of the overall power of the interface derives from the logical subsystem's flexible internal representation of queries and the sophisticated transformations it carries out.

5.1. Complex Selections

The second example query can be paraphrased in English as: "How many researchers between the ages of 20 and 30 not in rooms 120, 220, or 320 belong to the HCI group ?" Fig. 6 shows the Pasta-3 formulation.

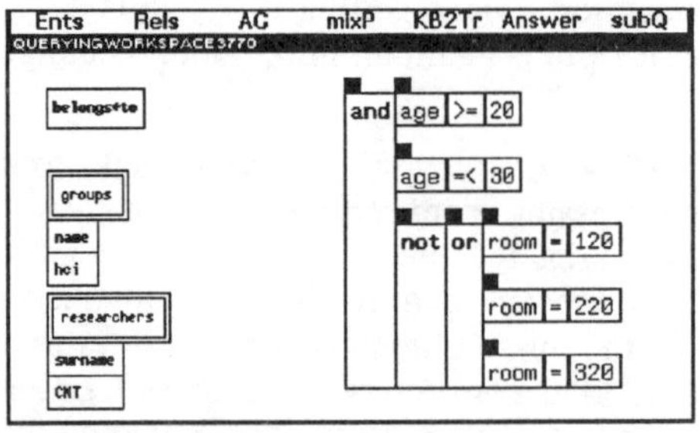

Figure 6: A second example query expression

Aside from the introduction of the count aggregation function (CNT), this example differs from the first one mainly in two ways:

- several E-R items are involved, and
- the selection is complex.

Users can bring Entities and Relationships into the workspace in the ways described for the previous example. Building a complex selection expression proceeds as follows. The user starts by constructing a single "atomic condition". This process begins with clicking on the "AC" item in the attached menu and leads to the state where an empty atomic condition icon has been created at the right of the workspace, ready to be instantiated.

To express that the "age" property is to be the left operand of this condition, the user brings up a menu of the properties of E-R items currently in the workspace and chooses "age". To specify the desired comparator, the user clicks in the middle box of the condition, pops up the comparator menu (rightmost in Fig.4), and opts for ">=". Specifying the right operand can in general be done in the three ways mentioned in Section 4.1 (menu selection, copy, type in).

The user now clicks on "AC" again to create a second empty condition box and fills it in analogously to the first. To add the conjunction connecting these two conditions, the user selects all the conditions to be connected by clicking in the small black squares at the top left corner of each condition. The "AND" connector can be obtained by clicking on the black square of the last condition to be chosen, getting the third pop-up menu in Fig. 4 and selecting "Insert AND". An AND logical operator icon is then put in place automatically by Pasta-3.

It is important to point out -- but impossible to show on paper -- that the structure of the condition tree can be modified at will by **dragging** the icons composing it. If an atomic condition or a logical operator icon is clicked, it (plus its entire subtree in the case of a logical operator) can be pulled away from its location in the tree and dropped elsewhere. Moving it away detaches it. Dropping it on a logical operator attaches it as the last operand of that operator. Dropping it in the background simply leaves it unattached at top level. In all cases, the graphical condition tree is automatically reshaped and redisplayed to take into account the changes in its logical structure. This GrDM capability provides an easy way to build up very complex logical conditions and thus removes one of the major difficulties encountered in using traditional, linear query languages: users get lost in the syntax of complex expressions before they manage to formulate the condition that they had in mind.

The user builds the rest of the query with similar operations.

5.2. Recursion, Quantification, Subqueries, Prolog

Four advanced features can now be presented, but not in detail.

Certain **recursive** queries[9] can be expressed through use of the "Duplicate" menu item, which adds another icon for the same Entity into the workspace and notes that there are multiple occurrences to be handled. Fig. 7 shows how to formulate the query "Which projects follow up the Outils project ?".

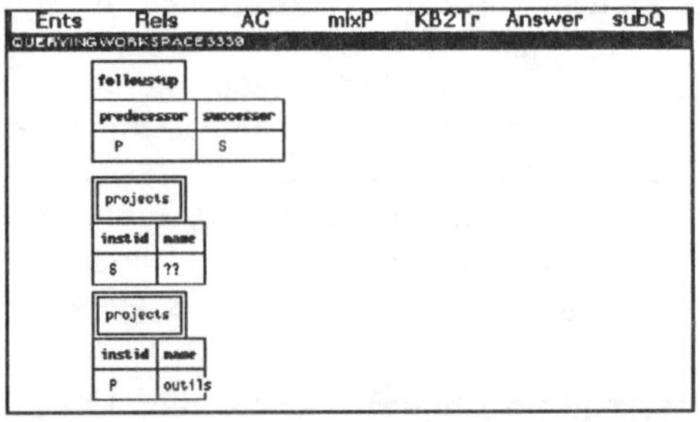

Figure 7: An example of a recursive query

In the schema used here, the "follows_up" Relationship is defined by means of a recursive rule, and the same Entity, "projects", is linked to it twice: once as the "predecessor" role and once as the "successor" role. This entity thus appears in the workspace twice, and the identical variable names ("P" in one case and "S" in the other) indicate which occurrence fills which role.

Logical Variables and Quantification: variable names can be typed in as property values, thus retaining an essential feature of the KB2 linear language. This pos-

[9]See [Cruz 87] for a graphical query language specializing in recursion.

sibility is also used in Fig. 7 where P and S are such variables. Quantification is always implicitly existential if not indicated otherwise. Explicit quantification is dealt with by means of the menu items, "universalQ" and "existentialQ". Choosing either displays the appropriate symbol at the left side of the E-R item icon. **Scope** is expressed by relative spatial position: the left-to-right ordering of quantifiers in the corresponding textual formulae is expressed by the top-to-bottom arrangement in the workspace of the icons representing the entities and relationships. Icons can be moved to wherever desired in the workspace simply by dragging them. Fig. 8 shows two queries (in two workspace fragments). On the left: "Is there a Sun that is used by all employees ?" (i.e. the same Sun is used by everyone); on the right: "Do all employees use a Sun ?" (i.e. each employee may or may not have his/her own individual Sun). The only difference between the two queries lies in the relatives scopes of the quantificrs thcy contain.

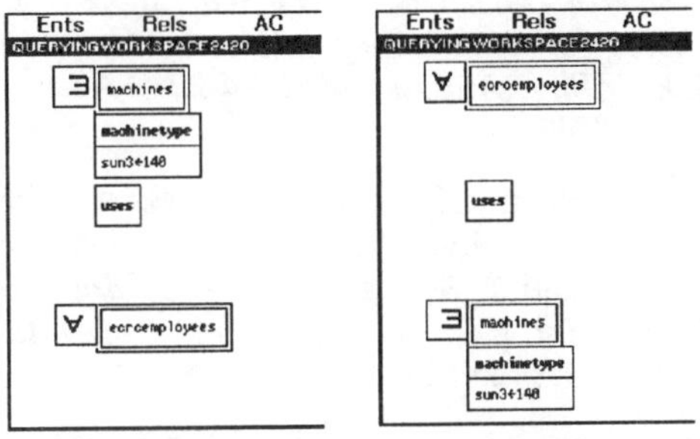

Figure 8: An example with two quantified queries

Subqueries[10]can be derived from workspaces by shrinking them to icons. With the "SubQ" command from the attached menu, icons so created can be copied into a query workspace and used as operands in both atomic and composite conditions, however correctness is checked only at evaluation and not on insertion. Fig. 9 shows an example. It supposes that there already existed an iconised query workspace called "KB_surnames" whose value is the set of the surnames of the members of the KB Group. The example expression asks "What are the ages of the researchers in the KB Group ?".

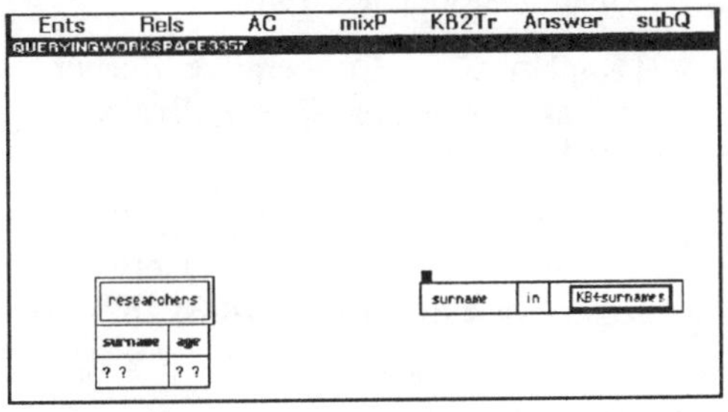

Figure 9: An example of a query containing a subquery

Mixing in Prolog -- creating an expression that contains both KB2 statements and Prolog predicates -- is done by using the "MixP" option in the attached menu. This pops up an editing window into which Prolog can be typed. The code thus entered is either prepended or appended to the KB2 translation of the query expression. Choice of appropriate variable names (thus enabling sharing between the two parts of the final query expression) is the user's responsibility. Variables instantiated during evaluation are treated like property names, and values for them are displayed along with other results. Fig. 10 includes three windows: a workspace, a KB2 translation window (corresponding to only the graphical part of the query), and an editing window for typing in Prolog. The complete query can be paraphrased as "How old would ecrc employees be if they aged by 10 % ?". A mix-in capability of this kind gives Pasta-3's query language a significant degree of on-the-fly extensibility which expert users can exploit to great advantage.

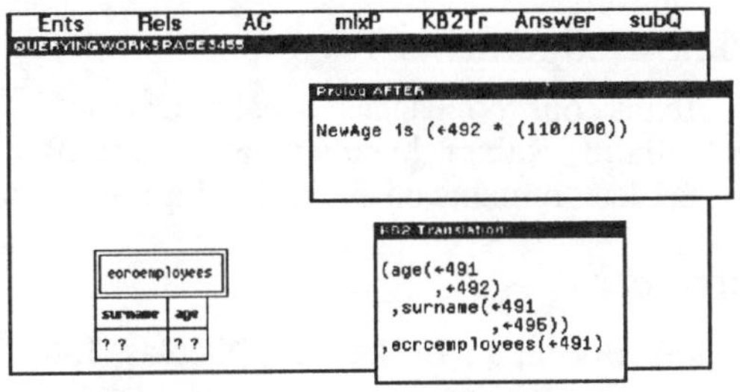

Figure 10: How Prolog code can be mixed in

6. Implementation and Conclusion

We built a throw-away prototype of a subset of the Pasta-3 query language in spring 1988, with the graphical subsystem running on Xerox 1108 hardware and the logical subsystem and the KB2 KBMS installed on a Sun 3-140 (the two subsystems being linked by a TCP connection over the local Ethernet). This version was extensively used for demos and gave rise to many critical observations and comments from people who had been users of KB2's textual language and from software ergonomists from ECRC shareholder companies.

A full-scale implementation begun in summer 1988 is nearing completion. On the graphical subsystem side, all functionality mentioned in this paper is up and running, except Prolog mix-in. The main implementation tool used is the TUBE User Interface Development Environment [Herrm 89, Hill 89]. On the logical subsystem side, coded in Prolog, the core functionalities are finished, however quantification and Prolog mix-in are not yet supported, and the implemented path completion algorithm is not as sophisticated as possible and not yet integrated with the other functionality.

We conclude by summarizing the most important aspects of the work described in this paper. The Pasta-3 query language is based on the **graphical Direct Manipulation** (GrDM) interaction paradigm, which relies

[10]in the same sense as for SQL

on a bit-mapped, multi-window screen and a mouse to implement clickable, draggable icons as the main representation of the query expression. Its purpose is not to invent an nth formal language, but rather to provide a new, sophisticated GrDM editing capability for the already existing formal constructs of the KB2 language. **Cooperative querying** is supported through: handy values, automatic path completion, and the edit-and-reevaluate loop. **Expressive power** is obtained by allowing: arbitrarily complex selections, recursive queries, logical variables and quantification, subqueries, and mixing in Prolog code.

These features of the Pasta-3 query language thus contribute to progress on the aspects of graphical querying where it was most needed. However a major open problem (already mentioned in Section 4.1) is the relationship between GrDM interfaces and DBMS back-ends: to what extent can DBMSs continue to be designed and developed without taking into consideration from the very beginning the requirements of GrDM interfaces ?

7. Acknowledgements

We thank our colleagues Johann Christoph Freytag, Herve Gallaire, Rainer Manthey, and Mark Wallace for their valuable comments on an earlier draft of this paper.

8. References

[Bocca 86] Bocca, J.; Decker, H.; Nicolas, J.-M.; Vieille, L.; Wallace, M. Some Steps Towards a DBMS Based KBMS. In Kugler H.-J. (editor), *Information Processing '86 - IFIP Congress Series*. IFIP, 1986.

[Bose 86] Bose, P.; Rajinikanth, M. An Intelligent Assistant to a Database System. *Texas Instruments Engineering Journal* 4(1):163-169, Jan/Feb., 1986.

[Bryce 86] Bryce, D.; Hull, R. SNAP: A Graphics-based Schema Manager. In *Proc. of the IEEE 2nd Int. Conf. in Data Engineering, Los Angeles*, pages 151-164. 1986.

[Campb 87] Campbell, D. M.; Embley, D. W.; Czejdo, B. Graphical Query Formulation for an Entity-Relationship Model. *Data & Knowledge Engineering* (2):89-121, 1987.

[Catarci 88] Catarci, T.; Santucci, G. Query By Diagram: A Graphic Query System. In *Proc. of the 7th Int. Conf. of Entity Relationship Approach, Rome, Italy*. The ER Institute, 1988.

[Chen 87] Chen,Tao; Wu, Jian-Kang. QPF -- A Versatile Query Language Based on ADTS. In *1987 Workshop on Visual Languages*, pages 199-205. Linkoeping University, August, 1987.

[Cruz 87] Cruz,Isabel;Mendelzon,Alberto;Wood,Peter. A Graphical Query Language Supporting Recursion. In *1987 SIGMOD Conference*, pages 323-330. ACM, May, 1987.

[Dart 88] Dart,Philip;Zobel,Justin. Conceptual Schemas Applied to Deductive Databases. *Information Systems* 13(3):273-287, 1988.

[Du 88] Du, H.; Manoochehr, A. GQL: A Graphical Database Language Using Pattern Images. In *Proc. of the Computer Graphics Int. Conf., Geneva, May*. 1988.

[Egenh 88] Egenhofer, M.J.; Frank, A.U. Towards a Spatial Query Language: User Interface Considerations. In *Proceedings of the 14th Conf. on Very Large Data Bases (VLDB'88), Los Angeles, California*, pages 124-133. 1988.

[Elmas 85] Elmasri, R.; Larson, J. A. A User-Friendly Interface for Specifying Hierarchichal Queries on an ER Graph Database . In *Proc. of the 4th Int. Conference on the Entity-Relationship Approach, Chicago*, pages 236-245. IEEE, October 28-30, 1985.

[Gimn 88] Gimnich, R. Implementing Direct Manipulation Query Languages Using an Adequate Data Model. In *Proc. of the 7th Interdisciplinary Workshop 'Informatics and Psychology': Visualization in Human-Computer Interaction, Schaerding, Austria, May 24-27*. Austrian Computer Society, 1988.

[Goldm 85] Goldman, K. J.; Goldman, S. A.; Kanellakis, P. C.; Zdonik, S. B. ISIS: Interface for a Semantic Information System. In *Proceedings of ACM-SIGMOD 1985 Int. Conf. of Management of Data, Austin, Texas*, pages 328-342. May, 1985.

[Herot 80] Herot, C. F. Spatial Management of Data. *ACM Transactions on Database Systems* 5(4):493-514, December, 1980.

[Herrm 89] Herrmann, M.; Hill, R. D. Abstraction and Declarativeness in User Interface Development -- The Methodological Basis of the Composite Object Architecture. In *IFIP Congress '89*. IFIP, 1989.

[Hill 89] Hill, R. D.; Herrmann, M. The Structure of TUBE -- A Tool for Implementing Advanced User Interfaces. In *EUROGRAPHICS '89*. 1989.

[Hutch 86] Hutchins, E.; Hollan, J.; Norman, D. Direct Manipulation Interfaces. *User Centered System Design*. Lawrence Erlbaum Associates, 1986.

[Kaplan 82] Kaplan, S.J. Cooperative Responses from a Portable Natural Language Query System. *Artificial Intelligence* 19:165-187, 1982.

[Kim 88] Kim, H.-J.; Korth, H. F.; Silberschatz, A. PICASSO: A Graphical Query Language. *Software-Practice and Experience* 18(3):169-203, March, 1988.

[King 84] King, R.; Melville, S. SKI: A Semantics-Knowledgeable Interface. In *Proc. of the Tenth Internat. Conf. on Very Large Databases (VLDB), Singapore*, pages 30-33. August, 1984.

[Kuntz 89a] Kuntz, Michel; Melchert, Rainer. Pasta-3: a Complete, Integrated Graphical Direct Manipulation Interface for Knowledge Base Management Systems. In *to appear in: Proc. 11th World Computer Congress*, pages . IFIP, August, 1989.

[Kuntz 89b] Kuntz, Michel; Melchert, Rainer. Pasta-3's Requirements, Design, and Implementation: A Case Study in Building a Large, Complex Direct Manipulation Interface. In *to appear in: Proc. Engineering for Human-Computer Interaction*, pages . IFIP, August, 1989.

[Kuntz 89c] Kuntz, Michel; Melchert, Rainer. Ergonomic Schema Design, Browsing with More Semantics, and a Graphical Direct Manipulation Query Language: the Pasta-3 End User Interface for E-R DBMSs. In *submitted to: 8th E-R Conf*, pages . E-R Institute, October, 1989.

[Lewis 83] Lewis, J. W. An Effective Graphics User Interface for Rules and Inference Mechanisms. In *Proc. of the CHI'83*, pages 139-143. December, 1983.

[Maier 83] Maier, D.; Ullman, J.D. Maximal Objects and the Semantics of Universal Relation Databases. *ACM Transactions on Database Systems* 8(1):1-14, March, 1983.

[McDon 75] McDonald, N.; Stonebraker, M. CUPID - The Friendly Query Language. In *Proc. of the 1975 ACM PACIFIC, San Francisco*, pages 127-131. ACM, April, 1975.

[Melch 87] Melchert, Rainer. Graphische Unterstuetzung beim Umgang mit Wissensbanken. In *Proceedings of Datenbanksysteme in Buero, Technik und Wissenschaft*, pages . BTN 87, April, 1987.

[Motro 88] Motro, Amihai; D'Atri, Alessandro; Tarantino, Laura. The Design of KIVIEW: An Object-Oriented Browser. In *Proceedings of the 2nd Int. Conf. on Expert Database Systems*, pages 17-31. George Mason University, April, 1988.

[Myers 86] Myers, Brad. Visual Programming, Programming by Example, and Program Visualization: A Taxonomy. In *CHI 86 Conference Proceedings*, pages 59-66. ACM SIGGRAPH, April, 1986.

[Pahwa 85] Pahwa, A.; Arora, A. K. Automatic Database Navigation: Towards a High Level User Interface. In *Proc. of the 4th Internat. Conf. on Entity-Relationship Approach, Chicago, Ill.*, pages 36-43. IEEE, 28-30 October, 1985.

[Poltr 86] Poltrock, S. E.; Steiner, D. D.; Tarlton, P. N. Graphic Interfaces for Knowledge-Based System Development. In *Proceedings of the CHI'86*, pages 9-14. ACM SIGCHI, April, 1986.

[Raeder 85] Raeder,Georg. A Survey of Current Graphical Programming Techniques. *IEEE Computer* 18(8):11-25, August, 1985.

[Reiser 88] Reiser,Brian et al. Visual Programming, Programming by Example, and Program Visualization: A Taxonomy. In *CHI 88 Conference Proceedings*, pages 39-44. ACM SIGGRAPH, May, 1988.

[Shnei 83] Shneiderman, B. Direct Manipulation: A Step Beyond Programming Languages. *IEEE Computers* (16):57-69, 1983.

[Teskey 84] Teskey, F. N.; Dixon, N.; Holden, S. C. Graphical Interfaces for Binary Relationship Data Bases. *Information Systems* 3(2):67-77, April, 1984.

[Udaga 82] Udagawa, Y.; Oshuga, S. Novel Technique to Interact with Relational Databases by Using a Graphics Display. *Journal of Information Processing* 5(4):256-264, 1982.

[Urspr 83] Ursprung, P.; Zehnder, C.A. An Interactive Query Language to Define and Use Hierarchies. *Entity-Relationship Approach to Software Engineering.* Elsevier Science Publishers B.V. (North Holland), 1983.

[Wahls 85] Wahlster, W. Cooperative Access Systems. In *Proc. of the Int. Conf. on Fifth Generation Computer Systems 1985*, pages Vol.1/ pp.103-111. 1985.

[Walla 86] Wallace, M. *KB2: A Knowledge Base System Embedded in Prolog*. Technical Report KB-12, ECRC GmbH, Munich, W. Germany, 1986.

[Willi 84] Williams, M. D. What makes RABBIT run ? *Int. Journal of Man-Machine Studies* 21:333-352, 1984.

[Wong 82] Wong, H. K. T.; Kuo, I. GUIDE: Graphic User Interface for Database Exploration. In *Proc. of the 8th Very Large Databases Conf., Mexico* , pages 22-32. 1982.

[Wu 86] Wu, C. T. A New Graphics User Interface for Accessing a Database. *Advanced Computer Graphics.* Springer-Verlag, Berlin, Heidelberg, New York, 1986.

[Zhang 83] Zhang, Z.-Q.; Mendelzon, A.O. A Graphical Query Language for Entity-Relationship Databases. *Entity-Relationship Approach to Software Engineering.* Elsevier Science Publishers B.V. (North Holland), 1983.

[Zloof 77] Zloof, M. M. Query-by-Example: a data base language. *IBM Systems Journal* (4):324-343, 1977.

ENIAM: A More Complete Conceptual Schema Language

Peter Creasy

Key Centre for Software Technology
Department of Computer Science
University of Queensland
St. Lucia Qld 4067
Australia

Abstract

The problem of knowledge representation has been extensively addressed in the information systems field and there has been a wide range of suggestions for capturing semantics. A number of systems and methodologies have been proposed. Most seem to suffer from a variety of problems which make representation difficult, e.g. mixing of type and instances, too many and complex structures. Representation languages which permit a graphical form are becoming more widely accepted under the motivation of being able to see the data structures which exist. However the semantics are somewhat limited with, for example, few integrity constraints being represented. In this paper we show how a conceptual schema language with a graphical form can be extended to permit a wider class of semantics to be represented.

1. Introduction

Conceptual schema languages with a graphical form have become increasingly significant in information systems (e.g. Harel 1988) as the importance of being able to see what the data structures are becomes accepted.

An essential part of any knowledge representation system is the representation of facts. As syntactic rules permit facts which are syntactically valid but semantically invalid within the universe of discourse, we additionally need constraint specifications. Correspondingly we distinguish two classes of data types - the fact encoding construct (a data structure type which is associated with a population of instances) and validation rules. It is the latter structures which are of interest in this paper.

NIAM (Nijssen 1977, 1986; Verheijn & van Bekkum 1982) is a methodology with an associated fact-based language with a graphical representation. It uses only a few validation constructs (e.g. mandatory role, uniqueness). While we can represent, what we regard as some of the more important constraints, there is still a large class which cannot be represented. In addition subtype definitions and derivation rules have no formal representation.

Most of the presentations in the integrity constraint area (e.g. Reiter 1988) use first order predicate logic (FOPL), a language unfamiliar to most analysts. Existential graphs (Peirce 1960; Roberts 1973) have been developed in a number of forms which correspond to propositional, first order and higher order logics. The graphs provide a very readable logic representation in graphical form which can be readily integrated with graphical knowledge representation languages. We discuss the first order form which is represented graphically using negation, conjunction and the existential quantifier.

We use the existential graphs to extensively extend the power of NIAM. This provides us with ENIAM, a graphical first order logic language which permits the subtype definitions, derivation rules and constraints to be formally graphically represented.

Finally, we discuss an extension to the NIAM methodology which guides analysts/users in the expression of constraints in ENIAM.

2. NIAM

The NIAM language is fact-based. It has a graphical representation which is the form generally used for conceptual schema design. The facts (fact type instances) are semantically irreducible associations between entity instances, with each entity instance playing a role in the fact (sentence). The methodology abstracts from facts to give fact types, the fact encoding construct.

Figure 1 shows a NIAM diagram which represents people holding passports and visiting countries. The main constructs are: entity types (circles), e.g. PERSON, COUNTRY; roles (rectangles), e.g. visited, was-visited-by, which correspond to verbs of a sentence; fact types (compound rectangles), e.g. VISIT, HOLD; label types (dotted circles), e.g. person-name; and constraints.

A relationship type between entity types is a fact type. Each fact or reference type consists of a number of roles. These describe the part (or role) each entity or label type (to which the role is joined) plays in the fact or reference type. For example the role played by PERSON in CALLED is "is-named".

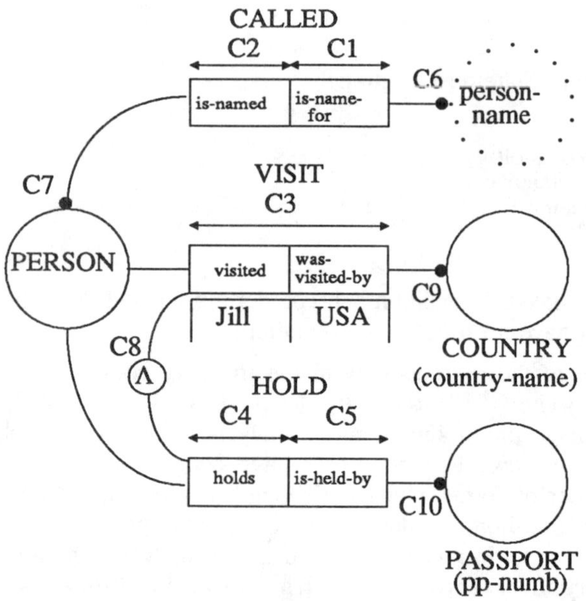

Figure 1:

Entities are non-lexical and are lexically identified by reference types (e.g. CALLED). These are special fact types which associate entity types with the identifying label type. The references can be abbreviated by writing the corresponding label type in parentheses following the entity. We have shown this for the entity type COUNTRY.

An example of a VISIT fact type instance is <Jill,USA>. We can show such populations as in figure 1, where the non-lexical enitites have been represented by the corresponding lexical representation. This fact can be read as *the person identified by person-name Jill visited the country identified by country-name USA.*

Some of the constraints we can represent in NIAM are intrafact (and intrareference), interfact uniqueness, mandatory role, equality, exclusion and subset constraints. The arrows above the roles represent intrafact or intrareference uniqueness constraints which are placed on the instances of the role or roles. They indicate that the role or roles uniquely identify a single fact instance, i.e. they specify functional dependencies. In this case constraints C4 and C5 indicate person ⟷ passport while C3 indicates person, country → ϕ. Constraints C1 and C2 indicate unique naming for the entity type PERSON.

The constraint C7 is a mandatory role constraint which indicates that if an entity (or label) takes part in any fact (or reference) instance then it must take part in an instance of the fact (or reference) type corresponding to the constraint. The subset constraint C8 indicates that anyone who visits a country must have a passport.

The fact types can be mapped to relational structures using, what is essentially, a synthesis algorithm. For figure 1 the resulting relations are VISIT(person , country) and HOLD(person , passport).

We have not mentioned all the features of the language. One which should be mentioned is subtyping, an important part of the language. An example is shown later. It should be noted that it is not necessary to write down the populations, fact type names, role names or constraint names unless they are needed for clarification. This results in a much less cluttered diagram.

3. Existential Graphs

Existential graphs were proposed by Peirce to represent logic and resulted from the work he did over a number of years. The graphs are simple with few concepts. They are one of many graphical representations for logic which have been proposed over the years, e.g. Venn diagrams (see Gardner 1983). The graphs use few operators (user symbols, lines and closures e.g. circles), are easy to read and write and offer the power of FOPL. Peirce proposed a number of what he called *parts* to the existential graphs. He called these the Alpha, Beta and Gamma parts. The parts of interest to us here are the Alpha part (equivalent to propositional logic) and the Beta part (equivalent to FOPL). Roberts (1973) has shown completeness and consistency for existential graphs. In particular Roberts uses Quine's (1955) ML logic system. Sowa (1984) has demonstrated the usefulness of existential graphs as a mechanism for representing FOPL.

The Alpha part has only three basic symbols: the sheet of assertion, the cut and the graph. The graphs are laid out on the *sheet of assertion* (SA), an arbitrarily large area, which equates to a model of the universe of discourse. A graph instance is something which asserts a possible state in the universe, viz. proposition. The sheet of assertion is also considered to be a graph, as the blank SA expresses whatever initially holds in the UoD.

Graph instances written on the SA are asserted to be true. By a graph is meant "any sign (symbol) which expresses in a proposition some state in the universe" (Roberts, p 32).

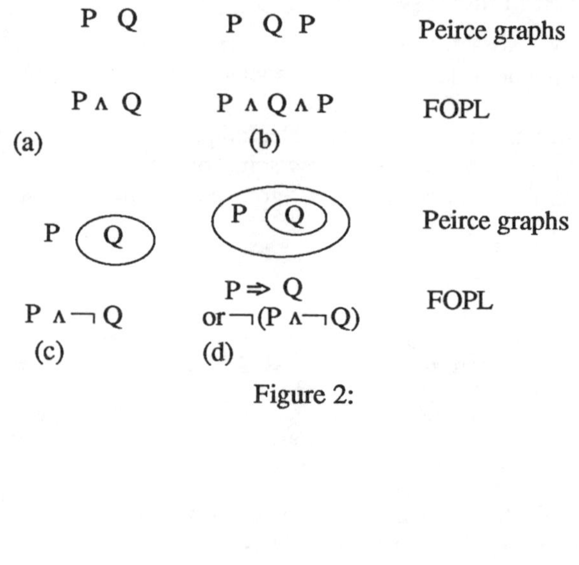

Figure 2:

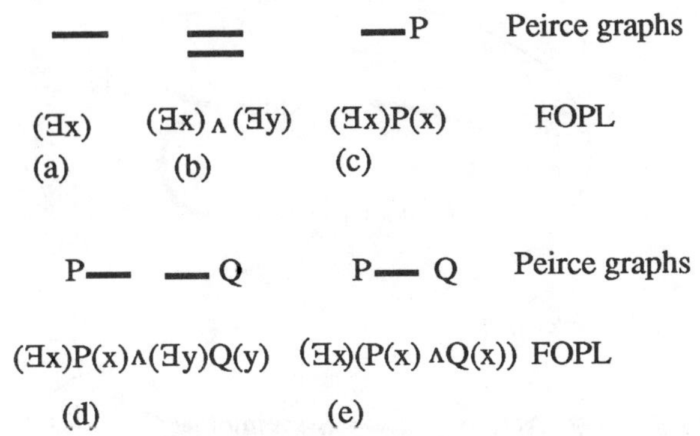

Figure 3:

Some examples are shown in figure 2. It should be noted that the shape of the graphs (e.g. whether lines are straight or curved or the shape of the closures) is not important. Figure 2(a) contains two graphs which together state the propositions represented by P and Q hold, i.e. the conjunction of these holds. Peirce distinguished graph and graph instance. In his terms there is strictly only one graph P of which there may be a number of occurrences (graph instances); e.g. we may have P repeated a number of times such as in figure 2(b). In the following we shall use the term graph rather than graph instance, except where we wish to make the distinction.

A single line, with no crossings, can be drawn around a graph (a cut) to indicate the negation of the graph. For example figure 2(c) is equivalent to P ∧ ¬Q. For material implication (e.g. P ⇒ Q, or "if P then Q") the term "scroll" is used. This example is shown in figure 2(d). The graphs are read from the outside inwards (*endoporeutic).* In this example (figure 2(d)) reading from the outside inwards we have the antecedent ("if P") and then the consequent ("then Q").

The Beta part additionally has a *line of identity*, used to represent an individual in the universe, and a *spot* which is equivalent to the FOPL predicate, e.g. P of figure 3(c). In the Alpha part the graph and the (propositional) symbol were one and the same. However in the Beta part a graph may consist of a number of symbols, hence the need to distinguish the symbol and the graph. The line of figure 3(a) indicates the existence of something, e,g, (∃x), while figure 3(b) indicates (∃x)(∃y), i.e. two individuals which may be the same.

We still have the convention that two graphs drawn on the same SA represent the conjunction of the graphs. Thus figure 3(d) represents (∃x)P(x) ∧ (∃y)Q(y). In figure 3(e) the line of identity between P and Q indicates that the two individuals represented by the ends are identical, i.e. (∃x)(∃y)(P(x) ∧ Q(y) ∧ x = y) or (∃x)(P(x) ∧ Q(x)). This can be extended to n individuals.

Just as a predicate can have more than one argument, more than one line may be attached to a predicate (symbol). The

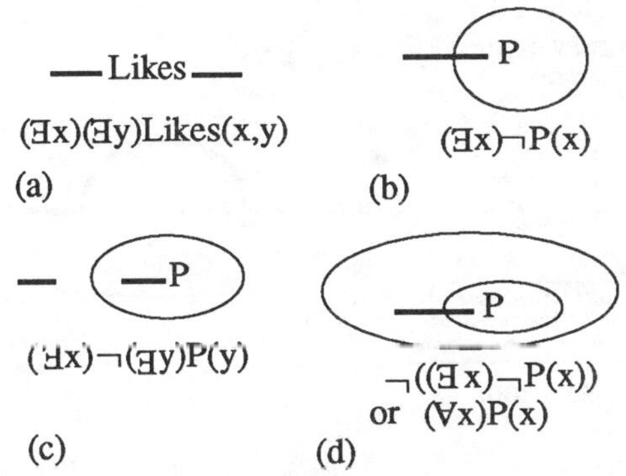

Figure 4:

places of attachment are called *hooks*, and the meaning of the hooks must be understood just as the positions of the predicate arguments in FOPL need to be understood. Figure 4(a) represents *someone likes someone*. In this case we arbitrarily decided to consider the hook in front of the predicate to be the person doing the liking and the other hook to be the person liked.

It is possible to have any nesting of cuts. Graphs are referred to as being evenly enclosed or oddly enclosed, depending on the depth of nesting. This is significant as those graphs evenly enclosed have been negated an even number of times and thus have a positive sign (no negation) in the expanded FOPL (or propositional logic) form, while those oddly enclosed have a negation sign. Anything not enclosed (i.e. on the SA) is considered to be evenly enclosed.

When a line (of identity) crosses a cut the nesting of the cuts determines the equivalent FOPL variable scoping, with the outermost extremity of the line determining the position of the existential quantifier (the endoporeutic view). For example figure 4(b) is equivalent to (∃x)¬P(x), i.e. the existence is at the outer level; figure 4(c) is equivalent to (∃x)¬(∃y)P(y); figure 4(d), where the outer part of the line of identity is oddly enclosed and P is evenly enclosed, is equivalent to (¬∃x)P(x), i.e. (∀x)P(x), in this case the existence is at level 1.

A line of identity passing through an empty cut indicates the non-identity of the individuals at the extremities. For example figure 5(a) is equivalent to (∃x)(∃y)(P(x) ∧ Q(y) ∧ x≠y). The negation of a ligature indicates the non-identity of the individuals denoted by the extremities of the ligature.

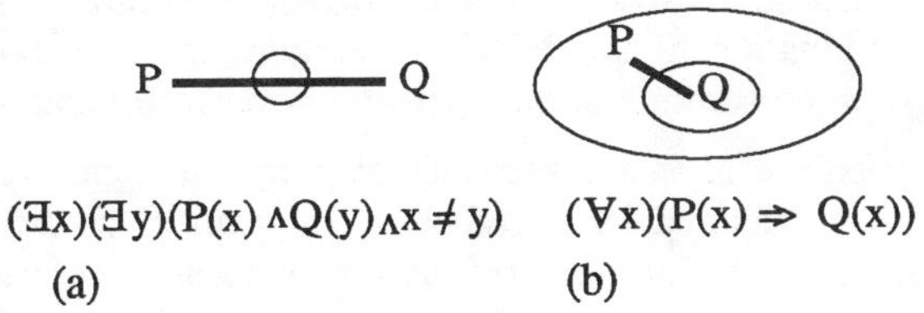

Figure 5:

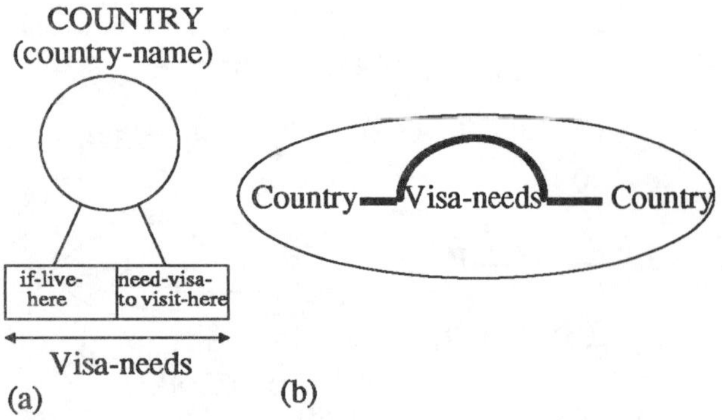

Figure 6:

We shall look at some examples to introduce a methodology for drawing the graphs and to compare the graphs with FOPL. As we have seen material implication is drawn with what Peirce called a scroll. Thus "if something is a P then it is a Q" is shown in figure 5(b). In this type of structure the antecedent is always oddly enclosed and the consequent evenly enclosed.

With only negation and conjunction we need these to represent disjunction. While this may appear awkward, we emphasize that the visual representation is important and one can quickly learn to recognize and draw this picture as a disjunction.

Constants are not treated differently from other symbols, which we have so far thought of as predicates. We treat a constant the same as a unary predicate. If necessary, we can distinguish constants by enclosing them in quotes. However for the usage with NIAM, as we shall see, this is not necessary as there is no ambiguity.

4. ENIAM

NIAM constraints are set oriented. In terms of sets, the existential graphs are "member oriented" in that we express in an exclusion constraint, say, that there is no element that is in both sets involved in the constraint. This potentially gives greater scope for expressing constraints. For example, given the NIAM schema of figure 6(a), we may wish to express that *one does not need a visa to visit ones own country*. We cannot express this with NIAM's set constraints as a country will generally appear in both roles, i.e. in both sets, and we thus can't use the exclusion constraint. However existential graphs do allow us to express "there cannot be a country which acts as a lives-in and a requires-visa-for in the same row". This is shown in figure 6(b). The combination of the NIAM constraints and the existential graph constraints is thus potentially powerful and simple.

To be useful as a constraints language the logic first needs to be extended by introducing "inequality" predicates ($<$, $\leq$, $\geq$, $>$, $\neq$) for numeric values. We can then use these predicates as in figure 7, which indicates that for *an active flight (one which we can make a booking on) the capacity must be less than the booked seats.* Bookings(flight,seats) and capacity(flight,seats) represent the bookings and capacity for an active flight. Either of the forms of figure 7 are acceptable. We interpret figure 7(a) as

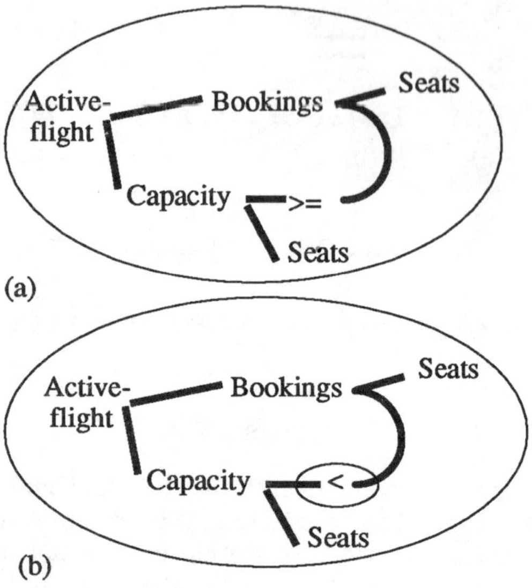

Figure 7:

$$\neg((\exists x)(\exists y)(\exists z)(\text{Active-flight}(x) \wedge \text{Seats}(y) \wedge \text{Seats}(z) \wedge \text{Bookings}(x,y) \wedge \text{Capacity}(x,z) \wedge y \leq z)).$$

One of the advantages of a graphical language is allowing a user to see immediately the structures (e.g. relationships between object types). Over-complication of a diagram defeats this advantage; thus it is important to keep the number of graphical symbols to a minimum. By using existential graphs it permits the full power of FOPL to be used to express constraints with few symbols. We could use set constraints of NIAM when appropriate, i.e. to express what we shall call our "specialty" constraints (uniqueness, mandatory role etc.), and existential graph constraints when membership constraints are required or NIAM does not have the appropriate symbol.

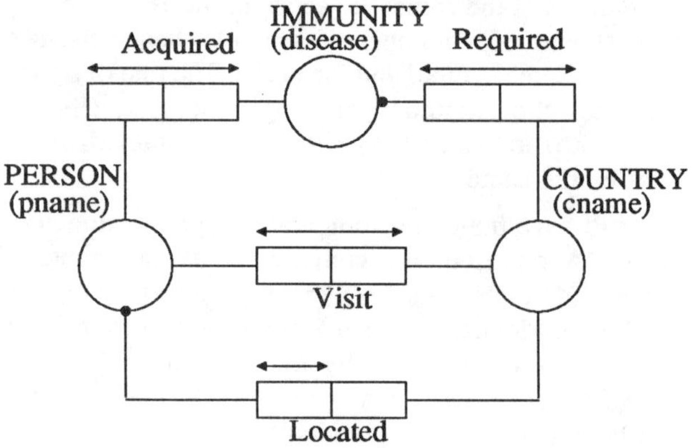

Figure 8:

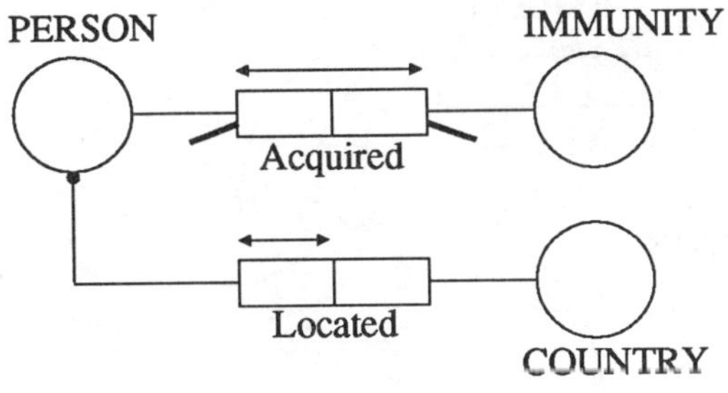

Figure 9:

4.1. The Extended Language

We shall use figure 8 as our example in demonstrating how we can expand NIAM's constraints. As most of the examples don't apply to all of figure 8 we shall only present the relevant sections in the examples. In figure 8 people are scheduled to visit countries. Each country has certain vaccination (immunisation) requirements and each person is recorded as having certain vaccinations. It is required to ensure that people have the necessary vaccinations to visit the countries they have been scheduled to visit.

If we wish to express the constraint that *there must be someone who has a vaccination* we use the fact type Acquired as an existential graph predicate as in figure 9.

As a further example suppose we have the subset constraint *anyone visiting a country must have a have a vaccination* we express this as the embedded existential graph in figure 10. However, if we use the fact types as predicates directly the diagram rapidly becomes unmanageable. For practical reasons we are excluded from using further instances of the predicates.

We thus also propose taking a "copy" or "image" of the relevant fact types and use those as the predicates in an existential graphs diagram. Thus the above constraint can be rewritten as in figure 11. The image A is an "image" of the fact population Acquired (what we have previously called a predicate instance). The dotted line indicates the image. The image is drawn in the same orientation as the fact type, thus the image role r is the PERSON hook. If it is required to draw the image in a different orientation to

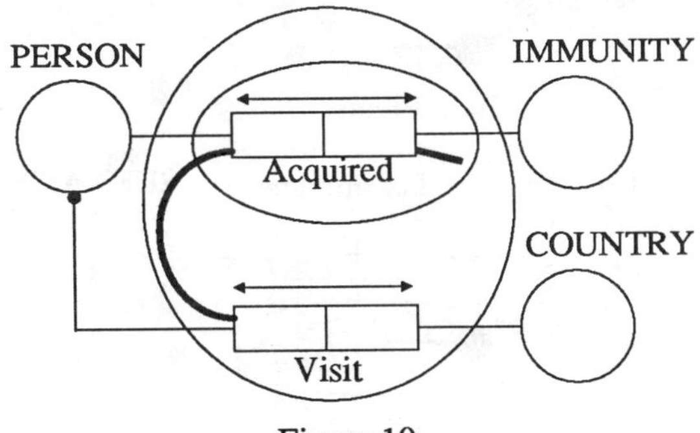

Figure 10:

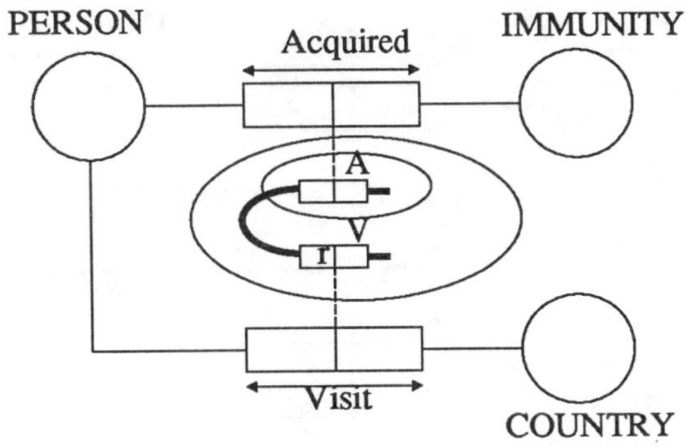

Figure 11:

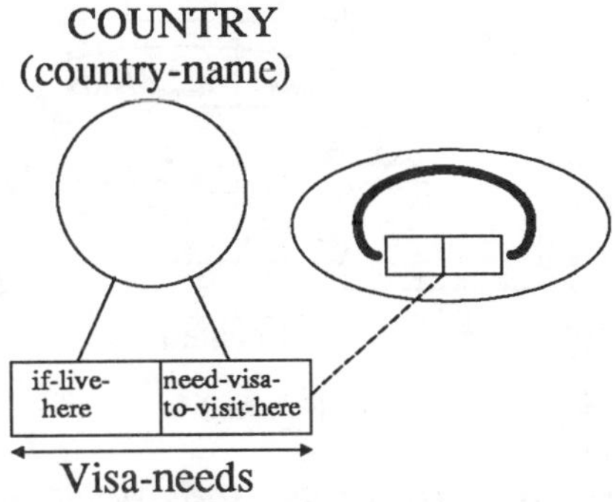

Figure 12:

the fact type we can always attach the image by a dotted line from a (non-central) role to the corresponding role in the fact type, although it would normally be obvious from the types over which the roles are defined. By a "non-central" role we mean a role other than the middle role of an n-ary for odd n. For example we express the constraint *no one needs a visa to visit their own country* as in figure 12, although in this case it is not really relevant which way the attachment is made.

In the embedded existential graphs we shall explicitly represent constants. For example *all visitors from Australia have a Polio vaccination* would be represented as in figure 13.

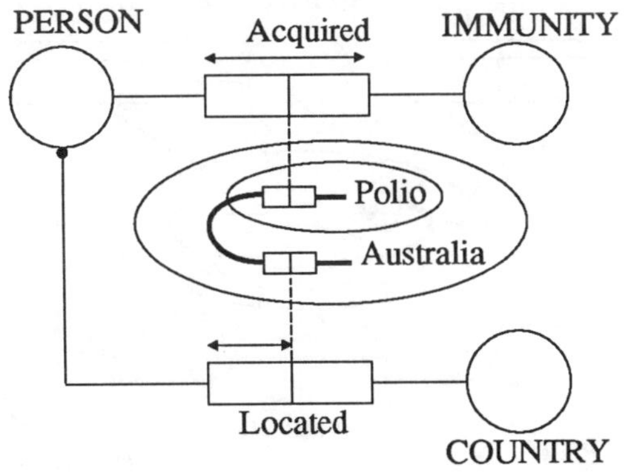

Figure 13:

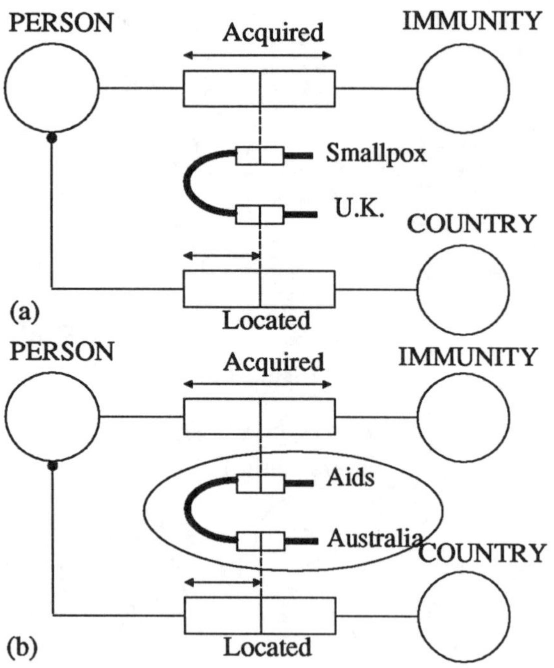

Figure 14:

The constraint *at least one U.K. visitor has a smallpox vaccination* is shown in figure 14(a), *no visitor from Australia has an Aids vaccination* is shown in figure 14(b).

Often we need constraints involving more than 2 fact types, in particular those involving a join. For example we may require that *people visiting a country must have the necessary vaccinations*, viz.

(Visit * Required)[person,immunity] $\subseteq$ Acquired.

We can express this in ENIAM in a more natural manner: *if a person is visiting a country which requires vaccinations then the person must have those vaccinations.* Using the scroll we express this in ENIAM as in figure 15.

We can also use the existential graphs to represent derivation rules. Where rules are of the form "consequent if antecedent",

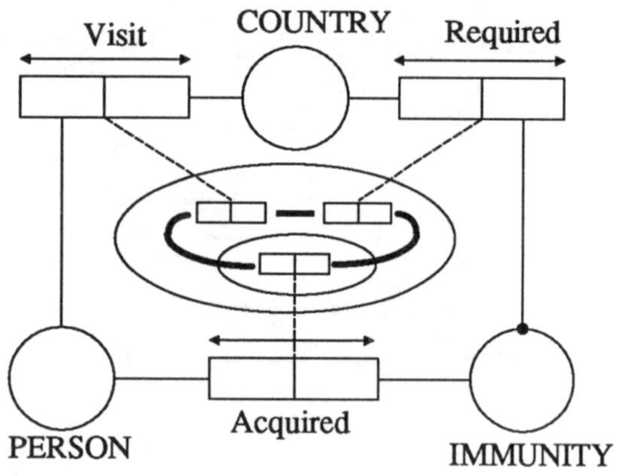

Figure 15:

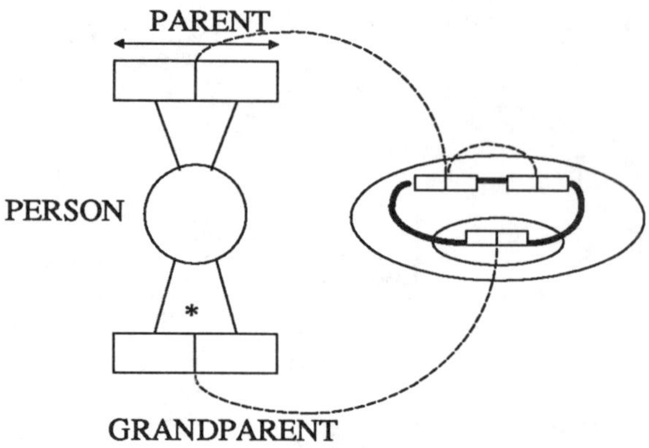

Figure 16:

e.g. *x is the grand-parent-of z if x is the parent-of y and y is the parent-of z* then we can use the scroll. This particular example is shown in figure 16.

It would be desirable to be able to represent an element of an entity type, i.e. to have the equivalent of unary type predicates. In this case we express the image of an entity type as a large blob. The constraint of figure 17 indicates that there exists at least one person in the information base (which does lead to problems in not permitting us to have an empty information base).

Given this construct we have a representation for the mandatory role. We represent *every person has a surname* as in figure 18.

We can use this construct to represent the subtype definition. The example of figure 19 is taken from Halpin (1988). We shall represent, say, the TVviewer definition as:

if Person viewing TV for PeriodRatio > 0 then TVviewer

Using this mechanism the subtype definition of figure 19 can be expressed as in figure 20.

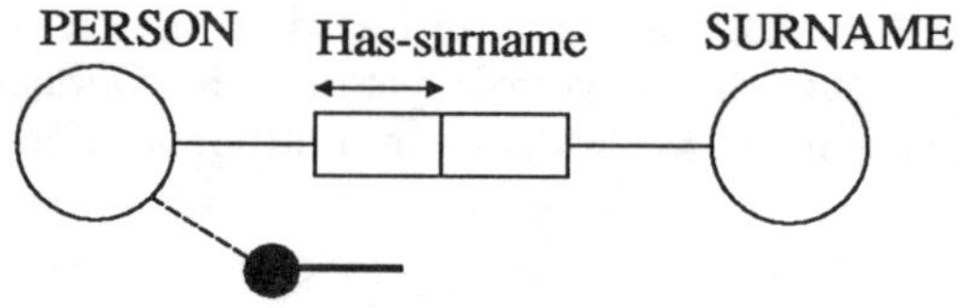

Figure 17:

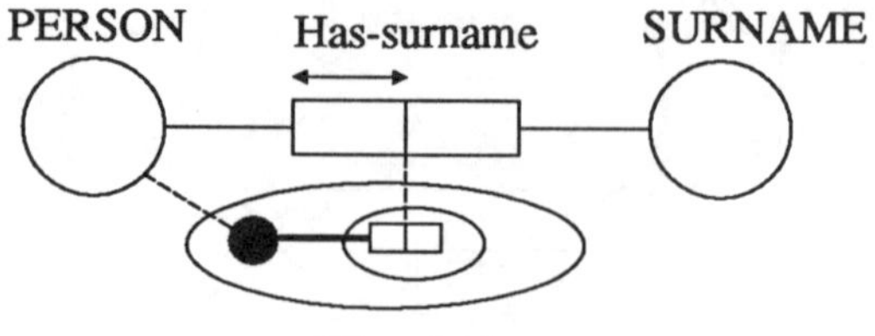

Figure 18:

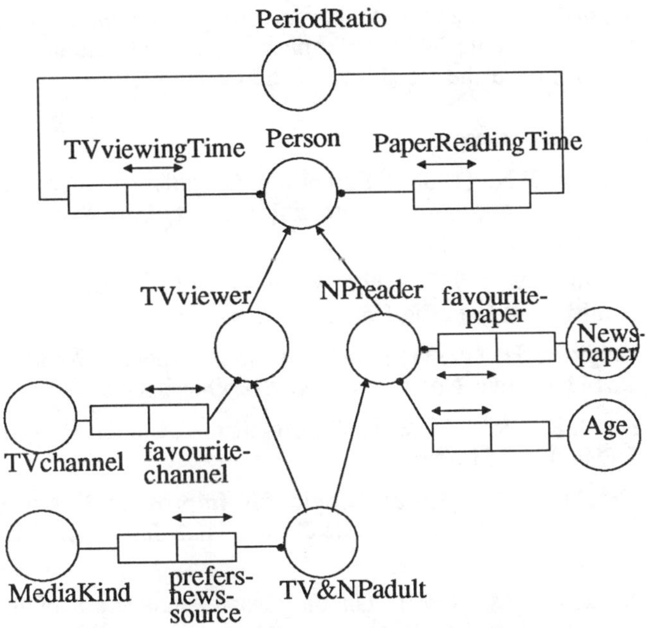

TVviewer = Person viewing TV for PeriodRatio > 0
NPreader = Person reading newspapers for PeriodRatio > 0
TV&NPadult = TVviewer ∩ NPreader having Age > 18

Figure 19:

4.2. The ENIAM Methodology

For most constraints we tend to use the "if then" construct (viz. the scroll). In our examples so far the only ones not using the scroll were existence or non-existence constraints. These constraints are fairly straight forward. The only problem anyone may have is in phrasing them as a negation. The existence constraints, e.g. *someone is booked on a flight*, are the simplest. However, if we start with an empty database then no facts are initially present and thus such constraints are initially violated. Nevertheless we shall consider such constraints in the following.

We claim that there are four basic structures which we can use to represent most constraints. These are:

(i) existence, e.g. *someone is to visit France*, figure 21(a).

(ii) non-existence

(a) of one element, e.g. *there is someone who is not to visit France*, figure 21(b).

(b) all elements, e.g. *no one is to visiting France*, e.g. figure 21(c).

(iii) scroll, e.g. *everyone is to visit France*, figure 21(d).

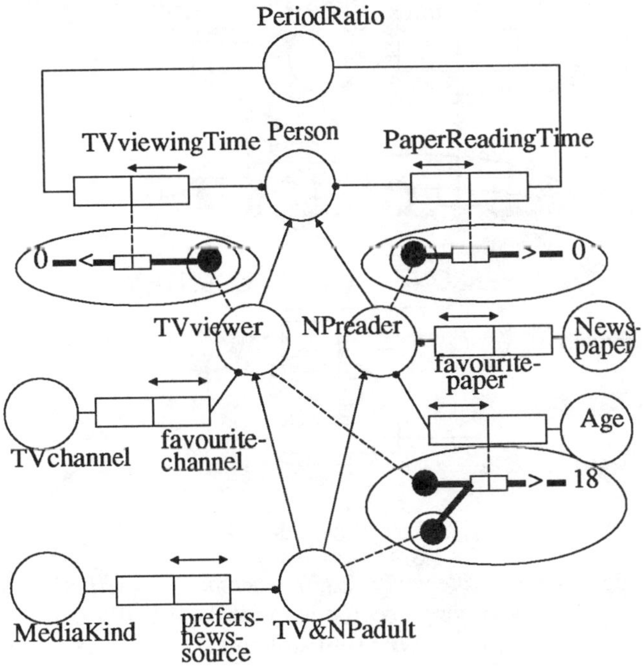

Figure 20:

Whenever we have constraints like *someone* . ., or *Jill* . . , i.e. those involving the existence of a fact we use a type (i) structure. For constraints that involve the existence of an unspecified individual for whom we wish to express a negative we use (iia). Other negatives without a conditional

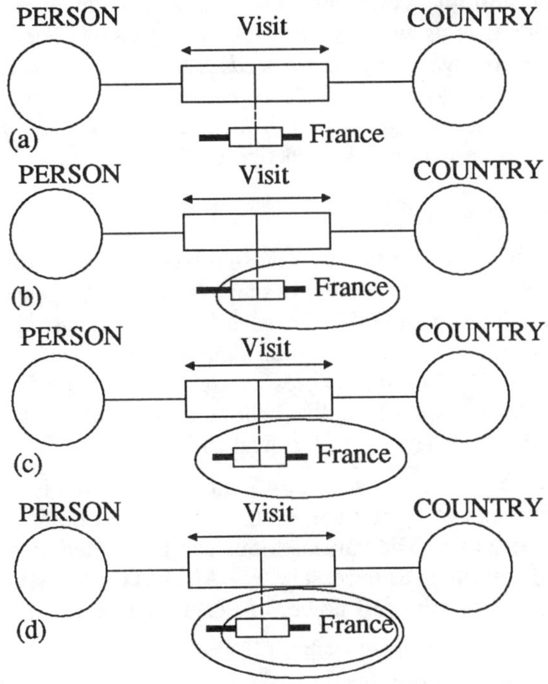

Figure 21:

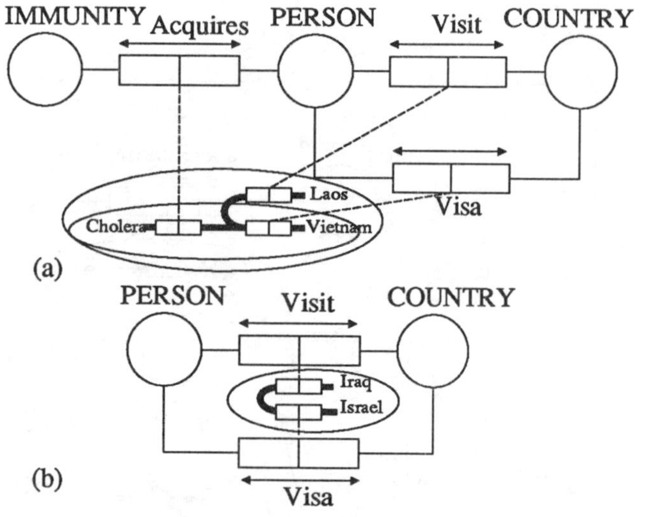

Figure 22:

such as *no <statement>*, e.g. *no person is to visit France* and those involving specific individuals, e.g. *Bill is not to visit France*, use a type (iib) structure.

Those involving (a) conditionals, such as *if <condition> then <statement>*, e.g. if someone visits a country that person needs a passport, (b) negative conditionals, e.g. *no one other than people visiting a country needs a passport*, and *no one may visit a country without a passport*, and (c) those involving all elements of a set, such as *any, every, when, whenever*, e.g. *everyone visiting a country needs a passport* use a type (iii) structure.

For multiple conditions and statements etc. we join the predicates as we have previously suggested. For example *everyone who visits Laos needs a cholera vaccination and a visa for Vietnam* would be expressed as in figure 22(a). While *no one may visit Iraq and have a visa to Israel* would be expressed as in figure 22(b).

5. Conclusions

We have shown that we can significantly extend the power of NIAM using a relatively simple graphical language which has the power of FOPL. We have demonstrated in a classroom environment that existential graphs can be easily learned and ENIAM used to express constraints, subtype definitions and deduction rules. We believe there is no necessity for analysts to have a background in logic to use ENIAM.

As we showed in section 4.2 most constraints involve the scroll and will not involve structures more complicated than those shown. The various forms can be learned by example and combined as necessary. ENIAM has the advantage that, if required, the full power of FOPL can be used.

We have shown elsewhere (Creasy 1988) that the graphs can be simply mapped to Prolog. We have also shown (loc. cit.) that the technique suggested by Nicolas (1982) can be applied to the graphs to permit efficient computation of most constraints. We have implemented the algorithms and demonstrated the feasibility of the technique.

6. References

Creasy, P.N. (1988) "Extending Graphical Conceptual Schema Languages", Internal Report, University of Queensland.

Gardner, M. (1983) *Logic machines and Diagrams*, (first published 1958), The Harvester Press, Brighton.

Halpin, T.H. (1988) *Introduction to Information Systems*, CS112 Lecture Notes, University of Queensland.

Harel, D. (1988) "On Visual Formalisms", *Comm ACM* **31**, 5 (May), pp 514-530.

Nicolas, J-M. (1982) "Logic for Improving Integrity Checking in Relational Data Bases," *Acta Informatica* **18**, pp 227-253.

Nijssen, G.M. (1977) "On the Gross Architecture of the Next Generation Data Base Management Systems," Proceedings 1977 IFIP Congress, Toronto. North Holland, Amsterdam.

Nijssen, G.M. (1986) "On Experience with Large-Scale Teaching and Use of Fact-based Conceptual Schemas in Industry and University," in *Proceedings IFIP Conference on Data Semantics (DS-1)*, R. Meersman and T.B. Steel Jr (Eds), Elsevier North-Holland, Amsterdam.

Peirce, C.S. (1960) *Collected Papers of Charles Sanders Peirce*, vol 4, A.W. Burks (Ed.), Harvard University Press, Cambridge, Mass.

Quine, W. V. (1955) *Mathematical Logic*, (revised edition) Harvard University Press, Cambridge, Mass.

Reiter, R. (1988) "On Integrity Constraints," *Proceedings 2nd Conference of Theoretical Aspects of Reasoning about Knowledge*, Pacific Grove, California, March 7-9, pp 97-111.

Roberts, D.D. (1973) *The Existential Graphs of Charles S. Peirce*, Mouton, The Hague.

Sowa, J.F. (1984) *Conceptual Structures: Information Processing in Mind and Machine*, Addison-Weseley, Reading, Massachusetts.

Verheijn, G. and van Bekkum, J. (1982) "NIAM: An Information Analysis Method" in *Information System Methodologies - A framework for understanding*, Olle et al (Eds), CRIS3 Task Group of IFIP WG8.1, IFIP.

FaceKit: A Database Interface Design Toolkit

Roger King
Michael Novak

Department of Computer Science
University of Colorado
Boulder, Colorado 80309

Abstract

FaceKit is an interface design toolkit for object-oriented databases. By combining techniques from the realm of User Interface Management Systems (UIMS) with a built-in knowledge about the specific kinds of techniques used by object-oriented databases, we have designed a system that allows a user to customize a database interface with minimal programming. Knowledge about object-oriented database constructs such as classes, groupings, etc., allows FaceKit to semi-automatically create graphical constructs appropriate to the object-oriented database environment. FaceKit is built on top of Cactis, an object-oriented database management system (DBMS), and is capable of creating interfaces that deal with Cactis both at the schema and data level.

Keywords: graphical interfaces, user interface management systems, object-oriented databases.

1. Introduction

FaceKit is a window based, interactive graphical system for designing (we do not support a formal design phase; some people might call FaceKit a tool for *building* an interface) graphics-based interfaces for object-oriented databases. Although more of a UIMS than a simple database interface, FaceKit is intended for designing a particular set of interfaces - those dealing with object-oriented databases. FaceKit knows about schemas, type-subtype hierarchies, methods, and database tools such as data definition languages (DDL). Therefore, knowledge about the types of interfaces being designed and the objects they manipulate is built into FaceKit. This knowledge allows default representations of database objects and representations defined or modified by the user. Both default and user-defined representations inherit properties in the same way as the database objects they represent. This allows the user to easily change any class or subclass of object representations and also leave the defaults in effect where desired. By taking advantage of the structure provided by the database, FaceKit makes designing a specific interface easier and faster, and forces the interface to stay consistent with the database.

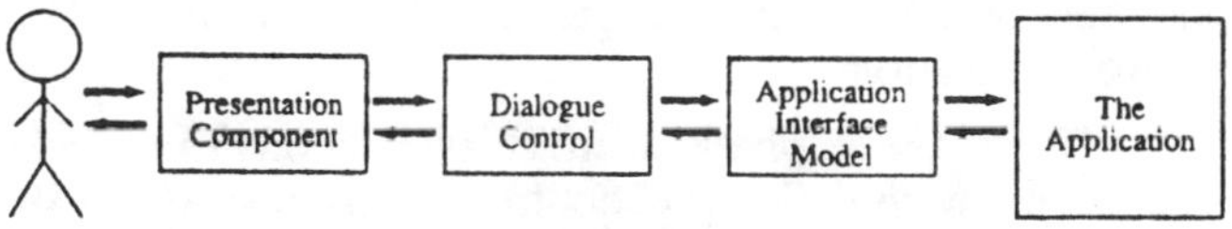

Figure 1

Database interfaces have undergone many changes in the last few years. Until a few years ago, most of the database interfaces we saw where non-graphical. Aside from programming language interfaces, they include relational algebras such as ISBL [Ull], relational calculi like QUEL [HSW75], and query languages such as SEQUEL [BoC74], which falls somewhere in between the two. These interfaces vary in many ways, but their general goal is to provide a general, fairly complete, interface to a database. Then came interfaces such as QBE [Zlo75], which provides a form style interface for specifying queries. In the last few years there has also been much interest in graphical database interfaces. Among these are interfaces such as ISIS [GGK85], Ski [KiM84], and SNAP [BrH86], which allow schema manipulation in an interactive graphical environment. They also include office forms systems such as FORMANAGER [YHS84], Freeform [KiN87], and SPECDOQ [KGM84]. These systems provide a non-expert interface for the storage and retrieval of office data.

Obviously, a single interface cannot be everything to everyone. Since many different database interfaces may prove useful, we feel that being able to quickly design a new or altered interface could prove very useful. UIMS's are designed for this very purpose. The question is, why not use an existing UIMS to rapidly generate interfaces for an object-oriented database? To answer this question, a little background on UIMS's is necessary.

We will use the Seeheim model of user interfaces [Gre84, Gre85] (figure 1) to examine and compare some of the existing UIMS systems. The Seeheim model is a logical model applicable to a wide variety of UIMS's. In this model the user interface is broken into three logical components: the presentation component, the application

This work was supported in part by ONR under contract numbers N00014-86-K-0054 and N0014-88-K-0559.

interface model, and the dialogue control. The presentation component is responsible for producing device output and gathering user input. The application interface model is responsible for representing application data and making it available to the interface, as well as providing the application with access to the interface. The dialogue component forms the bridge between the presentation component and the application interface model. It makes sure that the application carries out user requests and that the presentation component produces the output requested by the application.

The main emphasis in much of the UIMS research has been on the dialogue control component [Gre86]. Although there have been systems with dialogue models based on transition networks, grammers, and events, these system share a common perspective. UIMS's such as ADM [SRH85], Grins [ODR85], Menulay [BLS83], MIKE [Ols86], and Trillium [Hen86] all view an interface as a dialogue between the user and the application. The user makes a request and the application responds by performing some task and possibly providing some output. Similarly, the application makes input requests which the user responds to. The research emphasis in these systems has been on how to provide the user with the appropriate tools for specifying the dialogue.

Work done on the presentation component includes Peridot [Mye87] and Squeak [Car87] Both of these systems put heavy emphasis on allowing users to create new interaction techniques. Peridot users create these techniques by using examples to show how they should appear, while Squeak lets users apply direct manipulation principles [HHN, LeL83] to specify them.

There are also several systems which focus on the application interface model. Filters [Ege88] and Coral [SzM88] each provide a method of specifying relationships between application and interface objects. GWUIMS [SHB86] and Higgens [HuK88] both allow for sharing of data between the application and the interface. All of these systems use an object-oriented data model.

Another object-oriented system called GROW [Bar86] looks at some dialogue and presentation issues. GROW emphasizes building modifiable and reusable interfaces. The dialogue component uses messages for communication between the application and the interface. The presentation component provides a kernel of graphical objects arranged in a taxonomic hierarchy and it allows the user to specify inter-object relationships.

There are two main reasons why we don't wish to use any of the currently available UIMS's. First, even though we have seen systems which can communicate with application data in some manner, a general purpose UIMS has no knowledge of database schemas. Since FaceKit knows about database schemas, users may access database objects, methods, etc. and incorporate them directly into the interface. Similarly, interface objects are actually stored in the database and have the same structure as database objects. Thus, only one view of objects exists. Even those UIMS's which share some data between the interface and the application still require the user to view interface and application objects separately and to provide information about how the two are related. Having only one view of objects means that less information needs to be provided by the user. Furthermore, users do not need to keep track of two different types of objects.

By providing access to database tools such as query languages and methods, FaceKit also gives the user more ways of rapidly constructing an interface. Instead of writing code to generate an interface technique, the user may invoke a method in the database, or use a query language to define the technique. Interface objects previously created may also be used, because they too are stored in the database.

The second reason that we don't wish to use a currently available UIMS is that creating a new database interface often involves creating a new application. For example, an office forms interface involves much more than merely an interface to the database. New functionalities must be built into the interface. These may include new mathematical functions such as computing the city and state sales tax on a sales field, and utilities such as an inter-office memo system. Both of these new functionalities involve more than simple interaction with the DBMS. In order to support this type of interface, we wish to allow the user to interactively design both an interface and its corresponding application simultaneously [HuK88]. Unlike in a UIMS, our approach treats the interface being designed as a database environment, rather than a dialogue between a distinct user interface and an application. Instead of defining merely an interface, we define the visual, functional, and interactive aspects of the environment. Thus, certain otherwise hard to obtain functionalities such as binding representations to objects can be achieved. This also allows for "realistic" default interfaces and faster specification of representations. In fact, such an integration of database and UIMS technology has been suggested before [Gre87, Ols87].

The minus side of our approach is that neither interfaces outside the object-oriented database realm nor non-Cactis database applications are supported. However, many applications may be be placed into an object-oriented database. For example, much of the circuit board design software currently available uses simple files to store circuit information. Such systems could easily fit into the object-oriented database paradigm. One could argue that even aside from making it possible to use FaceKit with the application, storing the data in a database would make the application itself more manageable.

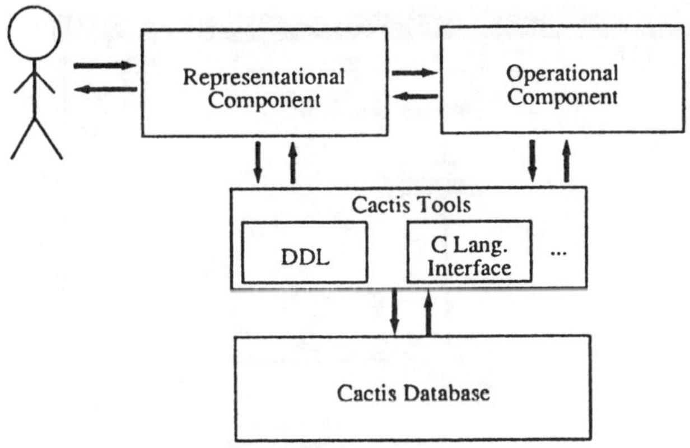

Figure 2

2. Modeling an Interface

FaceKit is built on top of an object-oriented DBMS named Cactis [HuK86, HuK87]. Since FaceKit has knowledge about both Cactis schemas constructs and the various schema manipulation tools (DDL, C language interface, etc.) available within Cactis, we will give a brief description of the Cactis data model before proceeding to describe the data modeling aspects of FaceKit.

Cactis views an application environment as a collection of *constructed objects*. An object may have *attributes* and *relationships*. Objects and relationships are typed. A constructed object's type is determined by two things: its attribute structure and its *connectors*. A connector allows a relationship to be applied to a certain object. An attribute is an atomic property of a constructed object. These atomic properties may be of any C data type, except pointer. A relationship is a directional mapping from one constructed object to one or more constructed objects. Restrictions such as non-null or unique may also be put on a relationship.

For example, a constructed object called person may have the attributes name, social_security_number, and age, which are all atomic and single-valued. It may also have a connector called children and a connector called parent. The directed relationships my_children and my_parent can be used to connect people to their immediate family. Relationships may be used to pass attributes from one object to another, in order to calculate derived attributes. In this way, the social security number of a person could be passed to a child over the my_children relationship, and used as the value of an attribute called my_parent's_social_security_number.

FaceKit uses the data model and the database management tools provided by Cactis for data and schema manipulation. An interface needs to communicate with these Cactis tools and it needs to manage visual representations and coordinate interface tasks. To accomplish this a FaceKit interface consists of two conceptual components: the *representational* component and the *operational* component. Both the representational and operational components communicate directly with Cactis (figure 2), but their tasks are quite different.

The primary responsibility of the representational component is managing the visual representations of database objects. For the purpose of this discussion, database object, or object, may refer to a constructed object, relationship, or attribute. Also, no distinction is made between schema and data objects at this time. The representational component builds, maintains and invokes the methods used to produce the visual representations of objects. Thus, if we build an interface where a data object called person is represented by a screen image of that person, the representational component makes sure that the method used to draw these screen images is invoked at the proper times, with the correct parameters.

The representational component is also responsible for the input and output associated with an interface. Input will generally consist of reading in some user data or command, while output will consist of either displaying new objects on the screen or invoking a different representation of objects already present. An object may have many different visual representations in the same interface. For example, a database may have a baby picture and an adult picture of the person in the example above.

The operational component is responsible for processing user queries and sending the results to the representational component so that the correct screen updates will be performed. The operational component performs a function much like that of the dialogue control in the Seeheim model. However, it has access to the Cactis database management tools. This means that no application interface model is necessary. Instead, the operational component talks directly to the application. This has several advantages. First, it allows FaceKit to store the operational description of an interface within Cactis. Therefore, Cactis manages both the data and the operational description. Not only does this make storage of interface descriptions convenient, but it also allows operations to be functionally dependent on anything present in the database. This allows an interface to change its behavior as the database is modified. Second, since these tools give the interface direct access to the application data, semantic feedback can be gathered merely by examining the relevant data. Similarly, constraints can be checked and adhered to.

Perhaps the most important effect of allowing the operational component access to these tools is utilized when designing interfaces. Instead of only being able to bind a user request to some application subroutine, the interface designer may construct a query using one of these tools. Since new interface functionality can be added quite rapidly in most cases, most of the functionality of Cactis can easily be plugged into an interface.

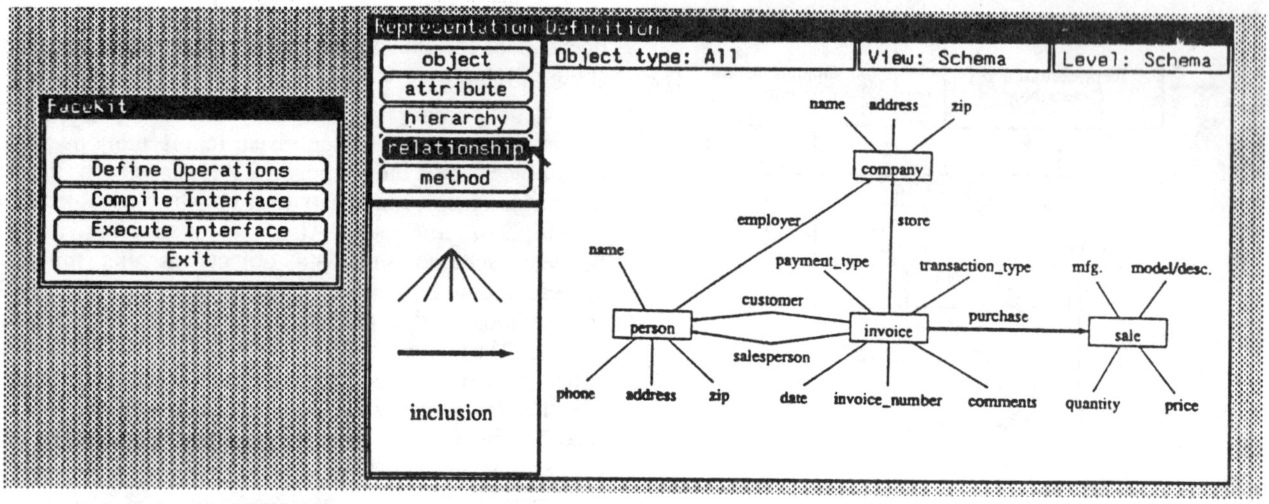

Figure 3

3. Designing an Interface

A very common UIMS approach to designing interfaces is to specify screen layout, then bind each possible user action to a specific application subroutine. Often, interface routines for the application program to call are also provided. FaceKit takes a different approach to designing interfaces. Our approach allows the user to define what we view as two somewhat different aspects of the interface: appearance, and functionality. When defining appearance we are really defining two kinds of visuals, interface constructs and database objects. By interface constructs we mean items such as menus, scrollbars, etc., as well as concerns like screen brightness, icon sizes, etc. Defining the appearance of database objects involves specifying representations for a class of objects. A representation may be identical for each data object in a class or it may be data dependent. In fact, it may be dependent on external data. For example, we may use the system clock to determine the brightness of a picture that represents the data object **sun**.

By defining the appearance of database objects separately, we need not worry about them when defining functionality. The type of the query result determines the screen appearance. Any type that has no user-defined representation will use a built-in default representation. For example, if a query results in an object of class **person**, the interface automatically uses the representation of person that has been defined. If one wishes to leave the screen alone and produce the query result elsewhere, this may also be specified.

Both appearance and functionality are defined interactively with FaceKit. In order to better explain how they are defined, we will show how one would go about defining an existing interface, an office forms system called Freeform [KiN87]. We will also show some sample applications of this interface. We will also show how an interface used to retrieve information about computer networks can be built using FaceKit.

3.1. Defining Representations with FaceKit

In figure 3, an object representation is being defined. The buttons on the left specify what kind of object is being defined and the line on top is a status line. In this case, the object type to be defined is **all** (the representation being defined is applicable to all types in the database). The current level and view are both **schema**. Two views are possible: **schema** and **class/subclass hierarchy**. The schema view is the standard graphical view of a database schema, while the hierarchy shows a forest of classes and subclasses. Two levels are also available: **schema** and **data**. The schema level lets us work with object types, while the data level lets us work with object instances. The commands for changing the view, level, and object type are available through a popup menu. The popup menu also has commands for placing restrictions on the representation being defined, moving around the schema, etc.

In the schema shown, the text enclosed in rectangles represents constructed objects and the unenclosed text represents the attributes of the constructed objects. The narrow arrows represent one-to-one relationships and the wide arrows represent one-to-many relationships.

Instead of representing a schema as a connected graph, we will represent relationships with inclusion. Thus, the object pointed to by an arrow will be included inside the object the arrow originates from. Also, attributes will be included within the constructed object they belong to. Once we choose **relationship** and **inclusion** (figure 3), the schema changes to look like figure 4. The label on top of each box is the name of the relationship

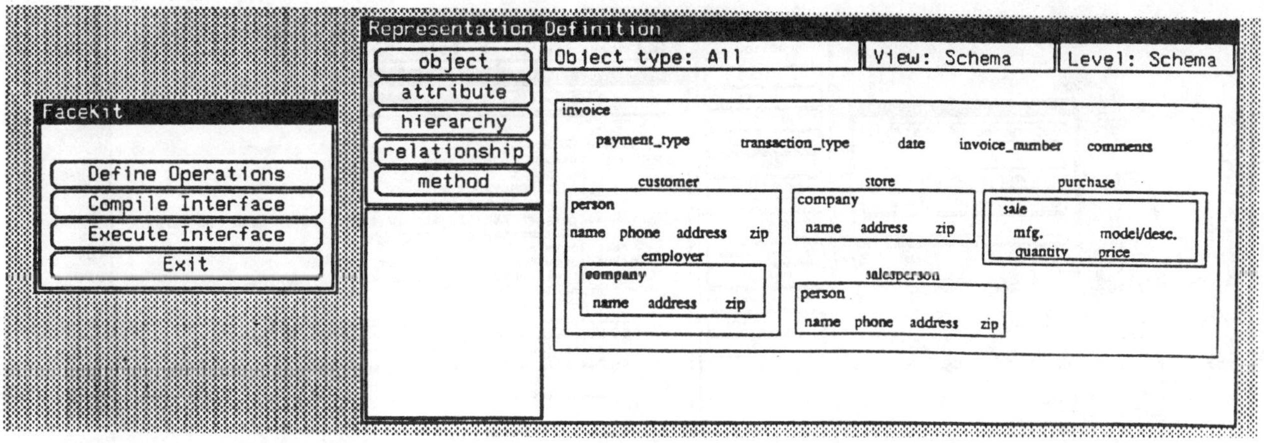

Figure 4

between the object represented by the box and the object it is included in. The double box around purchase denotes that it is a one-to-many relationship.

Although the representation looks fairly formlike now, some further refinement is still needed. Attributes need to have a line for filling in data behind them, representations need to be made different for various types of attributes, and multi-valued relationships such as purchase need to be represented with some appropriate structure like a table. Since the details of these changes are very similar to the examples above, we will not show them. Instead, we will give some examples of defining functionality.

3.2. Defining Operations with FaceKit

An operation definition window is shown in figure 5. The schema shown is a refined version of the schema in figure 4. Although there is no reason we cannot have both the representation definition and operation definition windows up simultaneously, we will only use one at a time in order to make our diagrams less crowded and confusing. A user building an actual interface would probably use them simultaneously.

By selecting an action (or sequence of actions) from the Action buttons and a result (or sequence of results) from the Result buttons, the user can bind the desired functionality to a set of actions. For example, suppose the sequence **pick**, **new action**, and **menu** is selected. This specifies the following sequence. The user picks an object.

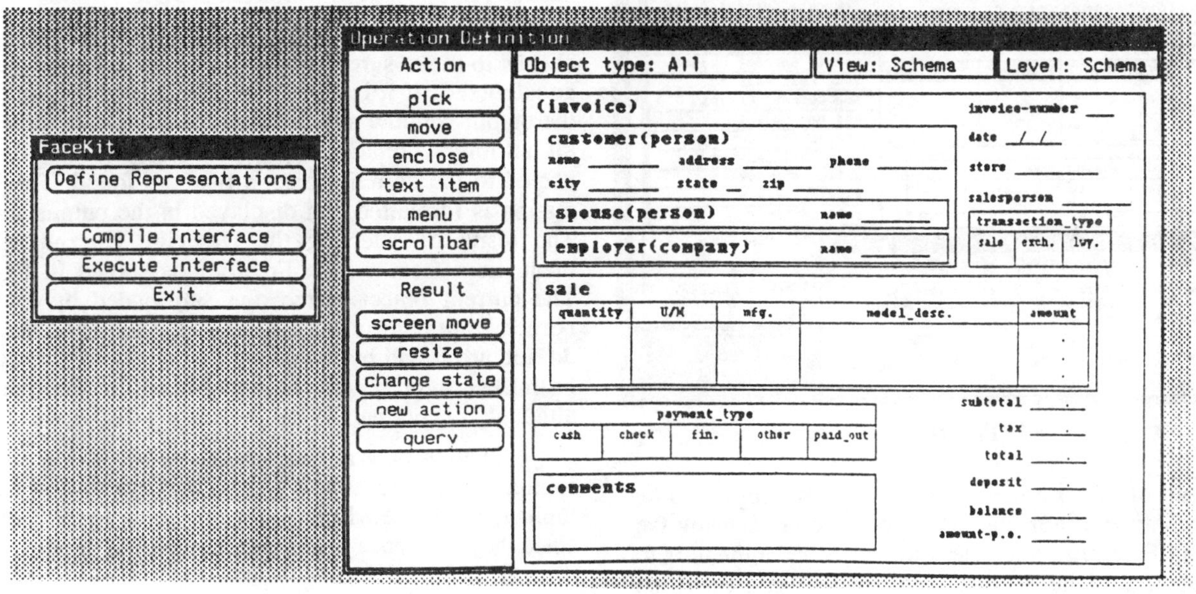

Figure 5

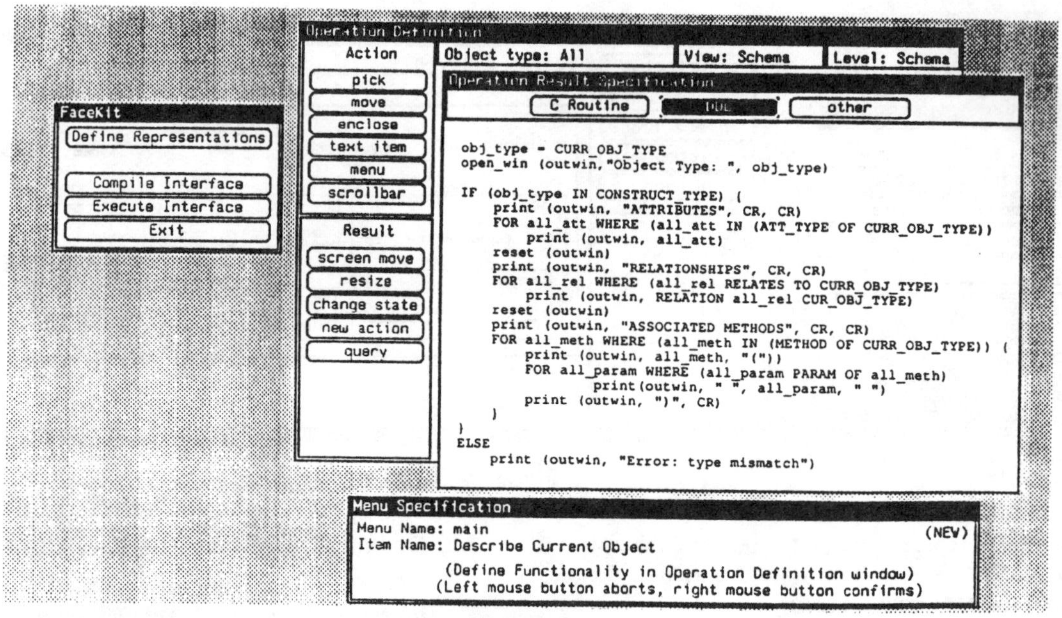

Figure 6

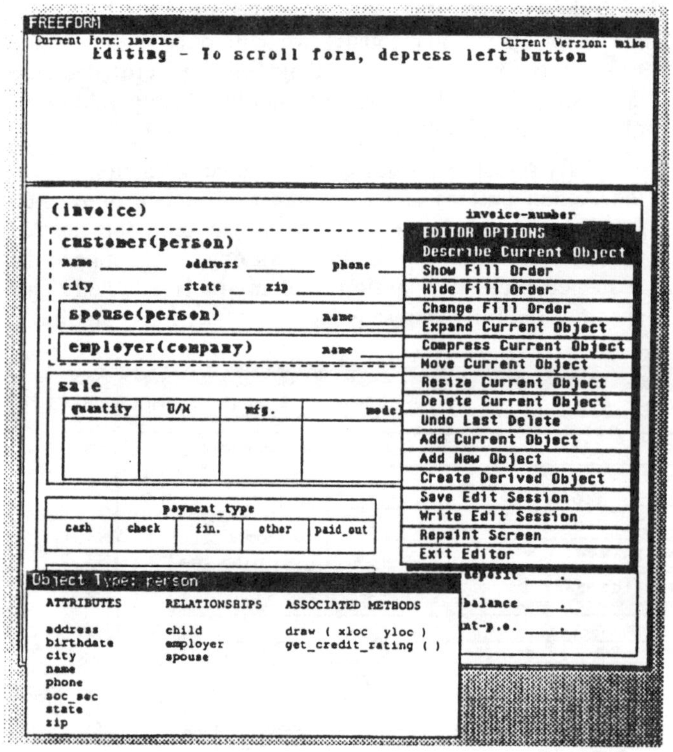

Figure 7

The result of this is a new action. This new action will be the appearance of a menu. Therefore, we are causing the a pick, in Freeform, to create a menu. Since a menu is to be created, a Menu Specification window is created, so that we may specify which menu to use. Figure 6 shows such a window, with a new menu being defined. We could have also chosen an already defined menu, but instead, we will demonstrate how one goes about defining a menu. This is done by defining each menu item and its corresponding functionality. The menu item being defined in figure 6 is **Describe Current Object**. The result of choosing this menu item will be a query, which will be defined in the Operation Result Specification window.

We will define this query using the DDL. The DDL being used is a simplified version of the Cactis DDL [SwS88], with some pseudo-code added. Open_win, print, and all variables in lower case are FaceKit constructs. All upper case words are DDL keywords.

First, the query opens an output window, then checks to make sure that the current object is of the class **constructed.** If it is not, an error message is displayed in the output window and the query is complete. Otherwise, all attributes, relationships, and methods directly connected to the current object's type are found. This information is formatted and displayed in the output window. The result of selecting the **Describe Current Object** menu item from within Freeform is shown in figure 7. The current object is the one surrounded by a dashed border. Both objects that appear on the form and those that do not, are found by the query. For example, the attribute birthdate is presented because it is in the database, although it is not part of the form being displayed.

Queries may also be produced by calling Cactis C routines. Currently, these are the only two query methods implemented. We will not give any examples of C queries, since they are conceptually similar to DDL queries.

3.3. A Software Engineering Example

Since most software engineers now have access to one or more computer networks, they need an easy way to

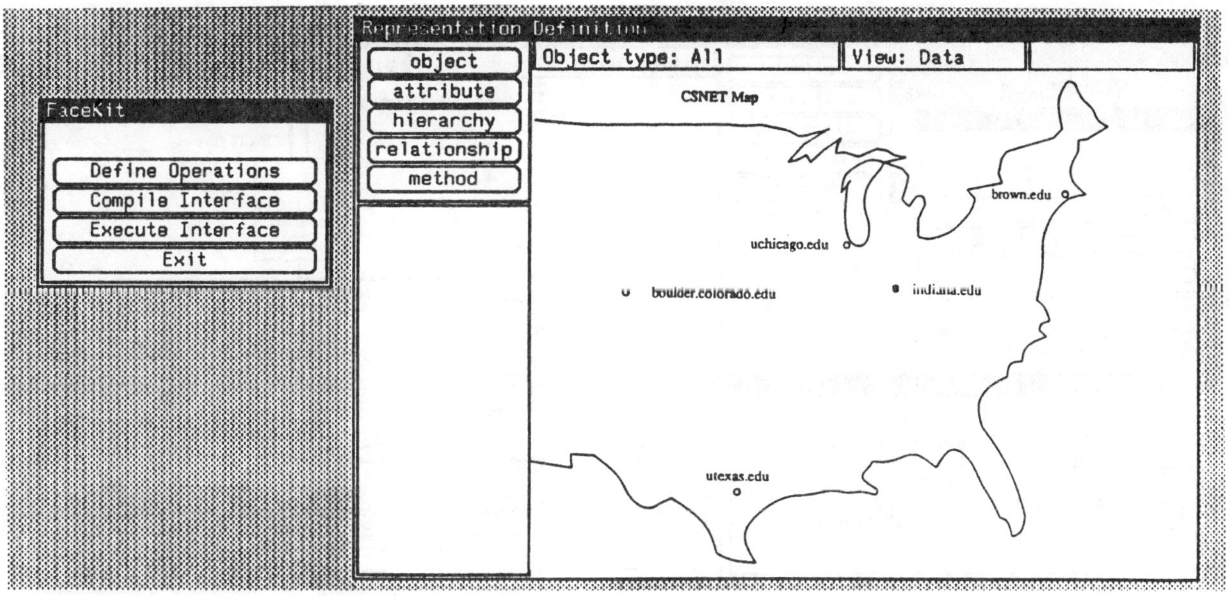

Figure 8

get information about these networks. With network configurations changing fairly often, storing the information in a database makes good sense. Having a graphical interface to access such information also makes good sense. Our next example shows part of such an interface being built with FaceKit. Since this interface does not look particularly "formlike", it is good for illustrating the variety of interfaces that can be built with FaceKit.

In Figure 8, we show a pictorial representation of a small subset of the CSNET sites in the United States. This representation was built by invoking a method (not shown) that plots database objects of type **CSNET_node**. Due to space constraints, only a small part of the map is shown on the screen. Note that actual data, not schema information, is being displayed.

Each **CSNET_node** contains a number of computers. Figure 9 shows a class hierarchy for the schema type **computer**. In this hierarchy, **computer** is the base type, **mainframe**, **workstation**, and **micro** are subtypes, and so on. Classes are represented by enclosed text and attributes are represented by unenclosed text. Attributes are connected with lines; subclasses are connected with arrows. Although this schema is not complete (for readability), we can see that computers such as a **Pyramid** and a **Sun 4** are different and contain different attributes. The database also contains methods which deal with this information. These methods are also accessible to FaceKit.

For example, a method which draws the representation of a computer may exist in the database (or one may be created specifically for this interface). FaceKit can access this information in order to create a graphical

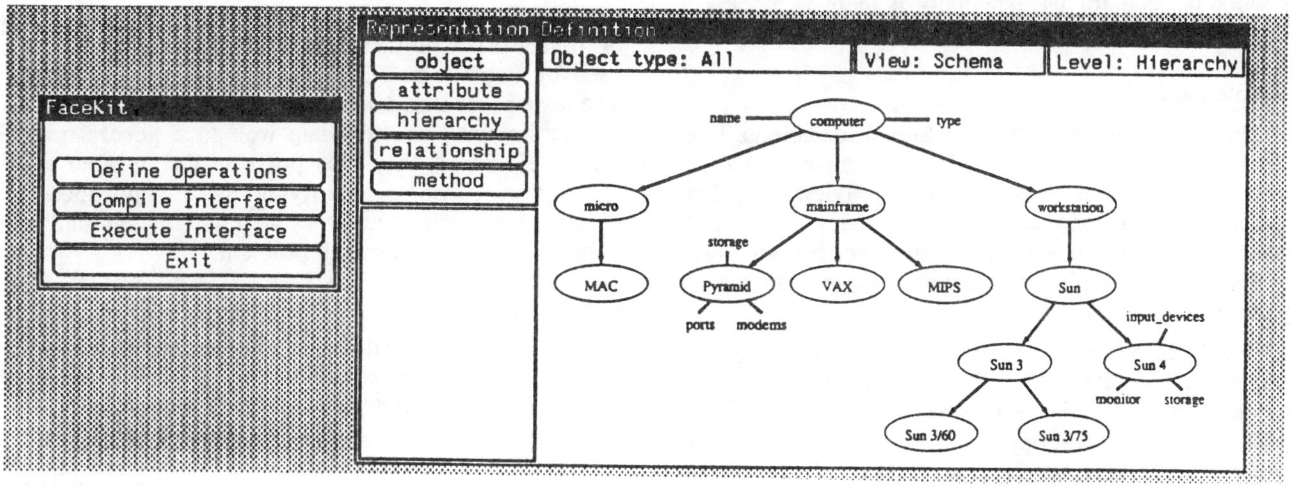

Figure 9

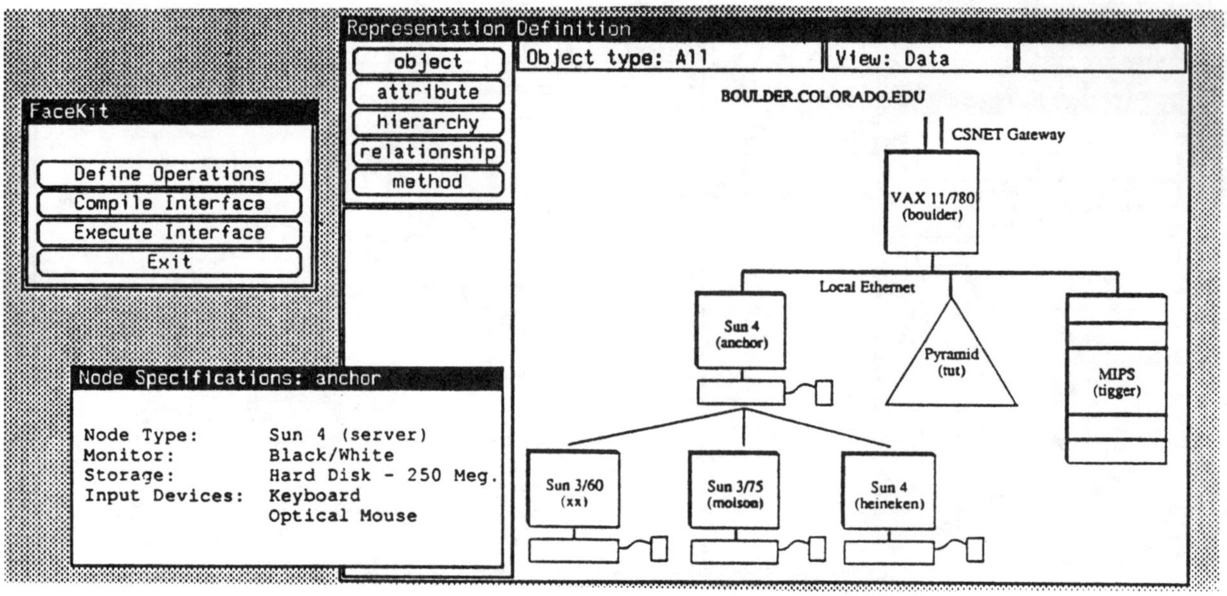

Figure 10

representation of a CSNET site (figure 10). Each computer at the selected site is shown along with its type and name. How they are connected to each other is also shown. In this example, the gateway to the CSNET is a **VAX 11/780** named boulder. FaceKit does not even need to know the specifics about each computer at that site. For each computer at that site, the database knows its type and the proper method for that type is invoked. To build this part of the interface, we only had to specify the screen relationships between the computers at a site.

Figure 10 also contains a popup window that displays attribute information about a selected computer. Since different computers have different attributes, the information given for a Sun will differ from the information given about a Pyramid. Once again, the database already knows these differences and therefore, much of our work is done for us, especially if there is already a built-in method for displaying attribute information.

4. Implementation

FaceKit is implemented on Sun workstations. It is written in C and runs in a UNIX environment. All the window management is done using Sunviews window package. This package also produces the graphics and handles user input. Since Sunviews manages the windows, they may be moved, hidden, resized, collapsed, etc. just as any other Sun window.

Methods for creating representations are stored in Cactis along with the database objects they represent. No distinction is made between them and regular Cactis objects. Operation descriptions are also in Cactis, but FaceKit creates separate database objects (distinct from regular Cactis objects) for storing them.

5. Future Directions

There are several areas of related research we would like to explore in the future. First, we want to look into giving FaceKit the ability to bootstrap interfaces. An interface designer could then design an interface using interfaces previously designed with FaceKit. For example, Freeform could be used to define representations and operations when building a new interface. Along the same line, we would like see if interfaces not built with FaceKit could be used as tools for building interfaces with FaceKit. This means that some methods for translating random interface queries into a form usable by FaceKit need to be developed. Whether this is feasible to any significant extent, remains to be seen.

There is also the issue of portability. Currently, FaceKit is built on top of Cactis. Is it possible to create a tool like FaceKit that can be easily connected to a wide set of object-oriented databases? Also, can the techniques used by FaceKit be successfully used by other UIMS's? The techniques that depend on extensive knowledge of the application may not translate well to a general purpose UIMS. However, they may prove useful in other special application UIMS's. Identifying these applications and seeing which techniques work well with them could prove to be an interesting research problem.

References

[Bar86] P. S. Barth, "An Object-Oriented Approach to Graphical Interfaces", *ACM Transactions on Graphics* 5, 2 (April 1986), 142-172.

[BoC74] R. Boyce and D. Chamberlin, "SEQUEL: A Structured English Query Language", *Proceedings of the ACM-SIGFIDET*

Workshop on Data Description, Access and Control, May 1974, 219 -261.

[BrH86] D. Bryce and R. Hull, "SNAP: A Graphics-based Schema Manager", *IEEE Conference On Data Engineering,* 1986, 151-164.

[BLS83] W. Buxton, M. R. Lamb, D. Sherman and K. C. Smith, "Towards a Comprehensive User Interface Management System", *Computer Graphics 17,* 3 (July 1983), 35-42.

[Car87] L. Cardelli, "Building User Interfaces by Direct Manipulation", *Digital Systems Research Center Tech. Report,* October 1987.

[Ege88] R. K. Ege, "Defining Constraint-Based User Interfaces", *Data Engineering 11,* 2 (June 1988), 54-63.

[GGK85] K. J. Goldman, S. A. Goldman, P. C. Kanellakis and S. B. Zdonik, "ISIS: Interface for a Semantic Information System", *SIGMOD Conference Proceedings* , May 1985, 328-342.

[Gre84] M. Green, "Report on Dialogue Specification Tools", *Computer Graphics Forum 3* (1984), 305-313.

[Gre85] M. Green, "The University of Alberta User Interface Management System", *Computer Graphics 19,* 3 (July 1985), 205-213.

[Gre86] M. Green, "A Survey of Three Dialogue Models", *ACM Transactions on Graphics 5,* 3 (July 1986), 244-275.

[Gre87] M. Green, "Directions for User Interface Management Systems Research", *Computer Graphics 21,* 2 (April 1987), 113-116.

[HSW75] G. Held, M. Stonebraker and E. Wong, "INGRES: A Relational Data Base Management System", *Proceedings of the AFIPS National Computer Conference 44* (May 1975), 409-416.

[Hen86] D. A. Henderson, "The Trillium User Interface Design Environment", *CHI 86 Proceedings,* April 1986, 221-227.

[HuK86] S. Hudson and R. King, "CACTIS: A Database System for Specifying Functionally-Defined Databases", *Proceedings of the International Workshop on Object-Oriented Databases,* Sept. 1986, 26-37.

[HuK87] S. Hudson and R. King, "Object-Oriented Database Support for Software Environments", *SIGMOD Conference Proceedings,* May 1987.

[HuK88] S. Hudson and R. King, "Semantic Feedback in the Higgens UIMS", *IEEE Transactions on Software Engineering 14,* 8 (August 1988).

[HHN] E. L. Hutchins, J. D. Hollan and D. A. Norman, "Direct Manipulation Interfaces", in *User Centered System Design* , 87-124 .

[KiM84] R. King and S. Melville, "Ski: A Semantic-Knowledgeable Interface", *VLDB Conference Proceedings,* Singapore, August 1984.

[KiN87] R. King and M. Novak, "Freeform: A User-Adaptable Form Management System", *VLDB Conference Proceedings,* Brighton, England, 1987.

[KGM84] H. Kitagawa, M. Gotoh, S. Misaka and M. Azuma, "Forms Document Management System SPECDOQ - Its Architecture and Implementation", *SIGOA Conference Proceedings,* June 1984, 132-142.

[LeL83] A. Lee and F. H. Lochovsky, "Enhancing The Usability of an Office Information System Through Direct Manipulation", *CHI Conference Proceedings,* 1983, 130-134.

[Mye87] B. A. Myers, "Creating Dynamic Interaction Techniques by Demonstration", *CHI + GI* , 1987, 271-278.

[ODR85] D. R. Olsen, E. P. Dempsey and R. Rogge, "Input/Output Linkage in a User Interface System", *Computer Graphics 19,* 3 (July 1985), 191-197.

[Ols86] D. R. Olsen, "MIKE: The Menu Interaction Kontrol Environment", *ACM Transactions on Graphics 5,* 4 (October 1986), 318-344.

[Ols87] D. R. Olsen, "Larger Issues in User Interface Management", *Computer Graphics (ACM SIGGRAPH Workshop on Software Tools for User Interface Management 21,* 2 (April 1987), 134-137.

[SRH85] A. J. Schulert, G. T. Rogers and J. A. Hamilton, "ADM - A Dialog Manager", *CHI 85,* April 1985, 177-183.

[SHB86] J. L. Sibert, W. D. Hurley and T. W. Bleser, "An Object-Oriented User Interface Management System", *Computer Graphics 20,* 4 (August 1986), 259-268.

[SwS88] M. Swain and C. Stepleton, "Design of a Data Definition Language for Cactis", *University of Colorado, Computer Science Dept. Tech. Report,* December 1988.

[SzM88] P. A. Szekely and B. A. Myers, "A User Interface Toolkit Based on Graphical Objects and Constraints", *OOPSLA Proceedings,* 1988, 36-45.

[Ull] J. Ullman, "Principles of Database Systems", Second Edition, Computer Science Press.

[YHS84] S. B. Yao, A. R. Hevner, Z. Shi and S. Luo, "FORMANAGER: An Office Forms Management System", *ACM Transactions on Office Information Systems 2,* 3 (July 1984), 235-262.

[Zlo75] M. M. Zloof, "Query By Example", *Proceedings of the National Computer Conference 44* (1975), 431-438.

A Low Communication Sort Algorithm for a Parallel Database Machine

Raymond A. Lorie and Honesty C. Young

IBM Almaden Research Center
San Jose, CA 95120-6099, USA

Abstract

The paper considers the problem of sorting a file in a distributed system. The file is originally distributed on many sites, and the result of the sort is needed at another site called the "host". The particular environment that we assume is a backend parallel database machine, but the work is applicable to distributed database systems as well.

After discussing the drawbacks of several existing algorithms, we propose a novel algorithm that exhibits complete parallelism during the sort, merge, and return-to-host phases. In addition, this algorithm decreases the amount of inter-processor communication compared to existing parallel sort algorithms. We describe an implementation of the algorithm, present performance measurements, and use a validated model to demonstrate its scalability. We also discuss the effect of an uneven distribution of data among the various processors.

1 Introduction

A database machine based on multiple nodes that share nothing is one way to provide the functionality of a conventional relational database system, with performance that goes far beyond what can be achieved with a single node. In particular, we consider a multiple node backend database machine that is connected to one or more host nodes. The interface to the database machine is SQL-like. This means that a complex query can be executed completely in the backend, and that only the request and the results are exchanged on the host–backend connection. For the sorting problem, however, we assume the host gathers sorted records in such a way that does not require sort key comparison. In other words, we do not carry out any comparison-based global sort or global merge at the host site. We also assume that the host extracts tuples from the communication messages and returns them to the application program.

The backend itself is a collection of n loosely-coupled nodes communicating by message passing on a network. Each node has its own processor(s), memory, channels, disks, and operating system. We assume that the hardware is symmetric in the sense that every node is equivalent to every other node. Besides the network which supports the n-to-n communication between the nodes, each node has a communication channel that it can use to receive a request from the host or send data back to the host. We refer to these communication channels as the n-to-1 network (see Fig. 1). We also assume that n-to-n communication shares the same media where each of the n-to-1 communication has its own media. One or more instances of the DBMS software may run on a single node, and each instance is called a *site*. The scheme of running multiple sites on a single node is called "virtualization" in [9].

We call data sites the sites where the records are originally stored and merge sites the sites where the merge operation is performed. In addition, there is a host site and a coordinator site. To simplify the discussion, we assume a site consists of a data site and a merge site, unless specified otherwise. P_i denotes the i-th site and P_{di} and P_{mi} denote the i-th data, and merge site, respectively. Depending on the algorithm, the sort operation may be performed on either the data sites or the merge sites.

Given a table T, with columns C1, C2, ... Ck, the sorting problem consists of returning to the host, the desired fields of the records that satisfy a predicate, ordered on the value of an expression computed on one or several columns. In order to simplify the analysis, we

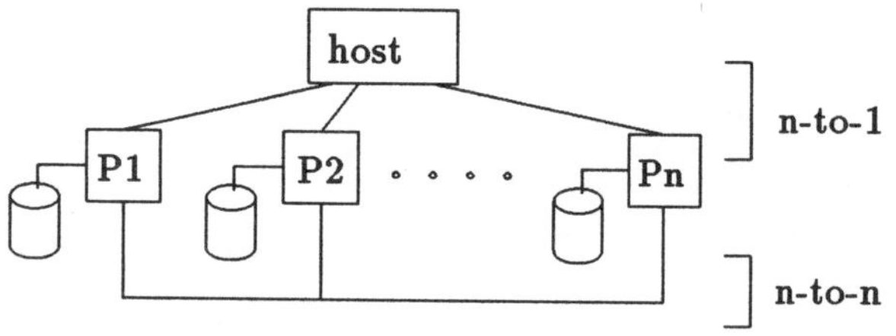

Figure 1: A Generic Backend Database Machine.

assume that the predicate is always true, that all fields are returned, and that the order is defined on a single column, as in the following SQL statement.

```
SELECT * FROM T ORDER BY C1
```

Barring data skew, the performance of a sort depends on the amount of CPU used at each site, the amount of disk I/O done at each site, the amount of n-to-1 communication, and the amount of n-to-n communication, and of course, the degree of parallelism. The cost complexity of n-to-n communication is typically superlinear: n^2 for a full crossbar switch or $n \log n$ for a multistage interconnection network. The cost complexity of n-to-1 communication in contrast is proportional to n[1]. So, a good sort algorithm for a distributed configuration is one that minimizes the amount of n-to-n communication, and exploits high degree of parallelism without substantially increasing the CPU and disk activity.

Section 2 discusses some related work. Section 3 proposes a novel, fully parallel algorithm (FPA) that decreases the amount of communication substantially. Section 4 describes briefly the ARBRE prototype built at the IBM Almaden Research Center, and provides some overall performance for a benchmark sort. Section 5 outlines a model to evaluate the new sort algorithm for a variety of parameter ranges. In section 6, we discuss the effect of an uneven distribution of data on the overall performance of the algorithm. In the conclusion, we summarize the benefits and applicability of the proposed sort algorithm.

2 Related Work

Researchers have studied the parallel sorting problem extensively (see the surveys in [2,8]). Most of the assumptions and results, however, are of theoretical interest (*e.g.,* the number of processors required is a function of the number of records to be sorted). Moreover, practical issues such as the times to read the source records from their data site(s) and to deliver the sorted records to a single host site are often ignored.

The sorting algorithms that are related to our current discussion are sometimes called "limited parallelism sorting algorithms" [2]. Four variations of such algorithms are worth mentioning.

Algorithm A: data sites send all records to a sort site S. S does the sort, then returns the sorted file to the host site. Although the scans are done in parallel, the sort itself is done serially.

Algorithm B: sort the data locally at each data site and merge the sorted file at a single merge site. This is similar to Tandem's FASTSORT [5] except that, in FASTSORT, the input data constitute a single file and a preliminary phase distributes the data before the local sorts take place.

Algorithm C: sort the data locally in each data site, organize the merge sites into a binary tree, and merge the sorted files in a pipelined fashion. This is a variation of the *parallel binary merge algorithm* in [3], which is itself an improved version of an algorithm described in [4] in that it pipelines the merge phase. The sort used in DBC/1012 [11] is similar to algorithm C in that each AMP does a local sort and sends the sorted stream through the Y-net, which carries out the merge operation in hardware.

Algorithm D: data sites use key-range partitioning to redistribute the data to sort sites; sort sites sort the redistributed data locally, and the host site concatenates the sorted file. This is a parallel version of the distributive partitioned sorting proposed in [7]. This algorithm computes, by some method, a partitioning of the key range R into subranges R1, R2, ..., Rn such that a roughly equal number of records falls in each subrange. Any well-known distributed random sampling algorithms may be used here. Each subrange Ri is assigned to one sort site in a one-to-one fashion; for example, Ri is assigned to P_{mi}. Then, each data site sends each of its records to the appropriate sort site that handles the range containing the record's sort key. When all records have been sent, all records with keys in the first range are all in P_{m1}. All records with keys in the second range are in P_{m2}, etc. Each site can then sort the records that it received. When done, P_{m1} returns its sorted records to the host. When it finishes, P_{m2} does the same, etc., until P_{mn}.

Each of the above algorithms has at least one phase where parallelism is not exploited. Algorithm A has a serial sort phase. Algorithm B has a serial (non-pipelined) merge phase. Algorithm C also has a serial (pipelined) merge phase. Algorithm D has the most parallelism, but it sends entire records on the n-to-n communication network, imposing a load of (number_of_records * record_length).

In algorithms B, C, and D, the return of the results

[1]n is the number of nodes in the system

to the host is serialized. There is a single site that sends the sorted records to the host in algorithms B and C. In algorithm D, the sorted records are sent first from P_1, then from P_2, etc. In our experiments, this is responsible for most of the response time.

In the next section, we propose a novel algorithm, similar to algorithm D, that eliminates the drawbacks mentioned above. The technique used to decrease the n-to-n communication is similar to the one used in a semi-join. To our knowledge, the only other research work that sends only the sort keys during an intermediate step of a sorting algorithm is the "key broadcasting" algorithm described in [10], where a parallel sort is executed in a common-bus local network environment.

3 A Fully Parallel Algorithm

Our sort algorithm has parallel sort, merge, and return-to-host phases. Thus, it is called a fully parallel algorithm (FPA). We will exploit the size difference between the sort key and the entire record by sending only sort keys from the data sites to the merge sites. The data sites send sorted files directly to the host which efficiently "merges" them without doing sort key comparisons. As explained below, the host's merge algorithm uses information computed during the merge sites' processing of the sort keys. The merge sites send only this information to the host; the sort keys are discarded as they are merged.

Before presenting the details of our sort algorithm, we describe it at a high level. Each data site sorts locally the portion of the file that it owns; in doing so, it computes the distribution of the keys that it encounters, and sends the distribution to a global coordinator. Based on such information from all data sites, the coordinator assigns each merge site a range of keys for which it is responsible. Each data site sends each of its record keys (in sorted order) to a certain merge site. For each record, the merge site is chosen as the site responsible for the key of that record. Each merge site merges the keys that it receives from all data sites, determining the order that the data sites deliver the next record to the host site.

3.1 Detailed Description

There are four kinds of tasks involved in the sort (see Fig. 2): the host task (always executed on the host machine), the coordinator (executed on any site), the scan tasks (one on each data site), and the merge tasks (one on each merge site). The tasks enclosed by the dotted line box are running on each site. The sort and the scan operations can be done by the same scan task because their operations do not overlap in time. In order to distinguish the work between merging the sort keys and returning the sorted records to the host, the data sites do not send sorted records to the host site until all the sort keys have been sent to the merge sites. Therefore, the scan task is also responsible for returning the sorted records to the host site.

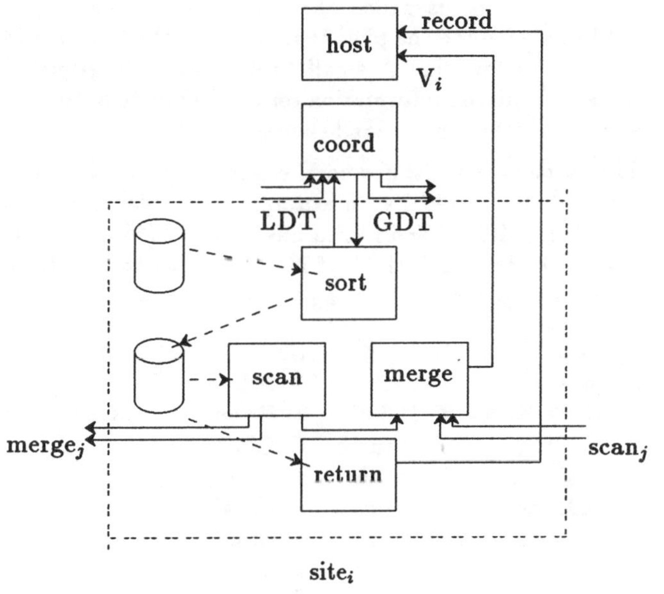

Figure 2: The Dataflow among Tasks.

The host task sends the sort request to the coordinator and waits until it starts receiving a stream of site Ids V_1 from the P_{m1}, etc. At that time, the host considers successively each value in V_1. A value i indicates that it needs to read the next record from the stream of records coming from data site P_{di}.

The coordinator waits for a request from the host. When it receives the request for a sort, it sends the request to all data sites and merge sites. It then waits for all data sites to send their distribution tables. Based on these distribution tables the coordinator constructs the GDT (Global Distribution Table). The GDT is then sent to all data sites. The coordinator needs to be involved no longer.

The sort itself has four phases–local sort, synchronization, merge, and return-to-host.

Phase 1: Local Sort
Phase 1 starts on any P_{di} as soon as it receives its sort request from the coordinator. P_{di} scans its local records and sorts the records using any algorithm. The result of the sort is a file S_i of sorted records. Each record in S_i contains a key and a record body. While the sort is taking place, two objects are being built besides the sorted output: (1) A list that contains the lowest key stored in a single page of S_i and a descriptor for that

page. This list is called the *index list*, because it is used to determine the page that contains the first record of a key range. (2) A small table, called distribution table, containing information on the key distribution; it is derived from the index list mentioned above.

This sorting algorithm and the data structures it uses are best illustrated by an example. Consider a system with three sites. Each record has two fields and the first (single byte) field is the sort key field. Let the contents of the sorted files S1, S2 and S3 be, respectively:

S1	S2	S3
A Alice	C Carl	A Adam
B Barbara	D Debbie	B Brian
G Greg	F Frances	C Christine
H Helen	F Frank	D Dan
H Henry	G Gloria	E Elizabeth
K Kevin	M Mary	E Eric
L Linda	N Nancy	J Jane
M Mark	P Peggy	J John
N Norman	R Ron	K Kathy
S Steve	S Sue	L Larry
	T Teresa	P Peter
	W Walter	R Ruth
		T Tom
		W Wendy

(a)

A particular sampling algorithm may generate the following distribution tables:

B	D	B
H	F	D
K	M	E
M	P	J
S	S	L
	W	R
		W

(b)

Phase 2: Synchronization

As soon as Phase 1 is completed on a P_{di}, this P_{di} sends its distribution table to the coordinator and waits for the GDT from the coordinator. Once the coordinator has received the distribution tables from all sites, it combines their contents and builds a single GDT. The GDT contains the boundaries of n subranges, such that the numbers of records with keys falling in each range are roughly equal. Each range is assigned to a merge site on a one-to-one basis (for example range R1 to site P_{m1}, etc). The coordinator then sends the GDT to all sites.

Let's use the previous example to show how the coordinator computes the GDT. The coordinator receives the data in (b), merges the distribution tables, and partitions the results into 3 roughly equal size groups, to obtain the range definitions that constitute the GDT:

```
B B D D E F   H J K L M M   P R S S W W

----------- F ----------- M -----------
lowest                          highest
```

Phase 3: Merge

During this phase, there are two tasks at each site, working concurrently: a producing scan task and a consuming merge task. Once a data site receives the GDT its producing scan task reads sequentially the records from its sorted file. For each record in the sorted file, the key is extracted and sent to the merge site P_j, where j is such that the key falls in the j-th range defined by GDT. It is received by the consuming merge task of P_j for merging.

A strictly sequential processing of file S_i is not optimal as the following scenario illustrates. Assume all sites start sending only records that fall in the first range. Therefore all traffic is destined for the merge site in charge of the first range, which becomes overloaded. When records in the second range are sent, another merge site becomes overloaded, and the process continues. Our solution to the problem consists of processing only a small portion of the records in the first range (for which the target is P_{m1}). The small portion may be a single page from each data site. While P_{m1} handles the corresponding merge, all producing tasks jump to the beginning of the second range, etc. After the first batch of Rn has been processed, each producing task sends its second batch of R1 and so on, in a round-robin fashion. Clearly, if one looks at a sufficiently large period of time, all data sites produce records that are sent to the appropriate merge sites. So all sites are kept busy, sending, receiving and merging records.

In the following example, we denote a key sent by data site P_{di} to merge site P_{mj} by <key, i>→ P_j[2] for illustrative purpose. P_i starts producing records, sending them to the indicated merge sites. If each data site sends 2 records in a batch, the following communication pattern occurs:

batch1:

<A,1> → P1 <C,2> → P1 <A,3> → P1
<B,1> → P1 <D,2> → P1 <B,3> → P1

<G,1> → P2 <G,2> → P2 <J,3> → P2
<H,1> → P2 <M,2> → P2 <J,3> → P2

<N,1> → P3 <N,2> → P3 <P,3> → P3
<S,1> → P3 <P,2> → P3 <R,3> → P3

[2] The sender's site identifier can be decoded easily from a message and is not included in the message body.

batch2:

<H,1> → P2	<F,2> → P1	<C,3> → P1
<K,1> → P2	<F,2> → P1	<D,3> → P1
...		

Since the data sites are not synchronized, the actual communication pattern will differ from the lock-step pattern shown here.

P_1 receives from all sites (including itself):

<A,1>	<C,2>	<A,3>
<B,1>	<D,2>	<B,3>
EOS	...	...
	EOS	EOS

Since the keys are sent to a different consuming merge task in each round, all merge tasks may start at about the same time. Clearly, the merge task of Pmk receives records that have keys that fall into the range for which the site P_{dk} is responsible, and all records received from a single site arrive in order. These ordered streams of keys can thus be merged to produce a single output stream. Each entry in the output stream contains only the identifier of the data site.

For example, consider the first merge site. It merges its records to obtain the following order:

<A,1> <A,3> <B,1> <B,3> <C,2> <C,3> ...

and produces the following V_1 stream:

```
V1      1 3 1 3 2 3 2 3 2 2 3 3
```

The other two merge sites produce their output streams in a similar fashion:

```
V2      1 2 1 1 3 3 1 3 1 3 1 2
V3      1 2 2 3 2 3 1 2 2 3 2 3
```

Phase 4: Returning Records to the Host

When Phase 3 is finished at site P_i, the corresponding stream V_i and sorted file S_i are sent as streams to the host. If the next entry in V is k, the host knows that it needs to get the next record from S_k sent by data site P_k.

In the example, the host receives a stream of site identifiers (V), which is the concatenation of V_1, V_2 and V_3. It also receives the sorted S's (a) from each site. Following the site id order of V, it retrieves records in that order:

A Alice	from P1
A Adam	from P3
B Barbara	from P1
B Brian	from P3
C Carl	from P2
C Christine	from P3
...	

3.2 Alternative Implementations

The algorithm as explained above is the one that we implemented and evaluated (see section 5 below). Some variations are worth discussing.

1. In the above description, the entire file S is read twice–once to send the keys to the merge tasks, and once to return the records to the host. It may be preferable to have the local sort write two files: S with the entire records and a much smaller file K with keys only. In phase 3, the smaller file K is read, instead of S.
2. When multiple merge sites are running on a single node (because of site virtualization), the merge phase can be divided into several sub-phases. The host site cannot start receiving records in the R2 ranges until it has received all records in R1, which cannot finish before P_{m1} finishes. If the merge phase has several sub-phases, the host can start receiving records in the ranges for which the corresponding merge operations have completed. That is, the host site receiving operation is pipelined with the merge site merging operation. The most important advantage of the pipelined scheduling is that the time to receive the records is expanded, thus, the resource requirements on the host can be smoothed out. Consequently, the same host can accommodate more nodes without incurring response time penalty.
3. Finally, other sampling algorithms may be used to compute the GDT.

4 Prototype

At IBM Almaden Research Center, we have built a prototype of a backend database machine, called ARBRE (Almaden Research Backend Relational Engine), in order to explore the use of parallelism within complex relational queries, and across many independent transactions [9].

A prototype of the database machine is now operational. It runs on three dyadic 4381 nodes providing 6 processors for the backend database machine; a fourth 4381 is used as a host. Each processor is approximately 3.5 MIPS (million instructions per second). The operating system is MVS and a 3088 is used for channel-to-channel communication. The database system is organized as follows: on each of the 4381's, four virtual sites are emulated by running four DBMS's in four address spaces, providing 12 sites in total. Each relation is split horizontally into partitions; each partition is stored in a different DBMS. The access method is borrowed from SQL/DS (DBSS) [6].

Since we have not implemented a query compiler, we

hand-code programs that specify the strategy that would normally be chosen by the optimizer. A strategy is expressed as a series of routines which are generally duplicated at each site. These routines produce or consume streams of data that are passed between disks, nodes, and the host. The execution is controlled by data flow and stream communication is controlled by a pacing mechanism.

We exercised the prototype for various sets of parameters. Here we report the results for one particular set of parameters which we use later to validate our model:

- number of sites : 12 (2 site per processor of 3.5 MIPS, 21 MIPS aggregate)
- number of records: 12 * 16000 records = 192,000 records
- record length : 64 bytes
- key length : 8 bytes (included in the 64)

For each of the four phases, we measured the CPU time, the number of I/O's, the number of messages, and the response time. Results on each site are shown below.

	sort	sync	merge	return	total
CPU (sec)	12.0	-	10.2	6.2	28.4
I/O	-	-	-	-	2884
Comm	2	2	159	509	672
Resp (sec)	44.7	1.6	22.4	23.6	92.3

Note that the CPU for the synchronization is too small to be measured and that we only show the total I/O because some pages logically modified in one phase are written later, due to the buffer page replacement algorithm, blurring the boundaries between phases.

5 Model

In this section, we model every phase separately, and immediately apply the model to the experiment just described.

Let's define the path length of various operations by the following symbols (approximate values from the prototype are given in parentheses):

- to extract a record from a list: Ex (= 950)
- to enqueue a record to a stream: Eq (= 150)
- to dequeue a record from a stream: Dq (= 150)
- to send or receive a message: Cm (= 4000)
- to perform an I/O: Io (= 8000)
- to carry out a key comparison: Cp (= 150)

The system parameters and their values are:

- the page size: P (= 4086[3])
- the message size: M (= 2048)
- the record length: Rl (= 64)
- the key field length: Kl (= 8)
- the number of sites: S (= 12)

The following symbols are also used, and all refer to one site only:

- the number of records: Ni
- the number of disk I/O's: Nio
- number of pages in the base table: Nb
- number of pages in the index list: Nl
- the number of communication messages: Nc

When two sites are sharing the same physical processor, we model the response time to be either the sum of the CPU and I/O times (when it is not CPU bound) or twice the CPU time (when it is CPU bound). Let us analyze what happens at one site.

Phase 1: Local Sort
We developed a model for the CPU and I/O times of local sorts and validated the model by running multiple experiments on our prototype. The detailed description of the local sort model is not relevant to this paper; thus, it will not be explained further. For the chosen parameters, the CPU time is 12 sec and the I/O time is 33 sec. Based on our response time model, the response time for 2 simultaneous sorts is (12 + 33) = 45 sec.

Phase 2: Synchronization
A message is sent from each site to the coordinator and from the coordinator to each site. This is a small linear overhead. Each site has an equal number of records to sort; so all sorts can be expected to terminate roughly at the same time; therefore we will neglect the time required for this phase but will return to the skew problem in a later section.

Phase 3: Merge
Using the symbols defined above, the first order[4] approximation gives us:

- $Nb = \text{ceiling}(Ni/\text{floor}(P/Rl))$
- $Nl = \text{ceiling}(Nb/\text{floor}(P/Kl + 4^5))$
- $Nio = Nb + Nl$
- $Nc = \text{ceiling}(Ni/\text{floor}(M/Kl))$

Consequently, we have

- $Nb = \text{ceiling}(16000/\text{floor}(4086/64)) = 254$
- $Nl = \text{ceiling}(254/\text{floor}(4086/12)) = 1$
- $Nio = 254 + 1 = 255$
- $Nc = \text{ceiling}(16000/\text{floor}(2048/8)) = 63$

The total path length and its breakdown are shown below:

[3] 4096 bytes less 10 bytes page header

[4] In particular, we ignore the path length related to the site identifier manipulation since each site identifier is very small (one to two bytes).

[5] The page descriptor has 4 bytes.

operation	formula	path length (M)
I/O	Nio * Io = 255 * 8000	2.04
Key Extraction	Ni * (Ex + Cp)) = 16000 * (950 + 150)	17.60
Enqueue	Ni * Eq = 16000 * 150	2.40
Send	Nc * Cm = 63 * 4000	0.25
Receive	Nc * Cm = 63 * 4000	0.25
Dequeue	Ni * Dq = 16000 * 150	2.40
Merge	Ni * (ceiling($\log_2$(S)*Cp) = 16000 * (4*150)	9.60
Total		34.5

The order of the operations shown above reflects the sequence for a "typical" record. The Cp term in the formula for key extraction is the path length associated with the partition range check. The CPU time on a 3.5 MIPS processor for 34.5 M instructions is 9.9 sec. Assuming each I/O takes 20 ms, the elapsed time for I/O is $255 * 0.020 = 5.1$ sec. The phase is strongly CPU bound. Therefore running simultaneously two such phases on one processor will yield an elapsed time of $(2 * 9.9) = 19.8$ sec.

These 19.8 sec are somewhat less than the measured 22.4 sec. Part of the difference is due to the fact that the merge needs always a record from a specific incoming stream in order to proceed. Therefore, some waiting time can be expected. A simulation has shown that the relative response time increase due to the deterministic merge varies very little with the increase in the number of sites: 3% for 4 sites to roughly 11% for 64 or 128 sites.

Phase 4: Return
For the return phase, we obtain the following path length:

operation	formula	path length (M)
I/O	Nio * Io = 254 * 8000	2.03
Record Extraction	Ni * Ex = 16000 * 950	15.20
Enqueue	Ni * Eq = 16000 * 150	2.40
Send	Nc * Cm = 500 * 4000	2.00
Total		21.6

The CPU time on a 3.5 MIPS processor for 21.6 M instructions is 6.2 sec. The I/O time is $254 * 0.020 = 5.1$ sec. Again, since the phase is CPU bound, the elapsed time for two such phases executed simultaneously is $2 * 6.2 = 12.4$ sec. This is true only if the host and the communication network are infinitely fast. In fact one can show that the host CPU requirement is 15.5 sec.

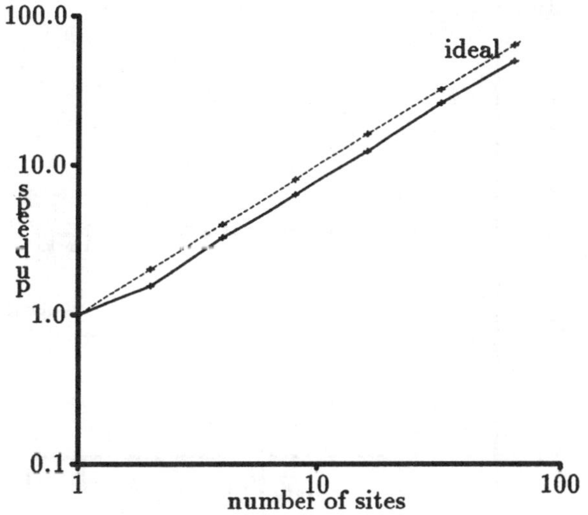

Figure 3: Response Time Reduction.

To summarize, let us compare the results of the model and the actual measurements:

		sort	merge	return
CPU		12.0	10.2	6.2
	(mod)	12.0	9.9	6.2
Resp.		44.7	22.4	23.6
	(mod)	45.0	19.8	>15.5

The response time of the return phase is longer than estimated because the channel-to-channel used in the prototype to implement the n-to-1 communication network becomes the bottleneck.

5.1 Exercising the Model

After having validated the model with actual measurements, we exercised it at will to examine response time reduction and horizontal growth. We adjusted some parameters to values that are more appropriate for high-performance relational DBMS, but difficult to change in our current prototype implementation. We assume that I/O streams read sixteen 4096 byte pages at a time with a latency of $(20 + 15 * 16) = 44$ ms, and message size of 4KB instead of 2KB. To eliminate the site virtualization effect, we run one site on a processor. We also consider 4 MIPS backend processors, and a 32 MIPS host processor.

We studied the response time reduction by sorting a total of 1 million 64 byte records that are evenly distributed among all sites. We also assume here that the GDT gives the exact partition. For the single site case, the synchronization, and merge phases are not executed. Fig. 3 shows the speed up with respect to the single site case for various configurations.

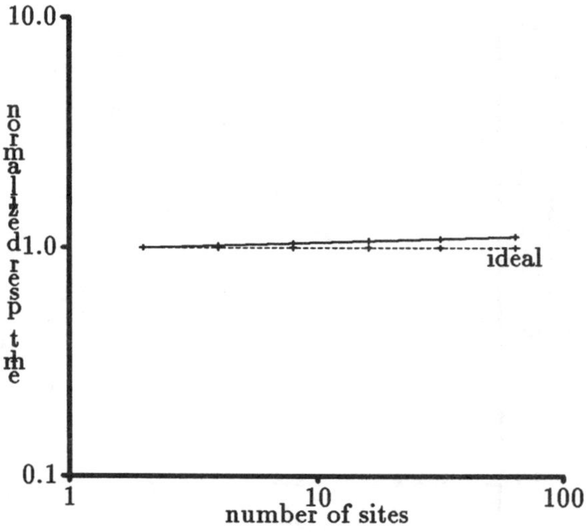

Figure 4: Horizontal Growth.

As the figure indicates, one pays for the distribution overhead when the number of sites goes from 1 to 2; afterwards the speed up is quasi-linear. This is true under the assumptions that the records are evenly distributed initially, and that the GDT ensures a pretty accurate partition. In Section 6, we will address the problem of uneven distribution of initial records.

We also studied horizontal growth by increasing the number of sites and the total number of records to be sorted, in the same proportion. That is, the number of records on each site remains unchanged. In particular, we sort 16K records per site. Since we are interested in multiple site cases, the response time in Fig. 4 is normalized to that of two sites. As expected, when the number of sites in the system increases, the response time increases slightly as well. (Remember that the number of key comparisons needed to produce a record in the merge phase varies with the logarithm of the number of sites.) For the chosen parameters, the response time is increased by about 2.4% when the number of sites doubles.

5.2 Comparison with Other Sort Algorithms

Let's use our model to compare the proposed fully parallel algorithm (FPA) with a variation of FPA (called FPA') without employing the "semi-sort" communication reduction technique and with the algorithms described in Section 2. The work load is to sort 1M 64 byte records using 64 4 MIPS processors. The communication CPU time is added wherever appropriate. The total CPU time is computed as the sum of all non-overlapped CPU times, and the total time is computed as the sum of the total CPU time and the I/O time. For all methods, the n-to-1 communication is 64 MB and is not discussed any further. In the table 5.2, two CPU times are shown for the return phases of both FPA and FPA': the smaller one is the CPU time on a backend processor while the larger one is that on the host processor. These two numbers are carried on in the subtotal and total. The real elapsed time should lie between the total CPU time and the total time.

The major reason for algorithm D performing much worse than FPA is that the sorted results are sent to the host in a serialized fashion. For algorithms B, C, and D, the scan time is not much smaller than the sort time because the particular database system used in the prototype has the property that reading a record for the internal sort is much cheaper (in terms of path length) than returning a record via the record-at-a-time interface. Algorithm C performs better than algorithm B because the merge fan-in of the former is much smaller.

Unless the n-to-n communication has a very high effective bandwidth, algorithms A, C, and D may experience a slow down because the n-to-n communication becomes a bottleneck when all sites exchange records among themselves. The bottleneck is caused by high aggregate communication, bursty communication or both, depending on the sort algorithm. For algorithm C, the amount of n-to-n communication is $64B * 1M * \log_2(S)$, where S is the number of sites. For a system with 64 sites, the total n-to-n communication is 384 MB. For all algorithms A, B, and D, the amount of n-to-n communication is $64B * 1M = 64$ MB. On the other hand, the amount of n-to-n communication for FPA is $8B * 1M = 8$ MB. Comparing FPA with FPA', we demonstrate that the "semi-sort" technique not only reduces the n-to-n communication but also decreases the total CPU path length.

For algorithms A and D, the n-to-n communication happens in bursts, even though the total response time is much longer than that of FPA. The n-to-n communication in algorithms B and C can be spread over the entire merge phase. However, the entire record is sent $\log_2(S)$ (6 under the chosen parameters) times for algorithm C.

In fact, the sorting algorithm reported in [1] (on which algorithms C is based) is limited by the network data transfer. Also, algorithms B and C were compared in [10], which concluded that algorithm B performs better than algorithm C in the common-bus local network for the same reason.

6 Sensitivity to Data Skews

The FPA automatically balances the load for both the merge and return phases, independently of the initial distribution of data. The local sort phase, however, depends upon this initial distribution: if one site has

algorithm		FPA	FPA'	A	B	C	D
CPU (sec)	scan			4.8	4.8	4.8	4.8
	sort	5.1	5.1	436.0	5.1	5.1	6.0
	merge	9.5	10.5		370.3	173.7	
	return	0.4/7.0	0.4/7.0	22.9			22.9
subtotal		15.0/21.6	16.0/22.6	464	380	184	32.7
I/O (sec)	scan			0.7	0.7	0.7	0.7
	sort	1.8	1.8	160.1	1.8	1.8	1.8
	merge	0.8	0.8				
	return	0.7	0.7	45.1			45.1
subtotal		3.3	3.3	206	2.5	2.5	47.6
CPU+I/O (sec)		18.3/24.9	19.3/25.9	670	383	187	80
n-to-n comm (MB)		8	64	64	64	384	64

Table 1: Algorithm Comparisons

more data than another, the local sort phase will take more time. The total elapsed time is thus a function of the maximum local sort time.

We propose the following technique to reduce the effect of skew. Suppose each data site periodically informs the other data sites where it is in the scan (in particular, if it has finished or not). Then, an overloaded site can send pages to other sites that have finished. The overloaded site still needs to read the pages and send them on the communication network; the receiving sites need to receive and extract the data from the page. If records are short, then scanning a page is time consuming; if records are long the path length for communication becomes more important and this load balancing technique is not as efficient. Of course having several sites being able to access the same disk would help in this case.

In order to study the effect of an uneven distribution of the data, let us consider one site that has A records while all other sites have B records. Assuming A > B, we define the skew as Z= A/B. It is easy to calculate the response time degradation of the initial scan as a function of Z. The corresponding curve is plotted as a dotted line in Fig. 5. The figure also shows how the proposed technique may improve the situation for different record lengths. Note again that the degradation applies only the pre-scan phase; its effect should be amortized over the whole duration of the global sort.

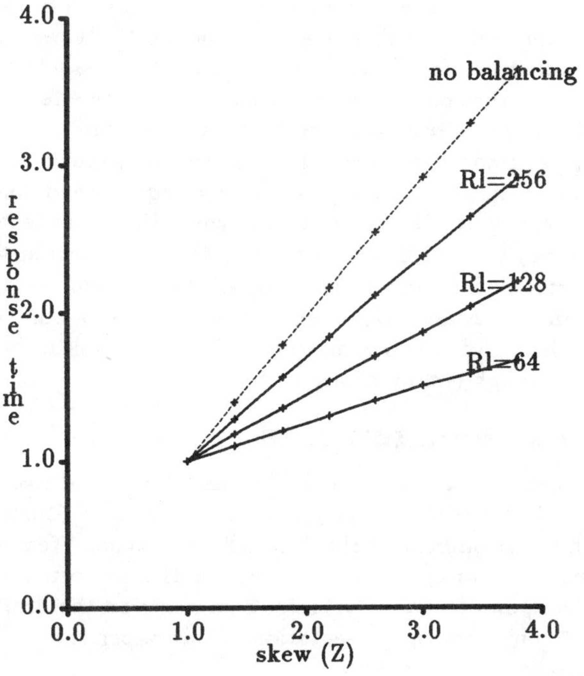

Figure 5: Effect of Redistributing Excess Records.

7 Conclusions

There is increased interest in using multiple nodes to improve the throughput and response time of database operations. Sort is just one of these operations; but it may be the most important one since it also plays a role in other operations such as duplicate elimination, grouping, joining, etc.

We have proposed an efficient parallel sort algorithm that assumes only a "shared nothing" architecture. In such an architecture, the nodes do not share main memory; nor do they share disks. Instead, they communicate among themselves, and with the host, by message passing.

The algorithm, evaluated by exercising a model validated by an actual prototype, exhibits the following nice properties:

1. All sites (except for very short intervals) are performed simultaneously, yielding high utilization of resources.
2. It scales very well. It decreases response time practically linearly when more processors are working on the same problem; it also supports horizontal growth, where 2n processors sort 2N records in roughly the same time that n processors sort N records. The primary reason for such good performance is that the sorted records are returned from all sites to the host site concurrently.
3. It reduces the amount of n-to-n communication and the path length to send/receive communication messages by using a "semi-sort" algorithm, where only keys have to be sent on the n-to-n communication network.

The efficiency of the algorithm is affected by the amount of skew in the data (the uneven distribution of data on the sites). Since skew only affects the initial scan of the data, and not the other phases of the sort, adverse effect is diluted. In any case, we discuss a load-balancing technique that can often be used to reduce the effect of skew on the initial scan itself. This technique is very efficient if the work needed to process a page is significantly larger than the work required to read and send/receive it. If this is not the case, the same technique can be used if several processors are connected to the same disk (which is generally needed in some small amount for availability, anyway); then the data can be read through the I/O channel with large prefetch rather than through communication.

Acknowledgements

We acknowledge the important contribution of Jean-Jacques Daudenarde, Gary Hallmark, and Jim Stamos to the development of the ARBRE prototype. We also thank Jim for useful discussions on the performance analysis, and for many valuable comments on the draft, which improved the presentation of this paper very substantially.

References

[1] M. Beck, D. Bitton, and W. K. Wilkinson. Sorting large files on a backend multiprocessor. *IEEE Transactions on Computers*, C-37(7):769–778, July 1988.

[2] D. Bitton, D. J. DeWitt, D. K. Hsiao, and J. Menon. A taxonomy of parallel sorting. *ACM Computing Surveys*, 16(3):287–318, September 1984.

[3] D. Bitton-Friedland. *Design, Analysis and Implementation of Parallel External Sorting Algorithms.* PhD thesis, Computer Sciences Department, University of Wisconsin—Madison, January 1982.

[4] S. Even. Parallelism in tape-sorting. *Commun. ACM*, 17(4):202–204, April 1974.

[5] J. Gray, M. Stewart, A. Tsukerman, S. Uren, and B. Vaughan. FASTSORT: an external sort using parallel processing. *Tandem Systems Review*, 2(3):40–47, December 1986.

[6] IBM Corporation. *SQL/Data System General Information.* IBM Form No. GH24-5012.

[7] P. J. Janus and E. A. Lamagna. An adaptive method for unknown distributions in distributive partitioned sorting. *IEEE Transactions on Computers*, C-34(4):367–372, April 1985.

[8] S. Lakshmivarahan, S. K. Dhall, and L. L. Miller. Parallel sorting algorithms. *Advances in Computers*, 23:295–354, 1984.

[9] R. Lorie, J. Daudenarde, G. Hallmark, J. Stamos, and H. Young. Adding intra-transaction parallelism to an existing dbms: early experience. *IEEE Data Engineering Bulletin*, 12(1):2–8, March 1989.

[10] K. P. Mikkilinent and S. Y. W. Su. An evaluation of sorting algorithms for common-bus local networks. *Journal of Parallel and Distributed Computing*, 5(1):59–81, February 1988.

[11] P. M. Neches. The anatomy of a data base computer system. In *Proceedings of the 2nd International Conference on Supercomputing*, pages 102–104, 1987.

Percentile Finding Algorithm for Multiple Sorted Runs

Balakrishna R. Iyer
Database Technology Institute
IBM Corporation
San Jose, CA 95161-9023

Gary R. Ricard
Applications Business Systems
IBM Corporation
Rochester, MN 55901

Peter J. Varman
Elec. & Comp. Eng. Department
Rice University
Houston, TX 77251-1892

Abstract External sorting is frequently used by relational database systems for building indexes on tables, ordered retrieval, duplicate elimination, joins, subqueries, grouping, and aggregation; it would be quite beneficial to parallelize this function. Previous parallel external sorting algorithms found in the database literature used a sequential merge as the final stage of the parallel sort. This reduces the speedup gained through parallelism in earlier stages of sort. The solution is to merge in parallel as well. Load balanced parallel two way merges and approximately load balanced parallel multi way merges are known. Measurements reported on parallel sorting that employs one of the approximate partitioning methods indicate that even if the sort keys are randomly distributed the load imbalance due to the approximation degrades speedup due to parallelism. Sort key value skews, known to occur in database workloads, can only exacerbate this problem. We give, prove and analyze an efficient exact method which can find any percentile of an arbitrary number of sorted runs. Application of our algorithm ensures load balance during the parallel merge. By removing the effect of skews of sort key values which caused loss of speed up in previous approaches our method can improve the speedup for parallel sorting on multiple processors. While we target our work to a parallel computer architecture of shared memory MIMD parallel processors, our results are also likely to be useful for other parallel computer architectures.

1 Introduction The need for database MIPS per database installation is outstripping the uniprocessor MIPS supplied by computer vendors. External sorting is frequently invoked by relational database systems for building indexes on tables, ordered retrieval, duplicate elimination, joins, sub-queries, grouping, and aggregation and is known to be a time consuming operation. External sorting on multiple processors is, therefore, an important and beneficial problem to be solved for relational database systems. For purposes of exposition and analysis we will assume a shared memory shared data computer architecture. Yet, the reader may find much of our work equally applicable to loosely coupled architectures.

1.1 Background We start by examining some of the parallel sort algorithms in the database literature. Perhaps the simplest technique is to partition the sort key range prior to accessing the database. Selected rows from the database are assigned to different buckets, each bucket corresponding to a key range. Rows in different buckets are sorted in parallel and then the result concatenated or used for further processing. This simple technique is not practical for the following two reasons: (a) tight upper and lower bounds for the sort key range are not easily determined before the rows from a database are selected, and (b) it is difficult to partition the sort key range into partitions so that about the same number of rows will have sort keys belonging to any one partition, load balancing is compromised.

Bitton et. al. (BIT83) propose two external parallel sort algorithms for use in database systems that they call parallel binary merge and block bitonic sort. Both algorithms employ the binary two way merge of sorted runs. In a typical external sort for database systems, sorted runs are first created as temporary relations on disk and then merged. If we use algorithms that rely on a two way merge the number of I/O's, temporary relation inserts and fetches per sorted row will be of the order of $\log_2(NR)$ where NR is the number of initial sorted runs created. Both I/O's and number of temporary relation interactions are expensive and they make algorithms based on a two way merge expensive, suggesting that the two algorithms are primarily useful for internal sort where there is no I/O or temporary relation interaction. Even for parallel internal sorting, Bitton's parallel binary merge algorithm suffers from the drawback that its last phase is essentially a sequential binary merge and it is therefore impossible to finish the sort earlier than the time for one sequential pass through the entire data regardless of the number of processors. While it may not be possible to avoid examining all the data for sorting, we must avoid examining it all serially if we are to obtain speedup linear in the number of processors (over a reasonable range of values for the number of processors). Valduriez and Gardarin (VAL84) generalized the parallel binary merge algorithm to a k-way merge algorithm. In their algorithm p processors will merge different sets of runs in parallel using a k-way merge. We then have p runs. These are merged (assuming $p < k$) sequentially on a single processor. A pipelined version of this algorithm has been prototyped (BEC88) and another implemented in a product (GRA88). The bottleneck due to a sequential merge is still present in all these algorithms.

The idea of employing multiple processors to merge the same runs appears in the literature in the context of merging two runs on two processors. Based on the pioneering work of Batcher (BAT68), an efficient algorithm was proposed (AKL87) to break each of two runs into two partitions so that key values in the first partition of each run were no greater than key values in the second partition of either run (magnitude requirement). Also, the sum of the number of elements in the first partition of the two runs was half the sum of the total number of elements in both runs (size requirement). The first partition of each run was merged on one processor and the second partition of each run is merged independently on another processor. Although the method meets the requirements for load balancing a parallel sort if there are two runs and two processors, it provides no obvious or intuitive algorithmic method for extension to an arbitrary number of equal sized partitions for an arbitrary number of runs. The idea of employing an arbitrary number of processors to merge two runs appears only recently (VAR88) (FRA88) (DEO88). A method to partition two runs into an arbitrary number of equal parts corresponding to percentile ranges is known. A multiprocessor multi-way merge can be implemented by a sequence of multiprocessor two-way merges. While this method gets around the bottleneck due to a final sequential merge, the I/O overheads incurred are proportional to $\log_2(NR)$, where NR is the number of sorted runs. The idea of employing an arbitrary number of processors to merge as many runs as there are processors (in one pass) appears to be first due to Quinn (QUI88). He proposed that m processors be used to merge m runs in parallel, each processor employing an m-way merge on $1/m$'th of the data.

1.2 Problem Statement How we obtain the m-way partitioning of an **arbitrary number** of runs for arbitrary m is the subject of this paper. We show how to solve the problem of partitioning an arbitrary number of m sorted runs into two parts such that one partition has an arbitrary fraction f of the total number of elements. Elements of one partition have keys greater than any element of the other partition. Recall that previous known exact algorithms can partition only **two runs** into multiple equal parts. Initially, we simplify the problem by making all runs equal in size, restricting the size to a power of two and pad the runs with $+\infty$ and $-\infty$ on either end (all other elements in the run must be strictly greater than $-\infty$ and strictly less than $+\infty$). The values taken by the fraction f are also restricted. An algorithm is given to find the partition of m such runs containing fraction $f = i/m$ of the data, where $1 \leq i \leq m$, and where the initialization (basis) step involves examining up to $O(m^2)$ elements.

1.3 Previous Solutions First we discuss previous solutions. One simple method to solve the partitioning problem approximately is to use a sampling technique (LOR89). This involves sampling the keys in the m runs, sequentially merging the samples, finding an exact partition of the samples (through a search of the merged runs of samples), and determining the exact m key ranges that partition the samples. The m keys that define the ranges are then used as indices into the runs to define the m partitions. Sampling techniques are discussed in (VIT85) and (OLK86).

An approximate technique that partitions the runs after they are formed has been proposed by Quinn (QUI88). Equally spaced apart keys from the first run are used to index into and partition the remaining runs. We illustrate this technique by means of an example of 3 runs and 3 partitions in Figure 1. The keys 9 and 15 from run 1 are selected since they partition it into three sequences of equal length. The two keys are used to partition the remaining two runs as shown in Figure 1. The first partitions of each run totalling to 11 keys will be merged on processor 1. The second partitions of each run totalling to 8 keys will be merged on processor 2, and the third partitions of each totalling to 8 keys will be merged on processor 3. This method does not exhibit exact load balancing. Experiments by Quinn show that the speed up of parallel sorting by employing this algorithm levels off beyond 10 processors for sorting a random permutation of 10,000 keys. Quinn attributes the loss of speed-up to the approximation and resulting loss of load balancing during the merge. Skewed distribution of sort key values in rows selected from a database are not uncommon in practice. In these cases the risk of the approximation is even greater. Quinn cites an example where the first run contains elements that are all smaller than the elements in all the other runs. Approximate partitioning leads to a merge time that is comparable to a sequential merge. Speedup would be severely compromised in other similar situations.

In order to ensure perfect load balance and prevent loss of speed up during the final parallel merge, an alternate and exact technique is proposed to partition the runs after they are formed. We give an algorithm that divides the m arbitrary sized sorted runs into two partitions such that one partition contains any given arbitrary fraction f of the total number of elements in the runs, and all elements in that partition are no greater than the elements in the other. By employing this algorithm repeatedly for $f = \frac{1}{m}, \frac{2}{m}, \ldots, \frac{(m-1)}{m}$

the m-way partitioning problem can be solved. It will be shown that as number of rows grows, the complexity of this algorithm (for a small number of processors) grows only logarithmically as the number of rows in each of the m runs. This is why we ignore the cost of partitioning the runs when estimating the elapsed time of the parallel sort algorithm. Recently, we became aware of an algorithm with asymptotically optimal number of comparisons for this problem (FRE82). The algorithm is quite complex and has a slower rate of convergence than the one proposed here, and requires significantly greater (about six times) the number of I/Os.

In the execution of complex queries in relational database systems we can visualize multiple processors employed in various ways. Specifically, for queries requiring ordering, e.g. sort/merge joins, multiple processors may be used for selecting rows and sorting different partitions of a relation or a composite produced as a result of previous steps (IYE88). As illustrated in Figure 2, the access of data from different partitions and the formation of runs is easily parallelized. Different partitions may be assigned to different processors. The assigned processor selects, sorts and merges rows belonging to a partition independent of other partitions. At the end, we need to merge (or merge/join) in parallel with multiple input runs. Our way to parallelize the merge (or merge/join) operation is to divide the sorted runs into equal parts based on key ranges and assign one part to each processor. Each processor merges (or joins) its partition of rows, roughly corresponding to different key ranges. (For joins it is possible to extend our problem and solution to ensure that there is no overlap among the key ranges. For sorting the smallest key in a range may appear as the largest key in the adjacent range.)

In the next section we state the algorithm for a restricted case of the problem, give its proof and an example. The generalized algorithm is discussed in section 3. We analyze the complexity of the algorithm in section 4. Concluding remarks are found in section 5.

Equal Sized Lists Algorithm

2.1 Problem The casual reader may skip the following subsections to the subsection containing the example. We start with a formal definition of the simplified problem. Let $A_r = a_{r,0}, a_{r,1}, a_{r,2}, \ldots, a_{r,N}$, $1 \leq r \leq m$, $N = 2^k$, be m ascending sequences, with $a_{r,0} = -\infty$, $a_{r,N} = +\infty$. , and $+\infty > a_{r,i} > -\infty$ for $i > 0$. The problem is to find for each r, $1 \leq r \leq m$, an index i_r that defines a prefix $\hat{A}_r$ of A_r with the properties 1 and 2 below. $\hat{A}_r = a_{r,0}, a_{r,1}, a_{r,2}, \ldots, a_{r,i_r}$ such that:

1. $\sum_{r=1}^{m} ||\hat{A}_r|| = f \sum_{r=1}^{m} ||A_r|| = fm(N+1)$ for $f = \frac{i}{m}$, $1 \leq i \leq m$ (formal statement of the partition size requirement in terms of the cardinalities $||.||$) and

2. $\forall x, x \in \hat{A}_r, 1 \leq r \leq m$, and $\forall y, y \in A_t, y \notin \hat{A}_t$, $1 \leq t \leq m$, $x \leq y$. (formal statement of the magnitude requirement).

We refer to $I = \{i_r, 1 \leq r \leq m\}$ as the partitioning function of $\{A_r, 1 \leq r \leq m\}$ and to $\{\hat{A}_r \mid 1 \leq r \leq m\}$ as the **leading partition** of $\{A_r \mid 1 \leq r \leq m\}$. The m sequences defined should be interpreted as the sequence of sort keys in the m runs. Element $a_{r,i}$ of the sequence is the sort key of the i'th row, of the r'th sorted run. The partitioning function is a set of indices into the m sequences that define where sequences must be split. The index i_r is the position of the sort key with the largest value in A_r and which belongs to the partition containing the smallest f fraction of the entire data. We refer to such an element of the sequence as a **boundary element** (for fixed length rows from a database sequentially stored on a disk or tape, it is straightforward to calculate the disk or tape address from the the index or the boundary element in the sequence of sort keys, for unequal sized rows we will need to use some form of indexing into the run).

Let A_r^p denote the subsequence of A_r consisting of $(2^p + 1)$ equally spaced apart elements, i.e.,

$$A_r^p = a_{r,0}, a_{r,N(p)}, a_{r,2N(p)}, \ldots, a_{r,N},$$

where $N(p) = N/2^p$. Let $I^p = \{i_1^p, i_2^p, \ldots, i_m^p\}$ denote the partitioning function of $\{A_r^p, 1 \leq r \leq m\}$. Recalling that $N = 2^k$, we note that $A_r^k = A_r$ and I^k is the partitioning function that we are looking for. We solve the problem by successively computing the partitioning functions I^{p+1} from I^p using O(m) I/O's and O($m \log_2 m$) comparisons.

2.2 Definitions Consider the subsequence A_r^p and its leading partition $\hat{A}_r^p$, $1 \leq r \leq m$. These consist of the elements
$A_r^p = a_{r,0}, a_{r,N(p)}, a_{r,2N(p)}, \ldots, a_{r,N}$
$\hat{A}_r^p = a_{r,0}, a_{r,N(p)}, a_{r,2N(p)}, \ldots, a_{r,i_r^p}$,
where $I^p = \{i_1^p, \ldots, i_m^p\}$ is the partitioning function of $\{A_r^p, 1 \leq r \leq m\}$.

For the set $\{A_r^p, 1 \leq r \leq m\}$ and partitioning function I^p, let a_{max}^p denote the largest of the m boundary elements, i.e., $a_{max}^p = \max_{1 \leq r \leq m}(a_{r,i_r^p})$. A consequence of the simplifications made in this subsection is that the leading partition of $\{A_r^p, 1 \leq r \leq m\}$ will always contain at least one element from each subsequence A_r^p. That is:

$$i_r^p \geq 0, \text{ for } 1 \leq r \leq m. \qquad (1)$$

This follows because of the restrictions we placed on the values that can be taken by the fractions f and the padding with $-\infty$'s and by the appropriate selection of the basis step to be described later. None of these assumptions will be needed in the generalized algorithm in the next subsection which embodies the same key idea as the algorithm we give next to solve the restricted problem.

At the start of the current iteration, we are given the partitioning function I^p of $\{A_r^p, 1 \leq r \leq m\}$ and $a^p_{\max}$. The former satisfies both the partition size and magnitude requirements. That is:

$$\sum_{r=1}^{m} ||\hat{A}_r^p|| = f \sum_{r=1}^{m} ||A_r^p|| \text{ (partition-size rqrmnt.) and}$$

$$a_{r,j} \leq a_{u,k} \text{ whenever} \qquad (2)$$

$$a_{r,j} \in A_r^p,\ 1 \leq r \leq m,\ 0 \leq j \leq i_r^p,$$
$$a_{u,k} \in A_u^p,\ 1 \leq u \leq m,\ i_u^p < k \leq N \text{ (magnitude rqrmnt.)}$$

The goal of the current iteration is to find from this partition of $\{A_r^p, 1 \leq r \leq m\}$ the partitioning for $\{A_r^{p+1}, 1 \leq r \leq m\}$ satisfying both partition size and magnitude requirements.

2.3 Algorithm and Proof First, note the cardinalities of the two subsequences A_r^p and A_r^{p+1} are related as $\sum_{r=1}^{m} ||A_r^{p+1}|| = \left(2 \sum_{r=1}^{m} ||A_r^p||\right) - m$ for $p \geq 1$, since A_r^{p+1} is obtained by inserting an element between every consecutive pair of elements of A_r^p and from (1). Let $X_r^{p+1} = a_{r,0}, a_{r,N(p+1)}, a_{r,2N(p+1)}, \ldots, a_{r,i_r^p}$ be the sequence obtained by taking the leading partition of A_r^p (i.e., $\hat{A}_r^p$) and adding to it the elements halfway between every two consecutive elements (note $N(p+1) = N(p)/2$). All the elements in X_r are equally spaced apart by $N(p+1)$. Its cardinality is easily computed as follows:

$$\sum_{r=1}^{m} ||X_r^{p+1}|| = \left(2 \sum_{r=1}^{m} ||\hat{A}_r^p||\right) - m = \left(2f \sum_{r=1}^{m} ||A_r^p||\right) - m$$
$$= f\left(\sum_{r=1}^{m} ||A_r^{p+1}|| + m\right) - m = \left(f \sum_{r=1}^{m} ||\hat{A}_r^{p+1}||\right) - m(1-f).$$

At least, in cardinality, we are close to the partitioning we want. We are only off by $m(1-f)$ elements.

For each r, $1 \leq r \leq m$, consider the new element of A_r^{p+1} that is $N(p+1)$ away from (greater than) the boundary element of A_r^p, i.e., the new element $a_{r,i_r^p+N(p+1)}$. Some of these could be smaller than $a^p_{\max}$, the largest boundary element of the previous iteration, and may have to be included in the leading partition of $\{A_r^{p+1}, 1 \leq r \leq m\}$ to satisfy the magnitude requirement. For each r, $1 \leq r \leq m$, for which this is true (i.e., $a_{r,i_r^p+N(p+1)} \leq a^p_{\max}$), we add the elements to the sequence X_r^{p+1} to obtain the sequence Y_r^{p+1}. If the condition is false, we let $Y_r^{p+1} = X_r^{p+1}$. Let the total number of elements so added be s. If we let j_r^{p+1} denote the index of the largest element in Y_r^{p+1} , then in the first case, $j_r^{p+1} = i_r^p + N(p+1)$, otherwise $j_r^{p+1} = i_r^p$. Its value will be further refined to give i_r^{p+1} next.

Let us compute the cardinality of the expanded sequence:

$$\sum_{r=1}^{m} ||Y_r^{p+1}|| = \left(\sum_{r=1}^{m} ||X_r^{p+1}||\right) + s \qquad (3)$$

$$= \left(f \sum_{r=1}^{m} ||A_r^{p+1}||\right) - m(1-f) + s$$

It should be clear from the construction that every element of Y_r^{p+1} is no greater than $a^p_{\max}$, and that every element of A_r^{p+1} not included in Y_r^{p+1} is at least as large as $a^p_{\max}$. In other words Y_r^{p+1} is a leading partition of $\{A_r^{p+1}, 1 \leq r \leq m\}$ that satisfies the magnitude requirement. A formal proof follows:
Claim: If $a_{r,j} \in Y_r^{p+1}$ then $a_{r,j} \leq a^p_{\max}$
Proof: If $Y_r^{p+1} \neq X_r^{p+1}$ then $a_{r,j} \leq a_{r,i_r^p+N(p+1)} \leq a^p_{\max}$
If $Y_r^{p+1} = X_r^{p+1}$ then $a_{r,j} \leq a_{r,i_r^p} \leq a^p_{\max}$.

Claim: If $a_{r,j} \in A_r^{p+1}$ and $a_{r,j} \notin Y_r^{p+1}$ then $a_{r,j} \geq a^p_{\max}$.
Proof: If $Y_r^{p+1} \neq X_r^{p+1}$ then
$a_{r,j} \geq a_{r,i_r^p+N(p+1)+N(p+1)} = a_{r,i_r^p+N(p)}$. From (2) $a_{r,i_r^p+N(p)} \geq a^p_{\max}$ thus $a_{r,j} \geq a^p_{\max}$. If $Y_r^{p+1} = X_r^{p+1}$ then $a_{r,j} \geq a_{r,i_r^p+N(p+1)} > a^p_{\max}$ by definition of Y_r^{p+1}.

Thus $\{Y_r^{p+1}, 1 \leq r \leq m\}$ defines a leading partition of $\{A_r^{p+1}, 1 \leq r \leq m\}$ which satisfies the magnitude requirement. The identification of $\{Y_r^{p+1}, 1 \leq r \leq p\}$ takes $O(m)$ I/O's and $O(m)$ comparisons.

Next we check if the partition-size requirement is satisfied; if not we adjust the number of elements to satisfy the size requirement.

Case 1: $s = m(1-f)$
From (3) it follows that $\sum_{r=1}^{m} ||Y_r^{p+1}|| = f \sum_{r=1}^{m} ||A_r^{p+1}||$, and hence $\hat{A}_r^{p+1} = Y_r^{p+1}$ and the partition-size requirement is satisfied. In this case: $i_r^{p+1} = j_r^{p+1}, 1 \leq r \leq m$.

Case 2: $s > m(1-f)$
From (3), we know $\{Y_r^{p+1}, 1 \leq r \leq m\}$ is close to being the correct leading partition of $\{A_r^{p+1}, 1 \leq r \leq m\}$

except that it has $s - m(1-f)$ too many elements. While preserving the magnitude requirement, we can satisfy the partition-size requirement for $\{A_r^{p+1}, 1 \le r \le m\}$ if we remove the $s - m(1-f)$ largest[1] elements from $\{Y_r^{p+1}, 1 \le r \le m\}$. The elements remaining in $\{Y_r^{p+1}, 1 \le r \le m\}$ after the largest $s - m(1-f)$ elements are discarded is a leading partition of $\{A_r^{p+1}, 1 \le r \le m\}$ that satisfies both requirements. For every element removed from Y_r^{p+1}, j_r^{p+1} is decremented by $N(p+1)$. Hence, if α_r elements are removed from Y_r^{p+1}, then $i_r^{p+1} = j_r^{p+1} - \alpha_r N(p+1)$, $1 \le r \le m$. Since, $0 \le s < m$ and $0 < f < 1$, the above computation can be bounded by $O(m)$ I/O's and $O(m \log m)$ comparisons. The $s - m(1-f) + 1$'th largest element in $\{Y_r^{p+1}, 1 \le r \le m\}$ is also found and set to a_{max}^{p+1}, for use in the next iteration. Note: $\sum_{r=1}^{m} \|\hat{A}_r^{p+1}\| = \left(\sum_{r=1}^{m} \|Y_r^{p+1}\|\right) - (s - m(1-f))$
$= f \sum_{r=1}^{m} \|A_r^{p+1}\|$ from (3),
as needed for the partition size requirement. Also, since the largest $s - m(1-f)$ elements are moved, the magnitude requirement remains satisfied.

Case 3: $s < m(1-f)$
From (3), we know $\{Y_r^{p+1}, 1 \le r \le m\}$ is close to being the correct leading partition of $\{A_r^{p+1}, 1 \le r \le m\}$ except that it contains $m(1-f) - s$ too few elements. While preserving the magnitude requirement, we can satisfy the partition-size requirement for $\{A_r^{p+1}, 1 \le r \le m\}$ by moving the $m(1-f) - s$ smallest[2] elements from $\{A_r^{p+1} - Y_r^{p+1}, 1 \le r \le m\}$ into $\{Y_r^{p+1}, 1 \le r \le m\}$. For every element moved into Y_r^{p+1}, j_r^{p+1} is incremented by $N(p+1)$. Hence, if α_r elements are moved into Y_r^{p+1}, then $i_r^{p+1} = j_r^{p+1} + \alpha_r N(p+1)$, $1 \le r \le m$. As above, the computation can be bounded by $O(m)$ I/O's and $O(m \log m)$ comparisons. The largest of the $m(1-f) - s$ elements moved is found and set to a_{max}^{p+1}, for use in the next iteration.
Note: $\sum_{r=1}^{m} \|\hat{A}_r^{p+1}\| = \left(\sum_{r=1}^{m} \|Y_r^{p+1}\|\right) + m(1-f) - s$
$= f \sum_{r=1}^{m} \|A_r^{p+1}\|$ from (3),
thus satisfying the partition size requirement. Since the smallest elements are moved into Y_r^{p+1}, the magnitude requirement remains satisfied.

If the induction is then applied, described as per the algorithm, then after $k = \log_2 N$ iterations, a leading partition of $\{A_r, 1 \le r \le m\}$ that satisfies both requirements is obtained.

2.4 Basis Step Let $p = \max\left(0, \mathrm{ceil}\left(\log_2\left(\frac{1-f}{f}\right)\right)\right)$.
By brute force, examine all the elements in A_r^p, determine $\hat{A}_r^p$, and set the initial values of i_r^p. Note that in the basis step we are considering at least $\frac{m}{f}$ elements because $\sum_{r=1}^{m} \|A_r^p\| = \frac{mN}{(N/2^p)} + m = m2^p + m$
$\ge \frac{m}{f} - m + m = \frac{m}{f}$. There are exactly $m - \infty$'s in our m sequences equal in number to $f\frac{m}{f}$,
the number of elements in the leading partition of $\{A_r^p, 1 \le r \le m\}$. Hence $i_r^p \ge 0$ as required by (1).

2.5 Example Consider as an example the four sequences given below, each consisiting of 9 elements. We would like to find a partition containing the 50'th percentile of the elements contained in the four sorted sequences.

```
  A1    A2    A3    A4

|-inf  -inf  -inf  -inf |
-------------------------
   x     x     x     x

   x     x     x     x

   x     x     x     x

   x     x     x     x

   x     x     x     x

   x     x     x     x

   x     x     x     x
-------------------------
|+inf  +inf  +inf  +inf |
```

In the basis step we examine only the first and last element of each sequence, and by brute force find the 50'th percentile of the 8 elements. For each sequence, each pair of elements considered are 8 elements apart. Because (in the simplified algorithm) we require the first element of each sequence to be $-\infty$ and the last to be $+\infty$ the boundary marking the 50'th percentile of the elements considered is as depicted above.

1 A tie between elements in the same sequence is broken by picking the element with the larger index. A tie between elements in different sequences is broken by picking the element from the larger indexed sequence.

2 A tie between elements in the same sequence is broken by picking the element with the smaller index. A tie between elements in different sequences is broken by picking the element from the smaller indexed sequence.

In the next iteration below, we consider elements 4 apart from each sequence. This introduces 4 new elements 27, 13, 5 and 10 from sequences A_1 , A_2, A_3 and A_4, respectively. The two largest among these 4 elements are assigned to the upper partition that comprises the 'smaller' 50'th percentile of elements considered and the remaining two elements are assigned to the lower partition or the 'larger' 50'th percentile as follows.

```
      A1    A2    A3    A4

 |-inf  -inf  -inf  -inf |
 --------------          |
     x     x  |  x     x |
              |          |
     x     x  |  x     x |
              |          |
     x     x  |  x     x |
 -------------|          |
 |  27    13 ||  5    10 |
 |            |-----------
 |   x     x  |  x     x
 |            |
 |   x     x  |  x     x
 |            |
 |   x     x  |  x     x
 |             -----------
 |+inf  +inf  +inf  +inf |
```

We keep track of the largest element belonging to the upper partition, 10. In the next iteration we consider elements 2 apart from each sequence. These elements as well as the partitioning from the previous iteration is given below.

```
      A1    A2    A3    A4

 |-inf  -inf  -inf  -inf |
 --------------          |
     x     x  |  x     x |
              |          |
    20     2  |  3     8 |
              |          |
     x     x  |  x     x |
 -------------|          |
 |  27    13 ||  5    10 |
 |            |-----------
 |   x     x  |  x     x
 |            |
 |  29    15  |  6    12
 |            |
 |   x     x  |  x     x
 |             -----------
 |+inf  +inf  +inf  +inf |
```

Without examining, we conclude that the elements 3 and 8 of sequences A_3 and A_4 are no greater than 10 from the knowledge that A_3 and A_4 are sorted sequences. Similarly without examination, elements 29 and 15 of sequences A_1 and A_2 are known to be no less than 10. The only in-doubt elements are 20, 2, 6 and 12 belonging to the the four sequences. We examine each and pick those less than 10 to be included in the upper (smaller half) partition and the others to the lower partition. Since there are exactly two out of four in-doubt elements less than 10, we have the new partition as follows.

```
      A1    A2    A3    A4

 |-inf  -inf  -inf  -inf |
 --------                |
     x  |  x     x     x |
 -------|                |
 |  20 ||  2     3     8 |
 |      |-------         |
 |   x  |  x  |  x     x |
 |       -------|        |
 |  27    13 ||  5    10 |
 |           ||    -------
 |   x     x ||  x |   x
 |           ||    | -----
 |  29    15 ||  6 | |12 |
 |            |------ |  |
 |   x     x  |  x   | x |
 |            ---------  |
 |+inf  +inf  +inf  +inf |
```

Note that of the elements considered so far we have found the 50'th percentile. The largest element in the upper partition continues to be 10. In the next step we consider all elements.

```
      A1    A2    A3    A4

 |-inf  -inf  -inf  -inf |
 --------                |
     4  |  1     2     7 |
 -------|                |
 |  20 ||  2     3     8 |
 |      |-------         |
 |  21  |  9  |  4     9 |
 |       -------|        |
 |  27    13 ||  5    10 |
 |           ||    -------
 |  28    14 ||  6 |  11
 |           ||    | -----
 |  29    15 ||  6 | |12 |
 |            |------ |  |
 |  31    19  |  7   |21 |
 |            ---------  |
 |+inf  +inf  +inf  +inf |
```

Again we can argue that there are only 4 in-doubt elements, viz., 4, 9, 7 and 11 from sequences A_1, A_2, A_3

and A_4, respectively. Three of these elements are less than 10 and are included in the upper partition and the other to the lower by using the method of the previous iteration. Such a partitioning has the deficiency that there are 19 elements in the upper partition and only 17 in the lower partition as given next.

```
     A1    A2    A3    A4

|-inf  -inf  -inf  -inf |
|                       |
|  4     1     2     7  |
-------                 |
  20  |  2     3     8  |
      |                 |
  21  |  9     4     9  |
      --------          |
  27    13  |  5    10  |
            |    -------
  28    14  |  6  |  11
            |     |
  29    15  |  6  |  12
            |     |
  31    19  |  7  |  21
            ------
 +inf  +inf  +inf  +inf
```

The imbalance is easily adjusted by picking the largest element from the upper partition, 10, and moving it to the lower. This solution of the 50'th percentile finding problem is as follows.

```
     A1    A2    A3    A4

|-inf  -inf  -inf  -inf |
|                       |
|  4     1     2     7  |
-------                 |
  20  |  2     3     8  |
      |                 |
  21  |  9     4     9  |
      --------    -------
  27    13  |  5  |  10
            |     |
  28    14  |  6  |  11
            |     |
  29    15  |  6  |  12
            |     |
  31    19  |  7  |  21
            ------
 +inf  +inf  +inf  +inf
```

3 Unequal Sized Lists Algorithm From a practical point of view, it is important to remove all the restrictions placed in the previous section. To accomplish this, the problem is restated and we derive a generalized algorithm from the simpler algorithm. The generalized algorithm handles arbitrary sized runs (not necessarily equal in size), arbitrary fraction of data f and does not assume infinity padding. The basis step is of order $O(m)$.

Let $A_r = a_{r,1}, a_{r,2}, a_{r,3}, \ldots, a_{r,N_r}$, $1 \le r \le m$, $N_r = ||A_r||$, be m ascending sequences. Analogous to Section 2.1 the problem is to find i_r , $1 \le r \le m$ for $\hat{A}_r = a_{r,1}, a_{r,2}, a_{r,3}, \ldots, a_{r,i_r}$ ($i_r = 0$ implies $\hat{A}_r$ is a null sequence.) such that:

1. $\left| \sum_{r=1}^{m} ||\hat{A}_r|| - f \sum_{r=1}^{m} ||A_r|| \right| \le 1, 0 \le f \le 1$ (partition-size requirement), and
2. $\forall x, x \in \hat{A}_r, 1 \le r \le m$, $\forall y, y \in A_t , y \notin \hat{A}_t,$ $1 \le t \le m, x \le y$ (magnitude requirement).

Since we allow for any arbitrary fraction f, the partition-size requirement can be met up to a difference of 1. We allow being off in cardinality by (up to) 1 from the exact fraction. Let us denote $I = \{i_r, 1 \le r \le m\}$ as the partitioning function of $\{A_r, 1 \le r \le m\}$.

We built on the algorithm that we discussed earlier for the constrained problem on equal sized lists. We will avoid padding the sequences with $+\infty$ or $-\infty$ by selectively including more and more of the m sequences during the iterations. Before the end of the last iteration all of the sequences will be included. Generally, we postpone the inclusion of smaller sequences until later in the iteration. Specifically, the problem is solved by obtaining a series of solutions $I^p = \{i_r^p, 1 \le r \le m\}$, successively for $p = 0, 1, 2, \ldots, k_{\max} - 1$ given a basis step at $p = 0$. The number of iterations $k_{\max} = \max_{1 \le r \le m}(k_r)$ where $k_r = \text{floor}(\log_2(N_r)) + 1$ and[3] I^p is the partitioning function of a selected subsequence of selected sequences. At the end of the p 'th iteration all sequences that have at least $2^{k_{\max} - p - 1}$ elements would have been included. The refined partitioning function I^{p+1} is found from the partitioning function I^p in the p'th iteration using $O(m)$ I/O's and $O(m \log_2 m)$ comparisons. The required definitions, the algorithm and an example are given in (IYE89).

4 Time Complexity Analysis All logarithms in the following discussion are base 2. Let p denote the

[3] Floor($\log_2(x)$) where x is a positive integer can be calculated using only integer arithmetic

number of processors. To partition m sequences into two partitions the number of iterations required is $\log(N_{max})$, where N_{max} is the number of elements in the longest sequence since elements are added in power-of-2 increments each iteration and, by definition, the algorithm must iterate $\log(N_{max})$ times to determine a partition of all elements. Each iteration involves $O(m)$ I/Os and no more than $O(m \log m)$ comparisons. This can be seen as follows. Of all the elements that are under consideration for partitioning during an iteration, the element immediately following the current partition boundary in each sequence must be inspected to determine if it must be moved into the leading partition. This requires at most $m - 1$ comparisons and $m - 1$ I/Os. The $m \log m$ term arises for the number of comparisons in Case 2 and Case 3 where at most m largest or smallest elements must be determined such that the partition-size requirement is satisfied. Note, this involves only $O(m)$ I/Os. Thus the worst case order of the number of I/Os is $O(m \log N_{max})$ and the number of comparisons is $O(m \log m \log N_{max})$. In the best case, the number of I/Os and comparisons for an iteration are both $O(m)$ and occurs if the iteration of the algorithm executes Case 1, thereby eliminating the need for determining up to m smallest or largest elements in Case 2 and Case 3. For database applications the number of rows to be sorted is much larger than the number of processors, and the cost of partitioning is far outweighed by the cost of merging the runs as discussed next.

Once the partitioning is found, each of the p processors can independently merge the portions of the m lists in its partition. (The following analysis assumes that the I/Os can also be parallelized, i.e., the I/O subsystem can support p concurrent accesses to m lists.) This can be done in a single pass using an m way merge using $O((N \log m)/p)$ comparisons per processor and $O(N/p)$ I/Os per processor, where N is the total number of elements to be merged. It is often the case in database applications that the number of I/Os primarily determines the execution time of a query. A sequential algorithm would require $O(N \log m)$ comparisons and $O(N)$ I/O's for the merge. For cases of practical interest for database systems the number of processors $p << N$ the number of rows to be sorted, and the number of runs to be partitioned $m = p$, since the processors can sort independently without interference until p runs are obtained (VAL84). We assume $m = p$ in the following discussion. The cost of partitioning itself is negligible; what matters is the merge cost.

There have previously been two general types of approaches to solving this partitioning problem. The first is to partition pairs of sequences exactly. This leads to workload balance and minimal contention for each 2-way merge but requires $\log(p)$ merge passes to complete. While any merging algorithm (AKL87, FRA88, VAR88, DEO88) which requires parallel 2-way merges can be done with $O((N \log m)/p)$ comparisons per processor, the number of I/Os per processor is $O((N \log p)/p)$, once again assuming that the I/Os can be parallelized, and the cost of partitioning for each iteration is negligible. I/O costs (these also represent the insertion and fetch costs into temporary tables that stores runs) need also to be minimized in a database system. The second partitioning method is to use a sampling method that divides the sequences based on a sort key (QUI88, see section 1). These methods perform a p-way merge of the partitions leading to a theoretical $O((N \log p)/p)$ for the expected number of comparisons per processor and $O(N/p)$ for number or I/Os per processors. However, since the original partitioning was only approximate, any skew in key values will lead to workload imbalance and a degradation in performance. In fact, Quinn (QUI88, see section 1) indicates that even with randomly distributed data, no speedup is obtained beyond $p = 10$ because of this workload imbalance.

In contrast, our algorithm will find an arbitrary exact partition for an arbitrary number of sequences. This leads to a nearly perfect workload balance and permits a single p-way merge phase. Thus the order of the merge is $O((N \log p)/p)$ for comparisons per processor and $O(N/p)$ for I/Os per processor, and solves the load balancing problem in parallel multi way merge. It is currently the best approach of which we are aware that will provide linear speedup in merge with respect to number of processors independent of key value distributions. Currently, implementations are underway to obtain empirical performance figures, and to optimize performance of the overall parallel sort by determining the optimal number of runs to merge in parallel given a fixed number of processors, I/O subsystem, and file organization.

5 Conclusions and Implementation Notes In this paper we examined the problems involved in employing multiple processors for parallel execution of a frequently used relational database function: external sorting. Previous solutions found in the database literature either rely on two way merges that makes the cost of merging expensive or use a final sequential merge that adversely impacts the speedup. We pick the best among three previously implemented sort algorithms for a shared memory parallel computer. Speedup due to this algorithm was shown to peak at about 10 processors. It failed to achieve efficiency and linear speedup beyond 10 processors because of approximations in the algorithm used to partition multiple sorted runs into equal percentile ranges. The approximations caused load imbalances among the multiple processors. We have given an algorithm that can solve the percentile finding problem exactly that works for an arbitrary number of runs and arbitrary percentiles eliminating the possibility of load imbalance. By employing our algorithm in parallel external sorting and eliminating load imbalances due to the previous approximations linear speed up is possible for the number of processors typically used for database related operations. We have also given an example to illustrate the working of our percentile finding algorithm. The percentile finding algorithm has been prototyped and tested successfully on various sets of sorted runs. A comparison between our algorithm and Quinn's parallel sorting method using random and skewed key values is necessary to determine the actual benefits provided by each approach.

From an implementation viewpoint we need to address the question of how to directly access an arbitrary *i*'th row of a run, as required by our percentile finding algorithm. The issue is for variable length rows. We can either use an index or estimate the location location based on an average row size. We suggest an alternate approach: to place load balancing requirements (partition size requirements) in terms of number of pages of rows merged by each processor, rather than in number or rows as per the exposition in this paper. The algorithm given in this paper can be extended to solve the run partitioning problem with such a load balancing requirement.

Acknowledgement We acknowledge help from Pat Selinger, Don Haderle, Dennis DeLorme, John Vriezen, John Palmer and Joel Wolf.

References

(AKL87) Akl, S. G., and Santroo, N., "Optimal Parallel Merging and Sorting Without Memory Conflicts" IEEE Trans. on Comp., C36, 11, 1987, 1367-69.

(BAT68) Batcher, K. E., "Sorting Networks and their Applications", Proc. AFIPS 1968 Spring Jt. Computer Conf., Vol 32, AFIPS Press, Arlington, VA.

(BEC88) Beck, M., Bitton, D. and Wilkinson, W. K., "Sorting Large Files on a Backend Multiprocessor", IEEE Transactions on Computers, 37, 7, July 1988.

(BIT83) Bitton, D., Boral, H., DeWitt, D. J., and Wilkinson, W. K., "Parallel Algorithms for Relational Database Operations", ACM Transactions on Database Systems, Vol. 8, No. 3, Sept. 1983

(DEO88) Deo, N. and Sarkar, D., "Optimal Parallel Algorithms for Merging and Sorting", Proc. Third Intl. Conf. on Supercomputing, Boston, May 15-20, 1988, 513-521.

(FRA88) Frances, R. S. and Mathieson, I. D., "A Benchmark Parallel Sort for Shared Memory Multiprocessors", IEEE Trans. on Computers, Vol C37, No. 12, Dec. 1988, pp. 1619-26.

(FRE82) Fredrickson, G. N. and Johnson, D. B., "The Complexity of Selection and Ranking in X + Y and Matrices with Sorted Columns:, J. Comp. and Sys. Sciences, 24, 197-208, 1982.

(GRA86) Gray, J., Stewart, M., Tsukerman, A., Uren, S. and Vaughen, B., "Fastsort: An external sort using Parallel Processing", Tandem Syst. Review, Vol. 2, Dec. 1986.

(IYE88) Iyer, B. R. and Dias, D. M., "System Issues in Parallel Sorting for Database Systems", IBM Research Report RJ 6585, San Jose, CA, November 30, 1988.

(IYE89) Iyer, B. R., Ricard, G. R. and Varman, P. J., "An Efficient Percentile Finding Algorithm for Multiple Sorted Runs", IBM Research Report, San Jose, CA, 1989.

(LOR88) Lorie, R. A. and Young, H. C., "A Low Communication Sort Algorithm for a Parallel Database Machine", IBM Research Report RJ 6669, San Jose, CA, February, 1989 (also to appear in VLDB 89).

(QUI88) Quinn, M. J., "Parallel sorting algorithms for tightly coupled multi-processors", Parallel Computing, 6, 1988, pp. 349-367.

(KNU73) Knuth, D. E., "The Art of Computer Programming: Volume 3", Addison-Wesley, Reading, MA, 1973.

(OLK86) Olken, F., Rotem, D., "Simple Random Sampling from Relational Databases", Proc. of the 12'th Intl. Conf. on Very Large Data Bases, Kyoto, Japan, August 1986, pp. 160-169.

(VAL84) Valduriez, P., Gardarin, G., "Join and Semijoin Algorithms for a Multiprocessor Database Machine", ACM Trans. on Data Base Systems, Vol. 9, No. 1, March 1984, pp. 133-161.

(VAR88) Varman, P. J., Iyer, B. R. and Haderle, D. J., "Parallel Merge on an arbitrary number of Processors", IBM Internal Report, August 1988 (available as IBM Research Report RJ 6632, San Jose, CA, December 30, 1988).

(VIT85) Vitter, J. F, "Random Sampling with a Reservoir", ACM Trans. on Math. Software, Vol. 11, No. 1, March 1985, pp. 37-57.

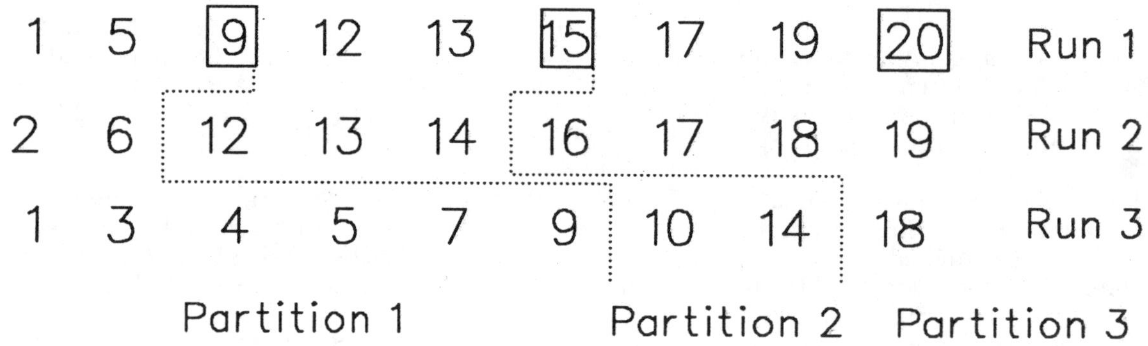

Figure 1

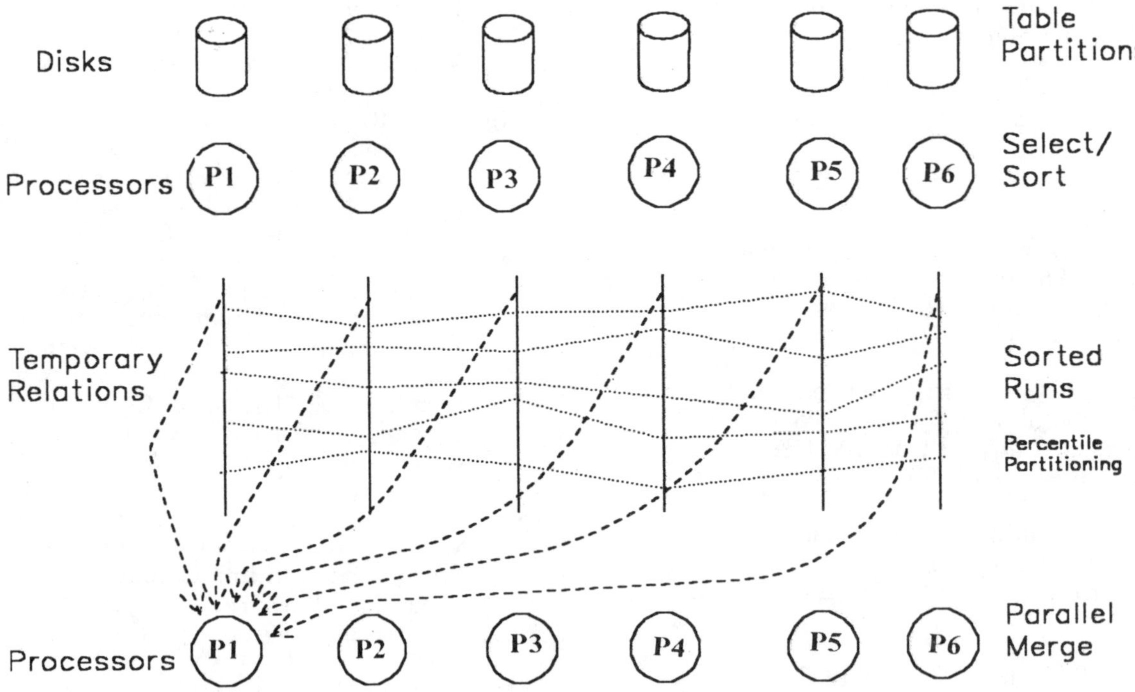

Figure 2

A Signature Access Method for the Starburst Database System

Walter W. Chang
Hans J. Schek[1]

IBM Almaden Research Center

Abstract

This paper describes a new signature generation method for constructing multi-level signature files to support both relational queries which contain multiple conjunctive (AND) predicates and generic document text queries. We describe the major problems with traditional multi-level signature files and then describe how to build multi-level signature files using a new composite method of parent signature generation. Performance of this signature generation scheme improves as more key fields are provided in the query. A combinatorial error problem common to all multi-level signature structures is identified and addressed. We show how a signature access method can provide query support for a large number of fields in a relation for which no index exists and can dramatically reduce the number of relation tuples that must be accessed during a normal scan. If one or more fields of the relation contain long field data such as text, the same signature mechanism can also provide query support for text search predicates.

1. Introduction

Signatures have been used extensively for information retrieval in electronic office and document filing systems and to some extent in database systems (Sacks-Davis87, Pfaltz80, Roberts79). Signatures are bit-vectors formed by a hash encoding of data objects such as long fields of text or relation tuples (records). Signatures are useful in filtering large quantities of data when queries are performed. In relational database systems, queries are traditionally answered by executing a plan generated by a query optimizer. The plan may require performing relatively expensive relation scans over a potentially large number of tuples, scanning one or more B-tree indexes which have been created and maintained over the key fields of the query, or by using some combination of the two.

In this paper we propose a signature access method within a relational database system to support conjunctive multi-field queries. If fields in a relation are used to store text data as externally defined abstract data types (Wilms88) we show how text search predicates can be supported by the same signature mechanism. We show that signature files can be incorporated into the Starburst extensible database architecture (Lindsay87) by extending an existing B-tree access method.

In Section 2 we present an overview, a definition of terminology, a discussion of single level signature files and the motivation for multi-level signature file structures. In Section 3 we describe the two main problems with multi-level signatures. Our proposal to solve these problems is presented in Section 4. In Section 5 we describe how an implementation of a multi-level signature file was achieved in the Starburst extensible DBMS. In Section 6. we discuss relevant query optimizer issues. In Section 7 we present concluding remarks.

2. A Summary of Signature Techniques

2.1. Signature Description and Terminology

Signatures compactly encode information about an object. For example, bits in the signature may represent field substrings of length k called $k - tuples$ (Harrison71). Examples of other signature encoding techniques are discussed in (Roberts79, Pfaltz80, and Deppisch86).

Data objects and query values are encoded using the same signature algorithm. When the bits of the query signature completely cover the signature bits of the data objects, the data object is a candidate that may satisfy the query. These data objects are then accessed and examined. Access of candidate objects that fail an exact match test are called *false drops*. False drops are due to hash collisions in the object signatures. Matches are called *hits*. Ideally, a query signature will reject all data signatures where the original data objects do not satisfy the specified query.

Signatures should be significantly smaller than the actual data object (10-20%), easy to compute, and provide a high degree of selectivity or data filtration. Prior work has shown that, optimal selectivity usually occurs when 50% of the bits in the signature pattern are set to 1s (Roberts79, Severance76). If too few or too many signature bits are set, the query signature cannot reject a large number of data object signatures and consequently, many additional objects must be accessed and examined. Under some conditions when duplicates are frequent, (Sacks-Davis83) and (Roberts79) show that values less than 50% are desirable.

[1] Current address: ETH Swiss Federal Institute of Technology, Switzerland.

All signature generation algorithms fall into one of two categories: superimposed or disjoint. Signature generation methods dependent on partitions within the data object (such as fields in a tuple) are called disjoint coding. For each partition or field of the object, different signature functions may be used. Each of these functions may yield signatures of different lengths. If one signature function is applied over an entire data object without regard to any internal object partitioning, the signature scheme is called superimposed coding. Superimposed coding schemes cannot directly preserve the relative ordering of attributes within an object.

The types of queries we want to support will determine whether disjoint or superimposed coding should be used. A Partial Match Query (PMQ) is defined to be a query with one or more predicates on which an exact string match is desired. Predicates are always connected conjunctively (by a Boolean AND operator.) An example of PMQ is shown below.

```
PMQ:    SELECT * FROM T1
          WHERE
           (COMPANY  = 'IBM') AND
           (DIVISION = 'Research');
```

When the match constraints are relaxed and substring matches are allowed, the queries are called Partial Partial Match Queries (PPMQ.) An example of PPMQ is shown below. Data and query values are regarded as strings, i.e. domains are handled as if they were of type CHAR and VARCHAR(). The "LIKE" operator used with the '%' wildcard symbol designate a substring match operation.

```
PPMQ:   SELECT * FROM T1
          WHERE
           (TITLE LIKE '%Engineer') AND
           (EMPLOYEE LIKE '%Walter');
```

PPMQs will always occur when text is a field type consisting of sentences or words. In this paper, we do not address using signatures to solve queries where the predicates may be connected by the logical OR operator or where predicates may be preceded by a logical NOT operator. These functions are not efficiently supported by signatures, although some discussion has addressed these issues (see Dadam83.)

An important characteristic of both PMQ and PPMQ is the concept of query weight. The query weight strongly influences the selectivity of a query signature. Let s_Q designate the signature of the query expression Q. The query weight is defined as the Hamming Weight (HW) of s_Q, i.e. $QueryWeight(s_Q) = HW(signature(Q))$ or simply the number of bits set to 1 in the query signature. As more bits are set in the query signature, the signature becomes more selective in filtering data object signatures. This property is true regardless of whether one or more predicates are entered and independent of the type (PMQ or PPMQ). Thus, queries with higher weights will cause fewer data objects to be accessed and will have fewer false drops since all bits of the query signature are less likely to be covered by bits in the object signature.

2.2. Single Level Signature Files

The simplest structure for a signature file is a single-level organization. Tuple signatures and Tuple-IDs (TIDs) are stored in a file in fixed sized units of storage called *pages*. A signature is computed for a query expression and then compared sequentially against all tuple signatures. Candidate tuples are retrieved using the TID and then examined. Single-level signature organizations may be suitable for applications where the number of data objects is small (e.g., less than 5K.) An example of a signature file and sample query are shown in figure 1. In this example a single predicate of (EMPLOYEE LIKE 'Chang') would have a signature of 01000010 and would qualify the first and fourth tuples but would reject all other tuples shown. After accessing and inspecting the first and fourth tuples, only the first would match the query.

Early in our work we conducted a series of single-level signature file experiments using records from internal IBM corporate telephone directory files. The results are summarized in the table in figure 2. These results and those of (Deppisch86) show that signature selectivity is very high as a function of the query weight. For query weights greater than 3, over 98% of the objects were immediately rejected by the signature. Other experiments, all with similar behavior convinced us that signatures would be a promising, simple, and generic method to provide coverage over several or all fields of each database record.

In the table in figure 2, our test relation required 270 pages and with no pages in memory took 45 seconds to scan on an IBM RT/PC workstation. A regular index over the LASTNAME field required 58 pages while a one-level signature file required 61 pages. This data indicated that access time was reduced by a factor of 2 to over a factor of 10. Combined signature/relation page IOs can be reduced by a factor of 2.5 when compared to a pure relation scan which would require all tuples on all 270 relation pages to be accessed. While this is a major improvement, a normal index still gives better search performance over signatures. However, one index file is required for each different key-column, while one signature file provides coverage for all columns of the relation. This

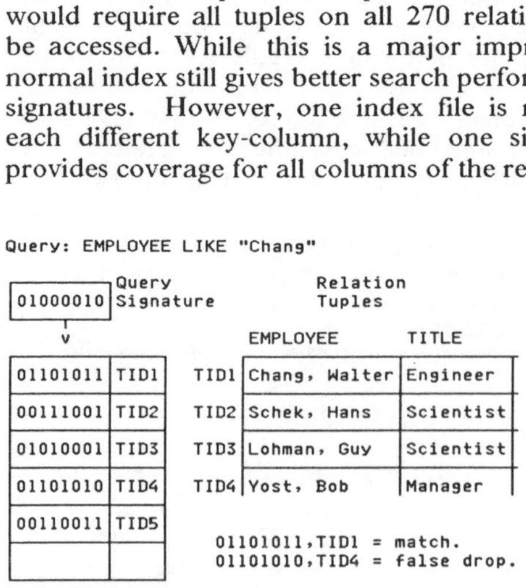

Figure 1: A Single Level Signature File.

has a significant impact on updates.

Relation contains 10000 tuples, Pages are 4KB.

Query Weight	Tuples Rejected by S1	Tuples Accessed by S1	False Drops	Matches	Time (Sec.)
1	7513.1	2486.8	2343.6	143.3	27.9
2	8051.9	1948.2	1503.7	444.5	21.4
3	9201.9	798.1	519.0	279.2	11.8
4	9860.8	139.2	136.4	2.8	4.8
5	9956.9	43.1	40.7	2.5	3.9
6	9988.6	11.4	10.3	1.1	3.4
7	9993.6	6.3	5.3	1.0	3.3
8	9998.4	1.6	0.6	1.0	3.0
9	9998.8	1.2	0.2	1.0	3.2
10	9999.0	1.0	0.0	1.0	3.1

Figure 2: Performance of A Single Level Signature File

To compare the search and update effort, we estimated the number of page accesses needed when a signature was used and when an index was used. We assumed each record contained F total fields and that a query specified f fields. For the signature access path, all 61 signature pages must be inspected during a query, independent of f. In case of the index, we take every value of the query and follow the path from the root to the leaf page where we fetch the tuple addresses (TIDs.) If each index contains l levels, we need $f \times l$ index page accesses.

Assume the number of different items to be put into an index is 10000 and the average number of different items within one record is 10. We utilize a B-tree for the items with 3 levels under these assumptions. Further we assume that the address list or TID-list of any item fits into one page. For many practical queries, $f=3$ or $f=4$ and therefore 9 to 12 index pages are accessed during a search as opposed to 61 in the signature case.

For insertion or deletion using the signature method one page access is needed. For indexes, we need at least as many (leaf) page accesses as we have values in the record, in our example $F=10$. If any of our index pages split, non-leaf pages must also be accessed. Thus, if the relation is fully indexed, each insert will require us to touch typically $F \times l$ (30 pages in our example). This discussion shows that signatures behave well in a dynamic environment and require substantially less update activity than full or even partial inversion schemes. We will show in the following discussion that multi-level signatures, if applied carefully, can meet this requirement.

2.3. Multi-Level Signature Organizations

In the previous example, if N = the number of tuples, a single-level signature file requires N comparisons between the query signature and the tuple signatures. The complexity is linear, and if N is large, we want to avoid the sequential signature scan, even if a single signature test is efficient.

One method of avoiding the linear scan is to cluster signatures together and create a group signature by superimposing or bit-ORing the individual signatures. A query signature is first compared to the group signature and if the query signature does not completely cover the group signature, the entire group can be discarded without further inspection. If the test with the group signature is positive, the query signature is tested against the individual signatures of the group. Shown below in figure 3 is a partial diagram of a multi-level signature structure.

Different multi-level signature organizations have already been investigated (Roberts79, Pfaltz80, Deppisch86, Sacks-Davis87). In this paper, B-tree terminology will be used. A "leaf signature" is stored at the lowest level, a "parent signature" is stored at higher levels.

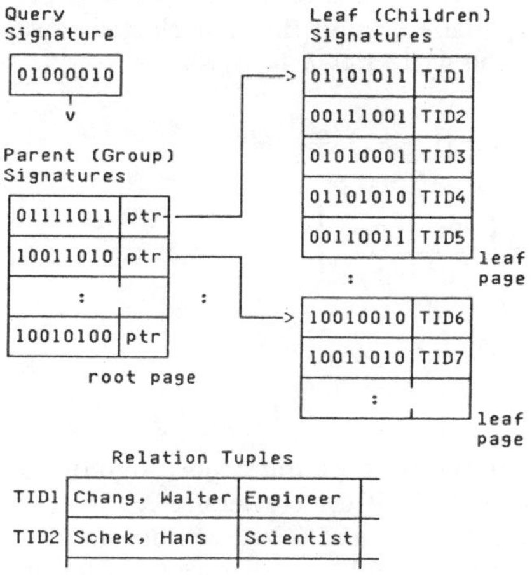

Figure 3: Multi-level signature file.

Each tuple signature and its TID is stored as a pair (SIG_i, TID_i) on leaf pages. The set of Signature-TID pairs on a leaf page forms a group. The parent signature of each group is denoted by s_P. Query signatures that reject the parent signature s_P reject all leaf signatures which belong in that group. In figure 3, since the bits of the query signature completely cover the bits of the first parent signature but not those of the second, the first leaf page would be searched but the second page would not. Bit-slice multi-level signature organizations exist but are not considered due to their significantly higher disk update costs (Sack-Davis87.) We now turn to the difficulties involved with multi-level signature organizations.

3. Problems with Multi-level Signatures

There are two major problems associated with multi-level signatures. One is the density of set bits in the parent signatures. The second problem is the rate of "combinatorial errors" caused by a large class of queries. These are key problems because both cause query signatures to unnecessarily qualify more object signatures thereby increasing the false drop rate. While the first problem has received some treatment, e.g., (Deppisch86), the second problem has received little attention until recently (Sacks-Davis87.)

Problem 1: Density of Set-1's in Parent Signatures

As more signatures are added into a group, the group signature saturates to an all 1's bit vector. For a signature s, the density of s is defined as $HW(s) \div m$ where m is the length of the signature in bits. This saturation effect unfortunately happens quickly for even a small number of signatures when leaf signatures have the "optimal" density of 0.5. Let N_L be the number of signatures/leaf page, α_L be the probability that the i'th bit of a leaf signature bit is 1, and α_P be the probability that the i'th bit of a parent signature bit is 1. The probability that the i'th bit in the parent is set to 1 is obtained by computing first the probability that the i'th parent bit is 0. For this to be true, each of the i'th bits of all N_L signatures in the leaf cluster must be 0. This yields the well-known formula:

(1) $$\alpha_P = 1 - (1 - \alpha_L)^{N_L} \text{, also}$$

(2) $$N_L = \left\lceil \frac{\ln(1 - \alpha_P)}{\ln(1 - \alpha_L)} \right\rceil$$

(3) $$\alpha_L = 1 - e^{\frac{1}{N_L}\ln(1 - \alpha_P)}$$

From equation (1), for optimal object signatures with $\alpha_L = 0.5$, α_P = 0.938 if the clustersize N_L is only 4. Experiments using 4-byte and 6-byte signatures confirmed these expectations. More advanced clustering techniques have been proposed and tested successfully by (Deppisch86) using S-Trees. Another solution to solve this problem is presented in (Sacks-Davis83) and uses large "segment" or block descriptors which are stored in a bit-slice organization. However, even if the parent level saturation problem is addressed, a second problem still remains.

Problem 2: The Combinatorial Error

The second serious problem is the combinatorial error. It appears in two seemingly different classes of signature applications: (1) conjunctive multi-predicate database queries consisting of PMQ and PPMQ, and (2) text search queries.

We consider the conjunctive multi-predicate case first. Assume we have N records with F fields each. Assume that we have already computed good selective signatures s_i for each of the N individual records. Such signatures either are obtained by a concatenation of the individual field signatures s_{ij} as in formula (4), or by superimposing the individual signature patterns for the $j'th$ field values of a record, as in formula (5).

(4) $$s_i = s_{i1} \mid s_{i2} \mid \ldots \mid s_{iF}$$

(5) $$s_i = s_{i1} \cup s_{i2} \cup \ldots \cup s_{iF}$$

For the following discussion, we may even assume that the s_i are perfect, i.e., there will be no false drops for whatever PMQ is posed. In reality, this can be achieved by the use of long signatures. We will now consider the group signature s_P formed by superimposing all single signatures s_i. Thus,

$$s_P = s_1 \cup s_2 \cup \ldots \cup s_{N_L}$$

Due to superimposing, any s_P signature represents not only the given N records but also all records which may be obtained by any Cartesian Product of field value combinations from the given field values of the N records. This causes our second problem. As an example, consider the simplest case of two records which contain only two fields. Here R_i designates a record with two fields and V_{jk} designates the $k'th$ value from the $j'th$ field.

$$R_1 = (V_{11}, V_{21}) \text{ ,} \qquad R_2 = (V_{12}, V_{22})$$

The group signature here would represent not only R_1 and R_2, but also records with combinations of these fields:

$$R_3 = (V_{11}, V_{22}) \text{ ,} \qquad R_4 = (V_{12}, V_{21})$$

With F fields, the group signature represents $(\mu_1 \times \mu_2 \times \ldots \times \mu_F)$ records where μ_j is the number of distinct V_{jk}'s or cardinality of the $j'th$ field. This product has a drastic influence on the "combinatorial error" which occurs when a query requests a record by specifying f fields. In the simple example above, by specifying two fields, four queries are possible, but only two of these queries will find matches. The other two queries will be satisfied by the group signature s_P but will be false drops since the corresponding tuples are not in the database. More generally, when field values are unique within the N records, the probability P_{Match} for finding a matching tuple given a query in which all F fields are constrained by a predicate, where the $j'th$ predicate consists of any value in field j, is given by:

(6) $$P_{Match} = \frac{N}{\prod_{j=1}^{F} \mu_j}$$

Therefore, the risk P_{Cerror} of having a combinatorial error is given by the following:

(7) $$P_{Cerror} = 1 - P_{Match}$$

This gives the probability for a combinatorial error when each of the f field values $V_1, V_2, \ldots, V_f$ from the query occur individually in at least one of the N records. More precisely, we assume that for each predicate for which the query values are V_{jk}, $k = 1, 2, \ldots, \mu_j$, there exists a record in the group which has value V_{jk} in the $j'th$ field.

Let $q_1...q_f$ represent the list of f fields for which predicates are specified in the query. If μ_{q_j} is the cardinality of a field in the query and is less than N, duplicate values will be present in this field. The total number of distinct records, considering only the $q_1...q_f$ fields, is represented by $N_{q_1...q_f}$. This value is a function of the cardinality and distribution of V_{q_j}'s. By generalizing (6) and (7), the match and combinatorial error probability for a specific query with f predicates is now given by:

$$(8) \qquad P_{Match} = \frac{N_{q_1...q_f}}{\prod_{j=1}^{f} \mu_{q_i}}$$

$$(9) \qquad P_{Cerror} = 1 - P_{Match}$$

Since the query contains values from the $N_{q_1...q_f}$ records, the query signature will necessarily cover the group signature. Thus, formula (9) gives an estimate for the probability that the group does not contain a matching record due to the combinatorial error. From our experiments, even for small values of N and f, this error is considerable. Consider a simple example in which field values are unique, i.e., $\mu_j = N$. Equations (8) and (9) reduce to the two special cases:

$$P_{Match} = \frac{1}{N^{f-1}}, \qquad P_{Cerror} = 1 - \frac{1}{N^{f-1}}$$

Suppose, for example, that $N=10$, $f=2$, and let $\mu_1 = \mu_2 = N$. Since there are no duplicates, $N_{q_1...q_f}$ is the total number of distinct records when two fields are considered and, is just N. For a query Q, the probability that the group of 10 records does not contain a match is already 0.9. As f increases, the combinatorial error also increases. If $f=3$ and $\mu_3 = N$, the probability that we do not have a match is already 0.99.

We now turn to the case of text. Substantial work has been done in order to find good signatures for text based on statistics and information theory (Schek78, Christodoulakis84, Faloutsos85a,85b,87). Since text is a set or sequence of words and a query is also a set or sequence of words, there is still a considerable combinatorial error problem. Consider the following text consisting of two phrases as an example:

```
t = <<database index extension>,
     <operating system kernel>>
```

Assume that a signature s has been assigned to t and that a query asks for all documents with the phrase < database kernal >. The signatures would indicate a match even though the query did not match the actual text. Text signatures built on a word basis represent not only the original text phrase. They also represent every phrase which could be formed by any combination of the words in the original phrase. Thus, we have to deal with the same combinatorial problem.

The probability for this error increases as more phrases in the text are clustered to compute s and as more words are specified in the query. Formula (9), shows this to be a problem with group signatures with only a few phrases and a two- or three-word query.

While various aspects of the 1's bit saturation problem have been addressed in the literature, the combinatorial error problem remains. (Schek78) has examined estimating frequent pairs or triples of words and more recently, (Sacks-Davis87) has proposed multilevel signatures using larger parent signatures and setting bits based on common word-pairs to address one aspect of this combinatorial problem.

4. A Solution to the Combinatorial Problem

In this section, we describe a general approach to building second level signatures that exhibit a significantly reduced false drop sensitivity due to the combinatorial error.

The Combinatorial Signature

We propose a new signature function that encodes combinations of values within a record rather than single values from a record. This is equivalent to computing a new signature based on a combination of multiple bits in the original object signature. First, a leaf signature S1 with m_1 bits is computed for each record. Techniques for finding the length and hash function for S1 are well-known (see Faloutsos87).

In our new technique, for each S1 leaf signature a larger *combinatorial* signature CS1 of m_{CS1} bits is computed. CS1 is formed by testing a complete set or subset of all r pairs of bit positions in S1 and setting one of the m_{CS1} bit positions in CS1 to 1 if 1's appear at both bit positions indicated in S1. This could be extended to bit triplets, quadruplets, etc. For now, we consider only bit-tuples consisting of bit pairs.

To form a group (parent) signature for a set of N_L leaf signatures, all of the CS1's are superimposed (bit-OR'ed). Assume signatures are computed for records R_1 and R_2 and that each field value V_{jk} sets one S1 signature bit when it is hashed. When a combinatorial signature CS1 is computed by selecting pairs of bits in S1, each CS1 bit indicates the presence of a combination of record attributes. If a query is posed which specifies combinations of values (such as (V_{11}, V_{22}) or (V_{12}, V_{21})), a conventional S2 signature formed by bit-ORing leaf-level S1 signatures will not reject S1 signature groups. This query is said to contain a *combinatorial error*. However, an CS2 signature formed by bit-ORing CS1 signatures will indicate the presence of combinations of values and have a high probability of rejecting lower level leaf signature groups. An example of how the new combinatorial signature CS1 works is shown in figure 4.

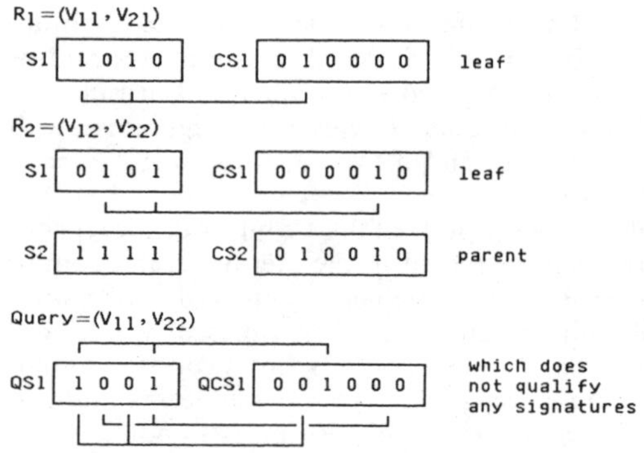

Figure 4: Rejection of query with combinatorial error

When a new object signature is incrementally added into the group covered by an CS2 signature, we compute the new object's CS1 signature from its S1 signature and then bit-OR the CS1 signature into the parent CS2.

The basic strategy for generating the combinatorial signature CS1 is to encode a full or partial subset of all combinations of S1 bit pairs. To compute CS1, we first identify a set of unique bit position pairs in S1. Each of these S1 bit pairs is bit-ANDed and each result is mapped to a specific bit position in CS1. When this is done systematically, the m_1 bits of S1 will form $m_1 - 1$ bit partitions in CS1 of size $(m_1 - 1)$, $(m_1 - 2)$,..., 1. For example: Bit 1 of CS1 is set by bit-ANDing bits 1 and 2 of S1. Bit 2 of CS1 is set by bit-ANDing bits 1 and 3 of S1. This is shown below in Figure 5.

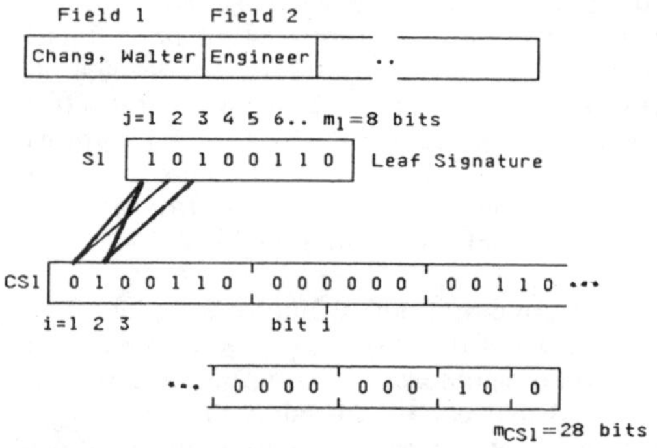

Figure 5: Computation of the combinatorial signature.

Using this method, we can determine the value of m_{CS1} for the required length of CS1. CS1 is partitioned into a set of bit fields of length $(m_1 - 1)$, $(m_1 - 2)$, ..., $(m_1 - (m_1 - 1))$. To find the total number of bits required for CS1, we sum the number of bits of all of the partitions:

(10) $$m_{CS1} = ((m_1 - 1)m_1) - \sum_{i=1}^{m_1 - 1} i, \quad \text{which gives:}$$

(11) $$m_{CS1} = m_1 \frac{(m_1 - 1)}{2}$$

Beyond a combinatorial signature, we can also use a second parent signature LS2 which is obtained by longer leaf signatures LS1 with m_{LS1} bits using other known methods (Sacks-Davis87). These longer LS1 signatures and the new combinatorial CS1 signature associated with S1 do not have to be stored in the leaf levels of the signature file since they are used only to build higher level signatures. In this sense, LS1 and CS1 signatures are "virtual".

Whether bit position pairs in S1 are selected at random or in some systematic way, for each CS1 bit, there is a small probability that selected S1 bit positions represent k-tuples from the same field. If this is the case, our new CS1 signature will not provide protection against combinatorial errors. Fortunately, this probability is low. If each record consists of F fields and we assume the signature encoding of the i'th field sets b_i signature bits in S1, the probability that a bit in CS1 was set by u S1 bits from the same record field i is given by:

$$P_{Same} = \left(\frac{b_i}{m_1} \right)^u$$

where $u = 2$ for bit pairs. For example, if a record consists of $F = 8$ fields and yields a signature S1 with $m_1 = 64$ total bits, and an average field requires $b_i = 8$ signature bits, $P_{same} = 0.016$. We expect this amount to be quite tolerable.

Thus, the key parameters in applying this technique are the selection of the CS1 generating algorithm, and an appropriate S1 signature. The selected density of S1 is crucial because it will determine the density in CS1 and the number of CS1 signatures which may be clustered to form CS2. We now address how signature parameters can be determined.

Determining Signature Densities

For each key to be stored in the index, we compute signatures S1, LS1, and CS1 from our data object, a record. S1 is a simple encoding using the algorithm given by (Harrison71) where we set k=3 and use k-tuples within fields of data objects. Leading and trailing blanks in the fields can be ignored. Our first goal will be to design signature functions such that parent signatures attain the desired density when they are formed by superimposing the N_L lower level signatures.

This can be achieved by using a general heuristic for determining the lengths, density, and hash functions for the signatures in our file. In Figure 6, only S1 signatures are stored in leaf pages, while longer LS2 and CS2 signatures are stored only at parent levels. When a group of virtual LS1 leaf signatures are superimposed, a parent signature LS2 of the same length is formed. CS2 is formed in a like way. The higher level signatures LS3 and CS3 are formed by superimposing groups of LS2 and CS2 signatures, respectively. We next discuss the LS1 virtual signature.

Determining Signature Lengths

To select the length m_{LS1} for a virtual leaf signature LS1, an estimate is made of the number of unique k-tuples contained in a record, since this determines the likely number of 1's set in the signature. Since a k-tuple is simply a substring sequence of k bytes, as k increases, the number of distinct k-tuples within the record decreases while for smaller k, k-tuples are more likely to be repeated. To simplify our design, we set k to 3 and assume a nearly uniform distribution of k-tuples within each record. For example, if the Harrison algorithm is used, an estimate is made for AVG_{kt}, the average number of unique k-tuples per record using equation (12). Equation (13) gives the leaf signature density α_{LS1}, given the target parent density α_{LS2} and leaf clustersize N_L derived from (3).

$$AVG_{kt} = \left\lceil \frac{Len(Record) - (k-d)}{d} \right\rceil \qquad (12)$$

$$m_{LS1} = \frac{AVG_{kt}}{\alpha_{LS1}} \qquad (13)$$

The determination of design parameters for signature CS1 is trickier. Since CS2 is formed by superimposing CS1's and CS1 is derived from S1, we work backwards. We first need the density α_{CS1} for CS1 given N_L and a selected parent target density of α_{CS2} (typically .5 or less.) Given the density that must occur in a parent signature CS2, α_{CS1} for a virtual leaf combinatorial signature CS1 can be directly determined by using equation (3). Since bits of CS1 are set by bit-ANDing pairs of bits in S1, the density α_1 for S1 can be derived from the product:

$$\alpha_{CS1} = \alpha_1^{\,2}. \qquad (14)$$

To determine the length m_1 of S1, we again use an estimate of the number of unique k-tuples present in each record and divide by α_1. By altering the k-tuple size and distance d between k-tuples, a high degree of flexibility is possible in deciding m_1. When $k = 2$ and $d = 1$, we have the simplest case of Harrison's algorithm. If k and d are made variable and k-tuples become equivalent to entire words, for large m_1 the signature generalizes into a Bloom filter. The final step is to determine the length m_{CS1} of signature CS1. This is done by using equation (11).

5. Incorporating Multi-Level Signatures in Starburst B-Trees

We now describe a generic organization for multi-level signature files using B-tree structures. Other hierarchical organizations are possible, but are not discussed here. (Korth82), (Prabhakar83), and others have attempted to integrate signatures into B-trees. We now generalize these efforts. In our approach, one or more signature fields are appended to leaf and parent key entries within the B-tree index. Signature entries in interior pages filter entire pages of lower level signatures while leaf signatures filter specific data objects during predicate evaluation.

For all non-leaf pages of a B-tree, page entries consist of ordered <Dividing-key, Pointer> pairs used to determine which child page must be accessed next. To incorporate signatures, for each non-leaf entry we also store one or more parent signatures which represent the aggregation of all lower level signatures in the child pages. We selected two different signature functions for each parent entry and designate these by LS2 and CS2.

The motivation for multiple signatures at non-leaf levels is as follows: If different group signatures are used, there is a lower chance that a group containing all false drops will be qualified by all the signatures generated from a query. The signature functions must be sufficiently different in the way object attributes are captured. Examples of these signature functions are those for which the k-tuple size is increased or the distance d between k-tuples is increased (Harrison71).

In figure 6 we show a partially constructed signature file with the constituent signature elements. The functions key(R_i), tid(R_i) indicate extraction of the key fields and TID of data record R_i. S1 indicates a short signature for each data record, LS1 indicates the longer "virtual" signature of the same object, and CS1 is the combinatorial signature derived from S1. ParKey indicates a regular B-Tree Parent Key used to navigate to the leaf pages.

At the leaf pages of the B-tree, entries now consist of the record key fields, the short S1 signatures, and TID. For each table record R_i, an LS1 and CS1 is computed and used only to form the appropriate LS2's and CS2's at the parent levels. LS1 and CS1 signatures are shown on the right of the leaf pages in figure 6 and are used to compute parent signatures by a bit-ORing operation.

Higher Signature Levels

To form higher signatures, we can repeat the idea of superimposing lower level parent signatures until the root level is reached. One method is to simply Bit-OR lower level parent signatures to form higher level parent signatures. By using properly designed leaf, "virtual", and combinatorial signatures, saturation can be delayed significantly as we move to higher levels of the index. Alternative hashing methods are possible by varying the Harrison parameters (k, m, d) or by using a class of universal hashing functions at the different levels of the index (Carter80).

Many of the concurrency problems and recovery issues which apply to index managers will also apply to our new B-tree signature files as well. The basic multi-record locking algorithms developed for the Starburst index component are extended for signature indexes, and are similar to the algorithms described in (Bayer77, Mohan89). Due to space constraints, we defer presenting the implementation details of the al-

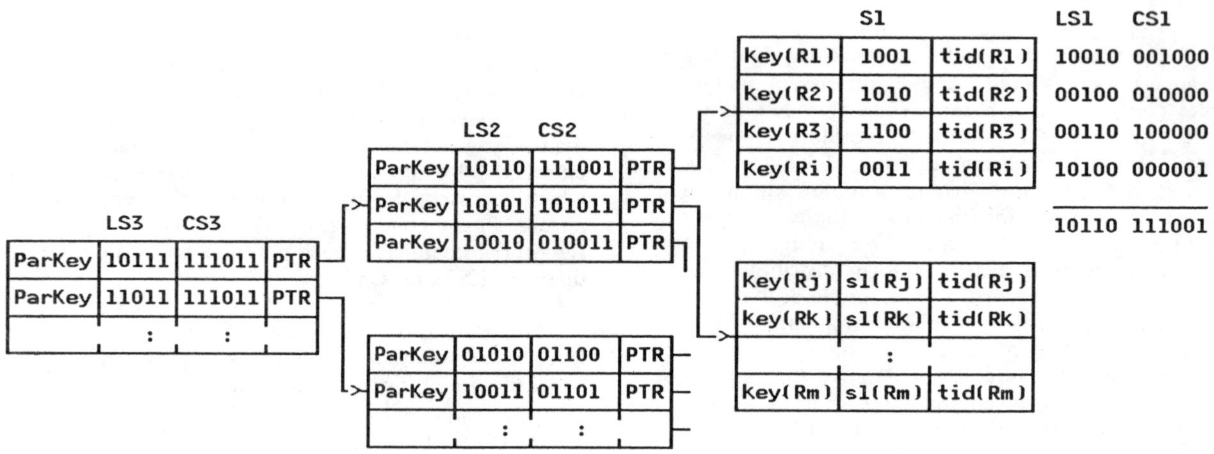

Figure 6: Multi-Level Signature Index incorporating S1, LS1, and CS1

gorithms for signature insert, delete, fetch, and scan operations as well as index locking and recovery to a future research report. We now address how the query optimizer would use a signature access method.

6. Starburst Query Optimizer Issues

For the query optimizer to decide whether or not to use an existing signature, it must know which relation columns are covered by the signature and then estimate the cardinality of field values, selectivity of a given query, and cost of using the signature file. To the optimizer, the signature access method or *attachment* has the same functional interface as other attachments in Starburst for tuple insert, delete, fetch, and scan (see Stonebraker80 and Lindsay87). Thus, the signature attachment will appear to the Starburst rule-based query optimizer as a scan LOLEPOP (Low-Level Plan Operator), described in (Haas88 and Lohman88).

Since the signature attachment will return an answer superset rather than a precise tuple set like a conventional index, a FILTER operator must be applied to eliminate tuples which may be false drops. The estimated cost of using a signature is based upon column cardinality, which is estimated by using a model of the probability any tuple will be accessed (selectivity), and a probabilistic model of page accesses.

The signature attachment could also be generalized for use in plans to solve text search predicates. If Harrison's k-tuples are set to be entire keywords in a document or a long field (Lehman89), than each $k-tuple$ represents a text keyword and precisely one bit per keyword will be set in signatures S1 and LS1. A signature index over text stored in a long field would be classified as an *exogenous* database attachment (Schwarz86) since the index may be bound to a long field, rather than a base table.

Our experimental results show that for conjunctive multi-predicate queries where f is large, signatures compete well with indexes. For queries with weights corresponding to two or more predicates (i.e., when the query weight exceeded 9), measurements showed that over 97% of the relation tuples were filtered and less than 20% of the signature file was accessed. Compared to relation scans, total page accesses (measured by page fixes) were reduced by over 92%. These preliminary tests indicate that signature files are a promising method for greatly accelerating relation scan performance for queries when normal indexes are not available.

7. Conclusions

We have identified an undesirable, naturally occurring false drop phenomena inherent to all multi-level signature structures which we have called the combinatorial error. We have presented a new combinatorial signature generation method which significantly reduces the negative effect of the combinatorial error. We have shown how to generically build multi-level signature structures from single level signature files by extending generic B-tree indexes such as the Starburst index component. Signature files can serve as a powerful extension access method to support not only database queries usually handled by relation or index scans, but also text applications.

Acknowledgments

We would like to express our thanks to the members of the Starburst project who provided us with many comments and suggestions during our work. We would also like to thank Uwe Deppisch at the Technical University of Darmstadt for his S-Tree experiments using our test data. Finally, we would like to thank Laura Haas, Guy Lohman, Pat Selinger, Bill Cody, and Bob Yost for helping to review this paper.

REFERENCES

[Bayer77] R. Bayer, M. Schkolnick, *Concurrency of Operations on B-Trees,* Acta Informatica 9, 1-21, (1977) (Springer-Verlag, 1977).

[Carter80] L. J. Carter, M. N. Wegman, *Universal Classes of Hash Functions,* IBM Research Report RC 6687 (#28796), IBM Thomas J. Watson Research Center (1977).

[Christodoulakis84] S. Christodoulakis, C. Faloutsos, *Design Considerations for a Message File Server,* IEEE Transact. on Software Engineering, Vol. SE-10, No. 2. (March 1984).

[Dadam83] P. Dadam, P. Pistor, H. J. Schek,. *A Predicate Oriented Locking Approach For Integrated Information Systems,* Information Processing 83: Proceedings of the IFIP 9th World Computer Congress (Paris, France, September 1983).

[Deppisch86] U. Deppisch, *S-Tree: A Dynamic Balanced Signature Index for Office Retrieval,* Proceedings of the 1986 ACM Conference "Research and Development in Informational Retrieval" (Pisa, Italy, September 1986).

[Faloutsos85a] C. Faloutsos, *Signature Files: Design and Performance Comparison of some Signature Extraction Methods,* Proceedings of SIGMOD, pp. 63-82. (1985).

[Faloutsos85b] C. Faloutsos, *Design of a Signature File Method that Accounts for Non-Uniform Occurence and Query Frequencies,* Proceedings of VLDB, pp. 165-170. (1985).

[Faloutsos87] C. Faloutsos, S. Christodoulakis, *Optimal Signature Extraction and Information Loss,* ACM Transactions on Database Systems, Vol.12, No.3 Sept. 87. (1987).

[Haas88] L. M. Haas, W. F. Cody, S. Finkelstein, J. C. Freytag, G. Lapis, B. Lindsay, G. Lohman, K. Ono, H. Pirahesh, *An Extensible Processor for an Extended Relational Query Language,* IBM Research Report RJ 6182, IBM Almaden Research Center, San Jose, CA. (1988).

[Harrison71] M. C. Harrison, *Implementation of the Substring Test by Hashing,* Communications of the ACM Vol 14, No. 21. (December 1971).

[Korth82] R. P. King, H. F. Korth, B. E. Willner, *Design of a Document Filing and Retrieval Service,* IBM Research Report RC 9696 (#42815) 11-18-82, Thomas J. Watson Research Center (1982).

[Lehman89] T. Lehman, B. Lindsay, *The Starburst Long-Field Manager,* 15th Proceedings of VLDB, 1989.

[Lindsay 87] B. Lindsay, J. McPherson, H. Pirahesh, *A Data Management Extension Architecture,* Proceedings of Association for Computing Machinery Special Interest Group on Management of Data, 1987 Annual Conference, May 27-29 (1987).

[Lohman88] G. Lohman, *Grammar-like Functional Rules for Representing Query Optimization Alternatives,* Proceedings of SIGMOD 1988.

[Mohan89] C. Mohan, F. Levine, *Aries-IX: An Efficient and High Concurrency Index Management Method Using Write-Ahead Logging,* IBM Research Report (In preparation), IBM Almaden Research Center (1989).

[Pfaltz80] J. L. Pfaltz, W. J. Berman, E. M. Cagley, *Partial-Match Retrieval Using Indexed Descriptor Files,* Communications of the ACM Vol. 23, No. 9 (September 1980).

[Prabhakar83] T. V. Prabhakar, H. V. Sahasrabuddhe, *Signature Trees - A Data Structure for Index Organization,* International Conference on Systems, Man, and Cybernetics Proceedings 1983 vol. 2 (1983).

[Roberts79] C. S. Roberts, *Partial-Match Retrieval via the Method of Superimposed Codes,* Proceedings of the IEEE Vol. 67, No. 12 (December 1979).

[Sacks-Davis83] R. Sacks-Davis, K. Ramamohanarao, *A Two Level Superimposed Coding Scheme For Partial Match Retrieval,* Information Systems Vol. 8, No. 4 (1983).

[Sacks-Davis87] R. Sacks-Davis, A. Kent, K. Ramamohanarao, *Multikey Access Methods Based on Superimposed Coding Techniques,* ACM Transactions on Database Systems Vol. 12, No. 4, December 1987 (1987).

[Schek78] H. J. Schek, *The Reference String Access Method and Partial Match Retrieval,* Proc. Information System Methodology 1978, Vol. 65., G. Bracchi, P. C. Lockemann, also see: IBM Scientific Center Report TR 77.12.008 (1978).

[Schwarz86] P. Schwarz, W. Chang, et. al., *Extensibility in the Starburst Database System,* IBM Research Report RJ 5311 (#54671) 09-23-86 (1986).

[Severance76] Severance, D. G., Lohman, G. M., *Differential Files: Their Application to the Maintenance of Large Databases,* ACM Transactions in Database Systems Vol. 1, No. 3 (September 1976)

[Stonebraker80] Stonebraker, M, *Retrospection on a Database System,* ACM Transactions in Database Systems Vol. 5, No. 2 (June 1980)

[Wilms88] P. Wilms, P. Schwarz, H. Schek, L. Haas, *Incorporating Data Types in an Extensible Database Architecture,* 3rd International Conference on Data and Knowledge Bases, June 28-30, 1988. Jerusalem, Israel (1988).

COMMUTATIVITY AND ITS ROLE IN THE PROCESSING OF LINEAR RECURSION

Yannis E. Ioannidis †

Computer Sciences Department
University of Wisconsin
Madison, WI 53706

Abstract

We investigate the role of commutativity in query processing of linear recursion. We give a sufficient condition for two linear, function-free, constant-free, and range-restricted rules to commute. The condition depends on the form of the rules themselves. For a restricted class of rules, we show that the condition is necessary and sufficient and can be tested in polynomial time in the size of the rules. Using the algebraic structure of such rules, we study the relationship of commutativity with several other properties of linear recursive rules. We show that commutativity is in the center of several important special classes of linear recursion, i.e., separable recursion and recursion with recursively redundant predicates.

1. INTRODUCTION

Several general algorithms have been proposed for the processing of recursive programs in database systems (DBMSs). Recursive query processing is recognized as an expensive operation, and all the proposed algorithms incur some significant cost. Thus, the importance of identifying special cases of recursion on which specialized and more efficient algorithms are applicable is obvious. Such special cases of recursion include bounded recursion (uniform and otherwise), transitive closure, separable recursion, and one-sided recursion. In this paper, we elaborate on another special case of recursion, where participating operators (or rules) commute with each other. When this happens, recursive queries can be decomposed into smaller queries, which are expected to have a lower total execution cost than the original query.

† Supported by the National Science Foundation under Grant IRI-8703592.

Proceedings of the Fifteenth International Conference on Very Large Data Bases

Commutativity has already been identified as a significant special case of recursion [Ioan88a]. It has been shown how several types of queries are affected by the applicability of commutativity and how the general algorithms for recursive queries are affected. Constants can also be used in conjunction with commutativity to reduce the amount of data the system has to look at to answer a query with selections. This earlier work on commutativity was done within the algebraic framework of linear recursive operators (rules) [Ioan86a, Ioan88a]. In this paper, we use the first order logic representation of rules to give necessary and sufficient conditions for two linear recursive rules of a restricted form to commute with each other. These conditions are based on the form of the rules themselves and make no direct use of the definition of commutativity, which requires composing the two rules in both ways and examining the two composites for equivalence. We also use the algebraic formulation of recursion to compare commutativity with other special classes of recursion, in particular, separable recursion and recursion with recursively redundant predicates, and discuss the effects of commutativity on the algorithms proposed for them.

The paper is organized as follows. Section 1 is an introduction. Section 2 is a summary of the algebraic model for linear recursion, which has been introduced elsewhere [Ioan86a, Ioan88a]. In Section 3, we define commutativity in the algebraic model and we compare the notion of commutativity with separability and recursive redundancy. Section 4 manipulates rules in the logic model and gives a decision algorithm for commutativity for a restricted class of rules. In Section 5, separability and recursive redundancy are reexamined for the restricted class of rules studied in Section 4. Finally, Section 6 presents our conclusions and gives some directions for future work.

2. ALGEBRAIC MODEL

In this section, we provide a summary of the algebraic model for linear recursion [Ioan86a, Ioan88a]. Consider a linear recursive rule

$$\mathbf{P}(\underline{x}^{(0)}) \wedge \mathbf{Q}_1(\underline{x}^{(1)}) \wedge \cdots \wedge \mathbf{Q}_k(\underline{x}^{(k)}) \rightarrow \mathbf{P}(\underline{x}^{(k+1)}), \quad (2.1)$$

where for each i, $\underline{x}^{(i)}$ is a vector of variables. No restriction is imposed on the form of the rule, or on the finiteness of the relations corresponding to the various predicates in the rule. Thus, for example, the rules can contain functions. Each one of $\mathbf{P}(\underline{x}^{(0)})$ and the $\mathbf{Q}_i(\underline{x}^{(i)})$'s is a *(positive) literal*. In relational terms, if $\{\mathbf{q}_i\}$ is a set of relations ($\mathbf{q}_i$

Amsterdam, 1989

corresponding to the predicate $\mathbf{Q}_i$), and $f(\mathbf{P}, \{\mathbf{q}_i\})$ is a function that accepts as input and produces as output relations of the same schema as **P**, the above rule can be expressed as

$$f(\mathbf{P}, \{\mathbf{q}_i\}) \subseteq \mathbf{P},$$

or equivalently $\mathbf{P} \cup f(\mathbf{P}, \{\mathbf{q}_i\}) = \mathbf{P}$. Given the existence of an additional nonrecursive rule of the form

$$\mathbf{Q}(\underline{x}) \rightarrow \mathbf{P}(\underline{x}), \qquad (2.2)$$

which in relational terms, given **q** a relation corresponding to **Q**, is expressed as $\mathbf{q} \subseteq \mathbf{P}$, the problem of recursive inference can be stated as one of finding **P** such that

(a) $\mathbf{P} = f(\mathbf{P}, \{\mathbf{q}_i\}) \cup \mathbf{q}$,
(b) **P** is minimal with respect to (a),
i.e., $\mathbf{P}'$ satisfying (a) implies $\mathbf{P} \subseteq \mathbf{P}'$.

The function $f(\mathbf{P}, \{\mathbf{q}_i\})$, having $\{\mathbf{q}_i\}$ as parameters and **P** as input, can be thought of as a linear relational operator applied to the recursive relation **P** to produce another relation of the same schema. Let R be the set of all such operators. We can establish an algebraic framework in which we can define operations on relational operators as follows. *Multiplication* of operators is defined by $(A * B)\mathbf{P} = A(B\mathbf{P})$ and *addition* by $(A+B)\mathbf{P} = A\mathbf{P} \cup B\mathbf{P}$. † For notational convenience we omit the operator $*$. The multiplicative identity ($1\mathbf{P} = \mathbf{P}$) and the additive identity ($0\mathbf{P} = \emptyset$, $\emptyset$ the empty set) are defined in obvious ways. The n-th power of an operator A is inductively defined as: $A^0 = 1$, $A^n = A * A^{n-1} = A^{n-1} * A$. Equality of operators in R is defined as $A = B \Leftrightarrow \forall \mathbf{P}, A\mathbf{P} = B\mathbf{P}$. Finally, a partial order $\leq$ is defined on R as $A \leq B \Leftrightarrow \forall \mathbf{P}, A\mathbf{P} \subseteq B\mathbf{P}$. The set R with the operations defined above forms a *closed semiring* [Ioan88a].

Having embedded the linear relational operators in the above algebraic framework of the closed semiring, the set of Horn clauses (2.1) and (2.2), assuming that A is the operator that corresponds to (2.1), can be rewritten as

$$A\mathbf{P} \subseteq \mathbf{P},$$

$$\mathbf{q} \subseteq \mathbf{P}.$$

The minimal solution of the system is the minimal solution of the equation

$$\mathbf{P} = A\mathbf{P} \cup \mathbf{q}. \qquad (2.3)$$

† The above definitions are valid only if the operators involved are appropriately compatible, e.g., for +, the operators have to agree on the schema of their input and the schema of their output. Although in the rest of the paper we only deal with appropriately compatible operators, the general algebraic theory incorporates all operators [Ioan88a].

The solution is a function of **q**. Hence, **P** can be written as $\mathbf{P} = B\mathbf{q}$, and the problem becomes one of finding the operator B. Manipulation of (2.3) results in the elimination of **q**, so that the equation contains operators only. In this pure operator form, the recursion problem can be restated as follows. Given operator A, find B satisfying the following:

(a) $1 + AB = B$,
(b) B is minimal with respect to (a), (2.4)
i.e., $1 + AC = C \Rightarrow B \leq C$.

Theorem 2.1: [Ioan88a] The solution of equation (2.4a) with restriction (2.4b) is $A^* = \sum_{k=0}^{\infty} A^k$.

The operator A^* is called the *transitive closure* of A. Theorem 2.1 is originally due to Tarski [Tars55], and in the database context, it was first examined by Aho and Ullman [Aho79b]. It is the first time though that the solution of (2.4) is expressed in an explicit algebraic form within an algebraic structure like the closed semiring of the linear relational operators. The implications of the manipulative power thus afforded on the implementation of A^* are significant [Ioan86a, Ioan86b, Ioan87, Ioan88a]. In this paper, we shall concentrate on the implications of commutativity of operators in the implementation of A^*.

Note that, although an operator A may be derived from a recursive rule, the operator itself is nonrecursive, i.e., it corresponds to a conjunctive query [Chan77]. Also note that A^* represents an operator. The query answer is the result of applying A^* to a given relation **q**. This is only an abstraction, however, that allows us to study recursion within the closed semiring of relational operators. It poses no restriction whatsoever in the processing order of the query, i.e., it does not enforce that first A^* is computed and then it is applied to **q**. For example, assume that A^* can be decomposed into B^* and C^*, i.e., $A^* = B^* C^*$, so that the final computation is $B^* C^* \mathbf{q}$. The computation may proceed by first computing C^*, then applying it to **q**, and then applying *seminaive* [Banc85] with B as the basic operator and ($C^* \mathbf{q}$) as the initial relation. The significance of the algebraic formulation lies in the abstraction that it offers, within which the capability of the decomposition $A^* = B^* C^*$ can be exhibited.

3. COMMUTATIVITY VS. SEPARABILITY and RECURSIVE REDUNDANCY

3.1. Commutativity

We say that two operators B and C *commute* if $BC = CB$. Consider computing the transitive closure of A, A^*, where $A = B + C$. It has been shown that if $CB \leq B^k C^l$, for some k, l with $k \in \{0,1\}$ or $l \in \{0,1\}$, then $A^* = B^* C^*$ [Ioan88a]. Commutativity is a special case of this condition. The computation of A^* is decomposed into two smaller computations, those of B^* and C^* (plus an additional multiplication of them). The complexity of B and C is smaller than that of A. In general, this is expected to affect the total cost of the computation significantly. Hence, it is important to be able to identify when two operators commute. In Section 4, we present a sufficient condition for commutativity, which for rules of some restricted form is shown to be necessary and sufficient.

3.2. Commutativity vs. Separability

Separable recursions have been identified by Naughton as an important class of linear recursion where efficient algorithms can be applied [Naug88]. In this section, we shall show that the efficient separable algorithm is applicable to the class of commutative recursions. For the sake of simplicity, we shall concentrate on two operators A_1 and A_2. The extensions of the results to an arbitrary number of operators is straightforward.

Theorem 3.1: Given two operators A_1 and A_2 that commute, and a selection σ that commutes with one of them, the separable algorithm can be used for the computation of $\sigma(A_1+A_2)^*$.

Proof: Let $A_1A_2=A_2A_1$. The transitive closure of the sum of A_1 and A_2 is given by $(A_1+A_2)^* = A_1^* A_2^*$ [Ioan88a]. Given an initial relation **q** and a query with a selection σ that commutes with A_1, we have

$$\sigma(A_1+A_2)^* \mathbf{q} = A_1^* (\sigma A_2^*)\mathbf{q}. \qquad (3.1)$$

To take advantage of the selection, the following algorithm may be used to derive the query answer given in (3.1). The variables B and C contain operators whereas the variables R and S contain relations. Multiplication of operators is shown explicitly for readability.

```
B := σ;
C := σ;
repeat
    B := B * A2;
    C := B + C;
until C doesn't change
R := C q;
S := R;
repeat
    R := A1 R;
    S := S ∪ R;
until S doesn't change
```

The first loop actually involves manipulating relations that are parameters of the various operators. If that is taken into account, and some small optimizations are incorporated so that, in every application of an operator inside each loop, only the new tuples produced in the previous iteration are used, the above algorithm is precisely the one proposed for separable recursions with full selections † [Naug88]. □

In general, given a set of operators $\{A_i\}$, $1 \le i \le n$, that are mutually commutative, and a set of selections $\{\sigma_i\}$, $0 \le i \le n$, such that σ_i commutes with all operators except A_i, the following holds:

$$\sigma_0\sigma_1\sigma_2 \cdots \sigma_n (A_1+A_2+\cdots+A_n)^* =$$

$$(\sigma_1 A_1)^* (\sigma_2 A_2)^* \cdots (\sigma_n A_n)^* \sigma_0.$$

Usually, most of the selections will not be present. In the presense of multiple selections, it is an interesting optimization problem to choose the order in which the various operators should be computed and the time when an operator should be applied to the input relation.

3.3. Commutativity vs. Recursive Redundancy

The class of recursions that contain recursively redundant predicates was also identified by Naughton [Naug86]. Consider the operator A that is the product of a set of operators $\{A_i\}$, i.e., $A=A_1A_2\cdots A_n$. In this case, every term in the series $A^* = \sum_{k=0}^{\infty} A^k$ is a product of the A_i's some number of times. An operator A_i, $0 \le i \le n$, is *recursively redundant* if there is some N such that each term in the series of A^* needs A_i factored in less than N times. The nonrecursive predicates appearing in A_i are also called recursively redundant. Before stating the main result of this subsection we need the following definitions. An operator B is *uniformly bounded*, if there exist K and N, $K<N$, such that $B^N \le B^K$. An operator B is *torsion*, if there exist K and N, $K<N$, such that $B^N = B^K$. Clearly, every torsion is uniformly bounded, but the opposite is not true in general. The effect of the presense of recursively redundant operators on the query processing algorithm of an operator is given by the following result [Ioan88a].

Theorem 3.2: If B and C are mutually commutative and B is torsion, then B is recursively redundant in $(B\ C)^*$.

Sketch of proof: Consider an operator A, such that $A = B\ C = C\ B$ and there exist K and N, $K<N$, such that $B^N = B^K$. Then,

$$A^* = (B\ C)^* = \sum_{m=0}^{K-1} B^m C^m + (\sum_{m=K}^{N-1} B^m C^m)(C^{N-K})^*.$$

The details of the above derivation are given elsewhere [Ioan88a]. Note that B^{N-1} is the highest power of B used in any term of A^*, i.e., B is recursively redundant. Clearly, the above formula corresponds to a more efficient algorithm than processing A as a whole, since B is processed only for a fixed finite number of times, i.e., $N-1$, beyond which only C is processed. □

---

† The precise definition of full selections is given by Naughton [Naug88]. The key observation is that if $A_1A_2=A_2A_1$ and $\sigma A_1=A_1\sigma$, then σ is a full selection.

4. CHARACTERIZATION OF COMMUTATIVITY

We now turn our attention to commutativity as expressed in a logic framework. We restrict ourselves to linear, function-free, constant-free, and *range-restricted* recursive rules, i.e., every variable in the consequent appears at least once in the antecedent as well. Thus, for any finite database, the answer to any query is finite. If a variable appears in the consequent of a rule, it is called *distinguished*, otherwise it is called *nondistinguished*. We assume that the rules have the same consequent and share no nondistinguished variables. Moreover, repeated variables in the consequent are replaced by distinct ones, while adding the appropriate equality predicates in the antecedent. Finally, although the original task is to compute the transitive closure of two recursive rules with the same consequent, we are interested in the commutativity of the underlying nonrecursive rules, i.e., conjunctive queries. (Commutativity as defined in Section 3, is a property of nonrecursive rules (operators).) Given a recursive rule, the corresponding underlying nonrecursive one will be written with $\mathbf{P_O}$ as the (output) predicate in its consequent and $\mathbf{P_I}$ as the (input) predicate in its antecedent. Nevertheless, we shall still be referring to these two predicates as instances of the recursive predicate.

We say that two rules r_1 and r_2 with the same consequent *commute* if composing r_1 with r_2, i.e., resolving the consequent of r_2 with the literal of the recursive predicate in the antecedent of r_1, (denoted by $r_1 r_2$) and composing r_2 with r_1 (denoted by $r_2 r_1$) give equivalent rules, i.e., given any relations for the predicates in their antecedents produce the same output relation for the predicate in their consequent. This, in turn, is equivalent to the existence of homomorphisms from each composite to the other [Chan77, Aho79a]. Given two rules r and s, a *homomorphism* $f : r \rightarrow s$ is a mapping from the variables of r into those of s, such that (i) if x is a distinguished variable then $f(x)=x$, and (ii) if $Q(x_1,...,x_n)$ appears in the antecedent of r, then $Q(f(x_1),...,f(x_n))$ appears in the antecedent of s. Clearly, the definition of commutativity suggests a straightforward algorithm to test it for two rules r_1 and r_2. The algorithm involves checking for equivalence of the two composites $r_1 r_2$ and $r_2 r_1$. Therefore, a polynomial time implementation of this algorithm is unlikely to exist, since equivalence of conjunctive queries is known to be an NP-complete problem [Chan77, Aho79a].

4.1. A Sufficient Condition

In this section, we shall give a sufficient condition for commutativity that avoids producing the two composites. The condition can be tested in exponential time, because it potentially involves testing for equivalence of conjunctive queries. The test, however, is still more efficient than the one based on the definition of commutativity, because the exponential part is only occasionally applied on parts of the original rules as opposed to always being applied on the composites of the two rules.

As a notation vehicle, we shall use a version of the α-graph of a rule (which we shall also call α-graph), which was introduced for the study of uniform boundedness [Ioan85]. The α-graph of a rule is defined as follows.

(i) For every variable in the rule, a node is put in the graph.

(ii) If two variables x, y appear in two consecutive argument positions of some nonrecursive predicate $\mathbf{Q}$ in the rule, an undirected edge $(x-y)$ is put in the graph between the corresponding two nodes x, y. Also, if x appears in a unary nonrecursive predicate $\mathbf{Q}$ in the rule, an undirected edge $(x-x)$ is put in the graph. In both cases, the label of the edge is $\mathbf{Q}$.

(iii) If two variables x, y appear in the same position of the recursive predicate $\mathbf{P}$ in the antecedent and the consequent respectively, then a directed arc $(x \rightarrow y)$ is put in the graph from node x to node y.

The following definitions about the connected components of the (underlying undirected graph of the) α-graph of a rule are also necessary. A *persistent* component is one that contains exactly one variable, which is distinguished, and no nonrecursive predicates (undirected edges). Its variable is also called persistent. A *permutation* component is one that contains only distinguished variables (i.e., their positions in the antecedent is a permutation of their positions in the consequent) and no nonrecursive predicates (undirected edges). Its variables are also called permutation variables. Any other component is called a *general* component. Each one of its distinguished variables that is the head and the tail of a directed arc (i.e., it appears in the same position in the recursive predicate in the antecedent and the consequent) is called *semi-persistent*. Any other remaining distinguished variable is called general. Notice that every persistent component is a permutation component as well. Also, every persistent variable is both a semi-persistent and a permutation variable as well. Whenever it is necessary to exclude persistent components (variables) from the other classes, we shall qualify the terms by *nontrivial*, e.g., a nontrivial permutation component is a permutation component that is not persistent.

Example 4.1: The following is the α-graph of the rule

$$\mathbf{P}(u,v,w,x,y,z) :- \mathbf{P}(v,u,w,w,y,z) \wedge \mathbf{Q}(x,y).$$

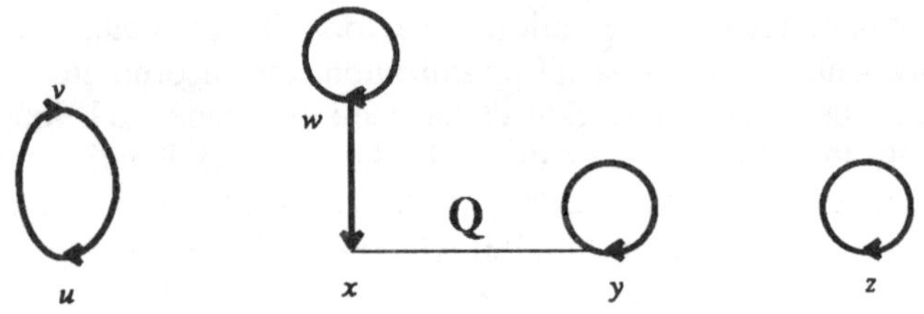

Figure 4.1: Example of an α-graph.

Variable z is persistent, variables w and y are semi-persistent, variables u and v are permutation, and variable x is general. □

For a rule r, we define the function h from the set of distinguished variables in r to the set of all variables in r. For a distinguished variable x, $h(x)$ is the variable that appears in the recursive predicate in the antecedent in the same position as x appears in the consequent. Since distinguished variables are assumed to appear exactly once in the consequents of rules (with the potential of repeated variables being realized by equalities in the antecedent), h is a function. Note that, if $h(x)=y$, then there exists a directed arc $(y \rightarrow x)$ in the α-graph of the rule. We also define powers of h as

$$h^1(x)=h(x), \text{ and}$$

$$h^n(x)=h(h^{n-1}(x)), \text{ if } h^{n-1}(x) \text{ is distinguished.}$$

For two rules r_1 and r_2, we define two more functions, g_{12} on the variables of r_2 and g_{21} on the variables of r_1. Since the two rules are assumed to share no nondistinguished variable, the former is defined as

$$g_{12}(z)=\begin{cases} z & z \text{ is nondistinguished} \\ h_1(z) & z \text{ is distinguished} \end{cases},$$

and similarly the latter. By definition, when $r_1 r_2$ is formed, a variable z in a predicate of r_2 is always replaced by $g_{12}(z)$.

Finally, two sets of permutation components are *consistent* if they are formed by the same (distinguished) variables, and for every variable x in them, it is $h_1(h_2(x)) = h_2(h_1(x))$. Clearly, consistent permutation components commute.

The following theorem gives a sufficient condition for commutativity of rules of the form specified in the beginning of Section 4. Its proof is omitted and can be found in the complete version of this paper [Ioan88b]. Another, less general sufficient condition for commutativity has been independently discovered and reported elsewhere [Rama89].

Theorem 4.1: Two rules r_1 and r_2 with the same consequent commute if

(a) the nontrivial semi-persistent variables in r_1 are semi-persistent in r_2 also (similarly for r_2),

(b) the nontrivial permutation variables in r_1 belong to consistent permutation components in r_2 (similarly for r_2),

(c) the general variables in r_1 either belong to an equivalent component in r_2, or they are persistent in r_2 (similarly for r_2).

Example 4.2: The following two rules, whose α-graphs are shown in Figure 4.2, commute with each other.

$$\mathbf{P_O}(x,y,z) :- \mathbf{P_I}(u,y,z) \wedge \mathbf{Q}(x,y)$$

$$\mathbf{P_O}(x,y,z) :- \mathbf{P_I}(x,y,v) \wedge \mathbf{R}(z,y)$$

Both composites are equal to the rule below.

$$\mathbf{P_O}(x,y,z) :- \mathbf{P_I}(u,y,v) \wedge \mathbf{Q}(z,y) \wedge \mathbf{R}(x,y)$$

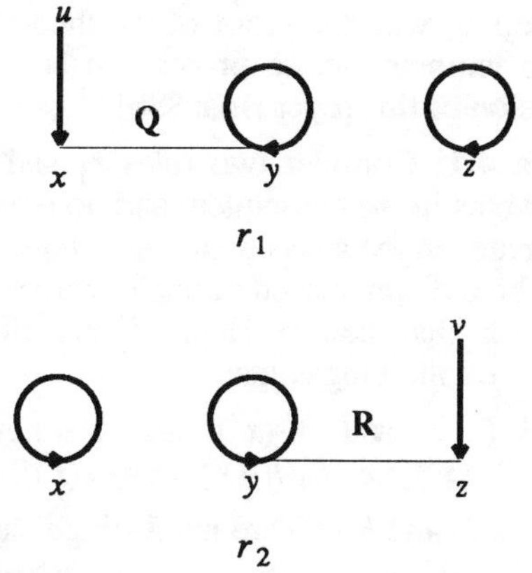

Figure 4.2: α-graphs of commuting rules satisfying the condition of Theorem 4.1.

Note that the condition of Theorem 4.1 is satisfied by the corresponding α-graphs. □

Unfortunately, as the following counter-example shows, the condition of Theorem 4.1 is not necessary for commutativity.

Example 4.3: The following two rules, whose α-graphs are shown in Figure 4.3, also commute with each other.

$$\mathbf{P_O}(x,y) :- \mathbf{P_I}(y,w) \wedge \mathbf{Q}(x)$$

$$\mathbf{P_O}(x,y) :- \mathbf{P_I}(u,v) \wedge \mathbf{Q}(x) \wedge \mathbf{Q}(y)$$

Both composites are isomorphic to the rule below.

$$\mathbf{P_O}(x,y) :- \mathbf{P_I}(u,v) \wedge \mathbf{Q}(y) \wedge \mathbf{Q}(w) \wedge \mathbf{Q}(x)$$

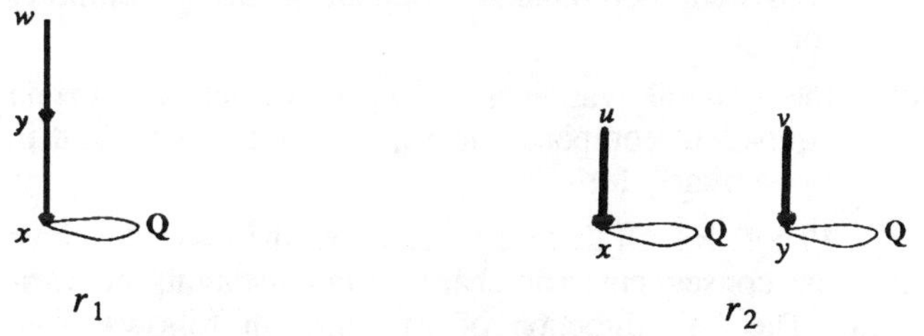

Figure 4.3: α-graphs of commuting rules not satisfying the condition of Theorem 4.1.

In this case, the condition of Theorem 4.1 is not satisfied. □

We are not aware of any necessary and sufficient condition for commutativity for rules of unrestricted form that is computationally or aesthetically better than the condition of the definition of commutativity. The following theorem shows that the condition of Theorem 4.1 is necessary and sufficient for commutativity if we restrict our attention to rules with *no repeated variables* in the consequent and *no repeated nonrecursive predicates* in the antecedent. The former restriction is enforced after all equalities have been eliminated from the antecedent. Before proceeding with the proof of the theorem, we need the following lemmas, whose proofs can be found in the complete version of this paper [Ioan88b].

Lemma 4.1: Consider two rules r_1 and r_2 with no repeated variables in the consequent and no repeated nonrecursive predicates in the antecedent that commute with each other. Let x be a distinguished variable, with $h_1(x)=x'$ and $h_2(x)=x''$, such that both x' and x'' are distinguished. Then, one of the following holds:

(a) both $h_1(x'')$ and $h_2(x')$ are distinguished and $h_1(x'')=h_2(x')$, i.e., $h_1(h_2(x))=h_2(h_1(x))$, or

(b) both $h_1(x'')$ and $h_2(x')$ are nondistinguished.

Lemma 4.2: Consider two rules r_1 and r_2 with no repeated variables in the consequent and no repeated nonrecursive predicates in the antecedent that commute with each other. Let $\{x_k\}$, $0 \le k \le n+1$, be a set of distinguished variables such that $h_1(x_k)=x_{k+1}$, i.e., $h_1^{k+1}(x_0)=x_{k+1}$, for $0 \le k \le n$, and x_0 appears in a nonrecursive predicate **Q**. Then, one of the following holds:

(a) $h_2(x_k)=x_k$, $0 \le k \le n+1$, or

(b) $h_2(x_k)=x_{k+1}$, i.e., $h_2^{k+1}(x_0)=x_{k+1}$, for $0 \le k \le n$, and x_0 appears in a nonrecursive predicate **Q** in r_2.

Theorem 4.2: Two rules r_1 and r_2 with the same consequent and no repeated variables in the consequent and no repeated nonrecursive predicates in the antecedent commute if and only if

(a) the nontrivial semi-persistent variables in r_1 are semi-persistent in r_2 also (similarly for r_2),

(b) the nontrivial permutation variables in r_1 belong to consistent permutation components in r_2 (similarly for r_2),

(c) the general variables in r_1 either belong to an equivalent component in r_2, or they are persistent in r_2 (similarly for r_2).

Proof: Recall that we assume that the two rules have the same consequents and share no nondistinguished variables. The "if" direction of the theorem follows from Theorem 4.1.

For the other direction of the theorem, assume that r_1 and r_2 commute. We shall show that for a distinguished variable x of r_1, one of (a), (b), or (c) holds in r_2, depending on the type of x. Since the theorem is symmetric in r_1 and r_2, the variables in r_2 are not examined. We shall always consider x being the first distinguished variable in the consequent, and we shall only be writing down the parts of the rules that are relevant to the proof. Also, unimportant variables will be denoted by _ .

(i) x is a nontrivial semi-persistent variable: In this case, x appears at least twice in the antecedent of r_1. This implies that it belongs to a general component, and that there exists a set of distinguished variables $\{x_k\}$, $0 \le k \le n+1$, such that $h_1(x_k)=x_{k+1}$, $0 \le k \le n$, with $x=x_n=x_{n+1}$, and x_0 appears in a nonrecursive predicate **Q**. If this is not true, then there must exist repeated variables in the consequent of r_1, which is a contradiction. Applying Lemma 4.2 for $x=x_n$ yields $h_2(x_n)=x_n$ or $h_2(x_n)=x_{n+1}$. Since $x=x_n=x_{n+1}$, this implies that in all cases $h_2(x)=x$, i.e., that x is semi-persistent in r_2.

(ii) x is a nontrivial permutation variable: If x is not a permutation variable in r_2 (i.e., it is a general variable or a nontrivial semi-persistent variable), then there exists a set of distinguished variables $\{y_k\}$, $0 \le k \le n+1$, such that $h_2(y_k)=y_{k+1}$, $0 \le k \le n$, with $x=y_{n+1}$, and y_0 appears in a nonrecursive predicate **Q** in r_2. By Lemma 4.2, this implies that either $x=h_1(x)$ or $x=h_1^{n+1}(y_0)$ and y_0 appears in a nonrecursive predicate **Q** in r_1. In the first case, x is a semi-persistent variable in r_1, and in the second case, x is a general variable in r_1. In both cases, this is a contradiction, since x is a nontrivial permutation variable in r_1. Hence, x must be a permutation variable in r_2 also.

Since x is a permutation variable in r_1 and r_2, $h_1(x)$ and $h_2(x)$ must also be permutation variables in r_1 and r_2 respectively ($h_i(x)$ is part of the same permutation as x in r_i). The argument in the previous paragraph can be applied in the case of $h_2(x)$ also and yield that $h_2(x)$ is a permutation variable in r_1 as well. Hence, $h_1(x)$, $h_2(x)$, and $h_1(h_2(x))$ are distinguished variables. By Lemma 4.1, $h_2(h_1(x))$ is also distinguished, and $h_2(h_1(x))=h_1(h_2(x))$. This implies that x belongs to consistent sets of permutation components in r_1 and r_2.

(iii) x is a general variable: This implies that there exists a set of distinguished variables $\{x_k\}$, $0 \le k \le n+1$, such that $h_1(x_k)=x_{k+1}$, $0 \le k \le n$, with $x=x_{n+1}$, and x_0 appears in a nonrecursive predicate **Q**. By Lemma 4.2, this implies that either $x=h_2(x)$, i.e., that x is a semi-persistent variable in r_2, or $x=h_2^{n+1}(x_0)$, and x_0 appears in a nonrecursive predicate **Q** in r_2, i.e., that x is a general variable in r_2. We shall examine the two cases separately.

If x is semi-persistent in r_2, we shall show that it cannot be nontrivial semi-persistent, i.e., it must be persistent. Assume to the contrary that x is nontrivial semi-persistent in r_2. From case *(i)* for r_2, we conclude that x is semi-persistent in r_1, which is a contradiction. Hence, x must be persistent.

If x is general in r_2, we shall show that it belongs to a component that is equivalent to its component in r_1. Recall that we examine the case where $h_2(x_k)=x_{k+1}$, for all $0 \le k \le n$, which implies that $h_1(x_k)=h_2(x_k)$. Since $x=x_{n+1}$ is an arbitrary variable in its component, we may conclude that for any distinguished variable z in that component, either both $h_1(z)$, $h_2(z)$ are distinguished and $h_1(z)=h_2(z)$, or both $h_1(z)$, $h_2(z)$ are nondistinguished, i.e., the structure of h for the components of z in r_1 and r_2 is the same. Hence, if we assume that the two components are not equivalent, there must be some nonrecursive predicate connected (through a series of nonrecursive predicates) to a distinguished variable in the component in r_1 that is not connected through the same series of nonrecursive predicates to the same distinguished variable in the component in r_2 (or vice-versa). Without loss of generality, assume that x is such a distinguished variable. Also without loss of generality, assume that $h_1(x)=h_2(x)=y$ is a distinguished variable, and that only nondistinguished variables appear in the nonrecursive predicates connected to x (except x). The other cases are treated similarly. This situation is depicted by the rules of Figure 4.4 and their composites. Clearly, since $y \ne x$ (x is general and not semi-persistent), the two composites are not equivalent, and r_1 and r_2 cannot commute, which is a contradiction. Hence, the assumption that the two components where x belongs in r_1 and r_2 are not equivalent is wrong. □

The complexity of the condition in Theorem 4.2 is given by the following theorem.

Theorem 4.3: Commutativity of two rules with no repeated variables in the consequent and no repeated variables in the antecedent can be tested in $O(a_r + n\, a_n + n^2)$ time, where a_r is the arity of the recursive predicate, a_n is the maximum arity of a nonrecursive predicate, and n is the maximum number of the nonrecursive predicates in the rules.

Sketch of proof: The algorithm has the following basic steps.

(a) Identify the connected components of the underlying undirected graphs of the α-graphs of the two rules, while identifying the type of every distinguished variable (i.e., persistent, nontrivial semi-persistent, nontrivial permutation, or general). The quantity $a_r + n\, a_n$ is an upper bound on both the nodes and the edges/arcs in the graph. So, this step can be done in $O(a_r + n\, a_n)$ time [Aho74].

(b) For every nontrivial semi-persistent variable in the one rule, check if it is semi-persistent in the other. This step takes $O(1)$ for every nontrivial semi-persistent variable.

(c) For every nontrivial permutation variable in the one rule, check if it belongs in a consistent permutation component in the other. This step takes $O(1)$ for every nontrivial permutation variable.

(d) For every general variable in the one rule, check if it is persistent in the other. If it is, do nothing. This step takes $O(1)$ for every such variable. If it is not, check if it belongs in an equivalent component in the other rule. Because the rules contain no repeated nonrecursive predicates in the antecedent, equivalence can be tested in polynomial time as follows (the components have to be isomorphic).

(d1) For every distinguished variable, check if the h functions of the two rules are compatible. This step takes $O(a_r)$ time.

(d2) For every nonrecursive predicate in the one rule, check if it exists in the other rule and if the mappings between the variables are consistent with the mappings of the other predicates. This step takes $O(n)$ time for the first check and $O(a_n)$ time for the second check for every nonrecursive predicate, for a total of $O(n\, a_n + n^2)$ time.

---

r_1: $\mathbf{P_O}(x,\cdots) :- \mathbf{P_I}(y,\cdots) \wedge \mathbf{R}_1(x,z_1) \wedge \mathbf{R}_2(z_1,z_2) \cdots \wedge \mathbf{R}_{m-1}(z_{m-2},z_{m-1}) \wedge \mathbf{R}_m(z_{m-1},z_m) \wedge \cdots$

r_2: $\mathbf{P_O}(x,\cdots) :- \mathbf{P_I}(y,\cdots) \wedge \mathbf{R}_1(x,z'_1) \wedge \mathbf{R}_2(z'_1,z'_2) \cdots \wedge \mathbf{R}_{m-1}(z'_{m-2},z'_{m-1}) \wedge \cdots$

$r_1 r_2$: $\mathbf{P_O}(x,\cdots) :- \mathbf{P_I}(\_,\cdots) \wedge \mathbf{R}_1(y,z'_1) \wedge \mathbf{R}_2(z'_1,z'_2) \cdots \wedge \mathbf{R}_{m-1}(z'_{m-2},z'_{m-1}) \wedge$
$\mathbf{R}_1(x,z_1) \wedge \mathbf{R}_2(z_1,z_2) \cdots \wedge \mathbf{R}_{m-1}(z_{m-2},z_{m-1}) \wedge \mathbf{R}_m(z_{m-1},z_m) \wedge \cdots$

$r_2 r_1$: $\mathbf{P_O}(x,\cdots) :- \mathbf{P_I}(\_,\cdots) \wedge \mathbf{R}_1(y,z_1) \wedge \mathbf{R}_2(z_1,z_2) \cdots \wedge \mathbf{R}_{m-1}(z_{m-2},z_{m-1}) \wedge \mathbf{R}_m(z_{m-1},z_m) \wedge$
$\mathbf{R}_1(x,z'_1) \wedge \mathbf{R}_2(z'_1,z'_2) \cdots \wedge \mathbf{R}_{m-1}(z'_{m-2},z'_{m-1}) \wedge \cdots$

Figure 4.4: Rules with a general variable in nonequivalent components and their composites.

The total time for steps (b), (c), and (d) without (d1) and (d2) is $O(a_r)$. If we add to that the time needed for (a), (d1), and (d2), we conclude that the total running time of the test is $O(a_r + n\, a_n + n^2)$. □

5. SEPARABILITY AND RECURSIVE REDUNDANCY REVISITED

In Section 3, we examined commutativity vs. separability and recursive redundancy as expressed in the abstract form of the algebra to obtain results that hold for any linear rules. In this section, we restrict ourselves to function-free, constant-free, and range-restricted rules and use our results from Section 4 to obtain more relationships of commutativity with separability and recursive redundancy for this class of rules.

5.1. Commutativity vs. Separability

Two rules r_1 and r_2 with the same consequent are separable [Naug88] if

(1) for any distinguished variable x, either $h_i(x)=x$ or $h_i(x)$ is nondistinguished, $i=1,2$,

(2) for any distinguished variable x, either both x and $h_i(x)$ appear under nonrecursiver predicates or none, $i=1,2$, and

(3) the sets of distinguished variables that appear under nonrecursive predicates in r_1 and r_2 are either equal or have an empty intersection.

We ignore a fourth clause that the original definition contained, since it was identified as nonessential for the correctness of the separable algorithm. For the case of two rules, one can take advantage of the efficient features of the separable algorithm only if in clause (3) the intersection of the sets of distinguished variables that appear under nonrecursive predicates in r_1 and r_2 is empty. With this assumption, we can prove the following lemma.

Lemma 5.1: If two rules r_1 and r_2 with the same consequent are separable, then they only contain semi-persistent and general variables, and any general or nontrivial semi-persistent variable in r_1 is persistent in r_2 (similarly for the variables of r_2).

Combining Lemma 5.1 with Theorem 4.1 yields the following theorem.

Theorem 5.1: If two rules are separable then they commute, but the opposite does not hold.

Proof: If two rules r_1 and r_2 are separable, by Lemma 5.1, the general variables of r_1 satisfy condition (c) of Theorem 4.1, and the nontrivial semi-persistent variables of r_1 satisfy condition (a) of the same theorem (similarly for the variables of r_2). Moreover, by the same lemma, there are no other types of variables in r_1 or r_2. Thus, by Theorem 4.1, the two rules commute.

The rules of Example 4.2 serve as examples of commutative rules that are not separable. They violate both conditions (2) and (3) of the separable definition. □

By Theorem 5.1, commutativity is a strictly more general notion than separability. Nevertheless, all the efficient processing algorithms for separable recursions are applicable for commutative rules as well (Theorem 3.1).

5.2. Commutativity vs. Recursive Redundancy

Given the α-graph of a rule, the *augmented* α-graph is produced by disconnecting all edges emanating from semi-persistent variables in the α-graph, and replacing the semi-persistent variable with a distinct nondistinguished variable for every such edge [Naug86]. A necessary and sufficient condition for a nonrecursive predicate in a rule of some restricted form to be redundant is given by the following theorem.

Theorem 5.2: [Naug86] A nonrecursive predicate in a rule with no repeated variables in the consequent and no repeated nonrecursive predicates in the antecedent is recursively redundant if and only if it appears in a bounded component of the augmented α-graph of the rule.

Theorems 3.2, 4.2, and 5.2 imply the following.

Theorem 5.3: Let $A=B\,C$ be an operator corresponding to a rule with no repeated variables in the consequent and no repeated nonrecursive predicates in the antecedent. Then, B is recursively redundant if and only if B and C commute and B is uniformly bounded.

Sketch of proof: From Theorem 4.2, we can prove that each component in the augmented α-graph, seen as an operator, commutes with the operators of all the other components. We can also show that, if the rules contain no repeated variables in the consequent and no repeated nonrecursive predicates in the antecedent, any uniformly bounded component of the augmented graph is torsion. Combining this with Theorem 5.2, yields that the condition of Theorem 3.2 is necessary and sufficient (in this case the properties of being torsion and uniformly bounded coincide). □

6. CONCLUSIONS

We have investigated the role of commutativity in query processing of linear recursive rules. Using the algebraic structure of such rules, we have identified commutativity as the essence of many properties that give rise to important classes of recursive rules, i.e., separable rules and rules with recursively redundant predicates. Focusing on range-restricted rules that contain no functions, and no constants, we have given a sufficient condition for such rules to commute. We have also shown that the condition is necessary and sufficient when the rules contain no repeated variables in the consequent and no repeated nonrecursive predicates in the antecedent. In that case, the condition can be tested in polynomial time in the size of the rules.

Commutativity emerges as a key property of linear recursive rules for which efficient algorithms can be applied. This paper is a first step in the investigation of its power. We believe that there is much more work to be done in this direction. Some problems we plan to study in the future are the following: characterize commutativity in more general classes of rules than the one studied in this paper; investigate the relationship of commutativity and one-sided recursion; investigate the relationship of commutativity and several optimizations proposed for the magic sets and counting algorithms (e.g., there seems to be a strong relationship between commutativity and the semijoin optimization [Beer87]); examine ways to take advantage of partial commutativity, i.e., when the transitive closure of a product of operators is to be computed, only a subset of which are mutually commutative; and examine ways to take advantage of commutativity appearing in some higher power of an operator.

7. REFERENCES

[Aho74]
Aho, A., J. Hopcroft, and J. Ullman, *The Design and Analysis of Computer Algorithms,* Addison Wesley, Reading, MA, 1974.

[Aho79a]
Aho, A., Y. Sagiv, and J. Ullman, "Equivalences Among Relational Expressions", *SIAM Computing Journal* **8**, 2 (May 1979), pp. 218-246.

[Aho79b]
Aho, A. and J. Ullman, "Universality of Data Retrieval Languages", in *Proc. of the 6th ACM Symposium on Principles of Programming Languages*, San Antonio, TX, January 1979, pp. 110-117.

[Banc85]
Bancilhon, F., "Naive Evaluation of Recursively Defined Relations", in *Proc. of the Islamorada Workshop on Large Scale Knowledge Base and Reasoning Systems*, Islamorada, FL, February 1985.

[Beer87]
Beeri, C. and R. Ramakrishnan, "On the Power of Magic", in *Proc. of the 6th ACM SIGMOD-SIGACT Symposium on Principles of Database Systems*, San Diego, CA, March 1987, pp. 269-283.

[Chan77]
Chandra, A. K. and P. M. Merlin, "Optimal Implementation of Conjunctive Queries in Relational Data Bases", in *Proc. 9th Annual ACM Symposium on Theory of Computing*, Boulder, CO, May 1977, pp. 77-90.

[Ioan85]
Ioannidis, Y. E., "A Time Bound on the Materialization of Some Recursively Defined Views", in *Proc. 11th International VLDB Conference*, Stockholm, Sweden, August 1985, pp. 219-226.

[Ioan86a]
Ioannidis, Y. E. and E. Wong, "An Algebraic Approach to Recursive Inference", in *Proc. of the 1st International Conference on Expert Database Systems*, Charleston, South Carolina, April 1986, pp. 209-223.

[Ioan86b]
Ioannidis, Y. E., "On the Computation of the Transitive Closure of Relational Operators", in *Proc. 12th International VLDB Conference*, Kyoto, Japan, August 1986, pp. 403-411.

[Ioan87]
Ioannidis, Y. E. and E. Wong, "Query Optimization by Simulated Annealing", in *Proc. of the 1987 ACM-SIGMOD Conference on the Management of Data*, San Francisco, CA, May 1987, pp. 9-22.

[Ioan88a]
Ioannidis, Y. E. and E. Wong, *"Towards an Algebraic Theory of Recursion"*, Technical Report No. 801, University of Wisconsin, Madison (submitted for publication), October 1988.

[Ioan88b]
Ioannidis, Y. E., *"Commutativity and its Role in the Processing of Linear Recursion"*, Technical Report No. 804, University of Wisconsin, Madison, November 1988.

[Naug86]
Naughton, J., "Redundancy in Function-Free Recursive Rules", in *Proc. of the 3rd Symposium on Logic Programming* , September 1986, pp. 236-245.

[Naug88]
Naughton, J., "Compiling Separable Recursions", in *Proc. of the 1988 ACM-SIGMOD Conference on the Management of Data*, Chicago, IL, June 1988, pp. 312-319.

[Rama89]
Ramakrishnan, R., Y. Sagiv, J. D. Ullman, and M. Vardi, *"Proof Tree Transformation Theorems and their Applications"*, Philadelfia, PA, Proc. of the 8th ACM SIGACT-SIGMOD-SIGART Symposium on Principles of Database Systems, March 1989, pp. 172-181.

[Tars55]
Tarski, A., "A Lattice Theoretical Fixpoint Theorem and its Applications", *Pacific Journal of Mathematics* **5** (1955), pp. 285-309.

Estimating the Size of Generalized Transitive Closures*

Richard J. Lipton and Jeffrey F. Naughton
Computer Science Department
Princeton University

Abstract

We present a framework for the estimation of the size of binary recursively defined relations. We show how the framework can be used to provide estimating algorithms for the size of the transitive closure and generalizations of the transitive closure, and also show that for bounded degree relations, the algorithm runs in linear time. Such estimating algorithms are essential if database systems that support recursive relations or fixpoints are to be able to optimize queries and avoid infeasible computations.

1 Introduction

Deductive database or knowledge base systems require estimates of relation sizes so that a good order of evaluation can be found for conjunctive queries [IW87,KBZ86,SG88,SG85]. Avoiding a bad choice of evaluation order is critical to the efficiency of the query evaluation procedure. For example, in the query

```
richCASenator(X) :- livesInCA(X) &
                    rich(X) &
                    senator(X).
```

first finding all senators (100) then finding those that live in California (2) then checking to see if those two are rich would be far more efficient than first finding all California residents (millions) then checking these to see if they are rich and senators.

*Work supported by DARPA and ONR contracts N00014-85-C-0456 and N00014-85-K-0465, and by NSF Cooperative Agreement DCR-8420948

Proceedings of the Fifteenth International
Conference on Very Large Data Bases

Some progress has been made toward estimating the size of relations that are defined by relational expressions rather than being stored in the database [Chr83,Dem80,HOT88,Lyn88,PSC84, Row83]. However, currently there are no known techniques for estimating the size of relations that are recursively defined. Such relations arise in not only in deductive database systems, but in any relational system that allows one to take the fixpoint of a relational expression. This paper considers a method for estimating the size of an important subset of such relations, which we term *generalized transitive closures.*

Previous work on generalized closures [AJ87,Ioa86, IR88,LMR87,Lu87,NSRU89,Nau87,RHDM86] has focussed on efficiently constructing the answer to the query rather than on answer size estimation. Of course, constructing the relation is one way to estimate its size; our goal is to get a rough estimate of the size much more cheaply.

The problem of estimating the size of a recursively defined relation is difficult. For concreteness, consider the transitive closure of a binary relation e. We may interpret e as being the edge relation of a digraph. Figure 1 shows a graph, with uniform out-degree 1, such that the closure of the graph is $O(n)$. Figure 2 shows another graph, also with uniform out-degree 1, such that the closure of the graph is $O(n^2)$. An estimation algorithm based on local (rather than global) properties of the graph will miss the difference between the two.

As another example of a hard case, consider the graph in Figure 3. Here, the graph again has uniform in- and out-degree 1. But while the transitive closure of the graph is $O(n^2)$, the closure of the graph formed by deleting the clique at the right is $O(n)$. Note that the portion of the graph that makes the difference between $O(n)$ and $O(n^2)$ here involves only $1/k$ of

Amsterdam, 1989

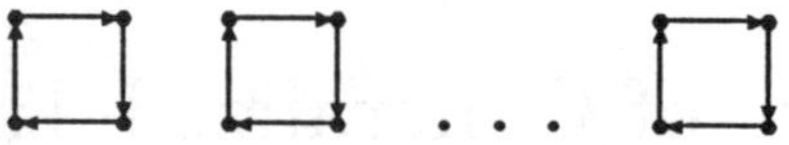

Figure 1: A bounded degree graph with closure $O(n)$.

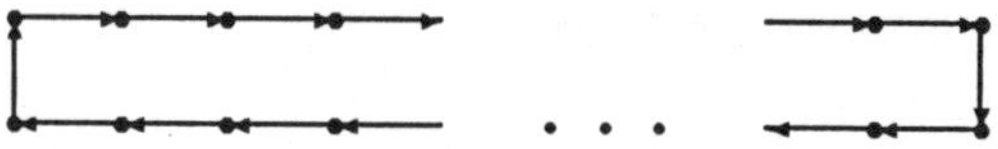

Figure 2: A bounded degree graph with closure $O(n^2)$.

nodes in the graph.

In this paper we give an adaptive sampling algorithm for estimating the size of generalized transitive closures. It is a sampling algorithm in that it estimates the size of the complete closure by sampling subsets of the closure; it is adaptive in that the number of samples made varies with the information obtained from each sample.

For the specific case of the transitive closure, if the underlying graph over which the transitive closure is being computed is of bounded degree, our sampling algorithm runs in linear time and, for $0 < \epsilon < 0.5$, estimates the size of the closure to within a factor of $1/\epsilon$ with probability $1-2\epsilon$. An algorithm for bounded degree graphs is especially important, as we expect that the bulk of relations encountered in practice are of bounded degree when interpreted as graphs.

When we move from the simple transitive closure to generalized closures, the error and accuracy of our estimation algorithm are unchanged. However, the running time will depend on the specific recursion in question.

The estimate produced by our algorithm is admittedly crude — by taking $\epsilon = 0.1$ we see that the algorithm estimates the closure to within a factor of 10 with 80% certainty. However, in optimizing queries, the key is to identify gross differences between the sizes of relations (e.g., the difference between the `livesInCA` relation and the `senator` relation in the example above.)

There is another reason why even rough estimates are essential in database systems that support recursion or fixpoints: even on modest sized databases, recursively defined relations can be too large to be computed. If a system is to be robust, it must detect such computations and produce a warning rather than attempting to compute the relation.

For example, suppose again that we wish to compute the transitive closure of a 100K tuple relation, and that each tuple of the relation takes 100 bytes. (Such a relation is small by database standards — it will easily fit in the memory of many workstations today — but it will serve to illustrate the point.) Then the transitive closure will fall somewhere between $\approx 10^5$ tuples ($\approx$ 10Mbyte) and $\approx 10^{10}$ tuples ($\approx$ a terabyte.) Computing a terabyte relation will almost certainly be infeasible, no matter how clever an algorithm we use to compute the closure. The estimating algorithm presented in this paper can determine quickly whether the proposed relation can sensibly be computed.

In Section 2 we present an urn model that establishes the theoretical underpinnings for the estimation algorithm. Section 3 discusses a general approach to estimating the size of binary recursively defined relations based on the results of Section 2. Section 4 considers the special case of the transitive closure. We conclude in Section 5.

2 An Urn Model

In this section we consider an urn model that abstracts the estimation problem.

Consider an urn U in which there are n *balls*. Associated with ball i is a number $1 \leq a_i \leq n$, for $1 \leq i \leq n$. The *cost* of sampling ball i is a_i; the cost of a set of samples is the sum of the cost of the samples in the set. The quantity we wish to estimate is $A = \sum_{i=1}^{n} a_i$. Consider doing so with the following procedure:

Algorithm 2.1 Repeatedly sample with replacement until S, the sum of the costs on the balls sampled, is greater than $2n$. Let the number of samples

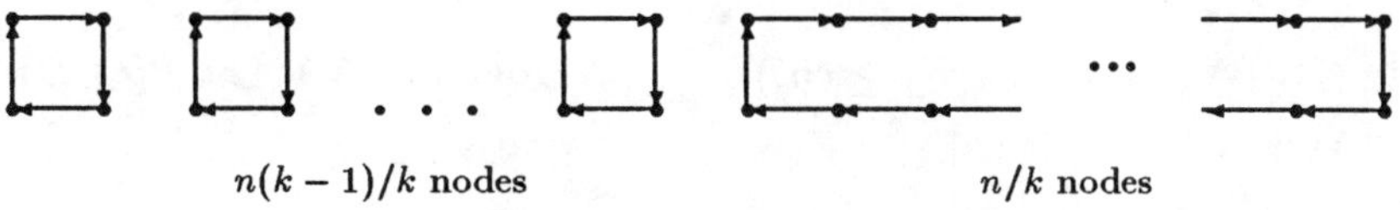

Figure 3: A hard example for size estimation.

this takes be m. Estimate that $\tilde{A} = nS/m$.

The main result of this section is the following theorem.

Theorem 2.1 *Let $\tilde{A}$ be the estimate of A in the above procedure, and let $A > 0$. Then for $0 < \epsilon < 0.5$, with probability $\geq 1 - 2\epsilon$ Algorithm 2.1 estimates A to within a factor of $1/\epsilon$.*

Before proving the theorem, we establish a series of lemmas. Let $\tilde{X}_i$ be a random sample from the urn.

Lemma 2.1 *Let $\tilde{X}_i$ be a random sample from the urn. Then*

1. $E[\tilde{X}_i] = A/n$.

2. $E[\tilde{X}_i^2] \leq A$.

3. $Var[\tilde{X}_i] \leq A$.

Proof:

1. $E[\tilde{X}_i] = \sum_{i=1}^{n} a_i/n = A/n$.

2. $E[\tilde{X}_i^2] = \sum_{i=1}^{n} a_i^2/n \leq \sum_1^n a_i = A$ (since $a_i/n \leq 1$).

3. $Var[\tilde{X}_i] = E[\tilde{X}_i^2] - E[\tilde{X}_i]^2 \leq A$.

□

The next lemma is useful in proving a lower bound on the expected number of samples made before stopping. We represent the probability of an event x by $Pr[x]$.

Lemma 2.2 *Let $m = \epsilon n^2/A$. Then*

$$Pr[\tilde{X}_1 + \ldots + \tilde{X}_m \geq cn] \leq \epsilon/(c-\epsilon)^2$$

Proof: Then

$$\begin{aligned} &Pr[\tilde{X}_1 + \ldots + \tilde{X}_m \geq cn] \\ &\quad = Pr[(\tilde{X}_1 - A/n) + \ldots + (\tilde{X}_m - A/n) \geq cn - \epsilon n] \\ &\quad \leq \frac{1}{(c-\epsilon)^2 n^2} m Var[\tilde{X}_i] \end{aligned}$$

But by Lemma 2.1, $Var[\tilde{X}_i] \leq A$, so

$$\begin{aligned} Pr[\tilde{X}_1 + \ldots + \tilde{X}_m \geq cn] &\leq \frac{1}{(c-\epsilon)^2 n^2} \frac{\epsilon n^2}{A} A \\ &\leq \frac{\epsilon}{(c-\epsilon)^2} \end{aligned}$$

□

We bound the error in the estimate with the following lemma.

Lemma 2.3 *Let $m \geq \epsilon n^2/A$ and $d > 0$. Then*

$$Pr\left[\left|\frac{n}{m}\left(\sum_{i=1}^{m} \tilde{X}_i\right) - A\right| \geq dA\right] \leq \frac{1}{\epsilon d^2}$$

Proof:

$$\begin{aligned} &Pr\left[\left|\frac{n}{m}\left(\sum_{i=1}^{m} \tilde{X}_i\right) - A\right| \geq t\right] \\ &\quad = Pr\left[\left|\left(\sum_{i=1}^{m} \tilde{X}_i\right) - \frac{m}{n}A\right| \geq mt/n\right] \\ &\quad = Pr\left[\left|\left(\sum_{i=1}^{m} \tilde{X}_i - \frac{1}{n}A\right)\right| \geq mt/n\right] \\ &\quad \leq \frac{n^2}{m^2 t^2} m Var[\tilde{X}_i] \end{aligned}$$

Let $m = \lambda n^2/A$, where $\lambda \geq \epsilon$. Then the above probability is less than or equal to

$$\frac{n^2}{m^2 t^2} m Var[\tilde{X}_i] \leq \frac{n^2}{\lambda \frac{n^2}{A} t^2} A$$

which is just $A^2/\lambda t^2$. Setting $Ad = t$ gives

$$\frac{1}{\lambda d^2} \leq \frac{1}{\epsilon d^2}$$

□

We can now prove Theorem 2.1.

Proof: (Theorem 2.1) There are two ways that the algorithm can fail — it can stop too early to guarantee a good error bound, or it can stop after enough samples but with a bad estimate.

First we claim that the procedure is unlikely to stop with $m \leq n^2/A$. We have that

$$Pr[((\exists j)(j \leq m) \wedge (\tilde{X}_1 + \ldots + \tilde{X}_j \geq cn))] \\ \leq Pr[\tilde{X}_1 + \ldots + \tilde{X}_m \geq cn]$$

where $m = \epsilon n^2/A$, because the event to the right of the inequality implies the event to the right. But by Lemma 2.2, the right side of this equation is at most $\epsilon/(c-\epsilon)^2$. Substituting $c = 2$ and noting that $0 < \epsilon < 0.5$, we get that this probability is less than ϵ.

Next we turn to the accuracy of the estimate. If $m = \epsilon n^2/A$, by Lemma 2.3 the estimate,

$$\tilde{A} = \frac{n}{m} \sum_{i=1} m\tilde{X}_i$$

is within dA of A with probability $\geq 1/(\epsilon d^2)$. Letting $d = 1/\epsilon$, this is just ϵ.

Putting the two ways of failure together, we get that the total probability of failure is less than $\epsilon + (1-\epsilon)\epsilon$, which is less than 2ϵ. Finally, note that because $A > 0$, there must be at least one i such that $a_i > 0$, so the algorithm will terminate. □

An interesting aspect of this urn theorem is that the sampling is adaptive: usually such sampling procedures perform a fixed number of samples. Here it is critical that the procedure adapt its behavior.

3 Binary Relation Estimation

First some notation. We let $R(Q, D)$ represent the binary relation defined by the query Q over the database D. The size of $R(Q, D)$ is just the number of tuples in D, and is represented by $|R(Q, D)|$. Using notation from relational algebra, the tuples in relation R with some constant c in column i is $\sigma_{i=c}(R(Q, D))$. In the remainder of this paper we will assume without loss of generality that $i = 1$. Such a binary query Q can be considered a generalization of the transitive closure, as it is expressing a "reachability" relationship between the constants of D.

We can map the urn problem of Section 2 to the problem of estimating $|R(Q, D)|$ as follows. We let the balls represent the constants appearing in D. If ball i represents a constant c_i, then $a_i = |\sigma_{1=c_i}(R(Q, D))|$. Sampling ball i corresponds to evaluating the query $\sigma_{1=c}(R(Q, D))$. Clearly, for any binary relation $R(Q, D)$, if there are n distinct constants in D, $|\sigma_{1=c}(R(Q, D))| \leq n$, and the relation $R(Q, D) = \cup_{c \in D} \sigma_{1=c}(R(Q, D))$. Note, however, that in the urn model we have $1 \leq a_i \leq n$, whereas here we have $0 \leq |\sigma_{1=c}(R(Q, D))| \leq n$.

We can now state the adaptive sampling algorithm for estimating $|R(Q, D)|$.

Algorithm 3.1 Let $R(Q, D)$ be a binary relation.

1. Set $s \leftarrow 0$.
2. Repeatedly choose a random constant c from D, and set $s \leftarrow s + \max(1, |\sigma_{1=c}(R(Q, D)|)$, until $s \geq 2n$.
3. Let m be the number of vertices chosen in Step 2. Estimate $|R(Q, D)| = ns/m$.

Theorem 3.1 *Let $0 < \epsilon < 0.5$. Then algorithm 3.1 estimates the size of $R(Q, D)$ to within $|R(Q, D)|/\epsilon + n(1 + 1/\epsilon)$ with probability $1 - 2\epsilon$.*

Proof: First, note that if $|\sigma_{1=c_i}(R(Q, D))| \geq 1$ for all i, then the correspondence between Algorithm 2.1 and Algorithm 3.1 is exact. Then by Theorem 2.1, Theorem 3.1 holds.

Next, note that for any sample $|\sigma_{1=c_i}(R(Q, D))|$, we have that

$$|\sigma_{1=c_i}(R(Q, D))| \leq \max(1, |\sigma_{1=c_i}(R(Q, D))|)$$

and

$$\max(1, |\sigma_{1=c_i}(R(Q, D))|) \leq |\sigma_{1=c_i}(R(Q, D))| + 1$$

This means that the quantity actually being estimated, say $|\tilde{R}(Q, D)|$, is such that

$$|R(Q, D)| \leq |\tilde{R}(Q, D)|$$

and

$$|\tilde{R}(Q, D)| \leq |R(Q, D)| + n$$

Since by Theorem 2.1 Algorithm 3.1 estimates $|\tilde{R}(Q, D)|$ to within $|\tilde{R}(Q, D)|/\epsilon$ with probability $1 - 2\epsilon$, Algorithm 3.1 estimates $|R(Q, D)|$ to within $(|R(Q, D)| + n)/\epsilon + n$, or $|R(Q, D)|/\epsilon + n(1 + 1/\epsilon)$. □

If $|R(Q, D)| > O(n)$, then the $O(n)$ term in the error of the estimate is asymptotically neglible.

We now turn to the running time of the algorithm. Because the algorithm is guaranteed to make at most $O(n)$ samples, if the time to perform an individual sample is $O(f(n))$, the time for the entire algorithm is $O(nf(n))$. However, if the problem $R(Q, D)$ has the property that the time to perform a sample is a well-behaved function of the size of the sample, we can prove a tighter bound.

Theorem 3.2 *Suppose that the running time of obtaining a sample $\max(1, |\sigma_{\$j=c_i}(R(Q, D))|)$ is bounded by $f(\max(1, |\sigma_{\$j=c_i}(R(Q, D))|))$, where $f(x)$ is such that for all $a, b > 0$, $f(a) + f(b) \leq f(a + b)$. Then Algorithm 3.1 runs in time $O(f(|R(Q, D)|))$.*

Proof: The proof follows from the observation that, for any function $f(x)$ such that $a, b > 0$ implies that $f(a) + f(b) \leq f(a+b)$, and any set x_i such that $x_i > 0$ for all i, we have $\sum_i f(x_i) = O(f(\sum_i x_i))$. □

The functions of interest in this paper (including n^k, where $k > 0$) are all well-behaved in this way.

Since in Algorithm 3.1 a sample is just a selection on the query, if we are to use our framework to estimate a query Q, we must have an efficient algorithm to answer selection queries on the relation defined by Q.

In the remainder of this section, we use Datalog notation. Informally, the Datalog rule

```
t(X,Y) :- t(X,W), e(W,Y).
```

may be read "the tuple `(X,Y)` is in `t` if there is some tuple `(X,W)` in `t` and some tuple `(W,Y)` in `e`." We consider only "safe" Datalog programs consisting of a linear recursive rule and a nonrecursive rule. Following Prolog conventions, we use uppercase letters to denote variables and lowercase to denote constants. Thus, the relational algebra expression $\sigma_{\$1=c}(t)$ may be rendered `t(c,Y)?`.

Consider estimating the size of the following trivial generalization of the transitive closure:

```
t(X,Y) :- e(X,W),t(W,Y).
t(X,Y) :- t0(X,Y).
```

(It is a generalization in that the body of the nonrecursive rule contains the predicate `t0` rather than `e`.) We might choose to sample the relation `t` based on its first argument, that is, we would evaluate `t(c,Y)` for constants `c` appearing in `t0` and `e`.

The straightforward way of evaluating the query `t(c,Y)` is to begin at the constant `c`, performing a breadth-first search through `e`. For each constant `a` encountered in this search, we check if there is a tuple `(a,b)` in `t0`, and if so, add `b` to the result.

This breadth-first search evaluates `t(c,Y)` in time linear in the sizes of `e` and `t0` hence can be considered "efficient." However, this cost of evaluating `t(c,Y)` is not any function of the answer size, because we could trace arbitrarily long paths through `e` such that there is no `t0` edge reachable from any node along that path.

On the other hand, if we sample on the second column of `t` instead, one can verify that if the number of answers returned by a sample is k, the time to evaluate the sample is $O(k^2)$.

Definition 3.1 Let r be a Datalog rule in which some predicate, say t, appears both in the head and in the body. Then a variable appearing in the same argument position both in the instance of t in the rule head and the instance of t in the rule body is a **`stable variable`**. An argument position of t is a *stable argument position* if it contains a stable variable.

First, we note that if the recursive predicate in a recursion has a stable argument position, a good heuristic is that the stable argument position is the column that should be sampled. This is because zero-size samples can be detected in $O(1)$ (a sample on a constant c is of zero size if and only if no tuple exists in the relation defined by the nonrecursive rule with the constant c in the sampled column.) Unfortunately, having a stable argument position is not sufficient to guarantee that the running time of a sample is a function of the size of the sample.

Example 3.1 Consider the recursion

```
t(X,Y) :- t(X,W),e(W,Y).
t(X,Y) :- a(X,W),b(W,Y).
```

and suppose that the database is

```
a(1,1).     b(1,1).
a(1,2).     b(2,1).
a(1,3).     b(3,1).
  .           .
  .           .
  .           .
a(1,n).     b(n,1).
```

where the relation `e` is empty. Then the sample `t(1,Y)?` will take time $O(n)$, but will only return a single answer. □

Identifying classes of recursions for which the time to compute a sample is a function of the sample size is an interesting open question. The following section shows that the standard transitive closure is one important example of a recursive query for which the running time of a sample is indeed a function of the sample size.

4 Transitive Closure

We will now use theorem 3.1 to provide an estimation result for the transitive closure of binary relations. For the purposes of this section, we consider the binary relation over which the closure is computed to be the edge relation of a digraph G. Hence we will refer to graphs rather than relations, edges rather than tuples, and vertices rather than constants.

Here "sampling the relation" just means choosing a vertex v, then counting the number of vertices reachable from v. If there are m edges in the graph, this clearly can be done in time $O(m)$. So, a trivial bound on the running time is $O(nm)$. Because $m < n^2$ this is better than the standard n^3 algorithm for computing the closure. However, a much better bound can be proven as a corollary to Theorem 3.2.

Corollary 4.1 *Let G be a digraph with n vertices. Then Algorithm 3.1 estimates the size of the transitive closure of G in time $O(n^2)$.*

Proof: Recall that a sample of a vertex v is the number of vertices in the connected component with v at the root. This quantity can be computed (via depth-first search, for example) in time bounded by the number of edges in the component. But the number of edges is at most the number of vertices squared. So the sample $|R_v(G,Q)|$ can be computed in time $O(|R_v(G,Q)|^2)$. The sum over all samples is $O(n)$, so we have by Theorem 3.2 that the total time is bounded by $O(n^2)$, as required. □

Note that the constant n here is not the size of the relation over which the closure is being performed — rather, it is the number of distinct constants that appear in that relation. In general, the size of the relation varies from n to n^2. For a relation of size n^2, the above bound is linear in the size of the input relation. For bounded degree graphs, we can prove a still better bound.

Corollary 4.2 *Let G be a bounded degree digraph with n vertices. Then Algorithm 3.1 estimates the size of the transitive closure of G in time $O(n)$.*

Proof: The proof follows that of Corollary 4.1, except that for a bounded degree graph there is a constant d such that no node has more than d out-edges. This in turn implies that the number of edges in a connected component is linear in the number of nodes in the component (instead of quadratic in the general case.) □

In the preceding proofs we discussed the sampling evaluation procedure in standard graph-processing terms. In particular, we mention the use of any standard search method, without reference to the number of disk I/O's necessary to perform the computation. If the relation is too big to fit in memory, specialized evaluation algorithms, which attempt to limit the number of disk I/O's, can be used to compute the $\sigma_{\$i=c}(R(Q,D))$. (See, for example, [IR88].)

A more detailed analysis of the special case of the transitive closure appears in [LN89]. The analysis shows that in general the algorithm is $\Theta(n\sqrt{m})$ on digraphs with n nodes and m edges, but is $O(m)$ if the graph is of almost uniform degree.

5 Conclusion

In this paper we have presented a framework for the estimation of the size of recursively defined relations. We have shown how the framework can be used to provide size estimation algorithms for the transitive closure and its generalizations. In a database system that supports recursion, such estimation algorithms are essential in order to optimize conjunctive queries and to avoid computations that are infeasible because the result is too large.

Many interesting questions remain. In general terms, the question is whether good estimation algorithms exist for arbitrary datalog recursions. While the work presented here represents a first step toward such an algorithm, it is possible that datalog is sufficiently powerful that no such general estimation procedure exists. Even if no such algorithm exists, the sampling algorithm presented here will be useful if we are correct in our conjecture that a significant fraction of the recursive relations encountered in practice will be generalized transitive closures.

References

[AJ87] Rakesh Agrawal and H. V. Jagadish. Direct algorithms for computing the transitive closure of database relations. In *Proceedings of the 13th VLDB Conference*, pages 255–266, Brighton, England, September 1987.

[Chr83] Stavros Christodoulakis. Estimating block transfers and join sizes. In *Proceedings of the ACM SIGMOD International Conference on Management of Data*, pages 40–54, San Jose, California, May 1983.

[Dem80] R. Demolombe. Estimation of the number of tuples satisfying a query expressed in predicate calculus language. In *Proceedings of the Sixth VLDB Conference*, pages 55–63, Montreal, Canada, 1980.

[HOT88] Wen-Chi Hou, Gultekin Ozsoyoglu, and Baldeao K. Taneja. Statistical estimators for relational algebra expressions. In *Proceedings of the Seventh ACM Symposium on Principles of Database Systems*, pages 276–287, Austin, Texas, March 1988.

[Ioa86] Yannis E. Ioannidis. On the computation of the transitive closure of relational operators. In *Proceedings of the Twelfth*

VLDB Conference, pages 403–411, Kyoto, Japan, August 1986.

[IR88] Yannis E. Ioannidis and Raghu Ramakrishnan. Efficent transitive closure algorithms. In *Proceedings of the Fourteenth VLDB Conference*, pages 382–394, Los Angeles, California, August 1988.

[IW87] Yannis E. Ioannidis and Eugene Wong. Query optimization by simulated annealing. In *Proceedings of the ACM-SIGMOD Conference on the Management of Data*, pages 9–22, San Fransisco, California, May 1987.

[KBZ86] Ravi Krishnamurthy, Haran Boral, and Carlo Zaniolo. Optimization of nonrecursive queries. In *Proceedings of the Twelfth VLDB Conference*, pages 128–137, Kyoto, Japan, August 1986.

[LMR87] Hongjun Lu, Krishna Mikkilineni, and James P. Richardson. Design and evaluation of algorithms to compute the transitive closure of a database relation. In *Proceedings of the Third International Conference on Data Engineering*, pages 112–119, 1987.

[LN89] Richard J. Lipton and Jeffrey F. Naughton. Estimating the size of the transitive closure of a digraph. Submitted for publication, May 1989.

[Lu87] H. Lu. New strategies for computing the transitive closure of a database relation. In *Proceedings of the Thirteenth VLDB Conference*, pages 267–274, Brighton, England, September 1987.

[Lyn88] Clifford A. Lynch. Selectivity estimation and query optimization in large databases with highly skewed distributions of column values. In *Proceedings of the Fourteenth VLDB Conference*, pages 240–251, Los Angeles, California, August 1988.

[Nau87] Jeffrey F. Naughton. One sided recursions. In *Proceedings of the ACM Symposium on Principles of Database Systems*, pages 340–348, San Diego, California, March 1987.

[NSRU89] Jeffrey F. Naughton, Yehoshua Sagiv, Raghu Ramakrishnan, and Jeffrey D. Ullman. Efficient evaluation of right-, left-, and combined-linear rules. In *Proceedings of the ACM SIGMOD International Conference on Management of Data*, Portland, Oregon, May 1989.

[PSC84] Gregory Piatetsky-Shapiro and Charles Connell. Accurate estimation of the number of tuples satisfying a condition. In *Proceedings of the ACM SIGMOD International Conference on Management of Data*, pages 256–276, Boston, Massachusets, June 1984.

[RHDM86] Arnon Rosenthal, Sandra Heiler, Umeshwar Dayal, and Frank Manola. Traversal recursion: A practical approach to supporting recursive applications. In *Proceedings of the ACM-SIGMOD International Conference on Management of Data*, Washington, D.C., June 1986.

[Row83] Neil C. Rowe. Top-down statistical estimation on a database. In *Proceedings of the ACM SIGMOD International Conference on Management of Data*, pages 135–144, San Jose, California, May 1983.

[SG85] David E. Smith and Michael R. Genesereth. Ordering conjunctive queries. *Artificial Intelligence*, 26:171–215, 1985.

[SG88] Arun Swami and Anoop Gupta. Optimization of large join queries. In *Proceedings of the ACM-SIGMOD Conference on Management of Data*, pages 8–17, Chicago, Illinois, May 1988.

Argument Reduction by Factoring

J. F. Naughton* R. Ramakrishnan† Y. Sagiv‡ J. D. Ullman§

Abstract

We identify a useful property of a program with respect to a predicate, called *factoring*. While we prove that detecting factorability is undecidable in general, we show that for a large class of programs, the program obtained by applying the Magic Sets transformation is factorable with respect to the recursive predicate. When the factoring property holds, a simple optimization of the program generated by the Magic Sets transformation results in a new program that is never less efficient than the Magic Sets program and is often dramatically more efficient, due to the reduction of the arity of the recursive predicate. We show that the concept of factoring generalizes some previously identified special cases of recursions, including separable recursions and right- and left-linear recursions, and that the specialized evaluation algorithms and rewriting strategies developed for those classes can be derived automatically by applying the Magic Sets transformation and then factoring the result.

*Department of Computer Science, Princeton University. Work supported by DARPA and ONR contracts NOOO14-85-C-0456 and N00014-85-K-0465, and by NSF Cooperative Agreement DCR-8420948

†Computer Sciences Department, University of Wisconsin, Madison, Wisconsin. Work supported in part by an IBM Faculty Development award and NSF grant IRI-8804319. Part of this work was done while visiting IBM Almaden Research Center.

‡Dept. of Computer Science, Hebrew University, Jerusalem, Israel. Work supported in part by grant 2545-2-87 from the Israeli National Council for Research and Development (ILNCRD).

§Stanford University. Work supported by NSF grant IRI-87-22886 and Air Force grant AFOSR-88-0266 and a grant of the IBM Corp.

Proceedings of the Fifteenth International Conference on Very Large Data Bases

1 Introduction

The Magic Sets transformation [BMSU86, BR87] is a rule rewriting technique that, given a query and a recursive program, produces a new program such that the semi-naive bottom-up evaluation of the new program constructs the answer to the query more efficiently than the original recursion. Magic Sets achieves its power by restricting the search of the underlying database to the portion of the database that is relevant to the query.

The Magic Sets transformation is conceptually simple and the potential savings gained by ignoring the irrelevant tuples in the database is large. However, for some important recursions much better algorithms are known. Intuitively, this is because Magic Sets does not reduce the arity (number of columns) of the recursive predicate. Since the size of the relation computed is bounded by n^k, where n is the number of distinct constants in the database and k is the arity of the recursive predicate, reducing the arity (k) can result in an order of magnitude increase in the efficiency of the algorithm.

In this paper we identify a useful property of a program with respect to a predicate, called *factoring*. If a program can be factored nontrivially with respect to a query, then the program can be rewritten to reduce the arity of the recursive predicate. Few programs and queries have the factoring property as written; however, in many important cases the Magic Sets transformation produces programs that do have the factoring property. While we prove that in general detecting factorable recursions is undecidable, we describe classes of recursions for which the Magic Sets transformation always produces a factorable recursion.

Amsterdam, 1989

Recently the separable recursions [Nau88] and the left- and right-linear recursions [NRSU89] have been identified as significant classes of recursion for which there are arity reducing evaluation algorithms. In this work we show that these classes of recursions are proper subsets of the class of recursions for which Magic Sets produces a factorable recursion. Furthermore, the special purpose evaluation algorithms of [Nau88] and the special purpose rewriting techniques of [NRSU89] can be derived automatically by simple optimizations applied to the factored Magic program.

We introduce the notion of factoring in Section 2, and show that in general it is undecidable. We describe classes of programs for which the corresponding "Magic" programs are factorable in Section 3. In Section 4, we summarize some simple optimizations that can be used in conjunction with factoring to refine a program. We discuss the connections between our approach, that is, Magic Sets followed by factoring, and the Counting transformation and the Separable, One-sided, and Right- and Left-linear classes of programs in Section 5. In Section 6, we present conclusions and directions for future work.

Example 1.1 As an example of the power of our approach, consider a definition of transitive closure including all three forms of the recursive rule, shown in Figure 1. We obtain the following program by first applying the Magic Sets transformation and then factoring:

$$\begin{aligned} m\_t^{bf}(W) &\text{ :- } ft(W).\\ m\_t^{bf}(5).&\\ ft(Y) &\text{ :- } m\_t^{bf}(X),\ e(X,Y).\\ query(Y) &\text{ :- } ft(Y). \end{aligned}$$

(This example is presented in detail in Section 3.) □

$$\begin{aligned} t(X,Y) &\text{ :- } t(X,W), t(W,Y).\\ t(X,Y) &\text{ :- } e(X,W), t(W,Y).\\ t(X,Y) &\text{ :- } t(X,W), e(W,Y).\\ t(X,Y) &\text{ :- } e(X,Y).\\ query(Y) &\text{ :- } t(5,Y). \end{aligned}$$

Figure 1: The three rule transitive closure.

2 The Factoring Property

Consider a program P, a query Q, and a predicate p that appears in P. Let P' be the program obtained by adding the following rules to P:

$$\begin{aligned} p_1(X_{i_1},\ldots,X_{i_k}) &\text{ :- } p(X_1,\ldots,X_n).\\ p_2(X_{j_1},\ldots,X_{j_l}) &\text{ :- } p(X_1,\ldots,X_n).\\ p(X_1,\ldots,X_n) &\text{ :- } p_1(X_{i_1},\ldots,X_{i_k}),\\ &\qquad p_2(X_{j_1},\ldots,X_{j_l}). \end{aligned}$$

where the X_i's are distinct variables. Here, $X_{i_1},\ldots,X_{i_k}$ and $X_{j_1},\ldots,X_{j_l}$ represent subsets of X_1 through X_n. We say that *(P, Q, p)* has the *factoring property* if P and P' compute the same answers to Q for all *EDB*s. More precisely, we say that $p(X_1,\ldots,X_n)$ can be factored into $p_1(X_i,\ldots,X_j)$ and $p_2(X_k,\ldots,X_l)$ in P with respect to Q. This holds trivially if either p_1 or p_2 contains all arguments of p. We say that p can be *non-trivially factored* if neither p_1 nor p_2 contains all arguments of p, and henceforth, we shall consider only non-trivial factoring of programs.

Note that factoring is defined for general logic programs, not just Datalog. The following theorem shows that factorability is undecidable even for Datalog programs.

Theorem 2.1 *It is undecidable whether a predicate in a given program is non-trivially factorable with respect to a given query.*

The proof of Theorem 2.1 is by reduction from the containment problem for Datalog queries, and assumes multiple IDB predicates. To our knowledge, the decidability of factoring for single IDB predicate programs is open.

We have the following simple observation, which suggests an equivalent definition of factoring.

Proposition 2.1 *Let P' be obtained from a given program P by the following transformation with respect to predicate p:*

- *Every body literal $p(t_1,\ldots,t_n)$ is replaced by the literals $p_1(t_{i_1},\ldots,t_{i_k})$ and $p_2(t_{j_1},\ldots,t_{j_l})$.*
- *Every rule with head $p(t_1,\ldots,t_n)$ is replaced by two rules with the same body, and with heads $p_1(t_{i_1},\ldots,t_{i_k})$ and $p_2(t_{j_1},\ldots,t_{j_l})$.*

P and P' compute the same answers to Q for all EDBs if and only if $p(X_1,\ldots,X_n)$, where the Xs are distinct variables, can be factored into $p_1(X_{i_1},\ldots,X_{i_k})$ and $p_2(X_{j_1},\ldots,X_{j_l})$ in P with respect to a query Q.

We refer to the transformation described in the above proposition as the *factoring transformation.* Note that applying this transformation results in a program that does not contain p, which is replaced by two predicates, p_1 and p_2, of strictly lower arity.

3 Classes of Efficiently Evaluable Programs

The Magic Sets transformation [BMSU86, BR87] rewrites a program with the objective of restricting the computation by propagating bindings in the query. We identify classes of programs for which the program produced by applying the Magic Sets transformation can be factored with respect to the recursive predicate.

3.1 Preliminary Definitions

We begin by introducing some terminology and conventions. We only consider programs in which there is a single (recursive) IDB predicate, say p, and there is a single reachable adornment, say p^α. We refer to such programs as *unit* programs.

A rule is said to be in *standard form* if every argument of p^α, in the head or the body, is a variable, and no variable appears in two arguments of the same p^α-literal. We require all rules to be in standard form, and we allow the use of a special predicates to ensure that this requirement does not entail a loss of generality. Thus, a literal $p^\alpha(X, X, 5, Y)$ could be replaced by $p^\alpha(X, U, V, Y), equal(V, 5), equal(X, U)$, while a literal $p^\alpha(X.Y, Z)$ must be replaced by the conjunct $p^\alpha(U, Z)$, $list(X, Y, U)$. Conceptually, *list* and *equal* are infinite EDB relations. Once this translation to standard form is done, the results in this paper can be used to test for factorability

We use $p^\alpha(\overline{X}, \overline{Y})$ to denote a p^α-literal, where $\overline{X}$ is the vector of variables in the bound argument positions of a p^α-literal, and $\overline{Y}$ is the vector of variables in the free argument positions.

Consider a rule in the adorned program with head literal $p^\alpha(\overline{X}, \overline{Y})$. A *left-linear* occurrence of p^α is a body literal $p^\alpha(\overline{X}, \overline{U})$, and a *right-linear* occurrence of p^α is a body literal $p^\alpha(\overline{V}, \overline{Y})$.

The following definitions generalize those in [NRSU89].

Definition 3.1 A rule is *left-linear* if it is of the form

$$p^\alpha(\overline{X},\overline{Y}) \;\text{:-}\; left(\overline{X}), p_1^\alpha(\overline{X},\overline{U}_1), p_2^\alpha(\overline{X},\overline{U}_2), \ldots, p_n^\alpha(\overline{X},\overline{U}_n), last(\overline{U}_1,\overline{U}_2,\ldots,\overline{U}_n,\overline{Y}).$$

where

- The rule is in standard form.
- $left(\overline{X})$ and $last(\overline{U}_1,\ldots,\overline{U}_n,\overline{Y})$ are disjoint conjunctions of EDB predicates.

Definition 3.2 A rule is *right-linear* if it is of the form

$$p^\alpha(\overline{X},\overline{Y}) \;\text{:-}\; first(\overline{X},\overline{V}), p^\alpha(\overline{V},\overline{Y}), right(\overline{Y}).$$

where

- The rule is in standard form.
- $first(\overline{X},\overline{V})$ and $right(\overline{Y})$ are disjoint conjunctions of EDB predicates.

Definition 3.3 A rule is a *combined rule* if it is of the form

$$p^\alpha(\overline{X},\overline{Y}) \;\text{:-}\; left(\overline{X}), p_1^\alpha(\overline{X},\overline{U}_1), p_2^\alpha(\overline{X},\overline{U}_2), \ldots, p_n^\alpha(\overline{X},\overline{U}_n), center(\overline{U},\overline{V}), p^\alpha(\overline{V},\overline{Y}), right(\overline{Y}).$$

where

- The rule is in standard form.
- $left(\overline{X})$, $center(\overline{U},\overline{V})$, and $right(\overline{Y})$ are disjoint conjunctions of EDB predicates.

We remark that some of the conjunctions of EDB predicates referred to in the above definitions may contain occurrences of the special EDB predicate *equal*. As a special case, a conjunction may contain only such occurrences.

3.2 Factorable Programs

We present theorems that identify classes of programs for which the corresponding Magic programs are factorable. The importance of these theorems lies in the technique that they exemplify: a two-step approach to optimizing programs in which the programs are rewritten using the Magic Sets transformation and subsequently factored if possible.

Let P be a program, Q a query, and P^{ad} the adorned program corresponding to a left-to-right evaluation of the rules of P. P^{mg} represents the program obtained by applying the Magic Sets transformation to P and Q.

Example 3.1 The rewriting algorithms presented in [NRSU89] were the first to derive automatically unary programs for single-selection queries for all three forms (left-linear, right-linear, non-linear) of the transitive closure. We achieve the same result here by first applying the Magic Sets transformation and

$$
\begin{array}{rcl}
m\_t^{bf}(W) & \text{:-} & m\_t^{bf}(X), t^{bf}(X,W). \\
m\_t^{bf}(W) & \text{:-} & m\_t^{bf}(X), e(X,W). \\
m\_t^{bf}(5). & & \\
t^{bf}(X,Y) & \text{:-} & m\_t^{bf}(X), t^{bf}(X,W), t^{bf}(W,Y). \\
t^{bf}(X,Y) & \text{:-} & m\_t^{bf}(X), e(X,W), t^{bf}(W,Y). \\
t^{bf}(X,Y) & \text{:-} & m\_t^{bf}(X), t^{bf}(X,W), e(W,Y). \\
t^{bf}(X,Y) & \text{:-} & m\_t^{bf}(X), e(X,Y). \\
query(Y) & \text{:-} & t^{bf}(5,Y).
\end{array}
$$

Figure 2: P^{mg} for the three-rule transitive closure.

then factoring the rewritten program. To illustrate the technique, we again consider the single program that includes all three forms of the recursive rule presented in Figure 1. The Magic Sets algorithm rewrites this program to produce the program in Figure 2.

If we identify m_t^{bf} tuples with goals in a top-down evaluation, we see that only the last occurrence of t^{bf} in a rule body generates new goals, and further, the answer to a new goal is also an answer to the goal that invoked the rule. In fact, every answer to a subgoal is also an answer to the query goal m_t^{bf}. Also, if c is generated as an answer to a subgoal, then a new subgoal $m\_t^{bf}(c)$ is also generated. These observations imply that it does not matter which subgoal an answer corresponds to; its role in the computation is the same in any case. That is, $t^{bf}(X,Y)$ can be factored into $bt(X)$ and $ft(Y)$ in the Magic program. This yields the program shown in Figure 3.

Applying further optimizations, discussed in Section 4, we finally obtain the following unary program:

$$
\begin{array}{rcl}
m\_t^{bf}(W) & \text{:-} & ft(W). \\
m\_t^{bf}(5). & & \\
ft(Y) & \text{:-} & m\_t^{bf}(X), e(X,Y). \\
query(Y) & \text{:-} & ft(Y).
\end{array}
$$

□

Definition 3.4 Let p be the only IDB predicate in a program P, and Q be a query on p. Then the combination of P and Q is an *RLC-stable program* if P consists only of right-linear, left-linear, and combined-linear rules plus one exit rule, and p^{α} is the only adorned version of p in P^{ad}.

We now define some auxiliary conjunctive queries that appear often later in this section.

$$
\begin{array}{rcl}
m\_t^{bf}(W) & \text{:-} & m\_t^{bf}(X), bt(X), ft(W). \\
m\_t^{bf}(W) & \text{:-} & m\_t^{bf}(X), e(X,W). \\
m\_t^{bf}(5). & & \\
bt(X) & \text{:-} & m\_t^{bf}(X), bt(X), ft(W), \\
 & & bt(W), ft(Y). \\
bt(X) & \text{:-} & m\_t^{bf}(X), e(X,W), \\
 & & bt(W), ft(Y). \\
bt(X) & \text{:-} & m\_t^{bf}(X), bt(X), \\
 & & ft(W), e(W,Y). \\
bt(X) & \text{:-} & m\_t^{bf}(X), e(X,Y). \\
ft(Y) & \text{:-} & m\_t^{bf}(X), bt(X), ft(W), \\
 & & bt(W), ft(Y). \\
ft(Y) & \text{:-} & m\_t^{bf}(X), e(X,W), \\
 & & bt(W), ft(Y). \\
ft(Y) & \text{:-} & m\_t^{bf}(X), bt(X), \\
 & & ft(W), e(W,Y). \\
ft(Y) & \text{:-} & m\_t^{bf}(X), e(X,Y). \\
query(Y) & \text{:-} & bt(5), ft(Y).
\end{array}
$$

Figure 3: The factored version of P^{mg}.

Definition 3.5 The conjunctive query *free_exit* is defined as follows:

$$free\_exit(\overline{Y}) \text{ :- } exit(\overline{X}, \overline{Y}).$$

where $exit(\overline{X}, \overline{Y})$ is the body of the exit rule.

The conjunctive query *bound_first* is defined for a given right-linear rule:

$$bound\_first(\overline{X}) \text{ :- } first(\overline{X}, \overline{U}).$$

where $first(\overline{X}, \overline{U})$ appears in the body of the rule.

The conjunctive query *bound* is defined for a given left-linear or combined rule:

$$bound(\overline{X}) \text{ :- } left(\overline{X}).$$

where $left(\overline{X})$ appears in the body of the rule.

The conjunctive query *bound* is defined for a given right-linear or combined rule:

$$free(\overline{Y}) \text{ :- } right(\overline{Y}).$$

where $right(\overline{Y})$ appears in the body of the rule.

The conjunctive query *middle* is defined for a given combined rule:

$$middle(\overline{U}, \overline{V}) \text{ :- } center(\overline{U}, \overline{V}).$$

where $center(\overline{U}, \overline{V})$. appears in the body of the rule.

Often by a slight abuse of notation we will refer to *left*, *right*, and *center* as conjunctive queries instead of using *bound*, *free* and *middle*.

Our first theorem essentially generalizes of the results in [NRSU89], although it must be used in conjunction with the additional optimizations described in Section 4 in order to do so. It uses the following definition.

Definition 3.6 Let P, Q be an RLC-stable program with IDB predicate p. Then P, Q is *selection-pushing* if the following conditions hold:

- For any combined or right-linear rule r in P, the conjunctive query "*free_exit*" must be contained in the conjunctive query "*free*" for r.
- For any pair of rules r_1 and r_2 in P, if both r_1 and r_2 contain a "*left*" conjunctive query, these must be equivalent. If one contains a "*left*" query, and the other a "*first*" query, the conjunctive query "*bound_first*" must be contained in the conjunctive query "*bound*".

Theorem 3.1 *Let P, Q be an RLC-stable program with IDB predicate p, and let $\overline{X}$ be the vector of variables appearing in bound arguments of p^α in the heads of the rules of P^{ad}, and let $\overline{Y}$ be the vector of variables appearing in free arguments of p^α in P^{ad}. If P, Q is selection-pushing then $p^\alpha(\overline{X}, \overline{Y})$ can be factored into $bp(\overline{X})$ and $fp(\overline{Y})$ in P^{mg} with respect to the query Q.*

Example 3.2 We illustrate the intuition behind selection-pushing and show that violating any of the associated conditions could destroy this property.

$$
\begin{array}{rcl}
p(X,Y) & :- & l1(X), p(X,U), c1(U,V), \\
 & & p(V,Y), r1(Y). \\
p(X,Y) & :- & l2(X), p(X,U), c2(U,V), \\
 & & p(V,Y), r2(Y). \\
p(X,Y) & :- & f(X,V), p(V,Y), r3(Y). \\
p(X,Y) & :- & e(X,Y). \\
query(Y) & :- & p(5,Y).
\end{array}
$$

The Magic Sets algorithm rewrites this to

$$
\begin{array}{rcl}
m\_p^{bf}(V) & :- & m\_p^{bf}(X), l1(X), \\
 & & p^{bf}(X,U), c1(U,V). \\
m\_p^{bf}(V) & :- & m\_p^{bf}(X), l2(X), \\
 & & p^{bf}(X,U), c2(U,V). \\
m\_p^{bf}(V) & :- & m\_p^{bf}(X), f(X,V). \\
m\_p^{bf}(5). & & \\
p^{bf}(X,Y) & :- & m\_p^{bf}(X), l1(X), p^{bf}(X,U), \\
 & & c1(U,V), p^{bf}(V,Y), r1(Y). \\
p^{bf}(X,Y) & :- & m\_p^{bf}(X), l2(X), p^{bf}(X,U), \\
 & & c2(U,V), p^{bf}(V,Y), r2(Y). \\
p^{bf}(X,Y) & :- & m\_p^{bf}(X), f(X,V), \\
 & & p^{bf}(V,Y), r3(Y). \\
p^{bf}(X,Y) & :- & m\_p^{bf}(X), e(X,Y). \\
query(Y) & :- & p^{bf}(5,Y).
\end{array}
$$

Factoring this program and applying further transformations described in detail in Section 4 yields

$$
\begin{array}{rcl}
m\_p^{bf}(V) & :- & bp(X), l1(X), \\
 & & fp(U), c1(U,V). \\
m\_p^{bf}(V) & :- & bp(X), l2(X), \\
 & & fp(U), c2(U,V). \\
m\_p^{bf}(V) & :- & m\_p^{bf}(X), f(X,V). \\
m\_p^{bf}(5). & & \\
bp(X) & :- & m\_p^{bf}(X), f(X,V), \\
 & & bp(V), fp(Y), r3(Y). \\
bp(X) & :- & m\_p^{bf}(X), e(X,Y). \\
fp(Y) & :- & m\_p^{bf}(X), e(X,Y). \\
query(Y) & :- & fp(Y).
\end{array}
$$

The transformations that produce the above program from the factored version of the Magic program preserve equivalence. We have applied these transformations in order to delete some unnecessary literals and rules in the factored program, thus making it easier to understand the essential ideas.

Consider the following EDB instance: $f(5,1)$, $e(5,6)$, $e(1,7)$, $e(2,8)$, $l1(1)$, $c1(6,2)$, $r1(7)$, $r1(8)$. Because the condition that *bound_first* should be a subset of $l1$ is violated by this EDB, 8 is incorrectly derived as an answer. Indeed, $m\_p^{bf}(1)$ is generated using $m\_p^{bf}(5)$ and $f(5,1)$. This generates $bp(1)$ using $e(1,7)$. Also, the tuple $e(5,6)$ gives us $fp(6)$. The critical step follows: the fact $fp(6)$ is used in the first rule with $bp(1)$, $l1(1)$ and $c1(6,2)$ to generate the fact $m\_p^{bf}(2)$. That is, the fact $fp(6)$, which is an answer to the goal $m\_p^{bf}(5)$, is incorrectly used where an answer to the goal $m\_p^{bf}(1)$ is required, thereby generating a spurious subgoal. One can verify that 8 is a valid answer if $l1(5)$ is added to the EDB. A similar example can be constructed if $l1$ and $l2$ are not identical, since the answer generated in response to a subgoal that satisfies $l1$ but not $l2$ can be used in the second rule to generate spurious subgoals.

Now consider the EDB instance: $f(5,1)$, $e(5,6)$, $e(1,7)$, $l1(5)$, $c1(6,1)$. The fact $fp(7)$ is incorrectly generated. The first rule is used to generate $m\_p^{bf}(1)$ from the query goal and the fact $e(5,6)$. The fact

$fp(7)$ is generated in response to this subgoal, but it cannot be an answer to $m\_p^{bf}(5)$ unless $r1(7)$ is true. The EDB instance violates the condition that *free_exit* should be contained in $r1$. This made it possible to generate subgoals whose answers are not answers to the original goal. □

Intuitively, we are separating the bound arguments from the free arguments, and we must ensure that every answer to a subquery (keeping in mind a top-down evaluation of the program) is also an answer to the original query. (We refer to the vector of values in the free arguments as the answer, corresponding to a query that is the vector of values in the bound arguments of a p^α-fact.) For this, we require that the *right* conjunctive queries be satisfied by every potential answer tuple, that is, *free_exit* is contained in every *right* conjunctive query. (Some answer tuples may be generated from left-linear rules, but these need not satisfy the *right* queries since there is a derivation of these answers to the original query that does not propagate these answers through right-linear occurrences of p^α.)

In addition, we must ensure that no spurious answers are generated. The main idea is that for every derivation of a fact using P^{mg}, there is an equivalent derivation in which the bound arguments of every left-linear p^α fact is identical to the bound arguments in the query. That is, in every recursive rule that contains a left-linear occurrence of p^α, we can replace the variables $X_1, \ldots, X_m$ in the bound arguments by the constants provided in the original query. This is in fact the motivation for the term "selection-pushing."

When a right-linear rule is applied to generate new subqueries, the answers to these subqueries could be used in left-linear occurrences of p^α. To justify this, we must ensure that a subquery invoking the right-linear rule is reachable from a subquery that satisfies the conditions on the bound arguments of the left-linear occurrences of p^α. Since every subquery is reachable from the initial goal, this is guaranteed if the initial query satisfies the (unique, for the given program) *left* conjunctive query. If the initial goal does not satisfy the *left* conjunctive query, then we cannot apply the right-linear rule, and the condition that the *bound_first* conjunctive queries should be contained in the *left* conjunctive query ensures this.

We can identify further classes of programs that can be factored.

Definition 3.7 Let P, Q be an RLC-stable program containing only combined recursive rules. Then P, Q is *symmetric* if the following conditions hold:

- Each rule contains exactly two occurrences of p^α in the body, and the "*middle*" conjunctive queries are all equivalent.
- For any recursive rule r in P, the conjunctive query "*free_exit*" must be contained in the conjunctive query "*right*" for r.

Theorem 3.2 *Let P, Q be an RLC-stable program with IDB predicate p, and let $\overline{X}$ be the vector of variables appearing in bound arguments of p^α in the heads of the rules of P^{ad}, and let $\overline{Y}$ be the vector of variables appearing in free arguments of p^α in P^{ad}. If P, Q is symmetric, then $p^\alpha(\overline{X}, \overline{Y})$ can be factored $bp(\overline{X})$ and $fp(\overline{Y})$ in P^{mg} with respect to the query Q.*

The Magic program for the following query can thus be factored:

$$
\begin{aligned}
p(X,Y) &\text{ :- } l1(X), p(X,U), c(U,V),\\
&\qquad p(V,Y), r1(Y).\\
p(X,Y) &\text{ :- } l2(X), p(X,U), c(U,V),\\
&\qquad p(V,Y), r2(Y).\\
p(X,Y) &\text{ :- } e(X,Y).\\
query(Y) &\text{ :- } p(5,Y).
\end{aligned}
$$

In summary, the results in this section are illustrative of a general approach to optimizing programs, in which we first apply the Magic Sets transformation and then factor. When we factor a Magic program and separate the bound and free arguments, we must establish two things:

- Every answer to a subquery is also an answer to the original query.
- No spurious subqueries or answers are generated.

Because testing for these classes of recursions in general requires testing for containment of conjunctive queries, and testing for conjunctive query containment is NP-complete [CM77, ASU79], testing for membership in these classes is also NP-complete. It is important that the measure of size here is the size of the recursion and query, not the database. An algorithm that is exponential in the size of the recursion and query (small) may be worth running during query planning in order to save time proportional to the size of the database (large) during query evaluation. Furthermore, in many cases, the conjunctive queries will be empty, in which case polynomial time algorithms for testing if a recursion satisfies Theorems 3.1 and 3.2 recursions exist.

4 Some Additional Optimizations

We use the following definitions.

Definition 4.1 A bound argument position of p^α is a *static argument position* if for every p^α-literal in the body of a rule, the variable in this argument position also appears in the same argument position in the head of the rule. (Recall that the head must also be a p^α literal, since we only consider unit programs.)

Definition 4.2 Let (P, Q) be a unit program — query pair, and let the ith argument of p^α be a static argument. Without loss of generality, let the variable in the ith argument of P^α always be X, and let the constant in the ith argument of the query Q be c. The program P is *reduced with respect to argument position i* as follows:

- Every rule r is replaced by $\sigma(r)$, where σ is the substitution $X \leftarrow c$.
- Every p^α-literal — in the head or the body of a rule — is replaced by a s^θ-literal with the same vector of arguments except for the ith argument, which is deleted. s is a new predicate with one fewer argument position, and θ is identical to the adornment α, but with the b corresponding to the ith argument deleted.

We begin with a result that augments the theorems presented in the previous section. Some programs that do not satisfy the conditions of these theorems can be transformed into programs that do by applying the following lemma.

Lemma 4.1 *Let (P, Q) be a unit program — query pair, let the ith argument of P^α be a static argument, and let P' be the reduced program. Then P and P' are equivalent with respect to Q.*

In the rest of this section, we summarize a few simple optimizations that are often applicable to factored programs.

If p^α is factored into bp and fp in P^{mg}, then the relation bp is contained in $magic\_p^\alpha$, since every rule defining bp contains $magic\_p^\alpha$ (with identical arguments) in the body. Further, for every rule defining fp (resp. bp) there is a rule with an identical body describing bp (resp. fp). Therefore, the goal $bp(\_)$, where $\_$ denotes an "anonymous" variable, succeeds if any fp goal succeeds, and vice-versa. These observations lead to the following propositions.

Proposition 4.1 *If a rule contains both bp and $magic\_p^\alpha$ in the body, with identical arguments, then we may delete the $magic\_p^\alpha$ literal.*

Proposition 4.2 *If a rule contains the literal $bp(\_)$ and also an fp literal, the literal $bp(\_)$ can be deleted.*

A symmetric proposition allows us to delete some $fp(\_)$ literals.

A similar observation is that if $m\_p^\alpha(\overline{c})$ is the original query goal, then $bp(\overline{c})$ is true if any fp goal succeeds. This is because every fp fact, in particular the successful fp goal, is an answer to the original query. However, note that in general, p^α may be factored but the original query may not be on predicate p^α.

Proposition 4.3 *Let the original query correspond to the fact $m\_p^\alpha(\overline{c})$. If a rule contains the literal $bp(\overline{c})$ and also an fp literal, then the literal $bp(\overline{c})$ can be deleted.*

Some additional simple observations that are useful are mentioned below.

Proposition 4.4 *We may delete a rule if the head literal also appears in the body, or if the head predicate is not reachable from the query predicate.*

This is a special case of deletion under uniform equivalence [Sag87].

Proposition 4.5 *We may introduce an "anonymous" variable in an argument position if the variable in it appears nowhere else in the rule.*

As shown in [RBK88], the preceding proposition can be strengthened to prove that an anonymous variable can be introduced in any existential argument position.

Example 4.1 Consider again the factored version of P^{mg} from the three-rule transitive closure (Figure 3.) We can delete the first and the third rules defining bt and the first two rules defining ft because the head literal also appears in the body. We can also delete the literal $m\_t^{bf}(X)$ from every rule that also contains $bt(X)$, and then replace all variables that only appear once in a rule by anonymous variables. This yields:

$$
\begin{aligned}
m\_t^{bf}(W) &\ \text{:-}\ bt(\_), ft(W).\\
m\_t^{bf}(W) &\ \text{:-}\ m\_t^{bf}(X), e(X,W).\\
m\_t^{bf}(5)&.\\
bt(X) &\ \text{:-}\ m\_t^{bf}(X), e(X,W),\\
&\qquad bt(W), ft(Y).\\
bt(X) &\ \text{:-}\ m\_t^{bf}(X), e(X,Y).\\
ft(Y) &\ \text{:-}\ bt(\_), ft(W), e(W,Y).\\
ft(Y) &\ \text{:-}\ m\_t^{bf}(X), e(X,Y).\\
query(Y) &\ \text{:-}\ bt(5), ft(Y).
\end{aligned}
$$

We can delete both body occurrences of $bt(\_)$ since the rules in which they appear also contain ft literals in the body. Similarly, we can delete the literal $bt(5)$ from the rule defining the query. This makes bt unreachable from the query, and we can delete all remaining rules for bt. This gives us:

$$\begin{array}{rcl} m\_t^{bf}(W) & :- & ft(W). \\ m\_t^{bf}(W) & :- & m\_t^{bf}(X), e(X, W). \\ m\_t^{bf}(5). & & \\ ft(Y) & :- & ft(W), e(W, Y). \\ ft(Y) & :- & m\_t^{bf}(X), e(X, Y). \\ query(Y) & :- & ft(Y). \end{array}$$

The second rule defining m_t^{bf} and the first rule defining ft can be deleted under uniform equivalence, and we finally obtain the following program:

$$\begin{array}{rcl} m\_t^{bf}(W) & :- & ft(W). \\ m\_t^{bf}(5). & & \\ ft(Y) & :- & m\_t^{bf}(X), e(X, Y). \\ query(Y) & :- & ft(Y). \end{array}$$

□

5 A Unifying Overview

We consider how the refinements of the Magic Sets transformation presented in this paper are related to some previously defined optimizations.

5.1 One-Sided Recursions

One-sided recursions were identified in [Nau87] as a class of recursions for which there are efficient evaluation algorithms. Here we restate the characterization of one-sided recursions.

Theorem 5.1 *(Theorem 3.1 from [Nau87]) Let D be a recursive definition with a single, linear recursive rule r. Then D is one-sided if and only if the full A/V graph for r has only one connected component with a cycle of nonzero weight, and that component has a cycle of weight 1.*

An important subset of the one-sided recursions are those such that the full A/V graph has one connected component with a cycle of nonzero weight, and that component contains exactly one cycle of nonzero weight, and that cycle is of weight 1. We call such a one-sided recursion a *simple one-sided recursion.* Any simple one-sided recursion can be "expanded" (by substituting the rule into itself some number of times) to produce a rule of the form

$$p(\overline{A}, \overline{B}) \ :- \ p(\overline{A}, \overline{C}), c(\overline{C}, \overline{D}, \overline{B}). \qquad (1)$$

where $\overline{A}$, $\overline{B}$, $\overline{C}$, and $\overline{D}$ are vectors of disjoint variables, and c is a conjunction of EDB predicates.

The preceding recursion is written in a form isomorphic to what we have called a left-linear recursive rule. However, the definition of left-linear is in terms of both the recursion and the specific query in question. By contrast, the one-sided recursions are defined independently of queries. Notice, however, that coupled with the query $p(\overline{c}, Y)$?, the preceding rule is left-linear; while coupled with the query $p(X, \overline{d})$? it is right-linear.

A selection that binds either every variable in $\overline{A}$ or $\overline{B}$ is a *full-selection.* With this definition, we can formalize the preceding discussion with the following theorem.

Theorem 5.2 *Let P be a simple one-sided recursion, expanded so that it is of the form of Equation 1. Let Q be a full-selection query on p, the recursive predicate of P. Also, let P^{mg} be the output of the Magic Sets algorithm on P and Q. Then P^{mg} and Q factor with respect to p.*

5.2 Separable Programs

Separable programs, defined in [Nau88], were defined to be class of recursions for which selection queries have efficient evaluation algorithms. Essentially, [Nau88] gave conditions that determine if a given recursion is separable and a schema for evaluating selection queries over separable recursions. Given a specific selection query on a recursion that is separable, the schema can automatically be instantiated to produce an evaluation algorithm for the query.

As was the case with one-sided recursions, the variables appearing in the heads of rules in separable recursions can be divided into equivalence classes (see [Nau88] for details.) A selection query that binds every variable of some equivalence class is a *full-selection query,* as before.

Theorem 5.3 *Let P be a separable recursion, let Q be a full-selection query on p (the recursive predicate of P), and let P^{mg} be the result of the Magic Sets transformation applied to P, Q. Then the pair P^{mg}, Q is factorable.*

The proof proceeds by showing that the conditions for separability given in [Nau88] guarantee that the pair P, Q will satisfy the conditions of Theorem 3.1.

To see that the converse is not true, that is, that there are factorable programs that cannot be viewed as full selections on separable recursions, note that separable recursions are all linear, whereas factorable programs need not be linear.

There is also a close connection between the instantiated separable recursion evaluation algorithm and the program resulting from Magic Sets followed by the factoring rewrite. Essentially, for a full selection on a separable recursion, the instantiated separable recursion evaluation schema represents the same computation as the semi-naive bottom-up evaluation of the output of the factoring rewrite applied to the Magic program.

5.3 Left- and Right-Linear Programs

In [NRSU89], recursions containing right-linear, left-linear, mixed-linear, and combined-linear recursions were identified and special rewriting algorithms in the spirit of the Magic Sets transformation were given. A simple check shows that the classes of programs defined in [NRSU89] are a proper subset of the programs satisfying the conditions of Theorem 3.1, and that Theorem 3.2 handles some additional programs. In addition, for the programs considered in that paper, the Magic Sets plus factoring transformation produces the same final program as the rewriting algorithms from that paper.

5.4 The Counting Transformation

The Counting transformation [BMSU86, BR87, SZ86] can be understood as a variant of the Magic Sets transformation. First, every derived predicate is augmented with some index fields, which, intuitively, encode the derivation of the fact. That is, the value of the index encodes the sequence of rule applications, and the literal that is expanded at each step, that was used to derive the fact. The program P^{mg} with these additional fields is then refined by deleting the fields corresponding to bound arguments in derived predicates.

When we describe Counting as reducing the arity of derived predicates, we ignore the new index fields that are introduced. The cost of computing the indices can be significant; in fact, this may make the Counting strategy more expensive than even Naive fixpoint evaluation, or cause non-termination.

There is an obvious parallel to factoring Magic programs, since the objective here is again to reduce the arity of derived predicates by separating the bound and free arguments. The connection is quite close — for the class of programs for which we have shown the Magic program to be factorable, the factored program (with some of the simple optimizations that we discussed in Section 4) is identical to the Counting program with all index fields deleted. In effect, this is a class of programs for which the benefits of the Counting strategy — reductions in predicate arity, and accompanying deletion of some literals and rules — can be obtained without the overhead of computing indices.

If a program contains left-linear or combined rules, the Counting program will not terminate since a rule is created that generates the same fact with an infinite number of values in the index fields. The following example is illustrative:

$$\begin{aligned} t^{bf}(X,Y) &\text{ :- } t^{bf}(X,Z), e(Z,Y). \\ t^{bf}(X,Y) &\text{ :- } e(Z,Y). \end{aligned}$$

The first rule generates the Magic rule:

$$magic\_t^{bf}(X) \text{ :- } magic\_t^{bf}(X).$$

With the indices added in the Counting transformation, this is:

$$cnt\_t^{bf}(X, I+1) \text{ :- } cnt\_t^{bf}(X, I).$$

This is a rule whose fixpoint evaluation does not terminate, given an initial cnt_t^{bf} fact, which is obtained from the query.

Theorem 5.4 *If a program satisfies the conditions of the factoring theorems in Section 3, and no rule contains a left-linear literal, then the factored Magic program, after deleting trivially redundant rules, is identical to the Counting program with all index fields deleted.*

The factoring approach allows us to reduce arities of some programs with left-linear literals, whereas the Counting program would never terminate in such cases. On the other hand, the well-known same-generation program is the canonical example of a program that cannot be factored, and in which the index fields introduced in Counting are necessary.

6 Conclusion and Directions for Future Work

We have shown that the Magic Sets transformation followed by factoring produces programs on which bottom-up evaluation efficiently produces the answer to the query.

The results presented in this paper motivate several interesting problems.

- We have identified classes of programs for which the corresponding Magic program can be factored. However, there are other interesting programs that can also be factored. Identifying broader classes of factorable programs is an interesting research direction.
- We showed that for the classes of factorable Magic programs identified in this paper, the indices in Counting were unnecessary. Can we show that the Counting indices are unnecessary in factorable Magic programs, independently of the sufficient conditions that we use to ensure factorability?
- Not all one-sided recursions have arity-reducing evaluation algorithms, and not all one-sided recursions produce factorable Magic programs. Does Theorem 3.1 cover all one-sided recursions that have arity-reducing evaluation algorithms?
- Suppose the program for p^α is factorable, but this predicate is not the query predicate. How can we identify cases in which p^α can be factored even though it is not the top-level query?
- Consider the various techniques for deleting rules and literals in Section 4 (additional optimizations). Does the order in which these are applied to a program affect the final result? If so, can we identify classes of programs for which the final result is unique?

References

[ASU79] Alfred V. Aho, Yehoshua Sagiv, and Jeffrey D. Ullman. Equivalence of relational expressions. *SIAM Journal of Computing*, 8(2):218–246, 1979.

[BMSU86] Francois Bancilhon, David Maier, Yehoshua Sagiv, and Jeffrey D. Ullman. Magic sets and other strange ways to implement logic programs. In *Proceedings of the ACM Symposium on Principles of Database Systems*, pages 1–15, Boston, Massachusetts, March 1986.

[BR87] Catriel Beeri and Raghu Ramakrishnan. On the power of magic. In *Proceedings of the ACM Symposium on Principles of Database Systems*, pages 269–283, San Diego, California, March 1987.

[CM77] Ashok K. Chandra and Philip M. Merlin. Optimal implementation of conjunctive queries in relational data bases. In *Conference Record of the Ninth Annual ACM Symposium on Theory of Computing*, pages 77–90, Boulder, Colorado, May 1977.

[Nau87] Jeffrey F. Naughton. One sided recursions. In *Proceedings of the ACM Symposium on Principles of Database Systems*, pages 340–348, San Diego, California, March 1987.

[Nau88] Jeffrey F. Naughton. Compiling separable recursions. In *Proceedings of the SIGMOD International Symposium on Management of Data*, pages 312–319, Chicago, Illinois, May 1988.

[NRSU89] Jeffrey F. Naughton, Yehoshua Sagiv, Raghu Ramakrishnan, and Jeffrey D. Ullman. Efficient evaluation of right-, left-, and combined-linear rules. In *Proceedings of the SIGMOD International Symposium on Management of Data*, pages 235–242, Portland, Oregon, May 1989.

[Ram87] Raghu Ramakrishnan. *Magic Templates: A Spellbinding Approach to Logic Programs.* In *Proceedings of the International Conference on Logic Programming*, pages 140-159, Seattle, Washington, August 1988.

[RBK88] Raghu Ramakrishnan, Catriel Beeri, and Ravi Krishnamurthy. Optimizing existential datalog queries. In *Proceedings of the ACM Symposium on Principles of Database Systems*, pages 89–102, Austin, Texas, March 1988.

[Sag87] Yehoshua Sagiv. Optimizing datalog programs. In *Proceedings of the ACM Symposium on Principles of Database Systems*, pages 349–362, Austin, TX, March 1987.

[SZ86] Domenico Sacca and Carlo Zaniolo. The generalized counting methods for recursive logic queries. In *Proceedings of the First International Conference on Database Theory*, 1986.

PANEL SESSION
Database Support for Hypertext

Panel chairman:
Nick Rossiter
University of Newcastle upon Tyne, UK

Panel Members:
Ted Nelson
Hector Garcia-Molina
Michael Heather
Eve Wilson
Peter Thanni

Hypertext is developing as a separate branch of computing but its aims and methods may to some extent be achieved within existing areas of computing science. Hypertext needs features like referential transparency and referential integrity. Obviously these and other desirable facilities are already well developed within database technology. The intention of this panel is to investigate the hypermedium as a very large database management system.

Present hypertext systems concentrate on the human-computer interface and the replication of physical addressing to represent links between one document and another but symbolic addressing is needed to cope with pre-existing forms of citation and for automated authoring of large quantities of text. Hypertext packages which rely on the whole document residing in main memory are very limited in the size of data that they can handle. Where they can page, data volumes are often restricted by authoring difficulties.

Hypertext systems are generally self-contained and cannot be easily integrated with other programs and data or used concurrently by a number of different authors. On hypermedia, they can store multimedia data in physical form but the semantic structure is not captured. There is limited opportunity for mapping and indirection between user views and storage structures: there has to be one fixed view - that of the author, with little scope for the preferences of individual readers. There is no independent level of control that can test or validate the data and track the navigation through documents: the design and construction of maintainable links is a major problem.

The main question is whether, to exploit the full power of hypertext, it is necessary to have some machine model expressing semantic detail of the documents held with a full abstract specification of the data-types involved and a multi-level architecture similar to that of a DBMS. A related question is the kind of model that is most suitable for representing the architecture of documents and multimedia data. Ideally gains in flexibility and information representation resulting from the use of database techniques should not be counterbalanced by a deterioration in the human-computer interface.

Besides being involved in the initial development, implementation and use of hypertext systems, members of the panel have experience in applying database techniques to textbases using E-R and relational models and object-oriented approaches such as Taxis and Postgres.

Finding Regular Simple Paths in Graph Databases

Alberto O. Mendelzon
Peter T. Wood†

Computer Systems Research Institute
University of Toronto
Toronto, Canada M5S 1A4

ABSTRACT

We consider the following problem: given a labelled directed graph G and a regular expression R, find all pairs of nodes connected by a simple path such that the concatenation of the labels along the path satisfies R. The problem is motivated by the observation that many recursive queries can be expressed in this form, and by the implementation of a query language, $\mathbf{G}^+$, based on this observation. We show that the problem is in general intractable, but present an algorithm than runs in polynomial time in the size of the graph when the regular expression and the graph are free of *conflicts*. We also present a class of languages whose expressions can always be evaluated in time polynomial in the size of both the database and the expression, and characterize syntactically the expressions for such languages.

1. INTRODUCTION

The design of our query language $\mathbf{G}^+$ [CMW87, CMW88] is based on the observation that many of the recursive queries that arise in practice—and in the literature—amount to graph traversals. For example, see [Agra87, GSS87, R*86]. In $\mathbf{G}^+$, we view the database as a directed, labelled graph, and pose queries which are graph patterns; the answer to a query is the set of subgraphs of the database that match the given pattern. In our prototype implementation, queries are drawn on a workstation screen and the database and query results are also displayed pictorially.

EXAMPLE 1.1: Let G be a graph representing airline flights: the nodes of G denote cities, and an edge labelled a from city b to city c means that there is a flight from b to c with airline a. Assume that we want to find all pairs of cities that are connected by a sequence of flights such that (a) at least one flight is with Air Canada (AC), and (b) no city is visited more than once.

†Current address: Department of Computer Science, University of Cape Town, South Africa.

This query can be expressed by the graph pattern in Figure 1.1.

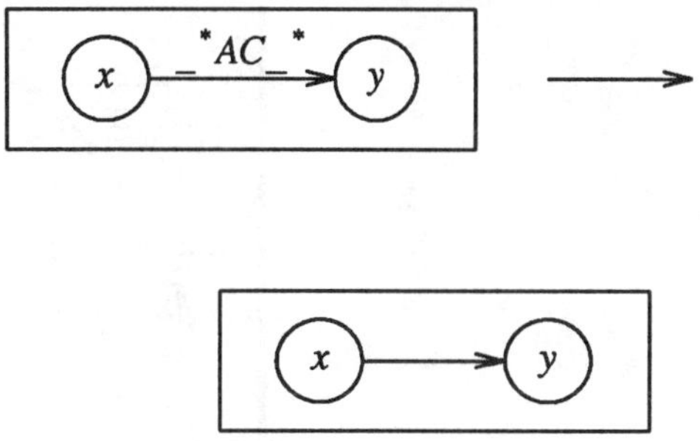

Figure 1.1. A graphical query.

The first box in the figure contains the pattern graph, while the second box contains the summary graph which specifies how the output is to be presented to the user. The edges of a pattern graph can be labelled with regular expressions; in this case the desired expression is $\_^*AC\_^*$ (where the underscore matches any edge label in G). This regular expression is used to match the edge labels along *simple* paths in G, thereby satisfying our original request. □

Figure 1.2 shows two screens from the G prototype. In the first screen, one window contains a database graph where the nodes are cities and the edges are flights between cities, labelled with airline names. The other window contains a query graph asking for all pairs of nodes connected by a simple path of a certain form. The second screen shows one of the answers, displayed by shading nodes of the database graph. Instead of using a summary graph, the user has chosen from a menu to display only the endpoints of each path.

Although queries in $\mathbf{G}^+$ can be a lot more general than exemplified above, this special case is challenging enough from an algorithmic point of view if we want to process queries efficiently. The problem addressed in this paper is: given a regular expression R and a graph G, find

Proceedings of the Fifteenth International Conference on Very Large Data Bases

Amsterdam, 1989

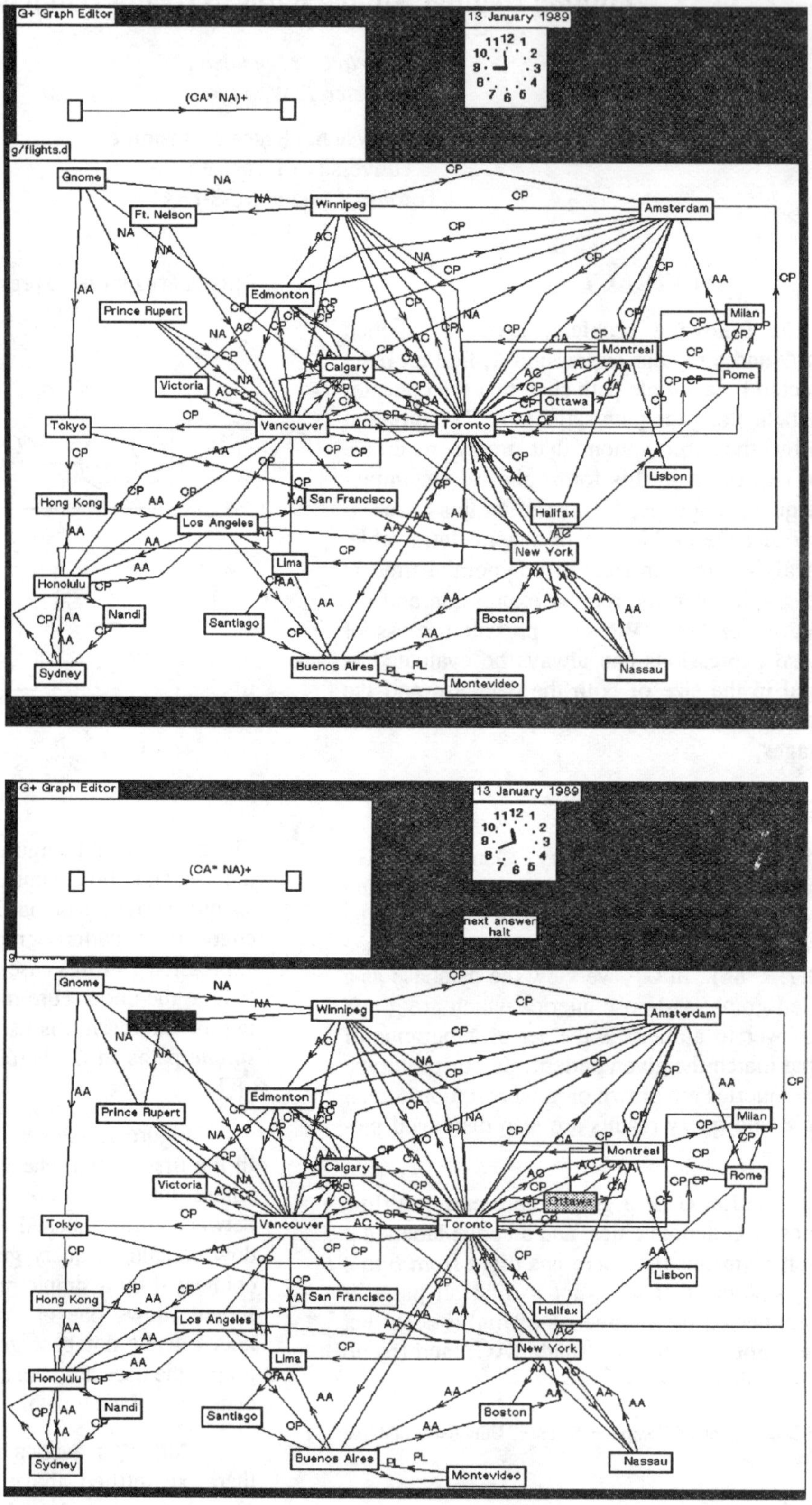

Figure 1.2. The G+ user interface

all pairs of nodes in G which are connected by a simple path p, where the concatenation of edge labels comprising p is in the language denoted by R.

When trying to find an efficient solution for this problem to incorporate in our implementation of G^+, we were somewhat surprised to discover that the query of Example 1.1 is in fact NP-complete. Using results in [FHW80, LaPa84], we show in Section 2 that for certain fixed regular expressions (such as R in the above example), the problem of deciding whether a pair of nodes is in the answer of a query is NP-complete, making the general problem NP-hard. We first attacked this problem by determining what it is in the language of R that makes the problem hard. In Section 3, we present a class of languages for which query evaluation is solvable in time polynomial in both the length of the regular expression and the size of the graph. We characterize these languages syntactically in terms of the regular expressions that denote them and the finite automata that recognize them. We then designed a general algorithm, presented in Section 4, which is correct for arbitrary graphs and queries, and is guaranteed to run in polynomial time in the size of the graph if the regular expression and graph are free of "conflicts", in a sense to be defined precisely in that section. As special cases, any query is free of conflicts with any acyclic database graph and any restricted expression query is free of conflicts with any arbitrary graph. Since we cannot restrict our prototype to work only on conflict-free queries and graphs, and it is expensive to test for conflict-freedom beforehand, it is quite convenient to have a single algorithm that works in all cases, and we have in fact incorporated the algorithm of Section 4 into our implementation.

2. INTRACTABILITY RESULTS

We begin by defining the graph structures as well as the class of queries over these structures in which we are interested.

DEFINITION: A *database graph* (*db-graph*, for short) $G=(N, E, \psi, \Sigma, \lambda)$ is a directed, labelled graph, where N is a set of *nodes*, E is a set of *edges*, and ψ is an *incidence function* mapping E to $N \times N$. So multiple edges between a pair of nodes are permitted in db-graphs. The labels of G are drawn from the finite set of symbols Σ, called the *alphabet*, and λ is an *edge labelling function* mapping E to Σ.

DEFINITION: Let Σ be a finite alphabet disjoint from $\{\varepsilon, \emptyset, (,)\}$. A *regular expression* R over Σ and the language $L(R)$ *denoted* by R are defined in the usual way. Let $G=(N, E, \psi, \Sigma, \lambda)$ be a db-graph and $p=(v_1, e_1, \ldots, e_{n-1}, v_n)$, where $v_i \in N$, $1 \le i \le n$, and $e_j \in E$, $1 \le j \le n-1$, be a path in G. We say p is a *simple path* if all the v_i's are distinct for $1<i<n$. We call the string $\lambda(e_1) \cdots \lambda(e_{n-1})$ the *path label* of p, denoted by $\lambda(p) \in \Sigma^*$. Let R be a regular expression over Σ. We say that the path p *satisfies* R if $\lambda(p) \in L(R)$. The *query* Q_R on db-graph G, denoted by $Q_R(G)$, is defined as the set of pairs (x,y) such that there is a simple path from x to y in G which satisfies R.
If $(x,y) \in Q_R(G)$, then (x,y) *satisfies* Q_R.

A naive method for evaluating a query Q_R on a db-graph G is to traverse every simple path satisfying R in G exactly once. The penalty for this is that such an algorithm takes exponential time when G has an exponential number of simple paths. Nevertheless, we will see below that in general we cannot expect an algorithm to perform much better, since we prove that, for particular regular expressions, the problem of deciding whether a pair of nodes is in the answer of a query is NP-complete. On the other hand, refinements can lead to guaranteed polynomial time evaluation under conditions studied in the following two sections.

Consider the following decision problem.

REGULAR SIMPLE PATH
Instance: Db-graph $G=(N, E, \psi, \Sigma, \lambda)$, nodes $x,y \in N$, regular expression R over Σ.
Question: Does G contain a directed simple path $p=(e_1, \ldots, e_k)$ from x to y such that p satisfies R, that is, $\lambda(e_1)\lambda(e_2) \cdots \lambda(e_k) \in L(R)$?

This is equivalent to asking "Is $(x,y) \in Q_R(G)$?". When the instance comprises only the db-graph, we refer to the problem as FIXED REGULAR PATH(R). That is, for FIXED REGULAR PATH(R) we measure the complexity only in terms of the size of the db-graph. We prove below that, for certain regular expressions R, FIXED REGULAR PATH(R) is NP-complete (which implies that REGULAR SIMPLE PATH is NP-hard).

THEOREM 2.1: *Let* 0 *and* 1 *be distinct symbols in* Σ. *FIXED REGULAR PATH(R), in which R is either (1)* $(00)^*$, *or (2)* $0^* 1 0^*$, *is NP-complete.*

PROOF: Consider the problem EVEN PATH, which is "Given a directed graph $G=(N,E)$ and nodes $x,y \in N$, is there a directed simple path of even length from x to y?", and DISJOINT PATHS, which is "Given a directed graph $G=(N,E)$ and pairs of distinct nodes $(w,x),(y,z) \in N \times N$, is there a pair of disjoint directed simple paths in G, one from w to x and the other from y to z?". For (1), the reduction is from EVEN PATH, which is shown to be NP-complete in [LaPa84]. For (2), the reduction is from DISJOINT PATHS, whose NP-completeness follows immediately from results in [FHW80]. □

Theorem 2.1 is a rather negative result, since it implies that queries might require time which is exponential in the size of the db-graph, not only the regular expression, for their evaluation. Thus, for regular expressions such as those in Theorem 2.1, we certainly would not expect an evaluation algorithm to run in polynomial time. This result, however, is not a function of the complexity of the particular regular expression but rather of the nature of the language denoted by the regular expression. A class of languages for which REGULAR SIMPLE PATH is in P is the subject of the next section.

3. RESTRICTED REGULAR EXPRESSIONS

We first introduce some terminology and definitions.

DEFINITION: A *nondeterministic finite automaton* (NDFA) M is a 5-tuple $(S, \Sigma, \delta, s_0, F)$, where S is a finite set of *states*, Σ is the *input alphabet*, δ is the *state transition function*, $s_0 \in S$ is the *initial state*, and $F \subseteq S$ is the set of *final states*. The concepts of the *extended* transition function δ^*, the language $L(M)$ *accepted* by M, and the *transition graph* associated with M are defined in the standard way, as is the notion of a *deterministic finite automaton* (DFA) [HoUl79].

DEFINITION: Given an NDFA $M=(S, \Sigma, \delta, s_0, F)$, for each pair of states $s,t \in S$, we define the *language from s to t*, denoted by L_{st}, as the set of strings that take M from state s to state t. In particular, for a state $s \in S$, the *suffix language* of s, denoted by L_{sF} (or $[s]$, for short), is the set of strings that take M from s to some final state. Clearly, $[s_0]=L(M)$. Similar definitions apply for a DFA.

Given a regular expression R over Σ, an ε-free NDFA $M=(S, \Sigma, \delta, s_0, F)$ that accepts $L(R)$ can be constructed in polynomial time [AHU74]. From now on, we will assume that all NDFAs are ε-free.

EXAMPLE 3.1: Figure 3.2 shows the transition graph T of a DFA M. State 0 is the initial state of M, while all states are final (denoted by a double circle). (We do not show (reject) states that are not on some path from the inital state to a final state.) $L(M)$ is denoted by the regular expression $0^*+1^*+0^*1$. The suffix language of state 1 is $[1]=0^*+0^*1$, while $[2]=\varepsilon$. □

Let R_1 and R_2 be regular expressions. In the subsequent analysis, it will be useful to refer to an NDFA which accepts the language $L(R_1 \cap R_2)$. The construction of such an NDFA is defined as follows.

DEFINITION: Let $M_1=(S_1,\Sigma,\delta_1,p_0,F_1)$ and $M_2=(S_2,\Sigma,\delta_2,q_0,F_2)$ be NDFAs. The NDFA for $M_1 \cap M_2$ is $I=(S_1 \times S_2, \Sigma, \delta, (p_0,q_0), F_1 \times F_2)$, where, for $a \in \Sigma$, $(p_2,q_2) \in \delta((p_1,q_1),a)$ if and only if $p_2 \in \delta_1(p_1,a)$ and $q_2 \in \delta_2(q_1,a)$. We call the transition graph of I the *intersection graph* of M_1 and M_2.

We saw in the previous section that, for certain regular expressions R, it is very unlikely that we will find an algorithm for evaluating Q_R on an arbitrary graph G that will always run in time polynomial in the size of G. One such regular expression is 0^*10^*. However, it turns out that if the regular expression $R=0^*10^*+0^*$ is specified instead, then Q_R is evaluable in polynomial time on any db-graph G. The reason is that if there is an arbitrary path from node x to node y in G that satisfies R, then there is a simple path from x to y satisfying R. In such a case, we need not restrict ourselves to looking only for simple paths in G, but can instead look for *any* path satisfying R. We define the corresponding decision problem below.

REGULAR PATH
Instance: Db-graph $G=(N, E, \psi, \Sigma, \lambda)$, nodes $x,y \in N$, regular expression R over Σ.
Question: Does G contain a directed path (not necessarily simple) $p=(e_1, \ldots, e_k)$ from x to y such that p satisfies R, that is, $\lambda(e_1)\lambda(e_2) \cdots \lambda(e_k) \in L(R)$?

LEMMA 3.1: *REGULAR PATH can be decided in polynomial time.*

PROOF: Given db-graph G along with nodes x and y in G, we can view G as an NDFA with initial state x and final state y. Construct the intersection graph I of G and $M=(S, \Sigma, \delta, s_0, F)$, an NDFA accepting $L(R)$. There is a path from x to y satisfying R if and only if there is a path in I from (x,s_0) to (y,s_f), for some $s_f \in F$. All this can be done in polynomial time [HRS76]. □

We are interested in conditions under which REGULAR SIMPLE PATH (which is appropriate because of our semantics) can be reduced to REGULAR PATH. The following lemma states one such condition.

LEMMA 3.2: *REGULAR SIMPLE PATH can be decided in polynomial time on acyclic db-graphs.*

PROOF: Follows immediately from Lemma 3.1 and the fact that every path in an acyclic graph is simple. □

Suppose that we want to characterize a class of regular expressions for which we can guarantee that REGULAR SIMPLE PATH reduces to REGULAR PATH and hence is solvable in polynomial time. If we assume that we know nothing about the structure of the db-graphs, we have to ensure that, for such a regular expression R, whenever string w is in $L(R)$, every string obtainable from w by removing one or more symbols must also be in $L(R)$. Otherwise, if $w=xay$ is in $L(R)$ but xy is not in $L(R)$ (where $a \in \Sigma$ and $x,y \in \Sigma^*$), we can construct a graph G comprising a single simple path from u to v and passing through z, in which there is a loop at z labelled a, the path from u to z is labelled x, and the path from z to v is labelled y (see Figure 3.1). There is a non-simple path from u to v in G that satisfies R but no simple path from u to v satisfying R.

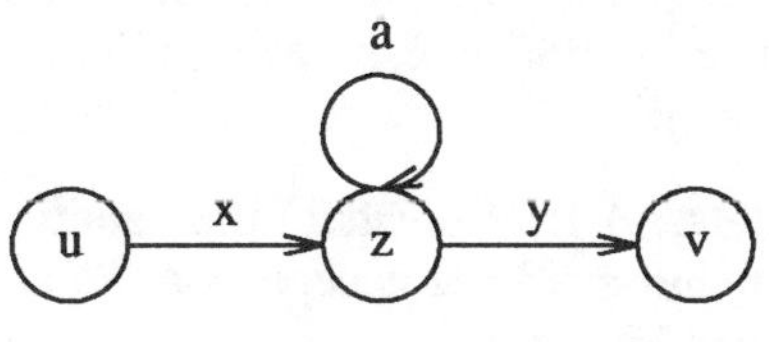

Figure 3.1.

DEFINITION: ([Carr79]) An *abbreviation* of a string w is any string that can be obtained from w by removing one or more symbols of w.

So we are looking for a class of regular expressions which denote languages that are closed under abbreviation. Now consider the following definition for the class of restricted regular expressions.

DEFINITION: For $a \in \Sigma$, denote the regular expression $(a+\varepsilon)$ by $(a?)$ (as is done in the *grep* utility of Unix[†], for example). Given a regular expression R, let R' be a regular expression obtained by replacing some occurrence of a symbol $a \in \Sigma$ in R by $(a?)$. R is *restricted* if and only if $R \equiv R'$, for any R' obtained from R as defined above.

EXAMPLE 3.2: The regular expression $R_1 = 0^* + 1^* + 0^* 1$ is restricted, since R_1 can be rewritten as $0^*(1?) + 1^*$, which is equivalent to $(0?)^*(1?) + (1?)^*$. Recall, from Theorem 3.1, that FIXED REGULAR PATH(R) is NP-complete for $R = 0^* 1 0^*$. R is not restricted, but $R' = 0^* 10^* + 0^*$ is restricted, since R' can be written as $0^*(1+\varepsilon)\, 0^*$, which is equivalent to $(0?)^*(1?)\,(0?)^*$. □

DEFINITION: A DFA $M = (S, \Sigma, \delta, s_0, F)$ exhibits the *Suffix Language Containment Property* (the *Containment Property*, for short) if, for each pair $s, t \in S$ such that s and t are on a path from s_0 to some final state and t is a successor of s, $[s] \supseteq [t]$ (that is, $L_{sF} \supseteq L_{tF}$).

EXAMPLE 3.3: Consider the regular expression $R_1 = 0^* + 1^* + 0^* 1$ from the previous example, and a DFA M accepting $L(R_1)$ whose transition graph T is given in Figure 3.2. We can verify that M exhibits the Containment Property by noting that [3] is denoted by 1^*, [2] by ε, [1] by $0^* + 0^* 1$, and [0] by $0^* + 1^* + 0^* 1$. Obviously, $[1] \supseteq [1]$ and $[3] \supseteq [3]$. It is easy to check that $[1] \supseteq [2]$, $[0] \supseteq [1]$ and $[0] \supseteq [3]$. The fact that M exhibits the Containment

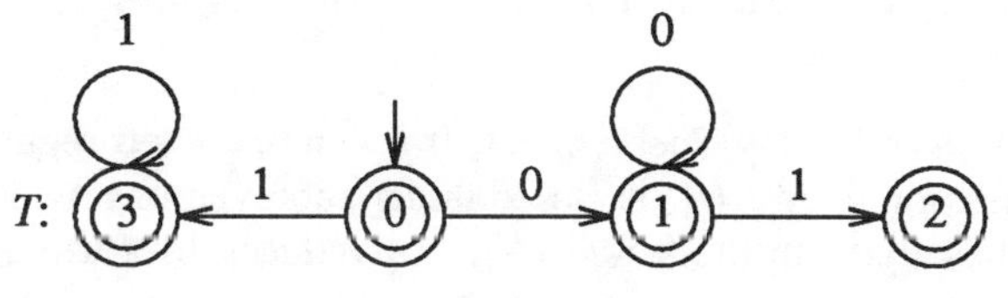

Figure 3.2. DFA for $0^* + 1^* + 0^* 1$.

Property and R_1 is restricted is no coincidence, as the following theorem states. □

THEOREM 3.1: *Let R be a regular expression over Σ, and $M = (S, \Sigma, \delta, s_0, F)$ be a DFA accepting $L(R)$. The following three statements are equivalent:*

(1) R is a restricted regular expression,

(2) $L(R)$ is closed under abbreviations, and

(3) M exhibits the Containment Property.

PROOF: (Sketch)

(1) => (2) Assume that R is restricted but that $L(R)$ is not closed under abbreviations. Then there is a symbol $a \in \Sigma$ and strings $x, y \in \Sigma^*$ such that $xay \in L(R)$ but $xy \notin L(R)$. Let $M_R = (T, \Sigma, \mu, t_0, E)$ be an NDFA that accepts the language $L(R)$. Let $T' = \mu^*(t_0, x)$, that is, the set of states M_R can be in after reading x. Since $L(M_R) = L(R)$ and $xy \notin L(R)$, for no $r \in T'$ can y be in $[r]$. On the other hand, $xay \in L(M_R)$, so there is a state $p \in T'$ such that $q \in \mu(p, a)$ and $y \in [q]$. Since R is restricted, adding an ε-transition from p to q leaves $L(M_R)$ unchanged. But if we do so, then $y \in [p]$, $xy \in L(M_R)$, and $L(M_R)$ is no longer equal to $L(R)$, which is a contradiction. We conclude that $L(R)$ is closed under abbreviations.

(2) => (3) Assume that $[s] \not\supseteq [t]$ for some pair s, t of reachable states in M such that $\delta(s, a) = t$, for some $a \in \Sigma$. That is, there is a string $y \in \Sigma^*$ for which $y \in [t]$ but $y \notin [s]$. Let $x \in \Sigma^*$ be a string for which $\delta^*(s_0, x) = s$. It follows that $xay \in L(M)$, but that $xy \notin L(M)$. Since $L(M) = L(R)$, we conclude that $L(R)$ is not closed under abbreviations.

(3) => (1) Assume that R is not restricted, and let M_R be an NDFA accepting $L(R)$ as above. Then there is an a-transition in M_R from s to t for which adding an ε-transition from s to t alters $L(M_R)$. Let $x \in \Sigma^*$ be a string for which $s \in \mu^*(t_0, x)$. That is, there is a string $y \in [t]$ such that $y \notin [r]$ for any $r \in \mu^*(t_0, x)$. Now consider the DFA M. Assume that $\delta^*(s_0, x) = p$. Since $L(M_R) = L(M)$ and $xay \in L(M_R)$, there must be a state q in M such that $\delta(p, a) = q$ and $y \in [q]$. However, $y \notin [p]$, for otherwise $xy \in L(M)$ which would mean that $L(M) \neq L(M_R)$. Hence, $[p] \not\supseteq [q]$, so M does not exhibit the Containment Property. □

[†] Unix is a trademark of AT&T.

COROLLARY 3.1: *REGULAR SIMPLE PATH can be decided in polynomial time for restricted regular expressions.*

PROOF: By Theorem 3.1, if R is a restricted regular expression, then $L(R)$ is closed under abbreviations, which implies that REGULAR SIMPLE PATH reduces to REGULAR PATH. By Lemma 3.1, REGULAR PATH can be decided in polynomial time. □

Note that we our results show that R being restricted is a sufficient condition for REGULAR SIMPLE PATH being solvable in polynomial time, but it is not necessary. For example, the expression 01^* is not restricted, but the corresponding query can be evaluated in polynomial time on arbitrary graphs using the algorithm we present in Section 5. It would be interesting to find other naturally defined classes of regular expressions that lead to a polynomial time solution for REGULAR SIMPLE PATH.

By adapting an algorithm to minimize the number of states of a DFA [HoUl79], we can compute the suffix language containment relation for all pairs of states in a DFA M. This relation will be used in the next section; it also provides an obvious method for testing whether or not a regular expression R is restricted (using Theorem 3.1). Since the construction of a DFA M accepting $L(R)$ may take exponential time (in the size of R), this test is not efficient. However, it is important to stress that we are trying to avoid the possibility of spending exponential time in the size of the db-graph in answering a query. Also, we have the following.

THEOREM 3.2: *Let R be a regular expression over alphabet $\{0\}$. Deciding whether R is not restricted is NP-complete.*

PROOF: By reduction from the problem of deciding whether a regular expression does not denote 0^*, which is shown to be NP-complete in [StMe73]. □

4. AN EVALUATION ALGORITHM

In this section, we describe an algorithm for evaluating a query Q_R on a db-graph G. As is to be expected from the results of Section 2, the algorithm does not run in polynomial time in general. It does, however, run in polynomial time under the sufficient conditions identified in Section 4, namely, when G is acyclic or R is restricted. In fact, we show that the algorithm runs in polynomial time if G and R are conflict-free, a condition implied whenever G is acyclic or R is restricted.

The evaluation algorithm traverses simple paths in G, using a DFA M accepting $L(R)$ to control the search by marking nodes as they are visited. We must record with which state of M a node is visited, since we must allow a node to be visited with different states (which correspond to distinct nodes in the intersection graph of G and M). In order to avoid visiting a node twice in the same state, we would like to retain the state markings on nodes as long as possible. Unfortunately, the following example shows that, in general, traversing only simple paths in G and retaining state markings can lead to incompleteness in query evaluation.

EXAMPLE 4.1: Consider the query Q_R, where $R = aaa$. An automaton M accepting $L(R)$ and a db-graph G are shown in Figure 4.1.

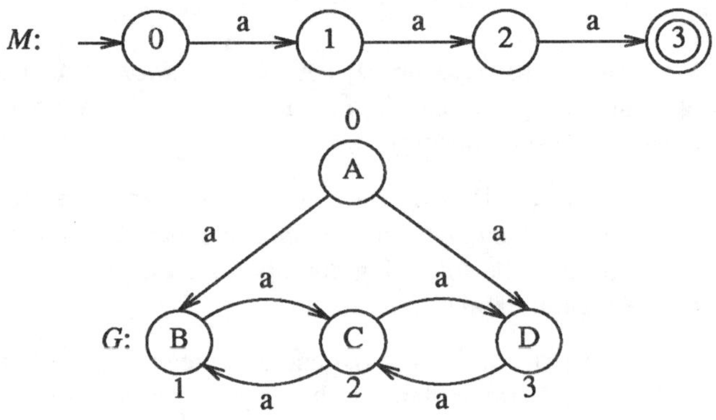

Figure 4.1. A DFA M and db-graph G.

Assume that we start traversal from node A in G, and follow the path to B, C and D. Nodes A, B, C and D are marked with states 0, 1, 2 and 3, respectively (as shown in Figure 4.1), and the answer (A,D) is found, since 3 is a final state. If we now backtrack to node C, we cannot mark B with state 3 because (A,B,C,B) is a non-simple path. So we backtrack to A, and visit D in state 1. If we have retained markings, however, we cannot visit node C as it is already marked with state 2. Consequently, the answer (A,B) is not found. □

As we will see below, it is safe to retain markings when G is acyclic or R is restricted . However, we can do better; because of the structure of a particular db-graph G, it may be the case that we can retain markings and evaluate Q_R in polynomial time even if G is not acyclic and R is not restricted.

DEFINITION: Let I be the intersection graph of a db-graph G and a DFA $M = (S, \Sigma, \delta, s_0, F)$ accepting $L(R)$. Assume that for nodes u and v in G and states s and t in M, there are paths p from (u, s_0) to (v, s) and q from (v, s) to (v, t) in I (that is, there is a cycle at v in G that satisfies L_{st}), such that no first component of a node on p or q repeats except for the endpoints of q. In other words, p and q correspond to a simple path and a simple cycle, respectively, in G. If $[s] \not\supseteq [t]$, then we say there is a *conflict* between s and t at v. If there are no conflicts in I, then I is said to be *conflict-free*, as are G and R.

It is obvious that if G is acyclic, then I is conflict-free no matter what regular expression R appears in Q_R. Also, if R is restricted, then, by Theorem 3.1, M exhibits the Containment Property; hence, I is conflict-free irrespective of the structure of G. We will show that Q_R can be evaluated in polynomial time if I is conflict-free. Hence, conflict-freedom is a third (weaker) sufficient condition for Q_R to be polynomial time evaluable. For example, if $R = (00)^*$ and G is a bipartite (cyclic) graph, then G and R are conflict-free and so Q_R can be evaluated in polynomial time on G (in contrast to Theorem 2.1).

The concept of conflict detection is embodied in the following query evaluation algorithm, Algorithm C. If no conflicts are detected, the algorithm retains markings, while whenever a conflict arises, it unmarks nodes so that no answers are lost.

EXAMPLE 4.2: Consider again the DFA M and the db-graph G of Example 4.1. Recall that, if markings were retained, the answer (A,B) would not be found. However, there is a conflict in the intersection graph of G and M. This is because node B in G can be marked with state 1 and there is a cycle at B which satisfies L_{13}, but $[1] \not\supseteq [3]$. Algorithm C detects such conflicts and unmarks node C on backtracking, enabling the answer (A,B) to be found. □

Algorithm C: Evaluation of a query on a db-graph.

Input: Db-graph $G = (N, E, \psi, \Sigma, \lambda)$, query Q_R.
Output: $Q_R(G)$, the value of Q_R on G.
Method:

1. Construct an DFA $M = (S, \Sigma, \delta, s_0, F)$ accepting $L(R)$.
2. Initialize $Q_R(G)$ to $\varnothing$.
3. For each node $v \in N$, set $CM[v]$ to *null* and $PM[v]$ to $\varnothing$.
4. Test $[s] \supseteq [t]$ for each pair of states s and t.
5. For each node $v \in N$,
   (a) call SEARCH-C(v, v, s_0, *conflict*)
   (see Figure 4.2)
   (b) reset $PM[w]$ to $\varnothing$ for any marked node $w \in N$.

The algorithm marks nodes in G with states of the DFA M when they are visited. Two types of marking are used for each node v: (1) a current marking ($CM[v]$), which (if non-null) tells us that v is already on the stack (Lines 7 and 14), and (2) a previous marking ($PM[v]$), which is a set of states and tells us earlier markings of v, excluding the current path (Line 15). Current markings are used to avoid following paths in G that are not simple, while previous markings are used where possible to

```
    procedure SEARCH-C (u, v, s, var conflict)
6.  conflict ← false
7.  CM[v] ← s
8.  if s ∈ F then Q_R(G) ← Q_R(G) ∪ {(u,v)} fi
9.  for each edge in G from v to w with label a do
10.   if δ(s,a) = t and t ∉ PM[w] then
11.     if CM[w] = q then conflict ← ([q] ⊉ [t]) fi
      else /* CM[w] is null */
12.       SEARCH-C (u, w, t, new-conflict)
13.       conflict ← conflict or new-conflict
      fi
    od
14. CM[v] ← null
15. if not conflict then PM[v] ← PM[v] ∪ {s} fi
    end SEARCH-C
```

Figure 4.2. Evaluation of a query on a db-graph.

prevent a node in G from being visited more than once in the same state during a single execution of Line 5(a). We do not describe the algorithm further here, but rely on a detailed example presented below to demonstrate how query evaluation, and in particular the detection of conflicts, is performed. □

The following theorem is proved in [Wood88].

THEOREM 4.1: *Algorithm C is correct; that is, given db-graph G and query Q_R, Algorithm C adds (x,y) to $Q_R(G)$ if and only if (x,y) satisfies Q_R.*

THEOREM 4.2: *In the absence of conflicts, Algorithm C runs in an amount of time which is bounded by a polynomial in the size of the db-graph.*

PROOF: Let G be a db-graph with n nodes and e edges. In the absence of conflicts, Algorithm C essentially performs n depth-first searches of the graph, and hence runs in $O(ne)$ time. □

EXAMPLE 4.3: Let $R = a((bc + \varepsilon)d + ec))$ be the regular expression for query Q_R. A DFA M accepting $L(R)$ and a db-graph G are shown in Figure 4.3. We demonstrate the execution of Algorithm C in evaluating Q_R on G. Assume that we start by marking node A of G with state 0 of M, after which we proceed to mark B with 1, C with 2, and D with 3. Since no edge labelled d leaves D, we backtrack to C and attempt to visit B in state 3. Although B already has a current marking ($CM[B] = 1$), this is not a conflict since $[1] \supseteq [3]$. The algorithm now backtracks to node B in state 1 and marks E with state 4. After backtracking again to B in state 1, the markings are given as in Figure 4.3. Next, the algorithm marks C with state 5 and D with 4. On backtracking to C and attempting to mark B with 4, a conflict is detected since $[1] \not\supseteq [4]$. So on back-

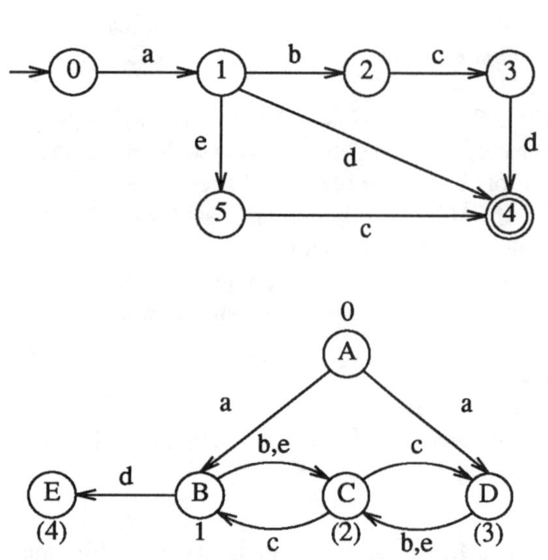

Figure 4.3. A DFA M and marked db-graph G (PMs are in parentheses).

tracking to A, the marking 5 is removed from C, 1 is removed from B, and neither is made a previous marking. Now D is marked with 1, but since C has a previous marking of 2, that marking is not repeated. So C is marked with 5 (which was previously removed), after which B can be marked with 4. When the algorithm backtracks to C and attempts to visit D, it discovers that D was previously marked with 4, so no conflict is registered. The markings are now given as in Figure 4.4. No more markings are generated by Algorithm C for the traversal rooted at A, so the set of answers with first component A is $\{(A,B),(A,D),(A,E)\}$. □

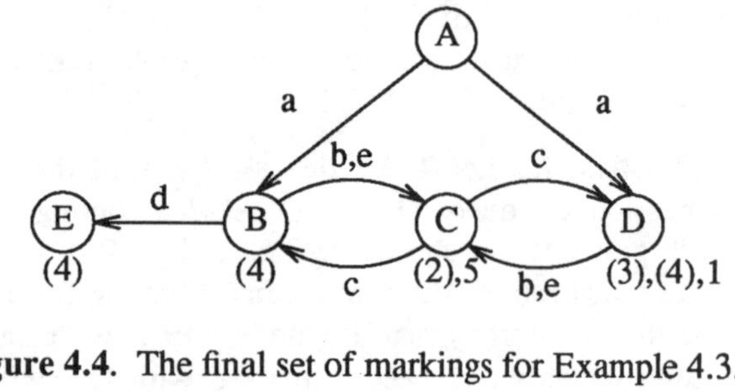

Figure 4.4. The final set of markings for Example 4.3.

If R is restricted or G is acyclic, no conflicts occur so Algorithm C runs in polynomial time. For the same reason, the algorithm also runs in polynomial time on the example of finding even paths in a bipartite graph. However, even in the presence of conflicts, Algorithm C can run in polynomial time in the size of G. This is the case, for example, if R is a (*)-free regular expression. In this case, $L(R)$ is finite; say the longest string in the language is of length q. Then there can be at most $O(n^q)$ simple paths in G, and in the worst case the algorithm traverses each one exactly once.

5. CONCLUSION

We have addressed the problem of finding nodes in a labelled, directed graph which are connected by a simple path satisfying a given regular expression. This study was motivated by the observation that many recursive queries can be expressed in this form, and by the implementation of a query language based on this observation.

We began by describing how a naive algorithm might evaluate such queries. Although this algorithm runs in exponential time in the worst case, we showed that we cannot expect to do better since the evaluation problem is in general NP-hard. Using the fact that the associated problem for paths in general (as opposed to simple paths) is solvable in polynomial time, we characterized the class of restricted regular expressions, whose associated queries can be evaluated in polynomial time. Finally, we presented an algorithm for evaluating arbitrary expressions on arbitrary graphs. This algorithm runs in polynomial time if either (a) the regular expression is restricted or closure-free, (b) the graph is acyclic, or (c) the regular expression and graph are conflict-free.

We should point out that the analysis in this paper, and the implementation itself, assume the graph can be entirely stored in main memory. This is a reasonable assumption in many cases, especially because in the intended applications of G^+ the graph is often only the fraction of the database that can be presented visually in a natural way. Relaxing this assumption provides an interesting area for further study. Other researchers, investigating similar algorithms for transitive closure, have claimed that they are amenable to efficient secondary storage implementation [IoRa88]. Our next version of the implementation will use the *HAM* hypertext package developed at Tektronix [DeSc86] as a graph storage server; this package maintains a graph on secondary storage, providing an efficient implementation of the graph-based operations that are the primitive steps of our algorithm.

Finally, we note that the problem of solving general G queries is more general than REGULAR SIMPLE PATH. For example, the language contains also aggregate operators that allow the expression of queries such as "find the length of the shortest path between two nodes." Queries with aggregates can be intractable independently of the properties of the regular expression; for example, one can also express the query "find the length of the *longest* simple path between two nodes, " which is NP-complete. Further discussion of aggregate operators can be found in [CrMe88]. A new language, that does not insist on simple path semantics and captures exactly the queries computable in logarithmic space, is presented in [Cons89].

ACKNOWLEDGEMENTS

This research was supported by the Natural Sciences and Engineering Research Council of Canada. We thank Mariano Consens and the anonymous reviewers for useful comments.

References

Agra87.
R. AGRAWAL, "Alpha: An Extension of Relational Algebra to Express a Class of Recursive Queries," *Proc. 3rd Int. Conf. on Data Engineering*, pp. 580-590, 1987.

AHU74.
A.V. AHO, J.E. HOPCROFT, AND J.D. ULLMAN, *The Design and Analysis of Computer Algorithms*, Addison-Wesley, Reading, MA, 1974.

Carr79.
B. CARRE, *Graphs and Networks*, Oxford University Press, Oxford, England, 1979.

Cons89.
M.P. CONSENS, "Graphlog: 'Real Life' Recursive Queries Using Graphs," M.Sc. thesis, Department of Computer Science, University of Toronto, 1989.

CrMe88.
I.F. CRUZ AND A.O. MENDELZON, "Summary Queries in the G+ Query Language," in *Office and Data Base Systems Research '88*, ed. F.H. Lochovsky, pp. 160-188, Tech. Report CSRI-212, Univ. of Toronto, 1988.

CMW87.
I.F. CRUZ, A.O. MENDELZON, AND P.T. WOOD, "A Graphical Query Language Supporting Recursion," *Proc. ACM SIGMOD Conf. on Management of Data*, pp. 323-330, 1987.

CMW88.
I.F. CRUZ, A.O. MENDELZON, AND P.T. WOOD, "G+: Recursive Queries Without Recursion," *Proc. 2nd Int. Conf. on Expert Database Systems*, pp. 355-368, 1988.

DeSc86.
N. DELISLE AND M. SCHWARTZ, "Neptune: a Hypertext System for CAD Applications," *Proc. ACM SIGMOD Conf. on Management of Data*, pp. 132-143, 1986.

FHW80.
S. FORTUNE, J. HOPCROFT, AND J. WYLLIE, "The Directed Subgraph Homeomorphism Problem," *Theor. Comput. Sci.*, vol. 10, pp. 111-121, 1980.

GSS87.
G. GRAHNE, S. SIPPU, AND E. SOISALON-SOININEN, "Efficient Evaluation for a Subset of Recursive Queries," *Proc. 6th ACM SIGACT-SIGMOD-SIGART Symp. on Principles of Database Systems*, pp. 284-293, 1987.

HoUl79.
J.E. HOPCROFT AND J.D. ULLMAN, *Introduction to Automata Theory, Languages, and Computation*, Addison-Wesley, Reading, MA, 1979.

HRS76.
H.B. HUNT, D.J. ROSENKRANTZ, AND T.G. SZYMANSKI, "On the Equivalence, Containment, and Covering Problems for the Regular and Context-free Languages," *J. Comput. Syst. Sci.*, vol. 12, pp. 222-268, 1976.

IoRa88.
Y.E. IOANNIDIS AND R. RAMAKRISHNAN, "Efficient Transitive Closure Algorithms," Computer Sciences Tech. Report #765, Univ. of Wisconsin-Madison, 1988.

LaPa84.
A.S. LAPAUGH AND C.H. PAPADIMITRIOU, "The Even-Path Problem for Graphs and Digraphs," *Networks*, vol. 14, pp. 507-513, 1984.

R*86. A. ROSENTHAL, S. HEILER, U. DAYAL, AND F. MANOLA, "Traversal Recursion: A Practical Approach to Supporting Recursive Applications," *Proc. ACM SIGMOD Conf. on Management of Data*, pp. 166-176, 1986.

StMe73.
L.J. STOCKMEYER AND A.R. MEYER, "Word Problems Requiring Exponential Time," *Proc. 5th Ann. ACM Symp. on Theory of Computing*, pp. 1-9, 1973.

Wood88.
P.T. WOOD, "Queries on Graphs," Ph.D. thesis, Department of Computer Science, University of Toronto, Tech. Report CSRI-223, 1988.

Towards an Open Architecture for $\mathcal{LDL}$

Danette Chimenti Ruben Gamboa Ravi Krishnamurthy

MCC, 3500 West Balcones Center Drive, Austin, Texas 78759

danette@mcc.com,ruben@mcc.com,ravi@mcc.com

Abstract

We extend $\mathcal{LDL}$ to allow programs to call external procedures and vice versa. This extension allows the modularization of $\mathcal{LDL}$, since external predicates are equivalent to external procedures written in $\mathcal{LDL}$. External predicates are viewed as infinite relations so that the traditional semantics of logic programs remain applicable. To avoid computing infinite relations, well-formedness conditions for programs in extended $\mathcal{LDL}$ are given. The traditional optimization framework can still be used; it is only necessary to define a new set of cost functions capable of handling the infinite relations. The problem of interfacing $\mathcal{LDL}$ programs with external procedures—exchanging complex objects and returning multiple solutions—is discussed. Thus, we provide a general framework to allow logic programs to interact with external procedures without sacrificing amenities such as optimization, safety, etc. This approach forms the basis for the implementation of externals and modules in the $\mathcal{LDL}$ compiler and optimizer at MCC.

1 Introduction

An $\mathcal{LDL}$[1] program calling a graphics routine or a windowing system calling an $\mathcal{LDL}$ program are obvious cases that motivate the incorporation of external predicates into $\mathcal{LDL}$. External predicates in $\mathcal{LDL}$ also allow a user to develop large programs by—what we call—*hot-spot refinement.* The user writes a large $\mathcal{LDL}$ program, validates its correctness and identifies the hot-spots; i.e., predicates in the program that are highly time consuming. Then, he can rewrite those hot-spots more efficiently in a procedural language such as *C*, maintaining the rest of the program in $\mathcal{LDL}$. Naturally, the correctness, optimization, safety and other amenities provided by the $\mathcal{LDL}$ environment should not be sacrificed. Hence, the optimizer must understand the nature of such external procedures (referred to as computed predicates), and optimize programs that use them.

Yet another reason for including external predicates in $\mathcal{LDL}$ comes from the optimizer. The time complexity of an optimization algorithm is typically exponential on the number of predicate occurrences. By modularizing the program and optimizing each module separately, compilation time can be reduced. Intuitively, the modularization allows the user to provide a unique binding pattern for the global (imported) predicate, thus simplifying the optimization task.

Computed relations were incorporated into relational databases in [MW81]. In this paper we present an extension of $\mathcal{LDL}$ to allow computed predicates by addressing the following three issues: 1) Providing $\mathcal{LDL}$ language extensions to allow external predicates; 2) Interfacing $\mathcal{LDL}$ programs with external predicates; 3) Optimizing $\mathcal{LDL}$ programs using external predicates. Since Datalog is a simpler subset of $\mathcal{LDL}$, we first discuss these issues in the context of Datalog and, subsequently, in the full context of $\mathcal{LDL}$.

We extend Datalog to Datalog^E, which allows the use of externals. We also provide well-formedness conditions for Datalog^E programs. By viewing external predicates as infinite relations, we show that the semantics of Datalog programs are preserved in the transition to Datalog^E. Then, we show that Datalog^E programs can be optimized in the traditional framework [Sell79, KBZ86], simply by redefining the cost functions to handle such infinite relations.

$\mathcal{LDL}$ is viewed as an extension of Datalog, where more complex operations and data types are allowed. We show that these extensions do not affect the semantics and optimization results presented in the context of Datalog^E. However, the presence of complex objects, e.g., sets and functor objects, does pose new problems, since the external predicates can manipulate them. We define enhanced cost functions, capable of handling complex objects as arguments to external predicates. We also provide protocols to interface $\mathcal{LDL}$ programs with external predicates and discuss implementation issues.

After overviewing the background in Section 2, the discussion in the context of Datalog is presented in Section 3. The issues in the context of $\mathcal{LDL}$ are discussed in Section 4, followed by the conclusion.

[1] $\mathcal{LDL}$, Logic Data Language, is a Horn clause based language with extensions for negation, set operations, updates, etc.

Proceedings of the Fifteenth International Conference on Very Large Data Bases

Amsterdam, 1989

2 Background

We briefly introduce Datalog, then $\mathcal{LDL}$ as an extension of Datalog, and, finally, the framework for optimization.

2.1 Datalog

We assume that the reader is familiar with Datalog and the associated definitions for rule, base/derived predicates, predicate occurrence and the mutually recursive property. A *rule base*, RB is a finite set of rules. The *extensional database*, EDB, is a finite set of finite *base* relations. A *query* is a predicate occurrence written as $?q(\ldots)$. For a given query, Q, we denote a *program* P by a triple (EDB, RB, Q). Given a program $P = (EDB, RB, Q)$, the answer to Q can be defined in the usual way.

Datalog is semantically defined using the bottom-up model, which coincides with the declarative model when the program is viewed as a first-order formula[Ll84]. However, the program execution does not have to be bottom-up. It is only required that the execution preserve the bottom-up semantics; i.e., the answers returned must be exactly the same as those returned under the bottom-up semantics. The use of sideways information passing [Ull85] and memo-ing (as is done in Prolog) results in many different executions (including top-down) that preserve the bottom-up semantics.

2.2 $\mathcal{LDL}$

Conceptually, we can view the execution of an $\mathcal{LDL}$ program as the manipulation of objects. Therefore, the $\mathcal{LDL}$ language can be viewed as an extension of Datalog in two categories: first, the addition of constructs such as negation, nondeterministic choice (i.e., declarative cut,!)[KN88a], updates[NK88], etc.; and second, the addition of functors, sets with their associated grouping constructor, etc. These extensions enhance the computational capability of $\mathcal{LDL}$ programs and the data model of the objects in the $\mathcal{LDL}$ programs, respectively. As we shall be interested mainly in the additions to the data model, only these extensions are reviewed below. We model $\mathcal{LDL}$ objects abstractly as follows:

1. An *atomic object* can be an integer, a string, etc.
2. A *functor object* is recursively defined as an object of the form $f(object_1, \ldots, object_n)$, where f is an n-ary functor symbol. A *tuple object* is a functor object with no functor name.
3. A *set object* is a (not necessarily homogeneous) collection of objects. For example, $\{(john, 10K), \ldots\}$ is a set object whose elements are tuple objects.

2.3 Optimization

The optimization of $\mathcal{LDL}$ (and therefore, Datalog) programs can be described abstractly as follows:

> Given a query Q, an execution space E, and a cost function defined over E, find an execution e in E that is of minimum cost; i.e.,
>
> $$\min_{e \in E}[Cost\ of\ e\ for\ Q]$$

Any solution to the above problem can be characterized by choosing: 1) an execution model and, therefore, an execution space; 2) a cost model; and 3) a search strategy.

The execution model encodes the decisions regarding the ordering of the joins, join methods, materialization strategy, etc. The cost model computes the execution cost. The search strategy is used to enumerate the search space, while the minimum cost execution is being discovered. These three choices are not independent; the choice of one can affect the others. For example, if a linear cost model is used, as in [KBZ86], then the search strategy can enumerate a quadratic space; on the other hand, an exponential space is enumerated if a general cost model is used, as in the case of commercial database management systems. Our discussion in this paper does not depend on either the execution space or the search strategy employed. However, it does depend on how the cost model is used in the optimization algorithm.

For brevity, we assume the exhaustive search as described in [KZ88]. As argued in that paper, the cost model for any execution can be composed of elementary cost functions for join, projection, and union, i.e., the traditional operations in relational queries. We can see how these cost functions are sufficient to model a recursive query by noticing that, in a bottom-up execution, recursion is implemented as a fixpoint iteration of a sequence of relational queries, which can be easily modelled using the traditional functions. Therefore, we limit the discussions in this paper to these cost functions.

The traditional cost functions use the description of the relations, e.g., cardinality or selectivity, to compute the cost. Observe that the operands for these functions may be intermediate relations, whose descriptions must be computed. We model the descriptor of a relation using the framework presented in [CP84], where it is called a profile. Such a descriptor encodes all the information about the relation that is needed for the cost functions.

The *descriptor* of an n-ary relation is a tuple containing $(b_i, c_i, i = 0, 1, \ldots, n)$, where b_i and c_i are the bag (the cardinality without removing duplicates) and cardinality of the ith attribute of the relation, and b_0 and c_0 are the bag and cardinality of the relation itself. The redundancy in the descriptors is assumed for ease of exposition. For example, consider the employee relation, $emp(Name, Age)$, containing 100 tuples with 50 distinctly named employees in 20 distinct age groups. The descriptor for this relation is (100,100, 100,50, 100,20).

Let the set of all descriptors be $\mathcal{D}$, and let $\mathcal{I}$ be the set of cost values denoted by integers. We are interested in two functions for each operation, σ, $\text{COST}_\sigma : \mathcal{D} \times \mathcal{D} \rightarrow \mathcal{I}$

and $\text{DESC}_\sigma : \mathcal{D} \times \mathcal{D} \rightarrow \mathcal{D}$. Intuitively, the COST_σ function computes the cost of applying the binary operation σ to two objects, and DESC_σ gives a descriptor for the resulting object. The functions for the unary operators are similarly defined. In this paper, we shall be mainly concerned with the DESC function.

3 Extending Datalog

In this section we incorporate the notions of modules and computed relations into Datalog to obtain Datalog^E and present well-formedness conditions for Datalog^E programs. Next, the semantics of Datalog^E programs is formalized. We then tackle the problem of optimizing Datalog^E programs by requiring schematic information for each computed predicate. It is shown that the traditional optimization framework can use this information to optimize Datalog^E programs.

3.1 Datalog^E

A Datalog^E program is characterized by the four-tuple (EDB, CDB, IDB, Q). The extensional database and the query are defined as before. The *computed database*, CDB, and the *intentional database*, IDB, are formalized below.

The *computed database* is a finite set of *infinite* relations. For example, $<, >$ and other arithmetic predicates fall into this category. Conceptually, any external procedure, e.g., a program written in *C*, can be viewed as an infinite relation with some constraints on the input and output arguments. Except for these constraints (to be discussed later), the usage of CDB predicates is identical to that of EDB predicates.

The *intentional database* is a finite set of pairs (RB, EQ), where RB is a rule base, and EQ is the set of exported predicates from RB. The exported predicates are query forms on the (derived) predicates defined in the RB. For convenience, we assume the exported predicates from any two RBs are disjoint. The exported predicates over all the RBs form the set of *global predicates* for the program. Any (derived) predicate in an RB that is not a base, computed or global predicate is called a *local predicate*. Without loss of generality, we assume that the set of predicates in the four categories, i.e., base, computed, global and local, partitions the set of all predicates in the entire program. Let us also redefine $DB = EDB \cup IDB \cup CDB$.

Global and computed predicates are referred to as *external predicates*, giving the rationale for Datalog^E.

3.2 Well–Formed Datalog^E Programs

We now introduce a class of Datalog^E programs. Our goal is to identify the programs that can be executed without generating infinite relations for external predicates and that can be compiled separately.

First, we recall the notion of *finiteness constraint* (FC) from [KRS88]. Intuitively, this captures the notion of the input/output requirements of an external predicate. A computed predicate $\underline{p}$ is said to satisfy a *finiteness constraint* of the form $\overline{X} \rightarrow \overline{Y}$ if and only if for each tuple T in p, the set of tuples $\{S[\overline{Y}] \mid S \in p \text{ and } S[\overline{X}] = T[\overline{X}]\}$, is finite.[2]

The finiteness constraint must be explicitly stated in the case of a computed predicate. This is done by specifying the input arguments $\overline{X}$ of the procedure corresponding to the computed predicate. It is assumed that this procedure can compute the (finite) set of values for $\overline{Y}$ in finite time, given a value for $\overline{X}$. It is the responsibility of the Datalog^E environment to invoke the procedure only after a value for $\overline{X}$ is known. This imposes an ordering on the predicate occurrences of a Datalog^E program. Consider the following example that uses the external predicates `mult` and `add`.

```
p(X,Y,Z)  <-  mult(X,X,XX), add(XX,YY,Z),
              mult(Y,Y,YY), b1(X), b2(Y).
```

We assume the FC $\{X, Y\} \rightarrow \{Z\}$ for both `mult(X,Y,Z)` and `add(X,Y,Z)`. Thus, none of the procedures can be invoked until some values are taken from the base relations, `b1` and `b2`. Moreover, the `add` procedure must be invoked after both occurrences of `mult`. The order of the `mult` invocations, however, can be chosen arbitrarily.

It is possible that the predicate occurrences of a Datalog^E program cannot be ordered in such a way that the input bindings required by an external predicate are supplied. Consider the following example:

```
p(X) <- add(X,Y,Z).
```

In such a case, the Datalog^E program is not well–formed. This can be recognized at compile time by using the following *covering condition.*

Given the query Q and an ordering per rule of the predicate occurrences, we construct an adorned version of the program as in [Ull85]. In an adorned program, every argument of each predicate occurrence (as well as the head predicate) is marked as bound or free. A program is *covered* if there exists an adorned program, i.e., an ordering per rule of the predicate occurrences, in which the input arguments of each external (both computed and global) predicate is marked as bound.

The FC need not be specified for global predicates, since it can be inferred during compilation. Briefly, the input bindings are specified in the query form for which the program is compiled. During compilation, safety analysis guarantees that the resulting execution will produce a finite result when applied to a finite database[KRS88]. In other words, if the arguments $x_1, x_2, \ldots, x_k$ are used for input and $y_1, y_2, \ldots, y_m$ are the output arguments specified, the compiler guarantees

[2] Note this notion is strictly weaker than that of functional dependency.

the FC $\{x_1, x_2, \ldots, x_k\} \rightarrow \{y_1, y_2, \ldots, y_m\}$. This leads to our earlier claim that the arguments of the global predicates in a Datalog^E program must also be marked as bound for the program to be covered.

In order to compile a global predicate separately, we must ensure that all information necessary in a global compilation scheme is available. In particular, we would like to guarantee the safety of the execution chosen by the system. Consider the following example involving two mutually recursive predicates.

```
p(X,Y)  ←  b(X,Y).
p(X,Y)  ←  q(X,Z),b(Z,Y).
q(X,Y)  ←  p(X,Z),b(Z,Y).
```

Clearly, a safe execution can be found if the predicates are defined in the same module. However, separate compilation could result in an unsafe execution, i.e., it could result in an infinite left–recursive descent.

Hence, we disallow mutually recursive global predicates in well–formed Datalog^E programs. A *well-partitioned* Datalog^E program is one in which any two mutually recursive global predicates are defined in the same RB.

We can now define the class of well–formed Datalog^E programs. A Datalog^E program $P = (EDB, CDB, IDB, Q)$ is *well-formed* if and only if it is covered and well-partitioned.

3.3 Semantics

In this section, we recall the declarative model semantics for Datalog programs and show that it is unaffected by the addition of external predicates. For the sake of brevity, our discussion in this section is informal.

A *substitution* is a non-empty finite set of ordered pairs $\{X_1/v_1, \ldots, X_n/v_n\}$ such that $(\forall 1 \leq i \leq n)$ X_i is a distinct variable, v_i is an atom. We view a substitution as a mapping on variables. If σ is a substitution and X a variable, the result of *applying* σ to X is defined as

$$X\sigma = \begin{cases} v, & \text{if } (X/v) \in \sigma \\ X, & \text{otherwise} \end{cases}$$

We informally explain the notion of the universe, U, of a Datalog (and Datalog^E) program P. Initially we take U_0 to be the set of all atomic objects in P. U_1 is defined as the set of all possible predicate occurrences that can be formed from the elements in U_0. U is then defined to be U_1 unioned with U_0.

We now define the satisfaction of a rule. Consider a rule of the form $p(\ldots) \leftarrow p_1(\overline{X_1}), \ldots, p_m(\overline{X_m})$ Let σ be a substitution and $I \subseteq U$. The rule is satisfied if either there is some $[p_i(\overline{X_i})\sigma] \notin I$, or if whenever $[p_i(\overline{X_i})\sigma] \in I, \forall i \in \{1, \ldots, m\}$ then $p(\ldots)\sigma$ is also in I.

Given a collection of rules RB, $I \subseteq U$ is a model of RB if I satisfies all the rules in RB. A model M of a given collection of rules RB is said to be minimal if no proper subset of M is also a model of RB. It has been shown that Datalog programs have a unique minimal model. This unique model is defined to be the meaning of the program.

Note that in the semantics defined above, finiteness of the base relations is irrelevant. Since we view external predicates as infinite relations, and use them interchangeably with base predicates in the programs, the same semantics defined above carry over to Datalog^E. Therefore, we use the same unique minimal model to be the meaning of Datalog^E programs.

The semantic definition is traditionally complemented with a constructed model semantics, that, in some sense, provides an operational semantics for the program. The constructed model semantics is not relevant to the topics discussed in the remainder of this paper and is, therefore, omitted.

3.4 Optimization of Datalog^E

The execution space and search strategy used in the optimization algorithm for Datalog can also be used for Datalog^E, since the usage and semantics of base and external predicates are very similar. We need only redefine the cost functions to be capable of evaluating the cost of operations on external predicates. Observe that the descriptor for a relation cannot model an infinite relation; therefore, we must define a new descriptor for the infinite relation. We then formalize the two functions to compute the cost and resulting descriptor for each basic operation, i.e., join, union and projection.

3.4.1 Descriptors for Infinite Relations

Let us denote the descriptor on infinite (finite) relations as *i-descriptors* (*f-descriptors*). An f-descriptor was previously defined. An i-descriptor is defined as a pair (SE,FE), where the SE and FE are *selectivity* and *fan-out expressions*. SE and FE are arithmetic expressions in which the operands are constants, cardinality, and bag of the input arguments[3]; and the operators are +, -, *, /, max, and min.[4] SE and FE are arithmetic expressions that evaluate to positive numbers less than and greater than 1, respectively. Intuitively, the SE represents the probability that a given binding for the input arguments will select a nonempty subset of the infinite relation corresponding to the computed predicate. The FE is the expected number of tuples in the selected subset of the infinite relation, conditioned on the fact that a nonempty subset was chosen.[5]

[3]We have assumed that *card* and *bag* are the only schematic information of interest. Any other information can be incorporated in the same way.

[4]Max and min are binary prefix operators, whereas the others are infix operators.

[5]Note that the traditional assumptions of uniformity and independence are implicit in the definition of the SE and FE.

Pred	Bound	SE	FE
$lt(X,Y)$	X,Y	$\frac{1}{3}$	1
$add(X,Y,Z)$	X,Y	1	1
$sqrt(X,Y)$	X	$\frac{1}{2}$	2
$eq(X,Y)$	X,Y	$\frac{1}{max(card(@1),card(@2))}$	1

Figure 1: SE and FE for some computed predicates.

A list of SE and FE for some computed predicates has been given in Figure 1. Note that the predicate *lt* is truly a predicate in the sense that it does not return any values, whereas *add* and *sqrt* are single-valued and multi-valued functions, respectively. The *lt* and *add* are very selective in the sense that all values for the input arguments choose a nonempty subset of their respective infinite relation. On the other hand, the *sqrt* predicate selects a nonempty subset only if the input argument is positive. As a result, the SE for this predicate is $\frac{1}{2}$. The use of *max* and *card* are exemplified in the descriptor for the *eq* predicate. The SE for this predicate is motivated by the desire to treat the following two rules identically.

```
p(X,Y)  ←  b1(X,X1),b2(X1,Y).
p(X,Y)  ←  b1(X,X1),b2(X2,Y),eq(X1,X2).
```

Note that card(@n) refers to the cardinality of the argument where the variable in the nth input argument is bound. Thus, the definition of card(@1) for *eq* is the cardinality of the second argument of *b*1, i.e., where the 1st input argument of *eq*, namely $X1$, is bound. Any references to the descriptor information of an input argument, e.g., *card* or *bag*, are resolved in this manner.

Note that in the above example for *eq*, the SE evaluates to a value depending upon the *card* of *b*1 and *b*2. This underscores the need to have an expression, not a constant, for SE and FE.

An i-descriptor can be used to model an f-descriptor with an FC restriction. Consider the *emp*(*Name*, *Age*) relation, containing 100 tuples with 50 distinct names and 20 distinct ages, i.e., the descriptor is (100, 100, 100, 50, 100, 20). Also consider the query ?*emp*(N, 30). Using the traditional assumptions regarding the uniformity of the distribution of values and the independence of attributes, we can estimate the result to contain 5 tuples with 5 distinct names and exactly one age, namely 30. Obviously, the result can be represented by an f-descriptor. We can also represent this result using the following i-descriptor, with finiteness constraint $Age \rightarrow Name$.

$$SE = min(1, \frac{20}{card(@2)}), \qquad FE = 5.$$

The selectivity expression is unity if the cardinality of the Age attribute, i.e., 20, is larger than the number of distinct values in the argument where the variable occurring in the Age attribute is bound. Otherwise, only 20 of those tuples will join with the *emp* relation, resulting in the selectivity $20/card(@2)$. The FE is five, since an age that exists in the relation has, on the average, five names associated with it. This example can be generalized to the following transformation.

f-to-i transformation:

Given a relation $r(A1, A2)$ with the f-descriptor $(b_0, c_0, b_1, c_1, b_2, c_2)$, the i-descriptor corresponding to r with respect to the FC $A1 \rightarrow A2$, is as follows:

$$SE = min(1, \frac{c_1}{card(@1)}), \qquad FE = \frac{c_0}{c_1}.$$

Intuitively, the SE and FE model the subset of the relation having a single value for the first argument. We shall use the f-to-i transformation in the definition of the cost and descriptor functions. Note that $A1$ and $A2$ could be vectors, and the above result could be extrapolated in the obvious way.

To completely model an f-descriptor of an n-ary relation with i-descriptors, we need to have an i-descriptor for each set of input arguments, i.e., 2^n i-descriptors. Therefore, an f-descriptor can be seen as a more concise representation of 2^n i-descriptors.

In summary, we have infinite and finite relations, with their respective descriptors. We redefine the set of all descriptors $\mathcal{D}$ to consist of the union of the set of all i-descriptors, $\mathcal{D}_i$, and the set of all f-descriptors, $\mathcal{D}_f$.

3.4.2 Cost and Descriptor Functions

In this subsection, we provide cost and descriptor functions capable of mapping from objects in $\mathcal{D}_i$. We assume that the schema has the following information for each computed predicate:

1. Name of the predicate;
2. FC, i.e., input arguments;
3. SE;
4. FE;
5. Cost of computing the answer for a value for each input argument.
6. Descriptor expression for output arguments.

As all of this information is needed to compute the cost and the resulting descriptor, we shall redefine the descriptor for an infinite object to consist of the above hextuple. The cost in the descriptor is the cost of computing the answer, using this external procedure, for a given value for each input argument. This cost is based on the unit cost which is used (by the optimizer) for the operations on non-external predicates. In $\mathcal{LDL}$ we use the cost of comparing a value as the unit cost. The database administrator who sets up the schema with the above information is expected to guesstimate the cost using the same units. The accuracy of this information is important only to an extent, as evident from the following desiderata used by most designers of query optimizers: *It is more important to avoid the worst execution than to obtain the best execution.* The rationale for this desiderata is that the spectrum of the cost of executions spans many orders of magnitude. Therefore, the cost supplied in the descriptor need only be a reasonable estimate. Note that we shall not need the descriptor expression for output arguments until the next section, when we extend the language to include complex objects such as sets, etc.

The operations are canonically viewed as follows:

1. Join of a and b:
$$p(X,Y,Z) \leftarrow a(X,Y), b(Y,Z).$$
2. Union of a and b:
$$p(X,Y) \leftarrow a(X,Y).$$
$$p(X,Y) \leftarrow b(X,Y).$$
3. Projection of c:
$$p(X,Y) \leftarrow c(X,Y,Z).$$

For each of the above operations, we describe the functions using the following finite and infinite relations:

1. infinite relations: $e1(X,Y)$ and $e2(X,Y)$ with $X \rightarrow Y$ as their FC; SE1 and SE2, FE1 and FE2, Cost1 and Cost2 as their SE, FE and Cost, respectively.
2. finite relations: $rx(X,Y)$ and $ry(X,Y)$ with (bx_0, cx_0, bx_1, cx_1, bx_2, cx_2) and (by_0, cy_0, by_1, cy_1, by_2, cy_2) as their descriptors.

The functions are defined below, using a generic cost function for operations on two finite relations, e.g., Jcost(rx,ry) as the cost of joining relations rx and ry. Even though these functions are for operations on the above binary predicates, the implementation of the $\mathcal{LDL}$ optimizer has successfully used this approach for predicates of arbitrary arity. For convenience, we define the following multiplication factors:
$$mf1 = SE1 * FE1; \qquad mf2 = SE2 * FE2.$$

1. Join:

- $\mathcal{D}_f \times \mathcal{D}_f$: Join of rx and ry:
  Assume Cost = Jcost(rx, ry).
  Assume Desc: Jdesc(rx, ry).

- $\mathcal{D}_i \times \mathcal{D}_i$: Join of $e1$ and $e2$:
  Cost = Cost1 + $mf1$*Cost2.
  Desc: SE = SE1 *ne; FE = FE1 *$mf2/ne$,
  where ne =(1 - $(1\text{-SE2})^{FE1}$) is the prob. of non-empty join of FE1 tuples with $e2$.

- $\mathcal{D}_i \times \mathcal{D}_f$: Join of $e1$ and ry:
  Cost = Cost1 + Jcost(ry, t),
  where t is a $mf1$-tuple relation.
  Desc: (B,C, B,1, B,mf1, B,min(mf1,cy_2)),
  where B = by_0*mf, C = bc_0*mf.

- $\mathcal{D}_f \times \mathcal{D}_i$: Join of rx and $e2$:
  Cost = bx_2 * Cost2 + Jcost(rx, t),
  where t is a $mf2$-tuple relation.
  Desc: similar to the $\mathcal{D}_i \times \mathcal{D}_f$ case.

2. Union

- $\mathcal{D}_f \times \mathcal{D}_f$: Union of rx and ry:
  Assume Cost = Ucost(rx, ry).
  Assume Desc: Udesc(rx, ry).

- $\mathcal{D}_i \times \mathcal{D}_i$: Union of $e1$ and $e2$:
  Cost = Cost1 + Cost2.
  Desc: SE = SE1 + SE2 - SE1 * SE2;
  FE = $(mf1 + mf2)$/SE.

- $\mathcal{D}_i \times \mathcal{D}_f$ and $\mathcal{D}_f \times \mathcal{D}_i$: Union of $e1$ and ry:
  Cost = Cost1 + Ucost(ry,t).
  Desc: SE = SE1 + min[1 , $(c_1/card(@1))$];
  FE = FE1 + (c_0/c_1).

3. Projection:

- $\mathcal{D}_f$: Pcost(rx), Pdesc(rx);

- $\mathcal{D}_i$: SE = SE1; FE = FE1; Cost = Pcost(t),
  where t is a $mf1$-tuple relation.

For brevity, we omit the detailed derivations of these formulae. Here we provide the intuitions for a few of the less obvious formulae and use these formulae to illustrate the impact on the optimizer.

The result of join is finite except in the join of $e1$ and $e2$. In this case, the SE for the result is the probability of selecting a nonempty result, which is the product of SE1 and the probability that at least one of the FE1 tuples from $e1$ join with $e2$; the FE reflects the fact that the fanout expression is conditional (i.e, the result is non-empty).

Union results in a finite relation only when both operands are also finite relations. The descriptor function for the union of $e1$ and ry uses the f-to-i transformation to convert the f-descriptor to an i-descriptor. The two i-descriptors are then unioned.

Since the result of the above functions may be an i-descriptor, some of the (local) derived predicates may also have to be modelled as i-descriptors. Therefore, these functions are applicable whenever two predicates with the appropriate descriptors are operated on.

The computation of the descriptor function that results in an i-descriptor is symbolic. Consider the case of joining $e1$ and $e2$, in which the resulting SE is an expression that is formed by multiplying the expressions symbolically. Such a symbolic manipulation of descriptor information was not needed in the traditional cost functions for optimizing relational queries. Such an expression is evaluated only when it is 'applied' to a base relation, as in the $\mathcal{D}_i \times \mathcal{D}_f$, and the $\mathcal{D}_f \times \mathcal{D}_i$ cases.

Given the above formulae for join, union and projection, it is natural to address the comprehensiveness of the descriptors for infinite relations. To do so, we must determine whether the information in the descriptors is sufficient for any Datalog program and whether the estimation is 'reasonably' accurate. The former is obvious from our observation that any bottom-up processing is composed of a sequence of these operations. The latter requires the validation of the formulae and has not yet been done. In this paper, the formulae have been provided simply to illustrate the approach. The derivations and validations will be addressed in a more complete paper.

In summary, we have proposed a model for the i-descriptor and shown that it can be used to define the cost and descriptor functions, which are then used in the global optimization algorithm. Note that the model is sufficiently general in the sense that the expressions can be extended to operate on any new schematic information that may be of interest. In that sense, we have presented a new framework for integrating external predicates into Datalog.

3.5 Implementation Issues

In order for the reader to understand the proposal in the proper context, we briefly outline the relevant implementation issues. The compiler translates each Datalog^E global predicate into a *C* procedure with references to the external procedures for all other external predicates. These references are resolved in the linking phase to produce an executable. An answer is produced when the executable is invoked with appropriate input parameters.

Some mechanism must be provided to convert objects between the Datalog^E and external representations. Moreover, an external procedure should be able to return (receive) one or more results to (from) a Datalog^E program. In the following section, we outline a more general implementation in the context of $\mathcal{LDL}$, which subsumes Datalog^E.

4 Extending $\mathcal{LDL}$

$\mathcal{LDL}$ has a more general data model than Datalog, in that it allows complex objects, such as functors, tuples and sets. The $\mathcal{LDL}$ semantics are presented in depth in [NT89]. As with Datalog^E, these semantics do not change with the addition of modules and externals. The syntax of modules and externals is presented in [CGK89a]. In the remainder of this paper, we concentrate on the implementation issues involved for the communication and optimization problems.

4.1 Implementation Issues

There are two orthogonal problems that must be solved in order to allow an $\mathcal{LDL}$ program to interact with a conventional language, such as *C*. There is an inherent incompatibility between the data types traditionally associated with procedural languages and the rich data types of $\mathcal{LDL}$. Moreover, procedural languages are ill-suited to the multiple solution paradigm basic to declarative languages, such as $\mathcal{LDL}$. In this section, we outline a solution which addresses both of these problems.

As mentioned in the background section, the complex data in $\mathcal{LDL}$ can be recursively defined using a set-tuple model. Basically, any object at the highest level is either a set, a tuple or an atom. So, we provide a set of primitives to compose and dissect any complex object partitioned along these three categories. The composing primitives include *create_atom, create_tuple, create_set* and *add_to_set* with appropriate parameters to define the constructed object. The *add_to_set* primitive is used to construct a set after it is initialized to the empty set using *create_set*. The dissecting primitives for atoms are conversion routines from the internal Datalog^E environment to the *C* environment. A tuple is dissected using the *get_ith_attr* and *get_name* primitives. Sets are dissected using the primitives *open_cursor* and *get_next_elem*, which can be used in a `while` loop of a *C* program.

In order to receive multiple solutions from a conventional subprogram, we must provide a facility to collect all the results with a single subroutine call. To do this, the calling $\mathcal{LDL}$ program initializes a temporary relation which is passed to the subprogram. The results are then stored into this relation by the subprogram. The calling program can read the tuples in this temporary relation as it would any other relation. In particular, the various $\mathcal{LDL}$ execution models [CGK89b], such as pipelined or materialized executions, apply, as do the various compile-time (local) optimizations, such as existential query optimization [RBK88] and intelligent backtracking. The following example shows how a subroutine can use these primitives to return multiple arguments.

procedure square_roots(*Rel*, x);
    Compute $\pm\sqrt{x}$ and *add_to_set*(*Rel*);
return;

$\mathcal{LDL}$ uses this same method when calling a global procedure. Hence, a conventional procedure can use this paradigm to call $\mathcal{LDL}$ by creating a temporary relation, calling the $\mathcal{LDL}$ compiled procedure, and, subsequently, reading the answers from the relation. We illustrate this in the following example.

procedure process_ancestors(x);
    create_set(*Rel*, 2);
    call $\mathcal{LDL}$ ancestor procedure with *Rel* and x;
    open_cursor(*Rel*);
    while (more tuples in *Rel*)
        let t be *get_next_elem*(*Rel*);
        /* Process t */
    end while;
return;

Note that the *Process t* above may include dissecting the tuple returned and processing the attributes of the tuple.

Observe that if a procedure returns a single value, then the above paradigm requires it to be circumscribed by a routine that puts the result in a relation and returns it to the $\mathcal{LDL}$ environment. We recognize that many *C* procedures that are single valued functions can be integrated in a simpler fashion; i.e., use them as functions to compute the answer. For these functions, the $\mathcal{LDL}$ environment takes care of the necessary conversions and totally avoids the intermediate step of putting the result in a relation. Note that the user is still required to declare this function in the schema and provide the descriptor information.

4.2 Optimization of $\mathcal{LDL}$

As in the case of Datalog^E, the addition of external predicates to $\mathcal{LDL}$ does not affect the execution space or the search strategy. The effect is limited to the cost functions. Even in the cost functions, only the computation of the descriptor function is affected.

The addition of complex objects to Datalog^E necessitates the extension of the i- and f-descriptors. The f-descriptor must be capable of describing the complex object. The i-descriptor is modified by adding information required to compute the descriptors for the output arguments. Note that the FC, SE and FE remain unchanged as these are not affected by the addition of complex objects to Datalog^E. In this section, we make the simplifying assumption that sets cannot have heterogeneous objects; e.g., a set cannot contain both f(X,Y) and g(X). Thus, we can model the complex object as nested sets. Note that the implementation of the $\mathcal{LDL}$ optimizer does not make this assumption.

We define the representation of a complex object using a grammar-like[6] formalism.

$$S \rightarrow \{\ S\ \} \mid [S, S, \ldots, S] \mid atom$$

Intuitively, the representation describes the nesting with {...} and [...] denoting a set and a tuple, respectively.

A nested set is associated with an f-descriptor similar to the one for flat relations. Each object/subobject is attributed with a bag and cardinality for a corresponding 'flat' relation. This relation is constructed as follows:

> First, identify the immediate set containing the subobject. For this set, append a unique number corresponding to each of the parents of this set to its elements and then union all the occurrences of this set. The bag and cardinality of this subobject are the corresponding values in the constructed set, i.e., relation.

Consider the following object:

{ [dno, loc, { [ename, { [cname, cage] }] }] }

This is a set of departments, each containing the department number (dno), location (loc) and the set of employees. Each employee has a name (ename) and a set of children. Each child has a name (cname) and age (cage). The bag and cardinality of the employee set is computed as if it were an atomic object in the department relation. The bag and cardinality for ename is computed for the set of all employees over all departments. Similarly, the bag and cardinality for the subobject, the children set, is computed over all departments. The bag and cardinality for cname is computed over all departments and over all employees. Intuitively, these numbers reflect the bag and cardinality of the universal relation corresponding to the above hierarchical schema. This allows the optimizer to estimate the result of joining two complex subobjects, especially at different levels.

Cost functions must be changed due to both the additional operations on complex objects and the revised set of f-descriptors. These changes are orthogonal to the problems incurred due to the addition of external predicates. Therefore, we avoid any discussion on this topic, except to note that the cost function extensions are based on the operations viewed as if they were on the conceptual universal relation mentioned above.

The addition of complex objects to the language also allows the external predicates to return complex objects for output arguments. Consequently, the optimizer must be provided with the structure of the output object. This may not be independent of the context, as evident from the following example. Consider the *eq* predicate, with only the first argument as input, i.e., an assignment predicate. The structure of the second argument may be atom, tuple or any complex object depending on the input argument. As another example, consider the predicate that returns the first argument of a tuple. The structure of this argument cannot be known a priori. So, once again, we use an expression for each output argument to describe its structure and to enable the optimizer to compute its descriptor correctly. The following grammar-like notation describes the expression syntax:

$$\begin{aligned} S \quad \rightarrow \quad & \{\ S\ \} \mid [S, S, \ldots, S] \mid atom \mid \\ & @I \mid element(S) \mid attr(S, I) \end{aligned}$$

where I is an integer; and *atom*, '@', *element*, *attr* are terminal symbols of the grammar. Note that the first three rules provide the mechanism to construct more complex objects. The fourth rule provides a reference to an input argument, i.e., I^{th} argument. This means that the descriptor is identical to the Ith argument, conditioned on the fact that a nonempty subset of the infinite relation is chosen. Applying this condition is useful in modulating the parameters of the descriptor for the input argument. The last two rules provide a way to dissect the complex object of an input argument.

For example, consider the output expression for the two predicates in Figure 2. The second argument is defined to be identical to the first for *eq* and to be an element of the set in *member*. In short, an expression is capable of referencing the descriptor of an input argument with the use of @I and then dissecting it (e.g., element, attr) or composing from it (e.g., {...}) to result

[6] Note that the we have not been precise in the rule for tuple, as it needs another recursive rule to represent the arbitrary number of attributes.

Predicate	Bound	Output exp
$eq(X,Y)$	X	@1
$member(X,S)$	X	$element$(@2)

Figure 2: Output for some computed predicates.

in any conceivable descriptor. Once again, note that the computation here is symbolic.

In summary, the descriptor function is enhanced so that it can compute the descriptor for each output argument of the external predicates. These descriptors will, in turn, be used in later cost computations.

5 Conclusions

We have extended $\mathcal{LDL}$ to include modules and external procedures, and have presented a framework for optimizing extended $\mathcal{LDL}$. The notion of well-formedness was introduced to identify the class of programs that could be separately compiled and could be executed without generating infinite relations. We addressed the inherent problems in merging a declarative language such as $\mathcal{LDL}$ with procedural languages. In particular we showed how an external procedure could manipulate complex $\mathcal{LDL}$ objects and handle the multiple solution paradigm of $\mathcal{LDL}$. We provided a set of functions to facilitate the above.

We viewed external procedures as infinite relations, and defined cost descriptors for these relations, called *i-descriptors*, based on the notions of fanout and selectivity. We extended the descriptor and cost functions to operate on *i-descriptors*. The traditional optimization scheme, using the new descriptors and extended functions, proved sufficient for optimizing extended $\mathcal{LDL}$ programs. The results reported here form the basis for the implementation of externals and modules in the $\mathcal{LDL}$ compiler and optimizer developed at MCC.

$\mathcal{LDL}$ contains constructs not addressed in this paper. Updates are of particular interest, since the $\mathcal{LDL}$ optimizer must preserve their relative ordering in a program[NK88]. The optimizer must be made aware of externals that have side-effects, as they must be treated in the same manner as updates. For example, this could be used to guarantee that a graphics device is initialized before anything is actually displayed.

References

[CGK89a] Chimenti, D., R. Gamboa, and R. Krishnamurthy. "Modules and Externals in $\mathcal{LDL}$," MCC Technical Report No. ACA-ST-036-89.

[CGK89b] Chimenti, D., R. Gamboa, and R. Krishnamurthy. "Execution Models in $\mathcal{LDL}$," MCC Technical Report in preparation.

[CP84] Ceri, S. and G. Pelagatti. *Distributed Databases: Principles & Systems*, McGraw-Hill Book Company, 1984.

[KBZ86] Krishnamurthy, R., H. Boral, and C. Zaniolo. "Optimization of Non-Recursive Queries," in *Proc. of the Conference on Very Large Data Bases (VLDB)*, Kyoto, Japan, 1986.

[KN88a] Krishnamurthy, R., and S.A. Naqvi. "Non-Deterministic Choice in Datalog Programs," in *International Conference on Databases*, Jerusalem, 1988.

[KRS88] Krishnamurthy, R., R. Ramakrishnan, O. Shmueli. "Framework for Testing Safety and Effective Computability of Extended Datalog," in *SIGMOD*, Chicago, 1988.

[KZ88] Krishnamurthy, R., and C. Zaniolo. "Optimization in a Logic Based Language for Knowledge and Data Intensive Applications," *Extending Data Base Technology*, Venice, 1988.

[Ll84] Lloyd, J.W. *Foundations of Logic Programming*, Springer Verlag, 1984.

[MW81] Maier, D., D. S. Warren "Incorporating Computed Relations in Relational Databases", in *Proc. SIGMOD*, Ann Arbor, MI, 1981.

[NK88] Naqvi, S. A., R. Krishnamurthy. "Database Updates in Logic Programming", in *Proc. SIGACT-SIGMOD Principles of Database Systems Conference (PODS)*, Austin, April 1988.

[NT89] Naqvi, S. A., S. Tsur. *A Language for Data and Knowledge Bases*, W.H.Freeman, 1989.

[RBK88] Ramakrishnan, R., C. Beeri, and R. Krishnamurthy. "Optimizing Existential Queries," in *Proc. SIGACT-SIGMOD Principles of Database Systems Conference (PODS)*, Austin, April 1988.

[Sell79] Sellinger, P.G. et al., "Access Path Selection in a Relational Database Management System," in *Proc. SIGMOD Intl. Conf on Mgt. of Data*, 1979.

[Ull85] Ullman, J. "Implementation of Logical Query Languages for Databases," in TODS, Vol. 10, No. 3, pp. 289-321, 1985.

EVENT-JOIN OPTIMIZATION IN TEMPORAL RELATIONAL DATABASES

Arie Segev and Himawan Gunadhi

School of Business Administration, University of California
and
Computer Science Research Department, Lawrence Berkeley Laboratory
Berkeley, CA 94720

ABSTRACT

An Event-Join combines temporal join and outer-join properties into a single operation. It is mostly used to group temporal attributes of an entity into a single relation. In this paper, we motivate the need to support the efficient processing of event-joins, and introduce several optimization algorithms, both for a general data organization and for specialized organizations (sorted and append-only databases). For the append-only database we introduce a data structure that can improve the performance of event-joins as well as other queries. Finally, we evaluate the performance of the proposed algorithms.

1. INTRODUCTION AND MOTIVATION

Temporal data models are designed to capture the complexities of many time-dependent phenomena, something that traditional approaches, like the relational model, were not intended to do. Many new operators are needed in order to exploit the full potential of temporal data models in enhancing the retrieval power of a database management system (DBMS). Many temporal operators have been introduced in the literature, (e.g. [Clifford & Tansel 85, Adiba & Quang 86, Clifford & Croker 87, Snodgrass 87]), yet with few exceptions, (e.g., [Lum et al 84, Rotem & Segev 87, Snodgrass & Ahn 87]), the issues of performance and optimization have not received as much attention. In a previous paper [Gunadhi & Segev 88], we identified a set of temporal joins and carried out preliminary investigation into their optimization.

In this paper, we study the optimization of *event−join* operations. The event-join operator was first introduced by [Segev & Shoshani 88a]; it is unique in that it combines temporal join and outerjoin components into a single operation. It is used primarily to group temporal attributes of an entity into a single relation; temporal attributes belonging to the same entity, but which are not synchronous in their event points, are likely to be stored in separate relations. Many queries require that they be grouped together as one relation, but differences in their behavior over time brings up the possibility that *null* values are involved in the operands and the join result.

This paper deals with optimizing event-joins in temporal relational databases. Its contributions are the following:

- Motivating and demonstrating the need to support the efficient processing of event-joins.
- As traditional processing cannot support event-joins, we have developed optimization algorithms for various situations, including static sorted databases, and dynamic databases with general data organization and append-only organization.
- In the context of the append-only database, we have developed a new data structures called the *AP*-Tree (*Append−Only* Tree). This tree is a variation of an *ISAM* and a B^+-tree combination, and is useful for other temporal queries besides event- joins.
- We compare the proposed algorithms by evaluating their costs and present some computational results.

The paper is organized as follows: in the next section, we discuss the relational representation of temporal data. In section 3, the event-join operator is defined and explained through an an example. Section 4 explores the optimization of event-joins for data that is sorted and data in a generalized setting; an algorithm for each is described in this section. Section 5 deals with the third main type of data: append-only databases, for which we propose two algorithms to optimize the event-join operator for such a database. The *AP*-tree is introduced in section 6. Section 7 presents an analysis of the costs and relative performance of the four algorithms developed. Conclusions and directions for further research are given in section 8.

This work was supported by the Applied Mathematical Sciences Research Program of the Office of Energy Research, U.S. Department of Energy under Contract DE-AC03-76SF00098.

2. RELATIONAL REPRESENTATION OF TEMPORAL DATA

A convenient way to look at temporal data is through the concepts of *Time Sequences* (*TS*) and *Time Sequence Collection* (*TSC*) [Segev & Shoshani 87]. A *TS* represents a history of a temporal attribute(s) associated with a particular instance of an entity or a relationship. The entity or the relationship are identified by a surrogate (or equivalently, the *time−invariant key* [Navathe & Ahmed 86]). For example, the salary history of an employee is a *TS*. In this paper, we are concerned with two types -- *stepwise constant* and *discrete*. Stepwise constant (*SWC*) data represents a state variable whose values are determined by events and remains the same between events; the salary attribute represents *SWC* data. Discrete data represents an attribute of the event itself, e.g. number of items sold. Time sequences of the same surrogate and attribute types can be grouped into a time sequence collection (*TSC*), e.g. the salary history of all employees forms a *TSC*. There are various ways to represent temporal data in the relational model; detailed discussion can be found in [Segev & Shoshani 88a]. In this paper we assume a representation as shown in Table 1.

MANAGER	E#	MGR	T_S	T_E
	E1	TOM	1	5
	E1	MARK	9	12
	E1	JAY	13	20
	E2	RON	1	18
	E3	RON	1	20

COMMISSION	E#	C_RATE	T_S	T_E
	E1	10%	2	7
	E1	12%	8	20
	E2	8%	2	7
	E2	10%	8	20

Table 1: Representing SWC Data with Lifespan = [1, 20]

We use the terms *surrogate*, *temporal attribute*, and *time attribute* when referring to attributes of a relation. For example, in Table 1, the surrogate of the MANAGER relation † is E#, MGR is a temporal attribute, and T_S and T_E are time attributes. We assume that all relations are in first temporal normal form (1TNF) [Segev & Shoshani 88a].

† We refer to the data construct as a 'relation', but we mean a 'temporal relation'. It is different from a standard relation because of the associated meta-data.

3. EVENT JOINS

An *Event−Join* groups several temporal attributes of an entity into a single relation. This operation is extremely important because due to normalization, temporal attributes are likely to reside in separate relations. To illustrate this point, consider an employee relation in a conventional database. If the database is normalized we are likely to find all the attributes of the employee entity in a single relation. If we now define a subset of the attributes to be temporal (e.g., salary, job-code, manager, commission-rate, etc.) and they are stored in a single relation, a tuple will be created whenever an event affects at least one of those attributes. Consequently, grouping temporal attributes into a single relation should be done if their event points are synchronized. Regardless of the nature of temporal attributes, however, a physical database design may lead to storing the temporal attributes of a given entity in several relations. The analogy in a conventional database is that the database designer may create 3NF tables, but obviously, the user is allowed to join them and create an unnormalized result.

Let $r_i(R_i)$ be a relation on scheme $R_i = \{S_i, A_{i1}, ..., A_{im}, T_S, T_E\}$. In many instances we illustrate the concepts using a single temporal attribute, that is, $m = 1$; all apply to any $m > 1$. Also, when the two surrogate *types* S_i of R_i and S_j of R_j are the same, we simply use S. Instances of surrogate S are denoted by $s1, s2, \cdots$. We use x_i to refer to an arbitrary tuple of r_i; $x_i(A)$ is the value of attribute A in tuple x_i. In order to describe the event-join between r_1 and r_2, we first present two basic operations *TE−JOIN* and *TE−OUTERJOIN*. TE-JOIN is the temporal equivalent of a standard equijoin; two tuples $x_1 \in r_1$ and $x_2 \in r_2$ are concatenated † if their join attribute's values are equal and the intersection of their time intervals is non-empty; the T_S and T_E of the result tuple correspond to the intersection interval. Semantically, this join condition is "where the join values are equal at the same time". In the case of event-joins, we are concerned only with a special case of TE-JOINs where the joining attribute is the surrogate. A TE-OUTERJOIN is a directional operation from r_1 to r_2 (or vice versa). For a given tuple $x_1 \in r_1$, outerjoin tuples are generated for all points $t \in [x_1(T_S), x_1(T_E)]$ where there does not exist $x_2 \in r_2$ such that

† It is not a standard concatenation since only one pair of T_S and T_E is part of the result tuples.

$x_2(S) = x_1(S)$ and $t \in [x_2(T_S), x_2(T_E))$. Note that all consecutive points t that satisfy the above condition generate a single outerjoin tuple. Using those operations the event-join, r_1 EVENT-JOIN r_2, is done as: templ $\leftarrow r_1$ TE-JOIN r_2 on S; temp2 $\leftarrow r_1$ TE-OUTERJOIN r_2 on S; temp3 $\leftarrow r_2$ TE-OUTERJOIN r_1 on S; result $\leftarrow$ templ $\cup$ temp2 $\cup$ temp3. Table 2 shows the result of an event-join performed between the MANAGER and COMMISSION relations of Table 1.

result	E#	MGR	C_RATE	T_S	T_E
	E1	TOM	∅	1	1
	E1	TOM	10%	2	5
	E1	∅	10%	6	7
	E1	∅	12%	8	8
	E1	MARK	12%	9	12
	E1	JAY	12%	13	20
	E2	RON	∅	1	1
	E2	RON	8%	2	7
	E2	RON	10%	8	18
	E2	∅	10%	19	20
	E3	RON	∅	1	20

Table 2: Result of Event-Join

The most troublesome components of the event-join are the outer-joins. The situation is further complicated by the time interval predicate associated with the TE-outerjoin, preventing the usage of non-temporal outerjoin procedures [Rosenthal & Reiner 84, Dayal 87]. An easy solution that comes to mind is to store all non-existence tuples explicitly, e.g., tuples like (1, ∅, 6, 8) are added to the MANAGER relation of Table 1. In that case the outerjoin components disappear, and the problem reduces to a TE-JOIN on S. Unfortunately, there are many situations where such a 'fix' will degrade overall performance rather than improve it. For example, if the whole S_i domain is represented in relation r_i, representing all non-existence data explicitly will in the worst case double the size of the table (this is the case of alternating state transitions between existence and non-existence). A much worse problem may arise when a relation contains only a fraction of the S-domain values, e.g., if on the average, only 5% of the employees of a large corporation earn commissions, adding to the non-existence data for the 95% other employees to the commission relation will add to storage cost, querying cost (including event-joins), and maintenance of the commission relation and any of its associated secondary indexes. Consequently, we divide event-joins into two types -- 'easy' and 'difficult'. Easy cases are those where the relations contain explicit tuples for all non-existence data and are sorted by (S, T_S) (the sorted case is detailed in the next section). Other cases are regarded difficult. In the remainder of the paper we are mostly concerned with the difficult cases.

4. EVENT-JOIN OPTIMIZATION

In this section we discuss the optimization of event-joins where the relations are either sorted or unsorted. Before we proceed with details of the algorithms, the important concept of tuple covering, which is used throughout the discussions, is presented first.

4.1. Concept of Tuple Covering

We first introduce the notion of *covering* which is used in all the event-join algorithms. To illustrate the concept, consider the example of Table 3.

r_1	r_2	Cover of x_1	Modified x_1
$s1,a,5,15$	$s1,b,1,2$	*None*	$s1,a,5,15$
	$s1,c,3,7$	$s1,a,c,5,7$	$s1,a,8,15$
	$s1,d,9,12$	$s1,a,∅,8,8$ $s1,a,d,9,12$	$s1,a,13,15$
	$s1,e,16,20$	$s1,a,∅,13,15$	*Full cover*

Table 3: Example of Tuple Covering

Relation r_1 has a scheme $R_1 = (S, A_1, T_S, T_E)$ and a single tuple <$s1, a, 5, 15$>. r_2 has a scheme $R_2 = (S, A_2, T_S, T_E)$ and four tuples as shown in the table. During the event-join, $x_1 \in r_1$ has to be compared with tuples $x_2 \in r_2$; assume that the order of comparisons is as shown in the table (top-down). A tuple x_2 contributes to the covering of x_1 if one or two result tuples $\{x_1(S), x_1(A_1), x_2(A_2), I_C\}$ can be derived, where $I_C \subseteq [x_1(T_S), x_1(T_E)]$. I_C can be viewed as a covered portion of x_1. The 'modified x_1' column in the table represents the uncovered portion of x_1. Note that in the covering process we have relied on the ordering of r_2 by time in deriving the outerjoin tuples (those with $x_2(A_2) = ∅$). Also, the covering column of the table contains only a subset of the final result since the covering of r_2's tuples is incomplete. The remaining result tuples should be derived from a TE-outerjoin from r_2 to r_1. In this particular example, the remaining result tuples are <$s1$, ∅, b, 1, 2>, <$s1$, ∅, c, 3, 4> and <$s1$, ∅, e, 16, 20>.

Determining and maintaining the information about the covered portion of a tuple is substantially different if the relations are not sorted by T_S. In the sorted case we can determine outerjoin tuples as the scanning progresses and the information about the covered portion of the tuple is maintained by simply modifying its T_S. In the general case, the covered subintervals can be encountered in a random order; moreover, an outerjoin result tuple associated with $x_1 \in r_1$ can be determined only when the scanning of r_2 is complete. We first present an algorithm for the case where r_1 and r_2 are sorted by S (primary order) and by T_S (secondary order). In the next subsection we discuss the general case. As can be seen from the above example, the particular values of A_1 and A_2 are immaterial as far as the logic of the event-join is concerned; we are only interested in existence or non-existence of these attributes. Consequently, in the remainder of the paper, whenever convenient, we use examples with relation schemas of (S_i, T_S, T_E), but the reader should keep in mind that at least one A_i attribute is part of the actual tuples. Also, the algorithms presented in this paper involve lots of housekeeping details. For lack of space we omit the details and provide only an outline of the algorithms. The logic of all algorithms is described ignoring blocking of tuples; it is trivially extended to handle blocking.

4.2. Event-Join Sort-Merge Algorithm

The *Sort−Merge* algorithm processes the event-join by taking advantage of the fact that both relations are in sort order. Unlike a conventional relation which requires only primary key order for sorting, the temporal relation needs to be sorted on S as the primary order and T_S as the secondary order. The event-join sort-merge algorithm, which will be referred to as Algorithm One, scans each relation just once in order to produce the result relation. At each iteration, two tuples (possibly with modified T_S), $x_1 \in r_1$ and $x_2 \in r_2$, are compared to each other and one or two result tuples will be produced based on the relationship between the tuples on their surrogate values and time intervals.

The first comparison in Algorithm One is on the surrogate value -- if they are unequal, it means that the tuple with the lower S value, say x_1, does not have any matching surrogates in the other relation; this implies that x_1 is fully covered, an outerjoin result tuple is generated, and the next x_1 tuple is read. If on the other hand $x_1(S) = x_2(S)$, there are many possible relationships that can exist between the time intervals of the two tuples; but there are just three distinct possibilities in terms of result tuples that have to be generated. The three cases are identified in Step 3 of Algorithm One.

Algorithm One

(1). Read x_1 and x_2. Repeat steps 2 to 4 until End-of-File (EOF). If EOF occurred for r_i, generate outerjoin tuples for the remainder of r_j's tuples (including the current tuple if not fully covered).

(2). If $x_i(S) < x_j(S)$, generate an outerjoin result tuple for x_i.

(3). For the situation where $x_1(S) = x_2(S)$, there are three cases to consider.

Case 1: $x_i(T_S) = x_j(T_S)$. Write an intersection result tuple.

Case 2: $x_i(T_S) < x_j(T_S)$ and $x_i(T_E) \geq x_j(T_S)$. Write one outerjoin tuple for x_i and one intersection tuple. Modify x_1 and x_2 and read next tuple(s).

Case 3: $x_i(T_E) < x_j(T_S)$. Write an outerjoin tuple for x_i.

(4). Modify x_1 and x_2 and read next tuple(s).

The next tuple of r_i is read only when the current tuple has been fully covered. Note that whenever we use the subscripts i and j in Algorithm One, $i = 1$ and $j = 2$ or $i = 2$ and $j = 1$. Also an intersection result tuple is equivalent to a TE-JOIN result tuple.

4.3. Event-Join Nested-Loop Algorithm

The Nested-Loop method described below does not assume any kind of ordering among the tuples in either relation. The event-join is achieved in two stages, the first of which is nested-loop with r_1 and r_2 being the inner and outer relations respectively. Tuples produced in the first stage are the result of either intersections or outerjoins from r_1 to r_2. In the second stage, the order of relations are now reversed for another nested-loop, but the only result tuples created here will be outerjoins from r_2 to r_1.

Unlike the sorted case, maintaining the information about the covered portion of x_i's time interval cannot be done by simply modifying T_S, and the following procedure is followed. In the first nested-loop, whenever a tuple x_1 from r_1 is first read, a list U is initialized with the pair of time-stamps associated with x_1. This list corresponds to the uncovered portions of x_1. For each tuple x_2, the algorithm applies the test of equality on the surrogate values and a non-null intersection over time. The second condition is needed because if two tuples share a common surrogate value but are disjoint over time, no conclusion can be derived (in contrast to the sorted case) as to whether an outerjoin is appropriate, unless the EOF for r_2 has been reached. Thus, while scanning r_2, the covering of x_1 is achieved only through interval intersections, and for each x_2, at most one intersection result tuple will be produced. Once this is accomplished, the uncovered subintervals associated with x_1 are determined, and appropriate outerjoin result tuples are generated. At the end of r_2's scan the interval of x_1 will either be completely covered, has one uncovered segment, or at most two segments. For each uncovered segment, the time pair representing them are inserted into U *in place* of the original entry. This ensures that U remains an *ordered* list; the ordering within U helps the search for the appropriate interval that is relevant for a TE-JOIN in subsequent iterations through r_2. Regardless of the number of entries in the list, any tuple x_2 can only intersect with *one* entry, otherwise it would mean that there are two or more tuples in r_2 having the same surrogate value and overlap in time. This implies that the condition of 1TNF has not been satisfied.

Unlike conventional nested-loop procedures, we need not retrieve all the tuples of the outer relation, since an empty U indicates that the original x_1 has been fully covered. In the event that the loop terminates because the end of file r_2 is reached, either the whole, or parts of x_1's time interval were left uncovered. An outerjoin result tuple is generated from each time pair in U; the time pair determines the time-start and time-end of the result tuple.

The second nested-loop differs from the first in that it produces only outerjoin tuples from r_2. Thus no result tuple duplicating a tuple already produced in the first stage is created. In order to reduce the number of unnecessary scans of r_i, the Algorithm uses a *hash-filter* [Bloom 70] created during the first stage as follows: when r_2 is scanned, each time an x_2 is found that participates in a TE-JOIN, the hash-filter is updated for that tuple. The hash-filter maintains H bits to represent N_{r_2} tuples, where $H \le N_{r_2}$. The hash-filter entries corresponding to $h(x_2)$, where h is the hash-function, are initialized to 0, and whenever an x_2 generates an intersection result tuple for the current x_1, $h(x_2)$ is set to 1. This table is kept in main memory, and in the best case scenario where there is sufficient memory to maintain one bit per tuple, the hash function is the count of x_2 tuples already accessed, and the table is a one dimensional array indexed by this count.

During the second stage, for each tuple in the inner relation r_2, if it hashes to a value of 0, then an outerjoin tuple is produced without scanning r_i. Otherwise, as in the first nested-loop, we carry out the same updates on the coverage of x_2, although no intersection tuples are produced. As before, outerjoin tuples are produced when it can be determined that no x_1 exists to cover the current x_2. Below we outline the steps of the algorithm, labeled as Algorithm Two. U_i denotes the list U for x_i, i =1, 2.

Algorithm Two

(1). [*Nested-Loop-1*] For each tuple in r_1: read r_2 and execute Step 2 until EOF for r_2 or x_1 is fully covered. If EOF for r_2, produce outerjoin tuples for x_1 based on U_1.

(2). If $x_1(S) = x_2(S)$ and the two time intervals intersect, then do: write an intersection result tuple. Update U_1. Set hash-filter entry for x_2 to 1.

(3). [*Nested-Loop-2*] For each tuple x_2 of r_2: if hash-filter bit = 0 produce outerjoin tuple immediately, and read next x_2. Otherwise read r_1 and execute Step 4 until EOF for r_1 or x_2 is fully covered.

(4). if $x_2(S) = x_1(S)$ and the two time intervals intersect then update U_2.

In the case of having space for a second bit for each of r_2's tuples, Algorithm Two can be further improved if a second filter is used. During the first stage, while covering x_1 it is possible that the time interval of x_2 contains that of x_1. In that case we set the corresponding filter entry to 1. Then, in Step 3 we also avoid the scan of r_1 if the first filter bit is 1 and the second filter bit is also 1.

5. APPEND-ONLY DATABASES

In the case of static history databases, one can store the data sorted by (S, T_S) and then apply Algorithm One; this provides the maximum efficiency for event-joins. For a dynamic temporal database, it may be too inefficient to keep the data sorted by (S, T_S), and consequently, either the operands are sorted prior to the application of Algorithm One, or Algorithm Two is used. If the database is append-only, the event-join algorithms can utilize this fact to enhance their efficiency.

There are several variations of append-only databases, some of which are not 'truly' append-only. As far as event-joins are concerned we view a database to be append-only if tuples are inserted at the end of the file and in order of the events that generated them. The tuples can have open-end or closed-end time intervals. To illustrate these points, consider Figure 1 that shows the time sequences for three surrogate instances with life-spans of [1, *NOW*]; each event point (X in the figure) corresponds to the generation of a new tuple for the surrogate (we are not concerned with the values of the temporal attributes). Let relation r_i represent that data; the states of that relation (for $t \ge 10$) are shown in Table 4. Note that such data is inappropriate for a WORM device since insertions also cause updates; for example, the event at time 10 led to updating ($s3$, 1, *NOW*) to ($s3$, 1, 19) and appending the tuple ($s3$, 20, *NOW*). If the representation of the data in this example uses time points instead of time intervals, it would be truly append-only.

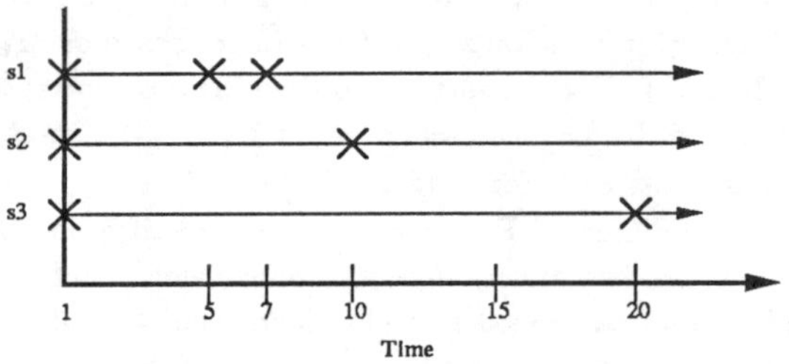

Figure 1: Time Sequences for 3 Surrogates with Lifespans = [1, *NOW*]

Snapshot at Time	State of r_i: $\{S_i, T_S, T_E\}$ (A_i is omitted)
$10 \le t < 20$	$s3, 1, NOW$ $s1, 5, 6$ $s1, 7, NOW$ $s2, 10, NOW$
$20 \le t <$ Next Event Pt	$s1, 1, 4$ $s2, 1, 9$ $s3, 1, 19$ $s1, 5, 6$ $s1, 7, NOW$ $s2, 10, NOW$ $s3, 20, NOW$

Table 4: Progression of an Append-Only Database States

Deletions in append-only temporal databases are significantly different than in conventional databases. In our case, they are storage management activities rather than user transactions. From a logical point of view deletions are a result of a change in the lifespan †, i.e. an increase in the value of *LS.START*. An example is a 'moving-window' lifespan $[NOW - l, NOW]$ where l is the length of history. In the case of step-wise constant sequences, deletion of data to reflect the new lifespan is not guaranteed to be contiguous; Table 5 illustrates this issue. The table shows the state of r_i at $t = 21$ (reproduced from Table 4) and the effect of changing the lifespan at $t = 22$ from $[1, NOW]$ to $[7, NOW]$. As can be seen from the table a new lifespan can cause updates and deletions at any point in the file. Although this example used open-end time intervals, the same problem occurs for any step-wise constant data regardless of its representation. It also demonstrates that maintaining the lifespan for an active database with small time granularity on a real-time basis can be prohibitively expensive. Fortunately, these updates and deletions can be done periodically without affecting the logical view of the data, that is, the physical lifespan can be different than the logical lifespan provided that the first contains the latter. For discrete data, the situation is much simpler and implementing a change in the lifespan can be done by simply updating a begin-of-file pointer to the first tuple whose time value is greater than or equal to the new *LS.START*.

† We use *LS.START* and *LS.END* to refer to the boundary points of the lifespan.

	State of r_i: $\{S_i, T_S, T_E\}$	
Tuple #	LS = [1, *NOW*] $t = 21$	LS = [7, *NOW*] $t = 22$
1	$s1, 1, 4$	*deleted*
2	$s2, 1, 9$	$s2, 7, 9$
3	$s3, 1, 19$	$s3, 7, 19$
4	$s1, 5, 6$	*deleted*
5	$s1, 7, NOW$	$s1, 7, NOW$
6	$s2, 10, NOW$	$s2, 10, NOW$
7	$s3, 20, NOW$	$s3, 20, NOW$

Table 5: Effect of Modifying the Lifespan of r_i at $t = 22$

If r_i is an append-only relation the order of its tuples corresponds to the order of their events, thus, they are ordered by T_S. Unfortunately, the event-join needs the primary order to be by S, and the surrogate instances of r_i can be in an arbitrary order. Nevertheless, we can take advantage of the ordering by T_S. We assume that if retroactive corrections to the history are necessary, they are done in batch mode offline and the file is reorganized to preserve the T_S-order; this is a reasonable course of action in most environments where the normal mode of operation is not error-correction. Another solution is to use an overflow area to store the 'correction records'; if their number is small (relative to the data file) they will not affect the performance of the event-join algorithms.

We present two event-join algorithms in this section. The first algorithm, stated as Algorithm Three below, follows the logic of the Nested-Loop algorithms, but is different in two important ways. First, when x_1 is compared against tuples of r_2 we do not necessarily have to complete r_2's scan -- since r_2 is append-only it follows that x_1 is fully covered if $x_1(S_1) = x_2(S_2)$ and x_2 fully covers x_1, or if $x_1(S_1) \neq x_2(S_2)$ and $x_2(T_S) > x_1(T_E)$. Second, as in the sorted case, the covered portions of x_1 are always contiguous and thus we can maintain that information by updating $x_1(T_S)$ as was done in Algorithm One. Unlike the sorted case we cannot write outerjoin tuples for x_2 when r_2 is scanned to cover x_1 (see Step 3 of Algorithm Three).

Algorithm Three

(1). [*Nested-Loop-1*] For each x_1: read r_2 and execute Step 2 until x_1 is fully covered or EOF for r_2 is reached. If EOF, generate outerjoin tuple for x_1.

(2). There are four cases to consider in this step.

Case 1: $x_1(T_S) > x_2(T_E)$ -- no result tuple is generated.

Case 2: $x_1(S) \neq x_2(S)$ and $x_2(T_S) > x_1(T_E)$ -- generate an outerjoin tuple for x_1.

Case 3: $x_1(S) \neq x_2(S)$ and $x_2(T_S) \leq x_1(T_E)$ -- no result tuple is generated.

Case 4: $x_1(S) = x_2(S)$ and $x_1(T_S) \leq x_2(T_E)$ -- do Step 3.

(3). Execute Step 3 of Algorithm One, except that no outerjoin tuple is written for x_2 if $x_2(T_S) < x_1(T_S)$, and the hash filter is updated whenever the time intervals of x_1 and x_2 intersect.

(4). [*Nested-Loop-2*] The procedure is similar to Steps 1 to 3, except: (i) If hash-filter entry for x_2 is 0, produce an outerjoin tuple without scanning r_1; (ii) Do not produce any intersection tuples; (iii) No filter updates occur and on EOF for r_2 the algorithm stops.

The second algorithm, stated as Algorithm Four below, avoids the final outerjoin from r_2 to r_1 by writing updated time-intervals for r_2's tuples while they are scanned for each x_1 tuple. This is achieved by creating a copy of r_2 which is updated during the first nested-loop. The benefit of this approach is that the second nested-loop is replaced by a single scan through r_2 in order to determine which tuples require outerjoins where no tuple has been found in r_1 with matching surrogates. The updating procedure for tuples in r_1 and r_2 is similar to that of Algorithm One.

Algorithm Four

(1). Create a working copy of r_2, and call it r_2'.

(2). [*Nested-Loop-1*] Procedure is the same as Steps 1 to 3 of Algorithm Three, except: (i) Step 3 is done exactly as in Algorithm One, that is, we write outerjoin tuples for x_2; (ii) x_2' is updated by writing in place its modified T_S; if x_2' is fully covered, its T_S is set to $T_E + 1$; (iii) No hash-filter is used.

(3). Read r_2' in a single scan, and for those tuples where $T_S \leq T_E$, produce an outerjoin result tuple.

Note that Step 1 of Algorithm Four can be done while scanning r_2 for the first x_1 tuple; subsequent x_1 tuples scan r_2'. Both of the above algorithms contains a nested-loop component to cover x_1 tuples by scanning r_2. This component is the most expensive part of the algorithms, and reducing the number of r_2's tuples scanned for each x_1 is very important. The append-only property helps in achieving that objective but we may further improve the performance by using a secondary index as described in the next section.

6. THE APPEND-ONLY TREE

Let r_1 and r_2 be append-only relations. We use a second subscript x_i whenever we need to identify specific tuples, that is, x_{ij} is the tuple x_i in location j (note that there is a one-to-one correspondence between tuple number and location number). We know that if $j_1 > j_2$, then $x_{ij_1}(T_S) \geq x_{ij_2}(T_S)$. Let x_1 be an arbitrary tuple of r_1 and assume we know the location of $x_{2\bar{j}}$, where $\bar{j}$ is the j that attains

$$\max_j \{x_{2j}(T_S) | x_{2j}(T_S) \leq x_1(T_E) \text{ and } x_2(S_2) = x_1(S_1)\}.$$

Then, we can start a backward scan of r_2 from location $\bar{j}$ until x_1 is covered. Location $\bar{j}$ can be identified using an index on (S, T_S). Such an index, however, if not available to support other queries, may be too expensive for a dynamic database. In this section we describe an index on T_S which is far cheaper to maintain compared to an S or (S, T_S) index. The index, referred to as *AP*-tree (Append-only Tree), can be viewed as a form of a load-only B^+-tree. Since the index points to records based on T_S, we omit the requirement that $x_{2\bar{j}}(S_2) = x_1(S_1)$, and thus start from the tuple which has the desired T_S and is the farthest (towards the end of the file). Figure 2 illustrates the process of covering x_1 when the *AP*-tree is used. As a specific example, consider the tuples of relation r_i in Table 4 at $t \geq 20$. Let a tuple of r_j be $(s1, 6, 7)$. To cover this tuple, only tuples of r_i with $T_S \leq 7$ should be examined. If we use an *AP*-tree, the tuple $(s1, 7, NOW)$ of r_i can be accessed directly, and following a backward scan the latest tuple to be read is $(s1, 5, 6)$. Without the index, we would have to scan r_i from the beginning and read 5 tuples (compared to two tuples with the index). In deciding whether or not to use the index, the cost of accessing it should also be taken into consideration. Using the index may be beneficial since the worst case of the backward scan is processing all the way to the beginning of the relation, e.g. if the first tuple of r_i in the above example would have been $(s1, 1, NOW)$. The main property that affects the usefulness of the index is the uniformity of event rate among surrogates of the outer relation.

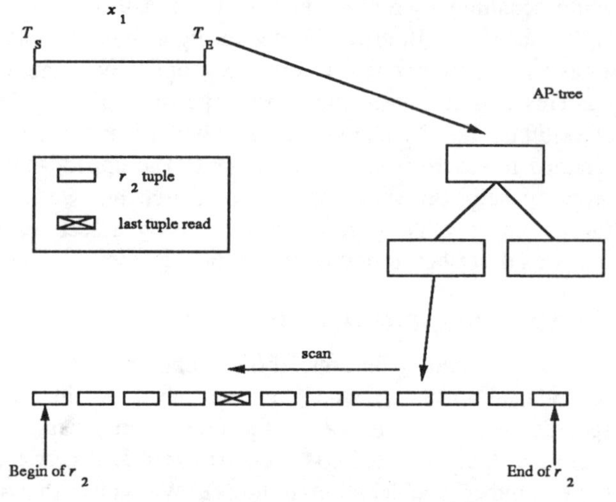

Figure 2: Covering Tuple x_1 Using *AP* -tree

Note that a uniform rate of events for an outer relation r_2 does not imply that the *AP*-tree need not be used for all $x_1 \in r_1$. Those x_1 tuples closer to the beginning of the file may benefit more from a forward scan. Currently, if the event rate is not uniform among the surrogates of r_2, an $x_1 \in r_1$ is likely to benefit from using the *AP*-tree if $x_1(S_1)$ is a very active surrogate in both r_1 and r_2.

We will now describe the basics of the *AP*-tree (more details can be found in [Gunadhi & Segev 89]). An *AP*-tree indexing r_i on T_S is shown in Figure 3. This tree is a hybrid of an *ISAM* index and a B^+-tree. The leaves of the tree contain all the T_S values in r_i; for each T_S value, the leaf points to the last (towards the end of the file) tuple with the specific T_S value. For example, tuples numbers 7, 8 and 9 have $T_S = 10$ (those tuples must belong to different surrogates since a given surrogate cannot have two tuples with the same T_S). Each non-leaf node indexes nodes at the next level. Note that the pointer associated with a non-leaf key value points to a node at the next level having this key value as the smallest node value. The significance of this decision is explained later on. Access to the tree is either through the root or through the right-most leaf. The *AP*-tree is different than the B^+-tree in several respects. First, if the tree is of degree $2d$, there is no constraint that a node must have at least d keys. Second, there is no node splitting when a node gets full. Third, the online maintenance of the tree is done by accessing the right-most leaf. Given the premise that deletions are treated as offline † storage management, only the right-hand side of the tree can be affected. The only online transactions that affects the T_S values in an append-only database is appending a new tuple. In most cases, just the right-most leaf is affected, either a pointer is updated or a new key-pointer pair is added, but if it is full a new leaf has to be created to its right, and in the worst case nodes are added along the path from the root to the right-most node and a new root node has to be created. In Figure 3 we show the effect of new tuples on the tree.

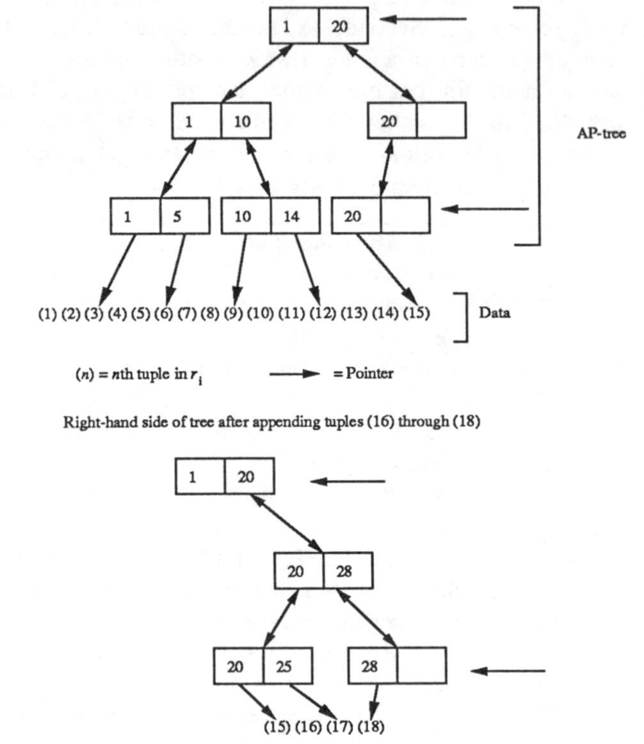

Figure 3: Example of *AP* -tree Before and After an Insertion

In the case of event-joins, an *AP*-tree on r_2 is used in the process of covering $x_1 \in r_1$. Therefore, we need to get to the leaf node pointing to $x_{2\bar{j}}$. The following procedure is adopted (v is a key value):

Procedure AP

1 Start at the root of *AP*-tree.

2 For each node visited, follow the pointer corresponding to $v^+ = \max\{v \mid v \le x_1(T_E)\}$.

† Reorganizing the tree to reflect deletions can be done during idle periods or low load periods. All the procedures function correctly regardless of the timing; the only issue is performance.

Several notes are in order. The fact that non-leaf nodes index lower level nodes based on the smallest rather than the largest key value assures that only one leaf node is visited, and that the maintenance of the tree is significantly cheaper than when the indexing is based on the smallest value. In step 2, we assumed that a v^+ exists. It is easy to see that a v^+ exists for all nodes except possibly for nodes on the path from the root to the left-most node. This case can be identified prior to accessing the *AP*-tree and thus prevents unnecessary index search.

7. COST ANALYSIS

In this section, we analyze the costs of the four algorithms. The following definitions are needed. W_{r_i} is the width (bytes) for each tuple in r_i. N_{r_i} is the number of tuples in r_i. B is the page size (bytes). P_{r_i} is the number of pages used for $r_i = \lceil (N_{r_i} \times W_{r_i}) / B \rceil$. Variables W_{r_i}, N_{r_i} and P_{r_i} apply to the operand relations and also r_{EJ}, which denotes the relation resulting from the event-join. M is the size (pages) of main memory available for an algorithm. $C_i(j)$ is the cost in disk I/Os of step j of algorithm i. α_i is the percentage of tuples in r_i that produce outerjoin tuples in r_{EJ}. β_i is the selectivity of the hash-filter on the tuples of r_i that require outerjoins. Finally, γ_i measures the average scan length through relation r_i when r_j is the inner relation for Algorithm Two, and γ_i' does the same for Algorithms Three and Four.

7.1. Algorithm One Costs

If the two relations are already sorted, the cost is $P_{r_1} + P_{r_2} + P_{EJ}$, which is the disk I/O time to join the two relations. For the case where the data need to be sorted first, each relation r_i is first sorted into F_{r_i} files, each M pages in size, where F_{r_i} is the number of files needed for the sort, and is equal to $\lceil P_{r_i} / M \rceil$. The F_{r_i} files are then merged together, and the total cost for the sorting/merging is $2(MF_{r_i} + P_{r_i})$. We are assuming that (1) $P_{r_i} \leq M$, and (2) the system allows F_{r_i} files to be opened simultaneously. If one or both of these assumptions are unsatisfied, the I/O costs will be greater. The total cost, $C_1(total)$, is $P_{r_1} + P_{r_2} + P_{r_{EJ}}$ if the relations are sorted, and $2M(F_{r_1} + F_{r_2}) + 3(P_{r_1} + P_{r_2}) + P_{EJ}$, if sorting is required.

7.2. Algorithm Two Costs

Assume that the hash-filter is kept in main memory and maintains one bit per tuple. This means that the selectivity factor β_i represents the portion of tuples in r_i with no matching surrogate values to be found in r_j. Take r_1 as the inner relation in the first nested-loop procedure. We present the cost of the algorithm in terms of its two nested-loop procedures which we label here as NL1 and NL2. $C_2(NL1) = P_{r_1} + \lceil (1 - \alpha_2)N_{EJ} / B \rceil + \gamma_2(1 - \alpha_1)\lceil P_{r_1} / M \rceil P_{r_2} + \alpha_1 \lceil P_{r_1} / M \rceil P_{r_2}$. The first term represents the cost of reading in r_1, the second term is the number of pages of result tuples written, the third term reflects the average number of reads in order to produce result tuples where x_1 is fully covered by r_2, and finally the last component is the cost of producing outerjoin tuples for r_1, which requires complete iteration through r_2 for every M pages of r_1.

As for NL2, $C_2(NL2) = P_{r_2} + \lceil \alpha_2 N_{EJ} / B \rceil + \alpha_2\beta_2 \lceil P_{r_2} / M \rceil + \gamma_1(1 - \alpha_2)\lceil P_{r_2} / M \rceil P_{r_1} + \alpha_2(1 - \beta_2)\lceil P_{r_2} / M \rceil P_{r_1}$. The first two components are the one time read cost of r_2 and the write cost for the outerjoin result tuples for r_2; the third subexpression is the cost of producing the outerjoin tuples with the help of the hash-filter; the fourth is the average cost of reads over the outer relation to determine that r_2 tuples are fully covered; and the last item is the cost of exhaustive search related to producing outerjoin tuples. The total cost is $C_2(total) = C_2(NL1) + C_2(NL2)$.

7.3. Algorithm Three Costs

For the first case of the append-only nested-loop, the hash filter is also employed; thus we assume that one bit per tuple is used. The difference in cost between Algorithms Three and Two are:(1) outerjoins can be performed on average as cheaply as covered tuples in terms of disk reads for Algorithm Three; (2) the average length of a scan through the outer relation, γ_i', is likely to be better than the γ_i of Algorithm Two, since there is a clustering of tuples on T_S. As before, $C_3(total) = C_3(NL1) + C_3(NL2)$. For the first nested-loop, $C_3(NL1) = P_{r_1} + \lceil (1 - \alpha_2)N_{EJ} / B \rceil + \gamma_2' \lceil P_{r_1} / M \rceil P_{r_2}$, where the second expression denotes the cost of iterating through r_2. For the second nested-loop, $C_3(NL2) = P_{r_2} + \lceil \alpha_2 N_{EJ} / B \rceil + \gamma_1'(1 - \alpha_2\beta_2)\lceil P_{r_2} / M \rceil P_{r_1} + \alpha_2\beta_2 \lceil P_{r_2} / M \rceil P_{r_1}$.

7.4. Algorithm Four Costs

The final algorithm differs further from the previous two nested-loop algorithms. The second part of the algorithm needs only a single scan through r_2. Although a temporary file needs to be created, it can be done during the first iteration through r_2 in order to save I/Os. Thus the total cost expression is: $C_4(total) = [P_{r_1} + 2P_{r_2}] + 2\lceil (1 - \alpha_2\beta_2)N_{EJ} / B \rceil + \lceil \alpha_2\beta_2 N_{EJ} / B \rceil + \lceil P_{r_1} / M \rceil \gamma_i' P_{r_2}$. The first expression (in brackets), represent the total cost of reading in the relations when they are in the inner loop, plus the additional overhead of creating r_2'. The second component is the write cost of event-join tuples during the first loop plus the cost of updating r_2'. The third component is the cost of generating the outerjoin result tuples during the second nested-loop. The fourth term is the cost of scanning through r_2 to produce the other result tuples.

7.5. Comparisons Among Algorithms

It is clear that Algorithm One is superior if the relations are already sorted. Also, the append-only algorithms dominate the algorithm for the general case. The interesting question is whether the relations, if not sorted, should be sorted, and then processed by Algorithm One. Figure 4 shows some preliminary results. It should be noted that we have assumed favorable conditions for the sorting, e.g., no limit on the number of files that can be opened simultaneously during a sort-merge execution; if this is not the case, the results will make Algorithms Three and Four more attractive.

Figure 4 shows the total I/O cost of the algorithms as a function of γ_i. We set the other parameters to be equal, i.e. P_{r_i} = 100,000 pages, $P_{r_{EJ}}$ = 200,000 pages, $\alpha_i = 0.1$, and $\beta_i = 0.5$. Additionally, we assumed that γ_i' is equal to γ_i. γ_i measures the percentage of blocks in the relation that have to be scanned. The graph in Figure 4(a) shows the performance of all four methods when γ_i was varied between 0.001 to 0.01. It shows that Algorithm Two does worst among the algorithms, while Algorithm Four's efficiency increases as the scan length gets shorter. It is better than Algorithm One at approximately $\gamma_i = 0.001$. Note that γ_i may be much more selective than 0.001 for an append-only database, since measured in disk I/Os, 0.001 is 100 blocks, which is still a large number. Figure 4 (b) highlights just the three best algorithms, so that a better comparison can be made at lower values of γ_i. The values of the above parameters reflect the filter selectivity and the number of tuples scanned for each inner relation tuple. The results validate our conjecture that one can do better than sorting in the append-only environment.

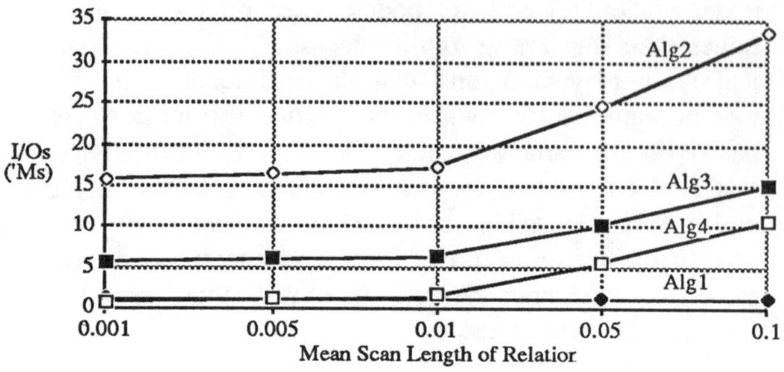

Figure 4(a)

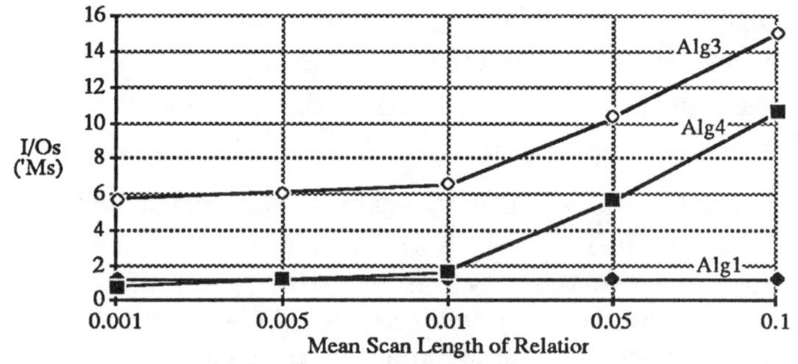

Figure 4(b)

Figure 4: Comparisons of Algorithms Over Gamma

8. Summary and Future Research

In this paper, we have addressed the problem of optimizing event-joins in a temporal relational database. Event-joins are important because normalization considerations are likely to split the temporal attributes of an entity among several relations. The event-join combines a temporal equijoin component and a temporal outerjoin component. Unlike a conventional outerjoin, the temporal counterpart consists of two asymmetric outerjoins, a fact that complicates its optimization. The complexity of processing event-join strategies depends on the nature of the data, its organization, and whether or not all non-existing data are represented explicitly. We addressed three cases of data organization; these are (in increasing order of complexity) data sorted by surrogate and time, append-only, and general optimization. For the sorted case (appropriate for static databases), the processing of an event-join is the most efficient since each relation has to be read only once. The append-only database is an appropriate organization for many dynamic temporal databases and an event-join algorithm can take advantage of the time ordering. For the append-only case we have introduced the AP-Tree, which is used to reduce the cost of scanning an outer relation in a nested-loop procedure.

In section 7, we have presented a cost analysis of the proposed algorithms. The algorithm for the sorted case (Alg. One) obviously dominates all others. The append-only algorithms (Algs. Three & Four) dominate the general nested-loop algorithm (Alg. Two); this is also expected. The interesting questions are whether, for the non-sorted case, the data should be sorted and then processed by Algorithm One. For the general case, the answer is yes (under the favorable sorting conditions that we assumed). For the append-only case the answer is dependent on the selectivity of the filter and the number of tuples scanned for each inner-loop tuple. Also, if the inner relation is significantly smaller than the outer relation, and the selectivity factors associated with the append-only algorithms are small, sorting will be less favorable. We are currently working on a comprehensive simulation test to validate our initial findings.

Finally, it should be noted that many of the concepts presented in this paper are applicable to other temporal queries; in particular other joins since the concept of covering is also relevant to them. In current and future research we try to devise more elaborate rules on when to use the AP-Tree. Also, as evident from the cost equations, estimation of several parameters are required.

References

Adiba, M, Quang, N.B., Historical Multi-Media Databases, *Proc. Int. Conf. on VLDB*, pp. 63-70, 1986.

Bloom, B.H., Space/Time Trade-offs in Hash Coding with Allowable Errors, *Comm. of the ACM*, 13, 7, 1970.

Clifford, J., Croker, A., The Historical Relational Data Model (HRDM) and Algebra Based on Lifespans, *Proc. Int. Conf. on Data Engineering*, pp. 528-537, 1987.

Clifford, J., Tansel, A., On an Algebra for Historical Relational Databases: Two Views, *Proc. ACM SIGMOD Int. Conf. on Mgt. of Data*, pp. 247-265, 1985.

Dayal, U., Of Nests and Trees: A Unified Approach to Processing Queries That Contain Nested Subqueries, Aggregates, and Quantifiers, *Pro. Int. Conf on VLDB*, pp.197-208, 1987.

Gunadhi, H., Segev, A. A Framework for Query Optimization in Temporal Databases, Lawrence Berkeley Lab Technical Report LBL-26417, 1988.

Gunadhi, H., Segev, A. Indexing Structures for Temporal Database, Lawrence Berkeley Lab Technical Report, 1989.

Lu, H., Carey, M.J., Some Experimental Results on Distributed Join Algorithms in a Local Network, *Proc. Int. Conf. on VLDB*, 1985.

Lum, V., Dadam, P., Erbe, R., Guenauer, J., Pistor, P., Walch, G., Werner, H., Woodfill, J., Designing DBMS Support for the Temporal Dimension, *Proc. ACM SIGMOD Int. Conf. on Mgt. of Data*, pp. 115-130, 1984.

Mackert, L.F., Lohman, G.M., "R" Validation and Performance Evaluation for Local Queries, *Proc. ACM SIGMOD Int. Conf. on Mgt. of Data*, pp. 84-95, 1986.

Navathe, S., Ahmed, R., A Temporal Relational Model and a Query Language, UF-CIS Technical Report TR-85-16, Univ. of Florida, 1986.

Rosenthal, A., Reiner, D., Extending the Algebraic Framework of Query Processing to Handle Outerjoins *Proc. Int. Conf. on VLDB*, pp. 334-343, 1984.

Rotem, D., Segev, A., Physical Organization of Temporal Data, *Proc. Int. Conf. on Data Engineering*, pp. 547-553, 1987.

Segev, A., Shoshani, A., Logical Modeling of Temporal Databases, *Proc. ACM SIGMOD Int. Conf. on Mgt. of Data*, pp. 454-466, 1987.

Segev, A., and Shoshani, A., The Representation of a Temporal Data Model in the Relational Environment, *Lecture Notes in Computer Science*, Vol. 339, M. Rafanelli, J.C. Klensin, and P. Svensson (eds.), Springer-Verlag, pp. 39-61, 1988a.

Segev, A., Shoshani, A., Functionality of Temporal Data Models and Physical Design Implementations, *IEEE Data Engineering*, 11, 4, pp. 38-45, 1988b.

Snodgrass, R., The Temporal Query Language TQuel, *ACM Transactions on Database Systems*, pp. 247-298, 1987.

Snodgrass, R., Ahn, I., Performance Analysis of Temporal Queries, TempIS Document No. 17, Dept. of Comp. Sci., Univ. of North Carolina, 1987.

ACHIEVING ZERO INFORMATION-LOSS IN A CLASSICAL DATABASE ENVIRONMENT*

Gautam Bhargava
Shashi K. Gadia

Computer Science Department
Iowa State University
Ames, IA 50011

Abstract. The research in temporal databases has, so far, concentrated on the history of an object as it exists in the real world. Instead, in this paper we view the history of an object as it is recorded in a database. Such a history is obtained by extrapolating the outcome of the updates (insert, modify, and delete) made to the object at discrete instants. Our model supports two kinds of query-users: the *system-user*, and the *classical-user*. For the classical-user, the interface to the database is identical to the usual interface in classical snapshot databases. We extend the classical relational model so that a transaction, i.e., an update or a (retrieval) query, is recorded in such a way that its effect can be determined at any time in the future; thus, our model is a *zero information-loss model* (Theorem 1). The logical structure imposed upon the model allows us to give a powerful algebra for the system-user to query the circumstantial information surrounding updates and queries. In addition, a single execution of a query can be identified with the relation it retrieves; thus, a user can query queries, query queries on queries, ad infinitum. The model represents an application of temporal databases to mainstream databases. It can be used in auditing, and as a foundation for building secure systems.

1. INTRODUCTION.

An update operation in classical databases is destructive - it not only destroys the environment in which it is executed, but also destroys the environment for queries. After an update is made, only the new database state is available, without even a clue to its past states.

A transaction in a database system is either an update, or a query. The activities in a database system consist of a sequence T of such transactions. We present a model, called the *zero information-loss model*, in which no information is ever lost - not even the circumstantial information surrounding the transactions. Transactions in our model are recorded by imposing a logical structure on T in such a way that the precise effect of any transaction, as well as the transaction itself, can be determined at any time in the future. In addition, the logical structure imposed upon transactions gives rise to a simple, yet powerful, algebra.

*This work was supported in part by the National Science Foundation under grant IRI-8810704.

Proceedings of the Fifteenth International Conference on Very Large Data Bases

Amsterdam, 1989

We use certain temporal relations, called transaction history relations, in which we associate a timestamp with every data value. Such relations form a foundation for the zero information-loss model. We use the time universe [0,NOW] to model the past and present (NOW) history of data in a uniform manner. As opposed to the usual approach of looking at the time dimension as the real world time, we consider it as the ***transaction time*** and use it to model the changing values of object attributes in the database.

1.1. Example. Consider the data value taken by the attribute COLOR, as shown in Figure 1.1. The semantics of this data value are: "from transaction time 15 to 20, the value of COLOR was known to be 'red' in the database, and from 21 to 25 the value was known to be 'blue'." Thus, for example, a user querying the color at transaction time 24 would retrieve 'blue' as its value. From this data value we can infer that the COLOR object was created at transaction time 15, modified at transaction time 21, and deleted at transaction time 26. This illustrates how the temporal relations in our model can be used to "store" update operations.

COLOR
[15,20] red [21,25] blue

Figure 1.1. A timestamped data value of COLOR

1.2. Definition of the zero information-loss model. Let $\mathcal{R}$ be a database scheme. Then, a *zero information-loss* model over $\mathcal{R}$ is a triple $\langle\langle\mathcal{D}$, Shadow$\rangle$, Q-log$\rangle$, where,

- $\mathcal{D}$ is the *basic model*. For every R $\in \mathcal{R}$ there is a transaction history relation r(R) $\in \mathcal{D}$, and a relation shadow(r) $\in$ Shadow.
- Q-log is a single relation.

The pair $\langle\mathcal{D}$,Shadow$\rangle$ can be viewed as a logically structured log of all updates. In a database system, the role of an update is to change the state of a database by modifying certain data values. Also, an update operation is associated with certain environmental parameters - the person authorizing the update, the reason for the update, the time of the update, etc. Along these lines, an update u to a relation r in our model can be decomposed into two parts: u^1, consisting of the new data value, and u^2, consisting of the circumstantial information surrounding the update u. The effects of u^1 and u^2 are reflected in r and shadow(r), respectively. This decomposition of the data and the circumstances of u is not lossy: the key of the updated tuple, and the time of the update can be used to glue them back together. The reason for separating r and shadow(r) is that their structures are very different. This separation allows us to maximally harness the potential of the relational approach for querying them.

The Q-log is a single relation used for storing a log of all the queries, together with the information about their execution environments. With Q-logs not only can we look at the details of (or, even re-execute) past queries, but also query queries, query queries on queries, ad infinitum. This extended capability has enormous potential in applications where it is important to monitor the accesses made to the database.

1.3. A comprehensive example. Suppose we have an emp transaction history relation with NAME, SALARY, and DEPT attributes, with NAME as the key, and that the shadow information corresponding to any update in the emp relation includes the transaction time (TT), the AUTHORIZER, the USER, and the REASON for the update. Now, consider the following sequence of transactions in the database:

T1: TT = 8; USER = Mark.
insert (NAME:John; SALARY:15K; DEPT:Toys) *in* emp
with (AUTHORIZER=Don; REASON=New Employee);

T2: TT = 40; USER = Ryne.
modify (NAME: John) *to* (DEPT: Shoes; SALARY: 20K) *in* emp
with (AUTHORIZER = Don; REASON = Reassignment);

T3: TT = 42; USER = Vance.
Q1: What is John's salary?

T4: TT = 48; USER = Rick.
insert (NAME: Doug; SALARY: 20K; DEPT: Auto) *in* emp
with (AUTHORIZER = Joe; REASON = New Employee);

T5: TT = 53; USER = Damon.
delete (NAME: John) *in* emp
with (AUTHORIZER = Don; REASON = Fired);

T6: TT = 54; USER = Andre.
Q1: What is John's salary?

T7: TT = 55; USER = Mitch.
Q2: What is John's department?

T8: TT = 56; USER = Don.
Q3: Who made inquiries about John's salary?

Corresponding to these transactions, the instance of our model $\langle\langle\mathcal{D}$, Shadow$\rangle$, Q-log$\rangle$ is shown in Figure 1.2. Note that the symbol ⊣ represents NOW.

Note that the values of NAME in emp and shadow(emp) set up a logical correspondence between tuples of emp and shadow(emp). By including the transaction time, this correspondence can be refined to a one-to-one correspondence between all updates to emp and the tuples in shadow(emp). As a result, the transactions {T1, T2, T4, T5} can be completely restored from the current state of emp and shadow(emp), and {T3, T6, T7, T8} can be completely restored from emp and the Q-log. Now, using TT, the transaction time, we can order {T1, T2, T4, T5} $\cup$ {T3, T6, T7, T8} to obtain the original sequence of transactions. Theorem 1 (the Zero-loss Theorem) says that this is true in general, i.e., a transaction log can be restored from the current state of the zero information-loss model; thus, indeed, there is no loss of information in our model.

The remainder of this paper is organized as follows. The basic model, $\mathcal{D}$, and a summary of a relational algebra for it, are briefly presented in Section 2. In Section 3 we formally define update operations, transactions, transaction logs, and state the Zero-loss Theorem. In Section 4 we

NAME	SALARY	DEPT
[8,52] John	[8,39] 15K [40,52] 20K	[8,39] Toys [40,52] Shoes
[48,→|] Doug	[48,→|] 20K	[48,→|] Auto

The emp transaction history relation

NAME	TT	AUTHORIZER	USER	REASON
John	8	Don	Mark	New Employee
John	40	Don	Ryne	Reassignment
Doug	48	Joe	Rick	New Employee
John	53	Don	Damon	Fired

The shadow(emp) relation

QUERY	TT	USER
Q1: John's SALARY	42	Vance
Q1: John's SALARY	54	Andre
Q2: John's DEPT	55	Mitch
Q3: USER ID of Q1	56	Don

The Q-log relation

Figure 1.2. The model ⟨⟨{emp}, {shadow(emp)}⟩, Q-log⟩

extend the basic algebra to query the whole zero information-loss model. We end this section with a list of assumptions made in this paper, a discussion of related papers, and some remarks about our notational conventions.

1.4. Assumptions. To simplify our presentation and avoid unnecessary distraction from the main theme of this paper, we make the following assumptions. These assumptions can be eliminated along the lines of [GB].

- The key of an object (tuple) in a transaction history relation $r \in \mathscr{D}$ does not change, and it is known correctly at all times. In the absence of this assumption, a new update operation, which allows keys of several objects to be changed simultaneously, becomes necessary [GB].
- At a given transaction time instant, there is only one transaction, i.e., the transactions are not batched together. This assumption is relaxed in [GB].
- No redundant updates are made, i.e., we do not change any attribute value of an object to itself. This assumption is also relaxed in [GB].

1.5. Related works. Algebras for temporal databases have appeared in [Ga1,Ta,LJ]. [Sn] introduces a temporal model with two time dimensions and gives a QUEL like query language for it. [GB], frequently referenced here, relaxes the assumptions made in this paper; a very preliminary version of [GB] appeared in [GY2].

1.6. Notations. We write a singleton {x} of any sort without braces. If R and S are schemes, RS denotes R∪S. If τ and τ' are tuples, then $\tau \circ \tau'$ denotes their concatenation; thus, if the schemes of τ and τ' are R and S, respectively, then the scheme of $\tau \circ \tau'$ is RS.

2. THE BASIC MODEL AND ITS ALGEBRA.

In this section we present a brief overview of the basic model $\mathscr{D}$. For a more detailed and generalized treatment of this basic model, the reader may refer to [GB].

2.1. The concept of time. We assume that we are given a universe of time instants [0,NOW] = {0, 1, 2, ···, NOW}, along with a linear order $\leq$ on it. NOW denotes the current time according to the system clock. A subset I of [0,NOW] is an *interval* if $\forall t_1 t_2 t_3 (t_1 \in I \wedge t_2 \in I \wedge t_1 \leq t_3 \leq t_2 \Rightarrow t_3 \in I)$.

2.2. Temporal elements. A *temporal element* is a finite union of intervals. An interval is obviously a temporal element. An instant t may be identified with the temporal element [t,t]; thus it may be regarded as a temporal element. The set of all temporal elements is closed under + (union), * (intersection) and - (complementation), and thus forms a Boolean algebra. Temporal elements make a striking simplification in query languages for temporal databases [GY1,Ga2].

2.3. Temporal assignments. To capture the time-variant properties of objects, we introduce the notion of a temporal assignment. A *temporal assignment* ξ to an attribute A, with a temporal element μ as its *temporal domain*, is a function from μ such that for each instant $t \in \mu$, $\xi(t)$ is an element of dom(A). Figure 1.1 shows a temporal assignment to COLOR with [15,25] as its temporal domain.

If ξ is a temporal assignment, then $[\![\xi]\!]$ denotes its temporal domain and $|\xi|$ denotes its range $\{\xi(t) : t \in [\![\xi]\!]\}$. We may denote a temporal assignment ξ as $\langle \nu_1 \mapsto a_1, \cdots, \nu_m \mapsto a_m \rangle$, where $\nu_1 \cdots, \nu_m$ are temporal elements, $\xi(t) = a_i$ if $t \in \nu_i$, $1 \leq i \leq m$. If ξ is a temporal assignment, and μ a temporal element, then $\xi \upharpoonright \mu$ denotes the restriction of ξ to the temporal domain $[\![\xi]\!] * \mu$.

2.4. θ-navigation. Suppose ξ_1 and ξ_2 are temporal assignments to θ-comparable attributes. We define $[\![\xi_1 \theta \xi_2]\!] = \{t \in [\![\xi_1]\!] * [\![\xi_2]\!] : \xi_1(t) \theta \xi_2(t)\}$. The construct $[\![\xi_1 \theta \xi_2]\!]$ is of fundamental importance in temporal databases. It evaluates to a time domain between $\emptyset$ and $[\![\xi_1]\!] * [\![\xi_2]\!]$. If the value is $\emptyset$, it implies that the two temporal assignments were never related. We define the Boolean $\xi_1 \theta \xi_2$ to be an abbreviation for $([\![\xi_1 \; \theta \; \xi_2]\!] = [\![\xi_1]\!] * [\![\xi_2]\!] \wedge [\![\xi_1]\!] * [\![\xi_2]\!] \neq \emptyset)$.

2.4.1. Example. If ξ_1 denotes the assignment to COLOR, as in Figure 1.1, and ξ_2 denotes $\langle$[0,NOW] $\mapsto$ blue$\rangle$, then $[\![\xi_1 = \xi_2]\!]$ = [21,25].

2.5. Transaction history tuples. A *transaction history tuple* τ, over a scheme R, is a function from R such that for each attribute A in R, $\tau(A)$ is a temporal assignment to A. Suppose τ is a transaction history tuple over R. If μ is a temporal element, then $\tau \upharpoonright \mu$ is the function from R such that $(\tau \upharpoonright \mu)(A) = \tau(A) \upharpoonright \mu$, for every $A \in R$. We say that τ is *homogeneous* if $[\![\tau(A)]\!]$ is the same for all attributes $A \in R$. In this paper we will only be interested in homogeneous tuples. We say that τ is *null* if $[\![\tau(A)]\!] = \emptyset$ for some $A \in R$.

2.6. Transaction history relations. A *transaction history relation* r over R, with $K \subseteq R$ as its *key*, is a finite set of non--null homogeneous transaction history tuples over R such that (i) $|\tau(A)|$ is a singleton for every $A \in K$, and (ii) if τ and τ' are tuples in r such that $\forall A \in K(|\tau(A)| = |\tau'(A)|)$, then $\tau = \tau'$. Figure 1.2 shows the emp transaction history relation over NAME SALARY DEPT with NAME as its key.

Note that our assumption that the value of a key attribute does not change with time has been used in the definition of a key of a transaction history relation. The behavior of a key in temporal databases is more complex than its classical counterpart; its detailed account would force the surfacing of several subtleties and auxiliary definitions [GY2,GB], and cause a substantial distraction from the main theme of the paper.

The algebra presented in this Section is limited in the following two respects: (i) We do not introduce certain restructuring operators which are necessary to change the key attributes of a transaction history relations. (ii) We give a limited definition of the projection and natural join operators. Both of these limitations are overcome in the detailed presentation provided in [GB]. An informal discussion of the problems involved is given in [GY2].

If r is a transaction history relation over R, then $[\![r]\!] = \cup_{\tau \in r} [\![\tau]\!]$. For example, for the state of the emp relation shown in Figure 1.2, $[\![emp]\!]$ = [8,NOW]. If μ is a temporal element then $r \upharpoonright \mu = \{\tau \upharpoonright \mu : \tau \in r\}$. If t is an instant, then $r \upharpoonright t$ is called a *temporal snapshot* of r at t.

2.7. An algebra for transaction history relations. We assume that a database $\mathscr{D}$, consisting of finitely many transaction history relations is given. The set of all algebraic expressions can be divided into three mutually exclusive groups: temporal expressions, Boolean expressions, and relational expressions. In the following definitions, a constant temporal assignment $\langle$[0,NOW] $\mapsto$ a$\rangle$ is abbreviated as a.

2.7.1. Temporal expressions. Temporal expressions, the syntactic counterpart of temporal elements are formed from $[\![r]\!]$, $[\![\xi_1 \theta \xi_2]\!]$, +, *, and -. If μ is a temporal expression and τ is a tuple, then $\mu(\tau)$, the result of substituting τ in μ, is defined in a natural manner.

2.7.1.1. Example. Let us consider the emp relation of Figure 1.2. Suppose τ denotes John's tuple. Then [[SALARY ≠ 20K]] is a temporal expression, and for τ its value is [8,39]. Similarly, [[DEPT = Shoes]](τ) = [40,52]. The result of substituting τ in the temporal expression [[SALARY ≠ 20K]] + [[DEPT = Shoes]] is [8,52].

2.7.2. Boolean expressions. Atomic Boolean expressions are of the form, TRUE, FALSE, $\xi_1\theta\xi_2$, and $\mu\subseteq\nu$. More complex expressions are formed by using the Boolean operators ∧, ∨, and ¬.

2.7.3. Relational expressions. Relational expressions are syntactic counterparts of transaction history relations. Every relation r ∈ $\mathscr{D}$ is a relational expression. In the following, we suppose r is a transaction history relation over R, with K⊆R as its key. The other relational expressions are defined as follows.

- If s is a relational expression over R, with K⊆R as its key, then r−s, r∪s, r∩s are transaction history relations over R, with K as their key. In giving the semantics for r∪s it is assumed that for the same object, the versions of their transaction histories in r and s do not differ at any instant; otherwise, r∪s is undefined. This is analogous to the classical case where, if r and s are snapshot relations with K as their key, then K may not be a key of r∪s.
- If K ⊆ X ⊆ R then Π_X(r) is a transaction history relation with K as its key. We recall that the condition K⊆X has been imposed in this paper to ensure that the key of Π_X(r) does not have to be changed.
- If f is a Boolean expression, and μ a temporal expression, then σ(r;f;μ) is a transaction history relation over R, with K as its key. σ is an important operator in temporal databases. It allows a tuple τ in r to be examined with respect to a Boolean criterion f, and then restricted to $\mu(\tau)$. It stands for the transaction history relation $\{\tau|\mu(\tau)$: $\tau\in r \wedge f(\tau)\}$.
- If s is a relational expression over S, with K' as its key, then r ⋈ s is a transaction history relation over RS with K∪K' as its key. The expression r ⋈ s evaluates to $\{(\tau\circ\tau')|([\![\tau]\!]*[\![\tau']\!])$: $\tau\in r \wedge \tau'\in s\}$. The restriction of $\tau\circ\tau'$ to $[\![\tau]\!]*[\![\tau']\!]$ is necessary to ensure homogeneity. In general, the key may be a proper subset of K∪K'.
- If A ∈ R and B ∉ R, then the transaction history relation $\rho_{A\leftarrow B}$(r) is obtained from r by renaming the attribute A to B. If A ∈ K, the new key is (K−A)∪B.

2.8. The query-users. As $\mathscr{D}$ encapsulates the entire database history of objects, it is very rich in content. Our model supports two kinds of query-users: (i) The *system-user*, who has access to the entire database $\mathscr{D}$ and can use the powerful operators to be introduced in Section 4. (ii) The *classical-user*, who can only access $\mathscr{D}$|NOW = {r|NOW: r∈$\mathscr{D}$}, the current state of the database. In $\mathscr{D}$|NOW, the only timestamps are NOW. Such timestamps are clearly redundant, and are hidden from the classical-user. This user has no use of temporal expressions; his/her needs are adequately covered by the Boolean expressions. Thus, the third argument of σ is not made available to him/her. In this way, the query interface for the classical-user of our model is similar to that for a user in a classical snapshot database.

2.9. Examples. We look at some queries that can be asked of transaction history relations by a system-user. All these queries refer to the emp relation of Figure 1.2.

2.9.1. Example. According to the database, at what transaction time was John's salary shown as 15K while he was working in Toys? This query is expressed as:

[[σ(emp; NAME = John; [[SALARY = 15K]] *
[[DEPT = Toys]])]]

2.9.2. Example. Give details about the employees for whom the database showed a salary greater than 24K while they worked in the Clothing or the Shoes department.

σ(emp; TRUE; [[SALARY > 24K]] * ([[DEPT = Shoes]] + [[DEPT = Clothing]]))

3. TRANSACTIONS AND ZERO INFORMATION-LOSS.

Informally, a transaction is either a single update or a single query. We impose a logical structure on the log of all transactions to obtain the model ⟨⟨$\mathscr{D}$, Shadow⟩, Q-log⟩ from which we can restore the log at any time. Thus, there is no loss of information in this model, and it is called a zero information-loss model. The advantage of the logical structure is that it lends itself to a powerful query language. The query language is the subject of study in the next section.

3.1. Shadow relations. Just as a transaction history relation records the update made in the value of an attribute, the shadow relation records the environment of the update. Consequently, every transaction history relation has a shadow relation associated with it, and for every update represented in the transaction history relation, there is a corresponding tuple in the shadow relation.

Suppose r is transaction history relation over R in $\mathscr{D}$, with $K \subseteq R$ as its key. Then the scheme of shadow(r) must contain {USER, TT} $\cup$ K. A USER is one who makes the update, and it can be a person, an algorithm, a machine, or any other decision making body. The transaction time TT is supplied by the system clock when an update becomes effective. The purpose of K is to establish a connection between an object in a transaction history relation, and the corresponding tuple in its shadow. Other attributes in shadow(R) depend upon the nature of r. For example, in an employee shadow relation it might be relevant to record who authorized the salary update, whereas in a weather record shadow relation it may make sense to record the space coordinates of the satellite reporting the information. Figure 1.2 shows the shadow(emp) relation, over NAME TT AUTHORIZER USER REASON.

3.2. Updates. Throughout this section we will assume that we are given a fixed but arbitrary transaction history relation $r \in \mathscr{D}$ over R, with $K \subseteq R$ as its key. If $\tau \in r$, then key(τ) denotes the snapshot tuple τ_1 over K, such that for every $A \in K$, $\tau_1(A) = |\tau(A)|$.

In the zero information-loss model, an ***update operation*** transforms a transaction history relation r, and its corresponding shadow relation, from one state to another. The update operations on $\langle \mathscr{D}$, Shadow$\rangle$, are as follows:

- *insert* τ *in* r *with* τ';
- *modify* key(τ) *to* new_τ *in* r *with* τ';
- *delete* key(τ) *with* τ';

where τ' is the part of the update operation that has the circumstantial or shadow information about the update. Several examples of these operations have already been given in Section 1. As mentioned in the introduction, an update operation u on $\langle \mathscr{D}$, Shadow$\rangle$ can be partitioned into two parts: u^1 and u^2, which modify $\mathscr{D}$ and Shadow, respectively. To make this partitioning more visible, we first make syntactic modifications in our update operations so that an update operation u can be expressed as $\langle u^1,u^2 \rangle$:

- $\langle$*insert* τ *in* r, *append* $\tau' \circ$key(τ) *to* shadow(r)$\rangle$
- $\langle$*modify* key(τ) *in* r *to* new_τ,
  append $\tau' \circ$key(τ) *to* shadow(r)$\rangle$
- $\langle$*delete* key(τ) *in* r, *append* $\tau' \circ$key(τ) *to* shadow(r)$\rangle$

If u is an update operator on $\langle$r,shadow(r)$\rangle$, and u^1 and u^2 are such that $u = \langle u^1,u^2 \rangle$, then u^1 and u^2 are called ***update operations*** on r and shadow(r), respectively. Note that "*append* $\tau' \circ$key(τ) *to* shadow(r)" is the only update operation for the shadow relation shadow(r). The semantics of an update operation on r can be defined and the following lemma can be proved easily.

LEMMA 1. If $\langle$r',shadow(r')$\rangle$ are obtained after update u to $\langle$r,shadow(r)$\rangle$ by the user x at transaction time t, then $\langle$r,shadow(r)$\rangle$, u, x, and t can be determined from $\langle$r',shadow(r')$\rangle$ alone.

3.3. Q-log. Q-log is essentially a log of all queries. A point of interest is that the same query may be made by different users at different times. For example, a compiled query program may be run at various times. The decision about how a query is to be stored - as a string, or as a parse tree, or as some other semantic entity - is left to the choice of the implementer. We only require that there be a decision procedure for deciding whether or not two queries are the same. This algorithm can be smart enough to try and deduce the logical equivalence of queries, or so simple as to consider textually different queries to be different.

We introduce a special attribute QUERY, and assume that dom(QUERY) consists of all queries. We define *Q-scheme* to be QUERY USER TT, and Q-log to be a relation over Q-scheme. Thus, a tuple τ in Q-log is of the form $\langle$Q, x, t$\rangle$, where Q is a query, x a user, and t a transaction time instant. The meaning of τ is that Q was executed by user x at transaction time t. Figure 1.2 shows an example of a Q-log relation.

Suppose s represents the current state of Q-log, the query q is asked by the user x, and the system answers it for the state of the zero information-loss model at transaction time t. Then the new state of Q-log becomes $s \cup \{q \circ x \circ t\}$.

LEMMA 2. Suppose the query q, by user x, at transaction time t, transforms the state s of a Q-log to the state s'. Then s, q, x, and t can be determined from s' alone.

3.4. Information about a transaction. The *information* associated with a transaction is the data which is deemed relevant for the application on hand. This definition places the burden of deciding relevance on the system designer. Thus, the system designer has to make a priori decisions about what kind of circumstantial information about transactions is of value and needs to be stored.

3.5. Transactions and transaction log. A *transaction* T in the zero information-loss model is of the form $\langle u/q,x,t \rangle$, where u/q is either an update or a query, x is a user, and t is a transaction time instant. If the transaction T, on the state Z of the zero information-loss model, results in the state Z', we write Z' as T(Z). The following follows from Lemmas 1 and 2.

LEMMA 3. If T is a transaction on Z, then from T(Z) alone, we can restore T and Z.

A *transaction log* is a sequence $T = \langle T_1, T_2, \cdots, T_n \rangle$ of transactions. Suppose $\emptyset$ denotes the initial empty state of the zero information-loss model. Then the *outcome* of the transaction log T is the state $T_n(T_{n-1}(\cdots(T_1(\emptyset))\cdots)$.

THEOREM 1 (The Zero-loss Theorem). A transaction log T can be restored from its outcome.

4. ALGEBRA FOR THE ZERO INFO.-LOSS MODEL.

In this Section we extend the algebra of Section 2 to query the complete zero information-loss model.

4.1. Classical operators for shadow relations. The operators for Shadow are as follows:

- Shadow relations are the classical 1nf relations, and we allow the classical relational operators on them.

4.1.1. Examples. Suppose emp(NAME SALARY DEPT) is a transaction history relation with NAME as its key, and management(DEPT MANAGER) is a transaction history relation with DEPT as its key. Suppose the scheme of shadow(emp) is as before, and the scheme of shadow(management) is TT AUTHORIZER DEPT USER REASON. The following queries are based on the shadow(emp) and shadow(management) relations.

4.1.1.1. Example. Give reasons for all the updates in emp.

$$\Pi_{\text{REASONS}}(\text{shadow(emp)})$$

4.1.1.2. Example. Give reasons for all the updates in emp that were authorized by Harry.

$$\Pi_{\text{REASONS}}(\sigma(\text{shadow(emp); AUTHORIZER = Harry}))$$

4.1.1.3. Example. Who made changes in both emp and management?

$$\Pi_{\text{shadow(emp).USER}}(\text{shadow(emp)} \bowtie_{\text{USER}} \text{shadow(management)})$$

4.1.1.4. Example. Who made the updates in emp between transaction time(TT) 10 and 20?

$$\Pi_{\text{USER}}(\sigma(\text{shadow(emp); TT} \geq 10 \ \wedge \ \text{TT} \leq 20))$$

The queries in the algebra for shadow relations can use temporal expressions involving relations in $\mathscr{D}$. This is illustrated by the following example.

4.1.1.5. Example. It is easily seen that the temporal expression $[\![\sigma(\text{management; MANAGER = Tom; [0,NOW]})]\!]$ retrieves the time when Tom is shown to be a manager in some department. Using this, the query "what were the reasons for updates in emp during the time Tom was a manager in some department?" is expressed as follows.

$$\Pi_{\text{REASONS}}(\sigma(\text{shadow(emp); TT} \subseteq [\![\sigma(\text{management; MANAGER = Tom; [0,NOW]})]\!]))$$

4.2. Navigation between $\mathscr{D}$ and Shadow. Suppose τ is a transaction history tuple over R. If $X \subseteq R$, we define the temporal expression $\delta(\tau(X)) = \{t\text{: for some } A \in X, \tau(A){\restriction}t \neq \tau(A){\restriction}(t-1)\}$. Clearly, $\delta(\tau(X))$ is the set of all transaction time instants when there is a change in $\tau(X)$.

4.2.1. Semijoins of relations in $\mathscr{D}$ and Shadow. Because the structures of a transaction history relation and a snapshot relation are quite different, we do not want to define their join. Instead we define semijoin operations, whose navigational nature is like a join, but instead of retrieving the concatenation of the tuples of the two operand relations, we only retrieve parts of tuples of one of the operand relations.

Suppose $r(R) \in \mathscr{D}$, e_1 and e_2 are expressions which evaluate to subsets of r and shadow(r), respectively, and f is a Boolean condition involving attributes of schemes of r and shadow(r). Then we introduce F_1 and F_2, called the ***filter operations***, as follows.

$$F_1(e_1,e_2,f) = \{\tau \in e_1\text{: } \exists \tau' \in e_2(\tau'(TT) \subseteq \delta(\tau(R)) \wedge f(\tau \circ \tau'))\},$$

$$F_2(e_1,e_2,f) = \{\tau' \in e_2\text{: } \exists \tau \in e_1(\tau'(TT) \subseteq \delta(\tau(R)) \wedge f(\tau \circ \tau'))\},$$

where $\tau'(TT) \subseteq \delta(\tau(R))$ expresses "τ' is a shadow tuple of τ corresponding to an update operation which modified at least one attribute in τ."

4.2.1.1. Example. Which tuples in emp were updated by Harry?

F_2(emp, shadow(emp), USER = Harry)

4.2.1.2. Example. Name the people who made changes in Tom's tuple during [10,20].

$\Pi_{USER}(F_1(\sigma(\text{emp; NAME = Tom; [0,NOW]}),$
$\text{shadow(emp)}), 10 \leq TT \wedge TT \leq 20))$

4.3. Operators for Q-log. Recall, a tuple in Q-log is of the form $\langle Q, x, t\rangle$, with the meaning that the query Q was executed by user x at transaction time t. If τ is the tuple $\langle Q, x, t\rangle$, then it is natural to identify it with the relation retrieved by Q when it was executed at transaction time t. (Note that from x, we can decode whether this user is a system-user or a classical-user.) We denote this relation as $[\tau]$, and define [Q-log] = $\{[\tau]: \tau \in$ Q-log$\}$. Thus, we arrive at the database Z = $\mathscr{D}$ ∪ Shadow ∪ {Q-log} ∪ [Q-log]. Z can be partitioned into two parts: Z^1 consisting of the classical 1-nf relations, and Z^2 consisting of transaction history relations. On Z^1 we allow the classical relational operators, and on Z^2 we allow the algebra described in Section 2. Additionally, in this section we have introduced operators on $\mathscr{D}$ ∪ Shadow. Clearly [Q-log] allows a query to be treated as a relation, which can be queried. Thus, we can query queries, query queries on queries, etc.

4.3.1. Example. Name the people who executed Q1.

$\Pi_{USER}(\sigma(\text{Q-log, QUERY = Q1}))$;

4.3.2. Example. What difference was observed in the execution of Q1 at TT = 3 and 10 by Harry?

[Q1,Harry,10] - [Q1,Harry,3].

4.3.3. Example. If Q is the query "what are all the naval bases," executed by Vance at transaction time t, then σ([Q,Vance,t]; LOCATION = Europe) can be used as an answer to the question "What naval base locations in Europe were revealed to Vance?" and the query $\Pi_{COMMANDER}$([Q,Vance,t]) can be used to answer the question "What COMMANDER information was released to Vance?"

The above example demonstrates the utility of querying the Q-log. This has great application in areas where it is necessary to monitor the accesses being made to the database. Such query capability allows the database administrator to analyze the information-flow patterns.

5. CONCLUSIONS.

We have presented a relational model which incorporates the information-rich transaction log. This is done in a manner which allows us to query the transactions and their environment. As a result our model is completely self-contained and has zero information-loss. We believe that it fulfills an important need for being able to examine the nature of transactions in a database system. It also applies to a knowledge based system where it is important to retain, both, a complete historical knowledge and the evolution process of such knowledge over time. We believe that this model can be used as a workbench for building very secure systems.

ACKNOWLEDGEMENT. We thank the anonymous referees for their helpful comments.

REFERENCES

[Ga1] Gadia, Shashi K. *A Homogeneous Relational Model and Query Languages for Temporal Databases.* ACM-Transactions on Database Systems, pp 418-448, vol 13, 1988.

[Ga2] Gadia, Shashi K. *The role of temporal elements in temporal databases.* Quarterly Bull. of the Computer Society of IEEE Technical Committee on Data Engineering, December, 1988.

[GB] Gadia, Shashi K. and Bhargava, Gautam. *A Formal Treatment of Updates and Errors in a Relational Database.* Submitted for publication.

[GY1] Gadia, Shashi K. and Yeung, Chuen-Sing. *Inadequacy of Interval Timestamps in Temporal Databases.* To appear in Information Sciences.

[GY2] Gadia, Shashi K. and Yeung, Chuen-Sing. *A Generalized Model for a Relational Temporal Database.* Proc ACM-SIGMOD Int. Conf. on Management of Data, June 1988, pp. 251-257.

[LJ] Lorentzos, Nikos A. and Johnson, Roger G. *Extending relational algebra to manipulate temporal data.* Information Systems, Vol 13, 1988, pp. 289-296.

[Sn] Snodgrass, Richard. *The Temporal Query Language TQuel.* ACM Transactions on Database Systems, Vol 12, 1987, pp 247-298.

[Ta] Tansel, A.U. *Adding Time Dimension to Relational Model Model and Extending Relational Algebra.* Information Systems, 1986.

Derived Data Update in Semantic Databases*

I-Min Amy Chen and Dennis McLeod
Computer Science Department
University of Southern California

Abstract

The derived data update problem involves the transformation of modifications of derived facts into corresponding changes to base facts and other derived facts. Using a database schema defined with a generic semantic database model which includes derived data specifications, techniques and algorithms are provided for appropriately modifying base data when derived data are changed. An experimental prototype system based upon this approach has been developed.

Keywords: derived data update, semantic databases, update propagation

1 Introduction

Derived data in a database is useful to accommodate multiple viewpoints on information, to maintain frequently referenced/computed data, and to support database protection/security [8, 9, 10, 11]. Consequently, a database is viewed here as consisting of base data, or explicitly stored facts, and derived data, which are computed from base data or derived data by derivation rules.

It is important for database users to be able to modify derived data directly, rather than issuing a less natural and possibly more complicated modification on the underlying base data. The *derived data update problem* [4, 5, 6, 7, 12, 13, 16] involves the transformation of modifications of derived data into corresponding changes to base data and other derived data. The transformation of modifications is usually termed *update propagation.*

*This research was supported, in part, by the USC AT&T Affiliates Program.

The derived data update problem is considered difficult because update propagation is not always unique; there may be no way or more than one way to transform modifications. For example, the object class A may be defined as derived by the union of classes B and C; inserting an instance to A can cause the instance to be inserted to B, to C, or to both. In this case, the update transaction is *ambiguous.*

The derived data update problem is similar to the *view update problem* [4, 5, 6, 7, 12, 13, 16] in the terminology of relational databases. Because it is difficult to support update propagation, most database systems permit view update in extremely limited cases, or allow view update only if update propagation procedures are explicitly specified by designers or users. The problem of supporting derived data update in semantic (structurally object-oriented) databases is largely open; most existing approaches restrict derived data to be not changeable (e.g., IFO [1], Galileo [3], SDM [9], and Taxis [14]). DAPLEX [15] supports derived data update, but users must specify the corresponding update on the base data; the user is responsible for assuring the correctness of such derived data update transactions.

Since conceptual schemas specified with semantic database models are richer than record-oriented (e.g., relational) schemas, it is possible to automatically decide update propagation in a more substantive manner. Our approach to the derived data update in a semantic database is termed *schema-based,* since update propagation is based on the schema definition, which includes class (type) and attribute definitions, constraints, and derivation rules. For example, changes to an attribute will have direct effects on its inverse, the attributes derived from this attribute, and the other attributes from which this attribute is derived. Changes to instances of a class will have direct effects on the immediate superclass and subclasses of the class. In our approach, schema information is used to yield update rules. As a simple example, there is an update-rule which states that if class A is the union of classes B and C, then insertion to either class B or class C will cause insertion to class A. Update propagation is based upon these update rules.

The organization of the remainder of this paper is as follows. A generic semantic database model upon which the schema-based approach is based is briefly introduced in section 2; this data model is a simplification of SDM [9]. In section 3, we describe derived data update rules based on schema information. An experimental prototype system embodying a derived data update algorithm based upon the rules is presented in section 4. Conclusions and directions for future research are given in section 5.

2 A Generic Semantic Database Model

A generic semantic database model (GSDM), which embodies the main features of prominent semantic database models [11], is used as a basis for our approach to schema-based derived data update. The GSDM is a simplification of SDM [9]. The **Appendix** contains an example database schema describing a university application environment specified using the GSDM.

A *database schema* in GSDM consists of a collection of *classes*. Each class has a set of instances (members), which are the objects belonging to the class. Every class has a class name, and an associated collection of attributes. There are two kinds of attributes: member and class attributes. Member attributes describe properties of instances of a class; for example, every instance of class *COURSE* has a course number, an instructor, a teaching assistant, and students enrolled in the course. Class attributes specify properties of the class as a whole, e.g., the total number of instances in the class *COURSE*. As shown in the **Appendix**, class *COURSE* has a class attribute *Total-courses*, and the four member attributes *Course-number*, *Taught-by*, *Has-ta* and *Students-currently-enrolled*. A class may have a cardinality limitation, which restricts the number of instances of this class. For example, the class *TEACHING-ASSISTANT* has the cardinality limitation: *with size between 0 and 25*.

Each member attribute has a name and a specification of the range (*value class*) of this attribute; for example, attribute *Course-number* of *COURSE* has value class *INTEGERS*. If the attribute is derived, then there is also a derivation rule; e.g., a person's monthly income can be derived by: *Annual-income / 12*. An attribute can also be specified as the *inverse* of an attribute of some other class.

An attribute can have associated constraints, e.g., single-valued or multi-valued. A multi-valued attribute may have an associated upper and lower bound. For example, attribute *Takes* of *STUDENT* has an associated constraint: *multi-valued with size between 2 and 5*. An attribute which is not allowed to have null values is specified by *may not be null*; e.g., the *Name* of a person is constrained in this manner. If the value of an attribute cannot be changed, then *not changeable* is specified. *Exhausts value class* requires that all members of the value class must be the value of the attribute for some instance(s). An attribute for which overlapping values are not allowed is specified with *no overlap in values*.

The specification of a class attribute is similar to that of a member attribute, except that class attribute does not have an inverse, and does not have constraints specific to member attributes such as *exhausts value class* or *no overlap in values*. A class attribute can have a derivation which involves properties of the class as a whole, such as the total number of instances in the class; e.g., class attribute *Total-courses* of class *COURSE* has the derivation: *number of members in this class*.

Classes in a GSDM database schema are organized as a collection of directed acyclic graphs (DAGs). A subclass can be derived from other classes or subclasses by one of the seven subclass constructors. A subclass can be (1) specified by attribute predicate; (2) explicitly specified (user-controlled); (3) the intersection of two classes; (4) the set difference of two classes; (5) the union of two classes; (6) the current set of values of a given attribute or (7) specified in certain format[1]. For example, subclass *COMPUTER-MAJOR-STUDENT* has the derivation rule: where Major = 'CS' or Major = 'CE' (a category 1 subclass). Subclass *STUDENT* is derived from *PERSON* with the derivation: specified (a category 2 subclass). Subclasses *BOTH-TA-RA* and *STUDENT-EMPLOYEE* are the intersection and the union of classes *TEACHING-ASSISTANT* and *RESEARCH-ASSISTANT* (category 3 and category 5 subclasses, respectively). Subclass *UNDERGRADUATE* is the set difference of classes *STUDENT* and *GRADUATE* (a category 4 subclass). *COURSES-TAKEN* has the derivation rule: where is a value of *Took* of *STUDENT* (a category 6 subclass).

A *database transaction* (or *transaction* for short) is a set of *database commands* (or *commands*) which after execution will not violate database integrity constraints. There are five kinds of modification commands: add instances to a class, remove instances from a class, change the value of a class attribute, add values to a member attribute, and delete values from a member attribute. Each of these five commands may cause update propagation if data changed is derived. In some cases, derived data update propagation rules can be constructed for individual commands, and in other cases the combination of commands within a transaction must be considered.

3 A Schema-Based Approach

We propose in this section an approach to derived data update for the GSDM. A list of update rules guiding derived data update is given. A derived data update transaction is allowed only if it follows update rules.

[1] Subclasses derived using data formatting specifications (e.g., for strings) are not explicitly considered here.

For example, "insertion to a class A which is the intersection of two classes B and C will cause the same instance to be inserted to B and C" is a sample update rule. An update rule may include default actions to resolve ambiguities. In the practical use of our approach, the database designer or user can override the defaults.

Update rules are gathered from the constructions of classes or attributes, and from the restrictions of classes or attributes. We call the former *structural information*, and the latter *constraints*. Structural information includes class definitions, subclass derivations, and derived attribute definitions. Constraints are specified in the class or the attribute definitions; examples of constraints include the cardinality limitation of attributes and classes, the existence of inverse of an attribute, and the restrictions that an attribute value cannot be null or cannot be changed. Update rules utilizing both structural information and constraints are presented here.

3.1 Update Rules From Structural Information

3.1.1 Derived Attributes

Derived attributes are those attributes which have associated derivation rules. For example, class *STUDENT-EMPLOYEE* has the derived attribute *Benefit* with the derivation rule: *Tuition-coverage* + *Stipend*. Changing the value of *Tuition-coverage* or the value of *Stipend* will cause the value of *Benefit* to change. However, direct update on the value of *Benefit* is not allowed since the update propagation is not unique.

Class *PERSON* has the derived attribute: *Monthly-income* with the derivation: *Annual-income* / 12. If a person's income changes, then the change can be made on *Annual-income*, and *Monthly-income* will change to the new value of *Annual-income* divided by 12. Changes can also be made to the value of *Monthly-income*, in which case the value of *Annual-income* is changed to be the new value of *Monthly-income* multiplied by 12.

Whether an update on a derived attribute is allowed depends upon how that attribute is derived. We have the following update rule for derived attribute update:

Update-Rule 1 (One-to-one Onto) *Update of a derived attribute is allowed if the derivation function f is a one-to-one onto total function. Update of underlying data will cause update of the derived attribute by using the function f. Update of the derived attribute will cause update of the underlying data by using the inverse of function f.*

3.1.2 Derived Subclasses

As stated in section 2, there are six kinds of subclasses in the GSDM considered here: (1) specified by attribute predicate, (2) explicitly specified, (3) the intersection of two classes, (4) the difference of two classes, (5) the union of two class, and (6) the current set of values of a given attribute. The following discussion is based on these subclass categories.

COMPUTER-MAJOR-STUDENT is a subclass of *STUDENT* with the derivation rule: *Major* of *STUDENT* is either 'CS' or 'CE'. Inserting a *COMPUTER-MAJOR-STUDENT* P1 with major 'CS' will propagate an insertion to class *STUDENT*. Inserting a *COMPUTER-MAJOR-STUDENT* P2 with major 'MATH' is an error. Inserting a *STUDENT* P3 with major 'CS' or 'CE' will cause the same student P3 to be inserted to class *COMPUTER-MAJOR-STUDENT*. Deleting an instance P4 from *STUDENT* will cause the deletion to be propagated to *COMPUTER-MAJOR-STUDENT*, providing P4's major is 'CS' or 'CE'. However, deleting a student from *COMPUTER-MAJOR-STUDENT* is ambiguous. This can be interpreted as a change of the student's major, or as a removal of that student from the class *STUDENT*.

Update-Rule 2 (Attr-Pred-Based-1) *Let T_c be a subclass of T_p which is derived by using an attribute predicate P. Inserting an instance A to T_c which satisfies P will cause the same instance to be inserted to T_p. However, if A does not satisfy P, then this is an error. Inserting an instance B to T_p, provided B satisfies P, will cause B to be inserted to T_c. Deleting an instance C from T_p will cause C to be deleted from T_c if C is also an instance of T_c.*

Update-Rule 3 (Attr-Pred-Based-2) *Let T_c be a subclass of T_p which is derived by $A = V$, where A is an attribute of T_p and V is a specified value. Inserting a new instance x to T_c will cause x to be inserted to T_p, and value V to be added to the attribute A of x.*

STUDENT is an explicitly specified subclass of class *PERSON*. Any insertion to *STUDENT* will be propagated to *PERSON*. Any deletion from *PERSON* will be propagated to *STUDENT*. New data inserted to *PERSON* can either be propagated to class *STUDENT* or not. To simplify the problem, the default action is not to propagate. Also by default, the deletion of an instance from *STUDENT* will not be propagated to *PERSON*.

Update-Rule 4 (Specified) *Let T_c be a subclass of T_p with derivation "specified." Inserting an instance A to T_c will cause the same instance to be inserted to T_p if it is not there already. Deleting an instance B from T_p will cause deletion from T_c if B is also an instance of T_c. Insertion to T_p or deletion from T_c will not be propagated (by default).*

Subclass *BOTH-TA-RA* is an intersection of class *TEACHING-ASSISTANT* (*TA* for short) and class

RESEARCH-ASSISTANT (*RA* for short), i.e., derived by "is in TA and is in RA." Inserting an instance P1 to class *TA*, providing P1 is also a research assistant, will cause P1 to be inserted to class *BOTH-TA-RA*. Inserting P2 to *BOTH-TA-RA* will cause P2 to be inserted to both classes *TA* and *RA*. Deleting P3 from *RA* will cause P3 to be deleted from *BOTH-TA-RA* if P3 is an instance of *BOTH-TA-RA*. Deleting P4 from *BOTH-TA-RA* is ambiguous; it can mean deleting P4 from *TA*, from *RA*, or from both.

Update-Rule 5 (Intersection) *Let T_a, T_b and T_c be subclasses of T_p, where T_c is derived by "is in T_a and is in T_b." Inserting x to T_a (or T_b), providing x is also in T_b (or T_a), will cause x to be inserted to T_c. Inserting x to T_c will cause x to be inserted to T_a and T_b (and T_p). Deleting an instance y from either T_a or T_b will cause y to be deleted from T_c if y is also an instance of T_c.*

Class *UNDERGRADUATE* and class *GRADUATE* are both subclasses of *STUDENT*. *UNDERGRADUATE* is a set difference of *STUDENT* and *GRADUATE* (i.e., derived by "is not in GRADUATE"). Inserting a student P1 to class *STUDENT* can cause P1 to be inserted to *GRADUATE* or *UNDERGRADUATE*. If the derivation of *GRADUATE* is satisfied (i.e., *Level* of P1 is equal to 'Graduate'), then the instance P1 will be inserted to *GRADUATE*. If the derivation is not satisfied, then P1 will be inserted to *UNDERGRADUATE*. Of course, there are cases where update propagation is ambiguous, e.g., P1's *Level* is unknown, or the derivation of *GRADUATE* is "specified." A reasonable default here is to insert the instance to class *UNDERGRADUATE*. Inserting an instance P2 to *GRADUATE* will cause P2 to be inserted to *STUDENT* if P2 is not already an instance of *STUDENT* (therefore, it is not an instance of *UNDERGRADUATE* either.) If P2 is an instance of class *STUDENT* and class *UNDERGRADUATE*, then P2 will be removed from class *UNDERGRADUATE*. Inserting a student P3 to class *UNDERGRADUATE* will cause the same instance P3 to be inserted to class *STUDENT* if it is not in *STUDENT* and is not in *GRADUATE*. However, it will be an error if P3 is an instance of *GRADUATE*. Changing a student from *GRADUATE* to *UNDERGRADUATE* must be done by first deleting that student from *GRADUATE*, and then inserting the same student to *UNDERGRADUATE*.

A transaction which deletes an instance P4 from *STUDENT* will delete P4 from *GRADUATE* or *UNDERGRADUATE* depending upon the nature of P4. Whether deleting an instance from *GRADUATE* is legal or not and how it will affect *STUDENT* depends upon how *GRADUATE* is derived. If deleting an instance from class *GRADUATE* is legal and it does not affect the same instance in *STUDENT*, then that instance will be added to *UNDERGRADUATE*. Deleting a student from *UNDERGRADUATE* is ambiguous; the update propagation can be either to delete that instance from *STUDENT*, or to add that instance to *GRADUATE*. Since it is not always possible to make the instance a member of class *GRADUATE*, the default action is to delete that instance from *STUDENT*.

Update-Rule 6 (Difference) *Let T_a and T_b be subclasses of T_p, and T_b is derived by "is not in T_a." Inserting an instance x to T_p will cause x to be inserted to T_a if the derivation of T_a is satisfied; it will cause x to be inserted to T_b if the derivation of T_a is not satisfied. If the derivation of T_a cannot be evaluated*[2]*, then x will be inserted to T_b by default.*

Deleting an instance y from T_p will cause y to be deleted from T_a or T_b, depending upon to which class y belongs. Deleting an instance y from T_a will cause y to be added to T_b if the deletion is legal and y is still an instance of T_p after it is deleted from T_a. Deleting an instance y from T_b will cause y to be deleted from T_p by default.

Classes *TEACHING-ASSISTANT* (*TA* for short), *RESEARCH-ASSISTANT* (*RA* for short), and *STUDENT-EMPLOYEE* are all subclasses of class *GRADUATE*. *STUDENT-EMPLOYEE* is the union of *TA* and *RA* (i.e., derived by "is in TA or is in RA"). Inserting an instance P1 to *TA* or to *RA* will cause P1 to be inserted to *STUDENT-EMPLOYEE*. Inserting a student to *STUDENT-EMPLOYEE* is ambiguous, because that student can be a teaching assistant or a research assistant. Deleting an instance from TA (or RA) will cause the instance to be deleted from *STUDENT-EMPLOYEE* if that instance is not in RA (or TA). Deleting an instance from *STUDENT-EMPLOYEE* will cause that instance to be deleted from both *TA* and *RA*.

Update-Rule 7 (Union) *Let T_a, T_b and T_c be subclasses of T_p; T_c is derived by "is in T_a or is in T_b." Inserting an instance x to T_a or T_b will cause x to be inserted to T_c if x is not already in T_c. Deleting an instance y from T_a (or T_b) providing y is not in T_b (or T_a) will cause y to be deleted from T_c. Deleting y from T_c will cause the same instance y to be deleted from both T_a and T_b (if it is there).*

Class *COURSES-TAKEN* is a subclass of *COURSE* with the derivation: is a value of *Took* of *STUDENT*; this means that the instances of *COURSES-TAKEN* are those instances of *COURSE* for which there exists student(s) who have taken that course. If the fact "P1 took CS102" is added to the database, then the corresponding fact associated with *COURSES-TAKEN*, viz., "CS102 is a course taken by P1" will be inserted to the database. If the fact "CS102 is a course taken by P1" is inserted to the database, then the corresponding fact "P1 took CS102" will be generated. Removing

[2] That is, the derivation rule is "specified," or values of attributes in the derivation are unknown.

the fact "P1 took CS102" will cause the fact "CS102 is a course taken by P1" to be deleted from the database, and vice versa. This example can be generalized as follows:

Update-Rule 8 (Attribute-Value-Based-Class) *Let T_p be a class with member attributes $A_1, \cdots, A_n$, and T_c is a class with derivation "is a value of A_i of T_p" (i is between 1 and n). Let t be an instance of T_p. Inserting a new value x to attribute A_i of instance t will generate corresponding data in T_c. Inserting new data to T_c will generate corresponding data in A_i of T_p. Deleting x from attribute A_i of instance t will remove corresponding data in T_c. Deleting data from T_c will cause the corresponding data in A_i of T_p to be removed.*

3.2 Update Rules From Constraints

There are cases when structural information is inadequate to decide a unique update propagation, but in which constraints embedded in the schema definition may be used to resolve the ambiguity. Examples of such constraints (on attributes) are *multi-valued* and *may not be null.* Value classes of attributes also provide information which is sometimes useful in resolving ambiguities.

3.2.1 Attributes and Their Value Classes

Class *STUDENT-EMPLOYEE* is the union of *TA* and *RA*. According to the discussion in the previous subsection, insertion to a union subclass such as *STUDENT-EMPLOYEE* is ambiguous. Information from member attributes and their value classes can however be used to decide a unique update propagation. For example, inserting an instance P1 to *STUDENT-EMPLOYEE*, providing that P1 is in charge of a course CS101, will cause this instance P1 to be added to *TA*, since *Is-in-charge-of* is an attribute of *TA*. However, P1 may also be an instance of *RA*, if classes *TA* and *RA* are not disjoint. A default action is to insert P1 to *TA* only.

Update-Rule 9 (Value-Class) *Suppose T_c is a subclass with derivation "is in T_a or is in T_b," T_a has an attribute Attr with value class T_d, and T_b does not have such an attribute. Inserting an instance x to T_c given Attr of x has a value from T_d will cause x to be inserted to T_a only (by default).*

3.2.2 Multi-valued with Size Restrictions

A multi-valued attribute may have a specified upper and lower bound, limiting the size of the attribute. A class can also have the size limitations. This information can be used to decide update propagation.

Class *STUDENT-EMPLOYEE* is the union of *TA* and *RA*. Suppose TA has a size limitation to be between 15 and 25. Insertion to *STUDENT-EMPLOYEE* may be ambiguous, but if there are already 25 TAs, then the new student employee can only be a RA, otherwise there will be a constraint violation. In this case, the system will suggest that the new data should be an instance of RA.

There are similar situations for intersection. *BOTH-TA-RA* is an intersection of *TA* and *RA*. Deleting an instance from *BOTH-TA-RA* providing there are 15 TAs will suggest the deletion should only be propagated to class *RA*.

Update-Rule 10 (Max-Size) *Let T_c be a subclass with the derivation "is in T_a or is in T_b"; T_a has a size limitation to be between Min and Max. Inserting a new instance x to T_c given T_a already has Max size will suggest that x should be inserted to T_b only.*

Update-Rule 11 (Min-Size) *Let T_c be a subclass with the derivation "is in T_a and is in T_b"; T_a has a size limitation to be between Min and Max. Deleting an instance x from T_c given T_a has Min size will suggest that x should be deleted from T_b only.*

3.2.3 Nullable

A class with an attribute specified as "may not be null" will prevent any insertion to that class unless the value of that attribute is given. Consider the class STUDENT-EMPLOYEE, and suppose attribute *Is-in-charge-of* of *TA* cannot have null values. Inserting an instance P1 to *STUDENT-EMPLOYEE* without giving a value to *Is-in-charge-of* will suggest that P1 is a research assistant.

Update-Rule 12 (Nullable) *Let T_c be a union of T_a and T_b; T_a has an attribute Attr which is specified as "may not be null." Inserting an instance x to T_c without giving a value to Attr will suggest that x should be inserted to T_b only.*

3.2.4 Changeable

In addition to the one-to-one case described previously, there are other cases in which a derived attribute can be updated. For example, attribute *Benefit* of class *STUDENT-EMPLOYEE* is defined to be: *Tuition-coverage* + *Stipend.* If *Stipend* is defined to be "not changeable," then a change to the attribute *Benefit* can result in a unique change to the attribute *Tuition-coverage.*

Update-Rule 13 (Changeable-1) *Let class T have attributes $A_1, \cdots, A_n$. Attribute A_n is derived by a function f with attribute names $A_1, \cdots, A_m$ $(m < n)$. If $A_1, \cdots, A_{m-1}$ are not changeable, then update on A_n will result in a unique update on A_m, which is* *an inverse of function f with values of $A_1, \cdots, A_{m-1}$ fixed.*

Suppose *MEN* is a subclass of *PEOPLE* with attribute *Sex* equal to "male," and attribute *Sex* is specified as "not changeable." In this case, deleting an instance from *MEN* can only mean deleting the corresponding instance from *PEOPLE*, but not changing the value of attribute *Sex*.

Update-Rule 14 (Changeable-2) *Let T_c be a subclass of T_p based on the attribute predicate P. If attributes used in P are all not changeable, then deleting an instance x from T_c will result in deleting x from T_p.*

3.2.5 Exhausts Value Class

Exhausts value class provides a global guide to deciding update propagation. Let *STUDENT*, *SCHOOL* and *COMPANY* be three classes; *ORGANIZATION* is the union of *SCHOOL* and *COMPANY*. Suppose *STUDENT* has an attribute *Attends* which has the constraint *exhausts value class* with value class *SCHOOL*. [3] Inserting "ABC" as an instance of *ORGANIZATION* will make "ABC" an instance of *COMPANY*.

Update-Rule 15 (Exhausts-Value-Class) *Let T_a, T_b, T_c and T_d be classes. Let T_d be defined as: "is in T_b or is in T_c." Suppose T_a has an attribute Attr with value class T_b with constraint "exhausts value class." Inserting an instance x to T_d without corresponding instances in Attr of T_a will insert x to T_c only.*

4 Experimental Prototype

An experimental prototype system based upon the approach described above has been designed and implemented. Currently this prototype is implemented in the C language and runs on SUN/UNIX. This prototype handles update propagation based upon the rules described above. The data representation and algorithms this prototype uses, along with a brief description of the implementation, are presented here. Two working examples are also included.

4.1 System Implementation

Figure 1 illustrates the overall structure of the experimental prototype. Here, a user first defines a database schema in the GSDM; this GSDM external schema is then translated into an internal representation called the *kernel database representation.* The internal representation consists a set of triplets with the format

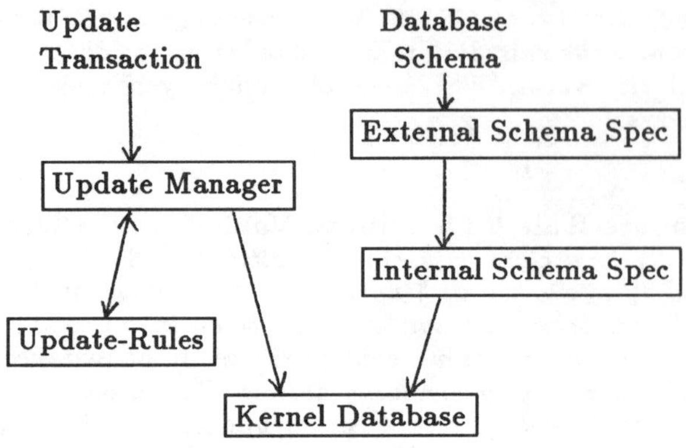

Figure 1: Block Diagram of Schema-Based Approach

(domain, mapping, range). For example, a schema description such as "STUDENT is a subclass of PERSON with derivation *specified*" is encoded into the internal triplets: (STUDENT, has-superclass, PERSON) and (STUDENT, has-derivation, "specified"). Database facts such as "P1 is a person with name John" are also encoded into internal triplets: (PERSON, has-instance, P1) and (P1, Name, 'John'). In this way, both data and meta-data are represented uniformly [2].

To modify the contents of a database, new data is either added to or deleted from the kernel database. Since this prototype is designed to test derived data update functions, only simple update commands are provided. Insertion is done by a command: INSERT (domain, mapping, range). For instance, inserting a fact that P2 is a person with name Mary is done by: INSERT (PERSON, has-instance, P2) and INSERT (P2, Name, 'Mary'). Deletion has a similar form: DELETE (domain, mapping, range). Update (or modification) is considered here as a deletion followed by an insertion with a connection between the two steps. A complete query language for this system, which includes retrieval commands, insertion commands, deletion with or without conditions, and update commands, is currently under development.

The update manager receives update commands from users, and uses an update algorithm, which embodies the update rules, to determine update propagation. The update algorithm is presented immediately below.

4.2 Derived Data Update Algorithm

The algorithm **Update-Propagation(X)** uses update tables to determine update propagation when command X is issued; the command X can be adding instances to or removing instances from a class, or

[3] Informally, this means that a school must be attended by some students.

setting or modifying the value of a member or class attribute. Generally speaking, X has format (op, d, m, r), where op is an operation, d is a domain, m is a mapping, and r is a range.

Update-Propagation(op, d, m, r):

- Case 1: m is an attribute name.

  1. **For each** triplet Y = (a, b, c)
     and (b is the inverse of m
     or b is an attribute derived from m
     or m is an attribute derived from b)
     { check the update table for
     derived attributes;
     if (a, b, c) is inserted to or
     deleted from the kernel database
     do **Update-Propagation**
     (op, a, b, c); }

  2. **For each** category 6 subclass z that is related to m
     { check the update table for
     category 6 subclasses;
     if (r, z, d) is inserted to or
     deleted from the kernel database
     do **Update-Propagation**
     (op, r, z, d); }

- Case 2: d is a base class, and m = "has-instance."

  For each y which is a category i subclass of d ($1 \leq i \leq 6$)
  { check the update table for category i
  subclasses;
  if (y, has-instance, r) is inserted to or
  deleted from the kernel database
  do **Update-Propagation**
  (op1, y, has-instance, r); }
  Whether op1 is insertion or deletion depends upon the entries in the update table.

- Case 3: d is a category k subclass ($1 \leq k \leq 6$), and m = "has-instance."

  1. **For each** y which is a superclass of d
     { check the update table for category k
     subclasses;
     if (y, has-instance, r) is inserted to or
     deleted from the kernel database
     do **Update-Propagation**
     (op1, y, has-instance, r); }
     Whether op1 is insertion or deletion depends upon the entries in the update table.

  2. **For each** z which is a category i subclass of d ($1 \leq i \leq 6$)
     { check the update table for
     category i subclasses;
     if (z, has-instance, r) is inserted to or
     deleted from the kernel database
     do **Update-Propagation**
     (op1, z, has-instance, r); }
     Whether op1 is insertion or deletion depends upon the entries in the update table.

  3. **If** d is a category 6 subclass
     and A is a related attribute of d
     { check the update table for
     derived attributes;
     if (r, A, d) is inserted to or
     deleted from the kernel database
     do **Update-Propagation**
     (op, r, A, d); }

4.2.1 Derived Attributes

Let A be an attribute, derived or not.

	Insert x to	Delete x from
A	Apply **One-to-One Onto and Changeable-1**; ambiguous, if not applicable.	Apply **One-to-One Onto and Changeable-1**; ambiguous, if not applicable.

4.2.2 Subclass Based on Attribute Predicates (Category 1)

Let T_c be a subclass of T_p based on an attribute predicate P.

	Insert x to	Delete x from
T_p	Insert x to T_c if P is satisfied.	Delete x from T_c.
T_c	Insert x to T_p if P is satisfied; Error if P is not satisfied; If P is with format A = V, then apply **Attr-Pred-Based-2** otherwise, ambiguous.	Apply **Changeable-2**. Ambiguous, if not applicable.

4.2.3 Specified Subclasses (Category 2)

Let T_c be a subclass of T_p with derivation "specified."

	Insert x to	Delete x from
T_p	No action (default).	Delete x from T_c.
T_c	Insert x to T_p.	No action (default).

4.2.4 Intersection Classes (Category 3)

Let T_c be a subclass with derivation "is in T_a and is in T_b."

	Insert x to	Delete x from
T_a	Insert x to T_c, if x is in T_b; no action, otherwise.	Delete x from T_c.
T_b	Insert x to T_c, if x is in T_a; no action, otherwise.	Delete x from T_c.
T_c	Insert x to T_a and T_b.	Apply **Min-Size**. Ambiguous, if not applicable.

4.2.5 Difference Classes (Category 4)

Let T_c be a subclass of T_p but not in T_a.

	Insert x to	Delete x from
T_p	Insert x to T_a, if derivation of T_a is satisfied; insert x to T_c, if derivation of T_a is not satisfied; insert x to T_c (default), if derivation of T_a cannot be evaluated.	Delete x from T_a and T_c.
T_a	Insert x to T_p, if x is not in T_p, T_c; delete x from T_c, otherwise.	Insert x to T_c, if the deletion is legal and x is still an instance of T_p.
T_c	Insert x to T_p, if x is not in T_a; error otherwise.	Delete x from T_p (default).

4.2.6 Union Class (Category 5)

Let T_c be a class with derivation "is in T_a or is in T_b."

	Insert x to	Delete x from
T_a	Insert x to T_c.	No action if x is in T_b; delete x from T_c, otherwise.
T_b	Insert x to T_c.	No action if x is in T_a; delete x from T_c, otherwise.
T_c	Check rules **Value-Class**, **Max-Size**, **Nullable**, **Exhausts-Value-Class**; ambiguous, if not applicable.	Delete x from T_a, T_b.

4.2.7 Attribute-Value-Based Subclasses (Category 6)

Let T_c be a subclass which is derived by: "is a value of Attr of class T_p."

	Insert x to	Delete x from
T_p	Insert corresponding instance to T_c.	Delete corresponding instances from T_c.
T_c	Insert corresponding instances to Attr of T_p.	Delete all instances in Attr of T_p corresponding to x.

4.3 Examples

We present in this subsection two illustrative working examples. For each example there is an initial database state, an update transaction, and a description of update propagation.

4.3.1 Example One

Consider the database transaction that inserts a graduate student P6 with name 'Tim' and student-id 789 to the database. This graduate student is taking the courses CS1 and EE1. He has an annual income of $18,000.

- **Initial Database (triplets):**

  (PERSON, has-instance, P1)

  (P1, Name, 'David')

  (COURSE, has-instance, CS1)

  (COURSE, has-instance, EE1)

  (CS1, Course-number, 123)

  (EE1, Course-number, 101)

  (COURSE, Total-courses, 2)

- **Transaction:**

  INSERT (STUDENT, has-instance, P6)

  INSERT (P6, Name, 'Tim')

  INSERT (P6, Student-id, 789)

  INSERT (P6, Takes, CS1)

  INSERT (P6, Takes, EE1)

  INSERT (P6, Level, 'Graduate')

  INSERT (P6, Annual-income, 18000)

- **Update Propagation:**

  INSERT (STUDENT, has-instance, P6) will insert the triplet (STUDENT, has-instance, P6) to the kernel database, and call **Update-Propagation** (insert, Student, has-instance, P6). Since STUDENT is a category 2 subclass of PERSON, the update table for category 2 subclasses will be referenced, and a new instance (PERSON, has-instance, P6) is generated. This new instance will result in the invocation of **Update-Propagation** (insert, PERSON, has-instance, P6); however, no changes are made as a result. STUDENT has a category 1 subclass GRADUATE; therefore, the update table for category 1 subclasses is also referenced, and a new instance (GRADUATE, has-instance, P6) is added to the database. **Update-Propagation**(insert, GRADUATE, has-instance, P6) is called, but no changes are made as a result.

  The command INSERT(P6, Takes, CS1) inserts an instance (P6, Takes, CS1) to the database, and calls **Update-Propagation** (insert, P6, Takes, CS1). The update table for derived attributes is checked, and (CS1, Students-currently-enrolled, P6) is generated because *Students-currently-enrolled* is the inverse of *Takes.* (EE1, Students-currently-enrolled, P6) is generated for similar reasons. (P6, Monthly-income, 1500) is generated from (P6, Annual-income, 18000).

4.3.2 Example Two

Now consider the database transaction which inserts a new teaching assistant into the database. This new teaching assistant has name 'John' and student-id 123. He took CS1 before, and is taking CS2 and EE1 now. He is in charge of a course CS1.

- **Initial Database (triplets):**

  (PERSON, has-instance, P1)

  (P1, Name, 'David')

  (COURSE, has-instance, CS1)

  (COURSE, has-instance, CS2)

  (COURSE, has-instance, EE1)

  (CS1, Course-number, 123)

  (CS2, Course-number, 222)

  (EE1, Course-number, 101)

  (COURSE, Total-courses, 3)

- **Transaction:**

  INSERT (TEACHING-ASSISTANT,
  has-instance, P2)

  INSERT (P2, Name, 'John')

  INSERT (P2, Student-id, 123)

  INSERT (P2, Took, CS1)

  INSERT (P2, Takes, CS2)

  INSERT (P2, Takes, EE1)

  INSERT (P2, Is-in-charge-of, CS1)

- **Update Propagation:**

  The command INSERT (TEACHING-ASSISTANT, has-instance, P2) adds the triplet (TEACHING-ASSISTANT, has-instance, P2) to the database, and calls **Update-Propagation**-(insert, TEACHING-ASSISTANT, has-instance, P2). Since TEACHING-ASSISTANT is a category 2 subclass of GRADUATE, the update table for category 2 subclasses is referenced, and the triplets (GRADUATE, has-instance, P2) and (P2, Level, 'Graduate') are generated. **Update-Propagation**(GRADUATE, has-instance, P2) will be executed, and the triplet (STUDENT, has-instance, P2) is generated. This new data in turn generates (PERSON, has-instance, P2).

  STUDENT has a category 3 subclass BOTH-TA-RA and a category 5 subclass STUDENT-EMPLOYEE; therefore, update tables for category 3 and 5 subclasses are checked, and (STUDENT-EMPLOYEE, has-instance, P2) is generated.

  (CS2, Students-currently-enrolled, P6) and (EE1, Students-currently-enrolled, P6) are generated from (P6, Takes, CS2) and (P6, Takes, EE1) respectively, because *Students-currently-enrolled* is the inverse of *Takes.* (CS1, Has-ta, P2) is the inverse of (P2, Is-in-charge-of, CS1). (CS1, COURSES-TAKEN, P2) is generated from (P2, Took, CS1), because class COURSES-TAKEN is derived by "is a value of Took of STUDENT."

5 Conclusions

In this paper, a schema-based approach to derived data update for semantic databases has been proposed. By contrast with view update in a relational database environment, this approach utilizes information in a semantic database schema, such as superclass and subclass relationships, value classes of attributes, inverses, etc., to propagate the effects of a modification of derived data. Some derived data updates which are ambiguous in a relational view update mechanism can be decided in this approach, e.g., insertion to class which is the union of two other classes. While the approach presented here is based upon the GSDM, it can be applied to other semantic database models as well.

An experimental prototype system based upon the approach described in this paper has been designed, implemented, and tested. It is important to note however that there are some cases in which derived data update operations have unique propagations but which our approach cannot detect. These difficulties result from complex derivation rules, particularly for derived attributes and category 1 subclasses.

Further research planned in this area includes: (1) extending the update rules for category 1 subclasses and for derived attributes, (2) completion of the experimental implementation of a query language and end-user interface for the prototype system, and (3) studying new approaches other than the schema-based to derived data update.

In particular, a second approach to derived data update termed the *rule-based approach* has also been studied and implemented in our research. This rule-based approach encodes all the schema-information into production rules, and then applies a rule inference mechanism to determine derived data update propagation. At present, we are conducting an analytic comparison of the schema-based and the rule-based approaches; this comparison includes the different meta-data roles and organizations in these two approaches, and their different mechanisms to determine derived data update. The study should provide further insight into the semantics of meta-data modification.

Acknowledgements

The authors would like to thank Richard Hull for his many comments and observations on the approach to derived data update described in this paper. The comments of the referees are also gratefully acknowledged.

References

[1] S. Abiteboul and R. Hull, *IFO : A Formal Semantic Database Model*, ACM Trans. on Database Systems, vol.12, no.4, Dec 1987

[2] H. Afsarmanesh and D. McLeod, *The 3DIS: An Extensible, Object-Oriented Information Management Environment*, ACM Trans. on Office Information Systems, 1989 (to appear)

[3] A. Albano, L. Cardelli and R. Orsini, *Galileo : A Strongly-Typed, Interactive Conceptual Language*, ACM Trans. on Database Systems, vol.10, no.2, June 1985, pp. 230-260

[4] F. Bancilhon and N. Spyratos, *Update Semantics of Relational Views*, ACM Trans. on Database Systems, vol.6, no.4, December 1981, pp. 557-575

[5] U. Dayal and P. A. Bernstein, *On the Correct Translation of Update Operations on Relational Views*, ACM Trans. on Database Systems, vol.8, no.3, September 1982, pp. 381-416

[6] A. L. Furtado, K. C. Sevcik and C. S. Dos Santos, *Permitting Updates Through Views of Data Bases*, Inform. Systems, vol.4, 1979, pp. 269-283

[7] G. Gottlob, P. Paolini, and R. Zicari, *Properties and Update Semantics of Consistent Views*, ACM Trans. on Database Systems, vol.13, no.4, December 1988, pp. 486-524

[8] P. Griffiths and B. Wade, *An Authorization Mechanism for a Relational Database System*, ACM Trans. on Database Systems, vol.1, no.3, September 1976, pp. 242-255

[9] M. Hammer and D. McLeod, *Database Description with SDM: A Semantic Database Model*, ACM Trans. on Database Systems, vol.6, no.3, September 1981, pp. 351-386

[10] S. Hudson and R. King, *An Adoptive Derived Data Manager for Distributed Databases*, Advances in Object-Oriented Database Systems, editor K.R. Dittrich, Springer-Verlag, 1988

[11] R. Hull and R. King, *Semantic Database Modeling: Survey, Applications, and Research Issues*, ACM Computing Surveys, vol.19, no.3, September 1987, pp. 201-260

[12] A. M. Keller, *The Role of Semantics in Translating View Updates*, IEEE Computer, January 1986, pp. 63-73

[13] A. M. Keller, *Choosing a View Update Translator by Dialog at View Definition Time*, Proc. Inter. Conf. on Very Large Data Bases, 1986, pp. 467-474

[14] J. Mylopoulos, P. A. Bernstein and H. K. T. Wong, *A Language Facility for Designing Database-Intensive Applications*, ACM Trans. on Database Systems, vol.5, no.2, June 1980, pp. 185-207

[15] D. Shipman, *The Functional Data Model and the Data Language DAPLEX*, ACM Trans. on Database Systems, vol.6, no.1, 1981, pp. 140-173

[16] A. Sheth, J. Larson, and E. Walkins, *Tailor: A Tool for Updating Views*, Proc. of Inter. Conf. on Extending Data Base Technology, Venice, Italy, March 1988

Appendix: (Example Database Schema)

```
PERSON
  identifiers :  Name
  member attributes :
    Name
      value class : STRINGS
      may not be null
      not changeable
    Age
      value class : INTEGERS
    Annual_income
      value class : REALS
    Monthly_income
      value class : REALS
      derivation : = Annual_income / 12

PROFESSOR
  interclass connection :
      subclass of PERSON where specified
  member attributes :
    Teaches
      value class : COURSE
      inverse : Taught_by

STUDENT
  interclass connection :
      subclass of PERSON where specified
  member attributes :
    Student_id
      value class : INTEGERS
    Major
      value class : STRINGS
    Level
      value class : STRINGS
    Takes
      value class : COURSE
      inverse : Students_currently_enrolled
      multi-valued with size between 2 and 5
    Took
      value class : COURSE
      multi-valued

COMPUTER_MAJOR_STUDENT
  interclass connection : subclass of STUDENT
        where Major = 'CS' or Major = 'CE'

GRADUATE
  interclass connection : subclass of STUDENT
        where Level = 'Graduate'

UNDERGRADUATE
  interclass connection : subclass of STUDENT
        where is not in GRADUATE

TEACHING_ASSISTANT
  interclass connection :
      subclass of GRADUATE where specified
  cardinality : with size between 0 and 25
  member attributes :
    Is_in_charge_of
      value class : COURSES_TAKEN
      inverse : Has_ta

RESEARCH_ASSISTANT
  interclass connection :
      subclass of GRADUATE where specified

STUDENT_EMPLOYEE
  interclass connection :
        subclass of GRADUATE
        where is in TEACHING_ASSISTANT
        or is in RESEARCH_ASSISTANT
  member attributes :
    Tuition_coverage
      value class : REALS
    Stipend
      value class : REALS
    Benefit
      value class : REALS
      derivation : = Tuition_coverage
                     + Stipend

BOTH_TA_RA
  interclass connection :
        subclass of GRADUATE
        where is in TEACHING_ASSISTANT
        and is in RESEARCH_ASSISTANT

COURSE
  identifiers :  Course_number
  member attributes :
    Course_number
      value class : INTEGERS
      single-valued
      may not be null
    Taught_by
      value class : PROFESSOR
      inverse : Teaches
    Has_ta
      value class : TEACHING_ASSISTANT
      inverse : Is_in_charge_of
    Students_currently_enrolled
      value class : STUDENT
      inverse : Takes
      multi-valued with size
                between 0 and 50
  class attributes :
    Total_courses
      value class : INTEGERS
      derivation : number of members
                   in this class

COURSES_TAKEN
  interclass connection :
        subclass of COURSE where is
        a value of Took of STUDENT
```

Using Integrity Constraints to Provide Intensional Answers to Relational Queries

Amihai Motro

Computer Science Department
University of Southern California
University Park, Los Angeles, CA 90089-0782

Abstract

An *intensional* answer to a query is a set of characterizations of the set of database values that satisfy the query (the *extensional* answer). Intensional answers provide users with additional insight into the nature of standard extensional answers. In this paper we describe a method that applies database *constraints* to generate intensional answers. These intensional answers characterize the extensional answers in two ways: (1) with constraints that are applicable to the extensional answer, and (2) with database views that are contained entirely in the extensional answer. Our method is to represent the definitions of constraints in special "meta-relations", and extend standard algebraic operators to these relations. When a query is presented to the database system, it is performed *both* on the actual relations, resulting in an extensional answer, and on the meta-relations, resulting in definitions of constraints that apply to the extensional answer, as well as database views that are contained entirely in the extensional answer. These definitions are translated into an intensional answer that accompanies the extensional answer.

1 Introduction

The information stored in databases is of two kinds. *Extensional* information (often called *data*) is information that applies to individual real world objects. *Intensional* information (often called *knowledge*) is information that applies to multitudes of real world objects. In the relational database model extensional information is expressed with *relations* over domains of *data values*, and intensional information is expressed with *constraints*, which are formulas in predicate logic that assert required relationships among the data values.

Relational database systems use constraints to enhance the integrity of the extensional information in the database by monitoring changes made to the relations. Assuming that initially the relations satisfy the constraints, thereafter update requests are accepted only if the resulting relations would not violate any of the constraints. For reasons of efficiency, most database systems do not implement constraints that are arbitrary predicate logic formulas. Usually, only very specific types of formulas are allowed, such as *range* constraints or *referential* constraints. But even a database system that implements only the notion of *relation key*, forbidding two tuples from having the same key value, is, in effect, enforcing a particular kind of constraint.

An *answer* to a query is a set of data values that satisfy a qualification specified in the query. Therefore, answers are derived entirely from the extensional information in the database. Indeed, the only intensional information that characterizes this set of values is the qualification specified in the query that generated it. Still, the intensional information in the database may include additional characterizations of the extensional answer. If this intensional information is retrieved, database answers may gain additional meaning.

In this paper we describe a method that provides such intensional answers. As an example, assume a personnel database that includes a relation EMPLOYEE with attributes NAME, TITLE, SALARY and DEPARTMENT, and two constraints: one states that all employees in the

This work was supported in part by NSF Grant No. IRI-8609912 and by an Amoco Foundation Engineering Faculty Grant.

design department are guaranteed a salary of at least $30,000, and the other states that all employees in research positions are in the design department. A query to retrieve all the employees in the design department may generate the extensional answer:

- Betty, Harry, Mary, Tom

and the intensional answer:

- All employees retrieved earn at least $30,000.
- All employees in researcher positions retrieved.

Thus, an intensional answer characterizes the extensional answer, providing more insight into the nature of the set of values retrieved.

In our approach, providing intensional answers from constraints is a specific case of the following more general problem: Given a query and set of database views that possess a particular property, what views of the answer possess this property?

In [6] we considered database views whose property is that they have guaranteed *integrity*. The problem then became: Given a query, what views of its answer have guaranteed integrity? Solving this problem enabled us to extend a database system so that every answer is accompanied by statements that define its integrity, resembling a *certification of quality*. In [5] we considered database views whose property is that they are *permitted* to a particular user. The problem then became: Given a query, what views of its answer should be permitted to this user? Solving this problem enabled us to extend a database system with an *authorization mechanism* that has several advantages over other known mechanisms.

In this paper we consider database views whose property is that they are always *null*. The problem then becomes: Given a query, what are the views of its answer that are always null? As we shall see, views of this type express constraints, and thus provide good intensional characterizations of answers.

In each case, our solution is to represent the definitions of the given database views in special relations, using the concept of *meta-tuples*. A meta-tuple defines a subview (i.e., a selection and a projection) of a single relation, and several meta-tuples can be used together to define general views. All meta-tuples that define subviews of the same relation are stored together in one *meta-relation*, whose structure mirrors the actual relation. Standard algebraic operators (product, selection and projection) are extended to these meta-relations.

When a query is presented to the database system, it is performed *both* on the actual relations, resulting in an answer, and on the meta-relations, resulting in definitions of views of the answer that inherit the particular property of the given views.

In the case of constraints, these views express constraints that apply to the answer. They provide the basis for the intensional answer that will accompany the extensional answer.

In the previous example, the two constraints may be restated as the following null database views: "employees in the design department who earn under $30,000" and "employees in researcher positions who are not in the design department".

When the employees in the design department are retrieved, we may infer from the first null view that the answer view "employees who earn under $30,000" is also null. The first component of the example intensional answer given earlier is simply a restatement of this conclusion.

Our method for inferring views of the answer that are always null may be extended to infer also views of the database that are contained entirely in the answer. When the employees in the design department are retrieved, we may infer from the second null view that the database view "employees in researcher positions" is contained in the answer in its entirety. This conclusion provided the second component of that intensional answer.

Altogether, the intensional answers characterize the extensional answers in two ways: (1) with *constraints*– views of the extensional answer that are null; and (2) with *containments*– views that are contained entirely in the extensional answer.

The remainder of this paper is organized as follows. In the next section we discuss related research works. Section 3 defines the language with which constraints are expressed, and Section 4 shows how constraints are stored. Section 5 defines the algebraic manipulations of constraints to yield constraints that are applicable to the answer. Section 6 shows how this process can be extended to infer containments. Section 7 demonstrates these method with several examples, and Section 8 concludes with a brief discussion of further refinements of the method. This representation and manipulation of views was first introduced in [6], and has been modified for the particular problem at hand.

2 Research on Intensional Answers

Recently, several research contributions have been concerned with related issues. While each adopts a different approach, all share a common goal, which is to answer queries more abstractly.

A logic database is composed of extensional predicates (*facts*) and intensional predicates (*rules*). Cholvy and Demolombe [1] are interested in providing answers to queries that are independent of a particular set of facts; i.e., answers that are derived only from the rules. The authors define a constructive derivation process that computes such answers. One of the problems they consider is how to avoid answers that are "irrelevant".

Also in the environment of logic databases, Imielinski [3] argues that rules should be allowed to occur in the answer to the query. This is shown to be beneficial both from the conceptual and the computational point of view. As an example, assume database predicates Teach(x,y), denoting that professor x can teach course y, and Group(x,z), denoting that professors x and z belong to the same research group. And assume a rule that requires that professors from the same group be able to teach the same courses. Then a query "who can teach the Database course" may be answered intensionally by "everybody in Smith's group". Exhaustive enumeration of this answer will be performed upon request.

Corella [2] notes that while research on knowledge representation produced much work on the derivation of taxonomies of concepts, concepts are also essential in responses to queries. The author defines a formal model of "semantic" retrieval (similar to intensional answers). Roughly, a concept is a unary predicate over a given domain, and a taxonomy is a finite tree of concepts, where the concept described by each node is subsumed by the concept described by its parent, and the union of sibling concepts is equal to their parent concept.

Shum and Muntz also note that answers that are exhaustive enumerations of individual objects are not always the most efficient or the most effective means of information exchange. In [7] they are concerned with implicit representation of answers through concise expressions that involve both concepts and individuals. For example, an acceptable answer to the query "Who earns more than $30,000?" is "all engineers except John Smith". In [8] they are concerned with providing aggregate responses, where preciseness is sacrificed for conciseness. An example would be an answer such as "90/120 engineers + 20/30 managers". In both cases they assume, like Corella, the availability of taxonomies that encompass all concepts and individuals.

3 Views, Queries and Constraints

We assume the following definition of a relational database [4]. A *relation scheme* R is a finite set of *attributes* $A_1, \ldots, A_m$. With each attribute A_i a set of values D_i, called the *domain* of A_i, is associated (domains are non-empty, finite or countably infinite sets). A *relation* on the relation scheme R is a subset of the product of the domains associated with the attributes of R. A *database scheme* $\mathcal{R}$ is a set of relation schemes $R_1, \ldots, R_n$. A database instance $\mathbf{D}$ of the database scheme $\mathcal{R}$ is a set of relations $R_1(\mathbf{D}), \ldots, R_n(\mathbf{D})$, where each $R_i(\mathbf{D})$ is a relation on the relation scheme R_i.

A *view* V is an expression in the relation schemes of $\mathcal{R}$ that defines a new relation scheme, and for each database instance $\mathbf{D}$ defines a unique relation on this scheme denoted $V(\mathbf{D})$. In this paper we consider views that are defined by *conjunctive relational calculus expressions* [9]. Using domain relational calculus, expressions from this family have the form:

$$\{a_1, \ldots, a_n \mid (\exists b_1) \ldots (\exists b_k)\, \psi_1 \wedge \ldots \wedge \psi_m\}$$

Where the ψ's may be of two kinds:

1. **membership:** $(c_1, \ldots, c_p) \in R$, where R is a database relation (of arity p), and the c's are either a's or b's or constants.

2. **comparative:** $d_1\, \theta\, d_2$, where d_1 is either an a or a b, d_2 is either an a or a b or a constant, and θ is a comparator (e.g., $<, \leq, >, \geq, =, \neq$).

In particular, each a and each b must appear at least once among the c's.

We shall refer to such views as *conjunctive views*. While this family is a strict subset of the relational calculus, it is a powerful subset. The family of conjunctive relational calculus expressions is precisely the family of relational algebra expressions with the operations *product*, *selection* and *projection* (where the selection expressions are conjunctive). The attributes that participate in the selection predicate will be called *selection attributes*, and the attributes that participate in the projection will be called *projection attributes*.

In this paper, conjunctive views serve to express both queries and constraints. When used as a query, a conjunctive view defines a relation that should be derived and delivered to the user. When used as a constraint, a conjunctive view defines a derived relation that should always be null. This follows from the fact that every constraint of the form $(\forall x_1) \ldots (\forall x_n)\, (\alpha(x_1, \ldots, x_n) \Longrightarrow \beta(x_1, \ldots, x_n))$, where x_i are domain variables and α and

β are safe relational calculus expressions with these free variables, may be rewritten as a null view: $\{x_1, \ldots, x_n \mid \alpha(x_1, \ldots, x_n) \wedge \neg\beta(x_1, \ldots, x_n)\} = \emptyset$.

As an example, consider a database with the following relation schemes:

$$\text{EMPLOYEE} = (\text{NAME}, \text{POSITION}, \text{SALARY}, \text{DEPARTMENT})$$
$$\text{DEPARTMENT} = (\text{D\_NAME}, \text{SUPERVISOR}, \text{DIVISION}, \text{BUDGET})$$

The following constraints state, respectively, that employees in the design department are guaranteed a salary of at least \$30,000, that employees in researcher positions are in the design department, that departments in the same division have the same budget, and that employees cannot earn more than their supervisors (all variables are quantified universally):

$$(x_1, x_2, x_3, x_4) \in \text{EMPLOYEE} \wedge (x_4 = \mathsf{design}) \Longrightarrow (x_3 \geq \$30,000)$$

$$(x_1, x_2, x_3, x_4) \in \text{EMPLOYEE} \wedge (x_2 = \mathsf{researcher}) \Longrightarrow (x_4 = \mathsf{design})$$

$$(y_1, y_2, y_3, y_4) \in \text{DEPARTMENT} \wedge (y_5, y_6, y_7, y_8) \in \text{DEPARTMENT} \wedge (y_3 = y_7) \Longrightarrow (y_4 = y_8)$$

$$(x_1, x_2, x_3, x_4) \in \text{EMPLOYEE} \wedge (x_5, x_6, x_7, x_8) \in \text{EMPLOYEE} \wedge (y_1, y_2, y_3, y_4) \in \text{DEPARTMENT} \wedge (x_4 = y_1) \wedge (x_5 = y_2) \Longrightarrow (x_3 \leq x_7)$$

These constraints may be restated as the following null conjunctive views:

$$\{x_1, \ldots, x_4 \mid (x_1, x_2, x_3, x_4) \in \text{EMPLOYEE} \wedge (x_4 = \mathsf{design}) \wedge (x_3 < \$30,000)\}$$

$$\{x_1, \ldots, x_4 \mid (x_1, x_2, x_3, x_4) \in \text{EMPLOYEE} \wedge (x_2 = \mathsf{researcher}) \wedge (x_4 \neq \mathsf{design})\}$$

$$\{y_1, \ldots, y_4, y_5, \ldots, y_8 \mid (y_1, y_2, y_3, y_4) \in \text{DEPARTMENT} \wedge (y_5, y_6, y_7, y_8) \in \text{DEPARTMENT} \wedge (y_3 = y_7) \wedge (y_4 \neq y_8)\}$$

$$\{x_1, \ldots, x_4, x_5, \ldots, x_8, y_1, \ldots, y_4 \mid (x_1, x_2, x_3, x_4) \in \text{EMPLOYEE} \wedge (x_5, x_6, x_7, x_8) \in \text{EMPLOYEE} \wedge (y_1, y_2, y_3, y_4) \in \text{DEPARTMENT} \wedge (x_4 = y_1) \wedge (x_5 = y_2) \wedge (x_3 > x_7)\}$$

For clarity, these constraints may be specified with an equivalent **disallow** statement:

disallow (EMPLOYEE)
where EMPLOYEE.DEPARTMENT = design
and EMPLOYEE.SALARY < \$30,000

disallow (EMPLOYEE)
where EMPLOYEE.POSITION = researcher
and EMPLOYEE.DEPARTMENT ≠ design

disallow (DEPARTMENT:1, DEPARTMENT:2)
where DEPARTMENT:1.DIVISION = DEPARTMENT:2.DIVISION
and DEPARTMENT:1.BUDGET ≠ DEPARTMENT:2.BUDGET

disallow (EMPLOYEE:1, EMPLOYEE:2, DEPARTMENT
where EMPLOYEE:1.DEPARTMENT = DEPARTMENT.D_NAME
and EMPLOYEE:2.NAME = DEPARTMENT.SUPERVISOR
and EMPLOYEE:1.SALARY > EMPLOYEE:2.SALARY

Thus, each **disallow** statement specifies a set of relations and a condition, disallowing tuples in these relations that satisfy the condition. The last two statements show how to handle cases where several membership subformulas reference the same relation. Note that constraints are stated on entire relations, not on particular attributes from these relations; disallowing a combination of tuples that satisfies a condition is equivalent to disallowing particular attributes from these relation that satisfy this condition. Thus, constraints do not apply projections, and are product-selection expressions.

As queries are also expressed with conjunctive views they may be specified with a similar **retrieve** statement as in the following query for the names and department of all employees of the Pacific division:

retrieve (EMPLOYEE.NAME, EMPLOYEE.DEPARTMEN
where EMPLOYEE.DEPARTMENT = DEPARTMENT.D_NAME
and DEPARTMENT.DIVISION = pacific

Finally, given a view, we define every view derived from it by a selection and a projection as its *subview*. In particular, every view is its own subview.

4 Storing Constraints

Constraints are stored in new relations that are added to the database. For each database relation R a *meta-relation* R' is added. The scheme of R' is identical to the scheme of R. Also, an auxiliary relation is defined: COMPARISON = (X, COMPARE, Y). In the previous example, the database is extended with the following relations:

$$\begin{aligned} \text{EMPLOYEE}' &= (\text{NAME}, \text{POSITION}, \text{SALARY}, \\ &\quad \text{DEPARTMENT}) \\ \text{DEPARTMENT}' &= (\text{D\_NAME}, \text{SUPERVISOR}, \text{DIVISION}, \\ &\quad \text{BUDGET}) \\ \text{COMPARISON} &= (\text{X}, \text{COMPARE}, \text{Y}) \end{aligned}$$

Relations EMPLOYEE′ and DEPARTMENT′ will be used to store membership subformulas of constraints. Comparative subformulas will be stored in relation COMPARISON.

Consider a constraint C,

$$C = \{a_1, \ldots, a_n \mid (\exists b_1) \ldots (\exists b_m)\, \psi_1 \wedge \ldots \wedge \psi_k\}$$

A subformula ψ of the kind $(c_1, \ldots, c_p) \in R$ is first modified so that the c's that are variables (i.e., a's or b's) that appear only once in the whole expression are replaced with $\sqcup$ (blank). Hence, each component of the modified subformula is either a constant (a value), or a variable, or a blank. This *meta-tuple* is stored in the meta-relation R'. A subformula ψ of the kind $d_1\, \theta\, d_2$, where θ is not $=$, is transformed into the tuple (d_1, θ, d_2) and is stored in the auxiliary relation COMPARISON. If θ is $=$, then all occurrences of d_1 in the other subformulas are substituted with d_2.

This representation of views in relations recalls the representation of queries in QBE [10]. As an example, Figure 1 shows an instance of the example database extended with the previous four constraints. For convenience of presentation, each pair of relations R, R' is shown as a single contiguous table.

Note that each individual meta-tuple may be regarded as defining a subview of the corresponding relation. The constants and variables specify the selection condition (and the entire relation is projected). For example, the meta-tuple $(\sqcup, \sqcup, x_1, \mathsf{design})$ stored in EMPLOYEE′ specifies a selection of all tuples of relation EMPLOYEE for which DEPARTMENT = design and SALARY = x_1. (A variable shared by another meta-tuple, such as x_1, specifies a selection condition which is satisfied by any value from a set of values defined elsewhere.)

EMPLOYEE

NAME	POSITION	SALARY	DEPARTMENT
Brown	engineer	35,000	manufacture
Green	technician	28,000	service
Jones	manager	51,000	design
King	manager	45,000	sales
Scott	manager	48,000	service
Smith	researcher	40,000	design
Wilson	manager	50,000	manufacture
Wood	salesperson	26,000	sales
		x_1	design
	researcher		x_2
		x_6	x_7
x_8		x_9	

DEPARTMENT

D_NAME	SUPERVISOR	DIVISION	BUDGET
design	Jones	pacific	275
manufacture	Wilson	pacific	275
sales	King	atlantic	160
service	Scott	atlantic	160
		x_3	x_4
		x_3	x_5
x_7	x_8		

COMPARISON

X	COMPARE	Y
x_1	$<$	30,000
x_2	$\neq$	design
x_4	$\neq$	x_5
x_6	$>$	x_9

Figure 1: Database Extended with Constraints

5 Manipulating Constraints

Assume a database $R_1, \ldots, R_n$ and a meta-database $R'_1, \ldots, R'_n$, COMPARISON. Let $C_1, \ldots, C_m$ be constraints (null views) defined on this database.

Let Q be a conjunctive query against this database. We are interested in constraints that are defined on its answer. That is, we are interested in subviews of Q that are null.

We describe a method that discovers such subviews. Basically, this method *generates* views of $C_1, \ldots, C_m$ that are subviews of Q, by manipulating the definitions of $C_1, \ldots, C_m$ algebraically. These manipulations mirror those that are necessary to implement Q. In effect, we *generalize* the standard product, selection and projection operations to manipulate also relations of view definitions. Clearly, every view of null views is also null, so the subviews of Q that are generated in this way are indeed null.

This method is illustrated by the commutative diagram shown in Figure 2. The solid lines describe the current situation: the null views R' define constraints on the database relations R, and the virtual relation A is derived from R to answer query Q. The dashed lines describe our method: query processing is extended to manipulate also R' to yield the null views A' that define constraints on the answer A.

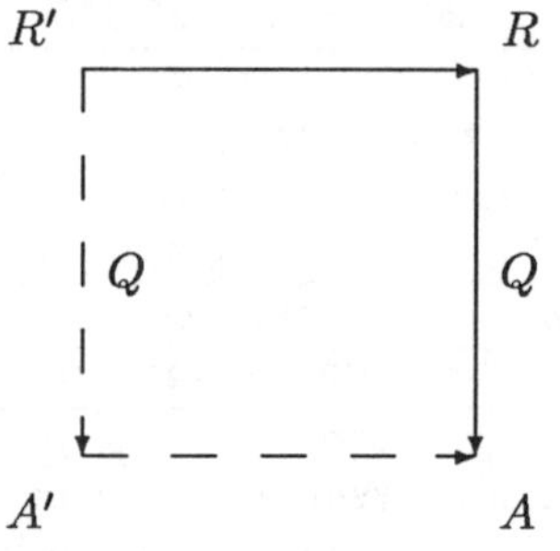

Figure 2: Extending Query Processing to Constraints

5.1 Meta-Relation Operations

Definition 1: Assume that R' and S' are meta-relations that define, respectively, views of R and S. The product of R' and S', denoted $R' \times S'$, is defined as follows. For every pair r and s of meta-tuples from R' and S', respectively,

$$\begin{aligned} r &= (a_1, \ldots, a_m) \\ s &= (b_1, \ldots, b_n) \end{aligned}$$

$R' \times S'$ includes the meta-tuple:

$$q = (a_1, \ldots, a_m, b_1, \ldots, b_n)$$

Proposition 1: Let $\mathbf{D}$ be an instance of this database, and let $r(\mathbf{D})$, $s(\mathbf{D})$ and $q(\mathbf{D})$ denote, respectively, the relations defined by r, s and q. Then $q(\mathbf{D}) = r(\mathbf{D}) \times s(\mathbf{D})$.
Proof: Let λ and μ denote, respectively, the selection predicates of r and s. Then $r(\mathbf{D})$, $s(\mathbf{D})$ and $q(\mathbf{D})$ can be expressed as the following product-selection expressions:

$$\begin{aligned} r(\mathbf{D}) &= \sigma_\lambda(R(\mathbf{D})) \\ s(\mathbf{D}) &= \sigma_\mu(S(\mathbf{D})) \\ q(\mathbf{D}) &= \sigma_{\lambda\wedge\mu}(R(\mathbf{D}) \times S(\mathbf{D})) \end{aligned}$$

We observe that $\sigma_{\lambda\wedge\mu}(R(\mathbf{D}) \times S(\mathbf{D})) = \sigma_\lambda(R(\mathbf{D})) \times \sigma_\mu(S(\mathbf{D}))$.

Definition 2: Assume that R' is a meta-relation that defines views of R. Let λ denote a primitive selection predicate (i.e., either $A'_i \,\theta\, c$, or $A'_i \,\theta\, A'_j$). The selection from R' by predicate λ, denoted $\sigma_\lambda(R')$, is defined as follows. Consider first the case $\lambda = A_i \,\theta\, c$, and let r be a meta-tuple from R',

$$r = (a_1, \ldots, a_i, \ldots, a_m)$$

Denote by μ the selection predicate expressed by a_i [1]. $\sigma_\lambda(R')$ includes the meta-tuple:

$$q = (a_1, \ldots, a'_i, \ldots, a_m)$$

where a_i represents $\lambda \wedge \mu$. Consider now the case $\lambda = A_i \,\theta\, A_j$, and let r be a meta-tuple from R',

$$r = (a_1, \ldots, a_i, \ldots, a_j, \ldots, a_m)$$

Denote by μ the selection predicate expressed by a_i and a_j. $\sigma_\lambda(R')$ includes the meta-tuple:

$$q = (a_1, \ldots, a'_i, \ldots, a'_j, \ldots, a_m)$$

where a'_i and a'_j represent $\lambda \wedge \mu$.

Proposition 2: Let $\mathbf{D}$ be an instance of this database, and let $r(\mathbf{D})$ and $q(\mathbf{D})$ denote, respectively, the relations defined by r and q. Then $q(\mathbf{D}) = \sigma_\lambda r(\mathbf{D})$.
Proof: $r(\mathbf{D})$ and $q(\mathbf{D})$ can be expressed as the following selection expressions:

$$\begin{aligned} r(\mathbf{D}) &= \sigma_\mu(R(\mathbf{D})) \\ q(\mathbf{D}) &= \sigma_{\mu\wedge\lambda}(R(\mathbf{D})) \end{aligned}$$

We observe that $\sigma_{\mu\wedge\lambda}(R(\mathbf{D})) = \sigma_\lambda\sigma_\mu(R(\mathbf{D}))$.

Definition 3: Assume that R' is a meta-relation that defines views of R. The projection of R' that removes its i'th attribute, denoted $\pi_{R-A_i}(R')$, is defined as follows. For every meta-tuple r from R',

$$r = (a_1, \ldots, a_m)$$

If a_i is $\sqcup$, then $\pi_{R-A_i}(R')$ includes the meta-tuple:

$$q = (a_1, \ldots, a_{i-1}, a_{i+1}, \ldots, a_m)$$

Proposition 3: Let $\mathbf{D}$ be an instance of this database, and let $r(\mathbf{D})$ and $q(\mathbf{D})$ denote, respectively, the relations defined by the meta-tuples r and q. Then $q(\mathbf{D}) = \pi_{R-A_i}(r(\mathbf{D}))$ [2].
Proof: Let λ denote the selection predicate of r. $r(\mathbf{D})$

[1] If a_i is blank, then μ is true.

[2] In general, we define $\pi_\alpha(R)$ as a projection on those attributes in α that are in R. Thus, if attribute A_i had already been removed, a projection on $R - A_i$ has no effect.

and $q(\mathbf{D})$ can be expressed as the following selection-projection expressions:

$$r(\mathbf{D}) = \sigma_\lambda(R(\mathbf{D}))$$
$$q(\mathbf{D}) = \sigma_\lambda \pi_{R-Ai}(R(\mathbf{D}))$$

We observe that if the i'th attribute of R does not participate in the predicate λ, then $\sigma_\lambda \pi_{R-Ai}(R(\mathbf{D})) = \pi_{R-Ai}\pi_\alpha\sigma_\lambda(R(\mathbf{D}))$.

Note that projection retains only meta-tuples whose projected attributes are disjoint from their selection attributes.

Propositions 1, 2 and 3 are summarized in the following theorem:

Theorem: Assume a database $R_1, \ldots, R_n$ and a meta-database $R'_1, \ldots, R'_n$, COMPARISON, with constraints (null views) $C_1, \ldots, C_m$. Let Q be a conjunctive query against this database. Let S be the relational algebra expression that implements Q. Let S' be the relational algebra expression obtained from S by substituting every reference to R with a reference to R'. S operates on the relations to yield the answer A. S' operates on the meta-relations to yield the meta-answer A'. Then, the meta-tuples in A' define subviews of A that are also views of $C_1, \ldots, C_m$.

The theorem guarantees that meta-tuples in A' define subviews of A that are views of $C_1, \ldots, C_m$ (and are therefore null). However, some meta-tuples may still contain references to meta-tuples outside A', and are therefore not expressible entirely within A'. Such subviews are avoided if S' is modified so that all products are performed first, and their result is pruned to retain only those meta-tuples that do not contain references to other meta-tuples. Also, as we explain below, it is advantageous to perform selections before projections. Altogether, S' is transformed to a sequence of products, followed by selections, and ending with projections. This simple strategy for implementing conjunctive queries is not necessarily optimal. However, we note that the optimality is not so essential for meta-relations, because they are relatively small. For the actual relations, where optimality is essential, a different strategy may be implemented.

5.2 Refinements

The theorem guarantees that the method for generating subviews is *sound*, but it does not guarantee that it is *complete*. That is, this method generates subviews of the result that are indeed null, but does not necessarily generate all such subviews. A method that would guarantee completeness would undoubtedly be of a different complexity altogether. Yet, with several simple refinements, it can be improved to generate additional desirable subviews. Two such refinements are sketched below. Both address the "loss of views" that occurs during projection, when meta-tuples are discarded if they restrict (with variables or constants) the attributes that are removed.

The first refinement modifies the product operation. Assume that Q is a product of R and S, followed by a projection that removes all the attributes of S. Obviously, Q is equivalent to R, and A' should retain all the meta-tuples of R'. However, these meta-tuples may be discarded by the projection, if they contain restrictions in the attributes contributed by S'. To handle this situation we may extend the product of meta-relations to include also these two tuples:

$$q_1 = (a_1, \ldots, a_m, \sqcup, \ldots, \sqcup)$$
$$q_2 = (\sqcup, \ldots, \sqcup, b_1, \ldots, b_n)$$

These tuples define all previous subviews of R and S as subviews of the product of R and S.

The second refinement modifies the selection operation. As defined, this operation requires conjuncting μ, the predicate expressed in the meta-tuple, with λ, the predicate expressed in the query. However, as all the tuples in the resulting relation satisfy λ, the expression $\mu \wedge \lambda$ is simply μ. Therefore, it appears that a simpler definition of the selection operation may be provided, which simply retains all meta-tuples without any modification. On the other hand, this simpler definition often would not generate the best definitions of views, nor would it detect views that are indeed irrelevant. As an example, assume a meta-tuple that selects the employees whose salaries are between \$30,000 and \$60,000, and consider the following four queries that select the employees whose salaries are (1) between \$20,000 and \$40,000, (2) between \$20,000 and \$70,000, (3) between \$40,000 and \$50,000, and (4) under \$30,000. In each case, the given view (employees whose salaries are between \$30,000 and \$60,000) could be retained as a view of the employees selected. However, it would be more desirable to handle this selection on a case by case basis, as follows. In the first query, modify the given view to define the employees whose salaries are between \$30,000 and \$40,000; in the second query, retain the given view without any modification; in the third query, modify the given view so it does not restrict the salary at all; and, in the fourth query, discard the given view altogether. In general, we observe four different cases: If λ *implies* μ, the meta-tuple is selected and the corresponding field is cleared (i.e., the variable or the constant is replaced by ⊔); if μ *implies* λ, the meta-tuple is selected without any modification; if λ and μ are *contradictory*, the meta-tuple is discarded; otherwise, the meta-tuple is selected, and is modified to represent $\mu \wedge \lambda$. Clearing selection predicates ensures that more meta-tuples will "survive" future projections. Determining the appropriate case for given μ and λ may require consulting relation COMPARISON, and, possibly, modifying it. While for most views and queries this task is quite simple, if an implementation chooses not to determine the case for predicates of certain form, then in those cases the relevant meta-tuple must not be selected. Note that the only other time where relation COMPARISON is used, is when the views in A' are described to the user.

6 Detecting Containments

Consider again the selection operation, and let $(a_1, \ldots, a_n)$ be a meta-tuple in the result of the product. This meta-tuple represents a *denial* of a conjunction of n primitive predicates. Denote by μ_i the predicate represented by a_i [3]. Then this meta-tuple represents the constraint:

$$\neg(\mu_1 \wedge \cdots \wedge \mu_n)$$

which may be formulated as an implication:

$$\mu_1 \wedge \cdots \wedge \mu_{j-1} \wedge \mu_{j+1} \wedge \cdots \wedge \mu_n \Longrightarrow \neg\mu_j$$

Let $\lambda_1 \wedge \cdots \wedge \lambda_m$ be the selection predicate expressed in the query. As discussed earlier, selection is performed with a *sequence* of steps, each comparing a μ_i with a λ_j. μ_i and λ_j are related in one of four different ways:

1. $\lambda_j \Longrightarrow \mu_i$
2. $\mu_j \Longrightarrow \lambda_i$
3. $\neg(\lambda_j \wedge \mu_i)$
4. Otherwise

In all but case 3, λ_j and μ_i are not contradictory, and the meta-tuple is retained (possibly modified). A meta-tuple that "survives" the entire selection sequence describes a constraint on the result.

Consider now case 3. It may be formulated as $\lambda_j \Longrightarrow \neg\mu_i$. Obviously, when this selection step terminates, the tuples in the result satisfy λ_j. Hence, they also satisfy $\neg\mu_i$. Recall that the meta-tuple may be represented as the implication:

$$\mu_1 \wedge \cdots \wedge \mu_{i-1} \wedge \mu_{i+1} \wedge \cdots \wedge \mu_n \Longrightarrow \neg\mu_i$$

[3] If $a_i = ⊔$, then $\mu_i = true$.

Thus, the tuples that satisfy $\mu_1 \wedge \cdots \wedge \mu_{i-1} \wedge \mu_{i+1} \wedge \cdots \wedge \mu_n$ must satisfy $\neg\mu_i$. Since the tuples of the result indeed satisfy $\neg\mu_i$, all the tuples that satisfy $\mu_1 \wedge \cdots \wedge \mu_{i-1} \wedge \mu_{i+1} \wedge \cdots \wedge \mu_n$ are included in the result. Therefore in case 3, the conjunction of all the constraint predicates except μ_i represents a view that is contained in the result.

We now extend the previous selection process. Consider a selection step that compares μ_i with λ_j. Case 1, 2 and 4 are treated as before. In case 3, the meta-tuple is not discarded; instead, a_i (which represents the predicate μ_i) is replaced by the symbol ¬, to indicate that it is never satisfied in the result. Future comparisons should leave this value unchanged. When an attribute that is removed by the projection contains ¬, the meta-tuple is retained, but this ¬ is "attached" to the meta-tuple.

7 Examples

When the entire processing ends, each meta-tuple in the result is examined. If it does not include any ¬ it is a constraint. If it includes exactly one ¬, it is a containment.

Constraints are translated into **disallow** statements; containments are translated into **contain** statements. The syntax of these statements is similar: the word **disallow** or **contain** are followed by an expression in the attributes of the result, similar to the **where** clause of a **retrieve** statement. This set of statements constitutes the *intensional* answer, and it accompanies the usual *extensional* answer.

Example 1: Consider the query of the introduction to retrieve the names, positions and salaries of employees in the design department:

retrieve (EMPLOYEE.NAME, EMPLOYEE.POSITION, EMPLOYEE.SALARY)
where EMPLOYEE.DEPARTMENT = design

This query may be implemented with the following sequence of algebraic operations:

1. A ← $\sigma_{(\text{DEPARTMENT}=\text{design})}(\text{EMPLOYEE})$
2. A ← $\pi_{\text{NAME,POSITION,SALARY}}(\text{A})$

The same operations that are applied to the database relations are applied to their meta-relations counterparts:

1. A′ ← $\sigma_{(\text{DEPARTMENT}=\text{design})}(\text{EMPLOYEE}')$
2. A′ ← $\pi_{\text{NAME,POSITION,SALARY}}(\text{A}')$

This sequence does not require products. Before se

lection, the meta-relation EMPLOYEE′ is pruned to include only those meta-tuples that do not reference other meta-tuples:

EMPLOYEE′

NAME	POSITION	SALARY	DEPARTMENT
		x_1	design
	researcher		x_2

The selection involves a single query predicate DEPARTMENT=design. In the first meta-tuple the constraint predicate is also DEPARTMENT=design. This corresponds to case 1, and the DEPARTMENT field is cleared. In the second meta-tuple the constraint predicate is DEPARTMENT≠design. This corresponds to case 3, and the DEPARTMENT field is replaced by ¬. We have:

A′

NAME	POSITION	SALARY	DEPARTMENT
		x_1	
	researcher		¬

Finally, the projection retains both meta-tuples:

A′

	NAME	POSITION	SALARY
			x_1
¬		researcher	

In response to a request to retrieve the names, positions and salaries of employees in the design department, the system issues this extensional answer:

NAME	POSITION	SALARY
Jones	manager	51,000
Smith	researcher	40,000

And this intensional answer:

disallow SALARY < $30,000
contain POSITION = researcher

These statements inform the user that all employees retrieved earn at least $30,000, and that all employees in researcher positions were retrieved.

Example 2: Consider a query to retrieve pairs of different departments that are in the same division, and their budgets:

retrieve (DEPARTMENT:1.D_NAME,
DEPARTMENT:1.BUDGET,
DEPARTMENT:2.D_NAME,
DEPARTMENT:2.BUDGET)
where DEPARTMENT:1.DIVISION =
DEPARTMENT:2.DIVISION
and DEPARTMENT:1.D_NAME <
DEPARTMENT:2.D_NAME

This query may be implemented with the following sequence of algebraic operations [4]:

1. A ← DEPARTMENT × DEPARTMENT
2. A ← $\sigma_{(\text{DIV:1=DIV:2})\wedge(\text{D\_NAME:1<D\_NAME:2})}(\text{A})$
3. A ← $\pi_{\text{D\_NAME:1,BUD:1,D\_NAME:2,BUD:2}}(\text{A})$

The same operations that are applied to the database relations are applied to their meta-relations counterparts:

1. A′ ← DEPARTMENT′ × DEPARTMENT′
2. A′ ← $\sigma_{(\text{DIV:1=DIV:2})\wedge(\text{D\_NAME:1>D\_NAME:2})}(\text{A}')$
3. A′ ← $\pi_{\text{D\_NAME:1,BUD:1,D\_NAME:2,BUD:2}}(\text{A}')$

The result of the product after outside references are removed is:

A′

D_N:1	S:1	D:1	B:1	D_N:2	S:2	D:2	B:2
		x_3	x_4			x_3	x_5
		x_3	x_5			x_3	x_4

The selection involves two query predicates. The first predicate is DIVISION:1=DIVISION:2. In both meta-tuples the constraint predicate is also DIVISION:1=DIVISION:2. This corresponds to case 1, and in each meta-tuple the DIVISION fields are cleared. The second predicate is D_NAME:1<D_NAME:2. In both meta-tuples the constraint predicate is *true*. Again, this corresponds to case 1, and in each meta-tuple the D_NAME fields remain clear. We have:

A′

D_N:1	S:1	D:1	B:1	D_N:2	S:2	D:2	B:2
			x_4				x_5
			x_5				x_4

Finally, the projection retains both meta-tuples:

A′

D_NAME:1	BUDGET:1	D_NAME:2	BUDGET:2
	x_4		x_5
	x_4		x_4

In response to a request to retrieve pairs of different departments that are in the same division, the system issues this extensional answer:

D_NAME:1	BUDGET:1	D_NAME:2	BUDGET:2
design	275	manufacture	275
sales	160	service	160

And this intensional answer:

disallow BUDGET:1 ≠ BUDGET:2

This statement (generated twice) informs the user that the budgets must be equal.

[4] When a relation has several attributes named A, then A : *i* denotes the *i*'th appearance of A.

8 Conclusion

We presented a method that addresses the important issue of providing more meaningful answers to queries. The relational model, which stores data in relations, is extended to store constraints in meta-relations, and the algebra of relations is extended to an algebra of meta-relations. Each query is processed against both the database and the meta-database, yielding an answer and a meta-answer. The meta-answer describes constraints and containments that apply to the answer.

Currently, the methods handle only conjunctive views (constraints, containments, and queries). Work is underway on extensions that will handle more general views; for example, views with *disjunctions* and views with *aggregate functions.* Since the methods do not necessarily detect all the constraints and containments that apply to the result, the intensional answers that are provided should be considered sound characterizations, that are not necessarily complete.

One of the difficult problems in providing intensional answers is how to identify the statements that are *relevant* to a query. For example, consider a query to list all employees and their salaries. While it is true that the set of employees listed satisfies the constraint that all those in researcher positions are in the design department, this information is probably irrelevant. It is possible to extend the model to include in the database additional knowledge that will assist in determining the relevant statements. A satisfactory implementation of this approach requires additional investigation.

An alternative solution is to define an intensional statement as relevant, if it can be expressed entirely with the attributes retrieved by the query. Thus, the constraint that all employees in researcher positions are in the design department is relevant only to queries that retrieve both POSITION and DEPARTMENT. While this solution may not be entirely satisfactory, it provides an extremely simple pruning mechanism, which is usually effective. Indeed, the projection operation implements this very pruning strategy, as it discards the meta-tuples whose projection attributes and selection attributes intersect, retaining only constraints and containments that are expressed entirely with the attributes retrieved by the query.

Another problem that remains to be solved is how to prune intensional answers, so they do not include statements that are implied by other statements. For example, assume a constraint that all female employees earn over $30,000, and a constraint that all employees over 35 years old earn over $40,000, and consider a query to list the female employees over 35 years old. The meta-answer will include two constraints: all employees listed earn over $30,000, and all employees listed earn over $40,000. Obviously, the former statement is redundant.

Work is also underway on a database "front-end" interface that will implement our methods and enable experimentation. The user will define constraints with **disallow** statements, and the system will insert automatically the appropriate meta-tuples into the meta-relations. In response to a **retrieve** statement, the user will receive the usual extensional answer, accompanied by inferred **disallow** and **contain** statements. Thus, the meta-relations and meta-tuple notation would be completely transparent, with all user-system communication done with customary query language statements.

Acknowledgement. The author is grateful to Alex Borgida for pointing out several relevant works.

References

[1] L. Cholvy and R. Demolombe. Querying a rule base. In *Proc. First Int. Conf. on Expert Database Systems*, April 1986, pages 365–371.

[2] F. Corella. Semantic retrieval and levels of absraction. In *Proc. First Int. Workshop on Expert Database Systems*, October 1984, pages 91–114.

[3] T. Imielinski. Intelligent query answering in rule based systems. *Journal of Logic Programming*, 4(3):229–257, September 1987.

[4] D. Maier. *The Theory of Relational Databases.* Computer Science Press, 1983.

[5] A. Motro. An access authorization model for relational databases based on algebraic manipulation of view definitions. In *Proc. Fifth Int. Conf. on Data Engineering*, February 1989, pages 339–347.

[6] A. Motro. Integrity = validity + completeness. (To appear in ACM Trans. on Database Systems.)

[7] C. D. Shum and R. Muntz. Implicit representation for extensional answers. In *Proc. Second Int. Conf. on Expert Database Systems*, April 1988, pages 257–273.

[8] C. D. Shum and R. Muntz. An information-theoretic study on aggregate responses. In *Proc. Fourteenth Int. Conf. on Very Large Data Bases*, August 1988, pages 479–490.

[9] J. D. Ullman. *Principles of Database Systems.* Computer Science Press, 1982.

[10] M. Zloof. Query-by-Example: a database language. *IBM Systems Journal*, 16(4):324–343, December 1977.

Integration of Buffer Management and Query Optimization In Relational Database Environment

Douglas W. Cornell **Philip S. Yu**

IBM Thomas J. Watson Research Center
Yorktown Heights, NY 10598

Abstract

In today's relational database systems, query optimization and buffer management are generally treated independently. However, the query access plan contains information on data access patterns which can be useful hints to the buffer manager. Furthermore, the optimal access plans under different buffer sizes can be quite different. In this paper, an integrated approach to buffer management and query optimization is proposed and analyzed. The query strategy of all transaction types is simultaneously considered together with the buffer allocation strategy so as to optimize overall system performance. As the buffer allocation depends upon the buffer holding time or transaction response time which is determined by the buffer allocation and query optimization strategies, an optimization method combining a integer programming model with a queueing model applied iteratively is developed to capture this effect. To reduce the size of the optimization problem, a methodology based on the concept of buffer consumption is proposed to pre-analyze the query and substantially cut down the access plans to be considered. A detailed simulation is used to demonstrate the superiority of the integrated strategy as compared to other buffer management strategies based on working set and hot set.

1. Introduction

Database systems have generally relied on memory buffers to reduce disk accesses. Even with the trend of ever increasing memory size, the memory buffer usually can not accommodate all the databases in the system, and some buffer management strategy is needed to make the best use of the buffer space. The traditional approach to memory management in a virtual memory environment uses a working set model where the popular replacement policy is LRU, which replaces the least recently used page by a new page [Denn68].

For a network or hierarchical database system, reference strings tend to be unpredictable except for batch processing. A study on network databases can be found in [Effe84]. These types of systems seem to fit reasonably well with the working set model. However, queries to relational databases [Codd70] imply a lot of information on data references. The discussion in this paper is based on a database system similar to system R [Astr76]. The query optimizer analyzes each query and generates an access plan which contains detailed information on how each relation is accessed. (Information related to access plans can be found in [Seli79].) These access plans assume that none of the referenced pages will be found already in the buffer. The buffer manager does not take hints on reference patterns from the query optimizer except prefetch for sequential scan. Although some variant of the LRU policy is often used for buffer management, it is not considered to be well suited for the reference patterns of relational databases [Ston81]. Previous research directed toward utilizing information available about reference patterns has been described in [Sacca82], [Chou85], and [Sacca86]. In [Sacca82], [Sacca86] a hot set model is proposed. The basic idea there is to determine a hot set for every query and allocate sufficient buffer space to cover the hot set before executing a query. Certainly, this is a local optimization for each query to provide it with sufficient buffer space to minimize disk IO accesses. The potential buffer contention among queries is not addressed in the buffer allocation strategy. As pointed out in [Saaca86], straightforward implementation of this idea can lead to problems such as infinite waits, long queries blocking short queries, etc.. Some ad hoc techniques to relieve these problems are also suggested. In [Chou85], active instances (due to references from different queries) of a file are given different buffer pools and are managed by different replacement disciplines. A DBMIN algorithm is proposed for estimating the size and discipline for each file instance of a query is described. To compare performance with the hot set strategy, a simulation is developed in which a query is described by the CPU usage, number of IO's, and hot set size. All these works investigate the "right" buffer allocation for a given query plan or access path selection without considering the effect of other queries.

In a transaction processing environment, the characteristics of all transactions/queries are known a priori. Thus, if the transaction rate for each type is also known, a buffer management strategy may be devised to optimize buffer allocation for all queries. Furthermore, the query optimizer may be integrated with buffer management to take into consideration the buffer availability in selecting an appropriate access plan for each query to optimize overall performance. In this paper, we examine the integration of buffer management and query optimization. A detailed simulation is developed to compare the integrated approach with buffer management schemes based on the LRU working set model and the hot set model, respectively. We discuss the issues of buffer management and query optimization that motivate the integrated approach in Section 2. The mathematical formulation of the optimization problem is given in Section 3. Section 4 shows the performance improvement through the integrated approach using a hypothetical example. To reduce the size of the optimization problem, Section 5 develops a methodology based on the concept of buffer consumption to substantially trim down the number of access plans to be considered in the

Proceedings of the Fifteenth International Conference on Very Large Data Bases **Amsterdam, 1989**

optimization procedure. Buffer consumption is defined to be the product of query response time and buffer allocation. An access plan with a small amount of buffer allocation is not desirable unless the buffer consumption is also small. Concluding remarks appear in Section 6.

2. Buffer Management and Query Optimization

The conventional approach to query optimization is to examine each query in isolation and select the execution plan with the minimal cost based on some predefined cost function of IO and CPU requirements to execute the query [Seli79]. The impact of buffer management generally is not reflected in the cost function. The number of IO's estimated in the cost analysis does not reflect the potential effect of other transactions concurrently under execution. On the other hand, it is also quite difficult to assess this effect without detailed knowledge on how other concurrent transactions are progressing.

For each query type, different execution plans show different sensitivities to the size of memory allocated for data and thus have different memory requirements. In this paper, we use the expression data buffer to refer to memory allocated for storing data from the database. The best access plan for a given buffer size is not necessarily the best plan for another buffer size. Furthermore, it is not even clear whether every query type needs to be given the same constraint on buffer allocation. Different query types may have different diminishing points of return on buffer size and face different penalties on suboptimal buffer size. An integrated buffer management and query optimization strategy is thus needed so that the access plans for all query types are considered together based on buffer availability.

Let's look at an example. Consider a query joining three relations, R_1, R_2, and R_3. The relations consist of 10, 20, and 40 pages with cardinalities (C_i, $i = 1, 2, 3$) of 50, 100, and 500, respectively. The join selectivities used to calculate the number of matching tuples from the join operations are taken to be .01 for all joins. Assume that no index is available. To simplify the discussion, nested-loop join is assumed to be the method of choice. The query optimizer still needs to decide the join order. Ignoring buffer availability, a conventional query optimizer would choose R_1, R_2, R_3 as the join order with R_3 the inner-most relation. The number of page fetches would be 3000 ($= 20 \times C_1 + 40 \times 0.01 \times C_1 \times C_2$) if none of the relations are kept in the buffer. The number of IO would be minimized if relations R_2 and R_3 can be kept in the buffer. A total of 61 ($=20+40+1$) buffer pages would be required. This strategy would be optimum if that amount of buffer space can be made available. If the join order is taken to be R_3, R_2, R_1 with R_1 the inner-most relation, the number of page fetches would be 15000 ($= 20 \times C_3 + 10 \times 0.01 \times C_3 \times C_2$) if none of the relations are kept in the buffer. However, with 31 ($=10+20+1$) buffer pages, the number of IO can be minimized to read in each relation exactly once. Of course, this access plan does need to perform more tuple comparisons. This illustrates that the optimum access plan depends upon the buffer allocation and that there are trade-offs between different hardware resource (CPU and IO) requirements. So far we have been looking at one query alone. In the presence of multiple queries, buffer space needs to be allocated among the different queries so as to optimize the overall performance. One thus needs to consider all query types together.

3. Methodology

The methodology to obtain a global optimization on buffer management and query optimization is now described in this section. We focus on join queries here, as select type queries have little buffer requirement. An integer programming formulation is used to select the best access plan and associated buffer allocation. The objective function is similar to the cost function used in a conventional query optimizer, but is the sum over all query types weighted by query frequencies. The constraints are that not only the buffer allocation for each query, but also the time-averaged buffer requirement over all queries must be less than the total buffer size. Note that the time-averaged buffer requirement for each query type is proportional to its response time and arrival frequencies in addition to the buffer allocation for each instance of its execution. The response time certainly depends upon access plan selection and buffer allocation.

To get around this inter-dependency problem, we can decompose the problem into two parts and take an iterative approach. The first part is the optimization just described which uses an assumed response time for each query type, and the second part is a queueing model to solve for the response time based on the access plan selections and buffer allocation from the first part (the optimization problem). The optimization problem then uses the response time from the queueing model to solve for an improved solution. Alternatively, we can use the average response time over all queries (derived in Section 3.2) as the objective function and employ a heuristic technique like simulated annealing [Kirk83] to solve the non-linear programming problem. Simulated annealing can be helpful to address very large size problems or optimize response times directly [Wolf88].

3.1 Optimization Problem Under the Integrated Strategy

Consider a set of join queries Q_i, $i = 1, ..., N_Q$, and relations R_k, $k = 1, ..., N_R$. For each query type Q_i, there is a set of different join plans, QS_{ij}, $j = 1, ..., n_i$. Under the integrated strategy, each join plan specifically indicates whether a joining relation is to be kept in the buffer. For example, in a two-way join of R_k and R_j, nested-loop join with R_k (or R_j) in the buffer as the inner relation and nested-loop join with R_k (or R_j) as the inner relation but not in the buffer are four potential join plans. Although there may be a large number of potential access plans for each query, we can eliminate most of them from consideration of the optimization procedure discussed below through some simple pre-analysis of the access plans. This is addressed in Section 5. Let X_{ij} be 1 if plan QS_{ij} is adopted for Q_i and 0 otherwise. That is to say, each join plan is identified by a (0,1) integer programming variable to indicate whether it is adopted or not. Let D_{ij} be the number of pages to be read from disk if plan QS_{ij} is adopted. D_{ij} can be calculated based on the join algorithms and buffer allocation strategies. Define λ_i to be the arrival frequency for Q_i.

Now we formulate the optimization problem. The objective function can be any cost function on IO and CPU over all query types. Here we use the aggregate IO rate as the cost function assuming IO is the performance bottleneck. Thus,

$$objective\ function = \sum_i \sum_j D_{ij} X_{ij} \lambda_i .$$

Constraints are included to guarantee that exactly one strategy is adopted for each query,

$$\sum_j X_{ij} = 1, j = 1,..., n_i .$$

Additional constraints are added to prevent the buffer from being overcommitted. First of all, no query can get a buffer allocation more than the total buffer size. Let B be the number of buffer pages available and F_{ij} be the number of buffer pages allocated for query Q_i under plan QS_{ij}. Thus F_{ij} is an input parameter derived from the access plan.

$$\sum_j X_{ij} F_{ij} \leq B, \qquad \text{for each } i,$$

Furthermore, the average buffer usage of each query type is estimated as the product of buffer allocation and response time times the arrival frequency. In order to accommodate the fluctuations in query workload, the average buffer usage needs to be less than some fraction, α, of the total buffer. The appropriate value for α is explored in the next section.

$$\sum_i \lambda_i \sum_j X_{ij} F_{ij} RT_{ij} \leq \alpha B$$

where RT_{ij} is the average response time of query Q_i under plan QS_{ij}. A starting value needs to be selected for each RT_{ij}, for $i = 1,..., N_Q$ and all plans $j = 1,..., n_i$ for query Q_i. The join plans selected by the solution of the resulting integer program are not yet optimum, since the initial response times are not chosen to be optimum. In Section 3.2, a queueing model is presented to estimate a new set of response times based upon the "optimal" strategy $\{QS_{i\beta_i}, i = 1,..., N_Q\}$ chosen by the optimization procedure. We iteratively solve the execution plan selection and buffer allocation from the integer programming problem and response time calculation from the queueing model until no further change in plan is observed or until the calculated average response time remains the same on successive iterations. If we choose a large response time for each query type to start the optimization problem, a very conservative buffer allocation will be made, as exaggerated response time leads to overestimates on the time average buffer requirement. The queueing analysis then provides an improved response time based on the strategy recommended from the optimization problem. With the improved response time or shortened buffer hold time, more buffer space can be allocated and lead to further improvement in the response time. It appears to converge after a few iterations.

3.2 The Queueing Model

So far we have been concentrated on buffer allocation. The concurrently executing queries compete not only for buffer but also CPU and IO. The contention on CPU and IO would increase the response time, thus the buffer hold time. The response time of the transaction processing system is determined using an open queueing network model. The system consists of a single CPU with speed *MIPS* and multiple (N_D) disks. Each database is assumed to be partitioned uniformly across the N_D disks based on its primary key. Under a given strategy, $\{QS_{i\beta_i}, i = 1,..., N_Q\}$, the overall system performance can be analyzed by assuming there is no reuse of the buffer between queries. The response time of each query can be calculated as its resource requirement on the CPU and disks can be predicted given the buffer allocation strategy. Define $U_{i\beta_i}$ to be the query-processing pathlength at the CPU for Q_i under $QS_{i\beta_i}$. Let I_1 be the CPU overhead for scanning/comparing a tuple, I_2 the overhead for returning a tuple to the application program, I_3 the overhead for an IO operation and I_{A_i} the application processing pathlength for type i query. $U_{i\beta_i}$ can be expressed in terms of these parameters as given in Appendix A. The total CPU processing cost is then

$$U_{CPU} = \sum_i \lambda_i U_{i\beta_i}$$

Let T_{IO} be the disk service time to perform an IO operation. The utilization of each disk assuming the load is spread uniformly across all disks for each relation can be shown to be

$$\rho_{IO} = \frac{T_{IO}}{N_D} \sum_i \lambda_i D_{i\beta_i} .$$

The average response time of Q_i is

$$RT_{i\beta_i} = \frac{U_{i\beta_i}}{MIPS - U_{CPU}} + \frac{D_{i\beta_i} T_{IO}}{1 - \rho_{IO}} ,$$

where the first component is the sum of the service time and waiting time at the CPU and the second component is the sum of those times at the disks. The average response time over all queries, under strategy, $\{QS_{i\beta_i}, i = 1,..., N_Q\}$, would then be $RT = \sum_i \lambda_i RT_{i\beta_i} / \sum_j \lambda_j$.

The following model parameters are used in the performance comparisons. The pathlength parameters are set at $I_1 = 70$, $I_2 = 5K$, $I_3 = 5K$ and $I_{A_i} = 20K$, for all i. The system consists of a 14 MIPS processor and 4 disks, each with a 20 millisec access time, i.e. $MIPS = 14$, $N_D = 4$, and $T_{IO} = 20$. During table scans, it is assumed that multiple pages are fetched together in each read IO, as in IBM DB2 [Teng84]. To estimate $D_{i\beta_i}$, a prefetch blocking factor of 10 is used here for a table scan.

4. Performance Comparison

4.1 Alternative Strategies

The proposed integrated buffer management and query optimization strategy is compared with two alternative buffer management strategies. One strategy, referred to as strategy W, is based on the working set model, where a global LRU replacement algorithm is used to manage the buffer. For strategy W, the access plans for all queries are decided without consideration of available buffering, i.e. by assuming that referenced pages are prefetched from disks on successive scans. The other strategy, referred to as strategy H, is based on an enhanced hot set model. The hot set in a multi-join query is defined here to be the subset of the joining relations (if table scan is used) or the non-leaf parts of the indexes (if index scan is used) that can fit into a given fraction y of the total buffer and result in the most I/O reduction. In choosing the hot set, we consider different access path selections, index vs segment, and select the one that requires the least IO. That is to say instead of selecting the hot set for an arbitrarily chosen access path, we choose the best access path and the corresponding hot set under a given buffer constraint. Still for strategy H, the join order is taken to be the same as that in strategy W to reduce the number of enumerations to be considered. Strategy H ensures that a buffer space of size equal to the query's hot set size can be allocated before a query gets executed. Otherwise the query has to be put into wait state. When additional buffer space becomes available, the longest waiting query whose hot set can be accommodated is scheduled. (An alternative is to schedule the query in FCFS order. For the example considered in this section, both scheduling schemes lead to very similar result.) For both strategies H and the integrated strategy, pages which are not in the hot set compete for buffer space, which are not pre-allocated to accommodate the hot sets, based on LRU. Note that for the nested-loop join example considered later in this section, the desired buffer allocation under DBMIN in [Chou85] for each relation based on the access path selected by Strategy H is similar to the buffer allocation Strategy H considered.

4.2 Simulation

A simulation program is developed to compare the three strategies under multi-way join queries. Although the integrated strategy can be applied to different join methods, to keep the comparison simple, nested-loop join is used as the join algorithm. Each join attribute is assumed to have an index. Even in this case, substantial differences in performance can be observed among the different strategies. The simulation attempts to capture the sequence in which tuples from different relations are scanned and compared, and in which pages are fetched from disks in a multi-way join. Two types of scan are simulated, table and index. In a table scan, under the uniform distribution assumption on attribute value, the number of tuples between matches has a geometrical distribution. (See [Chris84] for an analysis of the effects of assuming uniform distribution and attribute independence on performance projections.) In the simulation, instead of generating a random value for each tuple to carry out the join, we treat the number of tuples between matches as a random variable with geometric distribution and simulate the actions related to scanning and matching tuples. For example, scanning into a new page triggers an IO operation if the page is not in the buffer, finding a match in an outer relation initiates a scan into the inner relation, and reaching the end of scanning an inner relation returns the scan back to the outer relation. For index scans, the data page to be fetched is randomly selected as any one of the pages of the relation. Accessing of three index pages is required (assuming a three level index) as well as 1 data page. The simulation program generates the index page addresses assuming a B-tree index structure and captures the hit to re-referenced leave pages.

Transactions are placed on a wait list upon arrival. Each transaction has associated with it a value which is the least number of pages required on the free buffer list that will allow it to move to the CPU queue for execution. For the integrated strategy, this is the number of pages required to accommodate the relations to be kept in buffer, as determined by the optimization procedure. For strategy H, this is the number of pages in the hot set. There is no buffer requirement under strategy W. A transaction moves from the wait list to the CPU queue if it is the first in the wait list which has a buffer requirement less than the free page list and the multiprogramming level is not exceeded.

4.3 Example Problem

We illustrate the methodology using the following transaction processing environment with 14 transactions and 22 relations. Table 1 presents the cardinality and size of each relation. Table 2 describes the transaction characteristics. These include the number of joins and, for each join, the two relations involved and the selectivities on each join attribute, respectively. Also presented is the arrival frequency of each transaction. For example, transaction 1 has an arrival frequency of .07 Tx/sec and requires three joins. The join between relations R_2 and R_4, has a join selectivity on the join columns of both relations of .001.

Relation	Cardinality	No. of pages
1	500	90
2	1000	200
3	300	100
4	1200	50
5	500	70
6	1000	250
7	1000	60
8	100	100
9	500	400
10	200	100
11	4000	260
12	500	90
13	1000	200
14	300	100
15	1200	50
16	500	70
17	1000	250
18	1000	60
19	100	100
20	500	400
21	200	100
22	4000	260

Table 1. Relations for the example problem

The alternative access plans for a given join order are restricted for simplicity to using nested-loop join with the inner relations either kept in memory using table scan or not kept in memory using either table scan or index scan. If index scan is used, the non-leaf portion of the index will be kept in memory. The iterations required for the two step optimization is about six cycles to converge and consumes about 10 sec of CPU time on an IBM 3090 processor.

TX	Run Freq.	No. of Joins	Relations		Selectivities	
1	.07	3	2	4	.001	.001
			4	6	.001	.001
			6	8	.001	.01
2	.071	3	3	5	.004	.002
			5	7	.002	.001
			7	9	.001	.002
3	.11	3	1	4	.002	.001
			4	5	.001	.002
			5	7	.002	.001
4	.200	3	1	2	.002	.001
			2	8	.001	.010
			8	10	.010	.005
5	.06	4	3	5	.004	.002
			5	7	.002	.001
			7	9	.004	.002
			9	11	.002	.001
6	.210	1	1	4	.002	.001
7	.300	1	3	5	.004	.002
8	.072	3	13	15	.001	.001
			15	17	.001	.001
			17	19	.001	.01
9	.071	3	14	16	.004	.002
			16	18	.002	.001
			18	20	.001	.002
10	.11	3	12	15	.002	.001
			15	16	.001	.002
			16	18	.002	.001
11	.201	3	12	13	.002	.001
			13	19	.001	.010
			19	21	.010	.005
12	.062	4	14	16	.004	.002
			16	18	.002	.001
			18	20	.004	.002
			20	22	.002	.001
13	.210	1	12	15	.002	.001
14	.311	1	14	16	.004	.002

Table 2. Transactions for the example problem

In Figure 1, we plot the average response time from the simulation vs buffer size for the three strategies. We set $\alpha = 0.9$ for the integrated strategy and $\gamma = 0.65$ for strategy H, respectively. Note that these parameters, α, γ, are set at the optimum values for a 400 page buffer size. Sensitivities to these parameters are discussed later. The integrated strategy shows uniformly better performance over strategies W and H. This is especially the case when the buffer size is smaller (say, a 400 page buffer). Strategy H based on the hot set concept also shows significant improvement over strategy W using LRU, in agreement with [Sacca86]. This shows the importance of recognizing the semantics of the query and keeping the critical relations in the buffer. A simple LRU type strategy like strategy W, ignoring the query semantics, performs very badly. Still, strategy H is only a local optimization on each query. Imposing a uniform limit on hot set size over all queries can be suboptimal. The integrated strategy considers the behavior of all queries together to derive a decision on the amount of buffer to be allocated for the execution of each query.

Table 3 summarizes the access plan selection and buffer allocations for each of the three strategies when the buffer size is 400 pages. The optimization procedure to determine the access plan and buffer allocation requires about 6 iterations for this example. In the "Join Order" column, the first or leftmost relation is the outer most-relation in the join, and the last or the rightmost relation is the inner-most relation. In the "Scan Type" column, S and I represent table scan and index scan, respectively, for access of each of the relations presented according to the join order. In the "Buffer" column, y means the corresponding relation is kept in the buffer while n means that no effort is made to keep the relation in the buffer. For example, under the integrated strategy, the join order of transaction 1 is R_6, R_4, R_8, R_2, with R_2 the inner-most relation. All relations are scanned sequentially through table scan. All relations except R_6 are kept in the buffer. Note that under strategy W, no relation is pre-allocated any buffer space. A different join order may be chosen under the different strategies. Let us look specifically at transaction 2. Under the integrated strategy, R_7, R_5, and R_3 are kept in the buffer requiring a total of 230 pages, while R_9 with 400 pages is chosen to be the outer-most relation and is not kept in the buffer. Under strategy W, as no buffering of the relation is assumed, R_9 is instead chosen as the inner-most relation due to its low selectivity and the availability of an index. Under strategy H, the same join order is chosen as strategy W, but R_5 and R_7 are kept in the buffer. In Table 4, we compare some run time statistics between the integrated strategy and strategy H. Both the average IO's per transaction and the fraction of transactions delayed due to insufficient buffer size are substantially lower for the integrated strategy.

We observe that in the integrated strategy, the average response times determined from the queueing model in Section 3.2 and from the simulation are comfortably close to each other, differing by amounts within 20%. The difference is mainly due to the fact that the analysis does not capture some of the inter-transaction effect considered in the simulation. One is the IO reduction from buffer hit due to transactions referencing relations that were brought in by another transaction referencing the same relations. For the conditions of Figure 1, none of the strategies causes the transactions to read as many pages as it would if it ran alone on the system. The other is the delay due to temporary over-allocation of the buffer. In the integrated strategy, if a relation to be kept in the buffer for the execution of a transaction cannot be accommodated, the transaction execution is delayed. As shown in Table 4, this delay occurs quite infrequently. As the response time estimate in the integrated strategy is only used in the optimization constraint to measure the buffer requirement, the inaccuracy can be more or less absorbed in the factor α. We have examined the sensitivity of the integrated strategy to α and found that the response time remains the same over a wide range of α from 0.6 to 0.9 for the example shown.

Tx Type	Join Order	Strategy W Scan Type(s,i)	Buffer (y,n)
1	8,6,4,2	s,i,i,i	n,n,n,n
2	3,5,7,9	s,i,i,i	n,n,n,n
3	5,4,1,7	s,i,i,i	n,n,n,n
4	8,2,1,10	s,i,i,i	n,n,n,n
5	3,5,7,9,11	s,i,i,i,i	n,n,n,n,n
6	1,4	s,i	n,n
7	3,5	s,i	n,n
8	19,17,15,13	s,i,i,i	n,n,n,n
9	14,16,18,20	s,i,i,i	n,n,n,n
10	16,15,12,18	s,i,i,i	n,n,n,n
11	19,13,12,21	s,i,i,i	n,n,n,n
12	14,16,18,20,22	s,i,i,i,i	n,n,n,n,n
13	12,15	s,i	n,n
14	14,16	s,i	n,n

Tx Type	Join Order	Integrated Strategy Scan Type(s,i)	Buffer (y,n)
1	6,4,8,2	s,s,s,s	n,y,y,y
2	9,7,5,3	s,s,s,s	n,y,y,y
3	1,4,5,7	s,s,s,s	n,y,y,y
4	2,1,8,10	s,s,s,s	n,y,y,y
5	9,7,5,3,11	s,s,s,s,i	n,y,y,y,n
6	1,4	s,s	n,y
7	3,5	s,s	n,y
8	17,15,19,13	s,s,s,s	n,y,y,y
9	20,18,16,14	s,s,s,s	n,y,y,y
10	12,15,16,18	s,s,s,s	n,y,y,y
11	13,12,19,21	s,s,s,s	n,y,y,y
12	20,22,18,16,14	s,i,s,s,s	n,n,y,y,y
13	12,15	s,s	n,y
14	14,16	s,s	n,y

Tx Type	Join Order	Strategy H (γ = .65) Scan Type(s,i)	Buffer (y,n)
1	8,6,4,2	s,i,s,s	n,n,y,y
2	3,5,7,9	s,s,s,i	n,y,y,n
3	5,4,1,7	s,s,s,s	n,y,y,y
4	8,2,1,10	s,i,s,s	n,n,y,y
5	3,5,7,9,11	s,s,s,i,i	n,y,y,n,n
6	1,4	s,s	n,y
7	3,5	s,s	n,y
8	19,17,15,13	s,i,s,s	n,n,y,y
9	14,16,18,20	s,s,s,i	n,y,y,n
10	16,15,12,18	s,s,s,s	n,y,y,y
11	19,13,12,21	s,i,s,s	n,n,y,y
12	14,16,18,20,22	s,s,s,i,i	n,y,y,n,n
13	12,15	s,s	n,y
14	14,16	s,s	n,y

Table 3. Access plan selection and buffer allocation

(400 page buffer)

Strategy	I/O per transaction	Fraction of Tx with delayed starts
Integrated	50	.02
Strategy H (γ = .65)	438	.42

Table 4. Run statistics for 400 page buffer

Although the integrated strategy is found to be quite insensitive to the parameter α, strategy H is very sensitive to the parameter γ. Figure 2 shows the sensitivity of response time to γ under strategy H for a 400 page buffer. The variation in response time with the choice of γ is quite substantial. The variability in the response time curves points out the weakness of a local optimization strategy. Keeping a relation in memory can reduce the response time of the transaction referencing it, but may also create a detrimental effect for other transactions. This can occur if the reduction in response time is not large enough so that buffer consumption, which is the product of response time and buffer allocation, increases. Figure 3 shows the case for a 600 page buffer. A big zigzag in response time is observed. The peak at $\gamma = .5$ comes about because in going from $\gamma = .4$ to $\gamma = .5$, query Q_{12} has been allowed to keep relation R_{22} of 260 pages in the buffer. The simulation shows that although the number of page reads for query Q_{12} has been reduced from 5000 to 4020, the fraction of queries with delayed starts has increased from .03 to .32. At $\gamma = .6$, Q_{12} can keep both relations R_{18} and R_{22} with a reduction in IO to 1521. Because of the substantial reduction in its response time and hence buffer consumption for Q_{12}, the fraction of delayed starts is reduced to .25 even with larger buffer allocation. Note that the optimum γ depends upon not only the workload but also the buffer size. This would make selecting γ a nontrivial job for a real environment.

5. Reducing the Size of the Optimization Problem

In an environment with a large number of query types which require multiple joins over several relations, the number of variables, X_{ij}, can become very large. We can eliminate many of the access plans by pre-analyzing each query type separately before applying the global optimization. An important observation is that for a given query, some access plan is essentially inferior in the sense that there exists another plan that consumes less resources and offers better or similar performance. Inferior plans should be eliminated from consideration in the optimization procedure described in Section 4. For a given query, let us define buffer consumption, P_j^B, under access plan j to be B_jT_j where T_j and B_j are the execution time and buffer requirement of the query under plan j, respectively. Note that here we use execution time instead of response time. The additional queueing delay due to the impact of concurrently executing queries is ignored so we can examine each query separately. An access plan is feasible if the total buffer requirement is less than the total buffer size or some predefined maximum. We can state two simple rules to determine whether an access plan is inferior:

1. For any access plan i , if there exists another feasible

access plan j, such that $B_i = B_j$ and $T_i > T_j$, or $B_i > B_j$ and $T_i \geq T_j$, access plan i is regarded as an inferior plan.

2. For any access plan i, if there exists another feasible access plan j, such that $P_i^B > P_j^B$ and $T_i \geq T_j$, access plan i is regarded as an inferior plan.

The first rule indicates that a plan with a larger buffer allocation should achieve a smaller execution time. The second rule implies that a plan with a smaller buffer allocation and longer execution time is not necessarily desirable unless the buffer consumption is also smaller. Let access plan j_C using B_{j_C} buffer be the one with the minimum buffer consumption. Any plan with a larger buffer consumption than plan j_C should be rejected unless the extra buffer consumption leads to reduction in the execution time. Thus any plan using less than B_{j_C} buffer can be eliminated because it takes too little buffer so the long execution time results in large buffer consumption. Query strategies using more than B_{j_C} may achieve smaller response time at the expense of the buffer consumption. These strategies may be acceptable if there is sufficient buffer. This is to be determined by the global optimization procedure in Section 4. We need only to eliminate those where the response time improvement is small compared to the additional buffer used.

In Figure 4, we plot the execution time verses buffer allocation for transaction 1 of the Example problem described in Section 4. Several different access plan strategies correspond to each buffer size due to join-order permutations and the sum of the sizes of relations R_2 and R_8 being the same as the sum of the sizes of relations R_4 and R_6. Only the strategy with the smallest execution time for a given buffer allocation is plotted. Also, a buffer allocation is not considered unless a decrease in execution time can be realized compared with a smaller buffer allocation. That is to say, we only consider strategies which do not violate rule 1 stated above. Transaction 1 has 216 potential access plans, as there are 8 possible join orders (notice that not all permutations make sense, e.g. R_2, R_6, R_4, R_8) and each relation except the outer most can either be kept in buffer for table scan or not be kept in buffer while both table scan and index scan are possible alternative scanning strategies. The 216 candidate access plans of transaction 1 only result in five meaningful buffer allocations and access plans as follows:

1. 4 pages: join order R_8, R_6, R_4, R_2 with index scans on all except the outer most relation.
2. 50 pages: join order R_8, R_6, R_4, R_2 with R_4 kept in the buffer and index scans on R_6 and R_2
3. 150 pages: join order R_6, R_4, R_8, R_2 with R_4 and R_8 kept in the buffer and index scan on R_2
4. 250 pages: join order R_8, R_6, R_4, R_2 with R_2 and R_4 kept in the buffer and index scan on R_6.
5. 350 pages: join order R_6, R_4, R_8, R_2 with all pages of relations except the outer relation be kept in the buffer and accessed with table scans.

Figure 5 shows the plot of buffer consumption vs buffer size allocation for transaction 1. From rule 2, by comparing any two feasible strategies, the one with the higher buffer consumption can be eliminated unless it has the lower execution time. For a given buffer size of 400 pages, all five strategies are feasible as they all require less than 400 pages. Strategy 1 having 4 pages is the one with the minimum buffer consumption. By comparison with strategy 5, strategies 2, 3, and 4 are eliminated as they not only take longer to execute (Figure 4) but also have larger buffer consumption (Figure 5). Strategy 5, having 350 pages of allocated buffer, is still acceptable as its execution time is a lot smaller than strategy 1. Thus only strategies 1 and 5 need to be considered in the global optimization. The global optimization in Section 4 which considers all strategies has in fact picked out strategy 5. The number of strategies to be retained for global optimization from the other transactions of the illustrative example can be similarly reduced.

6. Summary

Existing query optimizers for relational databases attempt to minimize a linear combination of CPU and page IO costs for each query separately. The impact of buffer management is not reflected in the cost function. Although relational queries contain a lot of information on data references, it is generally not conveyed to the buffer manager by the query optimizer. In this paper, we propose and analyze a mathematical algorithm to integrate buffer management and query optimization. When making access plan selections, the integrated strategy explicitly considers whether a relation is kept in the buffer. The query strategies for all transactions in the workload are simultaneously determined to optimize an objective function on overall system performance with constraints on total buffer requirement. The mathematical formulation uses a two step approach combining an integer programming problem with a queueing network model to be solved iteratively. A detailed simulation is developed to study the performance improvement of the integrated strategy as compared with two other buffer management strategies based on the working set and hot set, respectively. The study shows that by ignoring all the reference information in the query, the LRU type strategy does very poorly. Explicitly taking into consideration the reference behavior from the query in buffer management, as in the hot set strategy, can improve the query performance. However, picking the right hot set size can be difficult. Integration of query optimization and buffer management to consider the requirement of all queries together gives substantial further improvement to the performance. This showed the importance of integrating of buffer management and query optimization strategies so that reference information from the queries can be captured to manage the buffer, and buffer availability can be reflected in the access path selection. Furthermore, each query can be pre-analyzed separately to reduce the number of access plans to be considered in the global optimization. We proposed an approach which eliminates an access plan if there exists another one which has less buffer consumption but similar or better performance. This pre-analysis was found to be able to greatly reduce the size of the optimization program.

Acknowledgement:

The authors would like to thank Guy Lohman for his suggestions on revising the paper.

References

[Astr76] M.M. Astrahan, M.W. Blasgen, D.D. Chamberlin, K.P. Eswaren, J.N. Gray, P.P. Griffith, W.F. King, R.A. Lorie, P.R. McJones, J.W. Mehl, G.F. Putzolu, I.L. Traiger, B.W. Wade, and V. Watson, "System R: Relational Approach to Database Management", ACM Transactions on Database Systems, Vol. 1, No. 2, June 1976, pp. 97-137.

[Chou85] H.-T. Chou and D. J. DeWitt, "An Evaluation of Buffer Management Strategies for Relational Database Systems", Proc. of the 11th International Conference on Very Large Data Bases, Stockholm, 1985, pp. 127-141.

[Chris84] S. Christodulakis, "Implications of certain assumptions in database performance evaluation", ACM Trans. Database Syst., Vol. 9, No. 2, June 1984, pp. 163-186.

[Codd70] E.F. Codd, "Relational Model of Data for Large Shared Data Banks", Comm. of the ACM, Vol. 13, No. 6, June 1970, pp. 377-387.

[Denn68] P. J. Denning "The Working Set Model for Program Behavior.", Comm. of the ACM, Vol. 11, No. 5, May 1968, pp. 323-333.

[Effe84] W. Effelberg and M. E. S. Loomis, "Logical, internal and physical reference behavior in CODASYL database systems", ACM Trans. Database Syst., Vol. 9, No.2, June 1984, pp. 187-213.

[Kirk83] S. Kirkpatrick, C.D. Gelatt, and M.P. Vecchi, "Optimization by Simulated Annealing", Science, Vol. 220, No. 4598, May 1983, pp. 671-680.

[Sacca82] G. Sacca and M. Schkolnick, "A Mechanism for Managing the Buffer Pool in a Relational Database System using the Hotset Model", Proc. of the 8th International Conference on Very Large Data Bases, Mexico City, Sept. 1982, pp. 257-262.

[Sacca86] G. Sacco and M. Schkolnick, " Buffer Management in Relational Database Systems", ACM Trans. on Database Syst., Vol. 11, No. 4, Dec. 1986, pp. 474-498.

[Seli79] P.G. Selinger, M.M. Astrahan, D.D. Chamberlin, R.A. Lorie, and T.G. Price, "Access Path Selection in a Relational Database Management System", Proc. of ACM SIGMOD 20, 1979, pp. 23-34.

[Ston81] M. Stonebraker, " Operating System Support for Database Management", Comm. of the ACM, Vol. 24, No. 7, July 1981, pp. 412-418.

[Teng84] T.Z. Teng and R.A. Gumaer, "Managing IBM Database 2 Buffers To Maximize Performance", IBM Systems Journal, Vol. 23, No. 2, 1984, pp. 211-218.

[Wolf88] J.L. Wolf, D.M. Dias, B.R. Iyer, and P.S. Yu, "A Hybrid Data Sharing-Data Partitioning Architecture for Transaction Processing", Proc. 4th Intl. Conference on Data Engineering, Los Angeles, CA, Feb. 1988, pp. 520-527.

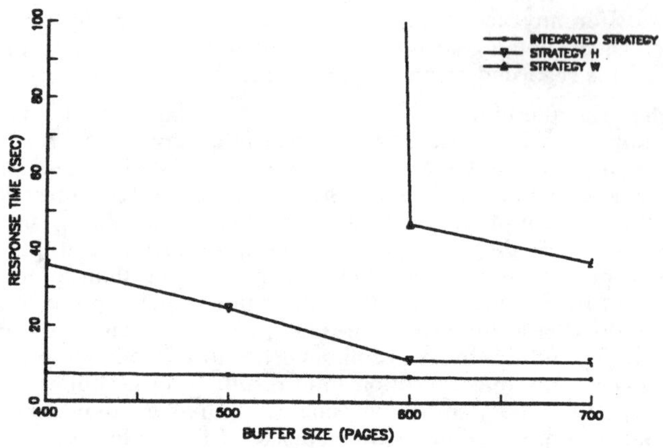

Fig. 1 Transaction response time vs buffer Size

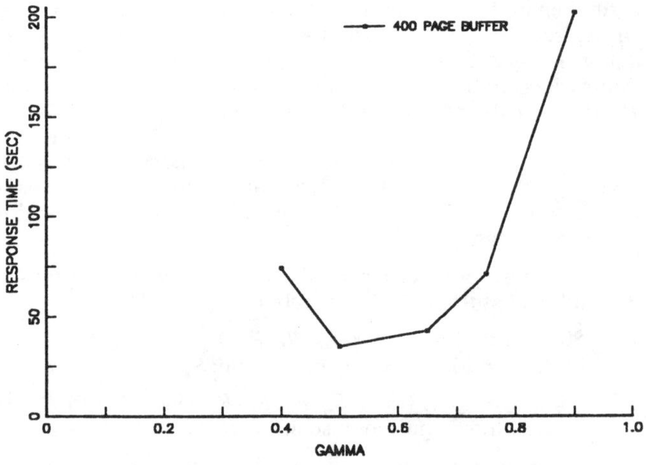

Fig. 2 Sensitivity of strategy H to γ (400 page buffer)

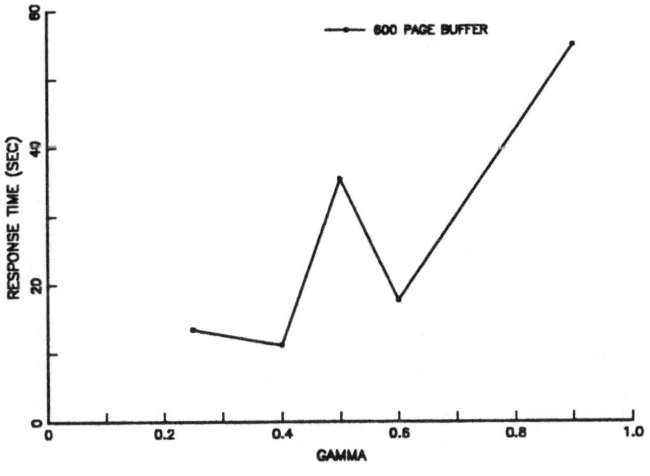

Fig. 3 Sensitivity of strategy II to γ (600 page buffer)

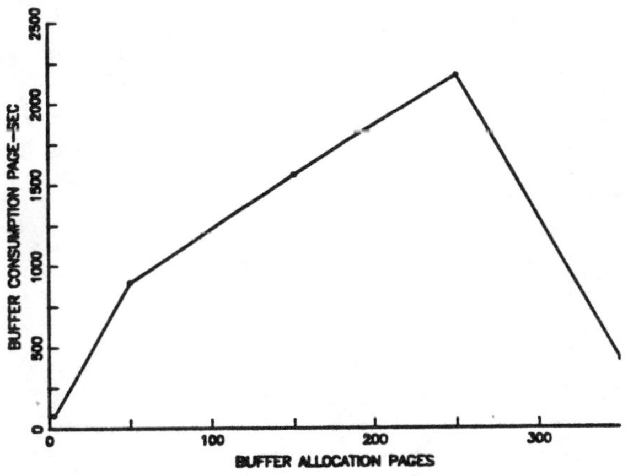

Fig. 5 Buffer consumption of transaction 1

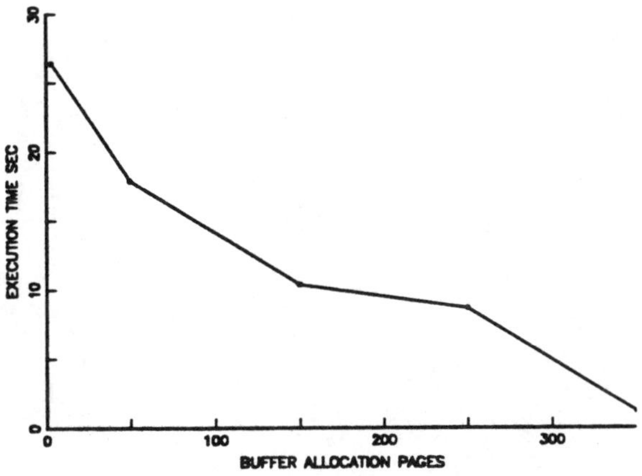

Fig. 4 Execution time of transaction 1

The Effect of Bucket Size Tuning in the Dynamic Hybrid GRACE Hash Join Method

Masaru KITSUREGAWA† **Masaya NAKAYAMA**‡ **Mikio TAKAGI**†

† Institute of Industrial Science, The University of Tokyo
7-22-1 Roppongi, Minato-ku, Tokyo 106, Japan

‡ Toyohashi University of Technology
1-1 Tempaku-cho, Toyohashi 440, Japan

Abstract

In this paper, we show detailed analysis and performance evaluation of the Dynamic Hybrid GRACE Hash Join Method (DHGH Method) when the tuple distribution in buckets is unbalanced.

The conventional Hash Join Methods specify the tuple distribution in buckets statically. However it may differ from estimation since join operations are applied with selection operations. When the tuple distribution in buckets is unbalanced, the processing cost of join operation becomes more costly than the ideal case when you use Hybrid Hash Join Method (HH Method). On the other hand, when you use the DHGH Method, the destaging buckets are selected dynamically, gives the same performance as the ideal case even if the tuple distribution in buckets is unbalanced such as Zipf-like distributions.

We analyze the total I/O cost of a join operation at various number of buckets. The result shows that we have to determine the number of buckets based on the tuple distribution in buckets rather than the size of the source relation. It is shown that we had better partition the source relation using a large number of small buckets instead of the smaller number of buckets almost filling the whole main memory adopted in the HH Method.

1 Introduction

The *join* is one of the most expensive relational algebra operations and many join methods have been proposed [Sto76,Kit83,Bra84,DeW84,Yam85]. Among them, '*split based hash-partitioned join methods*', such as GRACE Hash Join Method, present good performance for large scale database [Kit83,DeW84,Sha86]. In those join methods, the source relations are partitioned into number of "*buckets*" by a "*split function*" when the size of the smaller relation exceeds that of the available staging memory. The number of buckets is determined by the size of the source relation. In the Hybrid Hash Join Method (HH Method), which combines the good features of GRACE Hash Join Method and Simple Hash Join Method, the number of buckets is determined by the relative sizes of available staging memory and relation; it is computed so as to make the aggregated size of buckets are almost same to the size of staging memory, excepting for the first processing bucket called "R_1 *bucket*". So, if the split function does not suit for the tuple distributions in the source relations, some buckets exceed the memory size. Such buckets, named "*overflown buckets*" in this paper, diminish the performance of the HH Method and you cannot expect to achieve optimal performance [Nak88]. In order to minimize the overflown buckets, we divide the source relations into a large number of buckets. We name this approach as "*Dynamic Hybrid GRACE Hash Join Method*" (DHGH Method), because we dynamically select the first processed buckets and establish a scheduled processing sequence of buckets. In this join method, the unbalanced tuple distribution in buckets does not diminish the performance as shown in [Nak88].

The purpose of this paper is twofold. First, we evaluate the processing cost of the DHGH Method at various number of buckets to examine the minimal occurrence of overflown buckets and to identify the guideline for determining the number of buckets when the tuple distribution in buckets is unbalanced.

If we divide the source relations into a large number of buckets, the size of each bucket may become too small to fill up one page. We call such short-filled pages as "*fragment pages*". The I/O operations for them diminish the performance. Therefore, in our join algorithm, many of these short-filled buckets are put together to a larger bucket in accordance with the available main

Proceedings of the Fifteenth International Conference on Very Large Data Bases

Amsterdam, 1989

memory. This technique is called *"bucket size tuning"* as in [Kit83]. Second purpose of this paper is to evaluate this strategy. We show the detailed evaluation of this strategy and exhibit its effectiveness.

In section 2, we define the notations used in this paper and explain the environment of our performance evaluation.

In Section 3, we analyze the DHGH Method and show certain requirements for its algorithm. Roughly speaking, there are two conditions: the first condition is for eliminating overflown buckets, and the second one is for keeping the first processing bucket in the available staging memory.

Section 4 gives the performance results of three kinds of tuple distributions. Performance was evaluated by using various number of buckets. When we treat large sized relations, all results indicate that the elimination of overflown buckets is the key condition to attain good performance. They also show that large enough number of buckets is effective to minimize overflown bucket.

Section 5 shows the performance results for medium sized relations. In this section, we show the effect of bucket size tuning strategy. Bucket size tuning for the first processing buckets gives advantages when the number of buckets is small. On the other hand, bucket size tuning for other buckets is effective when the number of buckets is large because this tuning is also effective for I/O reduction of fragment pages.

Section 6 concludes this paper. The results of all evaluations show that we should divide relations into larger number of buckets than conventional suggestions. The conventional join method determines the number of buckets based on the relative size of the source relation and the available main memory, but we determine it based on the tuple distribution in buckets.

2 Overview of the Dynamic Hybrid GRACE Hash Join Method

Previous researches in hash-partitioned join methods did not concern themselves with the unbalanced tuple distribution in buckets and assumed optimal tuple distributions. Our proposed join method can dynamically handle most of distributions [1]. Our previous paper [Nak88] shows an evaluation for a triangular tuple distribution in buckets (Figure 1) as an example with good results. In this paper, we show a more detailed evaluation of the DHGH Method including the tuple distribution used in [Nak88]. In addition, Zipf-like distribution is selected for another example of unbalanced tuple distribution as is used in AS^{3}AP benchmark set [Tur87,Tur88].

[1] If there is a bucket which exceeds the available main memory size with same valued join attributes, we have to use nested loop manner for achieving join operation. This is an exceptional example for our algorithm.

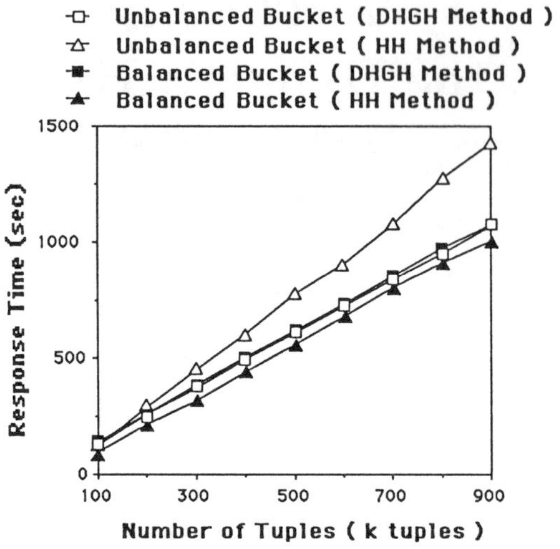

Figure 1: Comparison of processing time between our join method and conventional join method

2.1 The notations and definitions used in this paper

The two source relations of join operation are named R and S (R $\leq$ S). The result relation is named RES. When the size of relation R is smaller than that of the available main memory, Simple Hash Join Method [DeW84,DeW85] may be applied. While staging the whole relation R into the available main memory, we apply a **hash function** to each tuple and make a hash table. Then each tuple in relation S is staged and applied to the same hash function to achieve the join processing. In short, one scanning of each relation is enough to achieve join operation in this case.

On the other hand, when the size of relation R is larger than that of the available main memory, we have to divide it into sets of subrelations. We call these subrelations '**buckets**' and describe them as R_i ($1 \leq i \leq H_s$) in this paper. Here, H_s means the number of splitting buckets. The function to divide the source relations into buckets is called '**split function**'. When the bucket size is smaller than that of the available main memory, join operation for this bucket can be achieved by one scanning as described above. It means that join operation can be achieved by using two phase algorithms. The first phase is to divide the source relations into many buckets, and the second phase is to process the join operation for each bucket. The former phase is called '**split phase**' and the latter '**probe phase**'. Figure 2 shows the outline of join processing.

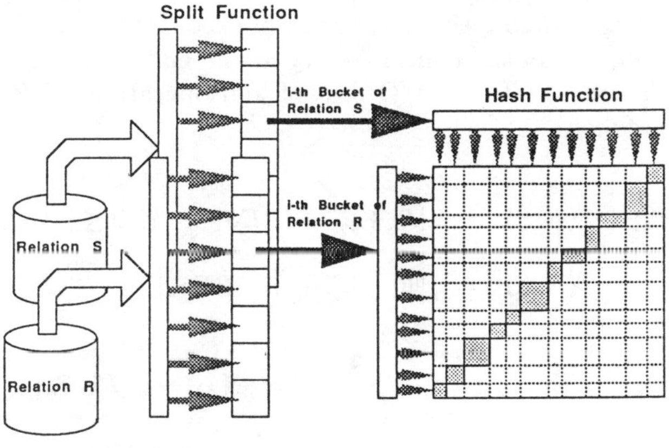

Figure 2: Two kinds of hash functions used by split based join methods

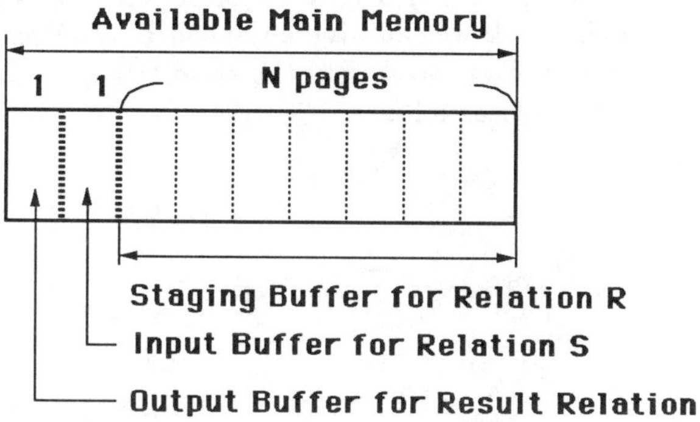

Figure 3: Memory usage for processing a join operation in our experiments

In the split phase, the CPU cost (the tuple splitting cost and the tuple movement cost) is negligible in comparison with I/O cost. So we treat buckets in main memory by page unit and tuples in the same bucket are placed closely together to save the I/O cost. As shown in Figure 3, main memory contains N pages for staging relations and 2 pages for input and output buffers.

When we partition relations into H_s buckets, $N - H_s$ pages are not used in the split phase if each phase is fully isolated (GRACE Hash Join Method). Instead of this method, the **"Hybrid Hash Join Method"** uses these free pages for staging R_1 bucket. This technique saves the I/O cost for R_1 bucket in join processing. In [DeW85,Sha86], it is shown that this method always gives the best performance among all of join methods when the following formula is satisfied.

$$H_s = \left\lceil \frac{|R| - |M|}{|M| - 1} \right\rceil + 1 \quad (1)$$

$$|R_1| = |M| - H_s + 1 \quad (2)$$

$$|R_i| = \frac{|R| - |R_1|}{H_s - 1} \quad (2 \le i \le H_s) \quad (3)$$

In the formula above, $|M|$ means the number of pages of available staging memory and $|R|$ means the number of pages of relation R. The notation $|M|$ is equivalent to N in our environment. In addition to this notation, we use $\{R\}$ to describe the number of tuples of relation R.

When the above formula is not satisfied for any reason, additional processing cost is required. If the size of R_1 bucket is small, the saving of the I/O cost is not significant. In such a case, other buckets get more tuples than the estimation and some buckets may exceed N pages (overflown buckets). These buckets must be recursively split into many **sub-buckets** and only then join operations can be performed. If the i-th bucket exceeds N pages and is partitioned into H'_s sub-buckets, we describe them as R_{ij} $(1 \le j \le H'_s)$. On the other hand, if the size of R_1 bucket exceeds the available staging memory, we have to re-split the R_1 bucket dynamically.

These additional overheads are necessary because of the statical decision of the R_1 bucket. In the DHGH Method, however, we dynamically determine the R_1 bucket and split the source relations into a large number of buckets to prevent overflown buckets. We discuss the method to determine the number of buckets in section 4.

When we divide the source relations into a large number of buckets, each bucket may have fewer tuples than $\{M\}$ (fragment pages). Such fragment pages can be put together into a bucket for further processing: i.e. a set of buckets, in effect, can be processed at a time. This strategy is called '**bucket size tuning**' and there are two kinds of size tuning. The size tuning for the buckets whose join processing are overlapped with split phase is called '**R_1 bucket size tuning**'. The other bucket size tunings is called '**R_i bucket size tunings**'. The advantages of size tuning are discussed in section 5.

2.2 Environment for performance evaluation

In this paper, we evaluate the performance for the Zipf-like distribution as is used in the AS^{3}AP benchmarks [Tur87,Tur88], This distribution is the generalization of the Zipf distribution [Knu73]. A power of the rank of an item is inversely proportional to its frequency:

$$i^z \times f_i = \frac{1}{constant^{(z)}} \quad (1 \le i \le n)$$

In this formula, z is the decay factor and $constant^{(z)}$ is the n-th harmonic number of order z. If $z = 0$, the distribution becomes uniform. When $z = 1$, it shows the Zipf distribution. And also, G. K. Zipf found the distribution of personal income when $z = 0.5$.

We assume that tuple distribution of each bucket follows this distribution. For instance, i-th bucket of relation R has such number of tuples as shown in the following formula:

$$\{R_i\} = \frac{\{R\}}{i^z \cdot \sum_{j=1}^{H_s} \frac{1}{j^z}} \qquad (1 \leq i \leq H_s) \qquad (4)$$

Though each bucket has an unbalanced distribution, we assume that each join attribute has a unique value in the relation in order to get the same total I/O cost for join operation for all distribution of buckets.

3 Performance analysis of the Dynamic Hybrid GRACE Hash Join Method

The algorithm of new join method was shown in [Nak88]. In this section, we show two conditions to ensure the good performance of the DHGH Method. As illustrated in the last section, the HH Method determines the number of buckets as small as possible. When the tuple distribution in buckets is unbalanced, performance becomes worse because of the overflown buckets. In the DHGH Method, we determine the number of bucket as large as we can for preventing overflown buckets; we divide the relation into number of buckets to satisfy the following formula:

$$\max |R_i| \leq |M| \qquad (5)$$

As for the size of R_1 bucket, it is determined by the formula (2) in the HH Method. In our method, however, the minimum sized bucket is dynamically selected for the R_1 bucket. So the condition is denoted as the following formula:

$$\min |R_i| \leq |M| - H_s + 1 \qquad (6)$$

Also, when we apply a join operation to small sized relations, we may collect a number of buckets to the first processing bucket. This schedule is called R_1 bucket size tuning. If you give the bucket identifier k in accordance with the size sequence, the previous condition (6) can by rewritten as follows. In this case, we assume that t buckets are selected as the first staging buckets.

$$\sum_{k=1}^{t} |R_k| \leq |M| - H_s + t \qquad (7)$$

In this paper, we analyze the I/O cost for join operation changing the number of buckets. If we can stage t buckets in the main memory at the end of split phase, we can denote the I/O cost of join operation, $C(R,S)$, as follows.

$$C(R,S) = |R| + |S| + |RES| \qquad (|R| \leq |M|) \qquad (8)$$

$$\begin{aligned} C(R,S) &= |R| + |S| + \sum_{k=1}^{t} |RES_k| \\ &\quad + \sum_{k=t+1}^{H_s} (|R_k| + |S_k| + C(R_k, S_k)) \\ &\qquad (|M| < |R|) \qquad (9) \end{aligned}$$

When the size of relation R is larger than that of available memory, total I/O cost of join operation is recursively defined. If we can divide the source relation into a number of buckets to satisfy the condition (5), $C(R,S)$ can be denoted as follows.

$$\begin{aligned} &C(R,S) \\ &= |R| + |S| + \sum_{k=1}^{t} |RES_k| \\ &\quad + \sum_{k=t+1}^{H_s} \{|R_k| + |S_k| \ + (|R_k| + |S_k| + |RES_k|)\} \\ &= |R| + |S| + |RES| \ + 2 \sum_{k=t+1}^{H_s} (|R_k| + |S_k|) \qquad (10) \end{aligned}$$

When all those buckets between $u+1$ and H_s are overflown buckets and thus further divided into $H_s^{(k)}$ sub-buckets satisfying the condition (5), total I/O cost becomes as follows.

$$\begin{aligned} &C(R,S) \\ &= |R| + |S| + \sum_{k=1}^{t} |RES_k| \ + \sum_{k=t+1}^{H_s} (|R_k| + |S_k|) \\ &\quad + \sum_{k=t+1}^{u} (|R_k| + |S_k| + |RES_k|) \\ &\quad + \sum_{k=u+1}^{H_s} \{|R_k| + |S_k| + |RES_k| \\ &\qquad +2 \sum_{l=t^{(k)}+1}^{H_s^{(k)}} (|R_{kl}| + |S_{kl}|)\} \\ &= |R| + |S| + |RES| \ + 2 \sum_{k=t+1}^{H_s} (|R_k| + |S_k|) \end{aligned}$$

$$+2\sum_{k=u+1}^{H_s}\sum_{l=t^{(k)}+1}^{H_s^{(k)}}(|R_{kl}|+|S_{kl}|) \qquad (11)$$

In this formula, we assume that $t^{(k)}$ sub-buckets can be collected into the first staging sub-bucket in split phase of k-th bucket. Both $t^{(k)}$ and $H_s^{(k)}$ are dependent on the tuple distribution in the k-th bucket and $t^{(k)}$ is also dependent on the size of k-th bucket. Here, we assume that the tuple distribution in sub-buckets is same distribution as in buckets and assume that $H_s^{(k)}$ ($u < k \leq H_s$) is equal to the H_s value.

In this remaining section, we analyze the I/O cost of join operations for the same two relations. As described before, each tuple has a unique value at join fields in the relation R and the amount of result relation becomes twice the source relation ($|RES| = 2|R|$): Formula (10) and (11) are rewritten as follows.

$$C(R,R) = 4|R| + 4\sum_{k=t+1}^{H_s}|R_k| \quad (formula\ 10) \qquad (12)$$

$$C(R,R) = 4|R| + 4\sum_{k=t+1}^{H_s}|R_k| + 4\sum_{k=u+1}^{H_s}\sum_{l=t^{(k)}+1}^{H_s}|R_{kl}| \quad (formula\ 11) \qquad (13)$$

Now, we can get information about extra I/O cost for overflown buckets. For detailed analysis, we have to fix the tuple distribution in buckets. We choose the triangular distributions for this analysis. The tuple distribution in buckets can be described as follows.

$$\{R_k\} = \frac{2\{R\}}{H_s(H_s+1)}\cdot k \qquad (1 \leq k \leq H_s)$$

Here T denote the minimum number of buckets to satisfy the condition (5). When t buckets can be staged into the staging memory at the same time, C_T can be described as follows.

$$C_T = 4|R| + 4\sum_{k=t+1}^{T}\frac{2|R|}{T(T+1)}\cdot k$$

When we use $T-1$ buckets in the split phase, the largest bucket (R_{T-1} bucket) becomes overflown bucket and the I/O cost of join operation is described as follows. (Appendix A shows the reason why we get such a formula.)

$$\begin{aligned} C_{T-1} &= 4|R| + 4\sum_{k=t+1}^{T-1}|R_k| + 4|R_{T-1,T-1}| \\ &= 4|R| + 4\sum_{k=t+1}^{T-1}\frac{2|R|}{T(T-1)}\cdot k + 4\frac{4|R|}{T^2} \end{aligned}$$

The extra I/O cost for processing the overflown bucket can be calculated as follows.

$$\begin{aligned} C_{extra} &= C_{T-1} - C_T \\ &= 8|R|\cdot\left(\frac{2}{T^2} - \frac{t(t+1)}{T(T-1)(T+1)}\right) \end{aligned}$$

This formula shows that an overflown bucket leads about $2/T^2$ of I/O cost disadvantages at most [2]. When we treat large relations, the occurrence of the overflown bucket is not so serious a problem because T is large. When we treat the small relations, however, we have to carefully determine the number of buckets to avoid the overflown bucket. The following two sections show these phenomena separately.

4 Performance results of three kinds of the bucket distributions

As shown in Figure 1, the HH Method does not suit for unbalanced tuple distribution in buckets. The processing cost for unbalanced buckets is about 1.4 times worse than that for balanced buckets. The reason for this phenomenon is not clearly discussed in [Nak88]. In this section, we evaluate the join performance of three kinds of tuple distribution in buckets as we change the number of buckets and the size of source relations.

We choose the triangular distribution, the Zipf-like distribution and the uniform distribution as the three kinds of tuple distribution in buckets. The former two distributions are used to evaluate the join performance for unbalanced buckets. The uniform distribution is used to get the base join performance.

In each distribution, we calculate the I/O cost during a join operation as we change the size of source relations and the number of buckets. We assume that the available staging memory has 100 pages and each page can contain 32 tuples in it. In this case, H_s can be changed between 2 and 99 ($= |M| - 1$). When the size of relation R is greater than that of the available staging memory, we need at least 2 buckets for achieving join operation. And the Dynamic Destaging Strategy, used in the DHGH Method, determines the upper bound of H_s. When we have no free pages in the staging memory at the split phase, destaging bucket is selected from the set of buckets which has more than one page. So the number of buckets must be less than $|M|$.

[2] For splitting the source relation into buckets, twice I/O operations (read the source relation and write to the temporal relation) are needed for each source relation. To probe join operation, another read and write operations are needed. Totally, 8 I/O operations are needed for basic join operation.

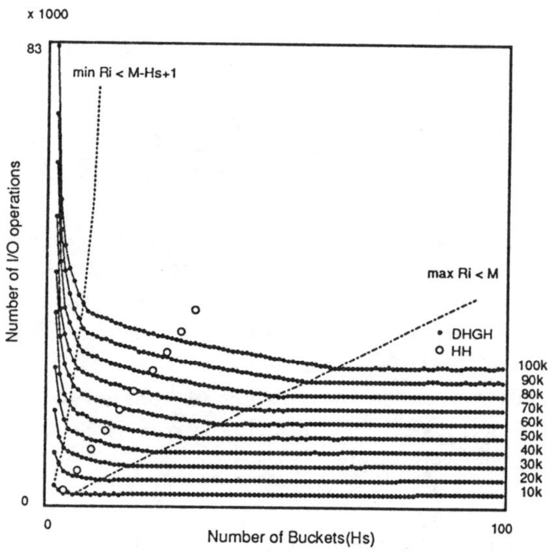

Figure 4: The performance results for triangular distribution

For all performance evaluations in this section, the number of tuples in source relations is changed every 10,000 tuples from 10,000 tuples to 100,000 tuples. As described above, at most 3,200 tuples can be staged in the available staging memory at a time. Most of the relations used for performance evaluation have more than five times larger tuples than the available staging memory contains. They are called "large sized relations" in [Nak88]. The performance results of "medium sized relations" are shown in the next section.

4.1 Performance results of the triangular distributions

First, we show the results of triangular distribution. This distribution is chosen in [Nak88] and we evaluate more detailed phenomena in this paper. Figure 4 shows the number of I/O operation for each number of bucket and for each size of the source relation.

In this figure, two different dotted lines show the boundary conditions of equation (5) and (7). The white points show the results of the HH Method and the black points show the results of the DHGH Method for each size of relation.

If the size of each bucket does not exceed the size of the available staging memory, H_s does not affect the performance that much. These phenomena are shown in the area whose H_s value is greater than the condition (5) line.

When some buckets exceed the available memory size, extra I/O cost is required to perform join operation for such overflown buckets. However, it is not so large on amount. These phenomena are shown in the area whose H_s value is between two condition lines.

When all the buckets exceed the size of available memory, the performance becomes extremely worse. These phenomena are shown in the area whose H_s value is smaller than the condition (7) line. In this situation, the maximum size of sub-bucket of $|R_{H_s}|$ bucket exceeds the size of $|R_1|$ bucket as follows.

$$\begin{aligned}
&\max|(\max|R_k|)_l| - \min|R_k| \\
&= |R_{H_s,H_s}| - |R_1| \\
&= \frac{2}{H_s+1}\cdot\frac{2|R|}{H_s+1} - \frac{2|R|}{H_s(H_s+1)} \\
&= 2|R|\cdot\frac{H_s-1}{H_s(H_s+1)^2} \quad > \quad 0
\end{aligned}$$

It means that sub-buckets whose size exceed the available staging memory need much more re-splitting operation to process the join operation. This is the reason why the performance of join operation becomes so much worse.

4.2 Performance results of the uniform distributions

In this subsection, we show the results of uniform distributions. Similar to the triangular distributions, we evaluate the I/O cost of join operation at various size of the source relation and the number of buckets. Figure 5 shows the results. Two different dotted lines, the white points and the black points have the same meaning as Figure 4.

Unlike the triangular distribution, the condition (5) is more severe than the condition (7). When all buckets satisfy both conditions, the I/O cost can be described by using formula (10) as follows:

$$\begin{aligned}
C(R,R) &= 4|R| + 4\sum_{k=t+1}^{H_s}|R_k| \\
&= 4|R| + 4\frac{|R|}{H_s}\cdot(H_s-t) \\
&= 8|R| - 4\frac{|R|}{H_s}\cdot t \qquad (14)
\end{aligned}$$

When T denotes the minimum number of the buckets which satisfies the condition (7), the I/O cost can be described as follows.

$$C_T = 8|R| - 4\frac{|R|}{T}$$

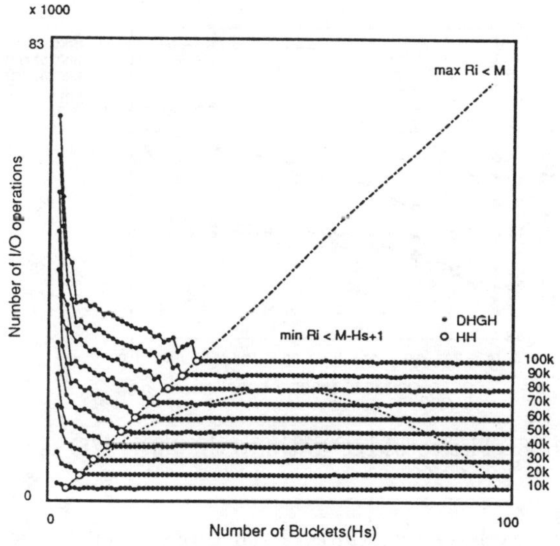

Figure 5: The performance results for uniform distribution

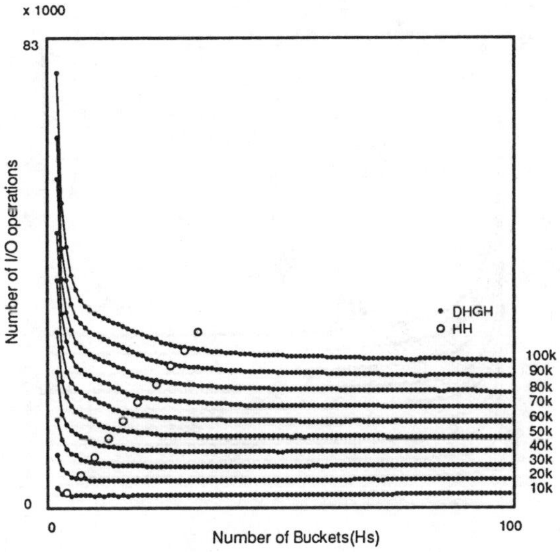

Figure 6: The performance results for Zipf-like distribution ($z = 0.5$)

When the condition (7) is not satisfied, the total I/O cost can be calculated as $t = 0$ in the formula (14).

$$C_{T-1} = 8|R|$$

The extra cost for the R_1 bucket overflown is described as follows. And the disadvantage becomes about $1/2T$.

$$C_{extra} = C_{T-1} - C_T = 8|R|\frac{1}{2T}$$

When the tuple distribution in buckets is balanced, each bucket has the same number of tuples. If the condition (5) is not satisfied, all destaging buckets become overflown buckets. The total I/O cost is described as follows.

$$\begin{aligned} C(R,R) &= 4|R| + 4\sum_{k=1}^{H_s}|R_k| + 4\sum_{k=1}^{H_s}\sum_{l=t+1}^{H_s}|R_{kl}| \\ &= 8|R| + 4\frac{|R|}{H_s}\cdot(H_s - t) \end{aligned}$$

In this formula, t is a constant value for all buckets. When T denotes the minimum number of the buckets to satisfy the condition (5), the I/O cost of join operation can be described as follows.

$$C_T = 8|R|$$

$$C_{T-1} = 8|R| + 4\frac{|R|}{T-1}\cdot(T-1-t)$$

And the extra I/O cost can be calculated as follows.

$$C_{extra} = C_{T-1} - C_T = 4|R|\cdot\frac{T-1-t}{T-1}$$

In this formula, both value of T and t are dependent on $|R|$. When $|R|$ is small, $(T-1-t)/(T-1)$ also becomes small and the extra I/O cost does not affect the performance. On the other hand, when $|R|$ is large, $(T-1-t)/(T-1)$ becomes large and the extra I/O cost will affect the performance.

4.3 Performance results of the Zipf-like distributions

To evaluate the performance for actual databases, we use the Zipf-like distributions in this subsection. The Zipf-like distribution has a parameter, the decay factor z. We evaluate the performance when z is 0.5 and 1.0. Figure 6 and Figure 7 show the results respectively. The white points show the results of the HH Method and the black points show the results of the DHGH Method.

We get similar results to those of the triangular distributions. However the results of Zipf-like distributions are not as good as that of the triangular distributions. Because Zipf-like distribution is more severe than the triangular distributions for the condition (5). In the triangular distributions, the largest bucket contains $2\{R\}/H_s$ tuples at most. However, in the Zipf-like distributions, it contains $\{R\}/(\sum_{j=1}^{H_s} 1/j^z)$ tuples. If we treat the Zipf dis-

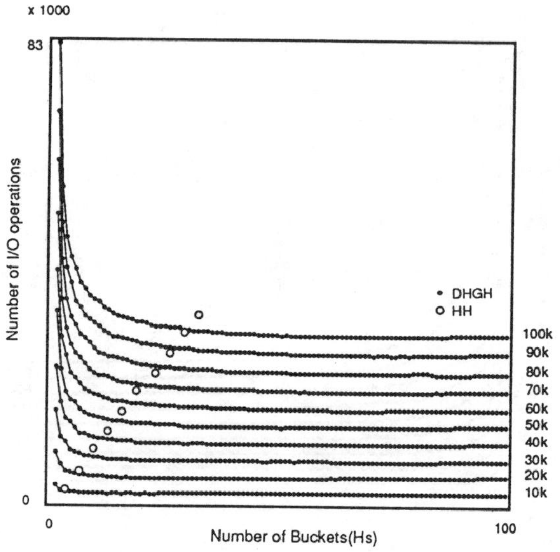

Figure 7: The performance results for Zipf-like distribution ($z = 1.0$)

tribution (z=1.0), sum of the inverse numbers becomes $\ln H_s$ and the largest bucket contains $\{R\}/\ln H_s$ tuples. It means that the source relation which contains more than 15k tuples produces overflown buckets even though we use dynamic destaging strategies. In this case, additional I/O cost for processing join operation is needed and the performance is degraded. This phenomenon is observed in the income distribution (z=0.5). If the relation contains more than 60k tuples, overflown bucket is occurred though the amount is little in such cases.

The following figure shows the total I/O cost ratio in each join method to the basic I/O cost for join operation ($8|R|$).

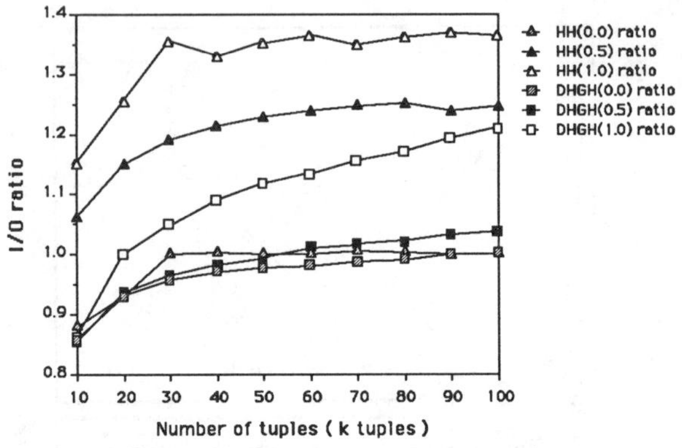

Figure 8: The total I/O ratio to the basic I/O operation

This figure clearly shows that we need additional I/O to apply join operation for Zipf-like distributions. It also shows that medium sized relation for all distribution does not have overflown buckets.

5 Discussion of bucket size tuning

We show the results for medium sized relations which has triangular distribution and discuss the effect of bucket size tuning in this section.

As described in [DeW84,DeW85,Sha86,Ger86], the advantage of R_1 bucket overlap processing is remarkable when the source relation is a small size as the available staging memory. In our environment, 3,200 tuples can be staged at a time. And we evaluate the performance of 5,000 tuples and 10,000 tuples relations. As illustrated before, we have two kinds of schedules about bucket size tuning, R_1 bucket size tuning and R_i bucket size tuning. To evaluate the detailed performance of bucket size tuning, we use following four kinds of algorithms.

- The algorithm 1 is applied neither to R_1 bucket size tuning and to R_i bucket size tuning.
- The algorithm 2 is applied only to R_1 bucket size tuning.
- The algorithm 3 is applied only to R_i bucket size tuning.
- The algorithm 4 is applied both to R_1 bucket size tuning and to R_i bucket size tuning.

Algorithm 1 includes the result of the HH Method. It determines the join processing sequence of buckets statically. The algorithm 2 only schedules a set of buckets in the R_1 bucket using the Dynamic Destaging Strategy. And the algorithm 3 schedules the join processing sequence of buckets by their sizes. A number of buckets are collected to fill up the available memory. This schedule can reduce the I/O operations for fragment pages. Lastly, the algorithm 4 denotes the DHGH Method itself. Figure 9 and Figure 10 show the result for 5,000 tuples relation and 10,000 tuples relation respectively.

These figures show that R_1 bucket size tuning is effective when the number of buckets is small. That is to say, this strategy is effective when the available staging areas for first processing bucket has a large number of pages. And the advantages of R_1 bucket size tuning decrease as the number of buckets increase. However, this overhead is much less than that of overflown buckets.

When the number of buckets is large enough to avoid the overflown buckets in this situation, each bucket has very small number of tuples in it. In this environment,

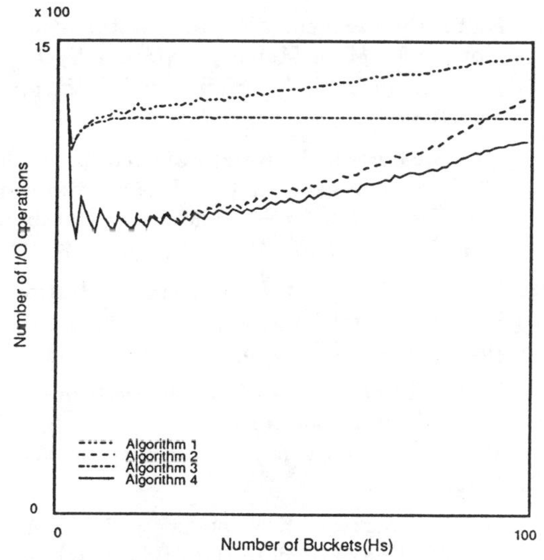

Figure 9: The I/O performance for 5000 tuples relation

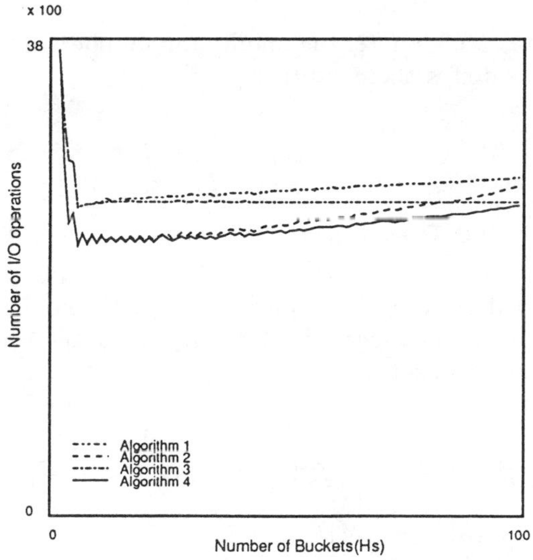

Figure 10: The I/O performance for 10000 tuples relation

the performance becomes worse because of the I/O operations for fragment pages. The result of algorithm 1 and algorithm 3 show these phenomena. So the R_i bucket size tuning is effective to reduce such overheads.

Difference between the result of 5,000 tuples relation and that of 10,000 tuples relation shows the fact that the effect of bucket size tuning is large when the source relation size is small.

6 Conclusions

Conventional split based hash-partitioned join methods determine the processing sequence of each bucket statically. They assume that the size of each bucket is almost same size. However, such cases are actually rare because join operations generally follow some selection/restriction operations and there is no guarantee that the split function is suitable for any kinds of tuple distribution in the source relation. Unbalanced tuple distribution in the buckets diminishs the performance in comparison with the estimation in the conventional join method. We extends the GRACE hash algorithms to obtain high performance than Hybrid Hash algorithms. We name it the Dynamic Hybrid GRACE Hash algorithms.

In this paper, we evaluate the join performance of three kinds of tuple distribution in buckets at various number of buckets and the size of source relations to investigate the phenomenon. The three distributions are; the uniform, the triangular and the Zipf-like distribution. Performance results show that we have to determine the number of buckets to minimize the overflown buckets. There are two conditions which describe the boundary conditions of overflown bucket:

$$\max |R_i| \leq |M|$$

$$\min |R_i| \leq |M| - H_s + 1$$

And the minimum number of buckets which satisfys such conditions depends on the distribution not the size of source relation. In our join method, if we cannot get information about tuple distribution in buckets before processing join operation, the number of buckets is chosen as the maximum number to minimize the overflown buckets.

Also, in this paper, we check the effect of bucket size tuning strategy using the medium sized relation. The R_1 bucket size tuning is efficient when the number of buckets is small and the amount of staging area for R_1 bucket can be taken largely. On the other hand, R_i bucket size tuning works together some fragment pages to reduce the I/O cost of them. So this tuning is efficient when the size of source relation is small and the number of buckets is large.

Appendix A How to lead the I/O cost formula with overflown bucket

When we partition the source relation R into H_s buckets and each bucket follows the triangular distributions, the tuple distribution in buckets becomes as follows.

$$\{R_k\} = \frac{2\{R\}}{H_s(H_s+1)} \cdot k \qquad (1 \le k \le H_s)$$

Here, we denote the minimum number of buckets which satisfys the condition (5) as T.

$$\begin{aligned} \max|R_k| &= \frac{2|R|}{T(T+1)} \cdot T \quad < \quad |M| \\ |M|T &> 2|R| - |M| \end{aligned} \qquad (15)$$

And if we use $T-1$ buckets to partition the source relation R, the largest bucket (R_{T-1} bucket) becomes overflown bucket in this case.

$$\begin{aligned} |R_{T-1}| &= \frac{2|R|}{T(T-1)} \cdot (T-1) \quad > \quad |M| \\ 2|R| &> |M|T \end{aligned} \qquad (16)$$

Using these conditions, we can lead the fact that the R_{T-2} bucket does not exceed the size of available staging memory.

$$\begin{aligned} & |M| - |R_{T-2}| \\ &= |M| - \frac{2|R|}{T(T-1)} \cdot (T-2) \\ &= \frac{1}{T(T-1)} [|M|T(T-1) - 2|R|(T-2)] \\ &> \frac{1}{T(T-1)} [(2|R| - |M|)(T-1) \; - 2|R|(T-2)] \\ & \qquad \qquad (\mathit{Condition}\ (15)) \\ &= \frac{1}{T(T-1)} [2|R| - |M|(T-1)] \\ &> \frac{1}{T(T-1)} [|M|T - |M|(T-1)] \\ & \qquad \qquad (\mathit{Condition}\ (16)) \\ &= \frac{|M|}{T(T-1)} \quad > \quad 0 \end{aligned} \qquad (17)$$

For all situations when T is larger than 2, only (T–1)-th bucket exceeds the size of available staging memory. It means that both u and $t^{(T-1)}$ in the formula (13) become $T-2$. And the I/O cost with the overflown bucket can be described as follows.

$$C(R,R) = 4|R| + 4 \sum_{k=t+1}^{H_s} |R_k| \; + 4 \sum_{k=u+1}^{H_s} \sum_{l=t^{(k)}+1}^{H_s} |R_{kl}|$$

$$= 4|R| + 4 \sum_{k=t+1}^{T-1} |R_k| \; + 4|R_{T-1,T-1}|$$

References

[Bra84] K. Bratbergsengen. "Hashing Methods and Techniques for Main Memory Database Systems". In *Proc. of the 10th Int. Conf. on VLDB*, pp. 323–333, 1984.

[DeW84] D. J. DeWitt, R. H. Katz, F. Olken, L. D. Shapiro, M. R. Stonebraker, and D. Wood. "Implementation Techniques for Main Memory Database Systems". In *ACM SIGMOD '84*, pp. 1–8, 1984.

[DeW85] D. J. DeWitt and R. Gerber. "Multiprocessor Hash-Based Join Algorithms". In *Proc. of the 11th Int. Conf. on VLDB*, pp. 151–164, 1985.

[Ger86] R. H. Gerber. "Dataflow Query Processing Using Multiprocessor Hash-partitioned Algorithms". Technical Report 672, University of Wisconsin, 1986.

[Kit83] M. Kitsuregawa, H. Tanaka, and T. Moto-oka. "Application of Hash to Data Base Machine and Its Architecture". *New Generation Computing*, Vol. 1, No. 1, pp. 66–74, 1983.

[Knu73] D. E. Knuth. *"The Art of Computer Programming"*, volume 3. Addison-Wesley Pub. Co., 1973.

[Nak88] M. Nakayama, M. Kitsuregawa, and M. Takagi. "Hash-Partitioned Join Method Using Dynamic Destaging Strategy". In *Proc. of the 14th Int. Conf. on VLDB*, pp. 468–478, 1988.

[Sha86] L. D. Shapiro. "Join Processing in Database Systems with Large Main Memories". *ACM Transactions on Database Systems*, Vol. 11, No. 3, pp. 239–264, 1986.

[Sto76] M. Stonebraker, E. Wong, P. Kreps, and G. Held. "The Design and Implementation of INGRES". *ACM Transactions on Database Systems*, Vol. 1, No. 3, pp. 189–222, 1976.

[Tur87] C. Turbyfill. *"Comparative Benchmarking of Relational Database Systems"*. PhD thesis, Cornell University, September 1987.

[Tur88] C. Turbyfill, C. Orji, and D. Bitton. "AS^{3}AP – An Ansi Sequel Standard Scalable and Portable, Benchmark for Relational Database Systems". University of Illinois Technical Report, December 1988.

[Yam85] Y. Yamane. "A Hash Join Technique for Relational Database Systems". In *Proc. on Foundations of Data Organization*, pp. 388–398, 1985.

PANEL SESSION
Building knowledge-based applications with cooperating databases

Panel chairmen:
Rakesh Agrawal
AT&T Bell Laboratories

Gerald A. Wilson
Quantum InKNOWvations Corp.

Both practical pilgrimages into issues involving the integration of AI and DBMS technologies has attracted attention in recent years. The promised "pot of gold" has always been that the combined technologies were "more" than the sum of the individual technologies. Database technology could bring performance perspective to AI; AI could bring enriching concepts to database systems. The result would be a cross-fertilization that would lead directly to a new class of functionally rich, yet computationally efficient, information systems.

Have practical, interesting, innovative, and/or useful results come from the attempts to marry or cross-fertilize these fields? Are the pragmatic results from the field encouraging or discouraging? What are the lessons learned?

We bring together in this panel a world-wide variety of expertise and experience to discuss this topic for the 90's. Panelists will describe their own experiences with both the development and implementation of end-user oriented applications in a variety of environments and with a variety of tools. They will describe both the rapture of happy outcomes and the agony of less than successfull attempts. The panel and the audience in concert will provide prescriptions for the future.

Random Sampling from B^+ trees *

Frank Olken and Doron Rotem
Computer Science Research Dept.
Information and Computing Sciences Div.
Lawrence Berkeley Laboratory
1 Cyclotron Road, Berkeley, CA 94720

Abstract

We consider the design and analysis of algorithms to retrieve simple random samples from databases. Specifically, we examine simple random sampling from B^+ tree files. Existing methods of sampling from B^+ trees, require the use of auxiliary rank information in the nodes of the tree. Such modified B^+ tree files are called "ranked B^+ trees". We compare sampling from ranked B^+ tree files, with new acceptance/rejection (A/R) sampling methods which sample directly from standard B^+ trees. Our new A/R sampling algorithm can easily be retrofit to existing DBMSs, and does not require the overhead of maintaining rank information. We consider both iterative and batch sampling methods.

1 Introduction

Virtually all database systems used to record financial transactions (accounting systems, inventory control systems, bank records, etc.) are subject to annual audit, usually involving random sampling of the records for corroboration. Yet commercial database management systems do not support queries to retrieve a random sample of some portion of the database. One reason is that previous proposals to support retrieval of random samples from databases have required the modification of standard access methods, and the maintenance of additional information in the indices. In particular, earlier authors have described sampling from modified B^+ trees, *called ranked B^+ trees*, which incorporate information which permits the computation of the rank of a record.

In this paper we discuss methods of sampling from regular B^+ trees, employing acceptance/rejection (A/R) methods. These new algorithms do not require any modification of the standard B^+ tree structures, nor do they require maintenance of any additional fields in the B^+ trees. Hence these new sampling methods can be more easily retrofit to existing DBMSs. While not quite as efficient as earlier proposals, it should be suitable for applications which only need sampling infrequently, e.g. for auditing. For B^+ tree files we discuss both iterative and batch sampling methods.

1.1 Why sample?

Random sampling is used on those occasions when processing the entire dataset is not necessary and is considered too expensive in terms of response time or resource usage. The savings generated by sampling may arise either from reductions in the cost in retrieving the data from the DBMS or from subsequent "post-processing" of the sample.

Retrieval costs are significant when dealing with large administrative or scientific databases. Post-processing of the sample may involve expensive statistical computations, or further physical examination of the real world entities described by the sample. Examples of the latter include physical inspection and/or testing of components for quality control [Mon85,LWW84], physical audits of financial records [Ark84,LTA79], and medical examinations of sampled patients for epidemiological studies. As noted above, most financial databases are subject to annual audits, which typically entail random sampling of records from the database for corroboration.

Sampling is useful for applications which are attempt-

*This is a condensed version of tech report LBL–25517. This work was supported by the Director, Office of Energy Research, Office of Basic Energy Sciences, Applied Mathematical Sciences Division of the U.S. Department of Energy under Contract DE-AC03-76SF00098. Authors electronic mail addresses: olken@csam.lbl.gov, olken@lbl.bitnet, rotem@csam.lbl.gov

ing to estimate some aggregate property of a set of records, such as the total number of records which satisfy some predicate. Thus random sampling is typ y used to support statistical analysis of a dataset, er to estimate parameters of interest [HOT88] or for hypothesis testing. See [Coc77] for a classic treatment of the statistical methodology. Applications include scientific investigations such as high energy particle physics experiments, quality control, and policy analyses. For example, one might sample a join of welfare recipient records with tax returns or social security records in order to estimate welfare fraud rates.

1.2 Simple Random Sampling

There are a variety of types of sampling which may be performed. In this paper we shall be concerned with *fixed size* samples, where the sample size has been specified by the user.

Another way of characterizing the sample is whether it is drawn with or without replacement. Let us assume that we are sampling from a distinct set of records. Samples drawn with replacement may contain duplicates, and are usually easier to obtain (as we shall see). However, samples drawn without replacement, which proscribe duplicates, generally provide more information for a given sample size.

If the inclusion probabilities for individual records are uniform we say that we have a *simple random sample (SRS)*.

In this paper we will deal with fixed size simple random samples. One can readily convert between simple random samples with and without replacement (denoted SRSWR and SRSWOR respectively). To convert from SRSWR to SRSWOR we merely remove duplicates (in time linear in the sample size via a hash table) (possibly we must increase the sample size to compensate for duplicate removals). To convert from SRSWOR to SRS we must generate synthetic duplicates, this also can be done in time linear with the sample size. If the sampling fraction (ratio of sample size to population size) is (as is typical) small, then there will be few duplicates in the SRSWR and the extra samples needed to compensate for duplicate removal will be insignificant. We shall ignore the costs incurred due to duplicate removal and replacement in this paper.

1.3 Notation and Efficiency Metric

In this paper we shall measure the efficiency of the sampling algorithm in terms of the number of disk blocks read. We have typically assumed that the desired sample size is much smaller than the number of records in the file, (hence the likelihood of duplicate records in the sample is small) so that we have ignored the difference between sampling with and without replacement. A table of frequently used notation follows.

2 Basic Techniques

2.1 Acceptance/Rejection Sampling

A basic tactic used in this paper is acceptance/rejection sampling. It can be used to construct *weighted samples* in which the inclusion probabilities of a record are proportional to some arbitrary weight (calculated from the record attributes). In this paper we will use it to compensate for algorithm or data structure induced variations in sample inclusion probabilities so as to finally obtain a *simple random sample*, i.e., one with uniform inclusion probabilities. A brief explanation of this classic sampling technique is included here for those in the database community who may be unfamiliar with it.

Suppose that we wish to draw a weighted random sample of size 1 from a file of N records, denoted r_j, with inclusion probability for record r_j proportional to the weight w_j. The maximum of the w_j is denoted w_{max}.

We can do this by generating a uniformly distributed random integer, j, between 1 and N, and then accepting the sampled record r_j with probability p_j:

$$p_j = \frac{w_j}{w_{max}} \tag{1}$$

The acceptance test is performed by generating another uniform random variate, u_j, between 0 and 1 and accepting r_j if $u_j < p_j$. If r_j is rejected, we repeat the process until some j is accepted.

The reason for dividing w_j by w_{max} is to assure that we have a proper probability (i.e., $p_j \leq 1$). If we do not know w_{max} we can use instead a bound Ω such that $\forall j, \Omega > w_j$,. The number of iterations required to accept a record r_j is geometrically distributed with a mean of $(E[p_j])^{-1}$. Hence using Ω in lieu of w_{max} results in a less efficient algorithm.

Acceptance/rejection sampling is well suited to sampling with *ad hoc* weights or when the weights are being frequently updated (since it does not require auxiliary indices).

2.2 A Review of Sampling from Files

Over the last 20 years there has been considerable work done on developing basic techniques for sampling from a single flat file (usually with fixed blocking). We employ some of these techniques in our work on query sampling. In Table 3 we list the major results, with citations to the relevant algorithms.

α_k	acceptance probability of record k
α	$= E(\alpha_k) =$ expected acceptance probability of record
β	$= (f_{avg}/f_{max}) =$ average acceptance prob. at of a node
b_{avg}	average blocking factor for variably blocked file
b_{max}	maximum blocking factor for variably blocked file
$C_{method}(s)$	cost of retrieving sample of size s, via specified method
f_i	fan-out from internal node i of B^+ tree, or number of records in leaf node i of B^+ tree
f_{avg}	average fan-out from internal node of B^+ tree, or average number of records in leaf node of B^+ tree
f_{max}	maximum fan-out from internal node of B^+ tree, or maximum number of records in leaf node of B^+ tree
h	height of B^+ tree(count root as height 1)
h'	height of ranked B^+ tree (usually same as h)
$i = 0$	denotes root node of B^+ tree
n	number of records in file
p_k	probability of inclusion of record k
$path_k$	path (node identifiers) from root to leaf containing record k
π	expected length of path in early abort algorithm
s	number of records desired in sample
s'	inflated sample size (to compensate for acceptance/rejection)
$thisnode$	pointer to a B^+ tree node (internal or leaf)
$Y(k, m)$	Cardenas's function for expected number of blocks referenced when retrieving k records from m block file
w_k	probability of sampling of record k on a simple random walk from root to leaf of B^+ tree

Table 1: Notation used in the paper.

ARHASH NI	A/R algorithm for hashed files Naive Iterative algorithm for B^+ tree files
EAI	Early Abort Iterative algorithm for B^+ tree files
RI	Iterative algorithm for ranked B^+ tree files
NB	Naive Batch algorithm for B^+ tree files
EAB	Early Abort Batch algorithm for B^+ tree files
RB	Batch algorithm for ranked B^+ tree files

Table 2: Algorithm abbreviations

Type of sampling	Citation	Expected Disk Accesses
SRSWR	[Yao77]	$O(s)$
SRSWR, variable blocking	this paper	$O(s(b_{max}/b_{avg}))$
SRSWOR	[EN82]	$O(s)$
Weighted RS	[WE80]	$O(s \log n)$
Sequential RS, known pop. size	[FMR62]	$O(n/b_{avg})$
	[Vit84]	$O(s)$
Sequential RS, unknown pop. size		$O(n/b_{avg})$
	[Vit85]	$O(s(1 + \log(n/s))$

Table 3: Basic Sampling Techniques from a single file. Assume each sample taken from a distinct disk page, i.e., $s \ll (n/b_{avg})$ For Vitter's algorithms assume random disk I/O.

3 Iterative Sampling from a B^+tree

3.1 A/R Sampling from a B^+ tree

In this section we discuss how to employ acceptance/rejection sampling to sample from a B^+ tree without requiring the storage of any additional information in the B^+ tree nodes. Although our new method has a higher retrieval cost than earlier methods based on ranked B^+ trees, it does not require any modification of existing access methods, nor any additional update costs. Hence this method will be preferred over earlier methods for applications where updates dominate sampling retrievals.

For expository reasons we commence with a discussion of the *naive method* (a random walk from root to leaf, followed by an acceptance/rejection test). Subsequently, we show that a modification of this method, known as *early abort*, dominates the *naive method.* In Section 4 we consider *batch* versions of each algorithm, which in turn dominate the original *iterative* algorithms discussed in this section.

We shall assume that the buffer pool is sufficiently large to cache one entire path (from root to leaf) of the B^+ tree. For simplicity of analysis, we neglect the minor effect of caching beyond the root page (for these iterative algorithms). This will not alter the relative performance of the various iterative algorithms.

3.1.1 The Problem

As with other file structures the problem is to produce uniform inclusion probabilities for the target data records. Simply choosing a random edge from each internal node will not suffice, because nodes reached from internal nodes with low fanout will be more likely to be sampled than those reached from nodes with high fanout.

3.1.2 Naive method

Basically, the naive method consists of performing acceptance/rejection sampling on complete random paths through the tree (from root to leaf). The acceptance/rejection sampling is used to correct the inclusion probability of each sampled path, so that every record (stored in the leaves of the B^+ tree) has the same inclusion probability. We dicuss this method primarily for expository reasons, since (as we shall show) it is dominated by the *early abort method* (described in Sect. 3.1.3).

In this method we select a random path from the root to a record in a leaf (i.e., at each internal node we choose a branch at random (equi-probably), at the leaf we select a record at random (equi-probably)). Upon reaching the leaf we perform an acceptance/rejection test to decide whether to keep this path. The acceptance probability is calculated as we traverse the path from root to leaf as the product of the ratios of actual fan-out to maximum fan-out at each node (except the root). We are in effect sampling from a full multi-way tree, discarding paths which do not actually exist.

We denote by B a B^+ tree of order m with height h (there exist h nodes on any path from root to leaf, including the root and leaf). Let f_i denote the fan-out of node i. This is the number of branches from an internal node, and the number of records in a leaf node. We designate the root node to be node zero.

Lemma 1 *The naive algorithm generates a simple random sample. The inclusion probability p_k for record r_k contained in leaf j for a single (root-to-leaf) path traversal is:*

$$p_k = {f_0}^{-1} {f_{max}}^{-h+1} \tag{2}$$

where $f_{max} = 2m+1$ is the maximum fanout of an internal node (and for simplicity also the maximum number of records on a leaf node).

Proof: Let $w_k = p(\text{sampling record } k \text{ on a random walk})$. Then

$$w_k = \prod_{i \in path_k} f_i^{-1} \tag{3}$$

where $path_k$ refers to the path from root to leaf node containing the record k. Let the acceptance probability for record k be α_k, defined:

$$\alpha_k = \prod_{i \in path_k, i \neq 0} (f_i / f_{max}) \tag{4}$$

Note that the product here excludes the root node, because it is common to all paths, hence it does not introduce any non-uniformity into the inclusion probabilities. Recall that all paths in the tree have the same height, h. Hence:

$$p_k = \alpha_k w_k = f_0^{-1} \prod_{i \in path_k, i \neq 0} (1/f_{max}) \tag{5}$$

$$p_k = {f_0}^{-1} {f_{max}}^{-h+1} \tag{6}$$

□

Theorem 1 *The expected cost of the naive method for a simple random sample with replacement of size s from a B^+ tree is approximately:*

$$E[C_{NI}(s)] \approx (\beta^{-1})^{h-1} s(h-1) + 1 \tag{7}$$

Proof: Caching is generally only effective for the root page, hence the $(h-1)$ factor, rather than h. The last term simply accounts for initially reading the root page. Hence:

$$E[C_{NI}(s)] \approx s'(h-1)+1 \tag{8}$$

where s' is the gross sample size necessary to produce a net sample of size s after acceptance/rejection sampling. By assuming that all the fan-outs are equal to f_{avg} we have:

$$E[s] \approx s'\beta^{h-1} \tag{9}$$

Inverting this equation we shall assume (ignoring stochastic variation) that the required gross sample size, s' is:

$$s' = (\beta^{-1})^{h-1}s \tag{10}$$

Substituting yields the theorem. □

3.1.3 Early abort method

The *early abort method* of sampling from a B^+ tree, derives from the *naive* method. Both are based on acceptance/rejection sampling of random paths in the tree. The difference is that the naive method traverses complete paths from root to leaf before deciding on acceptance or rejection while the *early abort method* performs an acceptance/rejection test at each node (except the root).

If the node is rejected, then we can abort searching this path, permitting early abortions of a path. We can in effect reject a leaf node while part way down the path to it, without requiring that we retrieve the entire path. Hence, this method clearly dominates the naive method in expectation.

One way of thinking about the algorithm is to imagine that we are sampling random paths from the full multiway tree. As soon as we go down a branch which is not present in the actual (partially full) tree, we abort that path.

At each node (except the root) along a path from root to leaf we perform an acceptance/rejection test with acceptance probability for node i denoted as β_i:

$$\beta_i = f_i/f_{max} \tag{11}$$

Recall

$$\beta = f_{avg}/f_{max} \tag{12}$$

The root node is accepted unconditionally, i.e., $\beta_0 = 1$.

Lemma 2 *For a single (root-to-leaf) path traversal, the early abort algorithm generates a simple random sample with inclusion probability p_k for record r_k, where:*

$$p_k = {f_0}^{-1}{f_{max}}^{-h+1} \tag{13}$$

Proof: Observe that the probability, b_k, of accepting a record k (in a leaf node) is simply the product of the β_i along the path:

$$b_k = \prod_{i \in path_k, i \neq 0} \beta_i = \alpha_k \tag{14}$$

i.e., the same as for the *naive* algorithm. Hence our result follows from the proof of the naive algorithm. □

In effect, the early abort algorithm searches the same paths as the naive algorithm, but it aborts the search at the first rejection. Thus for some paths, it accesses fewer pages. Hence its expected cost will be strictly less than that of the naive algorithm, unless all of the pages are full.

Theorem 2 *The expected cost of the early abort iterative method for a simple random sample with replacement of size s from a B^+ tree is approximately:*

$$E[C_{EAI}(s)] \approx (\beta^{-1})^{h-1}s\frac{(\beta^{h-1}-1)}{\beta-1}+1 \tag{15}$$

Proof: Here we have again assumed that all the fan-outs are equal to f_{avg}. As before, the gross sample size must be increased by a factor of $(\beta^{-1})^{h-1}$ to account for the losses due to A/R sampling.

Recall that for the naive algorithm the length of each path examined is h (with a one-path cache it will be approx. $h-1$). What is the expected length of a path searched in the early abort algorithm? Let π denote the average path length (assuming the root is cached). Then we ignore the root (because it is cached). Acceptance/rejection sampling at each node could cause an early abort - but this happens after the node has been read in - so we subtract one more from the exponent on the expected acceptance probability $\beta = (f_{avg}/f_{max})$ of a node. This gives:

$$\pi = \sum_{i=1}^{i=h-1} \beta^{i-1} \tag{16}$$

Summing we have:

$$\pi = \frac{(\beta^{h-1}-1)}{\beta-1} \tag{17}$$

The total cost is given by:

$$E[C_{EAI}(s)] \approx s'\pi + 1 \tag{18}$$

where the last term consists of reading the root and where again we have

$$s' = s(\beta^{-1})^{h-1} \tag{19}$$

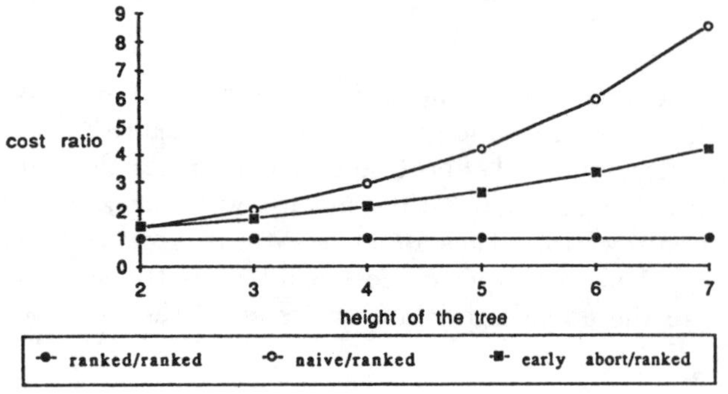

Figure 1: Iterative Algorithm Performance Graph

is the adjusted gross sample size to compensate for the attrition due to acceptance/rejection sampling. Substituting yields the theorem. □

Observe that from our definition of π that it must be less than $h-1$, hence we have:

Theorem 3 *The expected cost of the early abort iterative method is strictly less that the expected cost of the naive iterative method.*

Proof: Both methods must explore the same expected number of paths, as noted above the expected path length for early abort is less than that for the naive algorithm. □

We summarize the performance of the iterative algorithms in Figure 1.

3.2 Sampling from a ranked B^+ tree

In this section we discuss how to extract a fixed size simple random sample from a *ranked* B^+ tree file. A *ranked* B^+ *tree* file is one whose nodes have have been augmented with information which permits one to find the $j'th$ record in the file. We include this method as a benchmark against which to compare our new algorithms.

Suppose that we wish to sample from a file containing n records. We generate a uniformly distributed random number, j, between 1 and n, and then sample the $j'th$ record. To do this we must be able to identify the $j'th$ record. If our access method to the file is a B^+ tree, then we must be able to find the $j'th$ ranked record in the file. Hence we must store information in the tree which allows one to calculate the rank of each record.

This idea is discussed in [Knu73]. Similar ideas are used in [BK75] and [WE80]. Essentially we store in each node of the tree a count (partial sum) of the number of leaves in that subtree. In a binary tree [Knu73] the rank of each leaf can be calculated by suitable sums and differences of the count fields of all the nodes on the path from root to leaf. In a B^+ tree we promote the count fields one level in the tree so that each node stores not only the total count of leaves in its subtree, but also the counts for each child (alongside its key). Hence, while a rank access to the tree must still examine on average half the entries in each B^+ tree page, the number of disk pages which must be fetched is only equal to the height of the tree.

This matter is discussed in [SL88], and (for tries) in [Gho86].

For this algorithm, which we call the ranked iterative algorithm (denoted RI), we simply generate a random number, j, between 1 and n (the total number of records in the file) and then access the B^+ tree via the rank fields to retrieve the $j'th$ record. If there is no caching of disk pages, then each random probe retrieves a complete path from root to leaf, consisting of h' pages where h' is the height of the ranked B^+ tree.

Theorem 4 *The expected number of disk pages accessed for a simple random sample with replacement of size s from a ranked B^+ tree via the ranked iterative algorithm RI is:*

$$E(C_{RI}(s)) \approx s(h'-1)+1 \tag{20}$$

where h' is the height of the ranked B^+ tree, i.e., $h' = h/(1+log_{f_{avg}}(2/3))$, assuming that storing the rank field reduces the fan-out of internal nodes by $1/3$.

Proof: Proof omitted. See full paper. □

In practice, the ranked B^+ tree will have a height no more than one greater, than the corresponding B^+ tree.

As discussed earlier, duplicate removal can be performed in $O(s)$ memory. By checking online for duplicate random numbers, before fetching each record, we can obtain a SRS without replacement in the same number of disk accesses.

4 Batch Sampling from B^+*–trees*

In this section we consider batch methods of sampling from B^+ trees. Such methods are intended to reduce or eliminate the re-reading of disk blocks incurred by iterative sampling algorithms assuming that the buffer pool can hold one entire path through the B^+ tree. They process the entire sample as one batch, proceeding from one end of the file to the other. Batch sampling from ranked B^+ trees completely eliminates re-reading disk blocks. Batch sampling of regular B^+ trees via acceptance/rejection methods returns a sample of random (i.e., binomial) size. Hence, it may occassionally be necessary to repeat the process to obtain a sufficiently large sample.

4.1 Standard B^+ trees

4.1.1 Naive Batch

The naive batch sampling algorithm for standard B^+ trees simply applies the naive iterative algorithm in parallel. Instead of sampling paths through the tree one at a time, we process a batch of paths at once. The advantage is that we reduce the number of times pages must be rereferenced.

Suppose that we have an estimate s' of the required gross sample size needed to produce a net sample of size s. We start at the root with a gross sample size of s' and proceed with a depth first search of the tree.

At each internal node of the tree we allocate the incoming portion of the sample to the various subtrees by generating a multinomial random vector. A multinomial random vector from a population of size s in k cells is a vector $V = (v_1, v_2, ..., v_k)$ of length k, which records the number of balls v_i which fall in cell i, when s balls are thrown in the the cells at random (equiprobably). (More generally the probability of balls falling into each cell could vary, but we do not need this.) The uniform multinomial vector can be generated in two ways: by generating a random branch for each of s balls and incrementing the corresponding cell count, or alternatively by generating a Poisson random variable for each cell, and then adjusting the resulting variable with the first method, so that their sum is correct.

Only those branches with nonzero sample sizes allocated to them are pursued.

Upon reaching a leaf node we perform acceptance/rejection sampling on the portion of the gross sample allocated to this leaf. This we do by generating a binomial random variable, $x_k \sim B(s_k, \alpha_k)$ with parameters, s_k = gross sample size allocated to this leaf, and α_k =acceptance probability for this path k (as in naive iterative algorithm).

Having determined the net sample size for a particular leaf, we extract a simple random sample with replacement of this size from the records on the leaf (this is trivial).

We then continue with the depth first search of the tree until complete. The resulting sample may be the wrong size (because of acceptance/rejection). If the resulting sample is too large, we discard the excess (chosen at random). If the resulting sample is too small, we repeat the process (adding to our sample) until we have enough.

Lemma 3 Cardenas's Lemma *The expected number of disk blocks referenced, d, when sampling k records (with replacement) from a file of m equal size blocks is given by:*

$$d = m(1 - (1 - (1/m)^k)) \tag{21}$$

We will denote this function of m and k as $Y(k, m)$.

Proof: See [Yao77]. □

For the naive batch method, the effort to retrieve a gross sample of size s', is the same as the effort to perform batch searching on a B^+ tree, with the same batch size.

Theorem 5 *The expected cost in I/O to retrieve a sample of size s via the naive batch sampling method is approximately:*

$$E(C_{NB}(s)) \approx 1 + \sum_{j=1}^{j=h-1} Y(s', f_0 f^{j-1}) \tag{22}$$

where $f = f_{avg}$ and and $s' = s(\beta^{-1})^{h-1}$ (inflated gross sample size) as before.

Proof: As before, we use a gross sample size of s' to compensate for the losses due to acceptance rejection. This theorem is a derived by applying Cardenas's Lemma (Lemma 3) for each level of the tree. We have assumed that all nodes have fan-out $f = f_{avg}$, except the root which has fan-out f_0.

Let us number the levels 0, 1, 2, ..., h-1 from root to leaf, and let j denote the level number.

For level 0, we have s' records, 1 block (the root). For level 1, we still have s' records, and f_0 blocks. For level 2, we have s' records, and $f_0 f$ blocks. Thus for level j, $j > 1$, we have s' records and $f_0 f^{j-1}$ blocks, hence by Cardenas's Lemma the number of blocks accessed at level j will be $Y(s', f_0 f^{j-1})$. Summing yields the theorem. □

4.1.2 Early Abort Batch Method

The early abort batch method is simply the batch analog of the early abort iterative method. It is identical to the naive batch method, except that the acceptance/rejection sampling is performed (by computing binomial samples) at each node (except the root) as we search from root to leaf. As was the case with the iterative methods, the early abort batch method dominates the naive batch method.

Thus we commence our depth first search at the root with a gross sample size of s'. At the root, and each subsequent internal node, we allocate the incoming sample to the various branches by means of a multinomial random vector. Only those branches with nonzero sample sizes allocated to them are pursued.

At each level beyond the root, we perform acceptance/rejection sampling of the incoming sample by generating a binomial random variable, $x_i \sim B(s_i, \beta)$ with parameters s_i, (the incoming sample size), and acceptance probability β_i (as in the iterative early abort algorithm). The resulting net sample size for the node is then allocated the branches via a multinomial random vector. Only those branches with nonzero sample sizes allocated to them are pursued.

Upon reaching a leaf we perform one last acceptance/rejection test (by generating a binomial random variable, $x_i \sim B(s_i, \beta)$) to account for the variable loading of leaf pages. We then extract a simple random sample with replacement of size x_i of the records on the page.

Incorrect net sample sizes are dealt as described above for the naive batch sampling algorithm.

For the early abort batch method, the analysis is more complicated. The search paths which are aborted early do not generate as many page accesses. If we number the levels from 0 to h, denoting by *level i* the node at distance i from the root, then at level i, $i > 0$, we will access k_i records from $f_0 {f_{avg}}^{i-1}$ pages, where k_i is approximately a binomial random variable with expectation $s' \cdot \beta^{i-1}$, where we have ignored the variation in f_i, replacing it with f_{avg}.

Theorem 6 *The expected I/O cost to retrieve a simple random sample of size s' via the early abort batch sampling method is approximately:*

$$E(C_{EAB}(s)) \approx 1 + \sum_{j=1}^{j=h-1} Y(s'\beta^{j-1}, f_0 f^{j-1}) \tag{23}$$

where $f = f_{avg}$ and $s' = s(\beta^{-1})^{h-1}$ (inflated gross sample size) as before, and Y(k,m) is Cardenas's function defined above.

Proof: As before, we use a gross sample size of s' to compensate for the losses due to acceptance/rejection. This theorem is derived by applying Cardenas's Lemma (Lemma 3) for each level of the tree. We have assumed that all nodes have fan-out $f = f_{avg}$, except the root which has fan-out f_0.

Let us again number the levels 0, 1, 2, ..., h-1 from root to leaf, and let j denote the level number.

For level 0, we have s' records, 1 block (the root). For level 1, we still have s' records, and f_0 blocks. For level 2, we have $s'\beta$ records (because we have done acceptance/rejection at level 1), and $f_0 f$ blocks. Thus for level j, $j > 1$, we have $s'\beta^{j-1}$ records and $f_0 f^{j-1}$ blocks, hence by Cardenas's Lemma the number of blocks accessed at level j will be $Y(s'\beta^{j-1}, f_0 f^{j-1})$. Summing yields the theorem. □

4.2 Ranked B^+ trees

Finally, we consider batch sampling of ranked B^+ trees. We simply generate a simple random sample of the ranks, sort it, and then perform a batch search of the ranked B^+ treefile. Alternatively we could use Vitter's algorithm [Vit84] to generate the sequential skips required to determine the ranks of the sampled records. (Note that Vitter's algorithm generates a simple random sample without replacement, rather than with replacement. This is easily corrected.) In any case we are only concerned here with I/O, and we assume the sample of ranks can easily fit in memory.

We shall assume that we have a cache large enough to hold a complete path through the tree from root to leaf, so that reexamining pages along this path required to retrieve the sample is costless (in terms of disk I/O). Then retrieval of the sample is equivalent to batch searching of a B^+ tree, a classic problem treated by [Pal85] (among others). Essentially, the number of pages to be retrieved is simply the number of distinct pages in the union of all paths to sampled pages.

Theorem 7 *The expected I/O cost to retrieve a simple random sample of size s from a B^+ tree via the ranked batch sampling method is approximately:*

$$E(C_{NB}(s)) \approx 1 + \sum_{j=1}^{j=h-1} Y(s, f_0 f^{j-1}) \tag{24}$$

where $f = f_{avg}$, f_0 is the fan-out of the root, and $Y(k,m)$ is Cardenas's function defined above.

Proof: The proof is identical to that of Theorem 5, except that we do not need to inflate the sample size since we are not doing acceptance/rejection sampling. □

4.3 Comparisons

It is clear the batch algorithms will dominate the respective iterative algorithms, because they avoid rereading disk blocks which are used in more than one sample path.

Furthermore:

Theorem 8 *The early abort batch outperforms naive batch.*

$$E[C_{EAB}(s)] \leq E[C_{NB}(s)] \tag{25}$$

Proof: This result follows from the cost functions and the fact that Carndenas's functions $Y(k,m)$ is monotone increasing in k (the sample size). □

The relationship between the performance of the ranked batch algorithm and the early abort batch algorithm is more subtle. Because the ranked batch algorithm may increase the height of the B^+ tree it is possible that RB will perform worse than EAB. However, more typically RB will not increase the height of the B^+ tree and will out-perform EAB. How big is the difference between $E[C_{RB}(s)]$ and $E[C_{EAB}(s)]$? It is clearly less than a factor of $(\beta^{-1})^{h-1}$, the factor by which we inflate the gross sample size of EAB to compensate for the attrition due to acceptance/rejection sampling. For random B^+ trees β is known to be 0.7. Hence for a random B^+ tree of height 5, the methods differ by a factor of no more than 4.

5 Conclusions

Our most important results concern algorithms to retrieve simple random samples of B^+ trees, without any additional data structures or indices. These methods are based on acceptance/rejection sampling, and provide a simple, inexpensive way to add sampling to a relational database systems. They are appropriate for systems which only infrequently need to support sampling, e.g., for auditing. The batch early abort algorithm offers the best performance of algorithms of this class, and is to be generally preferred to the iterative algorithms.

Acknowledgements

The authors would like to thank Tekin and Meral Ozsoyoglu for their continuing encouragement. Jack Morgenstein first introduced us to the problem. We also wish to thank the referees for their comments.

References

[Ark84] Herbert Arkin. *Handbook of Sampling for Auditing and Accounting.* McGraw-Hill, 1984.

[BK75] B.T. Bennett and V.J. Kruskal. Lru stack processing. *IBM Journal of Research and Development,* 19(4):353–357, July 1975.

[Coc77] William G. Cochran. *Sampling Techniques.* Wiley, 1977.

[EN82] Jarmo Ernvall and Olli Nevalainen. An algorithm for unbiased random sampling. *The Computer Journal,* 25(1), 1982.

[FMR62] C.T. Fan, M.E. Muller, and I. Rezucha. Develoment of sampling plans by using sequential (item by item) selection techniques and digital computers. *Journal of the American Statistical Association,* 57:387–402, June 1962.

[Gho86] S. Ghosh. Siam: statistics information access method. In *Proceedings of the Third International Workshop on Statistical and Scientific Database Management,* pages 286–293, EUROSTAT, Luxembourg, 1986.

[HOT88] Wen-Chi Hou, Gultekin Ozsoyoglu, and Baldeo K. Taneja. Statistical estimators for relational algebra expressions. In *Proceedings of the Seventh ACM Conference on Principles of Database Systems,* pages 288–293, March 1988.

[Knu73] Donald Ervin Knuth. *The Art of Computer Programming: Vol. 3, Sorting and Searching.* Addison-Wesley, 1973.

[LTA79] Donald A. Leslie, Albert D. Teitlebaum, and Rodney J. Anderson. *Dollar Unit Sampling.* Copp Clark Pitmanan, 1979.

[LWW84] H.-J. Lenz, G.B. Wetherill, and P.-Th. Wilrich, editors. *Frontiers in Statistical Quality Control 2.* Physica-Verlag, Wurzburg, Germany, 1984.

[Mon85] Douglas C. Montogmery. *Introduction to Statistical Quality Control.* Wiley, 1985.

[Pal85] P. Palvia. Expressions for batched searching of sequential and hierarchical files. *ACM Transactions on Database Systems,* 10(1):97–106, March 1985.

[SL88] J. Srivastava and V.L. Lum. A tree based access method (tbsam) for fast processing of aggregate queries. In *Proceedings of the 4th International Conference on Data Engineering,* pages 504–510, IEEE Computer Scoeity, 1988.

[Vit84] Jeffrey Scott Vitter. Faster methods of random sampling. *Communications of the ACM,* 27(7):703–718, July 1984.

[Vit85] Jeffrey Scott Vitter. Random sampling with a reservoir. *ACM Transactions on Mathematical Software,* 11(1):37–57, March 1985.

[WE80] C.K. Wong and M.C. Easton. An efficient method for weighted sampling without replacement. *SIAM Journal on Computing,* 9(1):111–113, February 1980.

[Yao77] S. Bing Yao. Approximating the number of accesses in database organizations. *Communications of the ACM,* 20(4):260–261, April 1977.

AGGREGATE EVALUABILITY IN STATISTICAL DATABASES

F.M Malvestuto(†), M. Moscarini(‡)

(†) National Commission for Nuclear and Alternative Energy Sources, ENEA, Roma, Italy
(‡) Dipartimento di Matematica, Università di Roma "Tor Vergata", Roma, Italy

Abstract

Usually a statistical database contains many summary tables representing the distribution of the same statistical variable over the classes of as many partitions of a certain universe of objects. Existing query systems allow only queries on single tables. Indeed, in most cases additional queries can be evaluated by combining the information contained in similar tables in a suitable way.

In order to improve the responsiveness of the database and allow an integrated use of the stored information, we propose to inform the database system of the relationship among the partitions adopted in the tables. Such a relationship, called *intersection dependency*, states which classes of the partitions have a non-empty intersection and can be represented by a uniform multipartite hypergraph, called *intersection hypergraph*.

On the grounds of the algebraic properties of the intersection hypergraph and under the assumption of data additivity, we shall provide a characterization of evaluable queries, which allows us to define polynomial-time procedures both for testing evaluability and for evaluating queries.

1. Introduction

With the term *statistical database* we refer to "a numeric database containing statistics about classes of objects or individuals" [20] (social, economic, technological, environmental or demographic surveys are typical examples).

As usual, we shall assume that the unaggregated data from which the statistics have been computed is not available.

A statistical database is a collection of (*summary*) *tables*, each of which represents a distribution of a certain (*statistical*) *variable* (sometimes called "summary attribute" [14,20]) related to a given *universe* of objects or individuals, partitioned according to a set of (*category*) *attributes*, referred to as the *scheme* of the table.

Proceedings of the Fifteenth International Conference on Very Large Data Bases

Amsterdam, 1989

Example 1. *Universe:* Soviet people in the year 1959. *Variable:* Population (1000 individuals). *Scheme:* {**S**ex, **S**chooling, **P**arty-**M**embership} (the data is obtained by processing data from Bishop et al. [4]).

Table: Distribution of the soviet populatiion by schooling, sex and party (1000 individuals) 1959

Sex	Schooling	Party-Membership Yes	No
Male	< 4	0	13670
	4-7	1217	20568
	8-10	2140	21135
	> 10	672	2341
Female	< 4	0	34730
	4-7	1071	18115
	8-10	2441	24109
	> 10	696	2424

□

For any subset S of the scheme of a table, all possible combinations of values of attributes in S define a partition of the underlying universe into *classes*.

Example 2. The classes of the partition defined by {S, Sc} in Example 1 are the groups of individuals qualified by the following conditions

$$S=\text{Male} \wedge Sc< 4$$
$$S=\text{Male} \wedge 4 \leq Sc \leq 7$$
$$S=\text{Male} \wedge 8 \leq Sc \leq 10$$
$$S=\text{Male} \wedge Sc> 10$$
$$S=\text{Female} \wedge Sc< 4$$
$$S=\text{Female} \wedge 4 \leq Sc \leq 7$$
$$S=\text{Female} \wedge 8 \leq Sc \leq 10$$
$$S=\text{Female} \wedge Sc> 10$$

□

The partition defined by the whole scheme will be called a *base partition*; its classes correspond (one-to-one) to the entries in the table.

A typical query on a statistical database requires the evaluation of a variable for a subset (or *aggregate*) of objects in the underlying universe, qualified by a logical formula built up from specified values for given attributes by means of operators ∧, ∨ and ¬. An aggregate is *evaluable* if the corresponding value to be assigned to the variable is uniquely determined by the data stored in the database (note that even in the case that an aggregate is not evaluable, a query can be answered in an "approximate" way [11]).

In literature [6,7,9,10,15,16,18,19] only logical formulas with attributes from a single table scheme have been considered. This means that every aggregate is the union of classes of a base partition, that is, an element of the set field generated by that base partition.

Example 3. Referring to the table in Example 1, a user can ask for the population of the aggregate specified by the following formula

$$S = \text{Female} \quad \wedge \quad PM = \text{Yes} \quad \wedge \quad (4 \leq Sc \leq 7 \quad \vee \quad 8 \leq Sc \leq 10)$$

□

If the data is *additive under aggregation* [8], the evaluation of a variable for an aggregate can be carried out by summing the entries corresponding to the classes contained in that aggregate.

Example 3 (continued). The answer to the query in Example 3 is

$$\text{Population}(x) + \text{Population}(y) = = 1071 + 2441 = 3512 \text{ (1000 individuals)}$$

where

x: S = Female ∧ PM = Yes ∧ $4 \leq Sc \leq 7$
y: S = Female ∧ PM = Yes ∧ $8 \leq Sc \leq 10$ □

Indeed, a statistical database often contains "similar" tables, that is, distributions of the same variable over different partitions of the same universe of objects.

Similar tables may occur in the following typical cases:

- an organization ("data source") collects data on the objects of a given universe and then creates a database (called "abstract" in [17] and "partitioned database" in [5]) that is a collection of simple statistics;
- an organization creates a statistical database by putting together tables produced by different statistical agencies that refer to a common data source, but use different classification criteria.

Knowledge of relationships between attributes belonging to schemes of similar tables allows for additional aggregates to be evaluated [12,13]. The following example illustrates this statement in the case in which the *intersection dependency* between classes of the base partitions of similar tables is known, that is, we know which classes of the base partitions have a nonempty intersection.

Example 4. A statistical database contains the two following similar tables reporting the distributions of the employees of a research institute by department and by educational qualification. □

Table: Distribution of the employees by department

Department	Number
Information-Systems	30
Research	50
Administration	20

Table. Distribution of the employees by educational qualification

Educational-Qualification	Number
Ph.D.	15
High-School	20
Degree	65

Figure 1. Two similar tables.

Example 4 (continued). Suppose that we know the intersection dependency between the classes of the base partitions in Figure 1. We can represent this information with the bipartite graph in Figure 2, in which an edge represents a pair of nondisjoint classes of the two base partitions.

In this case, it is possible to evaluate aggregates which cannot be evaluated in traditional systems. An example of such an aggregate is

D = Information-Systems ∧ (E = High-School ∨ E = Degree) ∨ (D = Research ∧ E = Degree)

In fact, on examining the graph in Figure 2 we note that the aggregate qualified by the formula above refers to the employees who work in the Information-Systems or Research departments and are not Ph.Doctors. Consequently, the number of qualified employees is

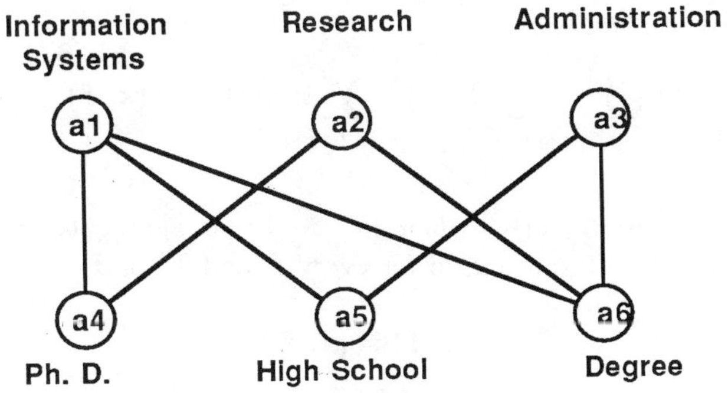

Figure 2. An intersection (hyper)graph

Number(a_1) + Number(a_2) - Number(a_4) = 30 + 50 - 15 = 65
where
a_1: D = Information-Systems
a_2: D = Research
a_4: E = Ph.D. □

The edges of the bipartite graph define the classes of a partition of the underlying universe that is finer than either of the two base partitions. Such a partition will be called *meet partition.*

Using formulas such as the one in the example (built up with attributes from the schemes of similar tables) we can express all the aggregates in the set field generated by the meet partition So, the following problem arises:

Problem. Is each aggregate in the set field generated by the meet partition evaluable ?

In this paper we approach this problem under the assumption that the data stored in similar tables is additive and that the underlying intersection dependency is known perfectly. In the general case in which the statistical database contains k similar tables, the intersection dependency can be represented by a uniform k-partite hypergraph [1] to be called *intersection hypergraph,* whose vertex set is in a one-to-one correspondence with the set obtained by union of the k base partitions, and whose edge set is in a one-to-one correspondence with the meet partition.

The paper is organized as follows. Section 2 contains the formal definition of an evaluable aggregate and the formulation of the problem of evaluability in the framework of linear equation systems. Section 3 contains an algebraic characterization of evaluable aggregates which leads to a quadratic procedure for testing evaluability, stated in Section 4. In Section 5 we shall show that evaluating an (evaluable) aggregate requires solving an equation system. Section 6 is dedicated to an illustrative example. Section 7 contains some basic properties of evaluable aggregates. In Section 8 we conclude by pointing out the advantages of an "informed" query system.

2. A formal framework

Let A be a partition of a *universe* Ω in *classes* of objects and A^+ be the *set field* generated by A, i.e., the family of subsets of Ω which are obtained as the union of a subset of classes of A. We call *aggregate* any element of A^+. A^+ turns out to be closed under set union, intersection and complementation.

Let F be a real-valued set function defined on a set field A^+; F is *additive* if for any two disjoint aggregates x and y in A^+, we have

$$F(x \cup y) = F(x) + F(y).$$

Observe that any additive function F on a set field A^+ is completely specified by the set of values $\{F(a) \mid a \in A\}$; in fact, by virtue of the additivity, for every $x \in A^+$ we have

$$F(x) = \sum_{\substack{a \in A \\ a \subseteq x}} F(a). \tag{1}$$

Consider now k partitions $A_1, \ldots, A_k$ of Ω. The *meet partition* $E = \{e_1, e_2, \ldots, e_m\}$ of $A_1, \ldots, A_k$ is the partition defined as the greatest lower bound of $\{A_1, \ldots, A_k\}$ [2], i.e., the partition of Ω whose classes are given by the nonempty intersections $a_1 \cap \cdots \cap a_k$ with $a_i \in A_i$ $(i = 1, \ldots, k)$.

The set field E^+ generated by the meet partition E is the least set field containing $A_1^+, \ldots, A_k^+$ [13].

Given the partitions $A_1, \ldots, A_k$ of Ω, let $F_1, \ldots, F_k$ be additive functions on the set fields $A_1^+, \ldots, A_k^+$, respectively. By $\mathcal{F}$ we denote the family of additive functions F on the set field E^+ which satisfy

$$F(a_i) = F_i(a_i) \quad \text{for all } a_i \in A_i \quad (i = 1, \ldots, k). \tag{2}$$

By (1) the condition (2) can be written as

$$\sum_{\substack{e_h \in E \\ e_h \subseteq a_i}} F(e_h) = F_i(a_i) \quad \text{for all } a_i \in A_i \quad (i = 1, \ldots, k). \tag{3}$$

This is a system of $n = |A_1| + \cdots + |A_k|$ equations with m unknowns $F(e_1), F(e_2), \ldots, F(e_m)$; we call *partitioned homogeneous system* the homogeneous system associated with (3).

If $\mathcal{F} \neq \emptyset$, we say that the set of additive functions $F_1, \ldots, F_k$ is consistent.

An aggregate x in E^+ is *evaluable* with respect to a consistent set of functions $F_1, \ldots, F_k$ if either $|\mathcal{F}| = 1$ or, for every pair of functions F and F' in $\mathcal{F}$ we have $F(x) = F'(x)$.

In what follows, we assume that a collection $A_1, \ldots, A_k$ of distinct partitions of a universe Ω, their meet partition $E = \{e_1, \ldots, e_m\}$ and a consistent set $\{F_1, \ldots, F_k\}$ of additive functions be given.

We conclude this section with some remarks relating the formal notions introduced in this section to the concepts discussed in the Introduction.

- $A_1, \ldots, A_k$ can be interpreted as the base partitions of k similar tables contained in the database;
- the values $F_i(a_i)$ for each $a_i \in A_i$ are the entries of the i-th table;
- the additivity of the functions $F_1, \ldots, F_k$ formalizes the intuitive concept of "additive data under aggregation" mentioned in the Introduction, which applies to a large number of statistical variables (count data, probability data, measurement data, etc.);
- as discussed in the Introduction, similar tables are assumed to come from a unique data source; this ensures the consistency of the set of functions $\{F_1, \ldots, F_k\}$ and, hence, guarantees that it is meaningful to use the k tables in an "integrated" way;
- the classes of the meet partition are in a one-to-one correspondence with the edges of the intersection hypergraph;
- the coefficient matrix of system (3) (as well as the coefficient matrix of the partitioned homogeneous system) is the incidence matrix of the intersection hypergraph.

3. Characterization of evaluable aggregates

To every aggregate x in E^+ we shall associate an m-tuple $\mathbf{x} = [x(1), \ldots, x(m)]$, where $x(h)$ is equal to 1 if class e_h is included in x and equal to 0 otherwise. We call $\mathbf{x}$ the *representative vector* of the aggregate x. Observe that the rows of the coefficient matrix $\mathbf{A}$ of the partitioned homogeneous system are the representative vectors of the classes in the partitions $A_1, \ldots, A_k$. We call *row space* the vector space $\mathcal{R}$ spanned by the rows of the matrix $\mathbf{A}$.

Analogously, we represent a function F on E^+ by the m-tuple $\mathbf{f} = [f(1), \ldots, f(m)]$ where $f(h) = F(e_h)$ for all e_h in E. We call $\mathbf{f}$ the *representative vector* of the function F. We call *solution space* the vector space $\mathcal{S}$ of the solutions of the partitioned homogeneous system.

Finally, by $(\mathbf{x}, \mathbf{y})$ we denote the scalar product of two m-tuples $\mathbf{x}$ and $\mathbf{y}$.

Thus, by additivity for every F in $\mathcal{F}$ and for every x in E^+ we have

$$F(x) = \sum_{\substack{e_h \in E \\ e_h \subseteq x}} F(e_h) = \sum_h x(h) f(h) = (\mathbf{x}, \mathbf{f}).$$

Consequently, when $|\mathcal{F}| > 1$, an aggregate x is evaluable if and only if for every F and F' in $\mathcal{F}$

$$(\mathbf{x}, \mathbf{f}) = (\mathbf{x}, \mathbf{f}'). \tag{4}$$

Lemma 1. *The solution space $\mathcal{S}$ and the row space $\mathcal{R}$ are complementary subspaces of $\Re^m$*

Proof. By definition, $\mathcal{S}$ is the set of all the m-tuples that are orthogonal to the rows of the coefficient matrix of the partitioned homogeneous system; hence, $\mathcal{S}$ and $\mathcal{R}$ are complementary [3]. □

Recalling that the dimension of a vector space is the maximum number of linearly independent vectors, we have that the dimension of the vector space $\mathcal{R}$ is equal to the rank r of the matrix $\mathbf{A}$ and, by Lemma 1, that the dimension of the vector space $\mathcal{S}$ is $m - r$.

Lemma 2. *An aggregate is evaluable if and only if its representative vector is orthogonal to the solution space $\mathcal{S}$.*

Proof. If $|\mathcal{F}| = 1$ then system (3) has a unique solution; consequently, the partitioned homogeneous system also has a unique solution, namely the zero vector, which is orthogonal to every vector.

If $|\mathcal{F}| > 1$, let $\mathbf{f}$ be a solution of (3). The set of the solutions of (3) coincides with the set of vectors $\mathbf{f}'$ such that

$$\mathbf{f}' = \mathbf{f} + \mathbf{s}$$

where $\mathbf{s} \in \mathcal{S}$.

Therefore, by exploiting the bilinearity of the scalar product, for every x in E^+ we have:

$$(\mathbf{x}, \mathbf{s}) = (\mathbf{x}, \mathbf{f}' - \mathbf{f}) = (\mathbf{x}, \mathbf{f}') - (\mathbf{x}, \mathbf{f})$$

and by (4) the lemma is proved. □

Recalling that the coefficient matrix of the partitioned homogeneous system coincides with the incidence matrix of the intersection hypergraph, from Lemma 1 and Lemma 2 we can derive the following:

Theorem 1. *An aggregate is evaluable if and only if its representative vector is linearly dependent on the*

a_4: E = Ph.D.
a_5: E = High-School
a_6: E = Degree

The meet partition E of A_1 and A_2 is formed by 7 classes corresponding to the edges of the bipartite intersection hypergraph given in Figure 2:

$$e_1 = a_1 \cap a_4$$
$$e_2 = a_1 \cap a_5$$
$$e_3 = a_1 \cap a_6$$
$$e_4 = a_2 \cap a_4$$
$$e_5 = a_2 \cap a_6$$
$$e_6 = a_3 \cap a_5$$
$$e_7 = a_3 \cap a_6$$

The incidence matrix $\mathbf{A}$ of the intersection hypergraph is:

$$\begin{pmatrix} 1 & 1 & 1 & 0 & 0 & 0 & 0 \\ 0 & 0 & 0 & 1 & 1 & 0 & 0 \\ 0 & 0 & 0 & 0 & 0 & 1 & 1 \\ 1 & 0 & 0 & 1 & 0 & 0 & 0 \\ 0 & 1 & 0 & 0 & 0 & 1 & 0 \\ 0 & 0 & 1 & 0 & 1 & 0 & 1 \end{pmatrix}$$

The reduced matrix $\mathbf{A}'$ is

$$\begin{pmatrix} 1 & 0 & 0 & 0 & -1 & 0 & 0 \\ 0 & 0 & 0 & 1 & 1 & 0 & 0 \\ 0 & 0 & 0 & 0 & 0 & 1 & 1 \\ 0 & 1 & 0 & 0 & 0 & 0 & -1 \\ 0 & 0 & 1 & 0 & 1 & 0 & 1 \\ 0 & 0 & 0 & 0 & 0 & 0 & 0 \end{pmatrix}$$

The partitioned homogeneous system in reduced echelon form looks like the following

$$\begin{array}{llllllll} s(1) & & & & & -s(5) & & = 0 \\ & s(2) & & & & & -s(7) & = 0 \\ & & s(3) & & & +s(5) & +s(7) & = 0 \\ & & & s(4) & & +s(5) & & = 0 \\ & & & & s(6) & & +s(7) & = 0 \end{array}$$

The corresponding basis of the solution space $\mathcal{S}$ is given by the vectors $\mathbf{s}_1$ and $\mathbf{s}_2$, the former of which is obtained by setting $s(5) = 1$ and $s(7) = 0$ and the latter by setting $s(5) = 0$ and $s(7) = 1$:

$$\mathbf{s}_1 = [1\ 0\ -1\ -1\ 1\ 0\ 0] \quad \text{and} \quad \mathbf{s}_2 = [0\ 1\ -1\ 0\ 0\ -1\ 1]$$

Consider now the two aggregates $x = e_2 \cup e_3 \cup e_5$ and $y = e_1 \cup e_5$ with representative vectors

$$\mathbf{x} = [0\ 1\ 1\ 0\ 1\ 0\ 0] \quad \text{and} \quad \mathbf{y} = [1\ 0\ 0\ 0\ 1\ 0\ 0]$$

Note that aggregate $\mathbf{x}$ is the aggregate considered in Example 4.

It is immediately seen that $(\mathbf{x},\mathbf{s}_1) = (\mathbf{x},\mathbf{s}_2) = 0$. So, x is an evaluable aggregate. But, $(\mathbf{y},\mathbf{s}_1) \neq 0$ and, therefore, y is not evaluable.

In order to express $\mathbf{x}$ as a linear combination of the rows of $\mathbf{A}$, take as a basis of row space $\mathcal{R}$ the set $\{\mathbf{a}_1, \mathbf{a}_2, \mathbf{a}_3, \mathbf{a}_4, \mathbf{a}_5\}$ of rows of $\mathbf{A}$ corresponding to the nonnull rows of the reduced matrix $\mathbf{A}'$. The coefficients λ_q of the required combination can be determined by solving the system

$$\mathbf{x} = \lambda_1 \mathbf{a}_1 + \lambda_2 \mathbf{a}_2 + \lambda_3 \mathbf{a}_3 + \lambda_4 \mathbf{a}_4 + \lambda_5 \mathbf{a}_5$$

by successive elimination of the unknowns; an admissible solution is given by:

$$\lambda_1 = 1$$
$$\lambda_2 = 1$$
$$\lambda_3 = 0$$
$$\lambda_4 = -1$$
$$\lambda_5 = 0$$

and, therefore, we have

$$\mathbf{x} = \mathbf{a}_1 + \mathbf{a}_2 - \mathbf{a}_4.$$

and

$$(\mathbf{x}, \mathbf{f}) = F_1(a_1) + F_1(a_2) - F_2(a_4).$$

7. Properties of evaluable aggregates

In this section we provide some basic properties of evaluable aggregates.

Proposition. *Let x and y be evaluable aggregates in E^+. We have:*

(i) *(disjoint-union property): if $x \cap y = \emptyset$, then $x \cup y$ is evaluable*
(ii) *(proper-difference property): if $x \supseteq y$, then $x - y$ is evaluable*
(iii) *(complement property): $x' = \Omega - x$ is evaluable.*

Proof. The statements can be easily proved by additivity. In particular for every F in $\mathcal{F}$ we have:

- in case (i), $F(x \cup y) = F(x) + F(y)$
- in case (ii), $F(x - y) = F(x) - F(y)$
- in case (iii), $F(x') = F(\Omega) - F(x) = \sum_{a_i \in A_i} F_i(a_i) - F(x)$ for any A_i. □

rows of the incidence matrix of the intersection hypergraph.

Proof. In fact, by Lemma 1 every vector orthogonal to the solution space $\mathcal{S}$ belongs to the row space $\mathcal{R}$. So, the theorem follows from Lemma 2. □

Corollary 1. *Each aggregate in E^+ is evaluable if and only if the dimension of the row space $\mathcal{R}$ equals the cardinality of the meet partition E.*

Corollary 2. *Each aggregate in E^+ is evaluable if and only if the solution space $\mathcal{S}$ contains only the zero vector* **0**.

4. Testing evaluability

Lemma 2 allows us to polynomially test the evaluability of an aggregate x in E^+ once a basis for $\mathcal{S}$ is given (recall that a basis of $\mathcal{S}$ is a set of $m-r$ independent vectors).

It is well-known [3] that a basis for $\mathcal{S}$ can be determined by transforming the partitioned homogeneous system in the so-called *reduced echelon form* [3], whose solution space coincides with $\mathcal{S}$. If the matrix **A** has rank r, we have

$$
\begin{aligned}
s(1) \quad & + c_{1,r+1}s(r+1) + \cdots + c_{1,m}s(m) = 0 \\
s(2) \quad & + c_{2,r+1}s(r+1) + \cdots + c_{2,m}s(m) = 0 \\
& \cdots \\
s(r) & + c_{r,r+1}s(r+1) + \cdots + c_{r,m}s(m) = 0
\end{aligned}
$$

This can be done [3] by first reducing the matrix **A** to a row-equivalent matrix **A**$'$ by repeatedly applying the following row operations:

(i) replace row $\mathbf{a}_i$ by $c\mathbf{a}_i$ for any scalar $c \neq 0$

(ii) replace row $\mathbf{a}_i$ by $\mathbf{a}_i + d\mathbf{a}_j$ for any $j \neq i$ and any scalar $d \neq 0$

and, then, rearranging the rows of **A**$'$.

Therefore, the task of transforming the partitioned homogeneous system into its reduced echelon form can be accomplished with $O(mn^2)$ elementary operations.

A basis $\{\mathbf{s}_1, \ldots, \mathbf{s}_{m-r}\}$ of $\mathcal{S}$ can be obtained by taking the $m-r$ **s**-solutions resulting from setting one of the parameters $s(h)$ $(h = r+1, \ldots, m)$ equal to 1 and the remaining parameters equal to 0. That is,

$$
\begin{aligned}
\mathbf{s}_1 &= [\ c_{1,r+1}, \ldots, -c_{r,r+1}, 1, 0, \ldots, 0] \\
\mathbf{s}_2 &= [-c_{1,r+2}, \ldots, -c_{r,r+2}, 0, 1, \ldots, 0] \\
& \cdots \\
\mathbf{s}_{m-r} &= [-c_{1,m}, \ldots, -c_{r,m}, 0, 0, \ldots, 1]
\end{aligned}
$$

At this point in order to decide whether x is evaluable or not, it is sufficient to verify that the scalar product of its representative vector **x** with each basis vector $\mathbf{s}_j$ $(j = 1, \ldots, m-r)$ of the solution space $\mathcal{S}$ vanishes.

It should be noticed that since a basis for $\mathcal{S}$ can be determined a priori on the grounds of the intersection dependency, the test for evaluability can be carried out without accessing the database.

From the foregoing discussion the following theorem follows.

Theorem 2. *Testing evaluability requires $O(m^2)$ elementary operations.*

5. Computing answers to evaluable queries

To answer a query related to an evaluable aggregate x, the database system needs to determine the coefficients of the linear combination of the rows of **A** which provides **x**.

Let $\mathbf{a}_1, \ldots, \mathbf{a}_r$ be the rows of **A** which give rise to the nonnull rows of the reduced matrix **A**$'$ mentioned in the previous section. The set $\{\mathbf{a}_1, \ldots, \mathbf{a}_r\}$ is a basis of the vector space $\mathcal{R}$ and, therefore, for each evaluable aggregate x we have

$$
\mathbf{x} = \sum_{q=1}^{r} \lambda_q \mathbf{a}_q \tag{7}
$$

which is a system of m equations with r unknowns, namely, $\lambda_1, \ldots, \lambda_r$, and can be solved using standard methods.

Once the coefficients λ_q have been determined, the answer to the query is given by

$$
F(x) = (\mathbf{x}, \mathbf{f}) = \sum_{q=1}^{r} \lambda_q (\mathbf{a}_q, \mathbf{f}) = \sum_{q=1}^{r} \lambda_q F_{i_q}(a_q)
$$

where a_q is the class in the partition A_{i_q} with representative vector $\mathbf{a}_q$ and F_{i_q} is the additive function on the set field generated by A_{i_q}.

6. An example

Consider the two base partitions $A_1 = \{a_1, a_2, a_3\}$ and $A_2 = \{a_4, a_5, a_6\}$ induced by the attributes Department and Educational-Qualification in Example 4, where

a_1: D = Information-Systems

a_2: D = Research

a_3: D = Administration

Let us consider, now the minimal evaluable aggregates, that is, those evaluable aggregates which contain no proper subsets which are evaluable

Theorem 3. *Every evaluable aggregate is a disjoint union of minimal evaluable aggregates.*

Proof. Let x be an evaluable aggregate which contains a minimal evaluable aggregate y. Then, by the proper-difference property the aggregate

$$z = x - y$$

is evaluable, too. So x is the disjoint union of y and z. If z is also minimal, then the theorem is proved; otherwise, repeat the same line of reasoning made for x to z. By finite induction, we prove that x is a disjoint union of minimal evaluable aggregates. □

8. Concluding remarks

In this paper we approached the problem of the evaluability of aggregates in a statistical database containing similar tables related by an intersection dependency.

We provided a characterization of evaluable aggregates. Our characterization allows us to define procedures both for testing evaluability and for evaluating queries.

These results are useful in designing an "informed" query system for statistical databases which promotes an integrated use of stored information. Such an "informed" query system allows the user to formulate a query involving attributes from several similar tables as if they were all contained in a single table ("universal" table), that is without knowing that these attributes have been extracted from different tables; for example, a user queries the database in Example 4 using the "universal" scheme {Department, Educational-Qualification}.

References

[1] C. Berge, *Graphs and Hypergraphs.* North Holland, 1973.

[2] G. Birkhoff, *Lattice Theory.* American Mathematical Society, 1967.

[3] G. Birkhoff and S. McLane, *A Survey of Modern Algebra.* Mc Millan, 1962.

[4] Y.M.M. Bishop, S.E. Fienberg and P.W. Holland, *Discrete Multivariate Analysis: Theory and Practice*, MIT Press, 1975.

[5] R. Brooks, M. Blattner, Z. Pawlak and E. Barret, "Using Partitioned Databases For Statistical Data Analysis", *AFIPS Conference Proc.* (1981), 453-457.

[6] E. Fortunato, M. Rafanelli, F.L. Ricci and A. Sebastio, "An Algebra for Statistical Data", *Proc. III Int. Workshop on "Statistical & Scientific Database Management"* (1986), 122-134.

[7] S.P. Ghosh, "Statistical Relational Tables for Statistical Database Management", **IEEE Software Engineering 12** (1986), 1106-1116.

[8] S.P. Ghosh, "Statistics Metadata", **Encyclopedia of Statistical Sciences 8** (by Kotz-Johnson). J. Wiley & Sons, 1988.

[9] G. Hebrail, "A Model of Summaries for Very Large Databases", *Proc. III Int. Workshop on "Statistical & Scientific Database Management"* (1986), 143-151.

[10] A. Klug, "Equivalence of Relational Algebra and Relational Calculus Query Languages Having Aggregate Functions", **JACM 29** (1982), 699-717.

[11] F.M. Malvestuto, "Answering Queries in Categorical Data Bases", *Proc. ACM 1987 Symp. on "Principles of Database Systems"*, 82-89.

[12] F.M. Malvestuto, "The Derivation Problem for Summary Data", *Proc. ACM SIGMOD 1988 Conf. on "Management of Data"*, 87-96.

[13] F.M. Malvestuto and C. Zuffada, " The Classification Problem with Semantically Heterogeneous Data", **Lecture Notes in Computer Science 339** (M. Rafanelli, J.C. Klensin and P. Svensson eds.), Springer Verlag 1989, 157-176.

[14] F. Olken, D. Rotem, A. Shoshani and H. Wong, "Scientific and Statistical Data Management Research at LBL", *Proc. III Int. Workshop on "Statistical & Scientific Database Management"* (1986), 1-20.

[15] Z.M. Özsoyoğlu and G. Özsoyoğlu, "An Extension of Relational Algebra for Summary Tables", *Proc. II Int. Workshop on "Statistical Database Management"* (1983), 202-211.

[16] G. Özsoyoğlu, Z.M. Özsoyoğlu and V. Matos, "Extending Relational Algebra and Relational Calculus with Set-valued Attributes and Aggregate Functions", **ACM Trans. on Database Systems 12** (1987), 566-592.

[17] N.C. Rowe, "Antisampling for Estimation: an Overview", **IEEE Trans. on Software Engineering 11** (1985), 1081-1091

[18] H. Sato, "Handling Summary Information in a Database: Derivability", *Proc. ACM-SIGMOD 1981 Conf. on "Management of Data"*, 98-107.

[19] H. Sato, "Fundamental Concepts of Social-Regional Summary Data and Inferences in their Databases", (thesis) Japan Economic Planning Agency, Tokyo 1982.

[20] A. Shoshani, "Statistical Databases: Characteristics, Problems and Some Solutions", *Proc. VIII Int. Conf. on "Very Large Data Bases"* (1982), 208-222.

Aggregates in Possibilistic Databases

Elke A. Rundensteiner and Lubomir Bic

Department of Information and Computer Science
University of California, Irvine

Abstract

Fuzzy set theory represents a uniform framework for extending the relational database model to handle imprecision of information found in the real world. None of the existing proposals for data models handling imprecision has dealt with queries involving aggregate operators. This paper presents a framework for handling aggregates in the context of imprecise information. Two kinds of aggregates, namely, scalar aggregates and aggregate functions, are being supported. We consider three cases: aggregates within approximate queries on precise data, aggregates within precisely specified queries on possibilistic data, and aggregates within vague queries on imprecise data. These extensions are based on fuzzy set-theoretical concepts such as the extension principle and the possibilistic expected value.

1 Introduction

It has been widely recognized that the uncertainty inherent in real world data has to be dealt with in database systems. Research addressing this problem has to a large extend been based on fuzzy set theory [11]. The two major objectives of these efforts are (1) enhancements to existing data models for representing incomplete and uncertain data, and (2) the development of new retrieval techniques for such data. The first issue addresses the limitation of conventional data models to allow attributes to take but one constant value from a domain [1, 8, 7]. The second problem is the inadequacy of current relational query languages (RQLs) to cope with different types of fuzzy representations of data. Most attempts to design an enhanced RQL, here called a FRQL [7, 12], are based on some form of the generally accepted relational algebra or relational calculus [2].

Database systems featuring only retrieval and update operations [2] are inadequate for many important applications [4]. Virtually all real-world problems need query capabilities involving aggregate and statistical functions [3]. This strongly suggests that the evaluation of aggregates has to be dealt with in the context of the extended relational models. However, most research on extending relational query languages has neglected this issue; the only notable exception is the work of Oezsoyoglu et al [5].

This paper investigates the incorporation of aggregate operators commonly found in traditional query languages like SQL or QUEL [3] into FRQLs. There are three general cases we consider. First, we handle approximate queries on a relational database containing precise information. An example of this situation is the query "What is the maximum salary of all *old* professors at UCI?" with the age and salary being precisely known and *old* being a fuzzy set defined on the Age attribute. The second case occurs when the query is exact but the information stored in the database is of possibilistic nature. An example of this case is the query "What is the maximum salary of professors at UCI?" where the salary is specified imprecisely in the database. The third case is obtained by combining the two approaches to handle queries with approximate restrictions against a database containing possibilistic specified data. The solutions proposed in this paper satisfy two principles, consistency and completeness.

The paper is structured in the following manner. First, we review the aggregate operators in the context of the relational data model (section 2). The basics of fuzzy set theory are presented in section 3. A possibilistic extension of the relational database model and the relational algebra is introduced in section 4, referred to as Possibilistic Relational Data Model (PRDM). Section 5 is devoted to our proposal for evaluating aggregates within the PRDM. Conclusions and future research are presented in the last section.

2 Aggregates in Relational Query Languages

In this section, we review how aggregates are handled within the classical relational data model [2, 4].

Proceedings of the Fifteenth International Conference on Very Large Data Bases **Amsterdam, 1989**

2.1 The Classical Relational Model

A relational data model consists of a set of *attribute names* A_i, a set of corresponding *domains* U_i, and a set of *relation schemas* R_i and relations r_i.

Definition 1 *A set of attributes* $\{A_1, ..., A_n\}$ *is called a* **relation schema** $R(A_1, ..., A_n)$, *or short R. Let* U_i *be the domain of the attribute* A_i, *with* $1 \leq i \leq n$. *A* **relation** *r on R is defined as a subset of the Cartesian product of the domains* U_i, *i.e.,* $r \subseteq U_1 \times ... \times U_n$. *Each tuple* $t_i \in r$ *has the form* $< a_{i1} \times ... \times a_{in} >$ *with* $a_{ij} \in U_j$. $t_i[A_j]$ *denotes the value of tuple* t_i *on attribute* A_j, *i.e.,* $t_i[A_j] = a_{ij}$.

Query languages for the relational database model (RQLs) are based on either the relational algebra or the relational calculus, both proposed by Codd [2]. In this paper, we use the relational algebra. We chose a notation similar to the one of QUEL [3] over a more formal mathematical one in order to make the queries more understandable. For a definition see [3].

2.2 Scalar Aggregates

Retrieval statements exclusively composed from retrieval operations are inadequate for most applications of database management systems [4]. Many real-world queries require the application of aggregate and statistical functions to database relations. Consequently, most commercially available systems provide a set of such aggregate operators [3]. In general, two types of aggregate operators are supported by RQLs; these are *scalar aggregates* and *aggregate functions* [3]. (For the latter see section 2.3).

Scalar aggregates take a set of tuples (a relation) as an argument and produce a single value as a result. The following describes their syntax and semantics.

Definition 2 *Let r be a relation on the relation schema* $R(A_1, ..., A_n)$. *The syntax for a scalar aggregate f on* A_i *of relation r is* $f((A_i)(r))$. *The semantics are defined by*

$$f((A_i)(r)) = y \; with \;\; y = f\{t[A_i] \mid t \in r\} \qquad (1)$$

A precise definition of the most common aggregate operators is provided next.

Definition 3 *Let the relation r defined on the relation schema R consist of n tuples, with* $n \geq 0$. *Let t be a tuple variable in r and A an attribute in R.*

1. $count((r.A)(r)) = n$
2. $sum((r.A)(r)) = \sum_{\forall t \in r} t[A] \; if \; n > 0$
3. $min((r.A)(r)) = \min_{\forall t \in r} t[A] \; if \; n > 0$
4. $max((r.A)(r)) = \max_{\forall t \in r} t[A] \; if \; n > 0$
5. $avg((r.A)(r)) = \frac{1}{n} \sum_{\forall t \in r} t[A] \; if \; n > 0$
6. $any((r.A)(r)) = sign(n) \;\; where \;\; sign(n) = +1 \; if \; n > 0 \; and \; sign(n) = 0 \; if \; n = 0$.

2. to 5 are assumed to result in 0 whenever n = 0.

A scalar aggregate can be calculated independently from the rest of the query and then simply replaced by its value.

2.3 Aggregate Functions

Aggregate functions compute aggregations over one or more subsets of a relation. They first partition the relation on the values of some attribute and then compute the aggregation separately for each partition. The result of an aggregate function is a relation whose number of tuples equals the number of initial partitions. The result tuples consist of the attribute values on which the partition has been formed and the corresponding aggregate value for each partition. The syntax and underlying semantics [4] are shown below:

Definition 4 *Let r be a relation on* $R(A_1, ..., A_n)$. *The application of an aggregate function f on the attribute* A_i *is specified by* $f\,((A_i)(r)\;BY\;A_j)$ *with* $1 \leq i,j \leq n$. *The semantics of this are given by*

$$f((A_i)(r)BY\,A_j) = \{t[A_j] \circ y \mid (t \in r) \wedge$$
$$(y = f(\{t'[A_i] \mid (t' \in r) \wedge t'[A_j] = t[A_j]\}))\}. \qquad (2)$$

Definition 4 can easily be extended to partition on more than one attribute by replacing the attribute A_j by a set of attributes X with $X \subseteq R$. An example is given next.

Name	Salary	Position	Department
Tom	3500	Assistant	ComputerScience
Jack	4500	Full	ComputerScience
Julie	4000	Full	ComputerScience
Mary	2500	Associate	ComputerScience
Frank	3500	Associate	Engineering

Figure 1: The Prof relation

Example 1 *Given the relation Prof in figure 1. The query "What is the average salary of professors in Computer Science for each rank?" can be expressed by averageSal = avg((Prof.Sal) (SELECT (Prof WHERE Prof.Department = 'ComputerScience')) BY Prof.Position) The result of this is depicted in figure 2.*

Position	Avg-Sal
Assistant	3500
Associate	2500
Full	4250

Figure 2: The Average-Sal relation

As shown below, the evaluation of an aggregate function can be performed in several stages. This decomposition not only facilitates an understanding of the underlying semantics, but it will also serve as a vehicle for the remainder of the paper.

Definition 5 *Let A_i and A_j be attributes of the relation r. An aggregate function, f((A$_i$)(r) BY A$_j$), can be evaluated in the following manner:*

1. *First, the relational expression denoted by r is evaluated in the usual manner.*
2. *Then, the tuples are partitioned by the distinct values of the attribute A_j.*
3. *The aggregate operator f is applied to each partition P generated in step 2, i.e., the aggregate function f is replaced by a set of scalar aggregates of the form $f((A_i)(P))$ which are evaluated by definition 3.*
4. *The result of the aggregate evaluation for each partition is associated with the attribute value from A_j based on which the partition was performed.*

The partition in step 2 of definition 5 is straightforward for precise (crisp) values of A_j. The partition of the relation r on the attribute A_j corresponds to a function from the values a_i of the domain of A_j to a set of tuples taken from r defined by

$$P_r^{A_j}(a_i) = \{t \mid t \in r \wedge t[A_j] = a_i\}. \tag{3}$$

This can easily be extended to a partition on a set of attributes instead of just one attribute A_j.

3 Basic Concepts of Fuzzy Set and Possibility Theory

This section introduces the basic concepts of fuzzy set and possibility theory as proposed by Zadeh [11].

Definition 6 *Let U be a universe of discourse. F is a* **fuzzy subset** *of U, if there is a membership function $\mu_F : U \rightarrow [0,1]$, which associates with each element u $\in$ U a grade of membership $\mu_F(u)$ in F. A fuzzy set F is denoted by $F = \{\mu_F(u_1)/u_1, \mu_F(u_2)/u_2, ..., \mu_F(u_n)/u_n\}$ where $u_i \in U$ for $1 \leq i \leq n$.*

A close connection between fuzzy sets and possibility theory has been established [11].

Definition 7 *A* **possibility distribution** Π_A *for A defined on U is represented by a* **fuzzy set F** *on U whose membership function μ_F is identical to the possibility distribution function π_A, i.e., ($\mu_F(u) = \pi_A(u)$) ($\forall u \in U$).*

Thus, a possibility distribution over a set U can be used to define a fuzzy set of U, or vice versa [7]. This explains why these two concepts are used interchangeably throughout this paper. Furthermore, possibility distributions subsume the conventional and set-value representation [10].

The notion of α-sets [11] allows us to get from fuzzy to crisp sets. This is useful, for example, if we want to make a binary decision based on imprecise information.

Definition 8 *Given a fuzzy set F over U. Then the α-level set of F, denoted by F_α, is defined by $F_\alpha = \{u \in U \mid \mu_F(u) \geq \alpha\}$.*

A fuzzy set F may be decomposed into its level sets through the resolution identity

$$F = \sum_\alpha \alpha F_\alpha \tag{4}$$

where αF_α is the product of α with the set F_α and $\sum_\alpha$ is the union with α ranging from 0 to 1. The product αF_α is calculated as one might expect, hence, $\alpha F_\alpha = \{\alpha/u_1, ..., \alpha/u_n\}$ with $u_i \in F_\alpha$ for all i. In this paper, we assume that the fuzzy sets are based on non-continuous possibility distributions, and hence the summation in equation (4) is finite.

Finally, the extension principle introduced by Zadeh [11] allows arithmetic operations based on numeric values to be extended to apply to possibility distributions.

Definition 9 *Given a binary operation $\circ$ defined on the elements of a universe of discourse U. Then, the operation $\circ$ can be extended to apply to any two possibility distributions Π_x and Π_y over U by the following:*

$$\begin{aligned} & \Pi_x \circ \Pi_y \\ &= \{\pi_x(u1)/u1 \mid u1 \in U\} \circ \{\pi_y(u2)/u2 \mid u2 \in U\} \\ &= \{min(\pi_x(u1), \pi_y(u2))/(u1 \circ u2) \mid u1, u2 \in U\}. \end{aligned}$$

This extension of arithmetic operations is well-defined, since, by assumption, the operation $(u1 \circ u2)$ is well-defined for u1, u2 $\in$ U and the minimum of two real numbers taken from [0,1] is also well-defined, resulting in a real number again from the [0,1] interval.

4 Extending the Relational Data Model

4.1 The Possibilistic Relational Data Model

Various attempts to enhance the relational database model by fuzzy extensions can be found in the literature [1, 7, 12, 10, 8, 5]. This section describes one of them, called the possibilistic relational data model (PRDM) [10], which serves as the basis for the work presented in this paper. A relation in PRDM is defined as follows:

Definition 10 *Let A_i for i from 1 to n be attributes defined on the domain sets U_i, respectively. Then a* **possibilistic relation** *r is defined on $R(A_1, A_2, ..., A_n)$ as a subset of the Cartesian product of a collection of possibility distributions:*

$$r \subseteq P(U_1) \times P(U_2) \times ... \times P(U_n)$$

where $P(U_i)$ denotes the collection of all possibility distributions on U_i.

Each tuple t of r consists of a Cartesian product of possibility distributions on the respective U_i's, i.e., t[A_i] = $\Pi(A_i)$ where $\Pi(A_i)$ is a possibility distribution on U_i.

PRDM allows for data values that can be modeled by possibility distributions. This includes multiple values (e.g. $\{23, 24, 25\}$), discrete possibility distributions (e.g. $\{0.7/130, 0.8/135\}$), linguistic terms as labels for fuzzy sets (e.g. *young*, *about* – 20, *light*), and single values (e.g. 140). For example, the fuzzy set *light* could be represented as $\{1.0/100, 0.9/110, 0.7/120\}$ as depicted in the possibilistic relation in figure 3.

Name	Age	Weight	Sex
Uwe	$\{23, 24, 25\}$	130	male
Anita	*about* – 20	*light*	female
Hans	*young*	$\{0.6/120, 1.0/130\}$	male
Mary	20	110	female

Figure 3: The person relation: a possibilistic relation

4.2 Relational Algebra for the PRDM

Several suggestions can be found in the literature on how to extend the relational algebra operations to deal with possibilistic data [7, 12, 10]. We limit our discussion here to the SELECT operation. (Other relational algebra operations may be found in [8, 10].) The SELECT operation allows for the specification of an approximate match.

Definition 11 *The syntax of the SELECT operation is*

$$SELECT(r\ WHERE\ r.A_i\ is\ F) \quad (5)$$

where F refers to a fuzzy set defined over the domain U_i of the attribute A_i. The query is evaluated by measuring the agreement of each tuple in the relation r with F. This possibility measure [7] is defined by

$$Poss(t[A_i]\ is\ F) = \max_{u \in A_i} \min(\pi_{A_i}(u), \mu_F(u)) \quad (\forall u \in U_i) \quad (6)$$

The result of a selection operation is a set of tuples, each with an associated measure of how it satisfies the query. It is, in general, useful to specify a threshold of acceptance $\alpha \in [0,1]$ to select all tuples that match the selection criteria to at least that degree α (see the α-level set concept of definition 8).

Lemma 1 *If the data is crisp, e.g., $t[A_i] = u = \{1.0/u\}$, then equation 6 simplifies to the following possibility measure for a tuple t of r:*

$$Poss(t[A_i] is F) = min(1.0, \mu_F(u)) = \mu_F(u). \quad (7)$$

For possibilistic data but a crisp selection equation 6 defaults to

$$Poss(t[A_i] = u) = min(\pi_{A_i}(u), 1.0) = \pi_{A_i}(u). \quad (8)$$

A selection condition comparing two imprecisely specified attribute values is evaluated similarly.

Definition 12 *Given the attributes A_i and A_j of relation r. To find all tuples with matching attribute values for A_i and A_j we write*

$$SELECT(r\ WHERE\ r.A_i\ =\ r.A_j) \quad (9)$$

This is evaluated by measuring the agreement of the two attribute values for each tuple in the relation:

$$Poss(t[A_i] = t[A_j]) = \quad (10)$$
$$\max_{u \in dom(A_i) \cup dom(A_j)} \min(\pi_{A_i}(u), \pi_{A_j}(u))).$$

This evaluation could be extended to incorporate the similarity between domain values [10], or, by applying additional fuzzy modifiers, such as, the modifier *very* to the fuzzy set *old* [12].

5 Extending FRQLs with Aggregates

In this section, it is investigated how the aggregate operators presented in section 2 can be redefined to cope with the possibilistic representation of data. The following three cases are to be considered:

- the data over which the aggregate is to be evaluated could be crisp or possibilistic;
- the selection of tuples considered for the aggregation could be precise or vague;
- the data on which the partition is to be based could be crisp or possibilistic.

The third case applied only to aggregate functions, while the others must be considered for both, scalar aggregates and aggregate functions. An important goal of this research in all of the above cases is to satisfy two principles: *consistency:* the generalized operations should default to the crisp operations for conventional data;

and *completeness:* the operations should be well-defined for the fuzzy case [1]. These principles require the generalized versions of aggregates to be natural extensions of their crisp counterparts.

5.1 Aggregate Evaluation of Vague Queries on Crisp Data

In the following, the aggregate operators are extended to cope with approximate queries (fuzzy predicates) on crisp data. An example is the query "What is the average of the *high* salaries of all professors?". More generally, the queries have the form

$$f((A_i)(SELECT(r\ WHERE\ \mathbf{r.A_j\ is\ F}))) \quad (11)$$

where F refers to a fuzzy set defined over the domain of the attribute A_j. It is not necessary that $i \neq j$ as the above example query indicates. F could be a complicated expression involving fuzzy modifiers or a conjunction/disjunction of several fuzzy sets.

According to lemma 1, the evaluation of "$t[A_j]$ *is F*" results in $\mu_F(u)$ if $t[A_j] = u$. This value indicates the degree to what the tuple matches the selection criteria. It implies that the different tuples should participate to different degrees in the evaluation of the aggregate. The number of tuples to be considered to be in agreement with the selection depends on the chosen level of acceptance, denoted by α. Recall that α is inversely proportional to the number of elements able to satisfy α. We introduce the notion of an α-level relation, based on which we then define how a scalar aggregate is to be evaluated in a fuzzy query.

Definition 13 *Let r be a relation defined over the relation schema $R(A_i, A_j, ...)$ and F be a fuzzy set over the domain of A_j. Let $\alpha \in [0,1]$. Then the α-level relation $A_j^r(\alpha)$ is defined by:*

$$A_j^r(\alpha) = \{t \mid t \in r \wedge \mu_F(t[A_j]) \geq \alpha\} \quad (12)$$

This is based on the concept of an α-level set F_α (definition 8). For a given $\alpha \in$ [0,1] all tuples which satisfy the proposition "r.A_j is F" to at least the degree α are collected in $A_j^r(\alpha)$.

Definition 14 *Given r, A_j, and F from definition 13. A query of the form shown in equation 11 is evaluated by calculating the aggregate f on $A_j^r(\alpha)$ for all α and associating α with the result. The semantics are:*

$$f((A_i)(SELECT(r\ WHERE\ r.A_j\ is\ F))) =$$

$$\{\mu_f(y)/y \mid \mu_f(y) = sup_\alpha\{f((A_i)(A_j^r(\alpha))) = y\}\}. \quad (13)$$

The fuzziness of F induces a fuzzy set of possible answers instead of one value. The following example is given to illustrate the above definition.

Position	$\mu_{Prestigious}$
Assistant	*0.5*
Associate	*0.8*
Full	*1.0*

Figure 4: The fuzzy set Prestigious

Example 2 *Let Prof be the relation defined in figure 1. Let Prestigious be a fuzzy set defined on the set of different positions shown in figure 4.*

The query "What is the average salary of employees holding prestigious positions?" is formally expressed by avg((Salary)(SELECT(Prof WHERE Prof.Position IS Prestigious))). This query is evaluated by constructing α-level relations (see definition 13 and 14). Let t1 be the first tuple in the Prof relation, t2 the second, etc.

For $\alpha = 1.0$, $Position^{Prof}(\alpha) = \{t2, t3\}$ by equation 12. Then, avg((Salary) ($Position^{Prof}$(1.0))) = avg(4500, 4000) = 4250. Similarly, for $\alpha = 0.8$, avg((Salary) ($Position^{Prof}$(0.8))) = 3625, and for $\alpha = 0.5$, avg((Salary) ($Position^{Prof}$(0.5))) = 3600. Thus the result of the query is the fuzzy set {1.0/4250, 0.8/3625, 0.5/3600}.

The threshold level α determines which values are taken into consideration for the evaluation. The smaller α is the more elements are included in the evaluation.

Lemma 2 *Let r be a relation defined on a relation schema containing the attributes A_i and A_j. If $\alpha_1 \leq \alpha_2$ then $A_j^r(\alpha_2) \subseteq A_j^r(\alpha_1)$. This implies that $max((A_i)A_j^r(\alpha_1)) \geq max((A_i)A_j^r(\alpha_2))$. Correspondingly, $min((A_i)A_j^r(\alpha_1)) \leq min((A_i)A_j^r(\alpha_2))$ and also $sum((A_i)A_j^r(\alpha_1)) \geq sum((A_i)A_j^r(\alpha_2))$.*

There is no monotonicity for the other aggregates [6]. It may be of interest to summarize the result of such a query in a more concise way. A simple approach is to give an α value which one considers to be a sufficient matching degree; then the result of applying the aggregate to $A^r(\alpha)$ would be returned. While the discussion in this section has concentrated only on scalar aggregates, the approach proposed here can be directly extended to aggregate functions. This is done by first partitioning the relation, which produces precise partitions, since the underlying data is crisp. Then, each partition can be treated as outlined above.

5.2 Evaluation of Scalar Aggregates on Possibilistic Data

The queries considered in this section correspond to the ones described in definition 3, however, possibility distributions are now allowed as attribute values instead of just simple constants. The proofs of consistency are omitted, but can be found in [9].

5.2.1 Generalized Count Aggregate

The count aggregate, defined in definition 3, returns the number of tuples of a relation. One can directly adopt this definition for the possibilistic case. This is referred to as *fcount*1. An alternative is the use of the sigma-count operation [11], which is defined as follows.

Definition 15 *Given a fuzzy subset F over U, i.e.,* $F = \{\mu(u_1)/u_1, \mu(u_2)/u_2, ..., \mu(u_n)/u_n\}$*. Then, the cardinality of F, called sigma-count, is the sum of the grades of memberships in F. Thus, sigma-count(F)* $= \sum_{\forall i} \mu(u_i)$.

We propose to use the sigma-count as an extended count aggregate, here called *fcount*2.

Definition 16 *Given a relation r defined on the relation schema R(..,A,...). Let A be defined on the domain* $U = \{u_1, ..., u_n\}$*. The relation r consists of tuples* t_i *with* $1 \leq i \leq m$*. Let* $t_i[A] = \{\mu_i(u_1)/u_1, ..., \mu_i(u_n)/u_n\}$ $(\forall i)$*. Then the fcount2 aggregate on the attribute A of r is defined by* $fcount2((A)(r)) = \sum_{i=1}^{m} \sum_{j=1}^{n} \mu_i(u_j)$.

The result of a *fcount*2 operation is a real number, but it is understood that the result may be rounded, if necessary. The *fcount*2 operation does not exhibit all features of the conventional count operation, i.e., it does not always return the same value for the different attributes of a relation. If this characteristic is required, then the designer will choose the *fcount*1 operator over the *fcount*2 aggregate. The *fcount*2 operator has been proven useful for defining the generalized average operator (see section 5.2.5).

Example 3 *The fcount1 for the Weight attribute of the person relation in figure 3 is 4, since there are four tuples. Whereas, fcount2((Weight)(person)) = 6.2.*

5.2.2 Generalized Sum Aggregate

The sum aggregate is is only defined for numeric domains. Necessary characteristics of the extended sum aggregate are commutativity and associativity, since the relational data model does not impose any order on the tuples of a relation [2]. The sum aggregate, here termed *fsum*, is based on the extension principle (definition 9).

Definition 17 *Given r, R, A, U and the tuples* t_i *from definition 16. The tuples* t_i *are assumed to take on possibilistic values for the attribute A, i.e., for i from 1 to m we have* $t_i[A] = \{\mu_i(u_{ki})/u_{ki} \mid 1 \leq ki \leq n\}$*. The fsum aggregate of the attribute A of a relation r is defined by* $fsum((A)(r)) = \{u/y \mid ((y = \sum_{ki=k1}^{km} u_{ki}) \wedge (u = \min_{i=1}^{m} \mu_i(u_{ki}))) \; (\forall k1, ..., km : 1 \leq k1, ..., km \leq n)\}$.

It is fairly straigthforward to see that the *fsum* operation is commutative as well as associative, since both the summation and the minimum operation are. The result of the *fsum* aggregate is in general a possibility distribution. An example of the fsum aggregate operator is given next.

Example 4 *The fsum aggregation of the Weight of the relation person of figure 3 is fsum((Weight)(person))* $= \{0.6/460, 1.0/470, 0.9/480, 0.7/490\}$.
This result states that the sum of all values is 470 with the possibility of 1.0, and that values close to 470 are also possible results.

5.2.3 Generalized Max Aggregate

The max aggregate is defined for any domain that has an order defined on it. The extension of the max aggregate to deal with possibility distributions, here called *fmax*, is again based on definition 9. Consequently, commutativity and associativity of the operation are guaranteed.

Definition 18 *Given r, A, U and* t_i*'s from the previous definition. The fmax scalar aggregate is defined by* $fmax((A)(r)) = \{u/y \mid ((y = \max_{ki=k1}^{km} u_{ki}) \wedge (u = \min_{i=1}^{m} \mu_i(u_{ki}))) \; (\forall k1, ..., km : 1 \leq k1, ..., km \leq n)\}$

This generalization of the maximum aggregate is well-defined for the same reasons that the extension principle is. The result of the *fmax* aggregate is in general a possibility distribution. An example of the *fmax* operation is given below.

Example 5 *The fmax of the Weight attribute of relation person of figure 3 is fmax((Weight)(person))* $= \{0.6/130, 0.6/130, 0.6/130, 1.0/130, 0.9/130, 0.7/130\} = 130$ *This is a realistic result, since 130 is indeed the maximum value.*

5.2.4 Generalized Min Aggregate

The min aggregate is extended for the possibilistic data models in the same manner as the max aggregate, except for replacing the symbol max in definition 18 by min.

Example 6 *The fmin of the Weight attribute of relation person of figure 3 is fmin((Weight)(person))* $= \{1.0/100, 0.9/110\}$*. This is intuitive, since 100 and 110 are among the lowest values of the Weight attribute.*

5.2.5 Generalized Avg Aggregate

One could define the generalized average aggregate in terms of the quotient of the generalized sum and count aggregates. We take an alternative route based on the possibilistic expected value, PEV, a concept introduced by Zemankova and Kandel [12].

Definition 19 *Let A be an attribute of a numeric domain U. Let $\pi(u_i)$ be the possibility distribution for value u_i, and n the number of $\pi(u_i)/u_i$ pairs for the attribute A. The PEV for attribute A is defined by*

$$PEV(A) = \frac{\sum_{i=1}^{n} \pi(u_i)*u_i}{n}$$

We propose to define the average aggregate *favg* in terms of the PEV operation by replacing the denominator n by *fcount2*.

Definition 20 *Given r, A, U and the tuples t_i from definition 16. Now, the favg operator is defined to be*

$$favg((A)(r)) = \frac{\sum_{i=1}^{m}\sum_{j=1}^{n} \mu_i(u_j)*u_j}{fcount2((A)(r))}.$$

The *favg* operator results in a real number, and may be rounded, if necessary. The *favg* operator is demonstrated by an example below.

Example 7 *Given the relation in figure 3. Since $fcount2((Weight)(person)) = 6.2$, we have* $favg((Weight)(person)) = (1*100+0.9*110+0.7*120 +1*130+0.6*120+1*130+1*110)/6.2 = 725/6.2 = 116.9$

Finally, the conventional *any* operator can be directly adopted from definition 3 since it tests whether there is a tuple in the relation or not, and thus does not consider the actual content of the relation.

5.3 Aggregate Evaluation of Aggregate Functions on Possibilistic Data

As was outlined in section 2.3, aggregate functions are based on the evaluation of scalar aggregates. This also holds for aggregate functions in the PRDM. If the attribute values on which the partition is based are precise values then the extensions to be made for the possibilistic relational database are straightforward as shown in the next section. The case where attribute values on which the partition is based are possibilistic is presented in section 5.3.2.

5.3.1 Partitioning on Precise Data

If the attribute values on which the partition is based consist exclusively of crisp values, then it is in fact possible to directly translate the procedure described in definition 5 by translating the individual operations of step 1, such as selection by WHERE clause, to their corresponding fuzzy counterparts. Thus, it may affect step 1 of the evaluation process described in definition 5. This is because the partition of tuples by the BY clause produces an exact partition. The only other change concerns the scalar aggregates applied in step 3 of the procedure which now are replaced by the generalized scalar aggregates as defined in section 5.2. The following example demonstrates the procedure of evaluating an aggregate function over possibilistic data by partitioning it on precise data.

Example 8 *Let person be the relation given in figure 3. The values for the Weight attribute are possibilistic, but the values for the Sex attribute are all precisely known. The query "What is the average Weight of the persons for each sex?" is formally expressed by favg((Weight)(person) BY Sex). This can be evaluated as follows:*

1. *Step one is not needed.*
2. *First, find the attribute values to partition on by*

   $$PROJECT_{[Sex]}(person) = \{male, female\}.$$

   Thus, by equation (3) there are the following two partitions:

   - $P^{Sex}_{person}(male) = \{(Uwe, ..., 130, male), (Hans, ..., \{0.6/120, 1.0/130\}, male)\};$
   - $P^{Sex}_{person}(female) = \{(Anita, ..., \{1.0/100, 0.9/110, 0.7/120\}, female), (Mary, ..., 110, female)\}.$
3. *Next, apply the scalar aggregate operator average to each partition:*
   - $favg((Weight)(P^{Sex}_{person}(male))) = favg(130, \{0.6/120, 1.0/130\}) = 127.7$
   - $favg((Weight)(P^{Sex}_{person}(female))) = favg(110, \{1.0/100, 0.9/110, 0.7/120\}) = 109.1$
4. *Associate the result obtained for each partition with the value on which that partition was based. The result relation is depicted in figure 5.*

Sex	*Avg-Weight*
male	*127.7*
female	*109.1*

Figure 5: The avg-weight-by-sex relation

5.3.2 Partitioning on Possibilistic Data

Below, we base the partitioning process on an attribute containing possibilistic data. Under these circumstances, one cannot find a real partition any more. The best one can hope for is to determine to what degree a tuple participates in a given partition. Furthermore, it is possible that tuples participate in and thus contribute to more than one partition. Note the similarity between this situation and the case discussed in section 5.1. For simplicity, we assume that the attribute over which the aggregate is to be evaluated is of crisp nature. The problem of partitioning on possibilistic data leads to the introduction of a new concept, the α-level partition, which is defined in definition 21. Definition 22, a variation of definition 14, describes how the aggregate function $f((A_i)(r)BY\,A_j)$ is to be evaluated when allowing possibilistic values for A_j while A_i consists of crisp values.

Definition 21 *Let r be a relation defined over the relation schema $R(A_i, A_j, ...)$. Let the attribute A_i be crisp, and the attribute A_j be possibilistic. Let D be the active domain of A_j, i.e. $D = \{a \mid (t \in r) \wedge \mu_{t[A_j]}(a) > 0\}$ denotes the set of values on which the partition is to be based. The partition function P as defined in equation 3 has to be modified to accommodate for membership values, thus, for all $a \in D$ we have:*

$$P_r^{A_j}(a) = \{t \mid (t \in r) \wedge (a \in D) \wedge (\mu_{t[A_j]}(a) > 0)\} \quad (14)$$

Let $\alpha \in [0,1]$. For each partition $P_r^{A_j}(a)$ define the α-level partition $L_r^{A_j}(\alpha, a)$, a function of two variables α and a, by:

$$L_r^{A_j}(\alpha, a) = \{\, t \mid (t \in P_r^{A_j}(a)) \wedge (\mu_{t[A_j]}(a) \geq \alpha)\} \quad (15)$$

Definition 22 *Given r, A_i, A_j and D from the previous definition. Then the query $f((A_i)(r)BY\,A_j)$ is evaluated by computing for each partition $P_r^{A_j}(a)$ the supremum of the aggregate evaluation of all associated α-level partitions $L_r^{A_j}(\alpha, a)$ for all α. More precisely,*
$f((A_i)(r)BY\,A_j) = \{a \circ F \mid (a \in D) \wedge$

$$F = \{\mu_f(y)/y \mid \mu_f(y) = sup_\alpha\{f((A_i)(L_r^{A_j}(\alpha, a))) = y\}\}. \quad (16)$$

An example is given next to illustrate the previous definition.

Name	Salary	Position
Tom	3500	$\{1.0/Assistant, 0.8/Associate\}$
Jack	4500	$\{1.0/Full\}$
Julie	4000	$\{1.0/Full, 0.5/Associate\}$
Mary	2500	$\{1.0/Associate, 0.7/Assistant\}$
Frank	3500	$\{1.0/Associate\}$

Figure 6: The Prof relation

Example 9 *Given the relation Prof from figure 6. The query "What is the maximum salary of professors per position" formally expressed by fmax((Salary)(Prof) BY Position) has to be evaluated according to definition 22.*

1. *Again, step 1 is not needed for this example.*
2. *First, find the attribute values to partition on; they correspond to the active domain D which is $D = \{Full, Associate, Assistant\}$. This results in three partitions according to equation (14). Each tuple in a partition has associated with it a membership value indicating to what degree it participates in that partition:*
   - $P_{Prof}^{Position}(Full) = \{(Jack, ...)\ with\ 1.0; (Julie, ...)\ with\ 1.0.\}$
   - $P_{Prof}^{Position}(Associate) = \{(Tom, ...)\ with\ 0.8; (Julie, ...)\ with\ 0.5; (Mary, ...)\ with\ 1.0; (Frank, ...)\ with\ 1.0\}$
   - $P_{Prof}^{Position}(Assistant) = \{(Tom, ...)\ with\ 1.0; (Mary, ...)\ with\ 0.7\}$
3. *Now, by equation 15, we can determine for all partitions $P_{Prof}^{Position}(a)$ the different α-level partitions $L_{Prof}^{Position}(\alpha, a)$ for each distinct α. Then, we apply the scalar aggregate operator, max, to those α-level partitions $L_{Prof}^{Position}(\alpha, a)$:*
   - *fmax((salary)($P_{Prof}^{Position}$(Full)) :*
     - *for $\alpha = 1.0$: fmax((Salary)($L_{Prof}^{Position}$(1.0, Full)) = max(4500, 4000) = 4500.*
   - *fmax((salary)($P_{Prof}^{Position}$(Associate)) :*
     - *for $\alpha = 1.0$: fmax((Salary)($L_{Prof}^{Position}$(1.0, Associate)) = 3500.*
     - *for $\alpha = 0.8$: fmax((Salary)($L_{Prof}^{Position}$(0.8, Associate)) = 3500.*
     - *for $\alpha = 0.5$: fmax((Salary)($L_{Prof}^{Position}$(0.5, Associate)) = 4000.*
   - *fmax((salary)($P_{Prof}^{Position}$(Assistant)) :*
     - *for $\alpha = 1.0$: fmax((Salary)($L_{Prof}^{Position}$(1.0, Assistant)) = 3500.*
     - *for $\alpha = 0.7$: fmax((Salary)($L_{Prof}^{Position}$(0.7, Assistant)) = 3500.*
4. *The result of the scalar aggregate evaluation for each α-level partition is associated with the respective level α. Finally, combine all these $\alpha/L_{Prof}^{Position}(\alpha, a)$ pairs for all α-level partitions of a given partition $P_{Prof}^{Position}(a)$ to form a possibility distribution. The resulting relation is depicted in figure 7.*

Position	Max-Salary
Full	$\{1.0/4500\}$
Associate	$\{1.0/3500, 0.5/4000\}$
Assistant	$\{1.0/3500\}$

Figure 7: The max-sal-per-position relation

Finally, if both attributes, i.e. the attribute to partition on and the attribute on which to evaluate the aggregate function are possibilistic, then according to definition 22 the result of the third step of the aggregate evaluation process would be of the following form: $0.7/\{1.0/2500, 0.8/3500, 0.5/4000\}$. This can be evaluated by taking the minimum of the membership values, which results in $\{0.7/2500, 0.7/3500, 0.5/4000\}$.

5.4 Vague Queries on Possibilistic Data

We conclude with a discussion of how to combine the aggregate evaluation approach dealing with vague queries (section 5.1) with that for handling possibilistic data (section 5.2 and section 5.3). At this point, we favor the simplification of the evaluation strategy as indicated at the end of section 5.1. More specifically, the user chooses an α value s/he considers to be an acceptable threshold value for an approximate selection. When the user specifies a query involving a vague selection clause, such as "Find all *old* people", the result will be the list of tuples which satisfy that selection criteria with at least the degree α. Hence, the result of a vague selection would be an α-level relation described in equation 12.

Introducing possibilistic data concerns us in as much as the data underlying the selection clause is possibilistic. Then the evaluation of a vague selection clause, such as "Age is *old*", has to be altered. Definition 14 is modified by evaluating the vague selection in accordance with equation 6 instead of lemma 1. This again produces a possibility measure for each tuple describing the degree of matching between the tuple and the selection clause. All tuples with a possibility measure above the threshold α are collected in the respective α-level relation, i.e., they will be considered further in the query evaluation process. In terms of the aggregate evaluation process outlined in definition 5, this means that the vague query part is dealt with in step 1 of the process. A non-base relation (α-level relation) is generated in step 1, based on which the aggregate evaluation process can continue with the remaining steps as discussed in section 5.2 and 5.3.

6 Conclusions

The PRDM [10] is a generalized version of the relational database model capable of capturing precise, as well as imprecise, data. By extending the query languages for the conventional relational database model, the model is able to handle vague queries on both types of data. However, to make use of the PRDM in real world applications, it became imperative to develop and incorporate suitable aggregate evaluation mechanisms, found in most existing relational database systems, such as, System R or Ingres [3].

The major contribution of this paper is the development of a framework of evaluation procedures for scalar aggregates as well as aggregate functions for the PRDM. In particular, our approach handles the application of aggregate operators within vague queries as well as the application of aggregates to possibilistic data found in the PRDM. We show that the handling of aggregate functions in the PRDM can be handled in a natural and intuitive manner. The usefulness of all new operators is demonstrated in a pragmatic manner by intuitive examples.

Possible extensions to this research are manifold. For instance, aggregate operations for fuzzy data models which are not based on the notion of discrete possibility distributions are needed, e.g., interval representations, continuous functional descriptions, or continuous possibility distributions. Also, the question of update operations in possibilistic databases is mostly unexplored. For example, it is not obvious how to perform an operation such as "Update the salaries of all *old* people".

References

[1] Buckles, B. P. and Petry, F. E. A Fuzzy Representation of Data for Relational Databases. *Fuzzy Sets and Systems.* 7, (1982), 213 - 226.

[2] Codd, E. F. A relational model of data for large shared data banks. *Communications of the ACM* 13, 6 (June 1970), 337 - 387.

[3] Date, C. J. *A Guide to Ingres.* Addison-Wesley Pub. Co., Reading, Mass, 1987.

[4] Klug, A. Equivalence of Relational Algebra and Relational Calculus Query languages Having Aggregate Functions. *Journal of ACM* 29, 3 (July 1982), 699 - 717.

[5] Oezsoyoglu, G., Oezsoyoglu, Z. M. and Matos, V. Extending Relational Algebra and Relational Calculus with Set-Valued Attributes and Aggregate Functions. *ACM Trans. on Database Systems* 12, 4 (Dec. 1987), 566 - 592.

[6] Prade, H. Global Evaluations of Fuzzy Sets of Items in Fuzzy Data Bases. *Int. Workshop on Fuzzy System Applications.* Fukuoka, Japan, (Aug 1988).

[7] Prade, H. and Testemale, C. Generalizing database relational algebra for the treatment of incomplete/uncertain information and vague queries. *Information Science* 34, (1984), 115 - 143.

[8] Raju, K. V. S. V. N. and Majumdar, A. K. Fuzzy Functional Dependencies and Lossless Join Decomposition of Fuzzy Relational Database Systems. *ACM Trans. on Database Systems* 13, 2 (June 1988), 129 - 166.

[9] Rundensteiner, E. A. and Bic, L. Evaluating Aggregate Functions on Possibilistic Data. University of California, Irvine, Information and Computer Science Department, Tech. Rep. 89-12, May 89.

[10] Rundensteiner, E. A. The Development of a Fuzzy Temporal Relational Database (FTRDB): An Artificial Intelligence Application. Master's thesis. Dept. of Computer Science. Florida State University. 1987.

[11] Zadeh, L. Fuzzy Sets. *Information and Control* 8, (1965), 338 - 353.

[12] Zemankova, M. and Kandel, A. *Fuzzy Relational Database - A Key to Expert Systems.* Koeln: Verlag TNV Rheinland, 1984.

Extending the Relational Algebra to Capture Complex Objects

Bernhard Mitschang
University of Kaiserslautern, 6750 Kaiserslautern, FRG

Abstract

An important direction in database research for non-standard applications (e.g. engineering or design applications) deals with adequate support for complex objects. Without doubt, the provision of network structures and shared subobjects as well as support for dynamic object definition and appropriate manipulation facilities is urgently needed for natural and accurate modeling as well as for efficient processing of the applications' objects. These concepts are the major concern of the molecule-atom data model (MAD model) and its molecule algebra which is introduced in this paper. They make the model stand out compared to the relational model and even to models limited to hierarchical and statically defined complex objects. By means of the molecule algebra a precise and complete specification of one conceivable kind of complex object processing and its inherent semantics is provided. Furthermore, this algebra is used as a sound basis to express the semantics of the high level query language MQL (molecule query language) that is able to deal with complex objects in a descriptive manner.

Keywords:

complex objects, structural object orientation, relational algebra extensions, formal specification

1. Motivation

Over the last few years the development of a new generation of database systems capable of supporting non-standard application areas such as engineering applications for CAD/CAM and VLSI design, knowledge-based systems, and office applications has emerged as an important direction in database system research.

One uniting characteristic over all these advanced applications is adequate support for complex objects, which is quite different to conventional data processing in business applications. Thus, most current research topics refer to some kind of object-orientation and extensibility reflected in the data model and in its implementation. There are different approaches which are distinguishable starting from only a few selected extensions of the relational model [RS87,LK84] and leading up to the integration and superposition of structures on relations [Da86, PSS-WD87,CDV88].

Though many of these proposals are interesting, they are mostly limited to hierarchical complex objects, which are in addition in most cases statically defined (i.e. fixed in the database schema). As stated in [BB84] *), the

- provision of shared subobjects, and
- support for dynamic object definition (of course in combination with powerful manipulation facilities)

is urgently needed for natural and accurate modeling and efficient processing of the applications' objects.

In this paper we attack the above mentioned problems by means of the molecule-atom data model (MAD model [Mi88b]), which is an advancement to the relational model and is based on a generalized notion of complex objects. These objects are called molecules and are dynamically constructed from atoms, which are used as basic building blocks. Since molecules can overlap having non-disjoint atom sets, the model allows in a natural way for the sharing of subobjects. With the relational model in mind, chapter two offers an easy to understand introduction to the relevant concepts underlying the MAD model. When dealing with these network-like structures it is necessary to rely on a sound and precise definition of the model's data structures and its operations. Since a formal specification of the data model seems to be indispensable, we introduce in chapter three the so-called *molecule algebra* capable of handling complex objects exhibiting network structures and shared subobjects. With that, a formal description of complex object processing (here, simply called molecule processing) and its inherent semantics is provided. Thus, the molecule algebra appears to be an extension to the relational algebra [Ul80] and also to the non-first-normal-form algebra [SS86] that supports only hierarchical complex objects without shared subobjects. In the fourth chapter, we show how such a formal specification (here, it is the molecule algebra) is used as a sound basis to define the semantics of a higher level query and manipulation language (here, it is the molecule query language MQL). Finally, we conclude with a short comparison to other models and an outlook to upcoming future research topics. As far as is known to the author, no satisfying complex-object algebra (capable of handling network and recursive structures as well as dynamic object definition) has been proposed until now.

*) [BB84] '... support for molecular objects should be an integral part of future DBMSs ...', where 'molecular' objects are classified according to their structure, leading to disjoint/non-disjoint and recursive/non-recursive complex objects.

Proceedings of the Fifteenth International Conference on Very Large Data Bases — Amsterdam, 1989

2. An Informal Introduction to the MAD Model

Throughout the paper we use a simple example from a geographic application area for further explanatory purposes. As shown in the upper part of fig.1, we concentrate on a cartographic view to our universe of discourse (here, Brasil, its states, cities, rivers etc.), which is also modeled by means of the well-known ER model. The corresponding ER diagram depicts a geographical model as part of the schema that is used to share all geographic structures (i.e. points to build up edges, which are in turn used to construct areas and nets) among all application objects (e.g. cities, states, rivers etc.) thereby avoiding any data redundancies: point-like objects (e.g. city), network-like objects (e.g. river, street, flight), and area-like objects (e.g. state) are modeled by means of this common geographical model. Thus, different complex objects are contained in one schema sharing common subobjects. For example, the river Parana shares with the states Minas Gerais, Sao Paulo, and Parana some edge and point tuples representing in one case the course of the river and in another case the border of the states. This sharing of components is expressed by the n:m relationship types. It is easy to imagine that a transformation to the relational model becomes quite cumbersome, since all n:m relationship types have to be modeled by some auxiliary relations. With this, the queries and their processing obviously become more complicated and perhaps less efficient. To overcome these problems and to provide an effective complex object concept, we have developed the MAD model as an extension of the relational model.

In the MAD model *atoms* are used as a kind of basic element to represent the real world entities. They play a similar role to tuples in the relational model. Each atom consists of attributes of various data types, is uniquely identifiable, and belongs to its corresponding *atom type*. Contrary to the foreign/primary key connections of the relational model, all relevant relationships between the entities are expressed by so-called *links* that are defined

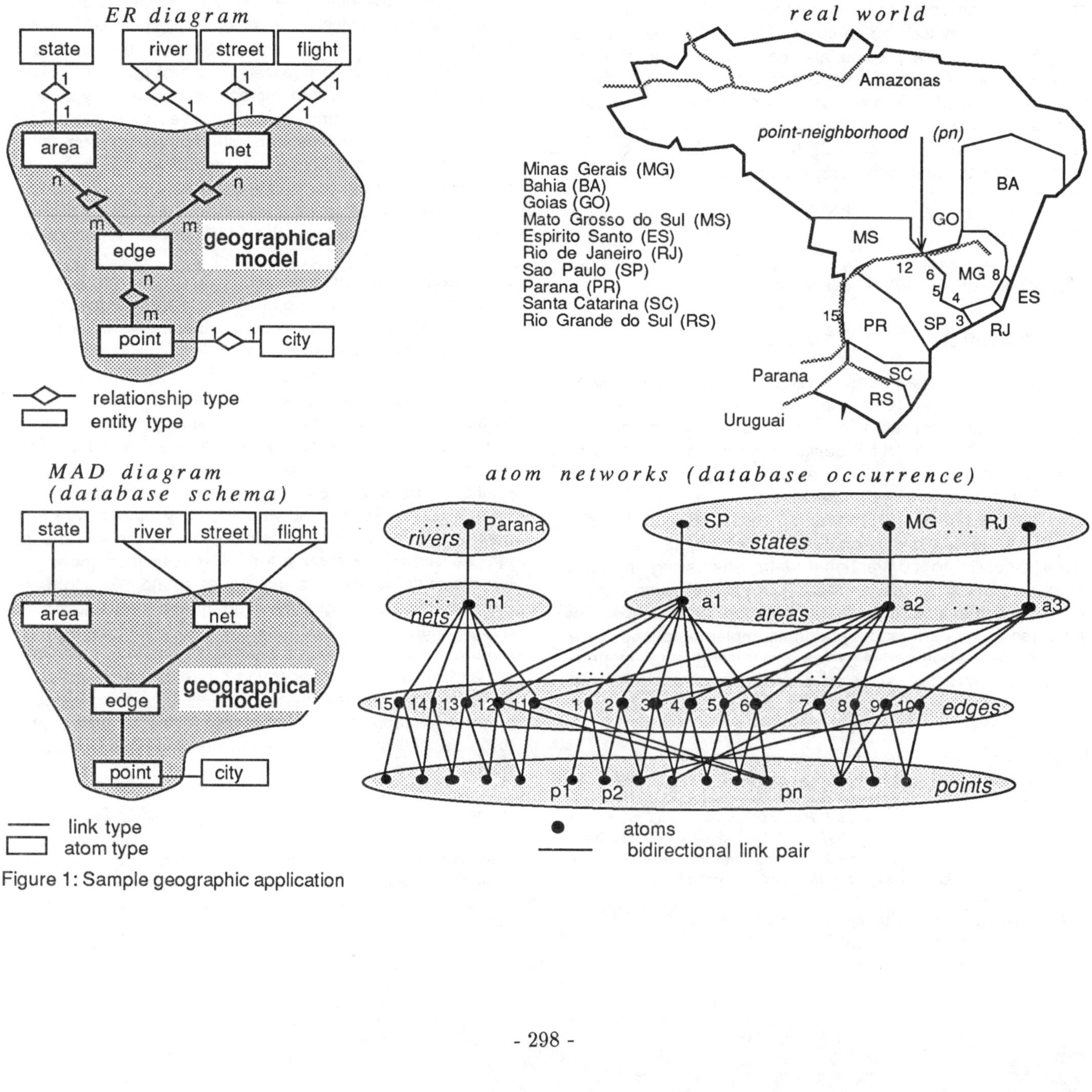

Figure 1: Sample geographic application

as *link types* between the corresponding atom types. Link types are used to accurately map all types of relationships (1:1, 1:n, and n:m). (Note, the link concept is a logical concept and shall not be confounded with physical storage structures). The direct representation and the consideration of bidirectional, i.e. symmetric links establish the basis of the model's flexibility. Hence, in the database all atoms connected by links form meshed structures, called *atom networks*. The corresponding *database schema* exhibits a set of atom types, whereof some are connected via link types establishing nondirectional graphs. The lower part of fig.1 shows these concepts applied to our cartographic application.

A closer look at the ER diagram and the corresponding MAD diagram in fig.1 reveals that there is a one-to-one mapping from the ER model to the MAD model associating each entity type with an atom type and each relationship type with a link type. Compared to the relational model, here we don't have to use any auxiliary structures.

Based on the atom networks the model's complex objects are dynamically definable as higher level objects seen as a structured (i.e. coherent) set of possibly heterogeneous atoms. In analogy to chemistry, we call these complex objects *molecules* to express more vividly that several atoms could 'combine' in different ways to build up the desired molecules. The 'formula' or the procedure to determine what kinds of molecules one is interested in is specified by means of the *molecule structure*. This description is a subgraph of the database schema, which establishes a coherent, directed and acyclic type graph, whose nodes are the atom types and whose edges are the link types. Each graph has a unique starting point called the root atom type. To each molecule structure, there exists a corresponding *molecule set*, which groups all molecules adhering to the specified structure. At least from the concept level point of view the *derivation of molecules* proceeds in a straight-forward way using the molecule structure as a kind of template, which is laid over the atom networks. Thus, for each atom of the root atom type one molecule is derived following all links determined by the link types of the molecule structure to the children, grandchildren atoms etc. till the leaves are reached. Derivation of the children atoms means performing the hierarchical join [LK84] along the specified branches. Both, the molecule structure together with its derived molecule set are denoted *molecule type*.

The flexibility of the MAD model stems from the fact that the same database (i.e. atom networks) can be used to derive totally different molecule types, just by specifying and deriving different molecule structures. This is shown in fig.2. The reason that it works well, firstly lies in the direct and bidirectional link concept allowing for a symmetric use of the database: e.g. looking at the 'point neighborhood', i.e. going from point to edge, and from edge to both area and state as well as to net and river, as exemplified in fig.2. Secondly, the database schema is primitive in the sense that it is not superposed by some static structures used, for complex object definition, as it is the case for example in [LK84,Da86]. Our complex object definition is defined on demand in the queries and not fixed in the schema. Therefore, we term our approach *dynamic object definition*. Since it is possible to combine molecules having non-disjoint atom sets, (which show a general graph structure, and not only a strict hierarchical one) the model allows in a natural way for the sharing of subobjects. This fact is especially depicted in fig.2. Of course, disjoint objects showing only hierarchical (graph) structures are just special cases thereof.

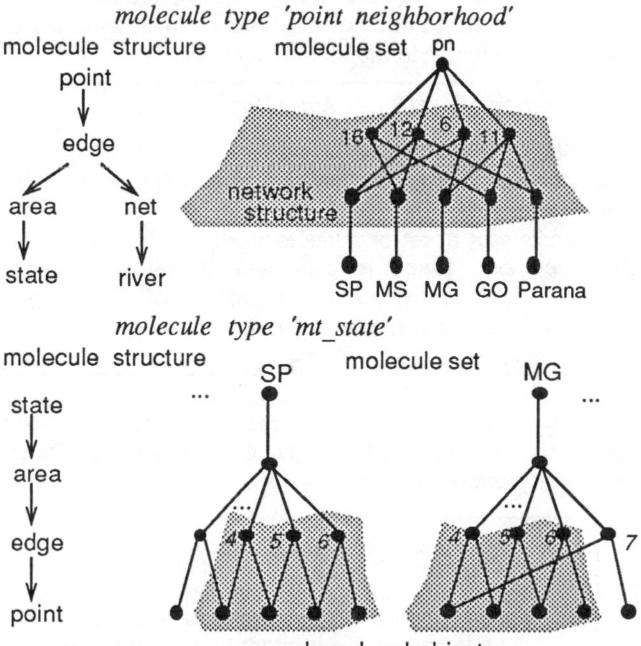

Figure 2: Some complex objects

3. The Molecule Algebra

In this chapter we provide a precise definition of the MAD model's data structures and of its algebra operations. These formal specifications will contain the relational model as well as the basic non-first-normal-form model (with some minor cuts) as degeneration. At a first glance, the reader might be astonished at the formalism introduced for more or less common facts. But in case of the MAD model and its inherent support for network structures and dynamic object definition, we cannot dispense with formal definition of basic concepts (e.g. type, structure or description, and occurrence information), because otherwise we could not define the effect of the algebra operators formally and we could not prove the closure of the molecule algebra. In the following we presume that the reader is familiar with the relational model and its formalization (e.g. [Ul80]) as well as with basic mathematical notations. Furthermore, we assume a given set N the elements of which will be used for naming purposes and the symbol ∇ indicating the end of definitions, theorems, and proofs.

3.1 Basic Concepts

Here we define the basic building blocks of the MAD model that is

- on the data structure side: atom and atom type, link and link type as well as atom network and database
- on the operational side: operations for projection, restriction, cartesian product, union, and difference, all working on atom types (and showing some effects on link types).

The data structures and the corresponding operations introduced here are very close to those known from the relational model. Therefore, we omit a detailed formal definition of well-known and common facts and concentrate on the aspects intrinsic to the MAD model.

> *Definition 1* *Atom Type*
>
> The triple at=<aname,ad,av> is an *atom type* iff aname$\in$N, ad is a valid atom-type description, and av is a valid atom-type occurrence. A valid atom-type description consists of a set of attribute descriptions, and a valid atom-type occurrence is a subset of the description's domain, which is the cartesian product of the attribute domains used. An element of the atom type occurrence is denoted *atom* ∇

Obviously, there is a direct association to the relational model. Fig. 3 compares the MAD concepts against the well known relational concepts.

relational concepts	MAD concepts
attribute	attribute
attribute domain	attribute domain
relation schema	atom-type description
tuple set	atom-type occurrence
tuple	atom
relation	atom type
database	database
-	link
-	link-type description
-	link-type occurence
-	link type
referential integrity (?)	referential integrity(!)
'relation domain'	database domain

Figure 3: Comparison of corresponding concepts

Each atom type is uniquely identified by its atom-type name. AT* denotes the set of all correctly defined atom types. To deal with atom types easily, we define for each at$\in$AT* the following auxiliary functions:

- nam(at) = aname provides the atom-type name,
- des(at) = ad provides the atom-type description,
- ext(at) = av provides the atom-type occurrence, and
- atyp(aname) = at is inverse to the function nam with atyp(nam(at))=at.

If C is a set of atom-type names, the function atyp is extended in the following way:

atyp(C) = {at | at=atyp(aname) and aname$\in$C}.

> *Definition 2* *Link Type*
>
> The triple lt=<lname,ld,lv> is a *link type* iff lname$\in$N, ld={$aname_1$, $aname_2$} with atyp($aname_1$)=at_1, atyp($aname_2$)=at_2, and at_1, $at_2 \in$ AT* as well as
> lv$\subseteq${l | l=<a_1,a_2> with $a_1 \in$ ext(at_1), $a_2 \in$ ext(at_2) and l is an unsorted pair}.
> Here lname is called the link-type name, ld denotes the link-type description, and lv marks the link-type occurrence consisting of a set of elements called *links* ∇

Each link type is uniquely identified by its link-type name. LT* specifies the set of all correctly defined link types. The use of nondirectional link types (and links) integrates the symmetry of the relationships of the MAD model into our formalization. It is allowed to define several link types using the same two atom types as well as using only one atom type (reflexive link type). For example, when modeling the bill-of-material application with its super-component and sub-component view, we just have to define one reflexive link type called 'composition' on the atom type 'parts'. Exploiting the link type's symmetry it is now easy to evaluate either the super-component view or only the sub-component view. Analogously to above, we define for each link type lt$\in$LT* the following auxiliary functions:

- nam(lt) = lname provides the link-type name,
- des(lt) = ld provides the link-type description, and
- ext(lt) = lv provides the link-type occurrence.

Using definition 1 and 2, we are now able to define the notion of a database:

> *Definition 3* *Database*
>
> Let AT$\subset$AT* be a set of different atom types and let LT$\subset$LT* be a set of corresponding link types. The pair DB=<AT,LT> specifies a *database* and DB*=<AT*,LT*> denotes the so-called database domain comprising all valid databases ∇

The definition of DB* is necessary, because all subsequently defined operations are closed under this database domain (see theorem 1). Each operation uses one or two atom types of a specified database and produces a new atom type with new link types that are all contained in a correspondingly enlarged database being part of the database domain. For the relational model its closure is defined in the same way: the result of each relational operation is a new relation that is contained in the 'relation domain' (see fig.3) comprising all valid relations.

Definition 4 *Atom-Type Operations*

Let DB=<AT,LT> be a database and let $at=<aname,ad,av>$, $at_1=<aname_1,ad_1,av_1>$, and $at_2=<aname_2,ad_2,av_2>$ be atom types within DB, and ad, ad_1 and ad_2 are in pairs disjoint. Now the following operations are defined:

- atom-type-projection
  $\pi[proj(ad)](at)=<aname_\pi,ad_\pi,av_\pi>=at_\pi$
  with $proj(ad)\subseteq ad$, $aname_\pi \in N$, $ad_\pi=proj(ad)$, and
  $av_\pi=\{a_\pi \mid a_\pi=a(proj(ad)) \text{ and } a\in av\}$
- atom-type-restriction
  $\sigma[restr(ad)](at)=<aname_\sigma,ad_\sigma,av_\sigma>=at_\sigma$
  with $restr(ad)\in$ qual-formulas(ad), $aname_\sigma\in N$, $ad_\sigma=ad$, and $av_\sigma=\{a \mid qual(restr(ad),a)) \text{ and } a\in av\}$; qual is a predicate that decides whether the atom at hand fulfills the qualification condition restr(ad).
- cartesian product
  $\chi(at_1,at_2)=<aname\chi,ad\chi,av\chi>=at\chi$
  with $aname\chi\in N$, $ad\chi=ad_1\cup ad_2$,
  $av\chi=\{a\chi \mid a\chi=a_1\&a_2 \text{ and } a_1\in ext(at_1),\ a_2\in ext(at_2)\}$
  and '&' denotes the concatenation of the two given tuples.
- atom-type-union
  $\omega(at_1,at_2)=<aname_\omega,ad_\omega,av_\omega>=at_\omega$
  with $name_\omega\in N$, $ad_\omega=ad_1=ad_2$, $av_\omega=av_1\cup av_2$
  and '$\cup$' denotes the union of two sets.
- atom-type-difference
  $\delta(at_1,at_2)=<aname_\delta,ad_\delta,av_\delta>=at_\delta$
  with $name_\delta\in N$, $ad_\delta=ad_1=ad_2$, $av_\delta=av_1-av_2$
  and '-' denotes the difference of two sets ∇

The detailed definition of these atom-type operations should be clear, so we decided to leave it out. All atom-type operations have indirectly some effects on the link types: the link types of the operand atom types are 'inherited' to the resulting atom type. Thus, the result atom type could be reused in subsequent operations. In particular this is necessary for the molecule operations (see next section), since the dynamic molecule derivation relies on the existence of link types. In order to save space, we omit the formal definition of the inheritance of link types. The interested reader is referred to [Mi88a].

Based on the above introduced constructs, we are now able to set up and prove the following theorem:

Theorem 1 *Atom-Type Algebra*

The set of the atom-type operations π, σ, χ, ω, δ forms an algebra on the database domain DB*, i.e. the result of each atom-type operation is representable in DB* ∇

Proof:

(1) To prove that each result atom type is valid:
This is already proved by the result atom type's construction (cf. def. 4)

(2) To prove that all inherited link types are well-defined:
This follows directly from their careful construction (cf. [Mi88a]) ∇

In fig.4 we show the formal specification of our geographic application of fig.1. Both the database schema and the atom networks (i.e. database occurrence) are defined within the database definition. Only the relevant data are shown.

Based on this database definition we are now able to raise some atom-type operations, whilst showing the corresponding relational algebra operations:

atom types

atom-type definition — *atom-type occurrence*

state = <state,{ name, hectare, ...}, {<BA, 1000, ...>,<MG, 500, ...>,<RJ, ...>, ..., <RS,...>}>∈ AT*

river = <river, {name, length, ...},{<Amazonas, 5875, ...>, <Parana, 3231, ...> ... <Uruguai,...>}>∈ AT*

area = <area, {area attributes},{a1, ...}>∈ AT*

net = <net, {net attributes},{n1, ...}>∈ AT*

edge = <edge,{edge attributes},{e1,e2,e3,...,e13,e14,e15}>∈ AT*

point = <point, {point attributes},{pn, ...}>∈ AT* . . .

atom

link types

link-type definition — *link-type occurrence*

state-area = <state-area,{ state,area} ,{<<SP,...>,a1>,<<MG,...>,a2>,<<RJ,...>,a3>,... }>∈ LT*

river-net = <river-net,{river,net},{<Parana,...>,n1>,<<Amazonas,...>,n2>,<<Uruguai,...>,n3>,...}>∈ LT*

area-edge = <area-edge,{area,edge},{<a1,e1>,<a1,e6>,<a2,e6>, <a2,e11> ,... }>∈ LT*

net-edge = <net-edge,{net,edge},{<n1,e11>,...}>∈ LT*

edge-point = <edge-point,{edge,point},{<e1,p1>,<e1,p2>,...}>∈ LT* . . .

link

database

atom types — *link types*

GEO_DB = <{ state, river, area, net, edge, point, ... },{ state-area, river-net, area-edge, net-edge, ... }>∈ DB*

Figure 4: Formal specification of the geographic database

• Cartesian product χ(area,edge) = border
with border =<border,{attributes of both area andedge},
{a1&e1,a1&e2,...}>

All link types having as component area or edge are inherited to the result atom type border. Analogously, the links are inherited to the result atoms.

The relational 'equivalent' looks like: border = area x edge with 'x' being the relational cartesian product.

In both cases the result border gets all attributes from its operands. On the relational side, the foreign keys indicating relationships (i.e. links) are implicitly inherited, too.

• Atom-type restriction σ[restr(hectare>1000)](border)
corresponds to the relational expression:
σ[hectare>1000](border).

Here it becomes obvious that the atom-type algebra exhibits the power of the relational algebra. Moreover, it avoids the problem of enforcing referential integrity, since the relevant relationships (i.e. the foreign/primary key concept of the relational model) are explicitly represented and maintained by means of the link concept. (There are no dangling references (i.e. links) and it is even possible to control cardinality restrictions specified in an extended link-type definition). In the next section we show how this explicit representation of the relationships serves to dynamically specify and derive the desired molecules.

3.2 Higher Level Structures and Operations

Analogously to the previous section, we introduce first the data structures and afterwards the operations for the higher level objects (i.e. complex objects, here called molecules). The central concept of the molecule algebra is the notion of the molecule type. According to the atom type, the definition of molecule type is based on both a molecule-type description and its corresponding molecule-type occurrence.

Definition 5 *Molecule-Type Description*

The pair md=<C,G> is a molecule-type description iff
- C = {$aname_1$,...,$aname_n$} and atyp(C)$\subset$AT*
- G = {dl_i | dl_i=<$lname_i$,$aname_{i1}$,$aname_{i2}$> with <$lname_i$,{$aname_{i1}$,$aname_{i2}$},lv>$\in$ LT* and i=1,...,n}
- md_graph(md) ∇

Def. 5 formalizes the notion of 'molecule structure' used in the previous chapter introducing a (type) graph whose nodes are atom types and whose edges are directed link types (cf. fig.2). Each triple dl has as first component its name, as second one the start node and as last component its end node. By means of the predicate md_graph we guarantee that this graph has the following properties: directed, acyclic, coherent, and having only one root node. At the same time each molecule-type description determines the domain of its corresponding molecule-type occurrence. This is expressed by the function m_dom:

Definition 6 *Function m_dom*

m_dom(md) = {m | m=<c,g> with
c$\subseteq${a | $\exists$at: a$\in$ext(at) and nam(at)$\in$C} and
g$\subseteq${l | $\exists$lt:l$\in$ext(lt) with ltyp(dl)=lt and dl$\in$G}
and mv_graph(m,md)}

The function ltyp is used to determine for each directed link type its associated nondirectional link type being an element of LT*. The elements of m_dom(md) are denoted *molecules* and the predicate mv_graph guarantees the correctness of the molecules with respect to the given molecule-type description md:

mv_graph(m,md) <=> md_graph(m) $\wedge$ total(m,md)

The already known predicate md_graph tests the above mentioned graph properties of the (molecule) graph whose nodes are atoms and whose edges are links. Of course we have to demand the same properties for both graphs (i.e. that of the molecule-type description and that of each molecule) in order to guarantee the correspondence of type and occurrences. The predicate total assures the molecule graph is maximally:

total(m,md) <=> $\forall$a$\in$c: contained(a,m,md) and
$\forall$a$\notin$c, but a$\in$ext(atyp(aname)) with
aname$\in$C: $\neg$contained(a,m,md)

contained(a,m,md)<=> (a$\in$ext(root(md))) $\vee$
($\forall$ <lname,$aname_i$,aname>$\in$G,
$\exists a_i\in$c with $a_i\in$ext(atyp($aname_i$)):
contained (a_i,m,md) $\wedge$ <a_i,a>$\in$g)

with a$\in$ext(atyp(aname)) with aname$\in$C,
and <a_i,a>$\in$ ext(ltyp(lname)).

The predicate root determines the root node of a given graph and the predicate contained is recursively defined on the molecule structure graph. It is used to derive all the atoms and links that are necessary to build up the desired molecules ∇

In some sense the above formalization defines the 'synthesis of molecules', that is a procedure explaining how to derive all molecules of a molecule-type occurrence according to a given molecule-type description. It is exactly the same procedure as already informally introduced in chapter 2: the derivation of the children atoms, i.e. hierarchical join along the specified branches is incorporated in the predicate contained. Based on def. 5 and 6 we are now able to make concrete our notion of molecule type:

Definition 7 *Molecule Type*

Given a database DB=<AT,LT>. The triple mt=<mname,md,mv> is a *molecule type* (defined over the database DB) iff the name of the molecule type mname$\in$N and md is a molecule-type description with md=<C,G> and atyp(C)$\subseteq$AT and ltyp(G)$\subseteq$LT as well as mv comprises the molecule-type occurrence with mv=m_dom(md) ∇

Definition 8 *Molecule-Type-Definition*

Given a database DB=<AT,LT>, C={$aname_1$,...,$aname_n$} with atyp(C)⊆AT, and G={dl_1,...,dl_m} with ltyp(G)⊆LT. Furthermore let be mname∈N and md_graph(<C,G>). Then the operator *molecule-type-definition* is defined as follows:

α[mname,G](C)=mt_α=<mname,md,mv>

with md=<C,G> and mv=m_dom(md) ∇

Obviously, the molecule type mt_α is valid according to def.7.

To prove the closure of the following molecule-type operations we have to introduce the function prop that materializes or 'propagates' the result of some molecule-type operations into a given database. Thereby, the database is enlarged by new atom types and link types in such a way that the previously propagated result set is now derivable as a corresponding molecule type over this extended database.

Definition 9 *Propagation of Result Sets*

Given a result set rst=<mname,rsd,rsv> with mname∈N, rsd=<C,G> with atyp(C)⊆AT*, ltyp(G)⊆LT*, and md_graph(rsd) as well as rsv={m|mv_graph(m,rsd)}. The function prop is defined as follows:

prop(rest,DB)=<mt,DB'> with

mt=<mname,md,mv> and DB'=<AT∪atyp(C'),LT∪ltyp(G')>

Furthermore, md=<C',G'> and mv=m_dom(md) with C' is the set of renamed atom types used in rsd that exhibit the same atom-type description but only a restricted atom-type occurrence: the corresponding atoms are selected only from the elements within rsv. The same happens to G' being the set of inherited link types that are used in rsd. Of course, md is built using only propagated atom types or inherited link types within C' or G' but it still shows the same graph structure as rsd ∇

Even without a formal proof (cf.[Mi88a]) it is obvious that mt is a valid molecule type over DB' such that mt=α[mname,ltyp(G')](atyp(C')). This is easily conceivable because md satisfies md_graph and all molecules that are constructible with respect to the molecule-type description md also belong to the molecule-type occurrence mv. Furthermore, mv is restricted in an appropriate way so that for each element within rsv there is exactly one equivalent molecule within mv and vice versa.

Figure 5 shows the approach used to define all molecule-type operations: the effect of each molecule-type operation is expressed in several portions starting with some operation-specific actions followed by the propagation of the specified result set or the execution of some atom-type operations. The last part to accomplish the molecule-type operation op_{mt} is to perform the corresponding molecule-type definition α.

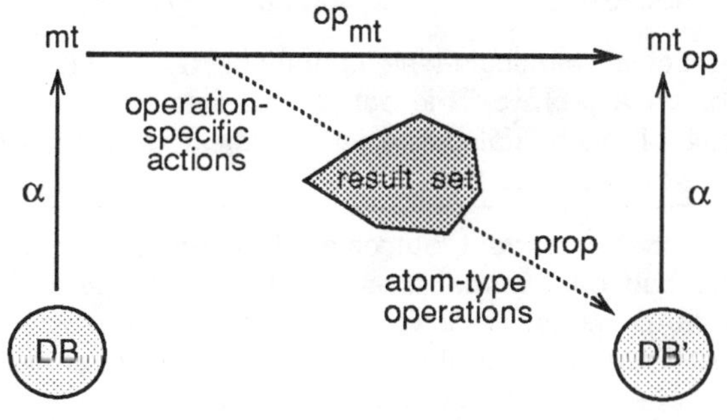

Figure 5: Diagrammatic view to the definition of molecule-type operations

Definition 10 *Molecule-Type-Restriction*

Given a database DB=<AT,LT> and the a molecule type mt=<mname,md,mv>. Furthermore let restr(md) be a qualification formula over md with qual(m,restr(md)) being a predicate that decides whether a molecule m∈mv fulfills the qualification conditions raised in restr(md). Then the operator *molecule-type-restriction* is defined as follows:

Σ[restr(md)](mt) = mt_Σ

with <mt_Σ,DB'> = prop(rst,DB)

and rst = <$mname_\Sigma$,rsd,rsv>

with $mname_\Sigma$∈N, rsd=md, and

rsv = {m | m∈mv and qual(m,restr(md))} ∇

Theorem 2

The molecule type mt_Σ is a valid molecule type over DB' ∇

Proof:

(1)To prove that rst is a valid result set, we only have to look at rsv, since $mname_\Sigma$∈N and rsd=md. By definition it is true that rsv⊆mv. Thus all elements m∈rsv fulfill mv_graph(m,rsd).

(2) The proof that mt_Σ is a valid molecule type over DB' follows straight-forwardly from the correctness of the propagate function prop (cf. def. 9) ∇

Due to space limitation we omit the definition of the following molecule type operations (again, the interested reader is referred to [Mi88a]):

- *molecule-type-projection* (Π),
- *molecule-type-cartesian-product* (X),
- *molecule-type-union* (Ω), and
- *molecule-type-difference* (Δ).

They are mostly defined using the molecule-type propagation and the atom-type operations of the previous section. Of course we can prove that the molecule types resulting from these 4 operations are valid over their correspondingly enlarged databases. Basically, these proofs are similar to that shown above for the operation molecule-type-restriction. Thus, the following theorem holds:

Theorem 3 *Molecule Algebra*

The set of molecule-type operations α, Σ, Π, X, Ω, Δ form an algebra on the set of molecule types. Thus, the result of each molecule-type operation is again describable as a molecule type ∇

The most important outcome of theorem 3 is that the described data model is closed under its operations, i.e. the MAD model is closed under its molecule-type operations. Thus, we are able to concatenate several molecule-type operations to build higher level and more complex operations and objects. For example, we can express the operation molecule-type-intersection Ψ by using the operation molecule-type-difference twice:

$$\Psi(mt_1,mt_2) = \Delta(mt_1, \Delta(mt_1,mt_2))$$

with mt_1, mt_2 being molecule types.

4. From an Algebraic Approach to a High Level Query Language

In this chapter we describe and exemplify a way how to exploit a model's algebra to define a high level as well as user-friendly query language for that model. Firstly, we show some examples of our molecule algebra using the geographic application described in fig.1, fig.2, and fig.4. Translating these algebra expressions into statements of our molecule query language (for short MQL) introduces the constructs and the semantics of this high level query language by means of the sound algebra definition provided in the previous chapter. As syntactical basis of MQL we have chosen an SQL-like formalism because of its widespread use, its syntactical simplicity, and its user-friendliness.

The first example shows the molecule-type-definition of the molecule type 'mt_state' (cf. fig.2):

α[mt_state,{<state-area,state,area>,<area-edge,area,edge>,
<edge-point,edge,point>}](state,area,edge,point)=mt_state

with

mt_state= <mt_state, <{state,area,edge,point},
{state-area,area-edge,edge-point>, mv>

molecule-type definition *molecule-type occurrence*

and

mv = {m1=<{<SP,...>,a1,e1,e2,e3,...,p1,p2,...},
{<<SP,...>,a1>,<a1,e1>,<a1,e2>,...} >, *molecule*

m2=<{<MG,..>, a2 ,e3,...},{<<MG,..>,a2>, <a2,e3>..},...}

component atoms *component links*

Some elements of this molecule-type occurrence are already shown in the lower part of fig.2 in a graphical representation.

The equivalent MQL statement looks as follows:

```
SELECT ALL
FROM mt_state(state-area-edge-point);
```

Obviously, the whole molecule-type definition is expressed in the FROM clause. (If there is only one link type defined between two atom types we can simplify the syntax of the molecule-type definition by using the symbol '-' instead of the link-type's name.)

The following MQL statement exploits the symmetry within the model's data structures:

```
SELECT ALL
FROM point-edge-(area-state,net-river)
WHERE point.name='pn';
```

The result of this query is shown in the upper part of fig.2. To express this in our molecule algebra, we firstly have to build the molecule-type 'point-neighborhood' using the operation molecule-type-definition:

point-neighborhood =

α[point-neighborhood,
{<point-edge,point,edge>,<edge-area,edge,area>,
<area-state,area,state>,<edge-net,edge,net>,
<net-river,net,river>}]
(point,edge,area,state,net,river)

In the second step we have to concatenate this molecule-type-definition with a molecule-type-restriction:

Σ[restr(point.name='pn')](point-neighborhood)

This example stresses the flexible and symmetric use of a link type to build up quite different molecule types. Molecule restriction in MQL is expressed within the WHERE clause, and molecule projection is accomplished within the SELECT clause.

5. Comparison and Outlook

After having sketched the MAD model and its molecule algebra as well as the high level language MQL, it is worthwhile to draw a comparison to other models and their algebras. A rough but expressive comparison can be done just by comparing the different complex object concepts supported: It is obvious that the MAD model with its provision for network structures and shared subobjects comprises all models that are based on flat or only hierarchically structured objects. Thus, the relational model, the extended relational model [LK84], and even the non-first-normal-form models [AB84,Da86] are just special cases thereof. A more detailed comparison is possible using the models' formalizations: comparing the molecule algebra with the NF^2 relational algebra [SS86] or other non-first-normal-form approaches [OY85,RKS85,BK86] results in the same outcome. Meanwhile, there are attempts to extend the basic NF^2 idea with sharing of subobjects, recursiveness, and dynamics in object definition (cf. [Oz88]).

As already mentioned above, the MAD model allows for reflexive link types and for other cycles in the database schema; e.g. for modeling a bill-of-material application. These cycles are normally queried in a recursive manner, for example asking for the parts explosion (i.e. sub-component view) of a given part. Therefore, we have started to extend the algebra as well as MQL to provide for recursive query facilities [Schö89] introducing recursive molecules types as data model objects.

Moreover, it is worth-while to recall the MAD model's capability for dynamic object definition: the complex

objects, i.e. the molecule types, to work with are defined in algebra expressions or high level language expressions and are not statically fixed in the database schema as it is the case in most of the above mentioned approaches. Our molecule algebra incorporates these facilities and, hence, offers a great flexibility in complex object management.
The comparison between the MAD model and the well-known (binary) ER model (without relationship attributes) is quite obvious. On one hand, the modeling power of the MAD model comprises the modeling capabilities of this type of ER model as already pointed out in chapter 2. On the other hand, it could also serve as a descriptive high-level 'ER language' with the molecule algebra serving as a sound 'ER algebra' (cf. [PS85,HG88,PRYS89]).
Meanwhile a first prototype implementation of MAD/MQL called PRIMA [HMMS87] has been finished. Its internal architecture shows two main components influenced by the construction of the molecule algebra: the basic component provides an atom-oriented interface (similar to the functionality of atom-type algebra) for the second component that performs molecule processing and implements an MQL interface (similar to the functionality of molecule algebra) to the application programs. Thus, this implementation endeavor exploits quite directly the theoretical experiences gained from our algebra definitions. Furthermore, we are confident that we can conveniently exploit the algebra to considerably simplify and enhance query transformation and query optimization as well as using it as a focal point for detailed investigations in query parallelism.

6. Acknowledgment

I'm deeply indebted to K. Meyer-Wegener who has considerably contributed to clarify and improve the presentation of important issues. T. Härder and H. Schöning have read and improved an earlier version, and M. Gaß, I. Littler, and H. Neu have helped in preparing the manuscript.

7. References

[AB84] Abiteboul, S., Bidoit, N.: Non First Normal Form Relations to Represent Hierarchically Organized Data, in: Proc. PODS Conference, 1984, pp. 191-200.

[BB84] Batory, D.S., Buchmann, A.P.: Molecular Objects, Abstract Data Types and Data Models: A Framework, in: Proc. 10th VLDB Conf., Singapore, 1984, pp. 172-184.

[BK86] Bancilhon, F., Koshafian, S.: A Calculus for Complex Objects, in: Proc. PODS Conference, 1986, pp. 53-59.

[CDV88] Carey, M., DeWitt, D.J., Vandenberg, S.: A Data Model and Query Language for EXODUS, in: Proc. ACM SIGMOD Conf., Chicago, 1988, pp. 413-423.

[Da86] Dadam, P. et al.: A DBMS Prototype to Support Extended NF^2-Relations: An Integrated View on Flat Tables and Hierarchies, in: Proc. ACM SIGMOD Conf., Washington, D.C. 1986, pp. 356-367.

[HG88] Hohenstein, U., Gogolla, M.: A Calculus for an Extended Entity Relationship Model Incorporating Arbitrary Data Operations and Aggregate Functions, in: Proc. 7th Int. Conf. on Entity-Relationship Approach, Rome, 1988, pp. 1-20.

[HMMS87] Härder, T., Meyer-Wegener, K., Mitschang, B., Sikeler, A.: PRIMA - A DBMS Prototype Supporting Engineering Applications, in: Proc. 13th VLDB Conf., Brighton, UK, 1987, pp. 433-442.

[LK84] Lorie, R., Kim, W., et al.: Supporting Complex Objects in a Relational System for Engineering Databases, IBM Res. Lab., San Jose, CA, 1984.

[Mi88a] Mitschang, B.: A molecule-atom data model for non-standard applications - requirements engineering, data-model design, and implementation concepts (in German), PhD Thesis, University Kaiserslautern, 1988, published by Springer Publ. Co. in the series IFB 185.

[Mi88b] Mitschang, B.: Towards a unified view to design data and knowledge representation, in: Proc. 2nd Int. Conf. on Expert Database Systems, Tysons Corner VA, 1988, pp. 33-49 (further publication by Benjamin/Cummings Publ. Co.).

[OY85] Ozsoyoglu, Z.M., Yuan, L.-Y.: A Normal Form for Nested Relations, in: Proc. PODS Conference, 1985, pp. 251-260.

[Oz88] Ozsoyoglu, Z.M. (ed.): Special Issue On Nested Relations, in: Data Engineering, IEEE, Vol. 11, No. 3, Sept. 1988.

[PRYS89] Parent, C., Rolin, H., Yetongnon, K., Spaccapietra, S.: An ER calculus for the entity-relationship complete model, Research Report 890401, Dep. d'Informatique, Ecole Polytechnique Federal de Lausanne, 1989.

[PS85] Parent, C., Spaccapietra, S.: An Algebra for a General Entity Relationship Model, in: IEEE Transactions on SE, Vol. 11, No. 7, July 1985, pp. 634-643.

[PSSWD87] Paul, H.-B., Schek, H.-J., Scholl, M.H., Weikum, G., Deppisch, U.: Architecture and Implementation of the Darmstadt Database Kernel System, in: ACM SIGMOD Conf., San Francisco, 1987, pp. 196-207.

[RKS85] Roth, M.A., Korth, H.F., Silberschatz, A.: Extended Algebra and Calculus for ¬1NF Relational Databases, Technical Report TR-84-36 of Univ. of Texas at Austin, 1985.

[RS87] Rowe, L.A., Stonebraker, M.R.: The POSTGRES Data Model, in: Proc. 13th VLDB Conf., Brighton, 1987, pp. 83-96.

[Schö89] Schöning, H.: Recursion in the MAD Model - Recursive Molecules as Objects of the Data Model (in German), in: IFB 204, Springer Publ. Co, 1989, pp 389-407.

[SS86] Schek, H.J., Scholl. M.H.: The Relational Model with Relation-Valued Attributes, in: Information Systems, Vol. 2, No. 2, 1986, pp. 137-147.

[Ul80] Ullman, J.D.: Principles of Database Systems, Pitman Publishing Limited, 1980.

Sorting, Grouping and Duplicate Elimination in the Advanced Information Management Prototype

G. Saake
Technische Universität Braunschweig, Abteilung Datenbanken, D-3300 Braunschweig, West Germany

V. Linnemann, P. Pistor
IBM Scientific Center Heidelberg, Tiergartenstrasse 15, D-6900 Heidelberg, West Germany

L. Wegner
Universität Kassel, D-3500 Kassel, West Germany

Abstract

Sorting, duplicate suppression and grouping are important operations in relational database management systems. This paper is devoted to the related language features and their implementation in the Advanced Information Management Prototype AIM-P. The query language HDBL is an SQL-like database language supporting the extended NF^2 data model. The proposed language extensions follow the classical SQL approach for sorting and duplicate elimination by extending the SFW construct with appropriate clauses. For the grouping operation we chose a new syntactical construct because the implicit structure transformation of grouping differs from the sorting and duplicate suppression operations. Finally, the integration into the query evaluation of the AIM prototype is described.

1. Introduction

Traditionally, sorting subsystems have played an important role in data base management systems. In relational systems like DB2, for example, they are the basis for producing sorted output, for eliminating duplicates, and for operations requiring an intermediate partitioning of input tables. They provide additional facilities for optimizing join operations and can be extremely advantageous for the creation of indexes.

This paper is devoted to the definition and implementation of sorting, grouping, and duplicate suppression in AIM-P /DaK86,ALPS86,LKD88/, a database system supporting an extended NF^2 data model. Some introductory information about the underlying data model is given in section 2. It illustrates how basic operations of the NF^2 algebra can be packaged in an SQL-like fashion, how their semantics are to be generalized to cope with the structures supported in AIM-P, and how they are related to sorting, grouping, and duplicate elimination.

With sorting (sect. 3) and duplicate elimination (sect. 4) we have chosen the classical SQL approach of providing these facilities via clauses added to the SFW construct. In comparison to initial austere proposals /PA86,PT86/, this approach allows for an intuitive and compact notation; in addition, this syntactical format seems to be better suited for translation into optimized sequences of lower level operations. These goals have also guided the design of the grouping operation (sect. 5). In this case however we had good reasons for developing syntactical solutions which deviate from classical SQL. Section 6 describes an evaluation machine which is able to exploit the facilities of an appropriately designed sorting subsystem in realizing these functions.

2. Extended NF^2 Data Model

2.1 Supported Structures

The eNF^2 data model of AIM-P /PT86,PA86,Pi87/ is based on the idea of non first normal form relations /AB84,SS84/. In terms of the extensions, the two models can be contrasted as follows:

- Top level objects are not necessarily relations. Scalars or tuples are admissible as well.
- The notion of relations is generalized to the notion of tables, i.e. collections of tuples.
- Collections may be ordered (lists) or unordered (sets or multisets). Collections may be free of duplicates ("unique") or not.

The work was done while the first author was at the IBM Scientific Center in Heidelberg on leave from Technische Universität Braunschweig

Proceedings of the Fifteenth International Conference on Very Large Data Bases

Amsterdam, 1989

- Collections need not necessarily be composed of tuples. Sets of integers or lists of tables are conceivable as well.
- Like the strict NF² model, tuples need not be composed of "atomic" items. Besides tables, however, fields may also contain tuples, or collections other than tables.

The following figure shows an eNF² structure which will also be used in the following chapters as an accompanying basic structure for language examples.

{ Literature }			
Title	< Authors >		{ Keywords }
	Surname	Firstname	
Designing a Generalized NF2 Model with an SQL-Type Interface	Pistor	Peter	Database Systems
	Andersen	Flemming	Nested Relations
			Query Languages
....			

The eNF² relation Literature is a multiset of tuples having two non-atomic attributes, namely an ordered list Authors and a multiset Keywords. In contrast to 'puritanical' relational and NF² database theory, the notions of sorting and duplicate elimination operators are well defined in the extended NF² data model because of the explicit handling of lists and multisets.

- The property 'is sorted' can only be stated for collections of type list, because for sets and multisets the ordering is irrelevant per definition.
- For collections other than lists, the property 'free of duplicates' makes only sense in data models which distinguish semantically between sets and multisets.
- The grouping operator involves nested structures, and therefore leaves the scope of 'flat' relational theory.

In contrast to the pure NF² data model, collections with duplicates are correct eNF² data structures. For this reason, duplicate suppression in AIM-P is exclusively done on demand in contrast to /KF88/ where the demand for duplicate suppression results implicitly from the semantics of the underlying pure NF² data model.

2.2 HDBL Language

For the eNF² setting, the AIM prototype provides an SQL-like language interface (Heidelberg Data Base Language HDBL). The concepts of this language /PHH83,PA86,PT86/ need not be outlined here in detail. Instead, a few comments and illustrative examples are sufficient to provide the background for the subsequent discussions.

As in other languages of the SQL family, the SELECT ... FROM ... WHERE construct (SFW construct) plays a central role in HDBL. It is an expression which maps collections into collections. Its general format is

```
(1.1) SELECT f(x1,..., xn)
(1.2) FROM   x1 IN Input1,..., xn IN Inputn
(1.3) WHERE  p(x1,..., xn)
```

The semantics of (1) can be understood quite naturally by analogous expressions of constructive set theory, viz.

```
(1')   {f(x1,..., xn) | x1 ∈ Input1 and ... and
                        xn ∈ Inputn and p(x1,..., xn) }
```

The term $f(x_1,\ldots, x_n)$ stands for an arbitrary expression, usually an expression returning a tuple; it provides the elements of the query result. The expression $p(x_1,\ldots, x_n)$ denotes a Boolean expression which controls whether a specific combination of elements $(x_1,\ldots, x_n)$ from the input collections $Input_1$ through $Input_n$ is rejected or not. The resulting collection is a list if at least one input collection is a list. In general, the result is finally assumed to be non-unique (i.e. the elements are not necessarily distinct from each other).

Expression f in (1.1) may be composed of any legal expressions of HDBL, especially SFW expressions. This is illustrated in the subsequent example:

```
(2.1) SELECT [FirstAuthor:  x.Authors[1],
(2.2)         ShortKeys   : (SELECT y FROM y IN x.Keywords
(2.3)                        WHERE  LENGTH(y) < 10)]
(2.4) FROM   x IN Literature
```

As this example demonstrates, the facilities of the SELECT clause in HDBL are far beyond plain projection in SQL for a flat setting (for similar facilities in NF² algebra see /SS84/). Rather than picking up the attributes Authors and Keywords, query (2) retrieves only the first author and short keywords.

Restriction, projection, and join are typical operations of relational algebra. HDBL supports them by SFW just like other SQL languages. In addition, SFW covers the typical NF² operations "nesting" and "unnesting". E.g.

```
(3.1) SELECT [x.Title, y.Surname, y.Firstname]
(3.2) FROM   x IN Literature, y IN Authors
```

provides Title and Authors information in an unnested ("flat") fashion. Nesting is demonstrated in the example (4) below. This example also gives further evidence that HDBL is not restricted to the table paradigm. The query

```
(4.1) SELECT [Rest      : x,
(4.2)         Partition: (SELECT y FROM y IN L
(4.3)                     WHERE x = y MOD 3)]
(4.4) FROM   x IN <0, 1, 2>
```

performs a modulo 3 partitioning of an integer list such that the partitions are displayed along with the grouping criterion.

The reader will have observed the key to this solution: the sequence (0, 1, 2) governing the partitioning is known in advance. A more general solution requires that a unique list (or set) of grouping values is extracted from the collections to be partitioned. In our case, (4) needs to be replaced by something like

```
(5.1) SELECT [Rest      : x,
(5.2)         Partition: (SELECT y FROM y IN L
(5.4)                     WHERE x = y MOD 3)]
(5.5) FROM    x IN MAKE_UNIQUE(SELECT z MOD 3 FROM z IN L )
```

Sofar we have addressed a wide range of operations that can be attributed to SFW expressions in a natural and intuitive fashion. With ordering we have an operation which is not covered by an interpretation of SFW as indicated in the analogy between (1) and (1'). Appropriate operations /PA86,RKB87/ like

```
(6)   ORDER /* every */ x IN L BY x MOD 3
```

are quite in the spirit of /Da84/: If a query result is to be ordered, this can be achieved by functional composition as in

```
(7)   ORDER x IN (SELECT ... FROM ... WHERE ...) BY ...
```

There might be situations where it is desirable to split operations into parts in order to control the sequence in which the parts are executed. On the other hand, there are good reasons for having compound operations for frequently used compositions. For example, users are quite happy to write

```
(8.1)  SELECT [x.Authors] FROM x IN Literature
(8.2)  WHERE  x.Title CONTAINS 'Darwin'
```

combining selection and projection in a dense notation rather than to write (quite correctly, by the way)

```
(9.1) SELECT [x.Authors]
(9.2) FROM    x IN (SELECT y FROM y IN Literature
(9.3)                    WHERE y.Title CONTAINS 'Darwin')
```

For that reason, the designers of SQL /CAE76, ISO86/ have overloaded the SFW construct by optional clauses for ordering, grouping, and duplicate suppression. We adopt this approach with ordering (sect. 3) and duplicate elimination (sect. 4). With grouping (sect. 5), however, we believe that a dedicated operation is justified.

3. Sorting

3.1 Ordering Relation

Sorting is an operation which takes collections and re-arranges them into a list which is ordered according to given sort criteria. In conventional DBMS simple sorting criteria (i.e. expressions yielding basic type results) are sufficient, since the objects to be sorted have a simple structure (usually flat records). In contrast, HDBL is orthogonal in that it provides ordering relations for any admissible eNF^2 value, be it atomic, a tuple, a set or a list. We have an intuitive, well defined ordering relation on tuples (proceeding the tuple fields from left to right) as well as on lists (using the lexicographic order known from text strings). An ordering relation between set or multi-set values is less obvious. E.g., is the set {1,5} less than {2,3,4} or not ?

A discussion on different choices for generalized ordering relations is presented in /KSW89/. The most preferable one defines an ordering relation on sets and multisets according to 'cardinality first, then iteratively comparing the minimum'.

3.2 Justification of SQL-like Sorting

As already indicated in section 2.2, a less austere facility for sorting is desirable. Of course, one could enrich the ORDER operation (see (6) or (7)) by facilities for rejecting input elements and for restructuring output elements. However, this would essentially be a duplication of SFW facilities. The SQL design - an optional order clause - is probably much more clever. Before giving evidence for this allegation, let's have a first impression of SQL-like sorting mechanisms within HDBL. The subsequent query gives an example of a seemingly simple ordering criterion. It lists all the Literature entries ordered by their titles. Rearrangement of Literature is based on the lexicographic ordering relation of strings.

```
(12.1)  SELECT l FROM l IN Literature
(12.2)  ORDER BY l.Title
```

The next query delivers *all* author entries in alphabetic order. Please note that both unnesting and ordering operations are expressed by a single SFW expression, and a complete tuple (see a in (13.2)) is used as sort criterion.

```
(13.1)  SELECT a FROM l IN Literature, a IN l.Authors
(13.2)  ORDER  BY a DESC
```

It should be noted that - *different from SQL* - the information on sort criteria need not be contained in the final output. This is illustrated in (14) which sorts the titles of the Literature table according to the number of authors:

```
(14.1)  SELECT l.Title FROM l IN Literature
(14.2)  ORDER  BY COUNT(l.Authors)
```

The SQL-like approach is advantageous for several reasons: First, it puts HDBL closer to established SQL standards /ISO86/. Second, a large class of queries can be formulated in a *more concise* fashion. For example, having nothing but austere sorting facilities, example (14) were to be rewritten as

```
(15.1)  SELECT l.Title
(15.2)  FROM   l IN (ORDER  l' IN Literature
(15.3)                      BY COUNT(l'.Authors))
```

or even worse:

```
(16.1)  SELECT l.Title
(16.2)  FROM   l IN
(16.3)     (ORDER l' IN
(16.4)            (SELECT [l".Title,
(16.5)                     SoMany: COUNT(l".Authors)]
(16.6)             FROM    l" IN Literature    )
(16.7)      BY l'.SoMany          )
```

It is not verbosity which is criticized here. It is the users' temptation to adopt the role of an optimizer, i.e. taking care of issues like

- compressing data before ordering (see (16.4-6)),
- arranging the sequence of SFW and ORDER operations according to whether order criteria are contained in the final output or not.

The preceeding arguments should not be understood as a case against subjecting the result of ordering operations to other operations which are applicable to lists, as in (17) below:

```
(17.1) SELECT [l.Title,
(17.2)        LexicographicallyFirstAuthor :
(17.3)              (SELECT a FROM a IN l.Authors
(17.4)               ORDER  BY a  ) [1] ]
(17.5) FROM    l IN Literature
```

3.3 Syntax and Semantics

The syntactical notation of the ordering clause is similar to the standard SQL proposal /ISO86/. A list of order criteria is added to the known SFW construct. The following rules in BNF syntax define the syntactical notation.

```
<SFW> :: SELECT <expr> FROM <from-list> [WHERE <expr>]
         [ORDER BY <order-expr> {"," <order-expr>}*]
<order-expr> :: <expr> [ ASC | DESC ]
```

An order expression is - in contrast to standard SQL - an arbitrary expression together with the ordering orientation (ascending or descending order). Scope and binding of variables in this expression are identical to scope and binding of the variables in the SELECT and WHERE clause of the SFW construct and are determined by the FROM list.

The following examples show ordering expressions with more complex expressions. Example (18) orders a list of points in the plane according to their distance to the origin.

```
(18.1)  SELECT p FROM p IN Points
(18.2)  ORDER BY p.x * p.x + p.y * p.y
```

Example (19) orders the authors by the number of their papers in descending order.

```
(19.1)  SELECT a FROM l IN Literature, a IN l.Authors
(19.2)  ORDER BY
(19.3)    COUNT( SELECT l' FROM l' in Literature
(19.4)           WHERE a ELEMENT_OF l'.Authors) DESC
```

As mentioned in chapter 3.1, we allow expressions yielding arbitrary eNF^2 type results as order criteria. As an example, query (20) orders the literature according to the set of authors derived from the author list.

```
(20.1)  SELECT l.Title FROM l IN Literature
(20.2)  ORDER BY MAKE_SET(l.Authors)
```

The semantics of an ordering expression does not seem to be complicated: The result of the SFW query is transformed into a list ordered by the specified sort criteria; this effect can be achieved by a *mental* evaluation model containing the following steps:

1. In a first step the cross product of all sets / lists occurring in the FROM-list is constructed (under consideration of the hierarchical variable binding). If lists are involved, the relative order of list elements is retained by following the rules given in /PT86,Da88/. If the FROM-list defines a join between two (or more) lists, nested loops may serve as a mental model for the construction of the relative order of the cross product lines.

   For the following steps, the variables are bound to the corresponding parts of the cross product lines.
2. All lines of the cross product are removed where the WHERE clause is evaluated to FALSE.
3. The sort expressions are evaluated for each remaining line.
4. The lines are sorted according to the sort criteria values computed in step 3.
5. The result is constructed out of the cross product lines as defined in the SELECT clause. This may imply a loss of the information needed for the preceding sort step.

The separation of the sort criteria computation phase (step 3) and the value comparisons (step 4) is necessary in this semantics definition, because HDBL expressions may have a nondeterministic semantics. An example is the conversion of a multiset into a list, where the order of the list elements can be arbitrarily chosen.

Before we finish the chapter and turn to the problem of duplicate elimination, we want to state a few requirements for sort algorithms to be used. Of course, the algorithms should be fast - but there are additional interesting requirements not appearing in the 'flat' case. The ordering relation on arbitrary extended NF^2 structures is recursively defined, so sorting on substructures may become a significant part of the total computation costs. Good sorting algorithms have to reduce these costs to a minimum by only partially sorting of substructures as proposed in /KSW89/.

4. Duplicate Elimination

4.1 Identification of Duplicates

The need for identification and deletion of duplicates arises in all data models offering sets and multisets as structuring facilities. On the one hand, set type structures should be free of duplicates per definition, and type conversion operations have to obey this property. On the other hand, the (relational algebra) operation 'projection' delivers usually results containing duplicates (not in relational theory, but in implemented database systems). Here we concentrate on explicit query features suppressing duplicates, and do not discuss implicit duplicate elimination due to set or uni-list type database structures.

What does 'free of duplicates' for extended NF^2 structures mean ? This property can only be stated for sets and lists, and there the intuitive meaning is 'containing no identical elements'. If we fix the term 'identical' as *representation independent equality* of NF^2 objects, we see that this requires a recursively defined equality on complex NF^2 structures, which has to take special care of set type structures. The formal background and efficient computation strategies for this property are again examined in the accompanying paper /KSW89/.

In many applications, the term 'identical' is used in a more restrictive fashion for duplicate identification. There, the identity of two structures is decided by inspecting only part of the structure. This occurs especially due to the use of semantic knowledge on the application area or due to optimizing duplicate elimination. In the case of relations, the use of *key attributes* is an example of such a restrictive identity definition - two tuples are assumed to be equal if their key attribute values are equal. In conventional relational databases, this notion is only used for database tables and not for result tables. We use these notions also for result structures and extend the concepts to *key expressions* where identity is defined by the equality of arbitrary expressions computed out of the set or list elements.

As an example for the need of key attributes (for duplicate elimination), we assume a personnel database consisting of personnel data like `Name`, `Age` etc., and an attribute `Picture` containing a 'bitmap' picture of the person. In a personnel relation, duplicate elimination should not compare the binary representation of the pictures to test identity of person records. On the other hand, in a set of fractional numbers we may fix equality by comparing the reduced fractions using a complex HDBL expression. If we use such an expression for determining the identity for duplicate suppression, we talk about *key expressions*. The usual key attributes are included as a special case.

As with the sort methods, the type 'list' needs special attention if key expressions are used for duplicate identifications. Structures with equal key values need not be equal on other parts of the structure. So the question arises, which of the occurring duplicates have to be eliminated. This problem can either be solved by the intuitively acceptable commitment of deleting the duplicate at the higher list position ('the first survives') or controlled by an additional parameter. In the following, we use the first variant.

In the evaluation of database queries, duplicate suppression makes sense in two evaluation phases. The most commonly used duplicate suppression is on the *result of the query*. Duplicate suppression on the result delivers always a result of type unique set (or unique list, if the input was of type list). For this reason, duplicate suppression on the query result can only have expressions over the result structure as key expressions. An example for duplicate suppression on the query result is a unique set of authors derived from the literature table.

The second, more subtle duplicate suppression is on the *input structures* of an SFW expression. This enables key expressions on non-selected data, for example if we want to project the titles out of the literature table only once unless associated with different author lists. This can be expressed by duplicate suppression with the keys `Title` and `Authors` on the input of the corresponding SFW construct (example (22)). It must be noted that the result of a query with duplicate suppression on the input may still contain duplicates.

4.2 Duplicate Suppression on Input

For duplicate suppression we have essentially the same design options for language features as for sorting. The duplicate suppression on input structures is closely related to sort expressions, since the mental model implies the removing of 'input lines' in the same evaluation phase where sorting is done (before result construction). Moreover, in this case legal key expressions are the same as legal sort expressions (without ASC or DESC keywords). We chose a descriptive notation analogous to the ORDER BY expressions :

```
(21.1)  SELECT l.Title FROM l IN Literature
(21.2)  IGNORE DUPLICATES ON l.Title
```

Example (21) shows that variables are bound to the input structures occurring in the FROM-list. This enables the formulation of the (above mentioned) query (22), where duplicate suppression is done with respect to non-selected information :

```
(22.1)  SELECT l.Title FROM l IN Literature
(22.2)  IGNORE DUPLICATES ON l.Title, l.Authors
```

Example (23) shows the use of arithmetic key expressions. The result contains only points with different distances from the origin (0,0).

```
(23.1)  SELECT p FROM p IN Points
(23.2)  IGNORE DUPLICATES ON p.x*p.x + p.y*p.y
```

The syntactical structure of an IGNORE clause is analogous to the order expression (without ASC and DESC). If an ignore and an order expression are combined in one SFW construct, their relative order is relevant due to the 'the first survives' rule for duplicate deletion. The following two SFW queries may deliver different results; in the second variant (25) the author with the 'lexicographic smallest' first name survives, whereas in the first variant (24) this depends on the order of the input lists.

```
(24.1) SELECT a                      (25.1) SELECT a
(24.2) FROM l in Literature,         (25.2) FROM l in Literature,
(24.3)        a in l.Authors         (25.3)        a in l.Authors
(24.4) IGNORE DUPLICATES             (25.4) ORDER BY a
(24.5)        ON a.Surname           (25.5) IGNORE DUPLICATES
(24.6) ORDER BY a                    (25.6)        ON a.Surname
```

4.3 Duplicate Suppression on Result

Now let us return to duplicate suppression in the query result. In the *mental evaluation model* this suppression has to be performed after the result construction phase. To guarantee the uniqueness property on the result structure, only expressions over result table elements are legal key expressions. Following the mental evaluation model, a duplicate suppression on the result structure is always performed *after* the processing of an ORDER or IGNORE clause.

Proposals /CAE76,PA86,PT86/ for relational as well as for extended NF^2 database languages describe several syntactic alternatives. In the language HDBL, the following three alternatives are candidates for an orthogonal language extension:

1. The *functional approach* /PA86,PT86,RKB87/ is similar to the notation of the sort function in example (16). Without keys it may be notated like 'make_unique(Literature)', with key parameters it resembles the functional order operator:

```
(26.1) UNIQUE x IN ( SELECT [l.Title, a]
(26.2)               FROM l IN Literature,
(26.3)                    a IN l.Authors)
(26.4)   ON x.Title
```

   The scope of the variables used in key expressions is bound to the result structure.

2. The *descriptive approach* is similar to the ignore clause described above. Legal key expressions are exclusively built on the basis of a system generated variable RESULT bound to the result structure.

```
(27.1) SELECT [l.Title, a]
(27.2) FROM l IN Literature, a IN l.Authors
(27.3) UNIQUE ON RESULT.Title
```

   The case of a desired unique result table without key restrictions is formulated by

```
(27.3') UNIQUE ON RESULT
```

3. The *SELECT clause modification approach* is derived from the SQL proposal and the use of key constraints in data definition languages. The SELECT clause defines the structure of the result, and duplicate suppression on the result may be formulated therein. The uniqueness of the result without key restrictions is expressed by the key word UNIQUE which directly follows SELECT, whereas key attributes are marked by KEY. The examples (28) and (29) show the use of both constructs.

```
(28.1) SELECT UNIQUE l.Title
(28.2) FROM l IN Literature

(29.1) SELECT [ KEY l.Title, l.Authors ]
(29.2) FROM l IN Literature
```

   It must be noted that in this approach only first level attributes are usable as key expressions. On the other hand, it may be orthogonal to a DDL extension introducing keys as integrity constraints into the schema definition.

The third alternative has a relevant lack on expressive power compared with the two other variants, so the final decision was between the functional and the descriptive approach. We opted for the second variant for the same reasons as with sorting. This decision also reduces the amount of additional concepts and language features.

4.4 Syntax and Semantics

The duplicate suppression on *input structures* is analogous to the order-by clause:

```
<SFW> :: SELECT <expr> FROM <from-list> [WHERE <bool-expr>]
         [IGNORE DUPLICATES ON <expr> {"," <expr>}*]
```

The mental evaluation model for duplicate suppression on input structures is very similar to the one for sort expression evaluation:

1. see order_by evaluation.
2. see order_by evaluation.
3. Computation of key expression values for the input lines.
4. If two lines have equal key expression values, one of them is removed. If the input structure is of type list, this removing has to obey the relative position stability.
5. Construction of result (see order_by evaluation).

In contrast to the IGNORE clause evaluation, duplicate suppression on the *result structure* allows only expressions over the variable RESULT (<result-expr>) to appear as key expressions.

```
<SFW> :: SELECT <expr> FROM <from-list> [WHERE <bool-expr>]
         [UNIQUE ON <result-expr> {"," <result-expr>}*]
```

One aim of the distinction of both duplicate suppression alternatives is that for duplicate suppression on the result structure we can syntactically derive the uniqueness of the result. To guarantee this property, we have to slightly modify our evaluation model.

1. see order_by evaluation.
2. see order_by evaluation.
3. see order_by evaluation.
4. Perform duplicate elimination for the result structure.
5. Computation of key expression values for the result elements.
6. If two lines have equal key expression values, one of them is removed.

The fourth step guarantees that the result itself is free of duplicates, and this property is not affected by the following steps. The duplicate suppression done wrt the key expressions alone is not sufficient because of the existence of nondeterministic expressions in the HDBL language.

If step 4 were omitted, a key expression with a nondeterministic function (for example conversion of a set into a list) may yield in step 5 different key expression values for identical result elements, and therefore uniqueness cannot be guaranteed.

5. The Grouping Operator

5.1 *Grouping in the eNF² Data Model*

In contrast to *pure* sorting or duplicate elimination, the 'grouping' or 'nesting' known from NF²-oriented languages *changes the structure of the result.* The fundamental idea is to assemble part of the result into a new created set (or list) such that all elements meet the same grouping condition. Because of this fundamental effect of changing the result structure, we decided to choose a functionally oriented language feature instead of a descriptive one (like the SQL proposal).

There are two alternatives for the structure of the result of a grouping operation. The first one is that the result structure contains only the constructed groups /PA86,PT86/. The new result structure is determined by the input structure, for example a 'set of input type' becomes a 'set of set of input type' after grouping. The disadvantage of this approach is the loss of the explicit grouping criterion information.

In the second alternative, the result is a binary relation between the groups and the group criterion. In terms of extended NF² structures, this structure is a set or list of tuples containing the groups and the grouping criterion values as field values. This structure implies the choice of attribute names as additional parameters of the group operator if we want to avoid system generated attribute names.

The result structure of the first variant can be computed from the result of the second variant by a simple projection on the group field of the result tuples. So we decided to choose the second variant because of its greater expressiveness. This choice implies that our grouping operator needs the following parameters :

1. The input structure, a set or list.
2. The grouping criterion (an expression with variables bound to the elements of the input structure).
3. The attribute identifiers for the resulting groups and the grouping values.
4. The structure description for the group elements (also an expression with variables bound to the elements of the input structure).
5. An (optional) predicate on the input structure (WHERE clause of the SFW construct).

Moreover, we decided to integrate the projection and selection operations into the group function to allow compact notations and to simplify query optimization. The following complex example shows the syntactical constructs chosen for the group function and the parameters:

```
(30.1)  GROUP [l.Title,l.Keywords]
(30.2)  INTO Publications
(30.3)  BY Author : a
(30.4)  FROM l IN Literature, a IN l.Authors
(30.5)  WHERE COUNT(l.Authors) < 4
```

The HDBL statement (30) delivers for each author all his (her) relevant publications (where publications are assumed to be irrelevant for an author if authored by four or more authors).

The result structure of group expression (30) is

```
(31.1)  LIST(TUPLE(Publications: LIST(TUPLE(
(31.4)                                  Title : ...,
(31.5)                                  Keywords : ...)),
(31.6)                  Author: TUPLE(Surname : ..., ... )  ))
```

The list types (instead of set types) follow from the list type structure Authors in the FROM list.

5.2 *Syntax and Semantics*

In more detail, the syntax of a group expression is fixed as follows:

```
<gexpr> ::=   GROUP <expr> INTO <id> BY <id> ":" <expr>
              FROM <from-list> [ WHERE <bool-expr> ]
```

The 'from-list' is built analogously to the SFW construct. To make the reader more familiar with this notation, we present a few further examples. As a first one, we group the complete literature by the number of authors :

```
(32.1)  GROUP l INTO LiteratureSet
(32.2)   BY NrAuthors : COUNT(l.Authors)
(32.3)   FROM l IN Literature
```

The second example (33) groups the literature titles by the set of their keywords.

```
(33.1)  GROUP l.Title INTO Titles
(33.2)  BY Keywords : l.Keywords
(33.3)  FROM l IN Literature
```

Please note that the type of the result of (33) is

```
(34)  SET(TUPLE( Titles:SET(TEXT), Keywords:SET(TEXT)))
```

As a last example, we group our points table by the distance to the origin (0,0) (assuming a function SQRT for computing the square root).

```
(35.1)  GROUP p INTO PointsOfEqualDistance
(35.2)   BY Distance : SQRT(p.x*p.x + p.y*p.y)
(35.3)   FROM p IN Points
```

The precise semantics of a grouping operation can readily be defined by an equivalent HDBL expression, using already defined language features. A group expression

```
(36.1) GROUP elementexpr INTO groupname
(36.2)  BY criterionname : criterionexpr
(36.3)  FROM fromlist WHERE predicate
```

is equivalent to the following HDBL expression, using functional composition and duplicate suppression.

```
(37.1) SELECT [ groupname :
(37.2)            (SELECT elementexpr FROM fromlist
(37.3)             WHERE predicate AND x = criterionexpr),
(37.4)          criterionname : x                        ]
(37.5) FROM x IN (SELECT criterionexpr FROM fromlist
(37.6)              WHERE predicate UNIQUE ON RESULT)
```

In this expression, the variable 'x' has as values all occurring group criterion values (in consideration of the WHERE clause). This equivalent HDBL expression fixes at the same time the result type of a grouping expression with respect to the from list. It is clear that an efficient implementation does not follow this semantics commitment.

6. Integration into Query Evaluation

6.1 The Sort Manager

The language features introduced in this paper are being implemented using a system component called *sort manager*. The sort manager offers at its call interface a sort function for arbitrary eNF^2 collections and allows for parameters controlling duplicate suppression and grouping during the sort process. For a discussion of appropriate sort algorithms for these functions see /KSW89/.

In detail, the parameters of a sort manager call contain the following data :

- The collection, an ordered or unordered relation, to be manipulated. The type of this collection is always set or list of tuples. The collection is an intermediate internal result structure, not a persistent database object.
- The order criteria. Allowed order criteria are top level attributes of the collection elements together with an ascending / descending flag.
- The duplicate elimination flag. If this flag is set, equality of the sort criteria values leads to elimination of one of the compared tuples.
- The grouping information. This parameter controls the grouping while performing duplicate elimination. It contains the (top level) group attribute, which must be of type set or list and must not be contained in the order criteria list. If a grouping attribute is defined, each time a duplicate elimination is performed the surviving tuple gets the union of both old grouping attribute values as new grouping attribute value.

The parameter restrictions give a less powerful functionality of the sort manager than introduced for the HDBL language extensions described in chapters 3, 4 and 5, but enable a straightforward implementation of the sort manager. The only critical point of the implementation are set-valued order criteria. In this case, the sort algorithm is recursively called to compute the order according to the 'cardinality first, then minimum' rule. The strategy to implement the full proposals of chapter 3 is described in the following subchapters.

6.2 Integration Strategy

We assume that duplicate elimination is done via sorting (for a discussion of other methods and a justification of this choice see /KSW89/). Later on, we will reduce grouping also to a slightly modified duplicate elimination (see 6.4). Due to these reasons, we concentrate mainly on the evaluation of sort expressions by stating that the used concepts carry over to duplicate suppression and grouping.

Of course, the mental model for sorting with 'computing of the cross product' is not an acceptable implementation idea. Instead of the complete cross product, only the result objects have to be computed along with additional attributes containing the information needed for key expressions or sort / grouping criteria. For example, consider the following expression :

```
(38.1) SELECT l.Title FROM l IN Literature
(38.2) ORDER BY COUNT(l.Authors)
```

For evaluation, we can compute the result of the extended query below (39), sort the intermediate result and as a last step remove the artificially added data needed for the sorting. Note that the sorted objects are considerably smaller than full lines of the input structure would be.

```
(39.1) SELECT [id1 : l.Title, id2 : COUNT(l.Authors)]
(39.2) FROM l IN Literature
```

This method is independent of the evaluation strategy implemented in a DBMS. It reduces the sort criteria internally to upper level attributes of a table and avoids multiple computations of sort expressions. For duplicate elimination, the result structure expansion enables the handling of both the ignore and the unique clause in an equal fashion.

Conceptually, we divide the query evaluation into the following phases:

1. Query analysis. Creation of internal query evaluation plan and catalog structure.
2. Optimization of the internal query representation.
3. Data retrieval out of the database (evaluation of retrieval part of the internal query representation).
4. Result modification phase (sorting etc).
5. Result processing (browsing, transfer to application software, inserting into database table etc.)

The implementation usually mixes phases 3 and 4 for efficiency reasons; for example, substructures are sorted directly after having been retrieved.

Here we concentrate on the added language features only. We subsume sorting, duplicate suppression and grouping under the notion of *result modification operations*. For generality, we talk about result modification expressions, which mean expressions modifying the result of a previous evaluation step (but not its internal result structure). Due to the definition in section 5.2, these operations are also sufficient to implement grouping. An arbitrary modification expression is composed using the following primitives :

1. **SORT:** This basic modification function has as parameters (among others) a reference to the input structure for the operation (which may be a substructure of a complex NF^2 object). In case of the SORT operator, a list of sort criteria is another parameter consisting of field identifiers with additional declarations like descending or ascending order. SORT also provides a facility for eliminating duplicates on the fly.
2. **Composition:** Modification expressions can be concatenated to sequences (for example, for the combination of sorting and duplicate elimination).
3. **Control:** As a final feature, we need loops over substructures (for their justification see also detailed discussion for optimization).

6.3 Details of the Evaluation Steps

Now we can take a closer look at the special activities needed for sorting and duplicate suppression in the single phases. The group operator is handled afterwards (see 6.4).

In the *query analysis phase*, the result catalog and the query evaluation plan have to be extended for the needed computed values of the sort / key expressions. The first step is the addition of the sort criteria to the result structure as additional attributes (see above example). The query evaluation plan is extended to retrieve the data into these attributes.

In the *optimization phase*, the original query evaluation plan constructed in the analysis phase is restructured for an efficient evaluation. Here we concentrate on optimizations related to the result modification. The following points can lead to optimized query evaluation plans.

- The use of indexes may influence the need for sorting, for example an index may deliver the elements already in the desired order making a sort operation superfluous.
- The arrangement of query modification expressions can be optimized. An important optimization technique is to avoid sorting of subcomponents which are removed afterwards due to duplicate elimination. An example is a query which delivers for each author his coauthors in alphabetic order.

```
(40.1)  SELECT [Author : a,
(40.2)          Coauthors :
(40.3)           SELECT ca
(40.4)           FROM l' IN Literature,
(40.5)                ca in l'.Authors
(40.6)           WHERE a ELEMENT OF
(40.7                   l'.Authors
(40.8)                AND NOT (a = ca)
(40.9)           ORDER BY ca          ]
(40.10) FROM l IN Literature,
(40.11)          a IN l.Authors
(40.12) UNIQUE ON RESULT.Author
```

  In this example it is better to first eliminate the duplicates and then do the sorting on the coauthors of the tuples which survive the duplicate elimination process.
- With key expressions for duplicate elimination, the relative order is irrelevant for the result. Thus, they can be re-arranged according to the costs of the equality tests, and according to their selectivity on the input structure.

In the *data retrieval phase*, we do not need additional activities because of the query evaluation plan modifications in the first phases.

The *result modification phase* contains the data manipulation algorithms needed for the new language features. The needed sort algorithms and order computations for extended NF^2 structures are discussed in /KSW89/ and are skipped in this paper. As a last step, the artificially added information is projected out of the result.

The result processing is not affected by the new language features.

6.4 Implementation of Grouping

As mentioned above, the grouping operator is defined by the use of duplicate elimination. The following algorithm sketch shows an elegant implementation of a grouping operator using the introduced formalism :

1. As a first step, the group expression is internally replaced by a SFW construct having the same result structure as the group expression itself. The example group expression

```
(41.1)  GROUP l INTO LiteratureSet
(41.2)   BY NrAuthors : COUNT(l.Authors)
(41.3)   FROM l IN Literature
(41.4)   WHERE NOT COUNT(l.Keywords) = 0
```

   is internally replaced by

```
(42.1)  SELECT [LiteratureSet : { l },
(42.2)          NrAuthors: COUNT(l.Authors) ]
(42.3)  FROM l IN Literature
(42.4)  WHERE NOT COUNT(l.Keywords) = 0
```

   After evaluation of this query, the attribute `LiteratureSet` of type multiset contains exactly one element (for each result object).

2. Now a modified duplicate elimination is executed for the result of the modified query (42) : Whenever duplicates on 'NrAuthors' are detected, the contents of the 'LiteratureSet' attribute are assembled and the surviving duplicate gets the assembled elements as new 'LiteratureSet' attribute (see chapter 4.1). For list type structures, the position stability of the grouped elements can be guaranteed using internally an artificially added position attribute.

7. Conclusions

In this paper we presented a language extension for the SQL-like NF^2 database query language HDBL, which offers powerful language features for sorting, duplicate suppression and grouping of extended NF^2 structures. Due to space limitations, some details had to be left out. They can be found in /SLPW89/.

Future work contains the implementation of this language extension as well as its improvement in practical applications. Interesting aspects for future work are the integration of sorting and duplicate elimination into query plan optimization and the requirements for an appropriate sorting algorithm to be included in the DBMS (for the latter see also /KSW89/).

The presented extensions affect only the DML part of the HDBL database language. Possible DDL part extensions and the enforcement of these constraints have to be investigated in the future.

Besides the proposed HDBL language extensions, the developed methods for duplicate elimination are also planned to be used for efficient evaluation of recursive queries as proposed in /Li88/. Here, special requirements like partially presorted input have also to be taken into consideration.

Acknowledgements

The authors would like to thank the colleagues in the AIM-P project, namely P. Dadam, R. Erbe, U. Herrmann, U. Keßler, K. Küspert, E. Roman and N. Südkamp for fruitful discussions and suggestions

References

AB84 S.Abiteboul, N.Bidoit: Non First Normal Form Relations: An Algebra Allowing Data Restructuring. Rapports de Recherche No 347, Institut de Recherche en Informatique et en Automatique, Rocquencourt, France, Nov. 1984.

ALPS86 Andersen, F., Linnemann, V., Pistor, P., Südkamp, N.: Advanced Information Management Prototype - User Manual of the On-Line Interface of the Heidelberg Data Base Language (HDBL). Techn. Note IBM Scientific Center Heidelberg TN86.01

CAE76 D.D.Chamberlin et al.: SEQUEL2: A Unified Approach to Data Definition, Manipulation and Control, IBM Journ. Res. Devel. 20 (1976), pp. 560-575.

Da84 C.J.Date: Some Principles of Good Language Design with Special Reference to the Design of Database Languages. ACM SIGMOD Record 14(3) (Nov. 1984), 1-7.

Da88 P. Dadam: Advanced Information Management (AIM): Research in Extended Nested Relations. IEEE Database Engineering, Special Issue on Nested Relations, Vol. 11, No. 3, 1988.

DaK86 P.Dadam, K.Küspert et al: A DBMS Prototype to Support Extended NF^2 Relations: An Integrated View of Flat Tables and Hierarchies. Proc. ACM SIGMOD 86, Washington D.C., May 1986, pp. 356-367.

ISO86 ISO International Organization for Standardization: Draft International Standard ISO / DIS 9075 : Information Processing Systems - Database Language SQL. 1986.

KF88 S. Khoshafian, D. Frank: Implementation Techniques for Object Oriented Databases. Proc. Advances in Object-Oriented Database Systems (K.R. Dittrich, ed.), LNCS 334, Springer-Verlag Berlin, 1988, pp. 60-79.

KSW89 K. Küspert, G. Saake, L. Wegner: Duplicate Detection and Deletion in the Extended NF^2 Data Model. Proceedings 3rd Intern. Conference on Foundations of Data Organization and Algorithms (FODO 1989), Paris, June 1989. (also Techn. Rep. IBM Scientific Center Heidelberg TR 88.11.012)

Li88 V. Linnemann: Functional Recursion and Complex Objects. Techn. Rep. IBM Scientific Center Heidelberg TR 88.12.017, Dec. 1988.

LKD88 V. Linnemann, K. Küspert, P. Dadam et al: Design and Implementation of an Extensible Database Management System Supporting User Defined Data Types and Functions. Proc. 14th Int. Conf. on Very Large Data Bases, Los Angeles 1988. (also Techn. Rep. IBM Scientific Center Heidelberg TR 87.12.011

PA86 P.Pistor, F.Andersen: Designing a Generalized NF^2 Model with an SQL-Type Interface. Proc. VLDB 86, Kyoto, Aug. 1986, pp. 278-288.

PHH83 P.Pistor, B.Hansen, M.Hansen: Eine sequelartige Schnittstelle für das NF^2 Modell. In J.W.Schmidt (ed.): Sprachen für Datenbanken. Informatik Fachberichte 72, Springer Verlag, Berlin-Heidelberg-New York, 1983. pp. 134-147.

Pi87 P. Pistor: The Advanced Information Management Prototype: Architecture and Language Interface Overview. IBM Scientific Center Heidelberg Techn. Rep. TR 87.06.004, June 1987.

PT86 P.Pistor, R.Traunmüller: A Database Language for Sets, Lists, and Tables. Information Systems, Vol. 11(4), 1986, pp. 323-336.

RKB87 M.A. Roth, H.F. Korth, D.S. Batory: SQL/NF: A Query Language for ¬NF Relational Databases. Information Systems Vol. 12, No. 1, 1987, pp. 99-114.

SLPW89 G.Saake, V.Linnemann, P.Pistor, L.Wegner: Sorting, Grouping and Duplicate Elimination in the Advanced Information Management Prototype. IBM Scientific Center Heidelberg Techn. Rep. TR 89.03.008, March 1989.

SS84 H.-J.Schek, M.Scholl: An Algebra for the Relational Model with Relation-Valued Attributes. TR DSVI-1984-T1, Techn. Univ. Darmstadt, 1984.

OPTIMIZATION OF RELATIONAL SCHEMAS CONTAINING INCLUSION DEPENDENCIES

Marco A. Casanova[1], Luiz Tucherman[1], Antonio L. Furtado[2], Anelise P. Braga[1]

[1]Rio Scientific Center - IBM Brazil - P.O Box 4624 - 22.071, Rio de Janeiro, RJ - Brazil

[2]Dept. de Informática - Pontifícia Universidade Católica do Rio de Janeiro
Rua Marquês de S. Vicente, 225 - 22.453, Rio de Janeiro, RJ - Brazil

ABSTRACT

A two-step optimization strategy for relational schemas that contains a class of inclusion dependencies is described. Both steps take into account additional information that indicates how to preserve each inclusion dependency in the presence of insertions and deletions. The first step eliminates inclusion dependencies which are redundant with respect to both the semantics of the data and the behavior of the transactions. The second step discards dependencies through a structural transformation that again preserves the semantics of the data and of the transactions and that applies both to 1NF and to NF² relational schemas.

1. INTRODUCTION

Among the classes of integrity constraints considered for the relational model, we find the *inclusion dependencies* [CFP], or INDs. For example, one may declare an IND between tables SECRETARY and EMPLOYEE to capture that all secretaries are employees. The specification of an IND is often complemented with insertion/deletion options [CFT,Da] that indicate how to preserve the dependency in the presence of insertions and deletions. For example, one specify that the deletion of an entry of table EMPLOYEE must propagate to the deletion of the corresponding entry of SECRETARY, if any, and that the insertion of an entry of SECRETARY must be blocked if there is no corresponding entry of EMPLOYEE.

Proceedings of the Fifteenth International Conference on Very Large Data Bases

This paper describes an optimization process that minimizes the set of INDs of a relational schema by eliminating redundant dependencies and by redesigning the schema, without modifying the basic semantics of the data and the behavior of the transactions. The process assumes that all INDs are *key-based* and that are qualified with insertion/deletion options. A key-based IND, or K-IND, is just an IND whose right-hand side is a key. This restricted class of INDs is sufficiently powerful to capture important semantic properties, such as hierarchies of sets of objects or relationships between sets of objects, in an information model.

The process of eliminating redundant inclusion dependencies is based on the inference rules for INDs described in [CFP], adapted to K-INDs and to the insertion/deletion options.

The process of discarding K-INDs through a structural transformation may work in the context of 1NF schemas or in the context of NF² schemas. For 1NF schemas, the transformation eliminates a K-IND by collapsing the two tables the dependency relates into a single table. For NF² schemas, the transformation also eliminates an inclusion dependency by nesting the two tables related. The transformation is used only when the dependency meets certain criteria that guarantee that the new table will not contain undue redundancies and that the old tables can be redefined as views of the new table. Dependencies satisfying these criteria are called *trivializable.*

From another perspective, the optimization process can be viewed as a design discipline for schemas containing a class of K-INDs, which is defined both for the traditional and the NF² variations of the relational model. In this aspect, the process differs substantially from the well-known design disciplines for the relational model [Ma], as well as from those defined for the NF² variation [AMM,FG,OY,RKS], since it is based on inclusion dependencies whereas the vast

Amsterdam, 1989

majority of published disciplines are based on functional or multivalued dependencies or their generalizations.

The organization of the paper is as follows. Section 2 discusses the elimination of redundant K-INDs. Section 3 describes the structural transformation assuming that all tables must be in first-normal form. Section 4 shows how to adapt the transformation for the NF^2 version of the relational model. Finally, section 5 contains the conclusions.

The reader is referred to [CTFB] for a more detailed presentation.

2. ELIMINATING REDUNDANT KEY-BASED INCLUSION DEPENDENCIES

2.1 Preliminary Definitions

We assume a scenario where the database designer specifies the conceptual schema of a database within the traditional relational model, that is, using relation schemes in first-normal form. We will be interested in three classes of integrity constraints, the null constraints, indicating when attributes do not admit null values, keys and a class of inclusion dependencies which is sufficiently powerful to capture hierarchies of sets of objects and to help define relationships over sets of objects.

More precisely, a *relation scheme* is an expression of the form $R[A_1,...,A_n]$ where R is the *name* and $A_1,...,A_n$ is the list of *attribute names* of the scheme, taken from a given set of identifiers. We will frequently use the term "attribute" instead of "attribute name".

Let $\boldsymbol{E}$ be a set of relation schemes with distinct names in the rest of this subsection. A *state* for $\boldsymbol{E}$ is a function σ that associates an *n*-ary relation $\sigma(R)$ to each name R of a scheme $R[A_1,...,A_n]$ in $\boldsymbol{E}$.

We assume familiarity with the relational algebra and denote the value of a relational expression Q in a state σ of $\boldsymbol{E}$ by $\sigma(Q)$.

Let λ denote both the null value and tuples, with arbitrary length, of null values. A *null constraint* over $\boldsymbol{E}$ is an expression of the form "R: $X \neq \lambda$" where R is the name of a relation scheme in $\boldsymbol{E}$ and X is a sequence of distinct attributes of R. We say that each attribute A in X *does not admit null values* in R. A state σ for $\boldsymbol{E}$ *satisfies* "R: $X \neq \lambda$" iff $t_A \neq \lambda$, for each tuple t in $\sigma(R)$ and for each attribute A in X.

A *key* over $\boldsymbol{E}$ is an expression of the form "R: K" where R is the name of a relation scheme in $\boldsymbol{E}$ and K is a sequence of distinct attributes of R. A state σ for $\boldsymbol{E}$ *satisfies* "R: K" iff, for each pair of tuples t,u in $\sigma(R)$, if $t_K = u_K$ then $t = u$ and, for each tuple v in $\sigma(R)$, $v_A \neq \lambda$, for each attribute A in K. Therefore, by definition, "R: K" implies "R: $K \neq \lambda$" and we also say that every attribute in a key "R: K" *does not admit null values* in R.

Let $\boldsymbol{K}$ be a set of keys over $\boldsymbol{E}$. We say that a key "R: K" in $\boldsymbol{K}$ is *minimal* iff there is no other key in $\boldsymbol{K}$ of the form "R: L" such that all attributes in L also occur in K. In the definition of a relational schema, to be soon introduced, we will indeed require that every key be minimal.

Let $\boldsymbol{K}$ be a set of keys and $\boldsymbol{N}$ be a set of null constraints over $\boldsymbol{E}$. A *key-based inclusion dependency* over $\boldsymbol{E}$, $\boldsymbol{K}$ and $\boldsymbol{N}$, or a *K-IND*, is an expression of the form $R_1[X_1] \subseteq R_2[X_2]:(\gamma,\delta)$ where, for i = 1,2,

- R_i is the name of a relation scheme in $\boldsymbol{E}$;
- X_i is a sequence of distinct attributes of R_i such that X_1 and X_2 have the same length and X_2 is defined as a key of R_2 in $\boldsymbol{K}$;
- γ is the *insertion option* and δ is the *deletion option* of the K-IND, taking values from the set $\{b^i,b^d,p^i,p^d\}$, with the following intended interpretation:

b^i	block immediately
p^i	propagate immediately
b^d	block deferredly
p^d	propagate deferredly

  with the restriction that γ can be equal to p^i or p^d only if all attributes of R_2, excluding those in X_2, admit null values, that is, no null constraint in $\boldsymbol{N}$ says the contrary.

The restrictions imposed on γ just avoid indeterminacies when propagating insertions into R_1 to insertions into R_2. Indeed, an insertion into R_1 determines only the value of the attributes in X_2. Hence, if some attribute A of R_2 did not admit a null value, the propagation of the insertion would have to arbitrate some non-null value for A, which is not reasonable.

The semantics of K-INDs has a *static* and a *dynamic* perspective. The static perspective is reflected by defining that a state σ for E *satisfies* $R_1[X_1] \subseteq R_2[X_2]:(\gamma,\delta)$ iff $\sigma(R_1[X_1]) \subseteq \sigma(R_2[X_2])$.

The dynamic perspective refers to the fact that the insertion and deletion options affect the behavior of the operations over R_1 and R_2 as defined in [CFT,Da]. For example, if the deletion option is b^i, for block immediately, then there is a test that rejects the deletion of a tuple t from R_2 in a state σ, if $t_{X_2} \in \sigma(R_1[X_1])$. Note that, since X_2 is a key of R_2, no other tuple t' in $\sigma(R_2)$ is such that $t'_{X_2} = t_{X_2}$. Moreover, the test need not consider updates on X_2 since we do not permit updates on keys. If the deletion option is b^d, for block deferredly, there is a test that aborts the transaction if the state φ at commit time is such that $\varphi(R_1[X_1]) \not\subseteq \varphi(R_2[X_2])$.

A *relational schema* is a pair $S=(E,I)$ where E is a set of relation schemes with distinct names and I is a set of null constraints, keys and K-INDs over E such that every key in I is minimal. A *state* of S is a state of E. A state of S is *consistent* iff it satisfies all constraints in I.

We recall that, given a set D of dependencies and a dependency d over a set of relation schemes E, we say that d is a *logical consequence* of D iff any state σ of E that satisfies all dependencies in D also satisfies d.

Let $S=(E,I)$ be a relational schema and T be a transaction over S. The *behavior* of T in S is the set of pairs (σ,σ') of states of E such that there is an execution of T, starting in σ and terminating in σ', that correctly fires all possible triggers and assertions, immediate or deferred, associated with the insertion/deletion options of K-INDs in I.

Let $S=(E,I)$ be a relational schema and d be a K-IND in I. We say that d is *redundant* in I iff

- d is a logical consequence of I-$\{d\}$;
- every transaction T over S has the same behavior in S and in $S'=(E,I$-$\{d\})$.

The notion of redundancy just extends the usual notion of logical consequence for dependencies to accommodate the behavior of transactions. For this reason, it was phrased in a non-traditional way.

2.2 A Motivating Example

Consider a database storing information about software modules, where a module can be in three stages: planned, specified and implemented. Assume that every implemented module must have been first specified and that every specified module must have been first planned. Then, a simplified relational schema for this database would contain the relation schemes

P[NAME,DESC]
S[NAME,SPEC]
I[NAME,CODE]

where P, S and I stands for 'planned module', 'specified module' and 'implemented module', respectively, and the following K-INDs:

(d^0_1) S[NAME]⊆P[NAME]: (b^i,p^i)
(d^0_2) I[NAME]⊆S[NAME]: (b^i,p^i)
(d^0_3) I[NAME]⊆P[NAME]: (b^i,b^i)

Let γ_i and δ_i be the insertion and deletion options of d^0_i. We intuitively prove that d^0_3 is only partially redundant or, more precisely, we argue that: (1) d^0_3 is a logical consequence of d^0_1 and d^0_2; (2) γ_3 is redundant; but (3) δ_3 is not redundant.

Indeed, (1) follows directly from the semantics of K-INDs. To prove (2), note that γ_3 is a consequence of γ_2 and the definition of d^0_1. Indeed, since γ_3 is b^i, γ_3 will block the insertion of a new implemented module m iff m is not yet planned. But if m is not yet planned, by d^0_1, then m is not yet specified. Hence, since γ_2 is b^i, γ_2 will also block the insertion of m, making it unnecessary to define γ_3. Finally, to prove (3), observe that, as δ_3 is b^i, δ_3 will reject the deletion of a planned module m, if m is already implemented, independently of any other assertion and trigger. On the other hand, if δ_3 were not specified, δ_1 and δ_2 alone would propagate the deletion of a planned module m to the deletion of its specification and implementation. Hence, d^0_3 is not redundant because its deletion option affects the behavior of transactions.

Suppose now that we replace d^0_1 by:

($d^0_1{}'$) S[NAME]⊆P[NAME]: (b^i,b^i)

Then, d^0_3 becomes redundant for $d^0_1{}'$ and d^0_2, which indicates that we may drop d^0_3 without affecting the semantics of the data and the behavior of the transactions. Intuitively, by d^0_2, all implemented modules must have been specified. Hence, if the deletion of a planned module m is blocked because m has already been implemented, the deletion will also be blocked because m must have already been specified.

In general, the database designer may specify a relational schema that contains any set of K-INDs. The goal of this section therefore is to analyse which of the K-INDs specified are redundant.

2.3 Conditions for Redundancy of K-INDs

To analyse when a K-IND is redundant in the presence of a set of K-INDs, we will adapt the inference rules for INDs given in [CFP] to take into account insertion and deletion options. In particular, one of the rules will use regular expressions to define sets of sequences of options. Thus, for example, the regular expression $(p^x \cup b^x)^+ b^x (p^x \cup b^x)^*$ denotes the set containing the sequences $p^x b^x$, $b^x b^x$, $p^x b^x p^x$, $p^x b^x b^x$, $b^x b^x p^x$, $b^x b^x b^x$, ...

The modified inference rules are:

(Extended Reflexivity)

$R[X] \subseteq R[X]:(p^i,p^i)$, if X is a key of R

(Extended Permutation)

if $S[B_1,...,B_m] \subseteq R[A_1,...,A_m]:(\gamma,\delta)$

then $S[B_{i_1},...,B_{i_m}] \subseteq R[A_{i_1},...,A_{i_m}]:(\gamma,\delta)$,

where $i_1,...,i_m$ is a permutation of 1,...,m

(Extended Transitivity)

if $R_{i-1}[X_{i-1}] \subseteq R_i[X_i]:(\gamma_i,\delta_i)$, for i = 1,...,k,

then $R_0[X_0] \subseteq R_k[X_k]:(\gamma,\delta)$

where γ and δ are as follows (with $x \in \{i,d\}$):

$\gamma_1 ... \gamma_k$	γ
$(p^x)^+$	p^x
$(b^x \cup p^x)^* b^x (b^x \cup p^x)^+$	p^x, b^x
$(p^x)^* b^x$	b^x

$\delta_1 ... \delta_k$	δ
$(p^x)^+$	p^x
$(b^x \cup p^x)^+ b^x (b^x \cup p^x)^*$	p^x, b^x
$b^x (p^x)^*$	b^x

Intuitively, the antecedents of the extended transitivity rule are ordered. Line 1 of the first table says that if all insertion options are of the form p^x then the insertion option of the conclusion must also be p^x; line 2 says that if all insertion options are immediate or all deferred (i.e., $x=i$ or $x=p$) and at least one insertion option *before the last one* is b^x then the insertion option of the conclusion can either be p^x or b^x; line 3 says that if all insertion options are p^x, except the last insertion which is b^x, then the insertion option of the conclusion must be b^x. The second table has a similar interpretation for deletion options.

For example, from

$R_0[X_0] \subseteq R_1[X_1]:(p^i,p^i)$
$R_1[X_1] \subseteq R_2[X_2]:(p^i,p^i)$
$R_2[X_2] \subseteq R_3[X_3]:(b^i,p^i)$

we may derive just $R_0[X_0] \subseteq R_3[X_3]:(b^i,p^i)$. Indeed, the insertion options of the K-INDs form the sequence $p^i p^i b^i$, which is in the set denoted by $(p^i)^* b^i$, but not in the sets denoted by $(p^i)^+$ or $(b^i \cup p^i)^* b^i (b^i \cup p^i)^+$, and hence the insertion option of the derived K-IND must be b^i. Moreover, the deletion options form the sequence $p^i p^i p^i$, which is in the set denoted by $(p^i)^+$, but not in the sets denoted by $(b^i \cup p^i)^+ b^i (b^i \cup p^i)^*$ or $b^i (p^i)^*$, and hence the deletion option of the derived K-IND must be p^i. Therefore, $R_0[X_0] \subseteq R_3[X_3]:(b^i,p^i)$ follows from the other three EK-INDs by the extended transitivity rule.

These rules are sound, which means that whenever a relational schema $\boldsymbol{S}=(\boldsymbol{E},\boldsymbol{I})$ has a K-IND $\boldsymbol{d}$ in $\boldsymbol{I}$ which can be obtained by the extended rules from the other dependencies in $\boldsymbol{I}$, we may optimize the database design by replacing $\boldsymbol{S}$ by a new schema $\boldsymbol{S}'=(\boldsymbol{E},\boldsymbol{I}')$, where $\boldsymbol{I}'=\boldsymbol{I}-\{\boldsymbol{d}\}$. This transformation will affect neither the semantics of the data nor the semantics of the transactions since a state σ of $\boldsymbol{E}$ satisfies $\boldsymbol{I}$ iff it satisfies $\boldsymbol{I}'$ and any transaction $\boldsymbol{T}$ has the same behavior in $\boldsymbol{S}$ and in $\boldsymbol{S}'$.

The reader is referred to [CTFB] for a fast algorithm testing redundancy for a special class of K-INDs.

3. ELIMINATING K-INDS IN THE 1NF RELATIONAL MODEL BY STRUCTURAL TRANSFORMATIONS

3.1 Preliminary Definitions

We now address an optimization process that minimizes the set of K-INDs of a relational schema by a structural transformation that preserves first-normal form. The transformation eliminates each K-IND that satisfies certain criteria by collapsing the two tables related by the dependency into a single table. The transformation is made transparent to users by redefining the original tables as views over the new table and by mapping their operations onto operations on the new table. The transformation is advantageous exactly because the view operations are much faster in most cases than the corresponding operations on the original relational schema, that is, on the schema contained the non-trivialized K-IND.

The definition of the transformation requires the following additional concepts. Let E be a set of relation schemes with distinct names. A *view definition* over E is a triple $(V[A_1,...,A_n],Q,\mathbf{T})$ where

- $V[A_1,...,A_n]$ is a relation scheme whose name is distinct from the names of the schemes in E;
- Q is an *n*-ary relational expression over E;
- **T** is a specification of correct translations for the operations over V into operations over the schemes in E (see [FC]).

When the specification of **T** is not of interest, we will denote the view definition by $V[A_1,...,A_n]=Q$. We also say that V is the *name* of the view and that Q is the *defining expression* of the view.

Let V be a set of view definitions over E. Suppose that the schemes in E and those defined in V have distinct names. We extend a state σ of E to V by assigning to each view V in V, whose defining expression is Q, the value $\sigma(Q)$.

Although a defining expression may be any valid relational expression, the structural transformation will produce only λ-restrictions, defined below, followed by projections.

Let λ again denote both the null value and tuples, with arbitrary length, of null values. We define a *λ-restriction* over a relation scheme R in E as a restriction of the form $R[A_{i_1}\neq\lambda]...[A_{i_k}\neq\lambda]$, where $A_{i_1},...,A_{i_k}$ are attributes of R. We will frequently use the following equivalences for relational expressions:

Original Expression	*Equivalent Expression*
$R[X][Y]$	$R[Y]$, if $Y\subseteq X$
$R[X\neq\lambda][Y\neq\lambda]$	$R[XY\neq\lambda]$
$R[X][Y\neq\lambda]$	$R[Y\neq\lambda][X]$, if $Y\subseteq X$

The transformation also requires extending the classes of integrity constraints previously introduced to relations defined by λ-restrictions.

Thus, an *extended null constraint* over E is an expression of the form "P: $X\neq\lambda$" and an *extended key* over E is an expression of the form "P: X" where P either is equal to the name R of a relation scheme in E or is a λ-restriction over R and X is a sequence of distinct attributes of R. The semantics of both classes of constraints is a direct generalization of the semantics of the original classes.

Let K and N be sets of extended keys and null constraints. An *extended key-based inclusion dependency* over E, K and N, or an *EK-IND*, is an expression of one of the forms

1) $R_1[X_1]\subseteq R_2[X_2]:(\gamma,\delta)$
2) $R_1[Y_1\neq\lambda][X_1]\subseteq R_2[X_2]:(\gamma,\delta)$,
3) $R_1[X_1]\subseteq R_2[Y_2\neq\lambda][X_2]:(\gamma,\delta)$ or
4) $R_1[Y_1\neq\lambda][X_1]\subseteq R_2[Y_2\neq\lambda][X_2]:(\gamma,\delta)$

where, for i = 1,2,

- R_i is the name of a relation scheme in E;
- Y_i is a sequence of distinct attributes of R_i such that $Y_2\cap X_2=\emptyset$;
- X_i is a sequence of distinct attributes of R_i such that X_1 and X_2 have the same length and X_2 is defined as a key of R_2 in K;
- γ is the *insertion option*, taking values from the set $\{b^i,b^d,p^i,p^d\}$. However, γ can be equal to p^i or p^d only if the dependency is of the forms (1) or (2) and all attributes of R_2, excluding those in X_2, admit null values, that is, no null constraint in N says the contrary.
- δ is the *deletion option*, taking values from the set $\{b^i,b^d,p^i,p^d,n^i,n^d\}$, where n^i stands for "propagate immediately by nullifying" and n^d stands for "propagate deferredly by nullifying". However, δ can be equal to n^i or n^d only if the dependency is of the forms (2) and (4) and all attributes in Y_1 admit null values in R_1.

The restrictions imposed on γ again just avoid indeterminacies when propagating insertions and the restrictions on δ just reflect when consistency can be restored, after a deletion from R_2, by nullifying attributes values.

We say that the EK-IND is *simple on the right-hand side* in cases (1) and (2) and that it is *simple on the left-hand side* in cases (1) and (3). We also say that the EK-IND is *from* R_1 *to* R_2.

The semantics of EK-INDs is a direct extension of the semantics of K-INDs and the new deletion options have the following interpretation. Consider a EK-IND of the forms (2) or (4). If the deletion option δ is n^i, there is a trigger associated with the

EK-IND that is fired right after each deletion affecting R_2 terminates. Let ***o*** be one such deletion and suppose that σ is the state right after ***o*** terminates and that ***D*** is the set of tuples that ***o*** deletes. The trigger sets to null the Y_1-values of every tuple u in $\sigma(R_1[Y_1 \neq \lambda])$ such that there is a tuple t in ***D*** such that $u_{X_1} = t_{X_2}$ and $t_{X_2} \in \sigma(R_1[Y_1 \neq \lambda][X_1])$. With option n^d, we associate a trigger, fired when each transaction commits data, that sets to null the Y_1-values of u, for each tuple u in $\varphi(R_1[Y_1 \neq \lambda])$ such that $u_{X_1} \notin \varphi(R_2[X_2])$, where φ is the commit state of the transaction.

The optimization method will work with more complex schemas, defined as follows. An *extended relational schema* is a pair ***SS*** = (***IS***, ***VS***) where

- ***IS*** = (***BS***, ***VD***, ***IC***), the *internal component* of ***SS***, is such that
  - ***BS*** is a set of relation schemes with distinct names, called *base schemes* of ***SS***;
  - ***VD*** is a set of view definitions over ***BS*** whose schemes have distinct names;
  - ***IC*** is a set of integrity constraints over the schemes in ***BS***, called *active constraints* of ***SS***, such that every key in ***IC*** is minimal.
- ***VS*** = (***E***, ***I***), the *visible component* of ***SS***, is a relational schema such that ***E*** $\subseteq$ ***BS*** $\cup$ *e*(***VD***), where *e*(***VD***) is the set of relation schemes introduced in view definitions in ***VD***.

An extended relational schema ***SS*** = ((***BS***, ***VD***, ***IC***), (***E***, ***I***)) is *correct* iff:

- every consistent state of (***BS***, ***IC***) induces, via ***VD***, a consistent state of (***E***, ***I***);
- the deletion and insertion options qualifying the inclusion dependencies in ***IC*** guarantee, again via ***VD***, the deletion and insertion options qualifying the inclusion dependencies in ***I***.

3.2 Description of the Transformation

This section describes a structural transformation for relational schemas, in first-normal form, that minimizes the number of EK-INDs. The transformation is called *trivialization.*

Let ***SS*** = ((***BS***, ***VD***, ***IC***), (***E***, ***I***)) be an extended schema. An active EK-IND ***d*** in ***IC*** is *trivializable* iff ***d*** is of the form $S[L] \subseteq R[K]:(\gamma,\delta)$ or of the form $S[L] \subseteq R[Z \neq \lambda][K]:(\gamma,\delta)$ where

1) S is a base scheme of ***SS*** with key L that has an attribute N that does not occur in L and does not admit null values;
2) R is a base scheme of ***SS*** with key K and Z is a list of attributes of R;
3) $\gamma, \delta \in \{b^i, p^i\}$;
4) there is no active EK-IND in ***IC*** from a scheme T to S such that the insertion option is p^i or p^d.

Condition (2) is a direct consequence of the definition of EK-IND and is repeated here just to highlight that K must be a key of R. Condition (3) indicates that the insertion and deletion options of ***d*** are either b^i or p^i, that is, they must be both immediate, but it does not require that the options be equal.

Conditions (1) and (2) permit, via the keys K and L, to collapse schemes S and R into a single scheme G without creating redundancies. Each tuple t in G always encodes exactly one tuple of R and, if $t_N \neq \lambda$, also encodes exactly one tuple of S. Condition (3) clarifies which are the insertion and deletion options compatible with the creation of G. Without any of these three conditions, collapsing R and S into G would produce a new relational schema that does not preserve the semantics of the original schema. Finally, condition (4) reflects the restrictions imposed on the insertion options that may qualify a EK-IND. Without it, step (7) of the trivialization process, described in what follows, would produce an invalid insertion option.

The trivialization algorithm takes as input a relational schema SS_0 and produces as output an extended schema SS_n. Each step of the algorithm transforms an extended schema SS_i into a new extended schema SS_{i+1} by trivializing an EK-IND d_i of SS_i as follows. Suppose that d_i is of the form $S[L] \subseteq R[K]:(\gamma,\delta)$ or of the form $S[L] \subseteq R[Z \neq \lambda][K]:(\gamma,\delta)$, where R[K,X] and S[L,Y] are base schemes of SS_i. Suppose that N is an attribute of S that does not belong to L and that does not admit null values. The new extended schema SS_{i+1} is obtained from SS_i through the following transformations:

1) Remove d_i from the set of active integrity constraints;
2) Add G[K,X,Y′] to the set of base schemes, where Y′ is a renaming of the attributes in Y to avoid conflicts with the attributes of K and X (we will denote by A′ the attribute of Y′ corresponding to an attribute A of Y).

3) Remove R[K,X] from the set of base schemes transforming it into a view by adding to the set of view definitions the triple (R[K,X],G[K,X],**T**), where **T** defines the translation of the operations over R to G as follows:

Operation	*Translation*	*Note*
insert into R (t)	insert into G ($K=t_K, X=t_X, Y'=\lambda$)	
delete from R where Q	delete from G where Q	(1)
delete from R where Q	delete from G where Q and $N'=\lambda$	(2,3)
update R set X=t where Q	update G set X=t where Q	

Notes:

(1) translation when $\delta = p^i$, that is, when deletions from R propagate immediately;

(2) translation when $\delta = b^i$, that is, when deletions from R block immediately;

(3) $N'=\lambda$ indicates that a tuple of G does not encode a tuple of S.

4) Remove S[L,Y] from the set of base schemes transforming it into a view by adding to the set of view definitions the triple (S[L,Y],G[$N' \neq \lambda$][K,Y'],**U**) where **U** is defined as:

Operation	*Translation*	*Note*
insert into S ($L=u_L, Y=u_Y$)	insert into G ($K=u_L, Y'=u_Y, X=\lambda$) *if* insert fails *then* update G set $Y'=u_Y$ where $K=u_L$ and $N'=\lambda$	(1)
insert into S ($L=u_L, Y=u_Y$)	update G set $Y'=u_Y$ where $K=u_L$ and $N'=\lambda$	(2)
insert into S ($L=u_L, Y=u_Y$)	update G set $Y'=u_Y$ where $K=u_L$ and $N'=\lambda$ and $Z \neq \lambda$	(3)
delete from S where Q	update G set $Y'=\lambda$ where Q′	(4)
update S set $Y=u_Y$ where Q	update G set $Y'=u_Y$ where Q′ and $N' \neq \lambda$	(5)

Notes:

(1) translation when $\gamma = p^i$, that is, when insertions into S propagate immediately. Note that, in this case, the trivialized EK-IND can only be of the form S[L]⊆R[K]:(γ,δ) because γ specifies propagation;

(2) translation when $\gamma = b^i$, that is, when insertions into S block immediately, and the EK-IND is of the form S[L]⊆R[K]:(γ,δ);

(3) translation when $\gamma = b^i$ and the EK-IND is of the form S[L]⊆R[$Z \neq \lambda$][K]:(γ,δ);

(4) Q′ is a renaming of Q reflecting the renaming of Y to Y′;

(5) the term $N' \neq \lambda$ avoids that the update becomes an insertion.

5) In each view defining expression, except those of R and S, replace R and S by their defining expressions, simplifying the result if possible.

6) In each translation of a view operation, except those associated with R and S, apply the translations specified in **T** and **U** to each operation over R and S.

7) Define K as a key of G. Each active integrity constraint C of SS_i, except the keys K of R and L of S, generates an active integrity constraint C′ of SS_{i+1} obtained by
   a) replacing R in C by its defining expression;
   b) replacing S in C by its defining expression and renaming the attributes of S to their new names in G;
   c) simplifying the result.

   Moreover, if C is a EK-IND, the deletion and insertion options of C′ remain the same as those of C, except if C is of the form S[X_1]⊆P[X_2]:(γ,δ) and the deletion option δ is p^i or p^d, in which case the deletion option of C′ becomes n^i or n^d, respectively.

8) Discard the view definitions whose schemes do not belong to the visible part.

The algorithm is correct in the sense that the visible component of the final extended schema SS_n is exactly the original relational schema SS_0 and that SS_n is correct (as defined at the end of section 3.1). In other words, the optimization is transparent to users since they will still see the original schema SS_0 and, moreover, the semantics derived from that of the internal component and from the view definitions of

SS_n is equivalent to the semantics originally specified. The correctness of the algorithm follows by induction on the number of iterations observing that, in each step, the dependencies over R and S, which are now views over G, become consequences of the dependencies over G. Indeed, by defining K as a key of G in step (2), we guarantee that K and L are keys of R and S and, through the generic transformation defined in step (7), we automatically map all null constraints, keys and EK-INDs involving R and S to equivalent dependencies over G.

As an example of the algorithm, consider a relational schema containing the relation schemes

P[NAME,DESC]   S[NAME,SPEC]
I[NAME,CODE]   C[E#]
M[NAME,E#]

standing for 'planned module', 'specified module', 'implemented module', 'chief-programmer' and 'manages', respectively, where keys are underlined and all attributes do not admit null values. Suppose also that the schema contains the following INDs:

(d^0_1) S[NAME]⊆P[NAME]: (b^i,p^i)

(d^0_2) I[NAME]⊆S[NAME]: (b^i,p^i)

(d^0_3) M[NAME]⊆P[NAME]: (b^i,p^i)

(d^0_4) M[E#]⊆C[E#]: (b^i,p^i)

Intuitively, in terms of the entity-relationship model, these dependencies say that I is a specialization of S, S is a specialization of P and M represents a relationship between P and C, which is n-1 with P on the "n" side.

The algorithm will transform this schema into a final extended relational schema with just two relation schemes and one EK-IND, thus trivializing the maintenance of three EK-INDs. For example, the trivialization of d^0_2 will produce one new scheme, SI[NAME,SPEC,CODE], and will transform S and I into views with defining expressions:

S[NAME,SPEC] = SI[NAME,SPEC]
I[NAME,CODE] = SI[CODE ≠ λ][NAME,CODE].

Furthermore, this trivialization will modified all constraints involving S and I. For example, d^0_1 becomes

SI[NAME]⊆P[NAME]: (b^i,p^i)

To conclude this section, we remark that the trivialization algorithm may be reinterpreted according to two opposite views. On one hand, if we understand a normalization process in the broad sense of a process that simplifies the treatment of dependencies by structural transformations, then the trivialization algorithm can be viewed as a normalization process. Indeed, define that an extended relational schema $SS_n = ((BS_n, VD_n, IC_n),(E,I))$ is in *extended key-based inclusion dependency normal form* (*EK-IND/NF*) iff IC_n has no trivializable EK-IND. We can then reinterpret the trivialization algorithm as a process to transform a given relational schema $S=(E,I)$ into an extended relational schema in EK-IND/NF. On the other hand, if we understand a denormalization process as any process that collapses into a single representation two logically separated concepts, then the trivialization algorithm can also be viewed as a denormalization process.

4. EXTENDING THE STRUCTURAL TRANSFORMATION TO THE NF2 RELATIONAL MODEL

This section briefly discusses, using an example, how to extend the structural transformation to NF^2 relational schemas. Without defining them explicitly, we shall use the basic notation of [Ja] and the counterpart for the NF^2 model of the concepts introduced in section 3. In particular, to indicate the operation that unnests an NF^2 relation, we shall use the symbol μ, subscripted with an expression E defining the part to be unnested. The unnest operation may be recursively applied to the result of other unnest operations. Note that μ automatically filters out the tuples whose unnested component is the empty set.

Let ***SS*** be an extended NF^2 relational schema. A EK-IND ***d*** in ***SS*** is *NF^2-trivializable* iff it is of the form S[L]⊆R[K] or of the form S[L]⊆μ_E(R)[K] where

1) S is a base scheme of ***SS*** with at least one attribute N not in L;

2,3,4) Identical to conditions (2),(3) and (4) of trivialization in the traditional model.

Note that, in the NF^2 model, L need not be a key of S and N may admit null values. The trivialization of ***d*** nests S into R, creating a single scheme G. Each tuple t of G always encodes the tuple of R with key t_K and the set of tuples u of S such that $u_L = t_K$. If this set is empty, the projection of t over the nested attributes of S will be simply the empty set, and not a tuple of nulls, as in the traditional model.

The trivialization algorithm can be extended to NF^2 relational schemas as the following example suggests. Let us consider a relational schema containing just two relation schemes:

E(SSN,NAME,DESC)
D(SSN,DNAME)

where E and D stands for 'employee' and 'dependent', respectively, and where the keys are underlined. Suppose also that the schema has one K-IND,

$(d^0{}_1)$ D[SSN] $\subseteq$ E[SSN]: (b^i,p^i)

Then, we may trivialize $d^0{}_1$ producing a single NF^2 relation scheme:

ED(SSN,NAME,DESC,D(DNAME))

and transforming E and D into views with defining expressions:

E(SSN,NAME,DESC) = ED[SSN,NAME,DESC]

D(SSN,DNAME) =
 $(\mu_{D;DNAME}(ED))$[SSN,DNAME]

In a general setting, it is convenient to carefully analyze the benefits of nesting a scheme S into another scheme when there are EK-INDs involving S that will not be trivialized because they will be transformed into EK-INDs over relations defined with the help of the unnest operator.

As for the traditional model, the definition of trivializable dependencies induces a normal form for the NF^2 model and the trivialization algorithm corresponds to a normalization process.

5. CONCLUSIONS

The optimization process described in this paper minimizes the set of K-INDs of a relational schema by eliminating redundant K-INDs and by trivializing EK-INDs without modifying the semantics of the application. As a conceptual design process, it differs from the strategies defined for the traditional as well as the NF^2 variations of the relational model because is based on inclusion dependencies and depends in a fundamental way on the insertion and deletion options to avoid altering the behavior of the transactions.

ACKNOWLEDGEMENT

We thank Prof. Claudia Bauzer Medeiros for the useful discussion concerning section 2.

REFERENCES

[AMM] H. Arisawa, K. Moriya, T. Miura, "Operations And The Properties On Non-First-Normal-Form Relational Databases", Proc. of the 9th International Conference on Very Large Data Bases, (1983), pp. 197-204.

[CFP] M.A. Casanova, R. Fagin, C. Papadimitriou, "Inclusion Dependencies and Their Interaction with Functional Dependencies", J. of Computer and System Sciences, Vol. 28, No. 1 (Feb. 1984), pp. 29-59.

[CFT] M.A. Casanova, A.L. Furtado and L. Tucherman, "Enforcing Inclusion Dependencies and Referential Integrity", Proc. 14th International Conference on Very Large Data Bases, Los Angeles (Aug. 1988), pp. 13-25.

[CTFB] M.A. Casanova, L. Tucherman, A.L. Furtado and A.P. Braga, "Optimization of Relational Schemas containing Inclusion Dependencies", Technical Report CCR078, Rio Scientific Center, IBM Brazil, Rio de Janeiro, Brazil (June 1989).

[Da] C.J. Date, "Referential Integrity" in *Relational Database: Selected Writings* Addison-Wesley Publishing Company, (1986), pp. 41-62.

[FC] A.L. Furtado and M.A. Casanova, "Updating Relational Views", in *Query Processing in Database Systems*, W. Kim, D.S. Reiner and D.S. Batory (eds.), Springer Verlag (1985), pp. 127-142.

[FG] P. Fisher, D. Van Gucht, "Determining When A Structure Is A Nested Relation", Proc. of the 11th International Conference on Very Large Data Bases, (1985), pp. 171-180.

[Ja] G. Jaeschke, "Recursive Algebra for Relations with Relation Valued Attributes", TR-85.03.002, IBM Heildeberg Scientific Center (Mar. 1985).

[Ma] D. Maier, *The Theory of Relational Databases*, Computer Science Press (1983).

[OY] Z. M. Ozsoyoglu, L. Yuan, "A New Normal Form for Nested Relations", ACM Trans. on Database Systems, Vol.12, No.1 (March 1987), pp. 111-136.

[RKS] M. A. Roth, H. F. Korth, A. Silberschatz, "Theory of non-first-normal form relational databases", TR-84-36, Dept. of Computer Science, Univ. of Texas at Austin, (1984).

The Case For Safe RAM

George Copeland, Tom Keller, Ravi Krishnamurthy and Marc Smith
MCC
3500 West Balcones Center Drive
Austin, Texas 78759
(512) 338-3473

Abstract

Battery-backed-up DRAM memories can be configured today to be almost as reliable as disk. This paper argues that it is cost-effective to employ Safe RAM in computer systems which support reliable updates.

Safe RAM allows systems that support reliable updates, such as database and transaction processing systems, to perform more efficiently. We show how a response time improvement can always be realized, and how a throughput improvement can be realized to the extent that a system has had to limit disk utilization to achieve adequate response time. We also show that Safe RAM is cost-effective today for most applications and will become increasingly cost-effective as more caching is used, and as DRAM standby power and disk active power decrease.

1 Introduction

This paper investigates whether it is cost-effective to employ in computer systems a memory that is made almost as reliable as disk, which we call *Safe RAM*. We define Safe RAM as a memory having enough backup power to keep both the memory and disk alive for long enough to copy the memory to disk, as well as adequate protection from runaway software. Some Safe RAM features are available in some systems today.† However, a convincing argument has not yet been made for the cost-effectiveness of Safe RAM. A major area of opportunity for Safe RAM is in support of reliable updates which must be made atomic and persistent. Although other benefits may exist for Safe RAM (e.g., improved availability), we base our cost-effectiveness arguments only on reliable updates.

Many computer systems, such as database and transaction processing systems, must support reliable updates. Such systems typically employ several conventional techniques for amortizing disk I/O across multiple transactions. The most common, caching (i.e., memory buffering for disk reads), amortizes reads. The addition of logging and checkpointing amortizes reliable data writes [Gra78, Hae83]. The further addition of group commit amortizes log writes [DeW84, Gaw85]. A common thread among these techniques is that disk is assumed to be a stable mass storage and RAM is not, so that a transaction must commit its reliable updates by forcing either the updated data pages or logs containing the updates out to disk. A performance limitation of this assumption is that a transaction must wait on at least one disk write for the log.

We can overcome this limitation by the strategy of spooling into Safe RAM all data and/or log pages containing reliable updates. A transaction can then safely commit after having written its updates to the Safe RAM, thus avoiding the need for forcing logs or pages containing reliable updates to disk during the critical path of a transaction. Pages are written from the Safe RAM to disk by a background task whenever the disk has no other I/O to perform. This strategy allows *all* disk writes for reliable updates to be removed from the critical path of *any* transaction. This gives Safe RAM a performance advantage for disk bottlenecked systems, which can be applied at different points along the throughput and response time continuum. We show that a response time improvement can always be realized, and that a throughput improvement can be realized to the extent a system has had to limit disk utilization to achieve adequate response time.

The performance advantages of Safe RAM increase in time as technology progresses. Safe RAM is of marginal advantage in processor-bottlenecked systems; however, processor MIPS are becoming cheaper at a much faster rate than disk arms. Also, DRAM standby power and disk active power, which contribute to the cost of Safe RAM, are decreasing at a rapid rate. Thus, *the reasons why Safe RAM may not have been a good idea in the past are rapidly diminishing*.

We first discuss the feasibility and cost of Safe RAM using conventional technologies in Section 2. In Section 3, we review conventional techniques for reliable updates and show how they can be easily extended to use Safe RAM. Sections 4 and 5 quantify the performance impact of this Safe RAM extension to the best of the conventional techniques and show when it is cost-effective. Throughout this comparison, we make assumptions that are pessimistic for Safe RAM. Section 6 discusses some related issues. Section 7 provides a summary.

2 Feasibility And Cost Of Safe RAM

In this section, we first describe how Safe RAM can be realized using conventional RAM, UPS (uninterruptible power supply) and memory-protection

† Tandem, Stratus, HP, DEC's Mira, IBM disk caches and Amdahl RAM disks have the capability of keeping memory alive during power failure. The IBM AS400 has an option that allows both the memory and disk to continue operation for several minutes.

Proceedings of the Fifteenth International Conference on Very Large Data Bases

Amsterdam, 1989

technologies using either of two physical organizations. Then, we describe how the additional cost involves only a few per cent of the system cost. We use this cost later as a basis for determining cost-effectiveness.

2.1 Realization With Conventional Technology

We view the storage hierarchy of the system as shown in Figure 2.1. It consists of 1) a main conventional RAM, 2) a Safe RAM which we call the *Safe*, and 3) a conventional disk. The main RAM contains code, system files, temporary results, transaction workspaces and a cache for the most frequently used pages of the database. The Safe is used as a stable intermediate storage for reliable updates to disk, such that the pages in the Safe are almost as resilient to system and power failures as pages in disk.† We use a foreground (i.e., high priority) disk queue for database reads and I/O for temporary data, and a background (i.e., low priority) queue to spool reliable data and log writes.

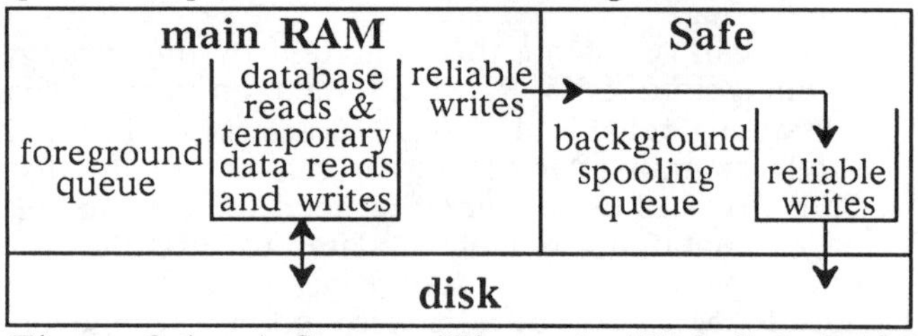

Figure 2.1: A Storage Hierarchy With Safe RAM

Although the Safe is built using off-the-shelf RAM technology, it must be made *almost as safe as disk.* A disk is nonvolatile and is protected by requiring access through a controller. Making RAM almost as safe as disk requires roughly comparable nonvolatility and protection.

Non-volatility is accomplished by using a UPS for the Safe, the disk and the disk controller. Upon UPS detection of power failure, these components are switched to UPS battery power and the Safe's contents are written to disk. When normal power is reinstated, recovery requires no special procedure. The UPS power requirement is the total active power of these components plus the standby power of the Safe's DRAM. The UPS energy (energy=power*time) requirement is dependent upon the time to write the Safe to disk.

Safe RAM is a storage medium. The physical failure rate of equally-priced disk and parity-based RAM subsystems are roughly equivalent [Bel88]. Any of the conventional recovery techniques used for disk media failure in conventional systems can be used to protect against Safe RAM media failure as well. Most database systems use a checkpoint and log technique for recovery from disk media failure, which writes a log of updates for each transaction before the transaction is allowed to commit and periodically writes a checkpoint of the database, both to a separate media than those containing the database [Gra78]. Some database systems offer an optional mirrored-disk technique for recovery from disk media failure, which maintains two copies of the database on separate disks by applying writes to both disk copies [Kat78]. The mirrored-disk technique improves availability because recovery is much faster than the checkpoint and log technique, but increases cost. The choice of whether to use this technique is based on the application's availability and cost requirements and the expected frequency of disk media failure. In systems that use the mirrored-disk technique, the checkpoint and log technique is usually also used as a backup for media recovery and for recovery from user and system errors which commit incorrect updates to the database. *A similar choice of whether to use a mirrored Safe for recovery from Safe media failure is possible.*

† [Gra78] (in Section 5.8.3.2) makes the historical note that database systems using core memories, which were non-volatile, assumed that the contents of in-core logs survived system crashes.

2.2 Physical Realizations

Figure 2.2 illustrates two physical realizations of the Safe. The *Separate Safe* realization employs a Safe which is separate from the main RAM and requires a separate mode of access (e.g., through a controller in a "silicon disk" implementation). The *Integrated Safe* realization makes the entire main RAM safe, so that the Safe pages can be distributed throughout the main RAM, intermixed with other pages (e.g., workspace and cache pages).

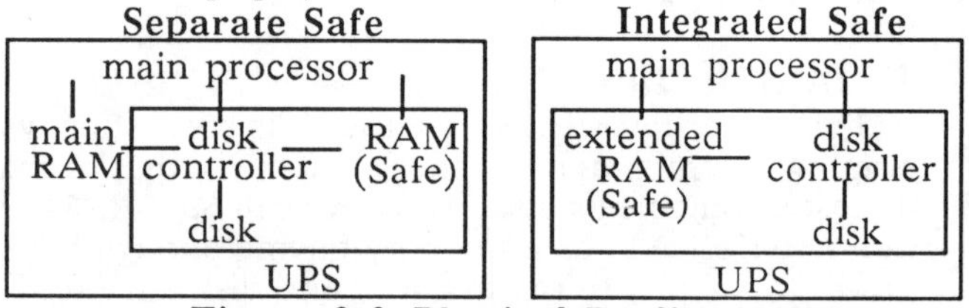

Figure 2.2: Physical Realizations

The advantages of the Separate Safe are:

- No additional protection requirements are needed, because the separate mode of access to the Separate Safe by the main processor protects the Safe from "runaway" system errors which result in undesired writes to memory. For the Integrated Safe, additional memory protection may be required to reduce the frequency of recovery. Conventional memory protection techniques (e.g., generous use of process virtual address spaces, write-protection mechanisms available in conventional MMUs, shadow memory pages or mirrored processor and memory pairs) can be used to ensure that only very limited types of system or user-program errors can damage Safe contents.
- Only the disk controller, the Safe and the disk require battery power and energy during the time required to write the Safe to disk. The Integrated Safe requires all of memory to be refreshed during this time.

The advantages of the Integrated Safe are:

- The size of the Safe in Integrated Safe can be dynamically varied up to the size of main memory. The size of the Separate Safe is fixed at a smaller moderate size and therefore requires the additional complexity of handling the case where the size of the Safe is exceeded (e.g., by forcing writes to the disk instead of the Safe until room becomes available in the Safe). Also, the cost of additional memory for the Integrated Safe to accommodate the Safe contents is limited to the average Safe size instead of a maximum Safe size as in the case of the Separate Safe.

- Copying pages between workspaces, the cache and the Safe is not necessary. Instead, pointers and protection status can be updated to reflect the move without physically copying the pages. The Separate Safe requires transfers between the main RAM and the Safe using a different, and probably slower, mode of access. This advantage is important for throughput if the processor is the bottleneck (since less work is required) and for response time (because this copying takes place during the critical path of a transaction). On the other hand, the copying required by the Separate Safe can reduce data contention because copying is considerably shorter than disk I/O and the page must be read locked during this time to ensure that a consistent copy of the page is written to disk.

For brevity, we will assume a Separate Safe for this paper. An analysis which includes the Integrated Safe may be found in [Cop88].

2.3 Assumptions About Technology

The technology estimates relevant to this report concern DRAM (dynamic RAM) cost, standby power and active power, processor active power, disk cost, speed and active power, and UPS cost, power and energy. In Figure 2.3, we provide estimates for both 1987 and 1993 for the sake of comparison. In particular, notice the rapid rate of decline of DRAM standby power [Fuj86, Ohs87, Saw88], a trend which favors a Safe RAM approach. We arrived at these estimates by simple extrapolation from historical trends and from various popular periodicals. We have pessimistically assumed that UPS costs will remain constant. The purpose of these numbers is not to make accurate predictions about technology. Instead, we use them to roughly calibrate various curves and provide ballpark estimates for several parameters. The conclusions that we draw in this paper are not highly sensitive to the accuracy of these estimates and our calculations can easily be redone by the reader with different technology characterizations and cost estimates as they become available.

2.4 Cost Of Safe RAM

In this section, we estimate the cost of the Separate Safe physical realization in both 1987 and 1993. We define the Safe as *cost-effective* if it can improve throughput by at least some *CostRatio* (ratio of system cost with Safe to without Safe) without increasing response time. For example, we determine later that the CostRatio is 1.043 for the Separate Safe realization using 1987 technology. This means that to be cost-effective, a Safe would have to increase throughput by more than 4.3%, without sacrificing response time.

The CostRatios that we calculate in this section are pessimistically high because of the following assumptions:

		1987	1993
DRAM	cost/Mbyte	$500/Mbyte	$25/Mbyte
	standby power	2watt/Mbyte	10watt/Gbyte
	active power	10watts	10watts
micro processor	active power	5watts	5watts
fast disk + controller	form factor	8"	5-1/4"
	unit cost	$5000	$2000
	unit capacity	500Mbytes	1000Mbytes
	cost/Mbyte	$10/Mbyte	$2/Mbyte
	track capacity	25Kbyte	50Kbyte
	avg seek time	0.018sec	0.010sec
	rotation time	0.015sec	0.010sec
	cost/access/sec for track for sector	 $165/access/sec $130/access/sec	 $40/access/sec $30/access/sec
	active power	100watts	50watts
UPS	cost/power	$1/watt	$1/watt
	time of power	20min (1200sec)	20min (1200sec)
	extra batteries	$0.5/Kwatt-sec	$0.5/Kwatt-sec

Figure 2.3: Technology Assumptions

- We assume a single disk system. The required Safe size for n disks is significantly smaller than n times the size required for one disk.
- We only include disk in the base cost. Including processor and memory in the base cost would reduce the CostRatio required to make Safe RAM cost-effective.
- We include a large memory size in the standby power requirements of the Integrated Safe. The UPS power requirements for RAM standby power is proportional to the RAM size.
- Only a very small percentage of the UPS energy is actually required to support Safe RAM, but we include the full cost of this unused energy in the CostRatio.

For the Separate Safe in 1987, the power required is about 116 watts, including a 100 watt disk, a 5 watt microprocessor, a DRAM with active power of 10 watts, and 100 2-Kbyte pages of Safe DRAM at 2 watt/Mbyte for standby power (0.4 watts). The time to write 100 sector-sized pages of Safe at 0.026 sec/access is 2.6 sec (assuming random writes to sectors). This is only about 0.2% of the energy capacity of the UPS, so that no additional battery energy is needed. This results in $116 worth of UPS. The additional RAM cost for the Safe would be $100. Thus, the total additional cost for the Safe is about $216. Because disk cost is $5,000, the CostRatio is 1.043.

For the Separate Safe in 1993, the power required is about 65 watts, including a 50 watt disk, a 5 watt microprocessor, a DRAM with active power of 10 watts, and 100 2-Kbyte pages of Safe DRAM at 10 watt/Gbyte for standby power (0.002 watts). The time to write 100 sector-sized pages of Safe at 0.016 sec/access is 1.6 sec (assuming random writes to sectors). This is only about 0.1% of the energy capacity of the UPS, so that no additional battery energy is needed. This results in $65 worth of UPS. The additional RAM cost for the Safe would be $5. Thus, the total additional cost for the

Safe is about $70. Because disk cost is $2,000, the CostRatio is 1.035.

Obviously, even for today's technology, only a negligible improvement in throughput is required to recover the dollar cost of Safe RAM because the dollar cost is so small.

3 History Of Reliable Updates Techniques

In this section, we summarize the historical evolution of conventional techniques for reliable updates and show how they can be easily extended to use Safe RAM.

In the simplest technique for reliable updates, a transaction reads pages from the public database and updates copies of the pages in the cache or a local workspace. When the transaction is ready to commit, a log of the updates is forced to disk, denoting the commit point. Then, the updated database pages are forced to disk to reflect the transaction-consistent state. The disk I/Os in the critical path of each transaction in this simple technique can be characterized by three variables:

- *Dr*: the number of read I/Os for pages that were not in the cache (and I/O for temporary data);
- *Dw*: the number of write I/Os for pages containing reliable updates;
- *Dl*: the number of log I/Os of reliable updates at commit time.

Some pages are read only once and discovered in cache by subsequent transactions, which amortizes reads. In the simple technique, a page may be overwritten on disk many times if that page is updated by many transactions. Many techniques have been proposed to amortize the writes over many transactions. Two such techniques are *checkpointing* (CP) which amortizes data writes, and *group commit* (GC) which amortizes log writes.

In CP techniques [Gra78, Hae83, Moh89], only the log is forced to disk at commit time of the transaction. A database page is written to disk only when that page is swapped out by the cache-replacement algorithm. As a result, a database page may be written to disk only once after many updates. To recover from failure, the system must reconstruct a consistent state of the database using the log. To avoid a recovery having to go arbitrarily far back in the log, checkpointing is periodically done, wherein all the updated pages still in the cache are forced to disk at the end of a checkpointing interval. This strategy incurs the following I/Os *in the critical path* of each transaction: 1) Dr, 2) Dw' (I/O for swapping out dirty pages for allocating memory for later transactions), 3) Dl, and 4) checkpointing overhead. Due to cache locality, Dw' is usually smaller than Dw (for the simple technique) and checkpointing overhead is amortized over many transactions. While checkpointing *reduces* the I/O for reliable updates, the data writes still lie in the critical path of some later transaction, because the cache-replacement algorithm will write pages out to allocate memory for later transactions. Thus, every write of an updated page is done in the critical path of *some* transaction.

After reducing the number of read and write I/Os per transaction via caching and checkpointing (perhaps close to zero for systems supporting memory-resident data), the log I/O cost becomes significant. In GC techniques [DeW84, Gaw85, Moh89], transactions are held up until a full log page can be written or a timeout occurs. Thus, the log write is amortized over a group of transactions. This strategy incurs the following I/Os in the critical path of each transaction: 1) Dr, 2) Dw', 3) Dl', and 4) checkpointing overhead. Dl' is usually less than Dl (in the simple and CP techniques), because the logs are amortized over several transactions. Further, a log I/O still lies in the critical path of each transaction, data I/Os still lie in the critical path of some later transaction, and each transaction is also held up for a period of time (waiting to commit) in its critical path. Consequently, the response time of each transaction reflects these I/O and waiting overheads.

Recall that in both the CP and GC techniques, dirty data pages are written out to disk by the cache-replacement algorithm to reallocate memory during the critical path of some later transaction. Alternatively, *spooling* can be used to migrate these dirty pages to disk. The advantage of spooling is that the dirty pages are written to disk as a background operation when no read I/O request is pending, instead of during the critical path of some transaction. The GC technique can be extended to use such a spooling mechanism to further improve performance. When this extension is made, checkpointing is usually also further optimized by making a log entry containing system status information (e.g., which pages are "dirty" and the status of transactions) instead of actually flushing the updated (dirty) pages, because spooling will keep the updated pages young enough so that a recovery does not have to go very far back in the log.

We consider Safe RAM a natural next step in this migration toward improved performance. Although Safe RAM may eventually influence the design of recovery algorithms, the use of Safe RAM does not necessarily require changing existing recovery algorithms provided that the Safe is treated the same as disk. Safe RAM can productively be used in conjunction with the most efficient conventional reliable update techniques (e.g., GC with spooling). In Sections 4 and 5, we compare the performance impact of Safe RAM on GC with spooling. In this context, the advantage of Safe RAM is that amortizing log writes via grouping does not cause a delay in commit times. This removes the log writes from the critical path of transactions. We quantify when this justifies the cost of Safe RAM.

4 Modeling

This section provides guidelines for the size of the Safe and provides equations for throughput and response time, which we use in Section 5 to show cost-effectiveness. We compare the best of the conventional reliable update techniques (e.g., GC with spooling) with and without the Safe. Throughout this section, we assume that the disk is the system bottleneck. We also make a number of assumptions, which we refer to as "pessimistic assumptions", that have a negative effect on the performance model for Safe RAM (e.g., overestimating the required size of the Safe and underestimating its improvements in response time and throughput). As a result, we get a *lower bound* for the improvement using Safe RAM, so that our cost-effectiveness argument is conservative.

4.1 Size of Safe

We are interested in determining the required size of the Safe given disk loading and how the load is made up of reads and writes. We will do this by defining two metrics, the mean time to overflow the Safe (MTTO) and the mean time to recover from the overflow (MTTR), and then, for given loads and Safe sizes, determining these means and seeing if they are acceptable. We show that even under extremely pessimistic assumptions, a moderate-sized Safe results in acceptable MTTO and MTTR.

We model the system by a single M/M/1 queue, making the following pessimistic assumptions in order to guarantee that the calculated MTTO will be far smaller than would be observed in an actual implementation:

- *All* disk I/Os are writes. The higher the proportion of reads to writes, the smaller the Safe needs to be. (If all disk I/Os were reads, then no Safe would be required.)
- Disk service times are exponentially distributed. This distribution has a variance higher than usually encountered in disk systems, resulting in longer queue length distributions (and thus a shorter MTTO).

For simplicity, we assume Poisson arrival rates. We assume a Safe of *k* pages (capable of holding k writes) and arrival and service rates characterized by means λ and μ. The probability of having exactly k writes in the queue is $(1-\rho)\rho^k$ where $\rho=\lambda/\mu$, the average utilization [Kle76]. The steady-state probability of having more than k writes in the queue (an overflowed Safe) is simply ρ^{k+1}. The probability of having k or fewer writes in the queue is $1-\rho^{k+1}$. The average rate at which the Safe overflows is

$$\lambda \text{ Prob}\left[\begin{array}{l}\text{a full Safe given that we}\\ \text{started with a non-full Safe}\end{array}\right]$$

$$= \lambda \text{ Prob}\left[\begin{array}{l}\text{k writes in the queue given that we}\\ \text{started with} \leq \text{k writes in the queue}\end{array}\right]$$

$$= \lambda \frac{(1-\rho)\,\rho^k}{1-\rho^{k+1}}$$

$$= \frac{(1-\rho)\,\rho^{k+1}\,\mu}{1-\rho^{k+1}} \quad .$$

The mean time between overflows is the inverse, or

$$\text{MTTO} = 1/\mu \,\frac{1-\rho^{k+1}}{(1-\rho)\rho^{k+1}} \quad .$$

Similarly, the mean rate at which the system exits the overflow state is

$$\mu \text{ Prob}\left[\begin{array}{l}\text{having k+1 writes in the queue given that}\\ \text{we started with k writes in the queue}\end{array}\right]$$

$$= \mu \frac{(1-\rho)\rho^{k+1}}{\rho^{k+1}}$$

$$= (1-\rho)\mu \; .$$

The inverse yields MTTR, or

$$\text{MTTR} = \frac{1/\mu}{1-\rho} \; .$$

Notice that MTTR is independent of k.

Figure 4.1 illustrates how MTTO (plotted on a log scale) depends exponentially on ρ for a given Safe size of k. We use a mean disk service time of $1/\mu$ = 26 msec for the plot. Note that for even (unrealistically) high disk utilizations (e.g., 0.90), the *pessimistic* bound of MTTO yields 10,800 sec = 3.0 hours for k=100. The bound of MTTR for this case is 0.26 sec. To summarize this extreme example, *if one implemented this system and ran an exclusively write workload on an overutilized disk with mean utilization of 90%, then one could expect a 100 page Safe to require writes to be forced to disk instead of the Safe for 0.26 sec once every 3 hours.*

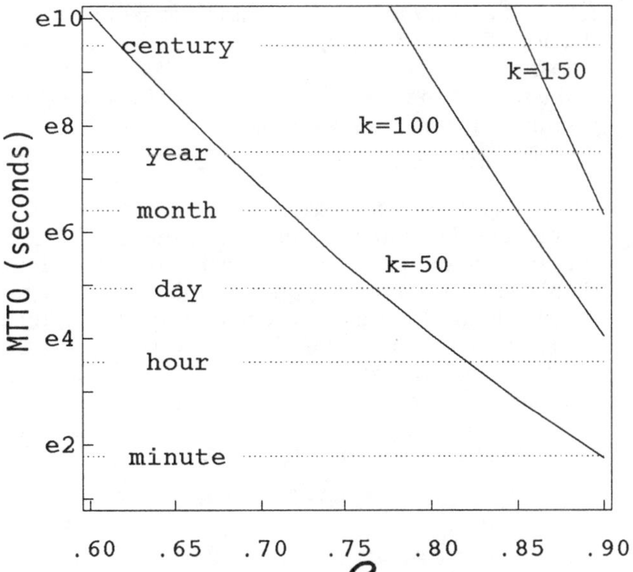

Fig 4.1: Mean Time to Overflow the Safe

4.2 Performance Model

This section develops the throughput and response time equations with and without Safe RAM. We assume the following:

- Only disk I/O is included in our throughput and response time equations. Processor delays are assumed to be negligible.
- The logs of G ($G \geq 1$) transactions are grouped into one page.
- Reliable writes are spooled for both the Safe and non-Safe case. We assume these spooled writes are removed from the critical path of any transaction.

This assumption is justified by the short length of the spooling queue described in Section 4.1 and by the fact that these pages will have recently-used cache status at the time they are put into the queue. These writes are nonpreemptive, so that the foreground queue must wait on any spooling I/Os which have already begun.

- There is a single database disk characterized by service rate μ and utilization ρ.
- There is a single log disk characterized only by service rate μ. Safe RAM never requires waiting for the log write, so that this response time delay is eliminated and a high log-disk utilization can be tolerated. Without Safe RAM, we optimistically assume that the log disk is never the system bottleneck and has low enough utilization so that its response time equals its service time of $1/\mu$. We also optimistically assume that a log write can be accomplished in a single I/O, regardless of how large Dw and G are. These optimistic assumptions are correspondingly pessimistic for Safe RAM.
- For simplicity, we ignore any checkpointing overhead, which is typically quite small.

When Safe RAM is not used for either disk, the commit time of multiple (G) transactions are delayed for a group commit, so that their logs can be written to the log disk with a single log write. We assume each transaction causes a total of D disk I/Os to the database disk, so that the maximum number of transactions per second without Safe RAM is

$$TPSo = \frac{\lambda_o}{D} = \frac{\rho_o \mu}{D},$$

where ρ_o is the database disk utilization without the Safe. The Dr reads are placed in the database disk's foreground queue, the Dw writes are placed in the database disk's spooling queue, and the log writes are placed in the log disk's foreground queue. Thus, the transaction response time has three terms.

First, each transaction must wait on Dr I/Os in the database disk's foreground queue. The average response time (queue wait plus service) for each of these I/Os is given in [Kle76] (Vol. 2, p. 121) as

$$\frac{\rho_o/\mu}{1-\rho_r} + 1/\mu ,$$

where $\rho_r = \frac{Dr}{D}\rho_o$ is the disk utilization due to the foreground queue.

Secondly, each transaction must wait on the other transactions in its group. This average group delay lies within the range:

$$\frac{1}{2}\frac{1}{TPSo}(G-1) = \frac{D}{2\rho_o\mu}(G-1) > \frac{D}{2\mu}(G-1).$$

The first expression assumes that each transaction is delayed for an average of half of the *average inter-arrival time* of G-1 transactions (we assume the last transaction does not have to wait). The second expression assumes that each transaction is delayed for an average of half of the *burst-service time* of G-1 transactions. We use the second expression for group delay because it is lower and thus is pessimistic for Safe RAM (i.e., it underestimates the savings due to Safe RAM).

Thirdly, each transaction must then wait on the log write in the log disk's foreground queue, which we assume is $1/\mu$. As a result, the average transaction response time without Safe RAM is

$$RTo = Dr\left(\frac{\rho_o/\mu}{1 - \frac{Dr}{D}\rho_o} + 1/\mu\right) + \frac{D}{2\mu}(G-1) + 1/\mu .$$

When Safe RAM is used for both data and log disks, the logs of multiple (G) transactions are grouped within the Safe containing the log disk's spooling queue, so that they can later be written to the log disk with a single log write, but without requiring the commit time of any transaction to be delayed. Each transaction causes a total of D I/Os to the database disk, so that maximum TPS with Safe RAM is

$$TPSs = \frac{\lambda_s}{D} = \frac{\rho_s \mu}{D},$$

where ρ_s is the database disk utilization with the Safe. The Dr reads are placed in the database disk's foreground queue, the Dw writes are placed in the database disk's spooling queue, and the log writes are placed in the log disk's spooling queue. Thus, each transaction must wait on Dr I/Os in the database disk's foreground queue. This is described by the first term in RTo, except that ρ_s is used instead of ρ_o. Safe RAM avoids the delays for group commit and log writes (second and third terms in RTo). As a result, RT with Safe RAM is

$$RTs = Dr\left(\frac{\rho_s/\mu}{1 - \frac{Dr}{D}\rho_s} + 1/\mu\right).$$

5 When Is Safe RAM Cost-Effective?

In this section, we quantify the performance improvements of Safe RAM and show when it is cost-effective.

5.1 Comparison Methodology

The performance improvement due to Safe RAM can be applied at different points along the throughput and response-time continuum. We examine throughput ratios (with Safe RAM over without Safe RAM) while constraining the response time with Safe RAM to be at least as good as without it. We also examine response time ratios (without Safe RAM over with Safe RAM) while constraining throughput to be equal. The numerators and denominators of these throughput and response time ratios are chosen to conveniently indicate that Safe RAM is better when either of the ratios is greater than 1. We define

$$TPSs/o \equiv \left.\frac{TPSs}{TPSo}\right|_{RTs \leq RTo} \text{and } RTo/s \equiv \left.\frac{RTo}{RTs}\right|_{TPSo=TPSs}.$$

We define *cost-effective* to mean that TPSs/o is greater than the CostRatio described in Section 2.4.

It is easy to prove that RTo/s>1 and TPSs/o$\geq$1 for all values of Dr, Dw, G and ρ_o, and that TPSs/o>1 when $\rho_o < \rho_m$, where ρ_m is the maximum allowable disk utilization even when the Safe is employed. The remaining questions are by how much these ratios exceed 1 and when Safe RAM is cost-effective (i.e., when TPSs/o > CostRatio). We derived TPSs/o and RTo/s analytically and plotted them using the following values:

Dr	Dw	D	Dr/D
3	4	7	0.43
0.3	0.4	0.7	0.43
6	1	7	0.86
0.6	0.1	0.7	0.86

x

G
1
4

x

ρ_m
0.9
0.7

The first case is the DebitCredit benchmark without caching [Ano85]. The second case constrains Dr/D to be the same as the first case while decreasing D via caching. The third case constrains D to be the same as the first case while increasing Dr/D. The fourth case constrains Dr/D to be the same as the third case while decreasing D via caching. We use G values of 1 (no grouping) and 4. We varied ρ_o from 0.2 to 0.9 and allowed ρ_s to go as high as 0.9 and 0.7. We included ρ_m=0.9 to represent the upper bound on disk utilization for any system. We also included ρ_m=0.7 to represent systems that require spare disk utilization for other purposes (e.g., disk mirroring†).

5.2 The Comparison

For the case when grouping is not used (G=1), Figure 5.1 illustrates the effect of Safe RAM on throughput when response time with Safe RAM is constrained to be at least as good as without it. The TPSs/o curves are shown in solid lines for ρ_m=0.9. TPSs/o = ρ_m/ρ_o except when ρ_s is forced to be below ρ_m by the response time constraint. The dashed line shows the 0.7/ρ_o curve. For ρ_m=0.7, the curves are the minimum of the 0.7/ρ_o curve and the solid curves. The higher of the two CostRatios (1.043) is also shown by the dotted line. Figure 5.3 illustrates TPSs/o when grouping is used (G=4). Figures 5.1 and 5.3 illustrate the following:

- As Dr is reduced (either by reducing Dr directly or by reducing D with Dr/D held constant), the curves come closer to the ρ_m/ρ_o curve, because the response time constraint becomes easier to meet. For G=1 and ρ_m=0.9, the (0.3, 0.4) curve is the same as the 0.9/ρ_o curve. For G=1 and ρ_m=0.7, the (0.3, 0.4) and (0.6, 0.1) curves are the same as the 0.7/ρ_o curve. For G=4 and ρ_m=0.9, all but the (6, 1) curve are the same as the 0.9/ρ_o curve. For G=4 and ρ_m=0.7, all of the curves are the same as the 0.7/ρ_o curve.
- As G is increased, the throughput improvement due to Safe RAM becomes much larger. This is due to the fact that Safe RAM eliminates the delay involved in group commit.
- As ρ_o approaches ρ_m, there is less room left for TPS improvement but the RT improvement increases.

For the case when grouping is not used (G=1), Figure 5.2 illustrates the effect of Safe RAM on response time with throughput held constant. The RTo/s curves are shown in solid lines for ρ_m=0.9. For ρ_m=0.7, the portion of the curves to the right of the dashed line do not apply. Figure 5.4 illustrates RTo/s when grouping is used (G=4).

The advantage of Safe RAM comes from the following:

- Log writes can be put in a spooling queue in addition to the data writes, so that the single log write is taken out of the response time's critical path. This advantage diminishes when Dr and ρ_o increase (the foreground queue wait time dominates) or when D and G increase (the group delay dominates).
- Response time is not delayed in order to exploit the grouped log writes. This advantage improves as G and D increase (the group delay dominates), and diminishes as Dr and ρ_o increase (the foreground queue wait time dominates). It has been argued that a group commit timeout value is typically much shorter (e.g., 50 to 100 msec) than the required response time (e.g., 1 sec), so that Safe RAM provides little advantage. However, in such systems, Safe RAM would allow a much larger G without impacting response time, so that more of the advantages of grouped log writes would be possible. Also, parallel systems often tend to centralize logging to exploit the maximum throughput advantage from group commit without increasing its response time penalty. Safe RAM allows more parallelism for logs, so that the maximum throughput advantage from group commit can be obtained without adding to transaction response time.

Our most significant point is: *Even with pessimistic assumptions, Safe RAM is cost-effective for systems that support reliable updates and have had to limit disk utilization to achieve adequate response times, provided that Dr is reasonably small.*

Safe RAM is expected to become increasingly cost-effective in the future because of the following:

- The CostRatio is expected to decrease in the future because of reduced DRAM standby power and cost per bit and reduced active disk power, even though we assumed that disk unit cost will decrease and UPS cost will be unchanged.
- The continuous improvements in DRAM cost per bit will cause caching to substantially increase in the future, so that Dr will decrease.

When viewing Figures 5.1 through 5.4, it is useful to recall the assumptions which we have made that are pessimistic for Safe RAM:

- The CostRatios were inflated because
  - we included the higher of the two CostRatios in the TPSs/o figures,
  - we assumed a single disk system,
  - we only included disk in the base cost,
  - we include a large memory size in the standby power requirements of the Integrated Safe,
  - we included the full UPS energy cost even though only a tiny percentage of the UPS energy is required to support Safe RAM, and

† For example, if disk mirroring were used with DebitCredit without caching (first case), then Dr, Dw and D become:

mode of operation	Dr	Dw	D
during normal operation	3	8	11
while one of the disks is out	6	8	14

To support the same arrival rate while one of the disks is out, maximum utilization during normal operation must be reduced to below 11/14=0.79 in order to keep utilization while one of the disks is out below 1.0.

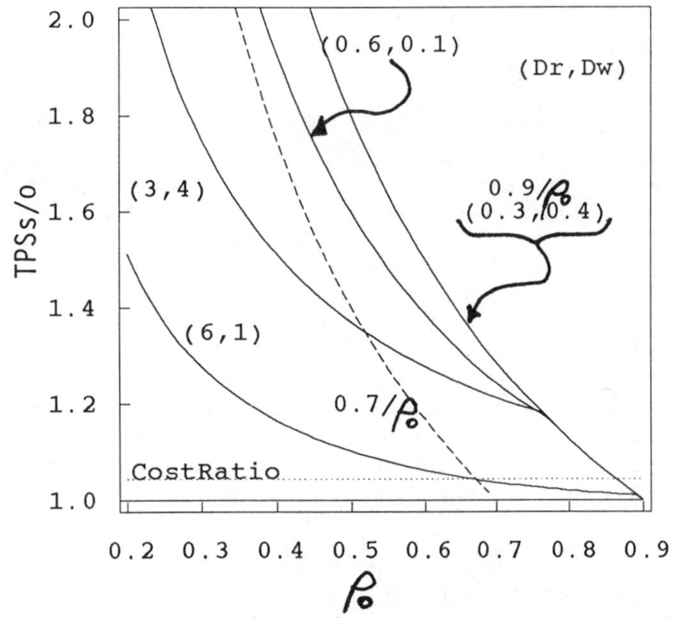

Fig 5.1: Throughput Improvement (G=1)

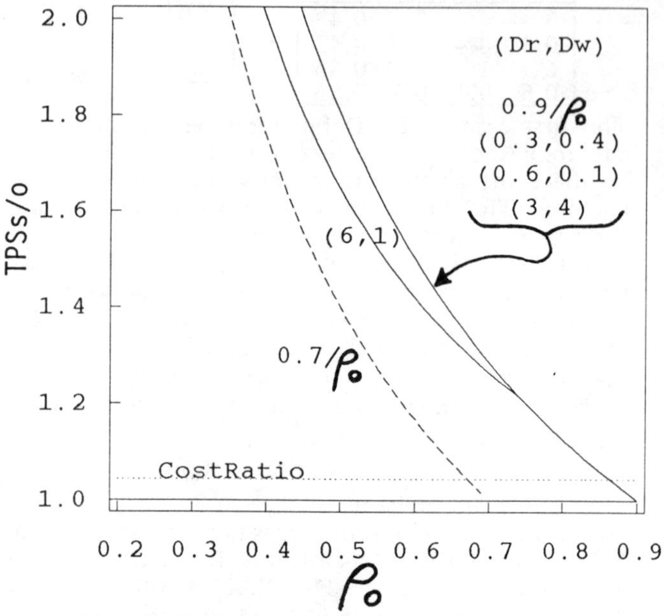

Fig 5.3: Throughput Improvement (G=4)

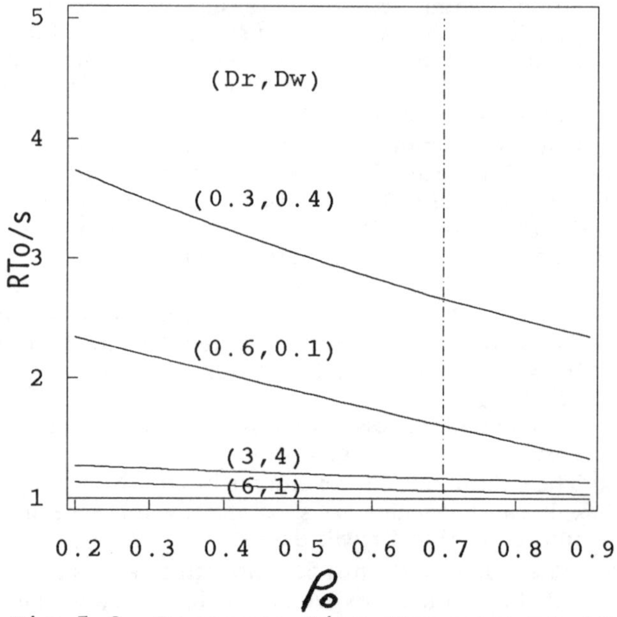

Fig 5.2: Response Time Improvement (G=1)

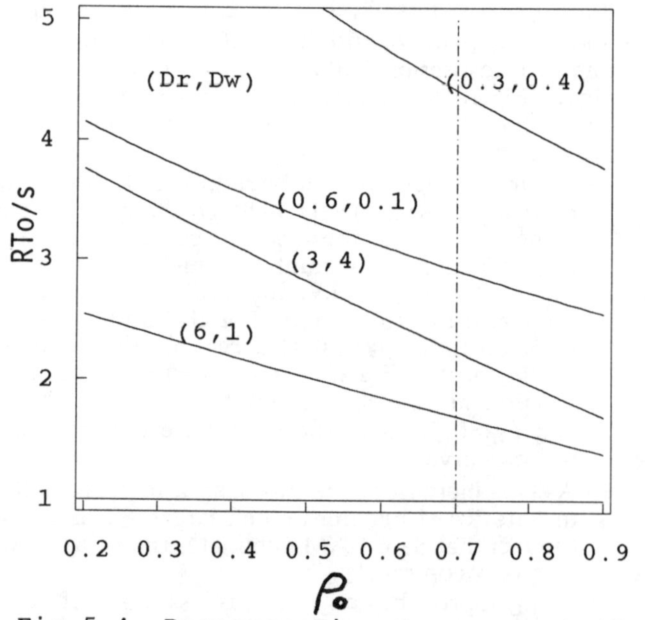

Fig 5.4: Response Time Improvement (G=4

- we overestimated the size of the Safe.
- We assumed that the log disk's delay without the Safe was a minimal $1/\mu$ due to low utilization (even though a log disk employing Safe RAM could tolerate a very high utilization) and regardless of how large G and D were (even though large G and/or D would cause multiple log writes).
- The group commit delay was based on burst service time instead of average inter-arrival time.

The accumulative effect of these assumptions can be quite significant.

6 Some Related Issues

This section describes several open issues concerning Safe RAM.

We showed that a typical existing UPS has considerably more energy capacity than is required to realize Safe RAM. (The energy requirements of Safe

RAM could be even further reduced by writing the pages in the Safe to contiguous disk locations during power failure and making the appropriate updates to disk page-mapping tables.) This suggests that a UPS with less energy and lower cost would suffice to realize Safe RAM. In fact, *a large-capacitance power supply might suffice*. Otherwise, the excess energy could be used to keep memory alive during power outage, so that subsequent disk reads of the data that was in memory at the time of power failure are unnecessary. The most cost-effective amount of energy for such higher availability could be calculated using an analysis similar to the 5-minute rule [Gra87] to trade off UPS energy cost for disk arm cost. Further, the excess energy could be used to keep the entire system alive during power outage to increase availability.

An interesting variation of the Separate Safe physical realization is the *Controller Safe*, which includes the Safe within the disk controller. This realization would allow today's systems which do not use group commit and which support reliable updates to plug in such a disk controller *transparently to the rest of the system* to achieve performance improvements.

For the Integrated Safe realization with a large cache, a separate spooling queue for cache-resident data (i.e., fixed in memory) with lower priority than the disk-resident data (i.e., subject to LRU swapping) has two advantages. One advantage is a significant reduction in the size of the spooling queue, because the separate spooling queue for cache-resident data can consist of pointers to cache-resident pages instead of the pages themselves. A second advantage is more efficient buffer management. Cache-resident data requires cache space anyway, whereas disk-resident data needs to be written out as early as possible to avoid being allocated by the LRU mechanism, which would cause a disk I/O during the critical path of some later transaction.

7 Summary

We argued the feasibility of Safe RAM using conventional technologies and estimated the additional cost required in both 1987 and 1993 using a physical realization called Separate Safe. We then described how Safe RAM can be used in conjunction with the most efficient conventional recovery techniques for improved performance. Finally, we quantified the performance effect of Safe RAM and described how, even with very pessimistic assumptions, Safe RAM is cost-effective *today* for systems that support reliable updates and have had to limit disk utilization to achieve adequate response times, provided that there are a reasonably small number of disk reads per transaction. Even though we assumed that disk unit cost will decrease and UPS cost will be unchanged, we showed how Safe RAM will be increasingly cost-effective in the future because caching will increase (due to DRAM cost per bit improvements) and because DRAM standby power and disk active power will decrease.

Acknowledgements

Thanks to Jim Gray for his encouragement and helpful suggestions.

References

[Ano85] Anon. et al, "A Measure of Transaction Processing Power," *Datamation* (April 1 1985).

[Bel88] J. Bell, private communications concerning experience as a Field Engineer and a Quality Assurance Manager (1988).

[Cop88] G. Copeland, R. Krishnamurthy and M. Smith, "The Case For Safe RAM," MCC Technical Report No. ACA-ST-080-88 (February 1988).

[DeW84] D.J. DeWitt, R.H. Katz, F. Olken, L.D. Shapiro, M.R. Stonebraker and D. Wood, "Implementation Techniques for Main Memory Database Systems," *Proceedings of the ACM SIGMOD Conference on Management of Data*, Boston (June 1984).

[Fuj86] S. Fujii et al, "A 50-μA Standby 1Mx1/256Kx4 CMOS DRAM With High-Speed Sense Amplifier," *IEEE Journal of Solid-State Circuits*, Vol. 21, No. 5 (October 1986).

[Gaw85] D. Gawlick and D. Kinkade, "Varieties of Concurrency Control in IMS/VS Fast Path," *IEEE Quarterly Bulletin on Database Engineering*, Vol. 8, No. 2 (June 1985).

[Gra78] J.N. Gray, "Notes on Database Operating Systems," in *Operating Systems: An Advanced Course*, Springer-Verlag, New York (1978).

[Gra87] J.N. Gray and F. Putzolu, "The 5 Minute Rule for Trading Memory for Disc Accesses and the 10 Byte Rule for Trading Memory for CPU Time," *Proceedings of the ACM SIGMOD Conference on Management of Data*, San Francisco (May 1987).

[Hae83] T. Haerder and A. Reuter, "Principles of Transaction-Oriented Database Recovery," *ACM Computing Surveys*, Vol. 15, No. 4 (December 1983).

[Kat78] J.A. Katzman, "A Fault-Tolerant Computing System," *Proceedings of the Eleventh Hawaii Conference on System Sciences* (January 1978).

[Kle76] L. Kleinrock, *Queueing Systems, Volume 1: Theory*, and *Queueing Systems, Volume 2: Computer Applications*, John Wiley & Sons, New York (1976).

[Moh89] C. Mohan, D. Haderle, B. Linsay, H. Pirahesh and P. Schwarz, "ARIES: A Transaction Recovery Method Supporting Fine-Granularity Locking And Partial Rollbacks Using Write-Ahead Logging," IBM Research Report RJ6649 (January 1989).

[Ohs87] T. Ohsawa et al, "A 60-ns 4-Mbit CMOS DRAM With Built-In Self-Test Function," *IEEE Journal of Solid-State Circuits*, Vol. 22, No. 5 (October 1987).

[Saw88] K. Sawada et al, "A 30-μA Data-Retention Pseudostatic RAM With Virtually Static RAM Mode," *IEEE Journal of Solid-State Circuits*, Vol. 23, No. 1 (February 1988).

ARIES/NT: A Recovery Method Based on Write-Ahead Logging for Nested Transactions

K. Rothermel, IBM European Networking Center, Tiergartenstrasse 15, D-6900 Heidelberg, W. Germany
rotherme@dhdibm1.bitnet

C. Mohan, Data Base Technology Institute, IBM Almaden Research Center, San Jose, CA 95120, USA
mohan@ibm.com

Abstract A simple and efficient recovery method for nested transactions, called ARIES/NT (Algorithm for Recovery and Isolation Exploiting Semantics for Nested Transactions), that uses write-ahead logging and supports semantically-rich modes of locking and operation logging is presented. This method applies to a very general model of nested transactions, which includes partial rollbacks of subtransactions, upward and downward inheritance of locks, and concurrent execution of ancestor and descendent subtransactions. The adopted system architecture encompasses aspects of distributed data base management also. ARIES/NT is an extension of the ARIES recovery and concurrency control method developed recently for the single-level transaction model by Mohan, et al. in the IBM Research Report RJ6649.

1. Introduction

The nested transaction concept was popularized by Moss [Moss81]. It has been implemented in several systems so far, such as ARGUS [LCJS87], Camelot [SpPB88], CLOUDS [DaLA88], LOCUS [MuMP83], and Eden [JNJBP82]. Nested transactions have at least three advantages over single-level transactions: First, they provide a means for controlling concurrency within transactions. Second, nested transactions can be used to protect a part of a transaction from failures of another part of the transaction - i.e., nested transactions can act as firewalls, preventing outside influences from affecting the internals. Third, nested transactions allow an easy and secure composition of transaction programs, by means of which the modularity of systems can be increased.

The goal of our work has been to devise an efficient and simple recovery method which guarantees the usual transaction atomicity properties for a very general model of nested transactions. Even though a number of papers have discussed the shadow-page recovery technique (as used in System R [GMBLL81]), write-ahead logging (WAL) is the method of choice in most commercial systems because of its efficiency, flexibility, and power (see [MHLPS89] for detailed comparisons). In WAL systems, an updated page is written back to the same nonvolatile storage (e.g., disk) location from where it was read. That is, **in-place updating** is done on nonvolatile storage. The ***WAL protocol*** asserts that the log records representing changes to some data must be physically written to stable storage **before** the changed data is updated on nonvolatile storage.

In this paper, a recovery method that uses WAL for nested transactions, called ARIES/NT (Algorithm for Recovery and Isolation Exploiting Semantics for Nested Transactions), is presented. ARIES/NT is an extension of the ARIES recovery and concurrency control method introduced by Mohan, et al. in [MHLPS89]. It retains all the features of the latter. ARIES tracks precisely the state of a page on nonvolatile storage as well as in main storage by associating a log sequence number (LSN) with every page. The LSN of a data base page is the address of the log record which describes the update most recently performed on the page. It is a monotonically increasing value that helps us relate the state of a page to logged updates of the page.

Like ARIES, ARIES/NT also logs all database changes, including even those performed during rollbacks of transactions. Updates performed during rollbacks are logged using what are called compensation log records (CLRs). During restart recovery, ARIES/NT guarantees that, for those transactions which were already aborting at the time of the system failure, only those actions that had not been undone ever before are rolled back. This means that the changes specified in a CLR are never undone (the changes are described in CLRs only for *redo* purposes during restart after a failure or during media recovery). This also causes a CLR to be used to keep track of how much of the transaction has already been rolled back and how much more remains to be rolled back. This tracking is done by recording, in a CLR, a set of pointers, each of which points to the next record to be dealt with in

Proceedings of the Fifteenth International Conference on Very Large Data Bases

Amsterdam, 1989

the chain of log records written by the transaction or one of its subtransactions (in ARIES, instead of a set of pointers, there will be only one pointer, called UndoNxtLSN, since ARIES was designed for the single-level transaction model).

The remainder of this paper is organized as follows. Section 2 introduces the notion of a nested transaction and describes the system architecture we assume. In section 3, the new recovery method ARIES/NT is presented. Section 4 describes other work on recovery algorithms for nested transactions. Finally, section 5 summarizes our work.

2. Transaction and System Models

In this section, we discuss the nested transaction model and the system architecture that we assume for ARIES/NT.

2.1. Nested Transactions

The transaction model that ARIES/NT is designed for is a generalization of the nested transaction model of [Moss81]. In this model, a transaction may contain any number of ***subtransactions***, and each subtransaction again may comprise any number of subtransactions.

A transaction which is not enclosed in another transaction is called a ***top-level transaction (TL-transaction)***. In the following, we use the term "transaction" to denote both TL-transactions and subtransactions. Transactions having subtransactions are called ***parents***, and their subtransactions are their ***children***. The ***ancestor*** (respectively, ***descendant***) relation is the reflexive transitive closure of the parent (child) relation. We use the term ***superior*** (respectively, ***inferior***) for the nonreflexive version of ancestor (descendant). The nesting hierarchy of a TL-transaction can be represented by a so-called ***transaction tree***, where the nodes of the tree represent transactions, and the edges represent the nesting relationship amongst the transactions.

Transactions can terminate either normally by committing or abnormally by aborting. Subtransactions appear atomic to the surrounding transactions and may commit or abort independently. A transaction is not allowed to commit until all its children have terminated. A subtransaction may abort without affecting the commit or abort outcome of the surrounding transaction. However, the commit of a subtransaction is relative; even if the subtransaction commits, aborting one of the subtransaction's superiors will undo its effects. Updates become permanent only when the enclosing TL-transaction commits.

The concurrency control scheme for nested transaction presented in [Moss81] is a locking approach. In this scheme, transactions may ***hold*** and ***retain*** locks. When a subtransaction commits, its locks are ***inherited*** by its parent, which then *retains* the locks. A transaction *holding* a lock for an object is allowed to access this object. The access is not allowed if it only *retains* the lock. Consequently, a retained lock is only a place holder indicating that transactions outside the hierarchy of the retainer cannot acquire the lock, but descendants of the retainer can.

The locking scheme introduced by Moss only allows for ***upward inheritance*** of locks, i.e., a parent may inherit locks of its children, but not vice versa. In [HaRo87b], a more general locking scheme that additionally supports ***downward inheritance*** of locks is proposed. ARIES/NT is flexible enough to be used together with concurrency control schemes enabling downward as well as upward lock inheritance [RoMo89]. Moreover, it accommodates the support of even semantically-rich lock modes like increment/decrement which permit multiple transactions to update the same data concurrently. This is the kind of feature that requires a recovery method to support operation logging (before and/or after image logging will not work) and to avoid the erroneous attempts to undo or redo some actions unnecessarily by precisely tracking the state of a page using the LSN and by writing CLRs.

Even concurrency control methods which are similar to locking, like the one described in [BaRa87], can be used with ARIES and ARIES/NT (see [MHLPS89]).

2.2. System Architecture

A distributed database system (DDBS) consists of a set of sites, which are interconnected via a communication network. Each site is comprised of a set of ***subsystems***, which cooperate with each other for the purpose of processing transactions. We assume that each subsystem has its own recovery component.

On each site of the DDBS, there exists a special subsystem called the ***bookkeeper***. Bookkeepers coordinate the initiation, migration and termination of transactions. Moreover, they maintain transaction state information, which is needed for recovery purposes, as well as information needed to build up transaction trees and to keep track of which transactions have visited which sites and which subsystems. The notations ***BK-subsystem*** and ***NBK-subsystem*** are used to denote a bookkeeper subsystem and a non-bookkeeper subsystem, respectively. An NBK-subsystem may be some component of a DDBS site, such as an index manager or a record manager.

A transaction may be distributed over several NBK-subsystems, possibly residing on different sites of the DDBS. A transaction is initiated at a subsystem either when the subsystem receives the first request to be performed within this transaction, or as soon as a child of this transaction commits at this subsystem, whichever happens first. From an NBK-subsystem's point of view, a subtransaction may be in two different ***states***, namely "unknown" and "active". At a subsystem, the initial state of a subtransaction is "unknown". The subtransaction enters the "active" state as soon as it gets initiated at the subsystem. It again becomes "unknown" when it commits or when one of its ancestors aborts.

TL-transactions alone may also reside in the "prepared" state. This state can be reached during the execution of a 2-Phase-Commit (2PC) protocol (e.g., the Presumed Abort protocol of [MoLO86]), which is used to terminate distributed TL-transactions. During the first phase of 2PC, the transaction enters the "prepared" state at a subsystem if updates were performed at that subsystem and that subsystem is willing to commit the transaction. After its termination (commit or abort) in the second phase, the transaction returns to the "unknown" state. Of course, it also becomes "unknown" when it is aborted in the "active" state.

Each BK in the DDBS guarantees the following:

- For each "active" subtransaction, an NBK-subsystem will eventually receive either a **commit** (for this subtransaction) or an **abort** (for an ancestor of this subtransaction) request from the local BK. A **commit** (respectively, **abort**) request contains the identifier of the subtransaction to be committed (aborted) as well as the identifier of the parent (inferiors, which are "active" at this subsystem) of this subtransaction. A commit request will not be issued until all the children of the committing subtransaction have terminated.
- For each "active" TL-transaction, an NBK-subsystem will eventually receive an **abort** or a **prepare** request from its local BK. While a **prepare** request contains the identifier of the TL-transaction to be prepared, an **abort** request includes the identifiers of the TL-transaction's "active" descendants at this subsystem. A prepare request will not be issued until all the children of the preparing TL-transaction have terminated.
- For each "prepared" TL-transaction, an NBK-subsystem will eventually receive a **commit** or an **abort** request from its local BK.
- For each "active" transaction, an NBK-subsystem can issue a query to the local BK asking for the transaction's inferiors residing in the "active" state at this subsystem. An NBK-subsystem needs this information when it decides to unilaterally abort a transaction.

3. ARIES/NT

In this section, first we give a brief overview of ARIES/NT, and then introduce the important data structures. Subsequently, we present the algorithm assuming normal operation, and finally describe how it operates during restart processing. Note that here we are primarily concentrating only on the ARIES/NT extensions, for the nested transaction model, to ARIES. [MHLPS89] should be consulted for detailed information and the rationale for the different basic design decisions.

3.1. Overview

In this section, we will first present the basic principles of ARIES (see also the section "1. Introduction") and then briefly describe the extensions that lead to ARIES/NT.

During restart after a failure, ARIES first scans the log, starting from the first record of the last complete checkpoint, up to the end of the log. During this **analysis pass**, information is gathered about (1) pages that were potentially more up to date in the buffers than in the nonvolatile storage version of the data base and (2) transactions that were in progress at the time of the crash. Then, ARIES ***repeats history***, with respect to those updates that were logged to stable storage but whose effects on data base pages did not get written to nonvolatile storage before the crash. This is done for ALL transactions, including for those transactions that were in progress at the time of the crash. This essentially reestablishes the state of the data base as of the time of the crash. No logging is done of the updates redone during this **redo pass**.

The next pass is the **undo pass** during which all in-progress transaction's updates are rolled back in reverse chronological order, in a single sweep of the log. Note that for those transactions which were already rolling back at the time of the crash, ARIES will only rollback those actions that had not already been undone. This is possible since history is repeated for such transactions and since the last CLR written for each already rolling back transaction points to the next nonCLR record, if any, that is to be undone.

Now, we will briefly describe the extensions to the above which lead to ARIES/NT. In both ARIES and ARIES/NT, all log records written by the same transaction are linked via a so-called **backward chain (BW-chain)**. In addition, in ARIES/NT, the BW-chains of *committed* subtransactions are linked to the BW-chains of their parents to reflect the transaction trees on the log. When a subtransaction T commits, a 'child committed record', which contains a pointer to the last record of T's chain, is written to the BW-chain of T's parent. Consequently, the BW-chain of an in-progress transaction together with the chains of its committed inferiors form a tree structure, which is called the transaction's **backward chain tree (BWC-tree)**. Since the parent/child relationships of committed subtransactions are stored on the log, subsystems can forget subtransactions after commit and the analysis pass need not collect information about committed subtransactions, which simplifies recovery.

When a transaction is aborted, the actions of that transaction and its (committed or active) inferiors are rolled back in reverse chronological order. Like ARIES, ARIES/NT logs database updates performed during rollback by means of CLRs. A CLR is also used to keep track how much of a transaction and its committed inferiors has already been rolled back and how much more remains to be undone. This is achieved by recording, in a CLR, a set of pointers, each of which points to the next log record to be processed in the BW-chain of the transaction or a committed inferior during undo.

As in ARIES, in ARIES/NT also, restart processing starts with an analysis pass, continues with a redo pass and ends with an undo pass. Redo processing of ARIES/NT works in exactly the same way as in ARIES, while the algorithms of the analysis and undo pass have been modified to support tree-structured log contents.

3.2. *Data Structures*

In this section, we describe some of the important data structures used by ARIES/NT at an NBK-subsystem.

Each record in the log of an NBK-subsystem belongs to a so-called ***backward chain (BW-chain)***. A BW-chain is associated with a transaction and connects the log records which are relevant for undoing and redoing this transaction at a particular subsystem.

The fields of a ***log record*** which are of interest in the subsequent discussions are:

- **PrevLSN:** Address of the preceding log record in the transaction's BW-chain. This value is NIL if this is the first log record.
- **LastLSN:** Present only in log records of the type C-Committed. When a subtransaction commits, a C-Committed record is added to the BW-chain of its parent. LastLSN contains the address of the last log record in the BW-chain of the committed subtransaction.
- **ChildId:** Present only in log records of the type *C-Committed.* It is the identifier of the child whose commit caused this log record to be written.
- **UndoNextSet:** Present only in log records of the type CLR. When written by a transaction T, it contains the address of the next log record of T that is to be processed during undo. Moreover, it also includes for each committed inferior of T which is only partially undone, the address of the next log record that is to be processed during undo in the inferior's BW-chain (for details see below).
- **PageId:** Present only in records of type Update (nonCLR) or CLR. The identifier of the data base page to which the updates of this record were applied.

In each NBK-subsystem, there exists a transaction table called ***TransTab***. For each transaction known by the subsystem, this table contains an entry consisting of **TransId, State,** and **LastLSN** (the address of the most recently written log record in the transaction's BW-chain).

A page in the buffer pool is said to be *dirty* if the buffer version of the page has some updates which are not yet reflected in the nonvolatile storage version of the same page. The table ***DirtyPages*** is used to represent the information about dirty pages. During normal processing, when a nondirty page is being fixed in the buffers for an update operation, the buffer manager inserts into the DirtyPages table a new entry containing the identifier of the page **(PageId)** and the current end-of-log address **(RecLSN** - recovery LSN), which is the address of the next log record to be written. Whenever a page is written back to disk, the corresponding entry is deleted from DirtyPages.

3.3. *Normal Processing*

The following two subsections describe the algorithms for database updates, transaction prepare and commit. The third subsection presents the algorithm for rollback, and finally, the fourth subsection describes how checkpoints are taken during normal processing.

3.3.1. Update

When an NBK-subsystem receives the first **work** request of a transaction, the recovery manager (RM) of this subsystem checks whether the transaction is already "active". The transaction would already be active, if a child of this transaction had executed at the subsystem and had committed before. If the transaction is still "unknown", the RM inserts an entry for the transaction in the RM's TransTab.

Whenever the execution of a work request causes a transaction to perform an update to an object in a page, this page is fixed in the buffer and latched in the exclusive (X) mode, the update is applied, an Update log record is added to the transaction's BW-chain, the LSN of the log record is placed in the page's LSN field, and the page is unlatched and unfixed (see [MHLPS89] for further explanations). The buffer manager uses the page LSN to enforce the WAL protocol. The page LSN is also used during recovery to determine the exact state of a page without having to examine any user data in the page. Before a transaction may update an object in a page, it must hold a lock for the object.

To support flexible storage management and to reduce the volume of log data, the changes to a page can be

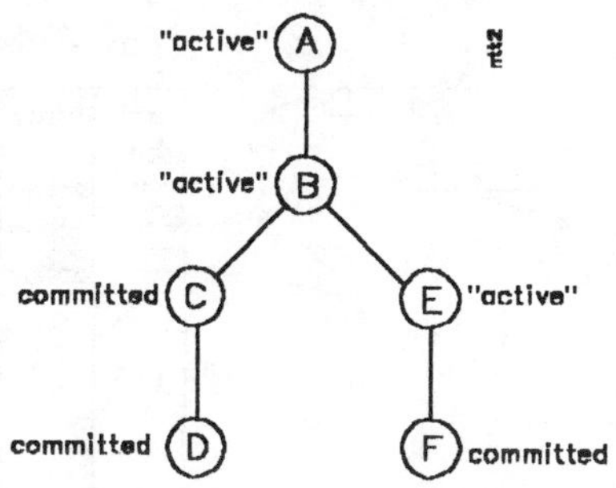

Figure 1: Transaction Tree with Transaction States

logged logically (logging operations, rather than before and after-images of modified data). This permits the semantics of the operations on the data to be exploited to permit additional concurrency. Moreover, since operations are logged, undo processing can affect a page totally different from the one affected during forward processing (see [MHLPS89, MoLe89]). Because the updates performed during rollbacks are logged, this permits us to support *page-oriented* redo processing and media recovery.

3.3.2. Prepare and Commit

Subtransactions: When an NBK-subsystem receives a **commit** request for a subtransaction T from its local BK, the RM of this subsystem checks whether there already exists an entry for T's parent in the local TransTab. If the entry does not exist, it changes the state of T's parent from "unknown" to "active" by inserting the corresponding entry in TransTab. Subsequently, it *asynchronously* writes a ***C-Committed*** (child committed) log record, which represents the first record in the BW-chain of T's parent. On the other hand, if T's parent is already "active", it only adds a C-Committed record to the BW-chain of T's parent. The C-Committed record contains, besides other information, the identifier of T (ChildId) as well as the log address of the last record in T's BW-Chain (LastLSN).

A BW-chain that represents the root of a BWC-tree is called a ***root chain***. An ancestor which represents a root chain is denoted as a ***root ancestor***. The BW-chains of TL-transactions as well as the BW-chains of aborted or "active" subtransactions always represent the root of a BWC-tree, whereas the BW-chains of committed subtransactions are always nonroot chains. Consequently, for each TL-transaction T, there might exist a forest of BWC-trees, which consists of T's BWC-tree and the BWC-trees of the "active" and aborted inferiors of T.

Figure 1 shows a transaction tree consisting of transactions A, B, C, D, E and F. Transactions A, B and E are in the "active" state, while the others are already committed. Figure 2 depicts a portion of the log of an NBK-subsystem S, which contains a BW chain for each transaction in this transaction tree. At time t1, there exists a forest comprised of three BWC-trees, which are associated with transactions A, B and E. To the BWC-tree of B, for instance, belong the BW-chains of B, C and D, where B's BW-chain is the root chain of this BWC-tree.

TL-Transactions: When an NBK-subsystem receives a **prepare** request for a TL-transaction during the first phase of 2PC, the RM of this subsystem adds a Prepare log record to the transaction's BW-chain and *synchronously* writes the record (and all log records preceding this log record) to the log on stable storage. The log record contains, besides other information, the identifier of the TL-transaction and a list of the update type locks (X, IX, etc.) held or retained by this transaction. The nonupdate type locks (S, IS, etc.) can be released at this time.

When an NBK-subsystem receives a **commit** request for a TL-transaction, it appends an End record to the transaction's BW-chain. Whether or not this record is written synchronously to stable storage depends on the kind of 2PC protocol used (for a detailed discussion, see [MoLO86]). After writing the log record and releasing the locks held or retained by the transaction, the subsystem deletes the transaction's entry in TransTab.

3.3.3. Rollback

When an NBK-subsystem receives an **abort** request for a transaction or it decides to unilaterally abort the transaction, it has to undo the effects of the transaction's "active" or "prepared" descendants (called ***known-descendants***). We define the ***BWC-forest*** of a transaction to consist of the BWC-trees of the known-descendants of this transaction. In order to rollback a transaction, the log records belonging to the transaction's BWC-forest have to be undone and compensated for.

When rollback for a transaction starts, the rollback process only knows the root chains of the transaction's BWC-forest; however, as rollback proceeds, it learns about all the nonroot chains of this forest. In order to keep track of which BW-chains of a transaction are currently known, for each of the transaction's known-descendants, a list called ***KnownChains*** is maintained by the rollback process. The KnownChains list of a known-descendant contains an UndoNext pointer for each BW-chain (of the known-descendant's BWC-tree) which has so far been encountered by the rollback process. The UndoNext pointer of a BW-chain points to the next log record in the chain to be read and processed.

When rollback of a transaction starts, the KnownChains list of a known-descendant, say X, of this transaction contains one UndoNext pointer, which

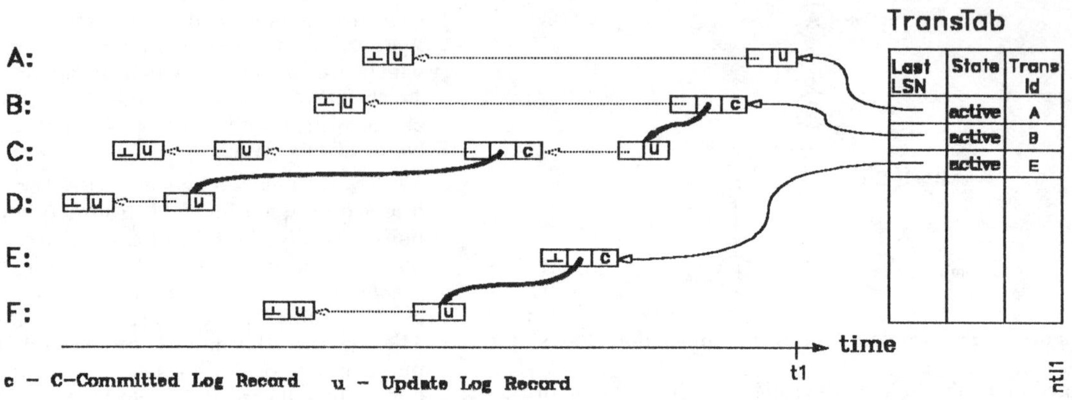

Figure 2: Log and TransTab at a Particular NBK-Subsystem

points to the last record of X's BW-chain. Whenever a C-Committed record belonging to X's BWC-tree is encountered in the log during rollback, a new UndoNext pointer is added to X's KnownChains list. Conversely, whenever the first log record of a BW-chain belonging to X's BWC-tree is read, the corresponding UndoNext pointer is removed from X's KnownChains list.

The CLRs compensating the log records belonging to a given BWC-tree are added to the root chain of this BWC-tree - i.e., CLRs can only occur in root chains (assuming there are no partial rollbacks - see [RoMo89]). If a subtransaction commits and later on it is aborted due to the rollback of one of its superiors, then that subtransaction's CLRs appear in the chain of log records of that superior.

Each CLR includes a set of log addresses: for each chain C′ which is a descendant of root chain C and which is known by the rollback process, the UndoNextSet field of a CLR added to C contains the address of the next log record in C′ to be processed. It is the information in the UndoNextSet that helps us avoid undoing the same nonCLR record more than once and also avoids having to undo CLRs.

Figure 3 shows the log of the NBK-subsystem S during the rollback of subtransaction B, which is a node of the transaction tree depicted in Figure 1. It is assumed that the rollback of B starts at time t1. The BWC-forest of B consists of two BWC-trees, the one of B and the one of E. In order to rollback B, the log records belonging to both these BWC-trees have to be undone and compensated for in reverse chronological order. The CLRs compensating log records belonging to the BWC-tree of B (E, respectively) are added to the BW-chain of B (E), which is the root chain of the BWC-tree of B (E). The UndoNextSet field of a CLR contains a set of log addresses. For example, CLR x6, which compensates Update log record u6, contains the LSNs of u3 and u4. Update record u3 (u4, respectively) is the next log record to be undone and compensated for in the BW-chain of D (C). At time t2, there exist two KnownChains lists, one for E and one for B. The KnownChains list for B contains two UndoNext pointers, which point to u4 and u3, whereas E's KnownChains list includes one pointer, pointing to u5. Of course, the next log record to be undone and compensated for is u5.

In order to describe how the rollback process acts on the different types of log records, we assume that it selects UndoNext pointer P residing in the KnownChains list of known-descendant X. Further, we assume that P points to log record R in BW-chain C. Depending on R's type, rollback acts as follows:

- *Update:* If R.PrevLSN equals NIL then R is the first log record in C - i.e., except R all undoable log records belonging to C have already been undone and compensated for. In this case, the rollback process can forget C by removing P from X's KnownChains list. If R.PrevLSN is not NIL, then P is updated to the value of R.PrevLSN. That is, after this update P points to the next log record of C to be read from the log.

  After updating X's KnownChains list, R is undone and compensated for if it is not a redo-only log record. The corresponding CLR is written to X's BW-chain, which represents C's root ancestor. The UndoNextSet field of this CLR contains the UndoNext pointers residing in X's KnownChains list. After writing the CLR, the page LSN and X's LastLSN in TransTab are updated to be the CLR's LSN.

  If X's KnownChains list is empty after removing P, then R is the last log record belonging to the BWC-tree of X that is yet to be processed. If this is the case, after processing R an End record is written to X's BW-chain, and then X is deleted from TransTab

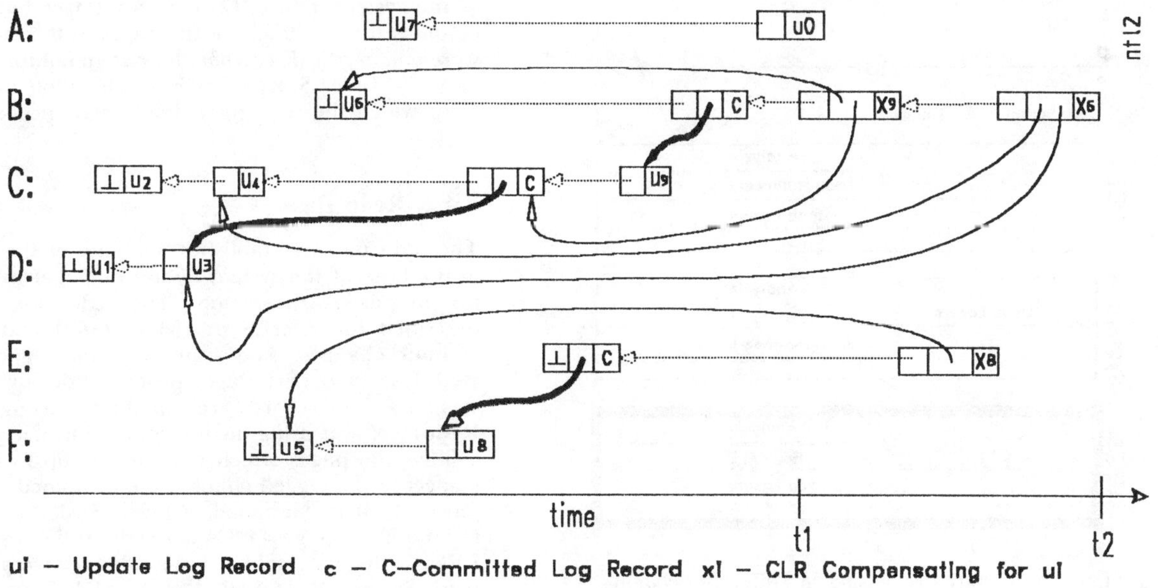

Figure 3: Logging Scenario

- ***C-Committed:*** The KnownChains list of X is updated by modifying P as described above. Moreover, a new UndoNext pointer P′ is added to X's KnownChains list, and is set to the value of R.LastLSN. Consequently, after this update, P′ points to the last record of a BW-chain, which is a child of C in X's BWC-tree.
- ***Prepare:*** P is changed to the value of R.PrevLSN - i.e., after this update P points to the next record in C to be processed.

The rollback algorithm described above only allows for *total* rollback of transactions. However, it has been extended to support *partial* rollbacks also (see [RoMo89]).

3.3.4. Checkpoints

For taking checkpoints periodically, the same *fuzzy checkpointing* mechanism as described in [MHLPS89] can be used without any changes. Taking a checkpoint is initiated by writing a ***BeginChpt*** log record. The checkpoint operation is completed by *synchronously* writing to stable storage an ***EndChpt*** record, which contains a copy of TransTab, and DirtyPages. The LSNs of the BeginChpt and EndChpt log records are stored in the ***Master*** record, which is in a well-known place on stable storage. Note that no dirty pages need to be forced to nonvolatile storage during the checkpoint operation. The assumption is that the buffer manager is periodically writing dirty pages to nonvolatile storage, as background activity, to keep the amount of redo work to be done at restart to a reasonable level.

3.4. Restart Processing

When a system failure occurs, the secondary storage version of the database may be left in an inconsistent state. It may contain uncommitted updates. Moreover, it may not contain some or all the updates of committed (or aborted) transactions. In order to reestablish a consistent database state, ***restart recovery*** has to be performed. Restart recovery is performed with three passes of the log, as outlined in the section "3.1. Overview". In the following, the three log passes are described in detail.

Figure 4 summarizes the recovery processing done by different systems. Note that, during restart, DB2, and System R do not redo any updates of the loser ("active" state) transactions. This is called ***selective redo***. In System R alone, the undo pass precedes the redo pass. There are a number of reasons for this and the correctness of this method relies on the shadow-page approach to recovery [GMBLL81]. The reader is referred to [MHLPS89] for detailed discussions concerning the problems caused by the recovery strategies used by the prior systems and how ARIES can exploit parallelism during the redo and undo passes to speed up restart. The algorithms for taking checkpoints during restart and for media recovery are the same as those described in [MHLPS89].

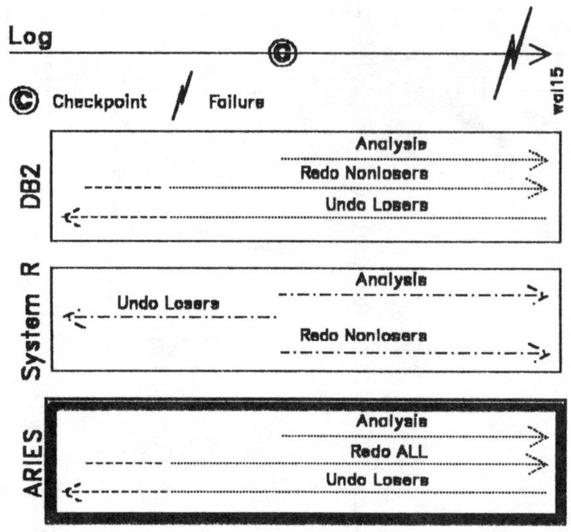

Figure 4: Restart Processing in Different Systems

3.4.1. Analysis Pass

The analysis pass begins by reading the Master record. The TransTab and DirtyPages tables are initialized from the EndChpt record. The log is scanned forward starting with the BeginChpt record until the end of the log is reached. During this scan, depending on its type, a log record R written by a transaction X is processed as follows:

- ***Update:*** If R represents the first log record in X's BW-chain, an entry is inserted into TransTab for X with State set to "active" and LastLSN set to R's LSN. Otherwise, in X's entry, LastLSN is set to R's LSN. If R.PageId is not in the DirtyPages table, a new entry is added to this table. The entry consists of R.PageId and R's LSN (as the RecLSN).
- ***CLR:*** In X's TransTab entry, LastLSN is updated. If R.PageId is not in the DirtyPages table, a new entry is added.
- ***Prepare:*** In X's TransTab entry, LastLSN is updated and State is set to "prepared".
- ***C-Committed:*** The entry in TransTab for R.ChildId is deleted.
- ***End:*** X is forgotten by removing X's entry from TransTab.

After the log is scanned until its end, **RedoLSN** is set to the minimum of the RecLSN values in DirtyPages. This is the log address from which the redo pass will start processing the log.

In summary, the output of the analysis pass is (1) the TransTab table, which contains the transactions that were in the "active" and "prepared" state at the time of the system failure, (2) the DirtyPages table, which contains the identifiers of the pages in the buffers that were *potentially* dirty when the system failure occurred, and (3) RedoLSN, which is the location on the log from which the redo pass should start processing the log.

3.4.2. Redo Pass

The redo pass reestablishes the state of the database at the time of the system failure and reacquires locks for "prepared" transactions. The redo pass algorithm described for ARIES in [MHLPS89] can be used without changes. The inputs to this algorithm are RedoLSN and DirtyPages produced during the analysis pass. Log records are examined starting from the RedoLSN and using the information in DirtyPages as a filter, the pages which may not be up to date with respect to the logged changes are examined. For each page which is examined, a log record's changes are redone if the page's LSN is less than the log record's LSN. Note that, unlike some other recovery methods (e.g., System R [GMBLL81]), ARIES redoes even the updates of transactions in the "active" state.

3.4.3. Undo Pass

During the undo pass, the effects of all transactions residing in the "active" state at the end of the analysis pass have to be undone. In the following, the set of BWC-trees of the these transactions are denoted by ***RestartUndo-Forest (RU-Forest)***. The fate of the transactions in the "prepared" state will be determined after contact is reestablished with the commit coordinator (see [MoLO86]).

Restart rollback is very similar to normal rollback. In reverse chronological order, the restart rollback process undoes and compensates for the log records which belong to the RU-Forest and which had not been undone ever before. It starts by initializing a KnownChains list for each transaction residing in the "active" state after the analysis pass. The KnownChains list of an "active" transaction contains one UndoNext pointer, which points to the last record of the transaction's BW-chain. Subsequently, it selects from these KnownChains lists the UndoNext pointer with the highest value. Then it reads the log record that the selected pointer points to and acts on it as described below. This is repeated until all the KnownChains lists are empty.

In order to describe how the restart rollback process acts on the different types of log records, we assume that it selects UndoNext pointer P residing in the KnownChains list of transaction X. Further, we assume that P points to log record R in BW-chain C. Depending on R's type the restart rollback process acts as follows:

- *Update, C-Committed:* Restart rollback acts on R in the same way as normal rollback does.
- *CLR:* P is removed from X's KnownChains list, and subsequently, all the UndoNext pointers in R.UndoNextSet are added to this list. That is, for each BW-chain which belongs to X's BWC-tree and which was known by a rollback process at the point of the system failure, an UndoNext pointer is added to X's KnownChains list. These pointers point to the log records to be undone next in the corresponding BW-chains.

4. Related Work

As far as we know, the only other recovery algorithm supporting WAL for nested transactions is the one presented in [Moss87]. Moss's WAL algorithm does not write CLRs and it does not seem to record LSNs in the nonvolatile storage versions of modified pages. That algorithm is not good enough to support fine-granularity of locking and avoid the numerous problems caused by not writing CLRs (the reader is referred to [MHLPS89] for detailed discussions concerning these topics and illustrations of the difficulties involved in supporting high concurrency efficiently using WAL). Especially with semantically-rich modes of locking (like increment/decrement) and operation logging, Moss's WAL approach of performing undo actions before redo actions during restart recovery and of not writing CLRs will not work.

Moreover, Moss does not discuss the concurrency control protocols and he assumes that the way the logging of changes is done guarantees idempotence. As discussed in [MHLPS89], the latter would make it impossible to support operation logging and efficient storage management for varying length objects. Use of LSNs on nonvolatile storage also is crucial for avoiding unnecessary and/or erroneous redo and undo work during restart recovery. Failures during restart processing will further compound the problems. ARIES and ARIES/NT support efficient restart and media recovery, and high concurrency by permitting operation logging, page-oriented redos and logical undos (see [MoLe89] for examples). For these reasons, we consider ARIES/NT to be fundamentally different from and more powerful than Moss's WAL algorithm.

Most recovery algorithms for nested transactions published so far are variations of the "version approach" described in [Moss81]. Briefly, this version algorithm goes as follows: When a transaction gets a write lock for an object, a new backup version of this object is created. This version is associated with the object and the transaction, and is used to restore the object should the transaction abort. The versions associated with an object are stored in a version stack, which is kept in volatile memory. When a subtransaction commits, its associated versions are offered to the parent. The parent accepts a version, if it does not already have an associated version for the same object. Otherwise, the offered version is discarded. If a transaction aborts, each of its associated versions is used to restore the objects directly or indirectly modified by the transaction. After that, the versions associated with the transaction can be destroyed. When a TL-transaction commits, the current state of each object directly or indirectly modified by this transaction is saved in stable storage. Then, all versions associated with this transaction are destroyed.

In most implementations of Moss's version algorithm (or variations of it), a version of an object is a complete copy of this object (e.g., see implementations in Eden [JNJBP82] and ARGUS [LCJS87]). An alternative approach is to store versions incrementally. In LOCUS [MuMP83], for example, file versions are stored incrementally - i.e., only those file pages that are new compared to the previous version need to be recorded in the new version.

Moss's version algorithm and its variations have several drawbacks:

If a transaction locks an object, a new version of the whole object has to be created. If the locked object is big, then the creation of new versions will be very expensive. On the other hand, if the objects to be locked are small, version stack management might become a substantial cost factor.

Those algorithms support only the **no-steal** buffer management policy, whereas ARIES and ARIES/NT support the no-steal as well as the **steal** policies (with *steal*, pages with uncommitted data may be written back to nonvolatile storage).

Partial rollbacks for nested transactions is not supported by them. Modelling savepoints [GMBLL81, HaRo87a, LHMW84] by means of subtransactions is too costly (inheritance of locks, generating new transaction identifier, etc.), since savepoints are used very frequently (e.g., to guarantee statement-level atomicity, which is required by ANS SQL).

5. Summary

A new recovery method for nested transactions, called ARIES/NT, has been presented. ARIES/NT is an extension of the ARIES recovery and concurrency control method that was introduced by Mohan, et al. in [MHLPS89] and that has been implemented to varying degrees in Starburst [Moha86], QuickSilver [HMSC88], the OS/2 Extended Edition[1] Database Manager [ChMy88], and DB2 V2R1. A high concurrency index management algorithm based on ARIES, and called ARIES/IM, has been implemented in the OS/2 EE Database Manager and is described, with extensions, in [MoLe89].

ARIES/NT is characterized by the following properties:

- It supports WAL for nested transactions, and, as far as we know, it is the only comprehensive WAL algorithm for nested transactions developed so far that permits semantically-rich modes of locking [BaRa87], operation logging, and efficient recovery.
- It allows arbitrary parallelism between related as well as unrelated transactions - i.e., a transaction may run concurrently with its superiors, inferiors, siblings and all other unrelated transactions.
- It supports concurrency control schemes allowing upward as well as downward inheritance of locks. That is, children may inherit locks to their parents, and vice versa.
- It supports savepoints at each transaction level - i.e., TL-transactions as well as subtransactions may establish savepoints.

Moreover, all the properties of ARIES described in [MHLPS89] hold for ARIES/NT also. No changes to fuzzy image copying (archive dumping), media recovery, buffer management, and deferred or selective restart algorithms of ARIES are necessitated by the introduction of the support for nested transactions. Some more details on ARIES/NT can be found in [RoMo89]. Additional discussions concerning BK-BK communications, BK recovery, BK data structures, and deadlocks arising from allowing concurrent execution of subtransactions and their ancestors will be presented in an expanded version of that paper [MoRo89].

6. References

BaRa87 Badrinath, B.R., Ramamritham, K. *Semantics-Based Concurrency Control: Beyond Commutativity*, **Proc. 3rd IEEE International Conference on Data Engineering**, February 1987.

ChMy88 Chang, P.Y., Myre, W.W. *OS/2 EE Database Manager Overview and Technical Highlights*, **IBM Systems Journal**, Vol. 27, No. 2, 1988.

DaLA88 Dasgupta, P., LeBlanc Jr., R., Appelbe, W. *The Clouds Distributed Operating System*, **Proc. 8th International Conference on Distributed Computing Systems**, San Jose, June 1988.

GMBLL81 Gray, J., McJones, P., Blasgen, M., Lindsay, B., Lorie, R., Price, T., Putzolu, F., Traiger, I. *The Recovery Manager of the System R Database Manager*, **ACM Computing Surveys**, Vol. 13, No. 2, June 1981.

HaRo87a Haerder, T., Rothermel, K. *Concepts for Transaction Recovery in Nested Transactions*, **Proc. ACM-SIGMOD International Conference on Management of Data**, San Francisco, May 1987.

HaRo87b Haerder, T., Rothermel, K. *Concurrency Control Issues in Nested Transactions*, **IBM Research Report RJ5803**, Almaden Research Center, August 1987.

HMSC88 Haskin, R., Malachi, Y., Sawdon, W., Chan, G. *Recovery Management in QuickSilver*, **ACM Transactions on Computer Systems**, Vol. 6, No. 1, p82-108, February 1988.

JNJBP82 Jessop, W.H., Noe, J., Jackobson, D.M., Baer, J.-L., Pu, C. *An Introduction to the Eden Transactional File System*, **Proc. 2nd IEEE Symp. on Reliability in Distributed Software and Database Systems**, Pittsburgh, July 1982.

LCJS87 Liskov, B., Curtis, D., Johnson, P., Scheifler, R. *Implementation of Argus*, **Proc. 11th ACM Symposium on Operating Systems Principles**, Austin, November 1987.

LHMW84 Lindsay, B., Haas, L., Mohan, C., Wilms, P., Yost, R. *Computation and Communication in R*: A Distributed Database Manager*, **ACM Transactions on Computer Systems**, Vol. 2, No. 1, February 1984.

MHLPS89 Mohan, C., Haderle, D., Lindsay, B., Pirahesh, H., Schwarz, P. *ARIES: A Transaction Recovery Method Supporting Fine-Granularity Locking and Partial Rollbacks Using Write-Ahead Logging*, **IBM Research Report RJ6649**, IBM Almaden Research Center, January 1989.

Moha86 Mohan, C. *An Overview of Starburst: An Extensible Relational DBMS*, **Proc. ACM-SIGMOD International Conference on Management of Data**, Washington, May 1986.

MoLe89 Mohan, C., Levine, F. *ARIES/IM: An Efficient and High Concurrency Index Management Method Using Write-Ahead Logging*, **IBM Research Report RJ6846**, IBM Almaden Research Center, June 1989.

MoLO86 Mohan, C., Lindsay, B., Obermarck, R. *Transaction Management in the R* Distributed Data Base Management System*, **ACM Transactions on Database Systems**, Vol. 11, No. 4, December 1986.

MoRo89 Mohan, C., Rothermel, K. *ARIES/NT: A Recovery Method Based on Write-Ahead Logging for a Very General Model of Nested Transactions*, **IBM Research Report**, IBM Almaden Research Center, Forthcoming.

Moss81 Moss, J.E.B. *Nested Transactions: An Approach to Reliable Distributed Computing*, **PhD Thesis**, Tech Rep MIT/LCS/TR-260, MIT, April 1981. Also as a modified version from MIT Press, 1985.

Moss87 Moss, E. *Log-Based Recovery for Nested Transactions*, **Proc. 13th International Conference on Very Large Data Bases**, Brighton, September 1987.

MuMP83 Mueller, E.T., Moore, J.D., Popek, G.J. *A Nested Transaction Mechanism for LOCUS*, **Proc. 9th ACM Symposium on Operating Systems Principles**, Bretton Woods, October 1983.

RoMo89 Rothermel, K., Mohan, C. *ARIES/NT: A Recovery Method Based on Write-Ahead Logging for Nested Transactions*, **IBM Research Report RJ 6650**, IBM Almaden Research Center, January 1989.

SpPB88 Spector, A., Pausch, R., Bruell, G. *Camelot: A Flexible, Distributed Transaction Processing System*, **Proc. IEEE Compcon Spring '88**, San Francisco, March 1988.

1 OS/2 is a trademark of the International Business Machines Corp.

Quasi Serializability: a Correctness Criterion for Global Concurrency Control in InterBase*

Weimin Du and Ahmed K. Elmagarmid
Computer Sciences Department
Purdue University
West Lafayette, IN 47907

Abstract

In this paper, we introduce Quasi Serializability, a correctness criterion for concurrency control in heterogeneous distributed database environments. A global history is quasi serializable if it is (conflict) equivalent to a quasi serial history in which global transactions are submitted serially. Quasi serializability theory is an extension of serializability. We study the relationships between serializability and quasi serializability and the reasons quasi serializability can be used as a correctness criterion in heterogeneous distributed database environments. We also use quasi serializability theory to give a correctness proof for an altruistic locking algorithm.

1 Introduction

The InterBase project in the Computer Sciences Department at Purdue University is an effort to investigate multidatabase management systems. The goal of this project is ultimately to build a heterogeneous distributed database system (HDDBS) that supports atomic updates across multiple databases.

An HDDBS integrates pre-existing database systems to support global applications accessing more than one element database. An HDDBS is different from a homogeneous distributed database system in that the element databases are autonomous and heterogeneous.

HDDBSs have become a very attractive research area recently. Within this research area, the transaction processing problem, especially that of concurrency control, has received considerable attention [GL84] [GP86] [Pu86] [BS88] [EH88] [LEM88]. Unfortunately, no satisfactory global concurrency control algorithm has been given yet. The lack of suitable theoretical tools for proving the correctness of these algorithms is one of the reasons.

Serializability has been generally used as the correctness criterion for the proposed concurrency control strategies. Unfortunately, serializability does not work well in heterogeneous distributed database environments. In [DELO88], we discussed the difficulties of maintaining global serializability in heterogeneous distributed database environments. The reason, we believe, is that serializability was originally introduced for centralized database environments and therefore is centralized in nature. Global concurrency control in heterogeneous distributed database environments, on the other hand, is hierarchical in nature due to the autonomy of the element databases. As a result, some of the proposed algorithms violate local autonomy (e.g., Sugihara's distributed cycle detection algorithm [Sug87]), while some allow low degree of concurrency (e.g., Breitbart's site graph testing protocol [BS88]), and others fail to maintain global serializability (e.g., [AGS87], [EH88] and [LEM88]).

The hierarchical nature of global concurrency control and the fact that local databases are autonomous make it very difficult to maintain global serializability in HDDBSs. On the other hand, the global concurrency controller (GCC) is relieved from some responsibilities (e.g., the correctness of local histories). This suggests that the correctness criteria for global concurrency control in HDDBSs should be based primarily on the behavior of global applications, with proper consideration of the effects of local applications on global

*This work is supported by a PYI Award from NSF under grant IRI-8857952 and grants from AT&T Foundation, Tektronix, and Mobil Oil.

ones. In this paper, we define such a criterion called quasi serializability (QSR). QSR is used as the correctness criterion for global concurrency control in InterBase.

In section 2, we define the terminology and assumptions used in this paper. We then introduce, in section 3, QSR as a correctness criterion for global concurrency control in HDDBSs. In section 4, we give a correctness proof, using QSR theory, of an altruistic locking algorithm [AGS87]. The correctness of this algorithm cannot be shown using serializability. Some concluding remarks are given in section 5.

2 Background

2.1 An HDDBS Model

A *heterogeneous distributed database system* consists of a set $D = \{D_1, D_2, ..., D_m\}$ of local database systems (LDBSs), a set $G = \{G_1, G_2, ..., G_n\}$ of global transactions, and a set $L = \bigcup_{i=1}^{m} L_i$ of local transactions, where $L_i = \{L_{i,1}, L_{i,2}, ..., L_{i,j_i}\}$. A *local transaction* is a transaction that accesses only one LDBS. The local transaction set L_i contains those local transactions that access LDBS D_i. A *global transaction* is a transaction that accesses more than one LDBS. The global transaction G_i consists of a set of global subtransactions, $\{G_{i,1}, G_{i,2}, ..., G_{i,m}\}$, where the subtransaction $G_{i,j}$ access LDBS D_j.

A *global history* H over $G \cup L$ in an HDDBS is a set of local histories $H = \{H_1, H_2, ..., H_m\}$, where the local history H_i (at LDBS D_i) is defined over global subtransactions $G_{1,i}, G_{2,i}, ..., G_{n,i}$, and local transactions $L_{i,1}, ..., L_{i,j_i}$.

Informally, an HDDBS is a collection of autonomous database systems wishing to cooperate. The element databases are connected and coordinated by a global database management system. Concurrency control is performed hierarchically. The local concurrency controller (LCC) at each LDBS is responsible for the correctness of the local history at that site. The GCC, which is added on the top of the existing LDBSs, coordinates the local histories at all local sites to guarantee the correctness of the global history. To simplify GCC's work, a global transaction is allowed to submit at most one subtransaction to each local site [GP86].

2.2 Assumptions

In this subsection, we list and explain the assumptions we have made in the study of QSR. Some of them are general in HDDBSs (e.g., local autonomy), while the rest are less general and are made to simplify the consistency problem of HDDBSs.

Local Autonomy

In an HDDBS, LDBSs are assumed to be autonomous. Autonomy is defined as follows [EV87] [DELO88].

- *Design autonomy:* Each of the LDBSs is free to use whatever data models and transaction management algorithms it wishes.
- *Communication autonomy:* Each of the LDBSs is free to make independent decisions as to what information to provide to the GCC or other LDBSs, and when to provide it.
- *Execution autonomy:* Each of the LDBSs is free to do anything (e.g., commit or restart) on any transactions running at its local site.

Basically, local autonomy reflects the fact that LDBSs were designed and implemented independently, and were totally unaware of the integration process. Local autonomy has significant effects on global concurrency control and the meaning of global consistency [DELO88].

Data Integrity Constraint

Another assumption we have made is that there is no data integrity constraint on data items at different sites. This assumption can be thought of as a consequence of local autonomy. Since each LDBS is designed and implemented independently, there is no relationship between data items at different LDBSs at all. In addition, it is also impossible to preserve these global data integrity constraints because neither the local users nor the LCCs of each LDBS are aware of the integration process and therefore the constraints[1].

As a result, global replication can only be in restrictive forms. In this paper, we assume that only those data items that are not directly updatable by local transactions are allowed to be replicated.

Intra Transaction Dependency

In this paper, we also assume that there is no value dependency between subtransactions of the same global transaction. This assumption is made because of the following consideration. In HDDBSs, it is very expensive (if not impossible) to support value dependency between subtransactions. Recall that LDBSs are

[1] Some people, however, believe that there should be some global integrity constraints on data items which are not identical to the local ones (see, e.g., [GP86]). However, they did not mention how these global constraints can be defined and preserved. We plan to investigate this further.

usually connected through a global database management system, and usually do not have direct communication. Since a global transaction can only submit one subtransaction to each site, the only way to implement value dependency is to submit those subtransactions with value dependency relations sequentially, even though they are executed at different sites.

We realized that this assumption is not necessary for QSR. However, the description of a less restrictive condition requires more concepts and work on the transaction model, and therefore is out of the scope of this paper (we refer readers to [ED89]). In addition, with this assumption, the appropriateness of QSR can be explained more easily and clearly.

3 Quasi Serializability

In this section we introduce Quasi Serializability. We first give its definition, and then discuss its various properties as a correctness criterion for concurrency control. We also discuss the relationships between QSR and other correctness criteria.

We assume that the reader is familiar with the basic theory and notations of serializability (see, e.g., [BHG87]).

3.1 Definitions

As mentioned earlier, QSR is used as a correctness criterion for global concurrency control. The basic idea is that, in order to preserve the global database consistency, global transactions should be executed in a serializable way, with proper consideration of the effects of local transactions. To understand this effect, let us expand the notion of conflict.

Let o_i and o_j be operations of two different transactions in a local history H, where o_i precedes o_j. We then say that o_i *directly conflicts* with o_j in H if they both access the same data item and at least one of them is a write operation. Whereas if o_i and o_j belong to the same local transaction (or global subtransaction), then we always say that o_i directly conflict with o_j in H. This, however, does not apply to operations of different subtransactions belonging to the same global transaction.

Let o_i and o_j be operations of two different transactions in a local history H. We say that o_i *indirectly conflicts* with o_j in H if there exist operations o_1, o_2, ..., o_k ($k \geq 1$) of other transactions such that o_i directly conflicts with o_1 in H, o_1 directly conflicts with o_2 in H, ..., and o_k directly conflicts with o_j in H.

Let G_i and G_j be global transactions in a global history H. We say that G_i *directly conflicts* with G_j in H, denoted $G_i \rightarrow_d^H G_j$, if one of G_i's operations directly conflicts with one of G_j's operations in a local history of H. We say that G_i *indirectly conflicts* with G_j in H, denoted $G_i \rightarrow_i^H G_j$, if one of G_i's operations indirectly conflicts with one of G_j's operations in a local history H. We also simply say that G_i conflicts with G_j in H, denoted $G_i \rightarrow^H G_j$, if either $G_i \rightarrow_d^H G_j$ or $G_i \rightarrow_i^H G_j$.

Notice that the conflict relation we defined here is history dependent and irreflexive. Therefore, an operation, o_i, (directly or indirectly) conflicts with another operation, o_j, in a history only if o_i precedes o_j in that history. Notice also that a global transaction might indirectly conflict with another global transaction in a history even if they do not access any common data items. The indirect conflicts are introduced by other transactions. It is the indirect conflicts that model the effect of local transactions on global transactions in a history.

Now let us introduce the notion of a *quasi serial history*. Unlike a serial history, only global transactions are required to execute serially in a quasi serial history. As we shall see later, this, together with the serializability of local histories, is sufficient to guarantee the correctness of global concurrency control in heterogeneous distributed database environments.

Definition 3.1 *A global history is quasi serial if*

1. *all local histories are (conflict) serializable; and*
2. *there exists a total order of all global transactions such that for every two global transactions G_i and G_j where G_i precedes G_j in the order, all G_i's operations precede G_j's operations in all local histories in which they both appear.*

Two global histories of an HDDBS are (conflict) *equivalent*, denoted $\equiv$, if their corresponding local histories are all (conflict) equivalent [BHG87].

Definition 3.2 *A history is quasi serializable if it is (conflict) equivalent to a quasi serial history.*

In a quasi serializable history, all local histories are serializable. In addition, global transactions are executed in a way that is serializable in terms of both direct and indirect conflicts. This kind of serializability is achieved by taking into account conflicts between both local and global transactions, although we are only interested in the behavior of global transactions.

Example 3.1 Consider an HDDBS consisting of two LDBSs, D_1 and D_2, where data items a and b are at D_1, and c, d and e are at D_2. The following global transactions are submitted to the HDDBS:

$$G_1 : w_{g_1}(a) r_{g_1}(d) \qquad G_2 : r_{g_2}(b) r_{g_2}(c) w_{g_2}(e)$$

Let L_1 and L_2 be some local transactions submitted at D_1 and D_2, respectively:

$L_1 : r_{l_1}(a)w_{l_1}(b) \qquad L_2 : w_{l_2}(d)r_{l_2}(e)$

Let H_1 and H_2 be local histories at D_1 and D_2, respectively:

$H_1 : w_{g_1}(a)r_{l_1}(a)w_{l_1}(b)r_{g_2}(b)$

$H_2 : r_{g_2}(c)w_{l_2}(d)r_{g_1}(d)w_{g_2}(e)r_{l_2}(e)$

Let $H = \{H_1, H_2\}$. Then H is quasi serializable. It is equivalent to the quasi serial history $H' = \{H'_1, H'_2\}$, where

$H'_1 : w_{g_1}(a)r_{l_1}(a)w_{l_1}(b)r_{g_2}(b)$

$H'_2 : w_{l_2}(d)r_{g_1}(d)r_{g_2}(c)w_{g_2}(e)r_{l_2}(e)$ □

3.2 The Quasi Serializability Theorem

There is a convenient graph-theoretic characterization of quasi serializability which is described in the following theorem. Let us first introduce the quasi serialization graph (QSG).

The *quasi serialization graph* of a global history H, denoted QSG(H), is a directed graph whose nodes are the global transactions in H, and whose edges are all the relations (G_i, G_j) $(i \neq j)$ such that $G_i \rightarrow^H G_j$.

QSG(H) is completely determined by the conflict relations of H. Therefore, QSG(H)=QSG(H') if H and H' are conflict equivalent. However, the reverse might not be true because QSG(H) loses some information of local transactions. Nevertheless, QSG(H) contains (in indirect conflict relations) enough information about local transactions for the testing of quasi serializability, as the following theorem indicates.

Theorem 3.1 *A global history H is quasi serializable if and only if all local histories are (conflict) serializable and QSG(H) is acyclic.*

Proof: (if) Suppose $H = \{H_1, H_2, ..., H_m\}$ is a global history over $G \cup L$, where G is a set of global transactions and L is a set of local transactions. Since QSG(H) is acyclic, it may be topologically sorted. Let $i_1, i_2, ..., i_n$ be a permutation of $1, 2, ..., n$ such that $G_{i_1}, G_{i_2}, ..., G_{i_n}$ is a topological sort of QSG(H). For each local history H_l $(1 \leq l \leq m)$, assume that $G_{i_1,l}, G_{i_2,l}, ..., G_{i_n,l}$ are the global subtransactions that appear in H_l. We show, below, that there is another serializable local history H'_l, equivalent to H_l, such that all of $G_{i_1,l}$'s operations precede $G_{i_2,l}$'s operations in H'_l, all of $G_{i_2,l}$'s operations precede $G_{i_3,l}$'s operations in H'_l, and so on.

In order to construct the equivalent history H'_l, let us group the operations in H_l into n operation sets based on global subtransactions.

For $p = 1$ **to** $n - 1$ **do**

$OP(i_p, l) = \{\ o \in H_l - \bigcup_{k=1}^{p-1} OP(i_k, l)$:
a. $o \in G_{i_p,l}$, or
b. o conflicts with one of $G_{i_p,l}$'s operations $\}$

$OP(i_n, l) = \{$ everything left $\}$

Informally, $OP(i_p, l)$ consists of all $G_{i_p,l}$'s operations and those operations in H_l that must precede some of $G_{i_p,l}$'s operations in any history which is conflict equivalent to H_l (but not in previous OP's). OPs are well defined and have the following properties:

1. $OP(i_p, l)$ contains all operations of global subtransaction $G_{i_p,l}$, but no operation of other global subtransactions, where $1 \leq p \leq n$.
2. $\bigcup_{k=1}^{n} OP(i_k, l)$ consists of all the operations in H_l.
3. $OP(i_p, l) \cap OP(i_q, l) = \emptyset$, where $1 \leq p \neq q \leq n$.

Let SH be the constructor that constructs from an OP a subhistory which has the same conflict relations as H_l as far as the operations in OP are concerned. This can be done by ordering all operations in OP in the same way as in H_l.

Now H'_l can be constructed as follows:

$$SH(OP(i_1, l)) \circ SH(OP(i_2, l)) \circ ... \circ SH(OP(i_n, l))$$

where $\circ$ stands for concatenation of subhistories. We claim that:

1. H'_l involves the same transactions as H_l;
2. H'_l is conflict equivalent to H_l; and
3. Global subtransactions in H_l are executed sequentially in H'_l.

The correctness of the first and the last statements are clear. We now show that the second statement is also true.

Let o_i and o_j be two operations in H_l and o_i conflicts with o_j. There exists integer p such that $o_j \in OP(i_p, l)$. If $p < n$, then o_j either belongs to $G_{i_p,l}$ or conflicts with one of $G_{i_p,l}$'s operations, and so is o_i. Therefore, either $o_i \in OP(i_q, l)$, where $q < p$, or $o_i \in OP(i_p, l)$ by the definition of $OP(i_p, l)$. In either case, o_i precedes o_j in H'_l. This is also true when $p = n$. So, o_i also conflicts with o_j in H'_l.

Let $H' = \{H'_1, H'_2, ..., H'_m\}$, then H' is quasi serial and equivalent to H. Therefore, H is quasi serializable.

(only if) Let H be a quasi serializable global history. Again, we assume that $G_1, G_2, ..., G_n$ are the global transactions in H. Let H' be a quasi serial global history which is equivalent to H. Then QSG(H) = QSG(H'). Since $G_1, G_2, ..., G_n$ are executed sequentially in H' and one operation can only conflict with subsequent operations in a history, QSG(H') must be acyclic and so does QSG(H). □

Example 3.2 The QSG of the global history H in example 3.1, as shown in Figure 1.(b), is acyclic. However, its serialization graph, as shown in Figure 1.(a), is cyclic. □

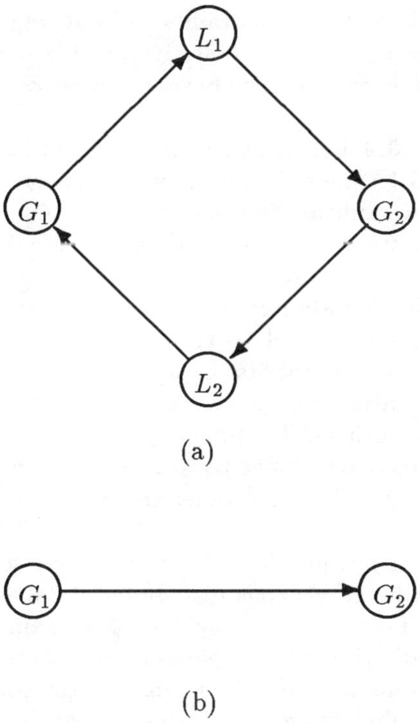

Figure 1: Serialization graph (a) and Quasi Serialization Graph (b) of H.

3.3 QSR as a Correctness Criterion for Concurrency Control

In this section, we show that quasi serializable histories are correct in terms of global concurrency control in heterogeneous distributed database environments. We do so by first discussing the database consistency problem, especially that of HDDBSs. We then discuss the ways that quasi serializable histories maintain the HDDBS consistency.

3.3.1 HDDBS Consistency Problem

It is generally accepted (see, e.g., [SLR76]) that a history is correct (or the database consistency is ensured) if

- Each transaction sees a consistent database.
- Each transaction eventually terminates.
- The final database after all transactions terminate is consistent.

Generally, a database is said to be consistent if it satisfies a set of consistency constraints. There are two types of consistency constraints associated with a database system:

1. Data integrity constraints
   These constraints are defined on data items and specify the real world restrictions on the values of data items. For example, the salary of an employee in a departmental database must always be greater than zero. These constraints are usually database dependent, but application independent. Therefore, they can be checked statically. In other words, whether a database satisfies the constraints can be checked entirely based on the values of data items.
2. Transaction consistency constraints
   These constraints specify the general restrictions on the interference between transactions. For example, two transactions should not mutually influence each other. These constraints can be database independent, but they cannot be checked statically. In other words, whether a database satisfies the constraints depends on all transactions executed and the interference among them.

Data integrity constraints can be expressed as a set of predicates, however they are not usually explicitly given. One way of preserving these constraints is to execute the involved transactions sequentially (assume that each transaction, when executing alone, will preserve these constraints.)

Transaction consistency constraints, on the other hand, usually cannot be expressed in predicate form. As with data integrity constraints, they are usually not explicitly given. They are generally required because of the conflict between user transparency and concurrent execution of transactions. Again, serial execution of the involved transactions is sufficient to preserve these constraints.

In an HDDBS, each LDBS has its own set of consistency constraints. When the LDBSs are integrated into an HDDBS, these consistency constraints are combined, together with some additional consistency constraints for the global database, to form the consistency constraints of the HDDBS.

In our HDDBS environment (as described in section 2), the consistency problem is different from what we have described above. The main difference is that there is no additional global data integrity constraint except the mutual consistency constraints for globally replicated data items.

On the other hand, there might be global transaction consistency constraints. This is possible because these constraints can be totally independent of any local transactions (e.g., constraints between two global transactions). We are currently investigating these and related problems. This is also necessary, as the following example shows.

Example 3.3 Consider an HDDBS consisting of two LDBSs, D_1 and D_2, where data item a is at D_1 and data items b and c are at D_2. The following global transactions are submitted to the HDDBS:

$G_1 : w_{g_1}(a)r_{g_1}(c)$ $\quad$ $G_2 : r_{g_2}(a)w_{g_2}(b)$

Let L be the local transaction submitted at D_2:

$L : r_l(b)w_l(c)$

Let H_1 and H_2 be local histories at D_1 and D_2, respectively:

$H_1 : w_{g_1}(a)r_{g_2}(a)$

$H_2 : w_{g_2}(b)r_l(b)w_l(c)r_{g_1}(c)$

In H_1, G_2 reads the value, a, written by G_1 directly. In H_2, however, G_1 might read the value, b, written by G_2 indirectly (e.g., local transaction L copies the value of b to c). Therefore, each global transaction sees only part of the effect of the other. Obviously, this kind of global inconsistent retrieval should not be allowed. However, it cannot be detected by LCCs. □

In summary, an HDDBS is consistent if and only if (1) it satisfies the global transaction consistency constraints (2) the globally replicated data items are consistent, and (3) all the LDBSs are consistent.

3.3.2 Correctness of Quasi Serializable Histories

Since a quasi serializable history is equivalent to a quasi serial history, we only need to show the correctness of quasi serial histories.

First, a quasi serial history preserves the mutual consistency of all globally replicated data items. This is true because they are updated only by global transactions that are executed serially.

Second, a quasi serial history preserves the global transaction consistency constraints. This is true because of the serializability of local histories, and the serial execution of global transactions. To see this, let us investigate various kinds of transaction consistency constraints in heterogeneous distributed database environments:

- Constraints on local transactions (or global subtransactions) at the same LDBS: They are preserved because of the serializability of local histories.
- Constraints on global transactions: They are preserved because the global transactions are executed serially.
- Constraints on local transactions (or global subtransactions) at different LDBSs: There is no direct interference between local transactions at different LDBSs because they do not access any common data item. Global transactions might introduce indirect interference between them. However, this is true only if there exist value dependency between subtransactions of the same global transaction, which is not the case in our environment. Therefore, constraints on local transactions at different LDBSs are always preserved because there is no interference between these local transactions.

Example 3.4 Let us consider the global history H in example 3.1. Since there is no value dependency between the two subtransactions of transaction G_2, the value of data item e written by G_2 at D_2 is not related to the value of data item b read by G_2 at D_1. Therefore, the value of data item e read by local transaction L_2 at D_2 is not related to the value of data item b written by local transaction L_1 at D_1. In other words, there is no relation between L_1 and L_2 (or they do not influence each other). The global constraints which can only be defined on global transactions are preserved because the global transactions are executed serially. □

It is worth noting that the above arguments are only true in heterogeneous distributed database environments. They may not hold in, for example, homogeneous distributed database environments. There are two reasons for this. First, in homogeneous environments, global integrity constraints can be defined on data items at different LDBSs. A quasi serializable history might not preserve these constraints. Second, it is possible for subtransactions of the same global transaction to have value dependency in homogeneous environments because they can communicate with each other. Therefore, two local transactions at different LDBSs might interact with each other. Unless all the transactions are executed in a serializable way, undesired interference between local transactions at two LDBSs might happen, and therefore constraints on these local transactions will not be preserved.

3.4 Comparison to Other Criteria

The most commonly used correctness criterion in general database environments (and also in heterogeneous distributed database environments) is serializability (see, e.g., [BS88]). Let G be a set of global transactions and $L = \{L_1, ..., L_m\}$ be sets of local transactions at various local sites. A global history over $G \cup L$ is serializable if it is computationally equivalent to a serial global history over $G \cup L$[2].

Two types of serializability exist, conflict serializability and view serializability [BHG87]. A global history over $G \cup L$ is conflict (or view) serializable if it is conflict (or view) equivalent to a serial history over $G \cup L$. In this section, we use CSR to denote the set of global histories which are conflict serializable, and

[2] L is treated as a set of "one site global transactions".

VSR for those which are view serializable. We use QSR to denote the set of global histories that are quasi serializable.

For a global history, its quasi serialization graph is a subgraph of its (conflict) serialization graph. Therefore, a conflict serializable global history is also quasi serializable. In other words, CSR $\subseteq$ QSR.

Theorem 3.2 *CSR $\subset$ QSR.*

Proof: We only need to show that CSR $\neq$ QSR. To see this, let us consider, for example, the global history H in example 3.1. It is quasi serializable, but not serializable. □

However, this is not true for VSR because the serialization graph for a view serializable history may not be acyclic. To see this, let us consider the following example.

Example 3.5 Consider an HDDBS consisting of two LDBSs, D_1 and D_2, where data items a, b and c are at D_1 and d is at D_2. The following global transaction is submitted to the HDDBS:

$G_1 : w_{g_1}(a)w_{g_1}(b)w_{g_1}(c)r_{g_1}(d)$

Let L_1 and L_2 be two local transactions submitted at D_1.

$L_1 : w_{l_1}(a)w_{l_1}(b)$ $\quad$ $L_2 : w_{l_2}(a)w_{l_2}(b)$

Let H_1 and H_2 be local histories at D_1 and D_2, respectively:

$H_1 : w_{g_1}(a)w_{l_1}(a)w_{l_1}(b)w_{g_1}(b)w_{l_2}(a)w_{l_2}(b)w_{g_1}(c)$

$H_2 : r_{g_1}(d)$

Let $H = \{H_1, H_2\}$. Then H is view serializable. Since H_1 is not conflict serializable (see [BHG87]), H is not quasi serializable. □

Theorem 3.3 *VSR $\not\subset$ QSR.*

The reverse is also not true. For example, the global history, H, in example 3.1 is quasi serializable but it is not view serializable.

Theorem 3.4 *QSR $\not\subset$ VSR.*

The relationships among QSR, CSR and VSR are illustrated in Figure 2.

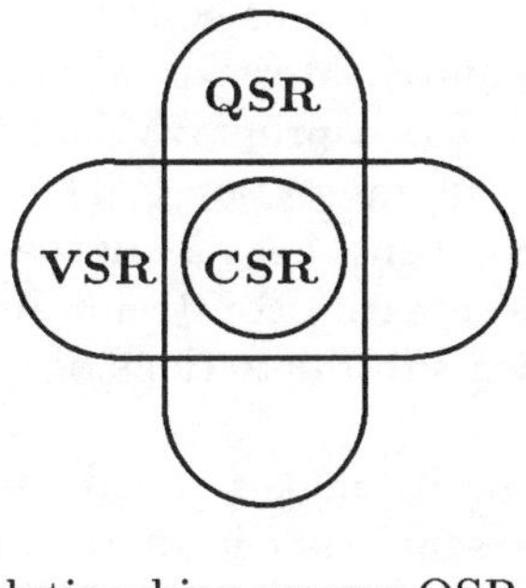

Figure 2: Relationships among QSR, CSR and VSR.

In [KS88], Korth proposed the use of predicatewise serializability (PWSR) as the correctness criterion for concurrency control in CAD database and office information systems. The basic idea of PWSR is that if the database consistency constraint is in conjunctive normal form, we can maintain the consistency constraint by enforcing serializability only with respect to data items which share a disjunctive clause. Clearly, PWSR concerns data integrity constraints only. In HDDBS environments, only the data items at the same LDBS can share a constraint clause. Therefore, a global history is predicatewise serializable if all the local histories are serializable. Since all the local histories in a quasi serializable history are serializable, each quasi serializable history is also predicatewise serializable. Let us also use PWSR to denote the set of histories that are predicatewise serializable. Then, we have QSR $\subseteq$ PWSR.

Theorem 3.5 *QSR $\subset$ PWSR.*

Proof: We need to show that QSR $\neq$ PWSR. This is true because the global history H in example 3.2 is predicatewise serializable, but not quasi serializable. □

4 A Correctness Proof of an Altruistic Locking Algorithm

It has been our goal in this paper to present a more flexible correctness criterion than serializability. This allows us to validate algorithms which provide a high degree of concurrency and do not violate local autonomies. In this section, we apply our quasi serializability theory to an altruistic locking algorithm [AGS87]. This algorithm was chosen for illustration because it was, we felt, correct, non-serializable and clearly stated.

4.1 How the Algorithm Works

The basic idea of the algorithm is to use global locking to coordinate the executions of global transactions at local sites. To improve the performance, the altruistic locking protocol provides a mechanism for global transactions to release locks before they finish.

Specifically, the altruistic locking protocol works as follows. A global transaction must lock a site before it can request work from that site. Once the global transaction's request has been processed, and if the global transaction will request no further work from that site, it can release its lock on the site. Other global transactions waiting to lock the released site may be able to do so if they are able to abide by the following restrictions. The set of sites that have been released by a global transaction constitutes the *wake* of that transaction. A global transaction is said to be in another global transaction's wake if it locks a site which is in that transaction's wake. The simplest altruistic locking protocol says that a global transaction running concurrently with another global transaction must either remain completely inside that transaction's wake, or completely outside its wake, until that transaction has finished.

4.2 A Non-Serializable Example

Although the authors of the altruistic locking algorithm believed that the algorithm ensures serializability of the global executions [AGS87], it actually does not. To see this, let us consider the following example.

Example 4.1 Consider an HDDBS consisting of two LDBSs, D_1 and D_2, where data items a and b are at D_1, and data items c and d are at D_2. The following global transactions are submitted to the system:

$G_1 : r_{g_1}(a)w_{g_1}(c) \qquad G_2 : w_{g_2}(b)r_{g_2}(d)$

Let L_1 , L_2 be the local transactions at D_1 and D_2, respectively:

$L_1 : w_{l_1}(a)r_{l_1}(b) \qquad L_2 : r_{l_2}(c)w_{l_2}(d)$

Let H_1 and H_2 be the local histories at D_1 and D_1, respectively:

$H_1 : w_{l_1}(a)r_{g_1}(a)w_{g_2}(b)r_{l_1}(b)$

$H_2 : w_{g_1}(c)r_{l_2}(c)w_{l_2}(d)r_{g_2}(d)$

Suppose G_1 locks D_1 before G_2 does. G_2 waits until G_1 finishes reading data item a and releases D_1. The same thing happens at D_2. In other words, G_1 gets the lock first. After updating data item c, G_1 releases the lock. G_2 then gets the lock and reads data item d. Since G_2 is completely in the wake of G_1, the history may be generated (or certified) by the algorithm. It is not hard to see that the global history, $H = \{H_1, H_2\}$, is not serializable. □

4.3 Correctness Proof

We prove that altruistic locking is a correct concurrency control algorithm (for HDDBSs defined in section 2) by proving that all global histories generated (or certified) by the algorithm are quasi serializable. We do so by constructing an acyclic QSG for every global history.

Let H be a global history over $G \cup L$, where $G = \{G_1, G_2, ..., G_n\}$ is a set of global transactions ($n > 1$). Suppose that H is generated (or certified) by the altruistic locking algorithm.

Let G_i and G_j be two global transactions in G which access a common database. If G_j gets any lock first, then either G_i is completely in the wake of G_j or G_i will not start until G_j has finished. Otherwise (if G_i gets any lock first), either G_j is completely in the wake of G_i or G_j will not start until G_i has finished.

Suppose that G_i conflicts with G_j. Then one of G_i's operations must conflict (directly or indirectly) with one of G_j's operations. Recall that an operation conflicts with another operation only if it precedes that operation in a history. One of G_i's operations must precede one of G_j's operations in a local history. Therefore, we have:

Lemma 4.1 *If the edge $G_i \rightarrow G_j$ $(1 < i, j < n)$ is in QSG(H), then all G_i's operations precede G_j's operations in all local histories.*

Theorem 4.1 *QSG(H) is acyclic.*

Proof: Suppose that there is a cycle in QSG(H): $G_{i_1} \rightarrow G_{i_2} \rightarrow \ldots \rightarrow G_{i_k} \rightarrow G_{i_1}$, where $1 \leq i_1, i_2, ..., i_k, k \leq n$. By lemma 4.1, all of G_{i_1}'s operations precede G_{i_2}'s operations at all local sites, all of G_{i_2}'s operations precede G_{i_3}'s operations at all local sites, etc., and all of G_{i_k}'s operations precede G_{i_1}'s operations at all local sites. In other words, all of G_{i_1}'s operations precede G_{i_1}'s operations at all local sites, a contradiction! □

5 Conclusion

We have extended serializability to verify the correctness of concurrency control algorithms for HDDBSs, resulting in *quasi serializability*. A global history in an HDDBS is quasi serializable if it is (conflict) equivalent to a *quasi serial* history in which all global transactions are submitted sequentially. We have shown that a global history is quasi serializable if and only if it has an acyclic *quasi serialization graph*. We have used this result to prove the correctness of an altruistic locking algorithm.

The main difference between quasi serializability and general serializability theories is that the latter treats global and local transactions equally while the former treats them differently. More specifically, quasi serializability theory is based primarily (not totally) on the behaviors of global transactions. This is appropriate in HDDBS environments because of the fact that there is no global data integrity constraint and no value dependency between subtransactions of the same global transaction.

Quasi serializability makes global concurrency control easier in the sense that it allows a higher concurrency than serializability. One immediate observation is that the global concurrency controller can ensure the correctness of the global history (i.e. quasi serializability) by simply controlling the submission of global transactions (e.g., serially). However, it is not obvious how to take advantage of quasi serializability to design global concurrency control protocols other than submitting global transactions serially. Further work is still needed in this area.

Another advantage of quasi serializable histories is the compatibility of (quasi) serialization and execution orders of the global transactions. This property is very useful for multi-level transaction management protocols where concurrency control is performed hierarchically and independently at different levels [BSW88].

Quasi serializability is intended to be used as a correctness criterion for global concurrency control in InterBase. It can also be used in any database system that meets the assumptions given in section 2. As a matter of fact, it can even be used in more general database systems (e.g., allowing restrictive value dependency between subtransactions of the same global transaction). We are now working on the problem of defining less restrictive transaction models suitable for quasi serializability and will report the result elsewhere.

Acknowledgements

The authors would like to acknowledge the contributions made by Yungho Leu and Shawn Ostermann. They have contributed immensely to the content and presentation of this paper.

References

[AGS87] R. Alonso, H. Garcia-Molina, and K. Salem. Concurrency control and recovery for global procedures in federated database systems. In *IEEE Data Engineering Bulletin*, pages 5–11, September 1987.

[BHG87] P. Bernstein, V. Hadzilacos, and N. Goodman. *Concurrency Control and Recovery in Databases Systems.* Addison-Wesley Publishing Co., 1987.

[BS88] Y. Breitbart and A. Silberschatz. Multibdatabase update issues. In *Proceeding of the International Conference on Management of Data*, pages 135–142, June 1988.

[BSW88] C. Beeri, H. Schek, and G. Weikum. Multilevel transaction management, theoretical art or practical need? In *Proceeding of the International Conference on Extending Database Technology*, pages 134–154, March 1988.

[DELO88] W. Du, A. Elmagarmid, Y. Leu, and S. Ostermann. *Effects of Autonomy on Maintaining Global Serializability in Heterogeneous Database Systems.* Technical Report CSD-TR-835, Purdue University, December 1988.

[ED89] A. Elmagarmid and W. Du. *Nested Transactions with Value Dependency Restrictions.* Technical Report CSD-TR-885, Purdue University, May 1989.

[EH88] A. Elmagarmid and A. Helal. Supporting updates in heterogeneous distributed database systems. In *Proceedings of the International Conference on Data Engineering*, pages 564–569, 1988.

[EV87] F. Eliassen and J. Veijalainen. Language support for mutidatabase transactions in a cooperative, autonomous environment. In *TENCON '87, IEEE Regional Conference*, Seoul, 1987.

[GL84] V. Gligor and G. Luckenbaugh. Interconnecting heterogeneous data base management systems. *IEEE Computer*, 17(1):33–43, January 1984.

[GP86] V. Gligor and R. Popescu-Zeletin. Transaction management in distributed heterogeneous database management systems. *Informtion Systems*, 11(4):287–297, 1986.

[KS88] H. Korth and G. Speegle. Formal model of correctness without serializability. In *Proceeding of the International Conference on Management of Data*, pages 379–386, June 1988.

[LEM88] Y. Leu, A. Elmagarmid, and D. Mannai. *A transaction management facility for InterBase.* Technical Report TR-88-064, Computer Engineering program, Pennsyvania State University, May 1988.

[Pu86] C. Pu. *Superdatabases for Composition of Heterogeneous Databases.* Technical Report CUCS-243-86, Columbia University, 1986.

[SLR76] R. Stearns, P. LewisII, and D. Rosenkrantz. Concurrency control for database systems. In *Proceeding of the Annual Symposium on Fundations of Computer Science*, pages 19–32, October 1976.

[Sug87] K. Sugihara. Concurrency control based on cycle detection. In *Proceeding of the International Conference on Data Engineering*, pages 267–274, February 1987.

The O_2 Object Manager: an Overview

Fernando Velez, Guy Bernard and Vineeta Darnis
Altaïr BP 105, 78153 Le Chesnay Cedex France

Abstract

O_2 is the Object-Oriented Database System being developed at Altaïr. The O_2 Object Manager is the layer of the system concerned with complex object access and manipulation, transaction management, persistence and disk management, and distribution in the server/workstation configuration. This paper describes the main choices made in the design of the Object Manager, and presents its overall architecture.

1 Introduction

The major objective of Altaïr is to build a new generation development environment for data intensive applications. The functionality of the system should include that of a DBMS, of a programming language, and of a programming environment. The target applications of the system are business applications, transactional applications (except very high performance transaction processing systems), office automation, and multimedia applications. The physical configuration we aim at consists of a server connected to a set of workstations (which may be heterogeneous). The server is the common repository for shared data. The target customers are application programmers and end users. Our main interest is in application programmers, since we consider the major problem to be the improvement of programmer productivity.

To meet these requirements, we decided to build an object-oriented database system and its programming environment. O_2 is both a Database System and an Object-Oriented System. As a Database System it provides support for accessing and updating large amounts of persistent, reliable, and shared data. As an Object-Oriented system, it supports features such as complex objects with identity, inheritance (of classes or types), encapsulation (of an object state by the methods defined on its class), overriding (redefining methods in classes), and run-time binding of methods to objects. Our interpretation of these notions are condensed in the definition of the O_2 data model [22], [21]. Section 2 presents a quick overview of the model.

Proceedings of the Fifteenth International Conference on Very Large Data Bases

O_2 has a complete programming environment. It provides tools such as editors, browsers, and debuggers. When developing an application, programmers specify a schema and write the code for methods. The schema is specified in the O_2 language. Methods are written using standard programming languages. In the current implementation of the system, methods are written in either CO_2 or $BasicO_2$, which are extensions of C and Basic respectively. The schema declarations are interpreted and handled by the Schema Manager. Methods are precompiled, then compiled by the "host" programming language compiler, and they may be loaded and executed.

A first throwaway prototype was built in 1987 and is reported in [3]. Since then, we redefined the system at all its levels: data model and language, programming environment, compilers, Schema Manager and Object Manager. The work reported here concerns only the last module.

The Object Manager (OM) is the piece of software that handles persistent and temporary complex objects with identity. Furthermore, objects are shared, reliable and move from a workstation to the server and vice versa. The OM is used by all the upper modules of the system.

We believe that the OM synthesizes techniques proposed in the database field and in the object-oriented programming field. A similar approach is found in GemStone [24], Orion [19], Encore [14], Vbase [2], and Iris [11]. However, it is is original in the following respects. Firstly, in the way it handles distribution: applications may run either entirely on the server, or entirely on the workstation, or the programmer may migrate control from one machine to another at his will when passing a message to an object. We consider this to be a powerful performance tuning facility, not found in the systems mentioned above and well adapted to a server/workstation configuration (the first three fall in this category).

Secondly, the system adapts itself to a wide range of application requirements ranging from development of applications to execution of data-intensive applications. We distinguish a *development mode* in which users are programmers developing applications from an *execution mode* in which previously built applications are executed and the main concern is performance. The systems mentioned above do not make such a difference and run in a single mode.

Lastly, the system implements a number of facilities that make the life easy for application programmers: (i) exceptional attributes for tuple objects may be attached

Amsterdam, 1989

at any time without performance degradation, (ii) persistence is implemented with a simple composition-based schema in which deletions are implicit (this obviously is not new in the programming language world), and (iii) clustering issues are clearly separated from the schema information and specified by the DBA in the form of a subset of the composition graph.

The rest of this paper is organized as follows: Section 2 discusses the functional requirements of the OM both with respect to the O_2 data model and with respect to the applications the system is intended to support. Section 3 presents the main design choices we faced when developing the OM. Section 4 presents an architectural overview of the system. Section 5 compares our system to other related work, and Section 6 concludes the paper, summarizes the differences with the first prototype reported in [3] and presents some future extensions.

2 Functional Requirements

2.1 Data Model Requirements

In what follows, we present the particularities of the model pertinent to the system design. In O_2, we distinguish between *values*, on which we can perform any of a set of predefined primitives, and *objects* which have identity and encapsulate values and user defined *methods*. Values can be *set, list* or *tuple* structured, or *atomic*. Each value has a *type* which describes its structure. Objects belong to *classes*. A class has a name, a type t (specifying the common structure of the set of objects of the class) and a set of methods M. Every object of a certain class has a value of type t and has the same set of methods M.

Classes and types are partially ordered according to an inheritance relationship. Multiple inheritance is supported. Our interpretation of tuple types follows the one proposed by Cardelli in [7], implying that inheritance has natural set inclusion semantics: if C is a subclass of C , then the set of all possible instances of C is included in the set of all possible instances of C . The prescriptive interpretation of tuple types implies that a tuple object or value may have "exceptional" attributes, i.e. attribute values not declared in its class or type. The O_2 code below shows this:

```
    class Person type tuple (name:string,
age:integer);

    O2 Person x ;
    *x = tuple(name: ''john'', age: 27,
my_opinion: ''nice fellow'') ;
```

Here, the expression "*x" refers to the value encapsulated in object x. Reference to values encapsulated in objects is allowed only in methods of class `Person` (or in other methods if the de-encapsulation method "*" is exported by the creator of class `Person` - see [23]). In order that the `my_opinion` attribute could be used while conserving a "strict" typing, it is necessary to have in the language a clause which permits to verify the existence of this exceptional attribute. This is scheduled for the next version of the language.

A *named* object or value is an object or value with a user-defined name. These objects or values are the roots of persistence. Persistence is *user controlled*: in order to make an object persistent, the user has to make it a component of an already persistent object.

An O_2 *schema* consists of classes, types, named objects and named values. A query language has been designed [4] which uses the distinction between objects and values and the existence, in the data manipulation language, of primitives to manipulate structured values.

2.2 Application Requirements

The system should support a wide range of application requirements. Consider the three following utilization scenarios:

(i) Users are programmers developing O_2 applications. They constantly define new classes and new methods, and methods are generally tested with few objects in main memory.

(ii) Users are "end users" (i.e., not necessarily programmers), and they execute applications which have been previously developed and tested. Here, we have a notion of "application" which contains a set of classes and methods. In this mode, the essential concern is execution speed. Data are massive, shared and have to be reliable. The schema is static.

(iii) Users are programmers developing (rapid) prototypes with O_2. Data would fit in main memory in many cases and is seldom shared. Execution speed is important and the schema may be flexible.

These three scenarios imply very different and contradictory requirements. In the first scenario, which we call *development mode*, flexibility is at a premium, and therefore, it is important to optimize method compilations and schema concurrency protocols. Execution speed is less important.

In the second scenario, the *execution mode*, we need all the functionality of a database system: persistence, disk management (i.e., indexing, clustering, and smart buffering), concurrency on objects (not necessarily on the schema), and recovery. Message passing should be optimized, referencing an attribute of a tuple value should be done in "compiled mode" (i.e., the system should not interpret an attribute name), and operations on large sets and lists should be done efficiently.

In the third scenario, the *resident mode*, the system could load all persistent data in main memory to enhance execution speed. This is in contrast with what we could call *object fault mode* which is typical of Data Managers.

2.3 Architectural Requirements

The OM should be as *canonical* as possible, i.e., other projects should be able to use the OM as a back-end in charge of persistence and transaction management for complex objects. This idea is not new: other projects such as Mneme [26] and ObServer [14] pursue this idea of a canonical object manager.

Another requirement is that it should be easy to downgrade the system to a single-site machine.

We tried to use existing products as much as possible. We use WiSS, the Wisconsin Storage System [10], as the low-level layer of the OM to provide persistence, disk management and concurrency control. It runs under Unix[1] System V but bypasses the Unix File System and does its own buffering.

3 Main Design Choices

3.1 Modes of Operation

3.1.1 Compiling Modes

The sharp differences in application requirements between the two modes (development mode vs. execution mode) mentioned in the previous section drove us to build two different "compiling modes". We designed the OM in such a way that these modes affect only the compilers. They do not imply building two different versions of the OM. In any case, both modes must share the same persistent data.

Execution mode applications are the result of "deeply compiling" some classes and methods developed by a programmer, to gain speed when executing these methods. The code is optimized in the following respects:

(i) Late binding is replaced by a function call whenever possible.

(ii) Access to tuple attributes by name is replaced by physical offsets.

(iii) No run-time method fault occurs: methods are loaded statically. In development mode, methods are loaded dynamically, as flexibility is required.

3.1.2 Transaction Modes

The OM interfaces either with a full-fledged execution mode application or with a development mode *session*. We shall denote both by the term *application*. An application is composed of one or more *transactions*.

Our notion of transaction differs from the classical database notion of transaction (namely, an atomic and serializable sequence of database commands) in the following respects:

(i) We distinguish concurrency on schema from concurrency on objects, and allow them to be activated or disactivated independently.

(ii) Recovery may be switched on or off. When running prototype applications, one may be willing to sacrifice safety for the sake of performance.

(iii) A transaction may choose to run in resident mode or object fault mode. Note that this is specified at execution time, not at compile time. A good candidate to run in resident mode is the Schema Manager. In the current version, the Schema Manager can run in object fault mode if it wishes to.

3.2 Handling Distribution

The OM has a *workstation version* and a *server version*. Both versions have (almost) the same interface. The main distinction is in the actual implementation: the workstation version is single user (as a workstation is single-user), memory based; while the server version is multi-user, disk based.

An important problem we faced while designing the system was: given an (execution mode) application program in O_2, how do we decide which part runs on the server and which part runs on the workstation? The main goal is to (i) make tasks run efficiently in the system configuration, and (ii) to program distribution as simply as possible. Recall that we are not dealing with a distributed query problem, but with general programs, since methods are programs.

Several approaches can be taken. The simplest is to make distribution transparent to the application programmer and make everything run on a single machine. Making distribution transparent and letting the system determine the "best site" is an unsolved problem up to now.

We decided to make the distributed architecture visible to the application programmer (but not to the end user). The programmer will be aware of the existence of two machines (the server and the workstation) and may explicitly specify on which of the machines a message passing expression is to be executed.

```
                                    ...
    /* programmer specifies that the message
''filter'', which tests */
    /* each element of set ''my_set'', should
be run on the server */

        result_object = [my_set filter() on
server] ;
                                    ...
```

In this case, if the selectivity ratio is high and the set is large, running the method on the server may result in better performance. On the contrary, methods displaying an object, editing a document object or performing computations on few objects should be run on

[1] Unix is a registered trademark of AT&T laboratories

the workstation to offload the server. Note that we associate location with message passing and not with methods.

For messages without site specification, the system decides the evaluation site. Possible strategies include naive ones, such as "execute in the machine in which you are currently executing". This is the one we have implemented. See Section 6 for foreseen improvements to this strategy.

Our design choices were made according to the criteria of *simplicity*, *transparency*, *performance*, and *reliability*.

Simplicity (and thus portability) is achieved by choosing standard and well-proven tools: the local area network which links the workstations and the server is Ethernet, the transport protocols are TCP/IP. *Transparency* results from the fact that the end user is never concerned with either the client/server task distribution or the heterogeneity of the machines. *Performance* is obtained by execution migration and by designing a communication protocol optimized for object moves, which are the main potential bottlenecks of any client/server architecture. Lastly, *reliability* is obtained by a failure detection mechanism which prevents indefinite resource holding which might result from an abnormal process termination.

3.3 Object Access

Objects are uniquely identified and accessed by object identifiers (*oids*). Object identifiers could be "logical", i.e., give no information about their location in secondary memory, as in GemStone [24], Orion [19] and ObServer [14]. With logical identifiers a correspondence table between oids and physical addresses is needed. Moving objects in secondary memory is straightforward, but the object table might be very big as it would contain one entry for each object in the database. One disk access is likely to be performed to retrieve the object table entry of the object, and a second to retrieve the object. We have chosen the (persistent) identifiers to be physical identifiers, i.e., reflecting their location on disk. The choice is largely motivated by the aforementioned performance reasons. Roughly speaking, an object will be stored in a WiSS record and the object identifier will be the record's identifier, i.e., an RID.

A major problem with physical identifiers is moving objects on disk without changing their identifiers[2]. The solution we adopted is to use forwarding markers (as in relational data managers such as System R). We modified WiSS in order to make the Object Manager have direct control over this mechanism.

[2]If we were to change the identifier of an object o we want to move on disk, we should be able to attain all objects referencing o to update their references, but this would imply using backward references in the composition hierarchy, and we consider these too heavy to maintain.

3.4 Object Representation

Recall that the O_2 model distinguishes objects from values (Section 2.1). In the OM, we deal only with objects and atomic values. Structured values will be given an identifier and are managed as "standard" objects. The system supports both the primitives for manipulating values as well as the message passing mechanism for objects. At the language level, objects are distinguished from values and errors such as passing a message to a value or applying a primitive to an object will be trapped at compile time. In the OM, however, there are primitives which distinguish oids denoting objects from oids denoting values.

3.4.1 Tuples

On disk, a tuple is represented as a record stored in a page. When a tuple outgrows a disk page, we switch to a different representation suitable for storing long records, the Long Data Item (or LDI) format provided by WISS. The oid of the tuple is unchanged (the RID of the original record).

In main memory, tuples are represented as a contiguous chunk containing the actual values. Only strings are stored away from the main chunk, which contains pointers to the proper locations. This way the strings may grow or shrink without requiring the entire object to change location. An exception to this rule comes from the fact that in O_2, a tuple object may have "exceptional" attributes, i.e., attribute values not declared in its class (Section 2.1). In such cases, the tuple object may grow in length. When a tuple grows, if an in-place extension is not possible, a level of indirection for the entire tuple value is generated.

3.4.2 Lists

Lists, which can be more accurately called *insertable arrays*, are represented as ordered trees as in [31] (with slight modifications to make fast scans). An ordered tree is a kind of B-tree in which each internal node contains a count of the nodes under it. The insertion and deletion procedures have to update node-counts (this is the essential difference from standard B-tree management). This structure is efficient to store small and large lists. In EXODUS [8], it has been used to implement long data items.

3.4.3 Sets

The representation for large sets of objects needs to be such that (i) membership tests are efficient, and (ii) scanning the elements of the set is also efficient. A set of objects is itself an object containing object identifiers of its members (the objects themselved are stored according to clustering strategies described in section 3.5). WiSS provides two kinds of indices: hash indices and B-trees. We use B-tree indices to represent large sets. However, using an index for a small set would be too costly. Therefore, there is a limit under which a set is represented as a WiSS record; a convenient value for this limit is the maximum record size in WiSS. Small sets are kept ordered. This decision was motivated by the fact that large sets are kept ordered, and binary operations on sets take advantage of this uniformity: unions, intersections and differences are programmed using "merge" algorithms.

3.4.4 Multimedia Objects

Two types of multimedia objects are implemented: unstructured text and Bitmap. From the user point of view, they are instances of the predefined classes `Text` and `Bitmap`. The predefined methods in these classes are `display` and `edit`. From the system point of view, texts are atomic objects of type `string` and bitmaps are atomic objects of type `bytes`, an unstructured byte string preceded by its length.

3.5 Persistence and Clustering on Disk

Recall that persistence is defined in the O_2 model as reachability from persistent root objects, which are named objects or values (Section 2.1). This is implemented by associating with each object a reference count. An object persists as long as this counter is greater than 0.

Newly created, persistent objects are given a persistent identifier when they are inserted in a file at transaction commit. The mapping of objects to files depends on control data given by the database administrator describing the placement of objects: the Placement Trees.

Placement trees are described extensively in [6]. The main idea is that if several objects are used together frequently, we should put them as close to one another as possible on disk. The main heuristic used here to postulate that two objects will be used together frequently is their relationship through composition structure. Roughly speaking, if class C' is a son of class C in a placement tree and an object o' of class C' is a component of object o of class C, we try in fact to store o' as close as possible to o. To do so, we use the elementary clustering facility offered by WiSS.

3.6 Handling concurrency at the O_2 level

Concurrency on objects is handled by WiSS at the server. However, the distributed architecture of the system does not make life easy. Consider a transaction executing on the workstation and updating shared objects. Before reaching the workstation, objects are read-locked. Note that it is difficult to anticipate which objects need to be write-locked, as objects are operated by methods of arbitrary complexity. There is no concurrency control on the workstation, as workstations are considered to be single-user. The problem is to ensure consistency, and this in a cheap way. Straightforward solutions are (i) each time the transaction wants to update an object in the workstation, migrate execution to the server and perform the update there: the concurrency control mechanism of WiSS will ensure consistency, (ii) update the objects locally and maintain an update log that will be executed in the server at transaction commit, (iii) run the transaction in the workstation and in the server in parallel (this is less straightforward!). Solution (i) has obvious performance problems, solution (ii) seems complicated to implement and has contention problems as all controls are performed at commit time, and solution (iii) is difficult to implement because the transaction itself may migrate execution site from one machine to another (see Section 3.2). None of these being all too satisfactory, we chose the solution now presented.

Concurrency in WiSS is handled by a two-phase locking algorithm on pages (and files). WiSS has an internal `lock_page` primitive which sets locks on pages. Our solution is to move this primitive to the programming interface, and ask for write locks explicitly from the workstation before an object is updated. The update proceeds in the workstation asynchronously without waiting for an answer. Unless the answer is `rollback`, the workstation is not informed. Otherwise, the process running the transaction in the workstation is informed and the transaction aborted. At transaction commit, the workstation process asks the server process if all requested write locks have been granted (we call this "precommit"), and only if this is the case, are the objects transferred. This way, consistency of shared objects in the server is preserved.

As mentioned previously, we handle concurrency on the schema differently from concurrency on objects. This is a consequence of our application requirements summarized in Section 2.2, but it is also due to the following observation: in development mode, the schema *is* the hot spot of the system. Even executing a message passing expression implies reading the schema information because methods belong to the schema. Therefore, a custom concurrency control for the schema is being designed which takes into account semantic information about schema updates to allow for increased parallelism. It just does not seem wise to put a read lock on a page each time a message passing expression is solved.

4 Overview of the System

4.1 Global Architecture

The OM is divided into four layers: (i) a layer which copes with the manipulation of O_2 objects and values and transaction control, (ii) a Memory Management layer, (iii) a Communication layer taking into account object transfers, execution migration, and application downloading, and (iv) on the server, a Storage layer devoted to persistence, disk management, and transaction support implemented by WiSS.

The following process layout has been adopted: on the workstation, an application and the workstation version of the OM form one unique process. There will be as many processes as running applications.

For each process running on a workstation, there will be a *mirror* process running on the server. In addition, there may be some *terminal* application processes running on the server which don't have any corresponding "partner" on a workstation. Given the large number of such concurrent applications on the server, the OM is compiled as re-entrant library modules to be shared among all applications. The lock table and the buffer managed by WiSS are shared by all processes. The OM Object Memory is also a global buffer in shared memory, as detailed below.

Both versions of the OM (workstation and server) are illustrated in Figure 1.

4.2 System Interface Module

This layer is the gateway to the OM. All object manipulations done by a transaction are treated by this module.

4.2.1 Object Manipulation

This submodule performs the following tasks:

(i) Creation and deletion of structured types (creation of classes and methods is the responsibility of the Schema Manager and not of the Object Manager).

(ii) Creation and deletion of objects.

(iii) Retrieval of objects by name.

(iv) Support for the predefined methods for objects. These are the methods of predefined class `Object`, namely: `value_equal`, `deep_equal`, `value_copy`, `deep_copy`, `display`, and `edit`.

(v) Support for set, list, and tuple objects.

For sets and lists, we have implemented a two-layer structure in which the upper layer takes care of logical operations (and reference counts) and the lower layer handles the data structures.

4.2.2 Support for Late Binding

This submodule exists both on the workstation and the server. In development mode, it provides support for late binding and also handles the application of the selected binary code to the receiver object. In execution mode, the message is replaced by a function call whenever possible and in the few cases which remain unsolved, method name resolution is done by the executing code in an ad-hoc manner.

4.3 Transaction Support

An O_2 transaction maps directly to a WiSS transaction. This submodule supports the functionality described in Section 3.1.2 and handles concurrency according to the strategy presented in Section 3.6.

When a transaction runs in memory resident mode, the system is given information about the persistent root objects it accesses, and fetches all objects accessible from these at transaction start. Once this is done, all inter-object references are changed to memory addresses.

At both sites, this submodule keeps track of the set of dirty objects (this includes newly created objects) and the set of persistent objects to be deleted from the database. When control is transferred from one site to another (at transaction commit or when migrating execution), these two sets, as well as the objects referenced to by these sets, are also transferred. In this way, transactions preserve their execution context. The coupling of the workstation version and the server version of the system is tight, as they communicate via "lower" system interfaces, as it is shown in the following sections (this has also been done in AIM-P [16]). This contrasts with (relational) distributed database systems, in which communication is done mostly via the high-level relational database interface.

4.4 Memory Management Module

This layer takes care of translating object identifiers into memory addresses. This includes handling object faults for objects requested by the application and not currently in memory. It is also responsible for managing the space occupied by objects in main memory.

As in Orion [19] and GemStone [24] (but unlike ObServer [14]), a dual buffer management scheme is implemented: a page buffer implemented by WiSS and an object buffer pool, the *object memory*. Objects in the page buffer are in their disk format. In the object memory, they are in their memory format.

On the server, an object fault implies reading a WiSS record and transferring it between the page buffer and the server object memory. Even though an object corresponds to one WiSS record, on every object fault, all the valid records on the same page as the object in question are transferred into the object memory. This "read-ahead" strategy is based on the fact that objects which have a strong correlation between them are clustered on the same or nearby pages and reading an entire page (which is anyway placed in the page buffer by WiSS) will accelerate further processing.

As some degree of sharing is expected among applications, the server object memory is implemented as a data segment shared by all the concurrent processes [15].

On the workstation, the object memory is private to each application. The memory allocation and deallocation tasks are left to the Unix virtual memory mechanism. An object fault is addressed to the Communication Manager, which in turn asks the server mirror process to send the object across the network. As before, all objects stored in the same page as the requested object are transferred to the workstation.

Both on the server and on the workstations, the memory address at which an object is stored never changes until the object migrates to another machine

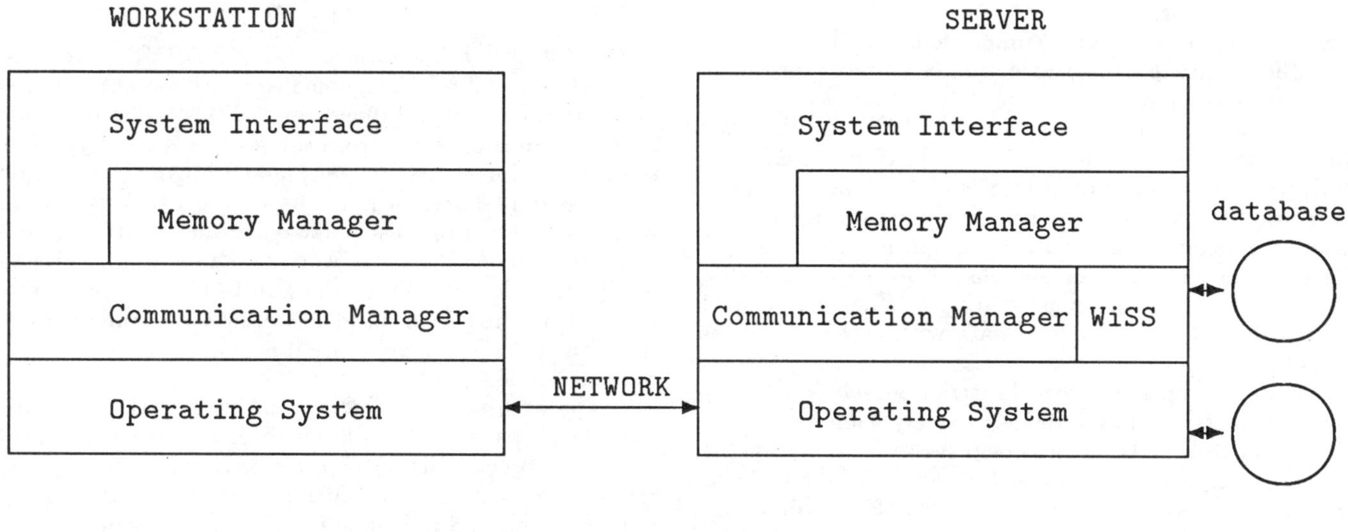

Figure 1: Architecture of the OM

or is written out to disk. While an object is in memory, in order to access it given its identifier, we use an object table. This table is hashed on identifier values and contains entries for resident objects only, as in LOOM [18].

4.5 Communication Module

4.5.1 Session Level

When an application is started on a workstation, a *mirror application process* must be started on the server in order to interact with the lower layers of the system.

The *workstation application process* and the *mirror application process* behave differently: whereas the workstation application process executes the code of the application as a transaction, the *mirror application process* is passive: it goes to sleep until receipt of an incoming message, processes it by *upcalling* the appropriate primitive and goes back to sleep until the next message arrives, or the workstation application process is ended.

4.5.2 Presentation Level

The heterogeneity between machines is handled by a presentation protocol, namely *XDR* [1], chosen for its simplicity. For object moves, the size of an object when received at a site depends on the machine parameters. In order to minimize extraneous memory-to-memory copying, this size is computed from the type of the transferred object. It is used to determine the place in memory where the object data will be copied, and data transfer is done directly from network to this memory address. Another optimization consists of grouping in a single message several objects to move (for instance at commit time). This minimizes the number of network accesses.

4.5.3 Execution Migration

In the current version of the system, the execution site for a message passing is specified statically by the application programmer. Embedded execution transfers may appear. For instance, if a selection operation results in displaying one tuple over 1000 in average, the programmer may specify that the selection operation should be done on the server but the display, of course, can be done only on the workstation. This way, the two sites may act alternatively as client or server. Our execution migration protocol is thus more than an unidirectional classical Remote Procedure Call mechanism.

5 Comparison with Related Work

We attempt to compare here the O_2 OM with other existing object managers. Caveat: as there is a lack of agreement on higher level object-oriented concepts for databases, object managers actually implement different kinds of functionality for different kinds of data models.

A major concern throughout the design has been to build a common kernel for different object oriented paradigms. Outside Altaïr, the OM is currently being used as a back-end bearing persistence to a frame-based expert system generator [27]. In fact, most of the semantics of the O_2 type system has been shifted to the upper layers, i.e., the Schema Manager and the Compilers. The part of the OM depending on classes and methods is essentially the message-passing mechanism and has been clearly isolated from the rest of the system, which handles shared and persistent typed objects with identity. Most object-oriented languages propose tuple objects and some have a notion of (ordered or unordered) collections. In this sense, the OM implements "state of the art" complex objects. The difference with nested relational kernels such as DASDBS [29] is that the latter are geared towards optimizing selections and joins, and don't support traversing the composition graph, nor important (object-oriented) functionality such as orthogonal persistence, garbage collection or support for methods and message passing.

Other projects such as Mneme [26] and ObServer [14] pursue this idea of a "canonical" object manager. Both provide untyped objects, the typed layer being implemented by a *client* of the object manager. In this case, traversing the composition graph is either done at the client level, with an obvious performance penalty, or the object manager must distinguish between object references and other kinds of data within the object, as in [26].

For space reasons, we will restrict ourselves to a subset of object-oriented DBMSs, namely GemStone [24] and Orion [19]. Other well known systems are Vbase [2], Iris [11]. Extensible DBMS such as Exodus [9] and Postgres [30] are close to object-oriented DBMS for their ability to deal with procedures and with an inheritance hierarchy.

5.1 GemStone

The original objective of GemStone [24] was to make Smalltalk a database system. Stone, the lower layer of GemStone has the same kind of functionnality as the OM : it provides persistence, secondary storage management, transactions, concurrency control, recovery, support for associative access, and authorization functions. However, the decomposition in processes is different than in O_2, as Stone is run as a single process on the server, and it communicates with different upper layer processes (Gem processes), one for each active user, also running on the server. Gem corresponds, roughly speaking, to the virtual image of the standard Smalltalk implementation. On the workstation, an interface to GemStone (PIM) has been developed and implements remote procedure calls to functions supplied by GemStone. So, the way the client/server architecture is handled is completely different as in our system.

Applications run exclusively in "development mode" in GemStone. Also, transaction management in GemStone bears little similarity to ours. An optimistic concurrency control scheme is implemented. Access conflicts are checked at commit time, rather than preventing them through locking. The main disadvantage with this scheme is that a long transaction can loose arbitrary amounts of work if prevented from committing. To alleviate this problem, a pessimistic concurrency control based on locking is explored in [28].

One of the major innovations of GemStone is its indexing mechanism to support associative access on large collections [25]. The main feature is that indexes can be defined on a *path* along the composition hierarchy, and that they index into sets rather than into classes.

5.2 Orion

Orion [5], [19] is an object-oriented DBMS developed at MCC extending Common-Lisp with object-oriented programming and database capabilities. The database type system is different from the host language type system, and Orion's persistence is not orthogonal. A version of Orion implements a multi-user, multi-task system in which a server provides persistence and sharing of objects on behalf of several workstations. Unlike O_2, an application does not have the ability to execute general methods in the server and the workstation while it runs. As with GemStone, the system is geared towards development mode.

The architecture of Orion consists of the following modules: (i) a *message handler* receives all messages sent to Orion objects, (ii) an *object subsystem* provides functions such as schema evolution, version control, query optimization, and multimedia information management, (iii) a *transaction subsystem* provides concurrency control and recovery mechanisms, and (iv) a *storage subsystem* provides access to objects on disk. The two last modules implement functionalities similar to those of the OM: The storage subsystem implements a dual buffer management scheme (page buffer and object buffer) and maintains a resident object table hashed by oid. However, as oids are logical, there is another hash-based oid-to-physical-id for all objects in the database. Buffer management for objects in the object buffer is far more complex than ours: fragmentation is dealt with by a garbage collection mechanism that reclaims space occupied by objects that can still be referenced to by the application, and thus an intermediate structure, the *resident object descriptor* or ROD, has to be introduced between the resident object table and the actual object. Applications point to RODs only, and these may not be swapped.

Transaction management [12] is rather "classical": concurrency control is based on a locking protocol and recovery is based on logging. However, a composite object (a hierarchy of exclusive and dependent composite objects) may be locked as a whole. To this end, the hierarchical locking protocol [13] has been extended. Also, a class-lattice locking protocol, which is needed to allow access to instances of a class while preventing changes to the definitions of the superclasses of the class has been proposed. Other important database features in Orion are indexing, authorization and version support. Indexes may be defined on a class or on subclasses of a class, and are used when evaluating queries.

6 Conclusions and Future Work

In this paper, we described the main choices involved in the design of the O_2 Object Manager, and presented its overall architecture. The OM is used by all the upper modules of the O_2 system. It combines well known techniques from the database field and the object-oriented programming field, and has the following original aspects:

(i) The programmer has control over distribution. This is done in an easy way when passing messages to objects. We consider this to be a powerful performance tuning facility.

(ii) The system adapts itself to a wide range of application requirements ranging from development of applications to execution of data-intensive applications.

(iii) The system implements a number of facilities that make life easier for application programmers: (i) exceptional attributes for tuple objects may be attached at any time without performance degradation, (ii) persistence is implemented with a simple composition-based schema in which deletions are implicit, and (iii) clustering issues are clearly separated from the schema information and specified by the DBA in the form of a subset of the composition graph.

Our first prototype [3] implemented an early version of the data model reported in [22]. There were no notion of values and lists were inexistent. All sets were implemented in the same way, regardless their size. Persistence was attached to classes, not to objects, so all objects of a persistent class were persistent. There were no clustering strategies, no transaction modes, no difference between execution mode and development mode. Finally, the Schema Manager was built in an ad-hoc way without using the object manager. For the next version of the system, the major thrust will be in the following directions :

(i) An ad hoc concurrency control for the schema, as mentioned in Section 3.6, is under design and will be implemented shortly.

(ii) Recovery and rollbacks are not implemented in the current version of WiSS. We plan to add these shortly. To support long transactions, we also plan to provide "savepoints" in order not to loose arbitrary amounts of work.

(iii) Indices will be provided at the O_2 level (WiSS provides already B+ trees and hash-based indices on files). The relationship of indices with both the inheritance hierarchy and the composition hierarchy (which has been explored separately in [20] and [25] respectively) is currently being investigated.

(iv) Our large objects are currently limited by the size of a WiSS long data item (1.6 Mbytes). We are looking into the possibility of modifying WiSS to be able to store large objects with virtually no size limitation.

(v) More work is needed towards determining when objects are "badly clustered" and should be moved on disk. Also, an incremental restructuring policy should be used.

(vi) For messages without site specification, the system currently implements message passing in the same site in which it runs. This strategy can clearly be improved by letting the system decide dynamically on the site of application of the method.

(vii) Versioning and authorization will be taken into account in the next version of the system.

Acknowledgements

We thank François Bancilhon for comments on an earlier draft of this paper and for the many discussions we had during the design of the system. We also thank the following persons involved in the development of the Object Manager: Véronique Benzaken, Constance Bullier, Philippe Futtersack, Gilbert Harrus, John Ioannidis, Jean-Marie Larchevêque, and Dominique Steve. The OM also benefit from discussions with Christophe Lécluse and Philippe Richard and especially from those with Sophie Gamerman and Claude Delobel. Paris Kanellakis, Michel Scholl and John Ioannidis read earlier drafts of this paper and suggested many improvements.

References

[1] *"External Data Representation Protocol Specification"*. SUN Microsystems, February 1986.

[2] T. Andrews and C. Harris. "Combining Language and Database Advances in an Object-Oriented Development Environment". In *Proceedings of the OOPSLA conference*, October 1987.

[3] F. Bancilhon, G. Barbedette, V. Benzaken, C. Delobel, S. Gamerman, C. Lécluse, P. Pfeffer, P. Richard, and F. Velez. "The Design and Implementation of O_2, an Object-Oriented Database System". In *Proceedings of the OODBS II Workshop*, Bad Munster, RFA, September 1988. Springer Verlag.

[4] Francois Bancilhon, Sophie Cluet, and Claude Delobel. *"Query Languages for Object-Oriented Database Systems: Analysis and a Proposal"*. Technical Report, Altaïr, 1989. In preparation.

[5] J. Banerjee et al. "Data Model Issues for Object-Oriented Applications". *ACM Transaction on Office Information Systems*, 5(1), April 1987.

[6] V. Benzaken and C. Delobel. *"Clustering Objects on Disk in an Object-Oriented Database"*. Technical Report, Altaïr, 1988. To appear.

[7] L. Cardelli. "A Semantics of Multiple Inheritance". *Computer Science*, Semantics of Data Types, Lecture Notes(173), 1984. Springer Verlag.

[8] M. Carey, D. DeWitt, J. Richardson, and E. Shekita. Object and file management in the exodus extensible database system. In *Proceedings of the 12th. VLDB Conference*, Kyoto, Japan, August 1986.

[9] Michael Carey et al. "The Architecture of the EXODUS Extensible DBMS". In *Proceedings of the International Workshop on Object-Oriented Database Systems*, Pacific Grove, California, September 1986.

[10] H.-T. Chou, David J. DeWitt, Randy H. Katz, and Anthony C. Klug. "Design and Implementation of the Wisconsin Storage System". *Software - Practice and Experience*, 15(10), October 1985.

[11] D. H. Fishman et al. "IRIS: an Object-Oriented DBMS". *ACM Transaction on Office Information Systems*, 5(1), January 1987.

[12] Jorge F. Garza and Won Kim. "Transaction Management in an Object-Oriented Database System". In *Proceedings of the ACM SIGMOD Conference*, Chicago, Illinois, May 1988.

[13] James N. Gray. "Notes on Database Operating Systems". In *Operating Systems: An Advanced Course*, Springer Verlag, 1978.

[14] Mark F. Hornick and Stanley B. Zdonik. "A Shared, Segmented Memory System for an Object-Oriented Database". *ACM Transaction on Office Information Systems*, 5(1), January 1987.

[15] John Ioannidis. *"Oh no!, we need Shared Memory"*. Technical Report, Altaïr, 1989. in preparation.

[16] k. Kuspert, P. Dadam, and J. Gumauer. Cooperative object buffer management in the advanced information management prototype. In *Proceedings of the 13th VLDB Conference*, Brigton, England, September 1987.

[17] T. Kaehler. "Virtual Memory on a Narrow Machine for an Object-Oriented Language". In *OOPSLA '86 Proceedings*, September 1986.

[18] T. Kaehler and G. Krasner. Loom - large object-oriented memory for smalltalk-80 systems. In *Smalltalk-80: Bits of History, Words of Advice*, Addison-Wesley, 1983.

[19] Won Kim, Nat Ballou, Hong-Tai Chou, Jorge F Garza, Darrel Woelk, and Jay Banerjee. "Integrating an Object-Oriented Programming System with a Database System". In *OOPSLA '88 Proceedings*, Los Angeles, California, September 1988.

[20] Won Kim, Kyung-Chang Kim, and Alfred Dale. *"Indexing Techniques for Object-Oriented Databases"*. Technical Report DB-134-87, MCC Technical Report, May 1987.

[21] C. Lécluse and P. Richard. "Modeling Complex Structures in Object-Oriented Databases". In *8th Symposium on Principles of Data Base Systems*, Philadelphia, Pennsylvania, March 1989. To appear.

[22] C. Lécluse, P. Richard, and F. Velez. "O_2, an Object-Oriented Data Model". In *Proceedings of the ACM SIGMOD Conference*, Chicago, Illinois, June 1988.

[23] Christophe Lécluse and Philippe Richard. *"The O_2 Database Programming Language"*. Technical Report, Altaïr, January 1989. Submitted for publication.

[24] David Maier and Jacob Stein. "Development and Implementation of an Object-Oriented DBMS". In Bruce Shriver and Peter Wegner, editors, *Research Directions in Object-Oriented Programming*, MIT Press, Cambridge, Massachussets, 1987.

[25] David Maier and Jacob Stein. "Indexing in an Object-Oriented DBMS". In *International Workshop on Object-Oriented Database Systems*, Pacific Grove, California, September 1986.

[26] J. Eliot B. Moss and S. Sinofsky. Managing persistent data with mneme: designing a reliable, shared object interface. In *Proceedings of the OODBS II Workshop*, Bad Munster, RFA, September 1988. Springer Verlag.

[27] Bertrand Neveu and Pierre Haren. "SMECI: an Expert System for Civil Engineering Design". In *First Int. Conference on Applications of Artifial Intelligence to Engineering Problems*, Southhampton, England, April 1986.

[28] Jason Penney, Jacob Stein, and David Maier. "Is the Disk Half Full or Half Empty?: Combining Optimistic and Pessimistic Concurrency Mechanisms in a Shared, Persistent Object Base". In *Workshop on Persistent Object Systems*, Appin, Scotland, August 1987.

[29] M. H. Scholl, H. B. Paul, and H. J. Schek. Supporting flat relations by a nested relational kernel. In *Proceedings of the 13th VLDB Conference*, Brigton, England, September 1987.

[30] M. Stonebraker. "The Design of the POSTGRES Storage System". In *Proceedings of the 13th VLDB conference*, Brighton, U.K., 1987.

[31] M. Stonebraker et al. "Document Processing in a Relational Database System". *ACM Transactions on Office Information Systems*, 1(2):143–158, April 1983.

ON CORRECTLY CONFIGURING VERSIONED OBJECTS

Rakesh Agrawal *and* *H. V. Jagadish*
AT&T Bell Laboratories, Murray Hill NJ

We consider the problem of configuring a system in software and design database domains, where a system comprises a version for each of its constituent objects. We present a syntactic characterization of a correct configuration, tied to a transaction model, that makes it possible to generate automatically all correct configurations of a system. One can also generate configurations that satisfy some selection criteria such as the absence and presence of specified features, or check whether a user-specified configuration is correct.

1. INTRODUCTION

Software and design databases invariably consist of versioned objects (*cf.* [2,4,5,7-9,14]). Versions of an object often represent alternatives or revisions. A system comprising a set of objects is *configured* by selecting a version for each of the objects that constitute the system. A configuration is also treated as a versioned object, so that more than one configuration can coexist.

An important issue in configuring a system is that the constituent versions must be compatible [12,15]. We all know that the version of a module compiled for Motorola 68000 should not be linked with the version of another module compiled for Intel 386. All the systems we know of provide very little by way of support for deciding what can constitute a correct configuration, leaving this decision to the user. In a system consisting of m objects, each with v versions, there can be up to v^m possible configurations. As systems become large, relying on user intuition to decide what constitutes a correct configuration is at the very least error-prone, if not altogether impossible.

This paper is an attempt to correct this deficiency by introducing a formal notion of the correctness of a configuration and mechanisms for generating and verifying correct configurations. We take inspiration from the rich database literature on transaction management (*cf.* [6]) and suitably extend the transaction model. Updates to the database and dependencies between versions are encapsulated in transactions that transform the database from one consistent state to another. Correct configurations are equated with consistent database states. However, consistent database states are not limited to the set of states that the database actually goes through during the execution of a set of transactions, but also includes states that could have been generated using different serialization orders consistent with dependencies in the transaction set. We do not require versions to be totally ordered and admit a much weaker notion of serializability than the 1-copy serializability [6] used in classical multiversion concurrency control theory.

We thus have a "syntactic" characterization of a correct configuration that is tied with the transactions that create and update versioned objects. Using this characterization, it is possible to generate automatically all correct configurations of a system. One can also generate those configurations that satisfy some selection criteria such as configurations that incorporate specified features, or check whether a user-specified configuration is correct.

There is often a distinction made between the *interface* of an object and its *implementation* [3]. Sometimes only the implementation is allowed to change and not the interface. At other times both change. These are often distinguished and sometimes given different names, one called a version, and the other a revision or an alternative. In this paper, we will not make this distinction, and will use the single term "version" to refer to all different implementations and interfaces of an object. We also do not differentiate between types of versions such as public, private, transit, working, etc. [7-9]. If necessary, the reader can imagine that we deal only with publicly released committed versions.

The idea of composite objects being obtained as configurations of primitive objects has been explored in [2]. In the terminology of [11] these primitive objects are *shared independent* constituents of a configuration. Whenever a new version of a primitive object is created, new configurations are also obtained, by means of *version percolation.* The number of composite object versions obtained grows in a combinatorial fashion, as pointed out in [7]. The solution proposed in [7], based on time-stamps, saves system storage by passing on to the user the burden of examining the new configurations possible and recording them in the database, if desired. Our approach, in this paper, avoids the need for storing configurations altogether by providing an efficient mechanism for putting configuration together on the fly.

Katz and Chang [10] also propose a system in which configurations can be composed from versioned objects. However, they rely upon a logic program (or rule base), provided by the user, to ensure correctness rather than using syntactic characteristics of the transaction model itself.

The organization of the rest of the paper is as follows. In Section 2, we introduce our transaction model and define what we mean by a correct configuration of object versions. In Section 3, we show how one can obtain all correct configurations from a given initial set of correct configurations. In Section 4, we discuss how to determine whether a user-specified configuration is correct. We also discuss how user requests for configurations containing (or not containing) certain specified features can be satisfied. We present our conclusions in Section 5. We assume that the reader is familiar with the basic notions of transactions and serializability; see [6] for a tutorial introduction.

2. BASIC CONCEPTS

We first introduce our transaction model, and then formally define what we mean by a correct configuration of object versions.

2.1 Transaction Model

A database is a collection of objects, each of which is independently versioned. For ease of exposition, we assume that the database starts with a "pre-creation" version of every object in the database. Subsequent versions are created by transactions that read and update database objects. Whenever an object is updated by a successful transaction, it results in the creation of a new version of the object. A transaction may update more than one object, but may create only one version of an object. All versions are created when the transaction commits. Deletions result in the creation of "post-deletion" versions.

An object version is created by exactly one transaction. Each object version is stamped with the signature of the transaction that created it. Given an object version x, the function $oid(x)$ returns the identity of the object of which x is a version, and the function $generator(x)$ returns the identity of the transaction that created x.

Associated with each transaction T_i is a *read set* R_i of object versions read by T_i and a *write set* W_i of object versions created by T_i. We differentiate between "reading" and "browsing " an object version. When a transaction T_i reads an object version x, it ensures that x, unless updated by T_i, will be configurable with whatever versions T_i creates. Browsing on the other hand is a "free" operation. When T_i browses an object version, it is not placed in R_i. A transaction can browse an object version x and then write object versions that are not compatible with x. Finally, transactions can read "pre-creation" versions but may not read "post-deletion" versions.

We put the following additional requirements on transaction behavior:

i. A transaction T_i that writes an object version must read at least one previous version of the same object, i.e., $\forall$ object versions $x \in W_i, \exists\, y \in R_i$ such that $oid(x) = oid(y)$.

ii. A transaction T_i reads no more than one version of an object that it updates, though it may read multiple versions of objects that it does not update, i.e., $\forall$ object versions $x, y \in R_i$, $\forall\, z \in W_i$: $oid(x) = oid(z), x \neq y \Rightarrow oid(x) \neq oid(y)$.

Thus, for each version of an object, one can identify a unique version of the same object from which it has been directly derived. We can, therefore, create a *version graph VG* for each object with one node for each version of the object and an edge from each version to the versions that are derived directly from it. More than one version can be derived directly from a version, and hence the version graph is a tree[1], as in most software and design databases.

We say that a version x of an object is *derived* from version y of the same object iff there is a path from y to x in the version graph of the object.

Simultaneous reads of the same object version by two different transactions do not conflict. Simultaneous updates to the same object in parallel by two different transactions do not conflict either, since they create different object versions.[2] Browsing also does not conflict with either simultaneous reading or writing of the same object version. However, a transaction cannot read any object version before it has been created. This requirement gives rise to a dependence between transactions, which can be represented in a *transaction dependence graph*.

The transaction dependence graph TG has a node corresponding to every transaction in the system, and there is an edge from a node corresponding to transaction T_i to a node corresponding to transaction T_j iff $R_i \cap W_j \neq \phi$. A transaction T_i is said to *depend* on

1. Allowing a transaction to read more than one version of an object that it updates, thereby allowing a version to be directly derived from more than one version, is equivalent to allowing the version graph to be a directed acyclic graph (DAG). We do not consider version graphs that are DAGs in this paper since trees are considered adequate in most practical applications [Won Kim, personal communication, 1988]. Note that we allow a transaction to browse multiple versions of an object that it updates. Thus, if a transaction has to peruse multiple versions of an object that it updates, it can "read" one from which branching will occur in the version graph and "browse" the remaining ones.

2. Note that in our transaction model, the system never aborts a transaction.

transaction T_j iff there is a path from T_i to T_j in *TG*.

Lemma (Acyclicity of the Transaction Dependence Graph): There is no cycle in the transaction dependence graph, as defined above.

Proof: Versions are created only after a transaction has committed. □

We can, therefore, topologically sort the transaction dependence graph, i.e., the transactions can be totally ordered in such a way that if T_i depends on T_j then T_i occurs later in the order than T_j. Each such total order constitutes a *serialization* of the transactions.

Read-only transactions are not significant in this model: while they do have a dependence on the transactions that write the object versions they read, they themselves write nothing and no other transactions depend on them. We, therefore, drop all read-only transactions from the transaction dependence graph for the rest of the paper.

2.2 Correct Configurations

A *configuration* is a set of object versions with no more than one version of any object, i.e., it is a set of object versions C with the property that $\forall\ x, y \in C : x \neq y \Rightarrow oid(x) \neq oid(y)$.

A *complete configuration* selects exactly one version of every object in the database, i.e., it is a configuration C such that for every object p in the database, $\exists\ x \in C$ such that $oid(x) = p$.

We are interested in *correct configurations*. We equate correct configurations with consistent states of the database. However, consistent database states are not limited to the set of states that the database actually goes through during the execution of a set of transactions, but also includes states that could have been generated using different serialization orders consistent with the dependencies in the transaction dependency graph. To this end, we introduce the notion of qualification of a configuration for a transaction.

A configuration C *qualifies* for a transaction T_i iff $\forall\ x \in R_i,\ \exists y \in C$ such that $y \in R_i \wedge oid(y) = oid(x)$. In other words, a configuration qualifies for a transaction if the following holds for all objects that the transaction reads: i) if the transaction reads only one version of an object, this version is included in the configuration; and ii) if the transaction reads multiple versions of an object, any one of the versions is included in the configuration. In addition, the configuration may contain versions of objects not in the read set of the transaction. For example, let p_i be a version of the object p, q_i a version of q, etc. Then, given a transaction with a read set $= \{p_1,p_2,p_3,q_1\}$, the configuration $\{p_1,q_1,r_1\}$ qualifies for it, while the configuration $\{p_2,q_2,r_2\}$ does not. Note that the set $\{p_1,p_2,q_1\}$ is not a configuration since it includes two versions of the object p.

If configuration C qualifies for transaction T_i, then T_i is said to be *applicable* to C. A new configuration D is said to be *generated* by *applying* T_i to C, as follows:

$$D = W_i \cup \{x \in C \mid (\forall\ y \in W_i : oid(x) \neq oid(y))\}$$

i.e., D is obtained by substituting in C the updated versions of the objects written by the transaction T_i.

We can now inductively define the set of correct configurations as follows:

1. The *initial complete correct configuration*, consisting of the "pre-creation" versions of all the objects, is a correct configuration.
2. Every subset of a correct configuration is also a correct configuration.
3. Given a transaction, T, and a qualifying correct configuration C, the configuration D generated by applying the T to C, is also correct.

Sometimes, we will find it necessary to work backwards from a configuration to determine how it could have been generated. To this end, we introduce the notion of the reverse transformation of a configuration. A configuration C is said to *reverse qualify* for a transaction T_i, iff the transaction's access set A_i is such that $\forall\ x \in A_i,\ \exists y \in C$ such that $y \in A_i \wedge oid(y) = oid(x)$. This definition of reverse qualification is the same as the definition of qualification with the substitution of the access set of the transaction for the read set. The *access set* A_i of transaction T_i is defined as:

$$A_i = W_i \cup \{x \in R_i \mid (\forall\ y \in W_i,\ oid(x) \neq oid(y))\}$$

i.e., the access set is the write set plus the versions of objects in the read set that do not occur in the write set. It can also be thought of as the read set with the updated versions replacing the old ones for objects that were updated by the transaction.

To *reverse apply* T_i to D, substitute in place of each element $x \in W_i$ in D, an element $y \in R_i$ such that $oid(y) = oid(x)$. That is, the resulting configuration C is obtained as

$$C = \{x \in D \mid x \notin W_i\} \cup \{x \in R_i \mid (\exists y \in W_i \text{ such that } oid(x) = oid(y))\}$$

This definition parallels the definition of applying a transaction. If configuration D is obtained by applying a transaction T to a qualifying configuration C, then D reverse qualifies for T, and C is obtained on reverse applying T to D, and vice versa. Note that the result of reverse application of a transaction to a reverse qualifying configuration is unique.

3. GENERATION OF ALL CORRECT CONFIGURATIONS

We now examine how to obtain all correct configurations from a given initial set of correct configurations. While this in itself may not be a problem of interest directly, the results developed in this section can readily be adapted for several problems of interest, some of which are discussed in Section 4.

3.1 A Naive Method

A straightforward technique to obtain all correct configurations is to create all possible serializations of the transaction dependence graph, and for each serialization to compute a set of correct configurations as each transaction "moves the database from one consistent state to another". However, a transaction can "access" not just the immediately preceding consistent state of the database, but any previous consistent state as well. When a transaction executes, the following procedures are executed to augment the set of correct configurations:

Transaction-Apply: Apply the given transaction to every qualifying correct configuration obtained previously from this serialization to generate new correct configurations.

Subset-Expand: For every new correct configuration, include every subset in the set of known correct configurations.

Thus, a transaction examines all possible preceding consistent states of the database, selects the ones that qualify for it, and from these generates new consistent states (subsets of which are also marked consistent).

For each serialization, the starting set of correct configurations consists of the initial complete correct configuration comprising of "pre-creation" version and all subsets thereof. These are all the correct configurations known before any transaction executes. The *processing* of a serialization is the execution of the procedure *Transaction-Apply* for each transaction in the serialization, in order, starting with a set of initial correct configurations. The *total processing* of a serialization is its processing with the procedure *Subset-Expand* executed between successive executions of the procedure *Transaction-Apply*. Thus, from the definition of correct configuration, the total processing of every serialization of the transaction graph, starting from the given initial set of correct configurations, yields the desired set of all correct configurations.

The problem with considering all possible serializations is that there are exponentially many of them. Moreover, each serialization generates an exponential number of correct configurations. This approach, therefore, is impractical. The rest of this section is devoted to finding techniques that will do better.

3.2 Incomplete Correct Configurations

Every subset of a correct configuration is a correct configuration, and new correct configurations are generated by applying transaction to qualifying correct configurations. To ensure that all correct configurations were found we had to execute procedure *Subset-Expand* between successive executions of procedure *Transaction-Apply*, thereby generating a large number of duplicates.

For example, consider a system in which there is a single transaction with a read set of $\{p_1\}$ and a write set of $\{p_2\}$, and an initial complete configuration of $p_1q_1r_1$. First applying *Subset-Expand*, we get the set of correct configurations $\{p_1q_1r_1, p_1q_1, p_1r_1, p_1, \phi, q_1, r_1, q_1r_1\}$. The first five of these qualify for the transaction[3]. Applying the transaction to each of these, we obtain as new correct configurations $\{p_2q_1r_1, p_2q_1, p_2r_1, p_2, \phi\}$. On applying *Subset-Expand* individually to each of these new configurations, we obtain 8, 4, 4, 2, and 1 correct configurations respectively, for a total of 19, whereas only 8 of them are distinct, and only 4 not included in the set we had before applying the transaction.

The lemma below helps us to eliminate this duplication.

Lemma (Subset Completion): Every correct configuration is a subset of a complete correct configuration[4].

Proof: If a correct configuration qualifies for a transaction, so does a complete correct configuration that is its superset, by definition of qualification. Since the only mechanism for generating correct configurations, beginning with subsets of a complete correct configuration, is to apply transactions for which it qualifies, we have a proof by induction. □

The consequence of this lemma is that it is sufficient to have a procedure to generate all complete correct configurations. All other correct configurations can be found directly as subsets of these.

Lemma (Complete Generation): A configuration, D, generated by applying a transaction T to a qualifying configuration C, is complete iff C is complete.

Proof: By definition of the application of a transaction and generation of a new configuration, B has all the

3. The empty set, ϕ, represents a configuration comprising no objects. It is included here for formal correctness.

4. This lemma is not as trivial as it may at first appear. It has an implication in the reverse direction from the one in the definition of a correct configuration, which said "Every subset of a correct configuration is a correct configuration". If we consider a slightly different definition of qualification, the lemma does not hold. For example, let a configuration C qualify for a transaction T_1 provided $\forall\ x \in R_i\ \forall\ y \in C\ oid(x)=oid(y) \Rightarrow x=y$. Consider a single transaction system with a read set of $\{p_1, q_2\}$ and a write set of $\{p_2, q_3\}$, and an initial complete correct configuration of p_1q_1. Clearly, this configuration does not qualify for the transaction so that the only complete correct configuration is the initial one. However, applying the procedure *Subset-Expand*, we get $\{p_1q_1, q_1, p_1\}$ all as correct configurations, the last of which now qualifies for the transaction according to our new rule, generating new correct configuration(s) that are not subsets of the initial complete correct configuration.

elements of A except that some objects have their versions updated. □

By virtue of the Complete Generation Lemma, we are guaranteed that we need only consider complete configurations if we wish to generate complete correct configurations. In conjunction with the Subset Completion Lemma, it implies that we need not totally process a serialization to obtain all correct configurations: it is enough to process a serialization beginning with the initial complete correct configuration, and then to apply the procedure *Subset-Expand* once at the end. Returning to the example just before the Subset Completion Lemma, we could first create $p_2q_1r_1$ by applying the transaction to the initial configuration $p_1q_1r_1$, and then run *Subset-Expand* at the end to generate all the correct configurations, with significantly fewer duplicates.

3.3 Regular Systems

We first introduce the notion of *weak dependence.* Transaction T_i *weakly depends* on transaction T_j iff

i. for every object in the read sets of T_i and T_j, at least one common version of the object exists in the two sets; and
ii. there is at least one common version that is read both by T_i and T_j, but is only updated by T_i and not T_j ; and
iii. there is no common version that is read both by T_i and T_j, which is updated by T_j; and
iv. T_j does not depend on T_i.

Formally, T_i weakly depends on T_j iff

$\forall\ x \in R_i,\ y \in R_j$ such that $oid(x) = oid(y)$
$\quad \exists\ z \in R_i \cap R_j$ such that $oid(z) = oid(x) = oid(y)\ \wedge$
$\exists\ x \in R_i \cap R_j,\ y \in W_i$ such that $oid(x) = oid(y)\ \wedge$
$\nexists\ u \in R_i \cap R_j,\ v \in W_j$ such that $oid(u) = oid(v)\ \wedge$
T_j does not depend on T_i.

The intuition behind weak dependence is that T_j can create a database state in which T_i can execute, while T_i cannot create a state in which T_j can execute. Therefore, even though there is no dependence between the transactions, in that T_i does not read something that T_j writes, we should execute T_j before T_i.

The weak dependence relation can be represented in a *transaction weak dependence graph* in which there is an edge from the node corresponding to the transaction T_i to the node corresponding to the transaction T_j iff T_i weakly depends on T_j. A set of transactions is said to constitute a *regular* system iff the transaction weak dependence graph is cycle-free.

The definitions of a regular system and weak dependence may appear to be rather arcane. We wanted to have the widest possible scope for our definition of a regular system. In a practical implementation, a simpler definition of weak dependence may be used, with some of the conjuncts removed, resulting in a narrower scope for what is a "regular system".

We argue that most systems in practice are likely to be regular systems. Consider why a transaction that updates one object (by creating a new version) would want to read versions of other objects. The most probable reason is because the updated object relies upon some properties of the objects read but not updated, of which only the versions read are guaranteed to possess the properties relied upon. One can then draw a "relies upon" graph over the objects. (In a UNIX® system, this graph could be obtained through a tool such as Cscope [13] where the objects are files and the "relies upon" relation is based upon functions called). If this "relies upon" graph is acyclic, which is often the case, then we have a regular system. Even if this graph has cycles in which pairs of objects mutually depend upon each other, we still have a regular system. In fact, it can be shown that if this graph has no simple cycles of length greater than 2, then we have a regular system. If we do have a situation in which object p relies upon object q which in turn relies upon r, and then r relies upon p without r relying on q or q on p, we may still have a regular system, but that is no longer guaranteed.

Define a *transaction graph* to be the union of the transaction dependence and weak dependence graphs so that the transaction graph has an edge from node T_i to node T_j iff there is an edge from T_i to T_j either in the transaction dependence graph (dependence edge) or the transaction weak dependence graph (weak dependence edge).

Lemma (Acyclicity of the Transaction Graph): The transaction graph in a regular system is cycle-free.

Proof: Suppose there is a dependence edge from T_i to T_j. There cannot be a dependence edge from T_j to T_i, since the transaction dependence graph is acyclic. There cannot be a weak dependence edge from T_j to T_i, since such weak dependence requires that T_i not depend on T_j. Now suppose that there is a weak dependence edge from T_i to T_j. There cannot be a weak dependence edge from T_j to T_i, since the transaction weak dependence graph in a regular system is acyclic. There cannot be a dependence edge from T_j to T_i by the definition of weak dependence once more. □

Thus, in a regular system, one can obtain one or more *regular serializations*, which are total orderings of all executable transactions such that if T_i depends or weakly depends on T_j then T_i occurs later in the order than T_j.

3.4 An Efficient Method for Generating All Correct Configurations

Lemma (Order Independence): The set of complete correct configurations generated by the processing of a serialization is not altered by switching the order in the serialization of two transactions that have no dependence and no weak dependence between them.

Proof: Consider two transactions, T_i and T_j, neither of which depends on the other. Let $r_k = \{oid(x) \mid x \in R_k\}$, and $w_k = \{oid(x) \mid x \in W_k\}$ be the sets of objects (not object versions) read and written by transaction T_k. Since T_i and T_j are not dependent, we know that $R_i \cap W_j = R_j \cap W_i = \phi$. (We use the absence of weak dependence later).

Consider a complete configuration C known to be correct before either transaction has been executed. We have four cases to consider:

1. C qualifies for neither T_i nor T_j.
   No new configurations are generated irrespective of the order of executing the transactions. Therefore the lemma holds for all such configurations, trivially.
2. C qualifies for T_i but not for T_j.
   Let C_i be generated by applying T_i to C. T_i does not write anything that T_j reads ($W_i \cap R_j = \phi$). Therefore, C_i cannot qualify for T_j, since C did not. Thus C_i is the only new complete configuration generated, irrespective of the order of transactions.
3. C qualifies for T_j but not for T_i.
   Using arguments similar to the previous case, we show that C_j is the only new composition generated irrespective of the order of execution of the transactions.
4. C qualifies for T_i and T_j, both.
   Now we have to consider several subcases.
   a. $r_i \cap w_j = r_j \cap w_i = \phi$.
      That is, neither transaction reads any objects written by the other. (However, both transactions may read some objects in common that they do not update). It follows that $w_i \cap w_j = \phi$. Let C_i be the configuration generated by applying T_i to C. C_i differs from C only in the objects included in w_i, none of which are in r_j. Therefore, since C qualifies for T_j, so does C_i. Let C_{ij} be generated by applying T_j to C_i. Similarly, we can obtain C_j, show that it qualifies for T_i, and obtain C_{ji}. Note, however, that since $w_i \cap w_j = \phi$, the changes made from C by the two transactions are in different sets of objects. Therefore, $C_{ij} = C_{ji}$. Thus, either order of application of the transactions to C and its derivatives, yields the same new configurations: C_i, C_j, and C_{ij}.
   b. $r_i \cap w_j \neq \phi \quad r_j \cap w_i \neq \phi$.
      In this case, clearly, C_i cannot qualify for T_j and C_j cannot qualify for T_i. Therefore the only new complete correct configurations produced are C_i and C_j.
   c. $r_i \cap w_j \neq \phi \quad r_j \cap w_i = \phi$.
      If $r_i \cap w_j \neq \phi$, then we must have $r_i \cap r_j \neq \phi$. Consider an object p read by both transactions and updated by T_j. Both transactions must read the same version of p, otherwise we would not have found a C that qualifies for both transactions. Furthermore, p is not updated by T_i; otherwise, we would have found p to be a member of $r_j \cap w_i$. Therefore, T_j weakly depends on T_i, a contradiction given the premise of the lemma.
   d. $r_i \cap w_j = \phi \quad r_j \cap w_i \neq \phi$.
      As in the previous case, we can show that T_i weakly depends on T_j, a contradiction. □

Lemma (Serialization Indifference): The processing of every regular serialization results in the same set of complete correct configurations.

Proof: Follows directly from the Order Independence Lemma and the definition of a regular serialization. (Recall that a regular serialization can be obtained only in a regular system). □

Lemma (Weak Dependence Ordering): In a regular system, given T_i weakly depends on T_j, the processing of a serialization with T_i occurring before T_j results in a subset of the complete correct configurations generated by a serialization that is identical except that T_j occurs before T_i.

Proof: Following arguments similar to those used in the proof of the Order Independence Lemma, we can dismiss the cases in which a configuration qualifies for no more than one of the two transactions. Since T_i weakly depends on T_j, we know that there are configurations that could qualify for both. Consider any such configuration C. Let C_j be generated by applying T_j to C. C_j differs from C in the objects that are updated by T_j. But we know that w_j includes no objects that are in $r_j \cap r_i$. Since $w_j \subseteq r_j$, it follows that $w_j \cap r_i = \phi$. Therefore, C_j qualifies for T_i as well. On the other hand, consider C_i generated by applying T_i to C. By definition of weak dependence, $w_i \cap r_j \neq \phi$. But since T_j does not depend on T_i, we must have $W_i \cap R_j = \phi$. In other words, C_i contains a version of at least one object that is different from the version(s) read by T_j. Therefore, C_i cannot qualify for T_j. Thus, executing T_i before T_j yields C_i and C_j from C, while executing T_j before T_i yields C_i, C_j, and C_{ij}. □

Theorem (Exhaustion): The processing of any regular serialization generates all possible complete correct configurations from a given initial complete correct configuration.

Proof: Follows from the Serialization Indifference Lemma, the Complete Generation Lemma, and the following facts:

i. If transaction T_j depends on T_i, then T_j must execute after T_i. Therefore only serializations need be considered. AND
ii. If transaction T_j weakly depends on T_i, then, from the Weak Dependence Ordering Lemma, we need only consider serializations in which T_i is executed before T_j. Therefore, only regular

serializations need be considered. □

Theorem (Uniqueness): In the processing of any serialization, each correct configuration is generated at most once.

Proof: Consider the generators of the object versions in the given configuration. There must be one that occurs last in the serialization. Call it T. The given configuration, C, is generated by applying T to some other configuration, D, which can be obtained by reverse-applying T to the given configuration. C cannot be generated by any previous transaction in the serialization, since T is the latest of the generators of the elements of C. For the same reason, C cannot be generated by any later transaction. If D can be obtained uniquely from C, then C is generated only once by transaction T. But D is unique by definition of reverse-application. □

In view of the Exhaustion Theorem and the Uniqueness Theorem, it follows that one can generate all complete correct configurations in a regular system by processing any one regular serialization starting from the given initial complete correct configuration. Moreover, during this procedure each complete correct configuration is generated exactly once, and one complete correct configuration is generated by every transaction application to an existing qualifying complete correct configuration. Therefore we have obtained a very efficient method of generating all complete correct configurations in a regular system.

4. GENERATION OF SELECTED CONFIGURATIONS

In the previous section, we discussed how to generate all possible correct configurations of a system. While the ability to do so is important, usually only a subset of all possible configurations need be generated, or a membership question answered. In this section, we look at these two types of queries. We first consider the problem of determining whether a user-specified configuration is correct. We then discuss how to satisfy user requests for configurations containing (or not containing) certain specified features.

4.1 Correctness Determination

Given a regular system, the following algorithm determines whether the given complete configuration is correct:

Choose any regular serialization of the transactions.

Repeatedly execute the following steps until the initial complete correct configuration is obtained, establishing the correctness of the complete configuration given, or until a configuration is obtained that does not reverse qualify for its last generator, establishing the incorrectness of the proposed complete configuration:

i. Find the generator of every object version in the configuration.

ii. Pick from this set of generators, the generator (transaction) that occurs last in the serialization.

iii. Reverse apply the transaction to the configuration, provided that the configuration reverse qualifies for the transaction, to yield the new configuration for the next iteration.

This algorithm is linear in the number of objects and in the number of transactions in the system. Its correctness follows directly from the exhaustion theorem, the fact that exactly one transaction writes an object version, and the uniqueness of the configuration generated by reverse applying a transaction to a configuration that reverse qualifies.

If the system is not regular, the above algorithm can be used with the modification that Step 2 must now be repeated for all possible serializations of the transaction dependence graph before declaring a configuration incorrect. The algorithm is terminated if the correctness of the given configuration is demonstrated in any serialization.

4.2 Configurations Containing Specified Features

Transactions encapsulate meaningful changes to the systems, and each transaction identifier can be mapped into a string that identifies corresponding changes. Thus, there could be a transaction called <add-feature-A> that possibly updates several objects. A system configuration ''has'' <add-feature-A>, if for each object updated by the transaction <add-feature-A>, the configuration uses a version that is reachable in the version-graph from the version generated by <add-feature-A>. The user can then ask for configurations that have <add-feature-A>, <add-bell-B>, <add-whistle-C>, etc. The following algorithm determines such configurations in a regular system:

Start with the initial complete correct configuration.

Choose any regular serialization of the transactions.

Repeatedly execute the following steps until all the transactions have been processed:

i. Take the next transaction in the serialization and apply it to all the qualifying configurations for this transaction to generate new complete correct configurations.

ii. If the transaction just applied corresponds to one of the features selected, delete all previously known complete correct configurations, and retain only those generated as a result of applying this transaction.

The set (possibly empty) of complete configurations finally left reflect the features added by all the specified transactions.

The correctness of this algorithm also follows directly from the exhaustion theorem and the fact that exactly one transaction writes an object version.

4.3 Configurations Not Containing Specified Features

If correct configurations are required that "do not include" certain features, the removal of the corresponding transactions from the serialization will ensure that no configuration is generated that includes these features. It may be more efficient not just to remove the undesirable transactions, but also all transactions that can be reached from them in the transaction dependence graph. The desired configurations can then be generated by starting with the initial correct configuration and applying all transactions, in order as per the truncated serialization, to the qualifying configurations.

5. CONCLUSIONS

We considered the problem of configuring a system comprising one version for each of its constituent objects. We presented a transaction model and a syntactic characterization of a correct configuration tied to this model, which equates correct configurations with consistent database states. We allowed version graphs that are trees. We also presented algorithms for generating all correct configurations and those configurations that satisfy some selection criteria such as absence and presence of specified features, and for checking whether a user-specified complete configuration is correct.

We feel that the theory and algorithms presented in this paper can provide the basis for developing automated system configuration tools, and help advance the current state of art in which very little is available to users by way of support in deciding and verifying what constitutes a correct configuration. Indeed, we plan to incorporate these algorithms in Ode [1] — an object-oriented database and environment being developed in our laboratory.

We intend to address in the future two limitations of the current work. One is the requirement that version graphs be trees. In some cases, especially in software engineering environments, DAGs are more appropriate. The second limitation is that the algorithm presented in Section 4.1 can be applied efficiently only to check whether a complete configuration is correct. Extension of the algorithm to incomplete configurations is non-trivial.

REFERENCES

[1] R. Agrawal and N. H. Gehani, "ODE (Object Database and Environment): The Language and the Data Model", *Proc. ACM-SIGMOD 1989 Int'l Conf. Management of Data*, Portland, Oregon, May-June 1989.

[2] T. M. Atwood, "An Object-Oriented DBMS for Design Support Applications", *Proc. IEEE 1st Int'l Conf. Computer-Aided Technologies*, Montreal, Canada, Sept. 1985, 299-307.

[3] D. Batory and W. Kim, "Supporting Version of VLSI CAD Objects", *IEEE Trans. Software Eng.*, 1986.

[4] D. Beech and B. Mahbod, "Generalized Version Control in an Object-Oriented Database", *Proc. IEEE 4th Int'l Conf. Data Engineering*, Los Angeles, California, Feb. 1988, 14-22.

[5] P. A. Bernstein, "Database Support for Software Engineering", *Proc. 9th IEEE Int'l Conf. Software Eng.*, March 1987, 166-178.

[6] P. A. Bernstein, V. Hadzilacos and N. Goodman, *Concurrency Control and Recovery in Database Systems*, Addison-Wesley, 1987.

[7] H. T. Chou and W. Kim, "A Unifying Framework for Version Control in a CAD Environment", *Proc. 12th Int'l Conf. on Very Large Databases*, Kyoto, Japan, Aug. 1986, 336-344.

[8] K. Dittrich and R. Lorie, "Version Support for Engineering Database Systems", Rep. RJ4769, IBM Research Lab., San Jose, California, July 1985.

[9] R. Katz, E. Chang and E. Bhateja, "Version Modeling Concepts for Computer-Aided Design Databases", *Proc. ACM-SIGMOD 1986 Int'l Conf. on Management of Data*, Washington D.C., May 1986.

[10] R. Katz and E. Chang, "Managing Change in a Computer-Aided Design Database", *Proc. of the 13th Int'l Conf. on Very Large Databases*, Brighton, England, Sep. 1987, 455-462.

[11] W. Kim, E. Bertino and J. Garza, "Composite Objects Revisited", *Proc. ACM-SIGMOD 1989 Int'l Conf. Management of Data*, Portland, Oregon, May-June 1989.

[12] D. E. Perry, "Version Control in the INSCAPE Environment", *Proc. 9th IEEE Int'l Conf. Software Eng.*, March 1987, 142-149.

[13] J. L. Steffin, "Interactive Examination of a C Program with Cscope", *Proc. USENIX Winter Conf.*, Dallas, TX, 1985, 170-175.

[14] W. Tichy, "Tools for Software Configuration Management", *Int'l Workshop Software Version and Configuration Control*, Grassau, FRG, Jan. 1988.

[15] J. F. H. Winkler, "Version Control in Families of Large Programs", *Proc. 9th IEEE Int'l Conf. Software Eng.*, March 1987, 150-161.

The Starburst Long Field Manager

Tobin J. Lehman
Bruce G. Lindsay

IBM Almaden Research Center

Abstract

Starburst is an experimental database management system prototype whose objectives include extensibility, support for knowledge databases, use of memory-resident database techniques, and support for large objects. We describe the structure of the Starburst long field manager, which was designed to manage large database objects such as voice, image, sound and video. The long field manager uses the buddy system for managing disk space, which allows it to allocate a range of small to very large disk extents (buddy segments) quickly and efficiently. When possible, a small number of large buddy segments are used to store long fields, which allows the long field manager to perform fewer disk seeks when transferring large objects to and from disk. The long field manager uses shadow-based recovery for long field data and write-ahead-log recovery for long field descriptor and allocation data. Internal space management synchronization is enforced by a combination of long-term and instantaneous locks.

Keywords: Starburst, relational database, long fields, storage methods, buddy system.

1 Introduction

The main goal of the Starburst experimental database management system (DBMS) is to provide a flexible platform on which research personnel can perform DBMS experimentation well into the 1990s. So far, this experimentation has been directed toward improving the structure and performance of the internals of the DBMS, *e.g.*, new recovery and logging mechanisms [Rothermel 89], extensible data management [Lindsay 87], new optimizer methods [Lee 88] [Lohman 88], new query processing methods [Haas 89], new methods for handling data types [Wilms 88], and new signature access methods [Chang 89].

As user-interface technology advances, user interfaces will make more use of multimedia presentations by manipulating voice, sound, image, video and animation. From the database perspective, each of these multimedia objects will require large amounts of storage. Furthermore, presenting images, playing sound, producing video signals, and displaying animation in real time provide significant challenges for a relational database.

The Starburst long field manager was designed with these applications in mind. The long field manager design goals include high data bandwidth with low CPU overhead and efficient storage management techniques. We have combined existing storage management techniques from database systems and file systems in a novel way to achieve these goals. The rest of the paper describes the design of the Starburst long field manager and is organized as follows.

Section 2 discusses the goals and the motivation behind the design of the Starburst long field manager, along with previous work in the area of data storage. Section 3 presents the storage design of the long field manager, including the space allocation data structures and algorithms. Section 4 discusses concurrency control and recovery for long fields. Section 5 summarizes the work and gives the project implementation status. Last, appendix A describes the SQL interface and long field operations.

For space considerations, this version of the long field paper has omitted some details and many diagrams. For more details, the interested reader should obtain the corresponding IBM technical report.

2 Starburst Long Field Manager Design Motivation

2.1 Design Goals

Three main goals drove the design of the Starburst Long Field Manager:

1. Storage allocation and deallocation must be efficient. We expect long field sizes to be large, on the order of 100 megabytes. It is important that the allocation mechanism minimize the time spent allocating and deallocating space to hold a long field.
2. The long field manager must have excellent I/O performance; long field read and write operations should achieve I/O rates near disk transfer speeds.
3. Long fields must be recoverable, but the recovery mechanism must not substantially slow down the long field operations.

To see how we might achieve these goals, we look at existing work in storage systems.

2.2 Previous Work in Storage Systems

Storage management of some kind is required by almost all large software systems. Overall, database systems and file systems have spent the most effort on designing fast and efficient storage mechanisms.

Database Systems

Past database systems were designed to manage simple facts — values that could be represented in fields of 255 bytes or less. Larger fields, such as those containing thousands or millions of bytes, presented problems to the database record manager, so the large fields were typically implemented *via* a separate long field mechanism.

The first SQL relational database system, System R [Astrahan 76], supported long fields with lengths up to 32,767 bytes. The System R long field manager divided long fields into a linked list of small manageable segments, each 255 bytes in length. Operations were restricted to reading and writing entire long fields; partial reads or updates were not supported. Later, an extension to SQL was proposed that provided operators for manipulating long fields [Haskin 82]. A new interface, the long field cursor, provided the ability for partial reading and updating of long fields. Along with the language extension, a storage mechanism was proposed that stored long fields as a sequence of 4 kilobyte data pages (rather than the previous scheme of a linked list of 255 byte records). The maximum length of a long field in extended SQL was about 2 gigabytes [SQL 85].

The Wisconsin Storage System (WiSS) [Chou 85] concurrently developed a similar mechanism for storing long fields. A WiSS long field was split into 4 kilobyte data pages, called *slices*. To reduce internal fragmentation, a *crumb*, a partially filled slice managed similarly to a database record, was used to hold the last segment of a long field if it did not occupy a full slice. A long field was represented by a directory of slices, plus a crumb. WiSS long fields had a size limit of 1.6 megabytes.

A more recent database system, EXODUS [Carey 86], stores all data objects in a general-purpose storage mechanism that can handle objects of any size — the limit is imposed by the amount of physical storage available. EXODUS uses a data structure that was inspired by the ordered relation data structure proposed for use in INGRES [Stonebraker 83]. The data structure is basically a B^+ Tree indexed on byte position within the object, with the B^+ Tree leaves as the data blocks. Objects smaller than a page are stored as single records (in a sense, a degenerate case of the $B+$ Tree structure — a single node). The EXODUS data structure is designed more for random access than for sequential access, although the leaves (the data blocks) may be configured to comprise several sequential disk pages if scan performance is being emphasized.

None of the database systems we've examined have emphasized high performance with respect to reading and writing large objects. Typically, performance issues are restricted to transaction processing systems involving "normal" database fields, those fields that are smaller than 255 bytes in length. The majority of the work involving the movement of large amounts of data at high speed has been in the area of file systems.

File Systems

File systems were used to manage large objects before database management systems existed. File systems associated with batch-oriented operating systems were designed to keep data stored either contiguously or clustered in physical extents, as data was read and written sequentially at high speed. Files with sequentially allocated disk blocks had excellent I/O performance, but sequential allocation caused problems with disk space due to external fragmentation. The next best thing to sequential allocation was physically clustered allocation in extents, or cylinder groups.

Early IBM operating systems such as DOS had file systems that left disk allocation up to the user; the user specified the number and location of disk extents required to hold the file. Later systems, such as IBM's OS/MVS, used hints from the user to plan for initial storage amounts as well as for future growth. IBM's CMS file system running under VM provided both automatic sizing and placement. Using the reasoning that blocks recently written will be read soon, the CMS file system allocated extents (disk tracks) that were close to the current position of the virtual disk arm.

The DEMOS file system [Powell 77], designed for CRAY computers, used the notion of physically clustered 4 kilobyte disk pages to enhance I/O performance. The inventors realized that sequential scanning of files, as opposed to random access, was the more frequent mode of access, so they designed accordingly. When a new disk block was added to the end of a file, it was allocated as physically close to the last block as possible. In order to prevent multiple files being written simultaneously from confusing the system, DEMOS also used a preallocation strategy to allocate a set of blocks for a file rather than single blocks.

The original UNIX file system [Ritchie 74] used 512 byte data pages that were allocated from random disk locations. The inventors seemed to follow the reasoning that, compared to batch systems, timesharing systems have less strict disk organization requirements. For example, with multiple users making disk requests, disk traffic can appear random. Even when two clients have contiguously allocated files, if their read requests are interleaved the disk arm can oscillate between the two file areas and eliminate any advantage of file clustering. The UNIX random page allocation scheme provided only medium-level performance at best, as virtually every disk page fetch operation resulted in a disk seek. Stonebraker noted that this results in poor performance for database systems that use the UNIX operating system [Stonebraker 81].

The UNIX Fast File System [McKusick 84] is an improvement over the original UNIX file system because it uses the idea of physical clustering of disk pages. The Fast File System uses larger data pages that are allocated from cylinder groups, thus sequential scan performance is improved significantly. For large files, a maximum of a megabyte of space is allocated from each cylinder group. The justification for this is that no single file should get all the space in a cylinder group, and that a megabyte of data between disk seek operations produces a respectably high transfer rate. To reduce internal fragmentation, the last large page can be split into smaller pages (fragments).[1]

The file system used by the Dartmouth Time Sharing System (DTSS) [Koch 87] uses the binary buddy system [Knuth

[1] Many page and fragment sizes are mentioned in [McKusick 84], but a common choice has been a 4 kilobyte page and a 1 kilobyte fragment.

75] for space allocation. The binary buddy system provides DTSS with variable size disk extents whose sizes are powers of two (units are disk pages).

When the file size is known in advance, DTSS allocates its disk extents in a manner that minimizes internal fragmentation. DTSS uses the bit representation of the size of the file (in pages) to determine the sizes of the extents needed to contain the file. Using N extents for file storage, the size of the file (in pages) is rounded up to the smallest binary number that contains N or fewer ones. For example, using 3 for N, 23_{10} (10111_2) would be rounded up to 24_{10} (11000_2) (one segment of 16 pages and one segment of 8 pages), 75_{10} (1001011_2) would be rounded up to 76_{10} (1001100_2) (one segment of 64 pages, one segment of 8 pages and one segment of four pages), and 231_{10} (11100101) would be rounded up to 256_{10} (100000000_2) (one segment of 256 pages). When the file size is not known in advance, DTSS allocates extents increasing in size until the file is contained. Later, the storage for the file is reallocated using the scheme described above once the file size is known.

Lessons Learned

Most database systems are quite primitive in their treatment of long fields. They treat the long fields as single values; although larger than simple scalar values, the long fields are read or written in a single operation. Those few database systems that handle long fields somewhat gracefully still do not pay particular attention to performance.

File system designers have addressed the I/O performance problem for large files and have found effective solutions. A file system using large disk extents can provide better performance than file systems using physically clustered disk pages, such as the UNIX Fast File System. A comparison of DTSS against the UNIX Fast File System shows that DTSS is much more efficient in allocating disk space, and through the use of large disk extents supplied by the buddy system, DTSS is able to sustain a high disk throughput with much less CPU usage. DTSS uses an average of three extents to hold a file, although it is able to allocate as many as twenty-four extents per file. The average number of extents per file is a system parameter, and is chosen to minimize the number of disk seeks per file scan, while providing a maximum of 3 percent disk space loss to fragmentation.

A survey of four systems using DTSS showed that the median static file size ranged between 6.4 kilobytes and 25 kilobytes, and that 98 percent of the files were under 64 kilobytes [Koch 87]. Using an average of three disk extents to hold files with these sizes was reasonable, but larger sizes, such as 100 megabytes, would probably require both larger extents and a larger number of them. The Starburst long field manager does exactly that.

3 The Long Field Manager Storage Design

3.1 Storage Design Overview

Figure 1 presents an overview of the data structures used for long field storage management. Given that it is not practical to store a long field directly in a relation[2], the next best thing is to store a long field descriptor in the relation. The long field descriptor is a *single level* directory of disk extents.

[2]IBM's Database 2 (DB2) actually does this, but it also restricts the size of a long field to 32 kilobytes—the size of the largest DB2 data page.

We chose a size limit of 255 bytes for the long field descriptor because it is stored in a tuple as a field value, and since tuples may not span pages, it is important to minimize the size of the descriptor.

The variable disk extents, called *buddy segments*, are taken from *buddy spaces* which are large fixed-size sections of disk that are reserved for long field use. Buddy spaces are taken from an even larger (variable size) portion of disk, labeled *DB Space* in figure 1. DB Space is a Starburst term for an area of disk typically used to hold a set of tables or indices. A buddy space comprises an allocation page and a data area. The allocation page describes the state and size of the individual buddy segments in the buddy space data area. Buddy segments contain only data—no control information—and therefore (given suitable system facilities) the data can be transferred directly from disk to an application's buffers.

3.2 The Buddy System

Discussion

The buddy system works well at managing block sizes that differ by several orders of magnitude [Knuth 75]. It is known for fast allocation and automatic coalescing of blocks on deallocation, but it has also been reported to suffer from poor space utilization due to fragmentation [Bromley 85, Chowdhury 87, Lloyd 85, Page 86, Peterson 77]. Many of the negative reports about the buddy system come from those trying to use the buddy system to allocate a *single* block of storage rather than a set of blocks. All buddy systems (binary, weighted, and fibonacci) have poor storage utilization due to external fragmentation when single extents are used to store data.

However, when multiple blocks are used, both internal and external fragmentation can be reduced greatly. The bounded fragmentation work done by Bryant and Franaszek [Bryant 85] at IBM's Yorktown Research Laboratory shows that the buddy system can be used effectively to address the problems of internal fragmentation and external fragmentation. Since the binary buddy system can supply buddy segment sizes in 2^N units (the units typically being disk pages), it is possible to allocate the correct sizes of buddy segments so that there is less than a single disk page lost to internal fragmentation for any long field.

Therefore, despite the Buddy System's reputation for poor space utilization, it is our choice as the space allocation method for the long field manager because we plan to use it effectively. We use a set of blocks in a range of sizes (including the smaller sizes) to reduce external fragmentation and trimming of blocks (into smaller buddy system blocks) to reduce internal fragmentation.

Description

We refer to a variable-size segment allocated with the buddy system as a *buddy segment*. In the Binary Buddy System, the size of a buddy segment is a power of two, where the units are disk pages. Segments of the same size sharing ancestors are considered buddies (*e.g.* segments 0000 and 0001 are buddies at the 1 unit level, segments 1000 and 1100 are buddies at the 4 unit level). Calculating the address of a buddy for a particular size is straightforward in the binary buddy system. The address of a segment XOR'd with its size gives the address of its buddy. For example, to find the

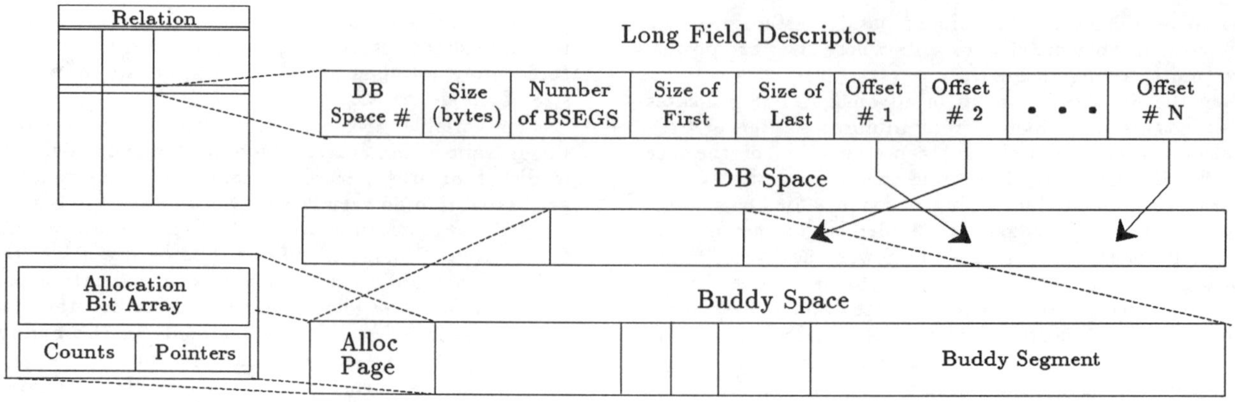

Figure 1: A Storage Overview of the Long Field Manager

size 4 buddy of block 1100, take the XOR of 1100 and 100, which gives 1000 — the size 4 buddy of block 1100 is 1000.

3.3 The Starburst Long Field Descriptor

A long field descriptor completely describes the storage of a long field. The descriptor is small, less than 255 bytes, yet it holds the disk storage information for long fields as large as 100 megabytes.[3] The descriptor contains a directory of buddy segment pointers, which allow direct access to long field data pages. The long field descriptor makes use of the property that the variable size segments are powers of two, thus segment sizes can be expressed as $log_2(size)$, rather than *size* which would require more storage.

3.4 Long Field Manager Algorithms

Creating a Long Field

Creating a long field involves: creating a long field descriptor, allocating buddy segments to hold the long field, and setting the array of offsets in the long field descriptor to point to the buddy segments. A long field may be allocated with or without prior knowledge of size. When the eventual size of a long field is not known *a priori*, successive segments allocated for storage double in size, (*e.g.* the first segment is a single page (1024 bytes), then the next segment is two pages (2048 bytes), then 4 pages, then 8 pages and so on until the maximum size of 2048 pages (2 megabytes) is reached). Once the segment size reaches the maximum, a sequence of maximum size segments is used until the entire long field is stored, up to the maximum long field size (120 megabytes). When the size of a long field is known before creation, maximum segments are used, rather than increasing sizes. The use of doubling sizes has three advantages:

1. It does not unnecessarily allocate large blocks that would then be broken into smaller blocks during the trimming process. (Long Field Trimming is explained in the next section.)
2. It uses the smaller sizes, thus reducing external fragmentation.
3. It avoids storing size information for each segment in the long field descriptor, since the segment size is implicit given the size of the first segment and the known pattern of growth.

[3]The design allows us to support up to approximately 400 megabytes if necessary. If even larger fields are needed, then the application must manage a set of maximum length long fields.

Sometimes, the exact size of a long field is not known, but an approximate size is available. Given a hint of approximate size from the long field append function, the long field manager will start with an appropriately sized first segment and allocate doubling segment sizes from there.

When allocating buddy segments for a long field, the allocator attempts to allocate buddy segments from the same buddy space, which is roughly similar to a contiguous group of cylinders. Therefore, we expect that disk seeks will be small even between buddy segments.

Long Field Trimming

To reduce *internal fragmentation*, the last long field segment is trimmed to the nearest page boundary whenever an application closes a long field (typically upon transaction commit). Consider how a segment of 16 pages would be trimmed to hold a string of bytes that occupy 11 pages. The sequence of segment sizes needed is straightforward; it is the binary representation of the number of pages needed to hold the long field. 11_{10} is 1011_2, thus there's a segment of size eight, a segment of size two, and a segment of size one. The remaining segments (size one and four) are released. After trimming, the last segment is logically a *set* of buddy segments, not a single segment. However, since they are contiguous, the last buddy segment pointer in the long field descriptor may still point to the first buddy segment in the set of segments. This representation may change if append operations (explained below) are performed on a long field that has been trimmed.

Appending to a Long Field

We believe that most long field applications, such as those involving voice, image, sound, or video, will read and write whole long fields and not perform update operations that affect only parts of long fields. There are some applications, however, that may find it necessary to perform partial updates to a long field. An application using the long field mechanism to contain a log file, for example, might perform append operations on a long field.

In most cases, the last segment of a long field will be trimmed to some fraction of its original size (to reduce in-

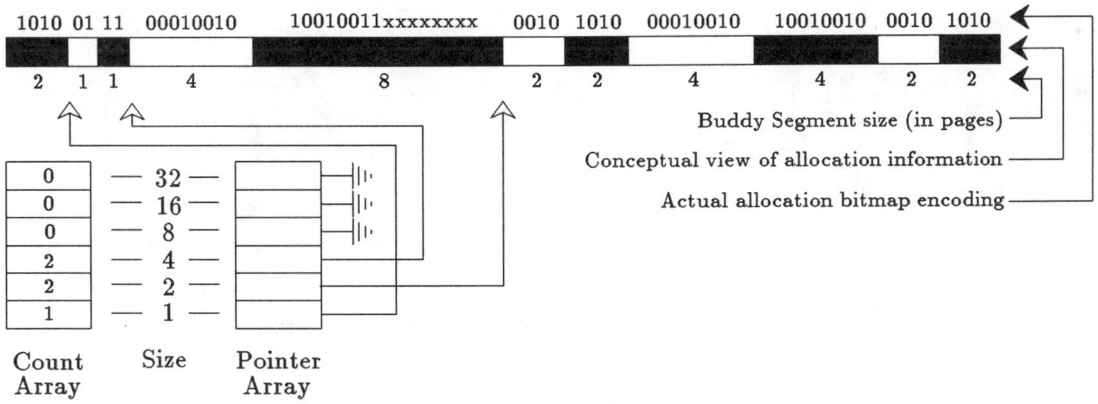

Figure 2: Allocation page structure

ternal fragmentation). In keeping with the design of the long field descriptor, an append operation that required more space than the existing trimmed segment would require a new segment with a size equal to the original size of the trimmed segment. The data from the trimmed segment would be copied to the new segment, and then the new data would be written to the new segment. If the new segment had extra space after the append operation, it would also be trimmed. Clearly, this would not be an efficient mechanism, as appending to very long fields would result in several megabytes of data being copied for each append operation.

Instead, when the long field manager receives an append request for a few bytes on a large object, it modifies the long field descriptor to handle appends more gracefully. The last segment offset in a long field descriptor typically points to a trimmed buddy segment, which consists of several smaller contiguous segments that are treated as a single segment. When an append operation is performed on a long field, the set of segments making up the trimmed segment is broken into separate segments, and the segment pointer for the trimmed segment becomes a directory for the new set of segments. Hence, just as the descriptor uses increasingly larger blocks for unknown sizes, it uses decreasingly smaller sizes for frequent append operations.

3.5 Long Field Manager Space Allocation

Long Field Descriptor Parameters

As we stated earlier, the long field descriptor must be minimized. The first five fields use 14 bytes, leaving about 240 bytes for the array of segment offsets (80 entries). Using a conservative estimate, a long field may have both increasing segment sizes, starting with a single page segment, and may also have decreasing segment sizes (because of appending). Using a 1 kilobyte data page, a maximum buddy segment size of 2 megabytes (2^{11} pages), would result in up to 11 slots being used for increasing sizes (the 11 slots hold a total of approximately 2 megabytes), up to 11 descriptor slots being used for decreasing sizes (also approximately 2 megabytes), and 58 slots being used for maximum size segments (116 megabytes), for a total of about 120 megabytes. (Notice that a long field composed of only maximum size buddy segments would hold more, but we chose to set the maximum long field length to the minimum of the possible maximum sizes.) Since the maximum length of the long field descriptor is fixed, the maximum length of a long field varies with the maximum buddy segment size. A maximum buddy segment size of 1 megabyte would result in a maximum long field size of 62 megabytes, a maximum buddy segment size of 4 megabytes would hold a long field of 232 megabytes and a maximum buddy segment size of 8 megabytes would hold a long field of 448 megabytes. Our current design, based on a 4 kilobyte allocation page, does not support buddy segments larger than 8 megabytes.

Page Allocation and Bitmap Encoding

Space in a long field DB Space is controlled by *allocation pages* stored in the long field DB Space. Recall from Figure 1 that the DB Space is divided into *buddy spaces*, and each buddy space contains an allocation page at the beginning of the space. Each allocation page controls about 16 megabytes of disk storage. Figure 2 shows the structure of an allocation page and the allocation bitmap representation. The allocation page controlling a buddy space has three parts: the allocation bitmap, the buddy segment Count Array, and the buddy segment Pointer Array.

The Count Array and the Pointer Array increase the efficiency of the search for a buddy segment. The Count Array shows the number of buddy segments of each size available in the space controlled by an allocation page, so that an allocator can determine immediately if it should look in a given allocation page for a given segment size. The Pointer Array provides a place to start looking for a segment of a particular size, so rather than starting each search from the beginning of the bitmap, the Pointer Array shows the first place where an available segment was last seen (by anyone updating the allocation page). On some occasions, the Pointer Array may point to a segment that is available, but on other occasions, the segment may have recently been allocated, hence the Pointer Array actually provides a *hint* rather than an absolute location. However, the pointer for a particular buddy size is guaranteed to be at least a correct starting point for a search for that size buddy segment. As segments are allocated, the Count Array is updated and the Pointer Array is updated to point to the location of the newly allocated segment (this is how the segment pointer moves forward). As segments are freed, the Count Array is updated and the Pointer Array is set to point to the segment closest to the beginning (this is how the segment pointer moves backward). Upon initialization, all of the pointers in the Pointer Array point to the first available segment of each size.

An obvious method for representing allocation informa-

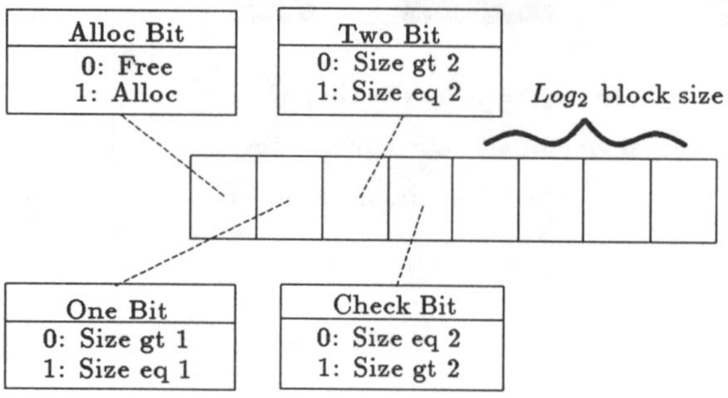

Examples

Bits	Meaning
11	A segment of 1 page (Allocated)
01	A segment of 1 page (Free)
1010	A segment of 2 pages (Allocated)
0010	A segment of 2 pages (Free)
10010100	A segment of 16 pages (Allocated)
00011000	A segment of 256 pages (Free)

Figure 3: Allocation bitmap encoding

tion in bitmap form would be to use a single bit for each page, where a '1' in a page's bit position means that the page is allocated and a '0' means that the page is free. With the buddy system, however, more information is needed: namely, the size of the buddy segment (in pages). We use two bits per disk page so that we can represent both the allocation status and the size of the buddy segment.

Two bits are sufficient for representing the size and allocation status of a single page buddy segment. When a buddy segment is larger than a single page, for every page in the buddy segment there are two additional bits available for size and status information. Eight bits are sufficient for representing the allocation status and size of segments that range from 1 page to 32768 pages.[4] Therefore, when eight bits are available, that is, when a segment is at least four pages long, the general purpose representation is used. When the segment size is one or two pages, then the specific representation for one or two pages is used. Figure 3 shows the bit encoding used to store the size and allocation information of various sizes of buddy segments. Notice that at most eight bits per segment are used to hold information, while any remaining bits are left unused.[5]

Encoding the size in the page allocation bitmap gives a considerable advantage when looking for a particular segment size. First, at most eight bits must be examined, regardless of the segment size, to determine the segment size and allocation status. Second, when a segment is freed, a simple arithmetic operation on the freed segment's address gives the address of the buddy which can then be checked immediately for possible coalescing.

Also, because of the buddy system, a segment of size 2^N can appear only on boundaries where the address is 0 modulo 2^N. Thus, looking at the size of a segment on those boundaries provides an immediate direction of where to look next.

With our allocation bitmap encoding, the bitmap *must* be read on *valid* segment boundaries. This invariant is maintained by the pointer array, which guarantees that the starting place for a search is valid. Without this invariant, an allocator searching the bitmap could easily become confused. Notice in Figure 2 that the segment of size 8 has 16 bits available, but uses only 8 bits. The trailing 8 bits do not carry any useful information, and could mislead an allocator that tried to read the bitmap on that nonsegment boundary.

4 Concurrency and Recovery for Long Fields

Concurrency control is managed by the database system at the record level, hence there is no user-level concurrency control mechanism needed for the long field manager. Locking the record that contains a long field descriptor also locks the long field associated with that record.

We examined the two main techniques for recovery of long fields: logging and shadow techniques. Shadowing [Lorie 77] is a recovery technique in which current page contents are never overwritten. Instead, new pages are allocated and written, while the pages whose values are being replaced are retained as shadow copies until they are no longer needed to support the restoration of the system state due to transaction rollback. Logging [Gray 81] is a well-known database recovery technique which supports state restoration and reconstruction of values that are updated in place by recording in a system log the old and new values of the updated data items. We chose the shadow technique for long field data because of its simplicity [Traiger 82] and its potential for causing fewer I/O operations. However, since the shadow technique does not provide protection against media failures some additional measures are necessary; the long field data can be stored on duplexed disks or logged. Duplexed disks allow higher throughput, but they are costly. Logging is less expensive, but it allows less throughput and has the additional problem of potentially degrading the performance of the entire system. We are currently investigating these alternatives.

When the Starburst long field manager updates (inserts or deletes) a long field, it creates a new copy of the updated long field data segments (which it transfers directly to disk) and logs the changes to the long field descriptor and the long field allocation page(s). The replaced long field data segments remain as shadow copies until the updating transaction commits. When a long field data value is deleted or replaced, the allocation information for the corresponding space in the long field file is marked free, but the allocation information itself is locked against *all* transactions (even the deleting transaction) until the deleting transaction completes, thus ensuring that deletes will be undoable. Should a delete operation need to be undone, the deleted space will be reclaimed and the long field descriptor will be reset to reference its original set of data segments. Locking the freed storage presents an interesting problem, since at the time the space is freed, the buddy system may auto-

[4] The number 32768 could be pushed to 262144, since we don't need to represent 2^0, 2^1, or 2^2, as they're done with a different encoding. However, we didn't need to add this extra bit of complexity because we plan on using a maximum of only 2048 pages in our design.

[5] Those bits would be used if the segment is split into smaller segments.

matically coalesce the released buddy segment into a larger buddy segment, which could possibly be coalesced into an even larger buddy segment, and so on. The newly freed (and possibly coalesced) buddy segment may have a different address from the originally freed buddy segment. As a result, simply locking the address of the released buddy segment is not adequate.

Our solution uses a set of exclusive-mode locks, intention-mode locks and instant conditional locks. Allocate (A) and intention allocate (IA) locks are *instant conditional* locks; they are not actually held like regular locks. Setting an instant conditional lock on a segment simply tests the lock status of the segment; it does not actually lock a segment, nor does it cause the requester to block in the event the segment is already locked. When an allocate lock is denied, the requester simply moves on to the next available buddy segment and tries again. An allocation bitmap structure is protected by an allocation page latch while a requestor attempts to set allocate locks, so mutual exclusion of multiple requestors is guaranteed.

When a buddy segment is freed, a release lock (R) is placed on its address and size, and an intention release lock (IR) is placed on all of its ancestors. These locks are held until transaction completion. When a transaction attempts to allocate a buddy segment, it sets instant conditional intention allocate (IA) locks on all ancestors of the intended buddy segment, and then sets an instant conditional allocate (A) lock on the intended buddy segment. The lock matrix is as follows:

	Lock Held			
		None	IR	R
Lock Requested	IR	OK	OK	n/a
	R	OK	n/a	n/a
	IA	OK	OK	NO
	A	OK	NO	NO

If an ancestor is locked with a release (R) lock, the intention allocate lock will be denied, as the ancestor must remain unallocated until its holding transaction completes. If the intended buddy segment is locked with either an intention release lock (IR) or a release lock (R) then all or part of the buddy segment must remain unallocated until its holding transaction completes.

An important aspect of any hierarchical locking scheme is the number of lock granularities. With 12 sizes of buddy segments (our current design), releasing a buddy segment of size 1 would require setting 1 release lock and 11 intention release locks. Rather than set a lock at each level, locks are set at only certain specified levels (granularities). For example, Figure 4 shows how we might set locks only at segment sizes 2 and 8 in a system with only 5 sizes. Setting locks at only certain levels trades concurrency for overall locking cost, as each lock may cover more segments than just the segment requiring a lock. In Figure 4 for example, segment 0011 is locked along with segment 0010, and pages 1000–1011 are locked along with segment 1100.

Related Work in Recovery

Numerous other methods have been used for supporting recoverable long fields or long sequential byte strings. An IBM database product, DB2, stores long fields directly in normal database records. DB2 records may not cross page boundaries, and since the largest page size in DB2 is 32 kilobytes, the maximum long field size in DB2 is 32 kilobytes. Recovery of long fields is the same as for other types — write-ahead logging. Another IBM database product, SQL/DS, has a long field manager that stores the long field value in a sequence of blocks, each block being small enough to fit into a database page. Each of these blocks has the format of a database record and the blocks are linked into a linear list whose head is in the database record which logically contains the long field. The values of the block of the long field are recovered using the "standard" database log techniques. As a result, the contents of an updated long field are written both to the log and to the database.

The remaining related work in recovery comes from the domain of "transactional file systems". A useful survey of several file systems which support transactions can be found in [Svobodova 84]. The predominant approach for these systems is the use of *shadow* copies which is similar to our approach. Most of the systems rely upon specialized techniques to recover control and allocation information following a failure. XDFS and CFS [Mitchell 82] both use shadow pages but depend upon *careful replacement* of control and allocation information to insure failure recovery. Both of these systems require extensive scanning of secondary storage to validate and reconstruct control information.

Paxton [Paxton 79] describes a shadow based file system which uses a form of careful replacement for control information known as *intentions lists*. Paxton's approach to storing long byte string values is similar to our method in that it uses shadow page replacement as the basic recovery mechanism. However, the use of intentions lists as the mechanism for transaction commit requires at least 3 extra write operations to reliably update the control information associated with the long byte string value. Our method requires only 1 extra (log) write of control information (the long field descriptor) to recoverably update a long field.

The Alpine file system [Brown 85] logs file content updates and copies file content from the log to the file following the commitment of the transaction modifying the file. This leads to 3 I/O operations per file page modified by the transaction. While file content changes in Alpine use a redo only log, file allocation operations are propagated immediately to secondary storage and use an undo log to support transaction abort. The disadvantages of the Alpine approach include the copying needed to propagate logged changes to the file proper and the need for synchronous I/O to log the update file control information. Our method writes long field content changes only to the long field file blocks and relies upon database recovery logging for control and allocation information recovery which implies that these changes do not need to be written synchronously and can be piggybacked on other log I/O operations.

The work closest to our method of long field recovery is the new file system for the Cedar Operating System [Hagmann 87]. The new Cedar file system uses the shadow technique to recover file content pages and uses a log to recover changes to file control information. File data pages are not logged and are vulnerable to media failure. Allocation data is not logged either, but it can be recovered by scanning the extent tables of all files. The logging and recovery granularity is the page and the log is a redo only log. Reservation of space for released pages is accomplished by postponing the allocation table updates until transaction commit. Our method uses record-level (sub-page) logging and recovery and uses the log for recovery of both control information

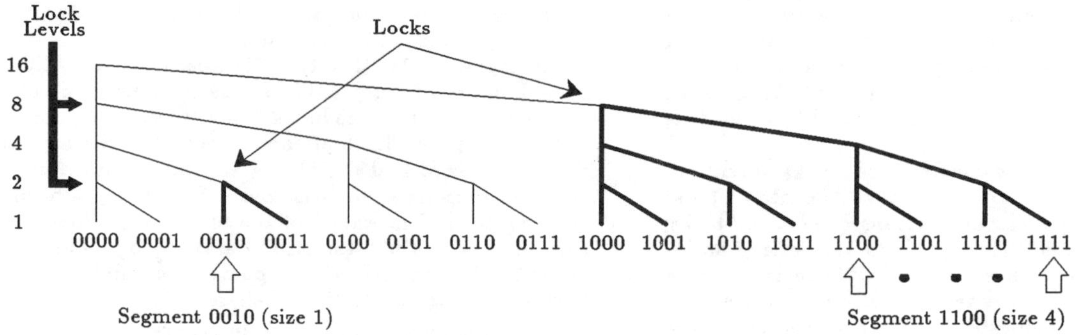

Figure 4: Sparse locking of buddy segment sizes.

and allocation data.

5 Summary

We designed the Starburst long field manager to manipulate long fields, with sizes in the 100 megabyte range, with great speed and efficiency. The long field manager uses the buddy system to manage disk space, which is compatible with growth and trimming algorithms that increase storage efficiency. Long field data segments contain only user data, which allows the long field manager to transfer data directly from disk storage to application buffers without copying data to an intermediate system buffer.

Recovery for long fields is based on shadows for long field data and write-ahead-log for long field allocation and descriptor information. The use of shadows prevents old and new values from both being saved — only the new pages need to be recorded. Locking of buddy segments for space reservation purposes uses exclusive-mode locks, intention-mode locks, and instant conditional locks in order to update the allocation information in place and perform concurrent space coalescing without allowing other transactions to see the updates to the allocation tables until the holding transaction completes.

Long Field Manager Status

The majority of the Starburst long field manager design and implementation took place during the summer of 1985. After some preliminary testing validated the low levels of the space allocation mechanism, further implementation on the Starburst prototype was postponed until several other necessary components were finished. We are just now finishing the implementation of the long field manager. We plan to run numerous benchmarks, comparing our long field manager performance to that of available file systems. One special benchmark in our plans involves storing a full length feature film (a Fred Astaire video — uncolorized) using the long field manager and then playing the video in its entirety in an application window. An uncompressed 2 hour video digitized for a 1 megabyte screen would be approximately 216 gigabytes, but using video compression hardware, we calculate that a 2 hour video will require approximately 500 megabytes.

The basic design of the Starburst long field manager was also implemented for IBM's Operating System/2 (OS/2) and incorporated in IBM's OS/2 Extended Edition Database Manager.

6 References

[Allchin 80] J. E. Allchin *et al.*, "FLASH: A Language-Independent, Portable File Access System," *Proc. ACM SIGMOD*, (May 1980).

[Astrahan 76] MM. Astrahan *et al.*, "System R: Relational Approach to Database Management," *ACM TODS*, Vol. 1, No. 2 (June 1976).

[Bromley 80] A. G. Bromley, "Memory Fragmentation in Buddy Methods for Dynamic Storage Allocation," *Acta Informatica*, Vol. 14, pp. 107–117 (1980).

[Brown 85] M. Brown *et al.*, "The Alpine File System," *ACM TOCS*, Vol. 3, No. 4 (November 1985).

[Bryant 85] R. Bryant and P. Franaszek, "Storage Allocation Policies via Bounded Fragmentation," *IBM Research Presentation*, Summer 1985.

[Carey 86] M. J. Carey *et al.*, "Object and File Management in the EXODUS Extensible Database System," *Proc. 12th VLDB*, (August 1986).

[Chang 89] W. Chang and H. Schek, "A Signature Access Method for the Starburst Database System," *Proc. 15th VLDB* (August 1989).

[Chou 85] H-T Chou *et al.*, "Design and Implementation of the Wisconsin Storage System," *Software Practice and Experience*, Vol. 15, No. 10 (October 1985).

[Chowdhury 87] S. K. Chowdhury and P. K. Srimani, "Worst Case Performance of Weighted Buddy Systems," *Acta Informatica*, Vol. 24, pp. 555–564 (1987).

[Eswaran 76] K. P. Eswaran *et al.*, "The Notions of Consistency and Predicate Locks in a Database System," *CACM*, Vol. 19, No. 11 (November 1976).

[Gray 75] J. Gray *et al.*, "Granularity of Locks in a Shared Data Base," *Proc. VLDB* (September 1975).

[Gray 81] J. Gray *et al.*, "The Recovery Manager of the System R Database Manager," *ACM Comp. Surveys*, Vol. 13, No. 2 (June 1981).

[Haas 88] L. Haas *et al.*, "Extensible Query Processing in Starburst," *Proc. ACM SIGMOD* (May 1989).

[Hagmann 87] R. Hagmann, "Reimplementing the Cedar File System Using Logging and Group Commit," *Proc. SOSP* (November 1987).

[Haskin 82] R. L. Haskin and R. A. Lorie, "On Extending the Functions of a Relational Database System," *Proc. ACM SIGMOD* (June 1982).

[Knuth 75] D. E. Knuth, *"The Art of Computer Programming. Vol 1, Fundamental Algorithms,"* Addison-Wesley, Reading Mass., 1969.

[Koch 87] P. D. L. Koch, "Disk File Allocation Based on the Buddy System," *ACM TOCS*, Vol. 5, No. 4 (November 1987).

[Lee 88] M. Lee *et al.*, "Implementing an Interpreter for Functional Rules in a Query Optimizer," *Proc. 14th VLDB* (August 1988).

[Lindsay 87] B. Lindsay *et al.*, "A Data Management Extension Architecture," *Proc. ACM SIGMOD* (May 1987).

[Lohman 88] G. Lohman, "Grammar-Like Functional Rules for Representing Query Optimization Alternatives," *Proc. ACM SIGMOD* (May 1988).

[Lorie 77] R. Lorie, "Physical Integrity in a Large Segmented Database," *ACM TODS*, Vol. 2, No. 1 (March 1977).

[Lloyd 85] E. L. Lloyd and M. C. Loui, "On the Worst Case Performance of Buddy Systems," *Acta Informatica*, Vol. 22, pp. 451–473 (1985).

[Mitchell 82] J. Mitchell and J. Dion, "A Comparison of Two Network-based File Servers," *CACM*, Vol. 25, No. 4 (April 1982).

[McKusick 84] M. K. McKusick, W. N. Joy, S. J. Leffler and R. S. Fabry, "A Fast File System for UNIX," *ACM TOCS*, Vol. 2, No. 3 (August 1984).

[Ousterhout 85] J. K. Ousterhout *et al.*, "A Trace-Driven Analysis of the UNIX 4.2 BSD File System," *Proc. 10th SOSP* (December 1985).

[Page 86] I. P. Page, "Improving the Performance of Buddy Systems," *IEEE Trans. on Computers*, Vol. C-35, No. 5 (May 1986).

[Paxton 1979] W. Paxton, "A Client-Based Transaction System to Maintain Data Integrity," *Proc. 7th SOSP* (December 1979).

[Peterson 77] J. L. Peterson, "Buddy Systems," *CACM*, Vol. 20, No. 6 (June 1977).

[Powell 77] M. L. Powell, "The DEMOS File System," *Proc. 6th SOSP* (November 1977).

[Ries 79] D. Ries and M Stonebraker, "Locking Granularity Revisited", *ACM TODS*, Vol. 4, No. 2 (June 1979).

[Ritchie 74] D. M. Ritchie and K. Thompson, "The UNIX Time-Sharing System," *CACM*, Vol. 17, No. 7 (July 1974).

[Rothermel 89] K. Rothermel, C. Mohan, "ARIES/NT: A Recovery Method Based on Write-Ahead Logging for Nested Transactions," *Proc. 15th VLDB* (August 1989).

[SQL 85] R. Lorie and J. Daudenarde, "Design System Extensions User's Guide," (April 1985).

[Smith 81] A. J. Smith, "Algorithms and Architectures for Enhanced File System Use," *Experimental Computer Performance Evaluation*, pp. 165–193 (1981).

[Stonebraker 81] M. Stonebraker, "Operating System Support for Database Management," *CACM*, Vol. 24, No. 7 (July 1981).

[Stonebraker 83] M. Stonebraker *et al.*, "Document Processing in a Relational Database System," *ACM TOFS*, Vol. , No. 2 (April 1983).

[Svobodova 84] L. Svobodova, "File Servers for Network Based Distributed Systems," *ACM Comp. Surveys*, Vol. 16, No. 4 (December 1984).

[Traiger 82] I. L. Traiger, "Virtual Memory Management for Database Systems," IBM Research Report RJ3489 (41346) (May 25, 1982).

[Wilms 88] P. Wilms *et al.*, "Incorporating Data Types in an Extensible Database Architecture," Procs. of 3rd Int. Conf. on Data and Knowledge Bases, Jerusalem, (June 1988).

[Verhofstad 78] J. S. M. Verhosfstad, "Recovery Techniques for Database Systems," *ACM Comp. Surveys*, Vol. 10, No. 2 (June 1978).

A The SQL Interface and Long Field Operations

In order to meet its extensibility goal, Starburst must accept the incorporation of new data types, new storage methods, and new access methods in a straightforward manner. The long field, as we have defined it, is a new storage method, and the "large unstructured byte stream" is a new data type. The creator of a new data type defines the interface to instances of the data type, defines the operations on instances of the data type, and specifies the storage method.

In the Starburst project, the large byte stream is the first data type to be implemented that does not fit into the standard mold for data types. Unlike scalar values such as integers, floating point numbers, or even the original System R long field, a byte stream value is not manipulated directly. The user of a byte stream type first selects a "handle" on it through an SQL SELECT statement or a CURSOR operation.

In general, an application may define a "handle" for any type instance. A handle for a type instance is similar to an open file descriptor; it is used to reference that type instance when performing operations on it. Functions can manipulate type instances *via* a handle, or in some cases, they can manipulate the type instance itself. Instances of scalar types, for example, can be manipulated directly, but byte streams may be manipulated *only* though a handle.

Defining the external interface to byte streams requires two steps. First, we must extend the standard SQL language with a special "apply" operator to provide a mechanism for calling the new operations. Second, we define a set of operations that manipulate byte streams.

A.1 The SQL Interface

In order to extend the System R interface to long fields, Haskin and Lorie proposed an extended definition of cursors to manipulate long fields [Haskin 82]. The proposed extension to SQL applied specifically to long fields, and was not usable by other data types. The APPLY function, as suggested here, is a general interface that allows an application to call operations defined on instances of any extended data types.

We extend SQL with an APPLY operation, in addition to its SELECT, UPDATE, INSERT, and DELETE operations. Although we have not yet determined the exact syntax of the APPLY operation, the general idea is that an application will use the APPLY operation when it asks the database system to manipulate instances of extended data types. Historically, database systems simply store data, and leave the manipulation up to the application. With long fields, however, it is more efficient to have the database perform some of the operations.

An example of the APPLY operation is:

```
EXEC SQL APPLY Read( :lf_handle, :position,
:count, :buffer );
```

A.2 Operations

When an application opens a byte stream through an SQL SELECT or CURSOR operation, it is given a handle that refers to the open long field. The application then supplies that handle when calling the operations defined for long fields. The supported byte stream operations are:

Clear(lf_handle) Clear the long field and set its value to NULL. (We make the distinction between a NULL value and zero length.)

Truncate(lf_handle, position) Delete the bytes from <position> to the end of the long field.

Read(lf_handle, position, count, BUFFER) Read <count> bytes from the long field at <position> and put them into <BUFFER>.

Append(lf_handle, count, BUFFER, hint) Append <count> bytes, taken from <BUFFER>, to the end of the long field. If the long field is zero length or NULL, then use <hint> as a value for estimated size. <hint> is used when the long field is being written over several append operations and the actual size of the long field is not known.

Replace(lf_handle, position, count, BUFFER) Overwrite <count> bytes at <position>, taken from <BUFFER>.

Length(lf_handle, length) Return the length of the long field in <length>.

Scheduling Real-Time Transactions with Disk Resident Data

Robert Abbott
Hector Garcia-Molina

Department of Computer Science
Princeton University
Princeton, NJ 08544

Abstract

Managing transactions with real-time requirements and disk resident data presents many new problems. In this paper we address several: How can we schedule transactions with deadlines? How do the real-time constraints affect concurrency control? How does the scheduling of IO requests affect the timeliness of transactions? How should exclusive and shared locking be handled? We describe a new group of algorithms for scheduling real-time transactions which produce serializable schedules. We present a model for scheduling transactions with deadlines on a single processor disk resident database system, and evaluate the scheduling algorithms through detailed simulation.

1. Introduction

A *real-time database system* (RTDBS) processes transactions with timing constraints such as deadlines. The system guarantees serializable executions while at the same time minimizing the number of transactions that miss their deadlines. Conventional database systems differ from RTDB ones in that the former do not take into account individual transaction timing constraints in making scheduling decisions. Conventional real-time systems, on the other hand, differ from RTDB systems in that they assume advance knowledge of the data requirements of programs and their goal is to guarantee *no* missed deadlines [14]. However, they do not guarantee data consistency. Such systems are called hard real-time. RTDB systems can be useful for applications which are both data-intensive and subject to real-time constraints. Such applications include computerized stock trading, command and control systems and computer-integrated manufacturing [1, 13].

There are many new and challenging problems in designing a RTDB. Two of these problems were studied in [1] : transaction scheduling and concurrency control. In particular, that paper presented several algorithms for resolving lock conflicts and for determining in what order to execute available transactions. The algorithms were studied via detailed simulations. Two major assumptions were made in that work: (a) the database was memory resident, and (b) only exclusive locks were available.

In this paper we continue our investigations of real-time scheduling and concurrency control. Assumptions (a) and (b) have been dropped, a new set of algorithms has been developed, and some additional issues and measures have been considered. The new results, we believe, provide substantial additional insights into the operation of RTDB systems.

Allowing the database to reside on disk, with a portion residing in a main memory buffer pool, introduces more interesting questions that one might initially imagine. For instance, the disk is now a resource that transactions must compete for. How are the disk requests to be scheduled? Do the same real-time priorities that worked for CPU scheduling work for disk scheduling? Some disk controllers do scheduling on their own (trying to minimize head movement). Does this interfere with the real-time scheduling? Since transactions now are suspended more frequently (lock waits and IO waits), there are more opportunities for CPU scheduling. How do the CPU scheduling algorithms respond? Finally, transaction commit must be considered. That is, transactions must flush their dirty pages to disk and write log records. What priorities should these operations receive? Should the log be placed on a separate disk?

Shared locks also introduce a new set of challenging questions. With exclusive locks only, conflicts always involve a pair of transactions, the holder and the requester. The conflict can be resolved by comparing the priorities (e.g., earliest deadline) of each. With shared locks, the holder can actually be a set of concurrently reading transactions, each with a different deadline. If the requester needs an exclusive lock, what is to be done? What priority does the group have? If the requester needs a shared lock, it could be granted immediately, but there may be other transactions already waiting for exclusive locks. How are the priorities of the waiting transactions compared against that of the new requester? Should the new requester be granted the shared lock or not?

We have extended the algorithms of [1] to cope with disk data and shared locks. In addition, we have studied concurrency control algorithms not considered initially, including one that *promotes* transactions that are blocking higher priority transactions. Finally, we have also considered two supplementary measures (in addition to mean number of missed deadlines). One is the mean tardiness of transactions, i.e., average time by which transactions miss their deadlines. The second is the response of the system to a batch of transactions that arrive at once. Such an "input step function" emulates a severe overload. Such overloads may not be frequent, but having algorithms that can cope with them gracefully is important.

2. Model and Assumptions

In this section we describe our basic assumptions and real-time transaction model. The system consists of a *single processor*, a disk-based database, and a main memory buffer pool. (The multiple processor case is also of interest, but we have not addressed it yet.)

The unit of database granularity we consider is the *page*. Transactions access a sequence of pages. If a page is not found in the buffer pool, a disk read is initiated to transfer the page to the pool. Modified pages are held in the pool until the transaction completes. At that time, the log is flushed and the transaction commits. Finally, the modified pages are written back to disk to free space in the buffer pool. (This buffer management strategy can be characterized as not ATOMIC, not STEAL and FORCE [7]) We assume that the buffer pool is large enough so that a transaction never has to write modified pages to disk until after commit. Thus, aborting a transaction involves no disk writes. We assume that the log is kept on a disk (or tape) separate from the database disks (the most common scenario in practice).

Each arriving transaction has a release time r, a deadline d, and a run time estimate E. The release time is the earliest time the transaction can be started and is usually the arrival time. The deadline is the desired maximum commit time. Estimate E approximates the duration of the transaction on an unloaded system. It takes into account both the CPU and disk read time involved. Parameters r, d, and E are known to the system as soon as the transaction arrives. However, the access pattern of the transaction is not known in advance. As the transaction is executed, it asks to read or write one page at a time. Our decision to assume knowledge of a run time estimate but no knowledge of data requirements is justified because it is easier to estimate the execution time of a transaction than to predict its data access pattern. In any case, E is simply an estimate that could be wrong or not given at all.

The RTDB system schedules transactions with the objective of minimizing the number of missed deadlines. As the system runs it may observe that some transactions have already missed their deadlines before committing. We assume that all transactions must be executed eventually, regardless of whether they are tardy or not. In this case, we will assume that the priority of tardy transactions increases, in order to limit the tardiness. Other options for handling tardy transactions and their performance are discussed in [1].

As discussed in the introduction, we assume that transaction executions must be serializable [5]. (If transaction semantics were known, other strategies could be used [6].) We assume that serializability is enforced via a locking protocol that allows for shared and exclusive locks. Deadlock detection is used to find and break deadlocks. We have selected a locking protocol because locking is widely used in practice; other strategies may be possible [2] but are not considered here.

3. Scheduling Algorithms

Our scheduling algorithms have three components: a policy for assigning priorities to tasks, a concurrency control mechanism, and a policy for scheduling IO requests. The priority policy controls how a priority is assigned to a real-time transaction. The concurrency control mechanism can be thought of as a policy for resolving conflicts between two (or more) transactions that want to lock the same data object. Some concurrency control mechanisms permit deadlocks to occur. For these a deadlock detection and resolution mechanism is needed. The third component controls how scheduling of the IO queue is done, i.e., whether a transactions's real-time constraints are used to decide which IO request is serviced next.

Each component may use only some of the available information about a transaction. In particular we distinguish between policies which do not make use of E, the runtime estimate, and those that do. A goal of our research is to understand how the accuracy of the runtime estimate affects the algorithms that use it.

3.1. Assigning Priorities

There are many ways to assign priorities to real-time tasks [9, 10]. We have studied three.

First Come First Serve. This policy assigns the highest priority to the transaction with the earliest release time. If release times equal arrival times then we have the traditional version of FCFS. The primary weakness of FCFS is that it does not make use of deadline information. FCFS will discriminate against a newly arrived task with an urgent deadline in favor of an older task which may not have such an urgent deadline. This is not desirable for real-time systems.

Earliest Deadline. The transaction with the earliest deadline has the highest priority. A major weakness of this policy is that it can assign the highest priority to a task that has already missed or is about to miss its deadline. When this is done, the system allocates resources to a transaction which cannot meet its deadline in favor of a transaction which could meet its deadline. One way to solve this problem is to use an overload management policy to screen out transactions that have missed or are about to miss their deadlines [1].

Least Slack. For a transaction T we define a slack time $S = d - (t + E - U)$, where t is the current time, and U is the amount of service time consumed by T so far. The slack time is an estimate of how long we can delay the execution of T and still meet its deadline. If $S \geq 0$ then we expect that if T is executed without interruption, it will finish at or before its deadline. A negative slack time is an estimate that it is impossible to make the deadline. A negative slack time results either when a transaction has already missed its deadline or when we estimate that it cannot meet its deadline.

Note that Least Slack is very different from Earliest Deadline in that the priority of a task depends on how much service time it has received. The slack of a transaction which is executing does not change. (Its service time and the clock time increase equally.) The slack time of a transaction which is not executing *decreases*. Hence the priority of that transaction *increases*.

A natural question to consider is how often to evaluate a transaction's slack. We consider two methods. With the first, *static evaluation*, the slack of a transaction is evaluated once when the transaction arrives. This value is the transaction's priority for as long as the transaction is in the system. (If a transaction is rolled back and restarted, the slack is recalculated. In effect, the transaction is re-entering the system as a new arrival.) Under the second method, *continuous evaluation*, the slack is recalculated whenever we wish to know the transaction's priority. This method yields more up-to-date information but also incurs more overhead.

Our performance studies have shown that sometimes it is better to use static evaluation and sometimes it is better to use continuous evaluation. (See Section 5.) The majority of our experimental results use static evaluation. We chose this because static evaluation performed better than continuous at higher load settings, which is where we performed many of our experiments.

3.2. Concurrency Control

If transactions are executed concurrently then we need a mechanism to order the updates to the database so that the final schedule is serializable. Our mechanisms allow shared and exclusive locks. Shared locks permit multiple concurrent readers. The priority of a data object O is defined to be the maximum priority of all transactions which hold a lock on object O. If O is not locked then its priority is undefined.

Let T be a transaction requesting a shared lock on object O which is already locked in shared mode by one or more transactions. Transaction T is allowed to join the read group only if the priority of T is greater than the maximum priority of all transactions, if any, which are waiting to lock O in exclusive mode. In other words, a reader can join a read group only if it has a higher priority then all waiting writers. Otherwise the reader must wait. Conflicts arise from incompatability of locking modes in the usual way.

We are particularly interested in conflicts that can lead to priority inversions. A *priority inversion* occurs when a transaction T of high priority requests and blocks on a lock for object O which has a lesser priority than T. This means that T has a higher priority then the transaction(s) which holds the lock on O. T must wait until the lock holder(s) releases its lock on O, either voluntarily or involuntarily. Conflicts which cannot lead to priority inversion, i.e., the priority of the requester is less than the priority of the object, are handled by having the requester wait. Of course a deadlock detection method must be employed to detect cycles of waiting transactions. We now discuss four techniques that we use to resolve conflicts.

In the following, let T_R be a transaction that is requesting a lock on a data object O that is already locked by transaction T_H. Furthermore, the lock modes are incompatible and T_R has a higher priority than the priority of O. Thus the priority of T_R is greater than T_H. Namely we have a priority inversion.

Wait. Under this policy, priority inverting conflicts are handled exactly as non-priority inverting conflicts. That is, the requesting transaction always blocks and waits for the data object to become free. This is the standard method for most DBMS which do not execute real-time transactions. All conflicts are handled indentically and the concurrency control mechanism makes no effective use of transaction priorities.

Wait Promote. Wait Promote handles conflicts as Wait does except when a priority inversion occurs. The high priority transaction T_R will block and wait but now we promote the priority of the lock holder T_H so it is as high as the priority of T_R. In other words, T_H inherits the priority of T_R. (Since locks are retained until commit time, T_H will keep its inherited priority until it commits or is restarted. In the event that T_H is restarted, e.g., because of deadlock, it assumes its normal priority. A pure implementation of priority inheritance would demote the priority of T_H if T_R is aborted before T_H finishes. We chose not to implement demotion. Our tests showed that it occurs so seldom that any difference in overall performance is not measurable.) This method for handling priority inversions was proposed in [12].

The reason for promoting T_H is that it is blocking the execution of T_R a higher priority transaction. Thus T_H should execute at an elevated priority so that it finishes sooner and T_R can resume execution sooner. Priority inheritance ensures that only a transaction with priority greater than T_R will be able to preempt T_H from the CPU. A transaction T_I of intermediate priority, a priority greater than T_H and less than T_R, would normally be able to preempt T_H. But with priority inheritance, T_I has a lesser priority than T_H which is now executing on behalf of T_R.

What if the object is locked by more than one transaction? In this event all transactions in the read group will inherit the priority of T_R. Note that a priority inversion can affect only some of the transactions in a read group. For example, the requesting transaction may have a priority that is greater than only some of the transactions in the read group. These transactions will inherit the greater priority of the requester. The priority of the other transactions in the read group remains unchanged. Thus every transaction holding a lock on object O has a priority that is at least as high as the highest priority transaction which is waiting for the lock.

Finally, the property of priority inheritance is transitive. If, for example, T_H is blocked by transaction T_{HH}, and the priority of T_{HH} is less than T_R then T_{HH} will inherit the priority of T_R. Note that when priority inheritance is combined with Least Slack, continuously evaluated, T_H inherits not a static priority but a priority function which evaluates the slack of T_R.

High Priority. The idea of this policy is to resolve a conflict in favor of the transaction with the higher priority. The favored transaction, the winner of the conflict, gets the lock on the contested object. We implement this policy by comparing transaction priorities at the time of the conflict. If the priority of T_R is greater than the priority of object O, and thus greater than every transaction holding a lock on O, then we abort the lock holders thereby freeing the lock for T_R. T_R can resume processing; the lock holders are rolled back and scheduled for restart. If the priority of T_R is less than or equal to the priority of O then T_R blocks to wait for O to become free.

An interesting problem arises when we use Least Slack to prioritize transactions. Recall that under this policy, a transaction's priority depends on the amount of service time that it has received. Rolling back a transaction to its beginning reduces its effective service time to 0 and raises its priority under the Least Slack policy. Thus a transaction T_H, which loses a conflict and is aborted to allow a higher priority transaction T_R to proceed, can have a higher priority than T_R immediately after the abort. The next time the scheduler is invoked, T_R will be preempted by T_H. T_H may again conflict with T_R initiating another abort and rollback.

Our solution to this problem is to compare the priority of T_R against that of each lock holder assuming that the lock holder were aborted. Using the notation $P(T_H)$ to denote the priority of T_H and $P(T_H^A)$ to denote the priority of T_H were it to be aborted, we can write this algorithm as follows:

High Priority Conflict Resolution Policy

IF for any T_H holding a lock on O
$P(T_H) < P(T_R)$ **AND** $P(T_H^A) < P(T_R)$
THEN Abort each lock holder
ELSE T_R blocks

Conditional Restart Sometimes High Priority may be too conservative. Let us assume that we have chosen the first branch of the algorithm, i.e., T_R has a greater priority than T_H and T_H^A. We would like to avoid aborting T_H because we lose all the service time that it has already consumed. We can be a little cleverer by using a conditional restart policy to resolve conflicts. The idea here is to estimate if T_H, the transaction holding the lock, can be finished within the amount of time that T_R, the lock requester, can afford to wait. Let S_R be the slack of T_R and let $E_H - U_H$ be the estimated remaining time of T_H. (Recall that U denotes the elapsed service time of a transaction.) If $S_R \geq E_H - U_H$ then we estimate that T_H can finish within the slack of T_R. If so, we let T_H proceed to completion, release its locks and then let T_R execute. This saves us from restarting T_H. If T_H cannot be finished in the slack time of T_R then we restart T_H (as in the previous algorithm). This modification yields the following algorithm:

Conditional Restart Conflict Resolution Policy

IF $P(T_H) < P(T_R)$ **AND** $P(T_H^A) < P(T_R)$
THEN
  IF $E_H - U_H \leq S_R$
  THEN T_R blocks
    T_H inherits the priority of T_R
  ELSE Abort T_H
ELSE
  T_R blocks

Note that if T_R blocks in the inner branch, then T_H inherits the priority of T_R. This inheritance is exactly the same as described in the Wait Promote algorithm.

We only implement Conditional Restart if the conflict is one-on-one, i.e., there is no read group involved. We chose this because it is NP-complete to choose a maximal subset of readers all of which can finish within the slack of the requester. Furthermore we do not consider chained blockings as we did in our earlier study [1]. That is, we only make the special Conditional Restart decision if the requester conflicts with exactly one lock holder and the lock holder is not blocked waiting for some other lock. Experience with our earlier simulations indicated that chained blockings were rare, so that the payoff for handling them in a clever way was limited.

3.3. IO Scheduling

In a non memory resident database system, the disk is an important resource which can be managed to optimize various performance criteria. In conventional systems the usual goal is to maximize the throughput of the IO system. One way that this is accomplished is by using a disk scheduling algorithm (e.g., SCAN [11]) to order the sequence of IO requests so that the mean seek time is minimized. While this may be good for maximizing throughput, it may be bad for a real-time system which is trying to meet transaction deadlines. For example, SCAN may order a batch of requests so that an IO request from a transaction with an early deadline is serviced last.

In this paper we study the consequences of using algorithms to schedule IO requests based on transaction priorities as opposed to minimization of disk head seek times. We have looked at two ways to schedule the IO queue.

FIFO. When FIFO is used to schedule the IO queue, requests are serviced in the order in which they are generated. This service order is somewhat related to transaction priorities because IO requests are generated by the

CPU, which is scheduled by priority. The ordering is essentially random with respect to track position on the disk.

Priority. Under this policy each IO request has a priority which is equal to the priority of the transaction which issued the request. The next IO request to service is the one with the highest priority. Thus a newly arrived request from a transaction with a high priority can leapfrog over other requests which have been waiting longer in the IO queue. We also expect this ordering to be random with respect to track positions on the disk.

In our model there are two types of IO requests: reads that are issued by unfinished transactions, and writes that are generated by committed transactions that are flushing their updates back to disk. (The log resides on a separate device, so it receives only log writes which are serviced FCFS.) Giving higher priority to reads over writes is desirable because it will speed the completion of transactions which are trying to meet their deadlines. Giving high priority to writes does not enhance performance directly because the transactions which issued the writes have already committed. In fact, as our studies have shown, giving high priority to writes can decrease performance if it excessively delays the servicing of read requests. The priority of writes cannot be too low however as writes must be completed in order to free space in the memory buffer pool.

In our experiments writes have the same priority as the transaction that issued them. For the priorities FCFS and Earliest Deadline, this means that writes are given a relatively high priority. (The arrival times and deadlines of committed transactions are usually earlier than those of uncommitted transactions.) Because we use static evaluation to implement Least Slack, the slack times of committed transactions are not necessarily larger or smaller than those of uncommitted transactions.

4. Simulation Model

Our program to simulate a RTDB was built using SIMPAS, a discrete system simulation language [3]. The names and meanings of the four parameters that control the configuration of the system resources are given in Table 1. The database log is maintained on a separate device which is of equal speed as the database disks. Each disk has its own queue of service requests.

Parameter	Meaning	Base Value
DBsize	# of pages in database	400
MemSize	# of pages in memory buffer	200
NumDisks	# of disks	2
IOtime	Time to access disk	25 ms.

Table 1. System Resource Parameters.

The database buffer pool is modeled as a set of pages each of which can contain a single database object. We do not model each buffer page individually, that is, we do not maintain a free list of pages, nor do we keep track of which pages have been modified. Instead we model the buffer pool as a collective set. When a transaction attempts to read an object, the system generates a random boolean variable which has the value true with probability *MemSize / DBsize*. If the value is true then the page is in memory and the transaction can continue processing. If the value is false then an IO service request is created and placed in the input queue of the appropriate disk. The database is partitioned equally over the disks and we use the function

$$D = \left\lceil \frac{i * NumDisks}{DBsize} \right\rceil$$

to map an object i to the disk where it is stored.

Transactions characteristics are controlled by the parameters listed in Table 2. Transactions enter the system with exponentially distributed inter-arrival times and they are ready to execute when they enter the system (i.e., release time equals arrival time). The number of objects accessed by a transaction is chosen from a normal distribution with mean *Pages* and the actual database items are chosen uniformly from the database. Each page is updated with with probability *Update*. Pages which are updated are locked exclusively, other pages are locked in shared mode. Updated pages are stored in the buffer pool until a transaction commits and then they are flushed out to disk.

Parameter	Meaning	Base Value
ArrRate	Mean arrival rate	7 jobs/sec
Pages	Mean # pages accessed per job	8
CompFactor	CPU service per page	15 ms.
Update	Probability page is updated	0.5
MinSlack	Min slack as fraction of R	2
MaxSlack	Max slack as fraction of R	8
EstErr	Error in E as fraction of R	0
Restart	Time to rollback transaction	5 ms.
MaxActive	Limit # of active jobs	25

Table 2. Transaction Parameters.

A transaction has an execution profile which alternates lock requests with equal size chunks of computation, one for each page accessed. Thus the total computation time is directly related to the number of items accessed. Let C denote the CPU requirement for a transaction; then $C = Pages' * CompFactor$. (We use $Pages'$ to denote the actual number of pages for a specific transaction rather than the mean.) The IO service requirement for a transaction has two components: IO requests to read pages from the disk into memory and IO requests to write the modified pages back to the disk. Since the writing of updates back to disk occurs after a transaction commits, this IO time is not included in the runtime estimate. Thus the total amount of pre-commit IO service time needed is $I = IOtime * Pages' * (1 - MemSize / DBsize)$ The total runtime service needed to commit a transaction executing in an unloaded system is $R = C + I$. The accuracy of a transaction's runtime estimate E with respect to R is controlled by the parameter *EstErr*, $E = R * (1 + EstErr)$. How we choose the value of *EstErr* is explained later when we discuss the experimental results.

The assignment of a deadline is controlled by two parameters *MinSlack* and *MaxSlack* which set a lower and upper bound respectively on a transaction's slack time. A deadline is assigned by choosing a slack time uniformly from the range specified by the bounds. The program does not account for time needed to execute the lock manager, conflict manager, and deadlock detection manager. These routines are executed on a per data object basis and we assume that the costs of these calls are included in the variable that states how much CPU time is needed per object that a transaction accesses. Context switching and the time to execute the scheduler is also ignored.

Deadlocks are detected by maintaining a wait-for graph and searching for cycles whenever a new arc is added to the graph. When a deadlock is detected a victim is selected by choosing the transaction with lesser priority of the two transactions which correspond to the arc which completed the cycle in the graph.

In the following sections we discuss some of the results of eight different experiments that we performed. Due to space considerations we cannot present all our results but have selected the graphs which best illustrate the differences and performance of the algorithms. For each experiment we ran the simulation with the same parameters for 20 different random number seeds. Each run, except for the input step function experiment, continued until at least 700 transactions were executed. For each algorithm tested, numerous performance statistics were collected and averaged over the 20 runs. In particular we measured the percentage of transactions which missed their deadlines, the average amount of time by which transactions missed their deadlines, and the number of restarts caused by lock conflicts. The percentage of missed deadlines is calculated with the following equation: $\%missed = 100 * (tardy\ jobs\ /\ 700)$.

Another metric that we use is the average tardy time of all committed transactions. A transaction that commits before or on its deadline has a tardy time of zero. It is these averages that are plotted in the following figures.

Note that we are not particularly interested in transaction response times as conventional performance evaluations of concurrency control mechanisms are. The reason is that response time is not critical as long as a transaction meets its deadline. We are interested in learning how the various strategies are affected by load, the percentage of updates per transaction, the size of the memory buffer, and error of the runtime estimate.

For many of the experiments the base values for parameters were as shown in tables 1 and 2. These values are not meant to model a specific real-time application but were chosen as reasonable values within a wide range of possible values. We chose the arrival rate so that the corresponding CPU utilization (an average 0.84 seconds of computation arrive per second) is high enough to test the algorithms. It is more interesting to test the algorithms in a heavily loaded rather than lightly loaded system. (We return to this issue in the conclusions section.)

Section 3 proposed three different methods for assigning priority and four methods for managing concurrency. Also IO scheduling can be done in two different ways. Taking the cross product yields 24 different algorithms. Table 3 summarizes the methods of Section 3 and provides the abbreviations that we will use when referring to them.

Priority	FCFS - First Come First Serve ED - Earliest Deadline LS - Least Slack
Concurrency	W - Wait WP - Wait Promote HP - High Priority CR - Conditional Restart
IO Scheduling	FIFO Priority

Table 3. Summary of Scheduling Policies.

5. Experimental Results

Effect of Increasing Load In this experiment we varied the arrival rate from 6 jobs/sec to 8 jobs/sec in increments of 0.5. The other parameters had the base values given in Tables 1 and 2. This set of parameters is designed to load the CPU more heavily than the IO system. The CPU utilization ranges from 0.72 to 0.96 seconds of computation arriving per second. The IO system experiences a range in utilization of 0.60 to 0.80 seconds of I/O service requests arriving per second.

Figures 1 and 2 graph the three priority policies for each of the four different concurrency control strategies. Our first observation is that FCFS consistently performs worse than ED and LS. Also, where the other priority algorithms perform better with the concurrency control strategies that use promotion, FCFS does not. This is understandable as FCFS is a poor notion of priority for real-time transactions.

The relative performance of ED and LS depends on the type of concurrency control that is used. Fortunately we can state some general rules as to when ED or LS may be better. The policy ED operates best under low levels of load and when the concurrency control strategy produces few restarts due to lock conflict resolution. At high load levels ED tends to assign high priorities to transactions which will miss their deadlines anyway. This can prevent transactions with later, but feasible, deadlines from finishing on time. Conflict resolution policies which can produce a lot of restarts (e.g., HP and CR) work against ED because the restarts effectively increase the arrival rate of new transactions into the system, thus increasing the load. Under HP and CR, LS produces fewer restarts than ED because LS makes use of an extra test that can prevent transactions from being restarted (see Section 3).

Examining the graphs, we see that ED is better than LS when the simple Wait strategy is used for concurrency control (Figure 1). When promotion is added to the Wait strategy, ED and LS perform comparably although ED is slightly better at lower loads and LS slightly better at higher loads. Both policies are much better than FCFS.

Under the two policies which restart transactions when conflicts occur, HP and CR, we note that LS is clearly the better policy when the load is high, Figure 2. Strategy ED performs well when the load is low but its performance degrades rapidly once the CPU utilization is greater than 0.84.

Having discussed the priority policies we now turn our attention to the concurrency control policies. The performance of a concurrency control strategy depends on the priority policy used, so it is hard to compare concurrency without including priority. We will compare combinations of priority (ED and LS) and the four concurrency control strategies. We ignore FCFS because of its poor performance. Figure 3 graphs the eight combinations.

The reader can see for himself that no one combination is always better. ED/WP is best when the load is light but LS/WP is better when the load is high. Thus we see that WP works well with both LS and ED, but CR works better with LS; HP is not very good with either priority policy.

Ideally we want our algorithms to schedule all transactions such that all deadlines are met. However, if this is not possible, then we would want to minimize the amount by which tardy transactions miss their deadlines. Figure 4 graphs the average overdue time in seconds against arrival rate. With the Wait Promote strategy, we observe that ED has the least average overdue time, then LS and finally FCFS. These results are not surprising for it is known that ED minimizes the maximum task tardiness and LS maximizes the minimum task tardiness [4]. The performance of the different concurrency strategies with respect to task tardiness parallels what we observed with missed deadlines, namely, W and WP are best with ED, and WP and CR are best with LS.

In the experiment described above, the priority policy LS was implemented using static evaluation. We are also interested in learning how LS performs when continuous evaluation is used.

Figure 5 compares LS/WP and LS/CR for both static and continuous evaluation. The reader can see that no algorithm is best under all load settings. Under low arrival rates, the continuously evaluated versions of LS work better than the versions that use static evaluation. However, under higher loads, the opposite is true. This is an interesting and counter-intuitive result. The explanation is that transactions executing under LS, continuously evaluated, can experience priority reversals. We illustrate with an example.

At time t let T_1 and T_2 be transactions with slack times of 5 and 6 seconds respectively. Since T_1 has the lesser slack and thus the greater priority, it gains the processor and begins to execute. Two seconds later a new transaction arrives. The scheduler is invoked and slack times are recalculated. The slack of T_1 is unchanged, the slack of T_2 has decreased by 2 to 4. So now T_2 has the greater priority and gains the processor. A similar event may occur 2 seconds later which would cause the priorities of T_1 and T_2 to reverse again. If T_1 and T_2 are the highest priority transactions in the system, this continuous method of priority evaluation virtually guarantees that the two transactions are executed concurrently in the manner just described. If T_1 and T_2 have overlapping read and write sets then this scheduling method increases the likelihood that a conflict will occur. It also increases the likelihood that a deadlock will occur. Both of these events are undesirable.

Biasing the Run Time Estimate The concurrency control strategy Conditional Restart uses the runtime estimate to decide if a low priority transaction holding a lock can finish within the slack time of the higher priority transaction requesting the lock. We can easily describe the behavior of CR for the extreme values of E. When E = 0, CR will always judge (assuming that the lock requester has a positive slack, this is true if the transaction has not missed its deadline) that the lock holder can finish within the slack of the lock requester. Thus the lock requester will always wait for the lock holder. Since promotion is used by CR, this behavior is exactly the same as WP.

At the other extreme, when E is much larger than R, the algorithm will nearly always judge that the lock holder cannot finish within the slack time of the lock requester. Thus the lock holder will be rolled back and restarted. This behavior is the same as the concurrency control strategy HP. When the value of E is not so extreme CR will behave somewhere between WP and HP.

The runtime estimate is also used by the priority policy Least Slack. We performed a number of experiments to study how biasing the runtime estimate affects the performance of LS. Our results showed that LS, under static evaluation, is surprisingly robust to inaccuracy in the estimate. To illustrate, we summarize a "worst case" experiment that yielded the largest performance degradation.

Half of the transactions were made to overestimate their runtime estimate $E = R * (1 + EstErr)$; the other half underestimated $E = R * (1 - EstErr)$. (Recall that R is the service time needed in an unloaded system. If *all* transactions over or underestimate, LS is not affected greatly.) The value of *EstErr* was varied from 0 to 5 in increments of 1. (Negative runtime estimates were converted to $E = 0$.) Figure 6 shows that even with this worst case bias, LS is not very sensitive to errors in the runtime estimate.

When continuous evaluation is used, the performance of LS is more sensitive to error in the runtime estimate. This is understandable since continuous evaluation means that the inaccurate runtime estimate will be used many more times to make scheduling decisons.

Effect of Increasing Memory In this experiment we varied the value of *MemSize* from 200 to 400 in increments of 50. The other parameters had the values shown in Tables 1 and 2. Thus the size of the memory resident fraction of the database varied from 0.5 to 1.0 of the total database.

Note that the proportion of memory resident database influences the assignment of deadlines to transactions. This happens because deadlines are assigned with respect to the total service time, both CPU and IO, required by a transaction. As the memory increases, the IO service requirement decreases (pages are more likely in memory), and the deadlines are shortened proportionately. In this way we can compare the scheduling of sets of transactions with similar task urgencies in system configurations of various memory sizes.

Our expectation, as memory size increases, is that all scheduling algorithms will perform better. The reason is that transactions will be doing much less IO and the time spent waiting for service in IO queues decreases. (CPU utilization and waiting time are unchanged.) Thus transactions will receive quicker service and are more likely to meet their deadlines.

Figure 7 graphs the results for the four algorithms which schedule all transactions and use LS to assign priority. It confirms our expectation that algorithms will perform better as the memory size increases. The results for other algorithms are similar.

Effect of Fraction of Pages Updated The value of *Update*, by controlling the size of the read and write sets, impacts the probability of conflict between two concurrent transactions. The value of *Update*, by determining the size of the writeset, also contributes to the IO load of the system. This is because updates are written to disk after a transaction commits.

In this experiment we varied the value of *Update* from 0.2 to 0.8 in increments of 0.1. When *Update* = 0.2, only 20% of a transaction's pages are locked exclusively. When *Update* = 0.8 80% of the pages accessed by a transaction are updated. The other parameters had the values shown in Tables 1 and 2. In this experiment the IO load varies from 0.49 to 0.91.

Figure 8 shows the four different types of concurrency control with the ED priority policy and with IO scheduling done by priority. The general shape of the curves is as expected: when *Update* is large, the performance is poor because the rate of conflict is high and because the overall IO load is high. As we saw earlier ED performs best with the concurrency strategies W and WP which cause fewer restarts than HP and CR. When *Update* is small, the conflict rate is lower and all algorithms perform equally well.

Figure 9 plots ED/W and ED/WP with priority IO scheduling against ED/W and ED/WP with FIFO IO scheduling. Somewhat surprising is the result that the versions with IO scheduling perform worse than the versions without IO scheduling in the high update (right side) and thus high IO load part of the scale. The result is explainable, however, when we note that at this end of the scale most of the IO, roughly 62%, consists of writing modified pages back to disk. As we discussed in Section 3, giving high priority to writes can delay the service of read requests with lower priorities. One conclusion that we can draw from this is that it may be advisable to give IO writes lower priorities than reads. We do not investigate this issue any further in this paper.

Performance Under a Sudden Load Increase In the previous experiments, we studied the performance of the various algorithms under an increasing but steady load. In this experiment we study the algorithms under a different type of load increase namely a batch of arrivals in an otherwise idle system. To simulate this input step function we programmed the simulator so that a set of transactions arrived all at the same time. The system then executed all the jobs in the set. The number of jobs in the set varied from 20 to 60 in steps of 10. The parameters controlling job characteristics were the same as shown in tables 1 and 2 except for the following differences: *Pages* = 10, *CompFactor* = 10, and *MaxSlack* = 30. The major difference here is the change in *MaxSlack*. This was necessary to produce a set of transactions with a large amount of variability in deadlines.

In reality, we would not expect all jobs in an overload to arrive at the same instant, nor would we expect the system to be completely idle at this time. However, our idealized step function model lets us study the impact of an overload without distracting second order effects. If an algorithm performs well under a step function, we could also expect it to do well in a system that has plenty of spare capacity under normal circumstances but is suddenly faced with a flurry of jobs (e.g., a radar system facing a sudden all-out enemy attack).

A good strategy for operation in this situation might be to employ an overload management policy to abort transactions which have missed their deadlines. If these transactions must be executed anyway then they should be restarted only after the spike has passed. We may also want to limit priority inversions from occurring. The reasoning is that if a high priority job blocks, it will almost certainly miss its deadline if forced to wait on a low priority job. Instead we may want to abort the low priority job and restart it later when the load has decreased.

Figure 10 shows the four different concurrency control policies for both the ED and LS priority assignment policies. IO scheduling is done according to job priority. It is easy to see that ED/HP is clearly the best algorithm. HP is also the best version of the LS algorithms The same results were observed when no overload management policy was used, although the effect was not as pronounced.

Effect of IO Scheduling In this experiment we compared the versions of scheduling algorithms which use transaction priorities to schedule IO requests against versions which service IO requests using FIFO. To better study the effects, we changed the base values of three parameters to create a configuration where the disks are much more loaded then the CPU. The parameters and the new values were: $Pages = 13$, $CompFactor = 5$, and $Update = 0.2$. (The other parameters had the values shown in Tables 1 and 2.) Thus the average transaction reads more pages and does less computation. The fraction of pages updated is less so the disks service more read requests than writes. The system load was varied by increasing the arrival rate from 6 jobs/sec to 8 jobs/sec in steps of 0.5. Hence, the CPU utilization varied from 0.39 to 0.52, and the IO utilization varied from 0.68 to 0.91.

Figure 11 plots the four algorithms which use LS and FIFO against the same four algorithms using LS and Priority IO scheduling. The results clearly show that using transaction priorities to schedule IO requests can yield significant performance improvements over scheduling using FIFO. This is especially true for the two concurrency algorithms which use promotion, i.e., WP and CR. This is understandable since a transaction which inherits the higher priority of a blocking transaction will receive better service from the disk and will execute faster. This unblocks the high priority transaction sooner, and it has a better chance of meetings its deadline.

Priority is better than FIFO but both methods ignore disk head position and do not minimize the average seek time. Would a disk scheduling algorithm which minimizes average seek time, (e.g., SCAN) but is ignorant of transaction time constraints, yield better performance in a real-time system than our algorithm which considers transaction time requirements? We cannot answer this question directly but we have performed an experiment which indicates how much faster the average seek time of a SCAN type algorithm would have to be for it to match the real-time performance of our Priority algorithm. In this experiment we fixed the parameter settings at the values they had in the previous experiment and set $ArrRate = 8$ job/sec. Then we increased the speed of the disks by decreasing the value of *IOtime* from the initial 25 ms. to 20 ms. in steps of 1. Figure 12 plots the performance of LS/WP/FIFO, which, like a SCAN type algorithm, does not use transaction priorities for IO scheduling. The horizontal line is the performance of LS/WP/Priority when $IOtime = 25$ ms. The two curves intersect at $IOtime = 22.8ms$ indicating that a SCAN type algorithm must have an average seek time roughly 9% less than Priority in order for it to yield comparable real-time performance. Of course it is possible to design a SCAN type algorithm which also uses transaction priorities, but we have not investigated this possibility.

6. Conclusions

In this paper we have presented various transaction scheduling options for a real-time database system. Our simulation results have illustrated the tradeoffs involved, at least under one representative database and transaction model. Before reaching some general conclusions, we would like to make two observations.

The first observation is that our base parameters represent a high load scenario. One could argue that such a scenario is "unrealistic." However, we believe that for designing real-time schedulers, one must look at precisely these high load situations. Even though they may not arise frequently, one would like to have a system that misses as few deadlines as possible when these peaks occur. In other words, when a "crisis" hits and the database system is under pressure is precisely when making a few extra deadlines could be most important [8].

It could also be argued that some of the differences between the various scheduling options is not striking. In many cases, the difference between one option and another one is a few percentage points. If we were discussing transaction response times, then a say 10 percent improvement would not be considered impressive by some. However, our graphs show missed deadlines (in most cases) and we believe that this is a very different situation. Again, the difference between missing even one deadline and not missing it could be significant. Thus, if we do know that some scheduling options reduce the number of missed deadlines, why not go with the best one?

And which are the best options? It is difficult to make any absolute statements, but we believe the following statements hold under most of the parameter ranges we tested. (All our additional results not shown in this paper also substantiate these statements.)

When the load is continuous and steady:

(a) Of the tested priority policies for real-time database systems, Least Slack (LS) is the best overall. It always performed better than FCFS, and was better than ED under the higher load ranges. Earliest deadline (ED) is the second choice for assigning priorities.

(b) Of the concurrency control policies we tested, Wait Promote (WP) is the best overall. It performs very well when combined with either LS or ED. Conditional Restart (CR) is the second choice for LS. CR also performs well with ED at low loads but poorly at high loads. Wait is the second choice for ED. High Priority (HP) does not do well with either LS or ED.

(c) Although LS performed better at minimizing the number of late transactions, ED is the best choice for minimizing the mean tardy time of commited transactions.

(d) When the IO system is heavily loaded, using real-time transacion priorities to schedule IO requests yields significant performance gains over scheduling IO requests in a FIFO manner.

(e) Using continuous evaluation to implement LS yields better performance when the load is low. Static evaluation achieves better performance when the load is high. Also, static evaluation is less sensitive to error in the runtime estimate than is continuous evaluation.

When the load is an "input step function":

(f) ED is the best priority policy to use. It performs better than LS at the higher load settings for three of the four concurrency control policies.

(g) HP is the best concurrency control policy. It is the best of the four when combined with either ED or LS. The second and third choices for concurrency are CR and WP.

Acknowledgments

The authors thank the referees for their helpful comments.

This research was supported by the Defense Advanced Research Projects Agency of the Department of Defense and by the Office of Naval Research under Contracts Nos. N00014-85-C-0456 and N00014-85-K-0465, and by the National Science Foundation under Cooperative Agreement No. DCR-8420948. The views and conclusions contained in this document are those of the authors and should not be interpreted as necessarily representing the official policies, either expressed or implied, of the Defense Advanced Research Projects Agency or the U.S. Government.

References

1. Abbott, Robert and Hector Garcia-Molina, "Scheduling Real-time Transactions: a Performance Evaluation," *Proceedings of the Conference on Very Large Database Systems*, pp. 1-12, VLDB, August 1988.
2. Bernstein, Philip A. and Nathan Goodman, "Timestamp-Based Algorithms For Concurrency Control In Distributed Database Systems," *Proceedings of the Conference on Very Large Databases*, pp. 285-300, IEEE, Montreal, Canada, Oct., 1980.
3. Bryant, R. M., *SIMPAS 5.0 User Manual*, University of Wisconsin-Madison.
4. Coffman, Edward G. Jr. and Peter J. Denning, *Operating Systems Theory*, Prentice Hall, 1973.
5. Eswaran, K. P., J. N. Gray, R. A. Lorie, and I. L. Traiger, "The Notions of Consistency and Predicate Locks in a Database System," *CACM*, vol. 19, no. 11, pp. 624-633, November 1976.
6. Garcia-Molina, Hector, "Using semantic knowledge for transaction processing in a distributed database," *Transactions on Database Systems*, vol. 8, pp. 186-213, ACM, June, 1983.
7. Haerder, Theo and Andreas Reuter, "Principles of Transaction-Oriented Database Recovery," *Computing Surveys*, vol. 15:4, pp. 287-317, ACM, 1983.
8. Jensen, E. Douglas, "The Future Reality of Real Time," *UNIX Review*, January 1989.
9. Jensen, E. Douglas, C. Douglass Locke, and Hideyuki Tokuda, "A time-driven scheduler for real-time operating systems," *Proceedings Real-time Systems Symposium*, pp. 112-122, IEEE, 1986.
10. Liu, C.L. and J.W. Wayland, "Scheduling algorithms for multiprogramming in hard real-time environment," *Journal of the ACM*, vol. 20, pp. 46-61, ACM, January, 1973.
11. Peterson, James L. and Abraham Silberschatz, *Operating System Concepts*, Addison Wesley, 1986.
12. Sha, Lui, Ragunathan Rajkumar, and John P. Lehoczky, "Priority Inheritance Protocols: An Approach to Real-Time Synchronization," *Tech. Report*, vol. CMU-CS-87-181, CMU, Pittsburgh, December 1987.
13. Stankovic, John A., "Misconceptions About Real-time Computing: A Serious Problem for Next-Generation Systems," *Computer*, vol. 21, pp. 10-19, IEEE, October 1988.
14. Zhao, W., K. Ramamritham, and J. A. Stankovic, "Preemptive Scheduling Under Time and Resource Constraints," *Transactions on Computers*, vol. C-36, pp. 949-960, IEEE, August 1987.

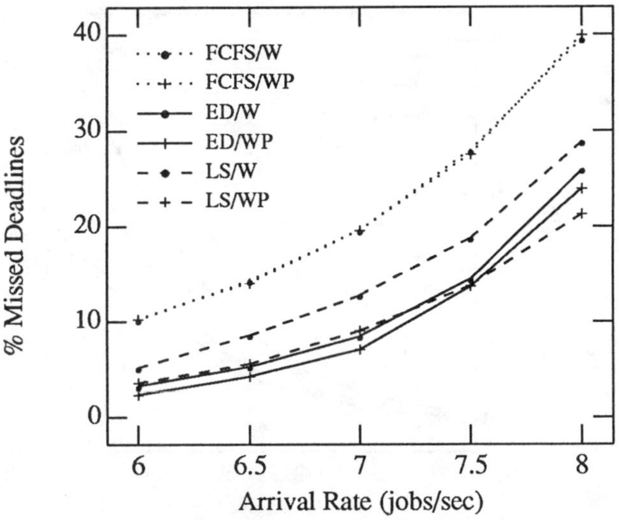

Figure 1. FCFS, ED, and LS each with W and WP.

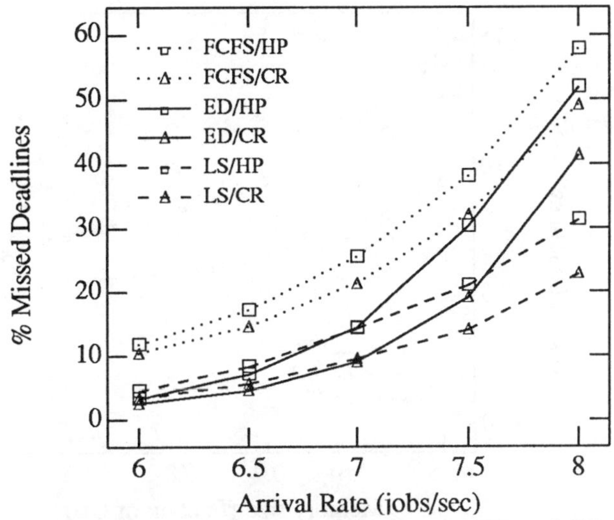

Figure 2. FCFS, ED, and LS each with HP and CR.

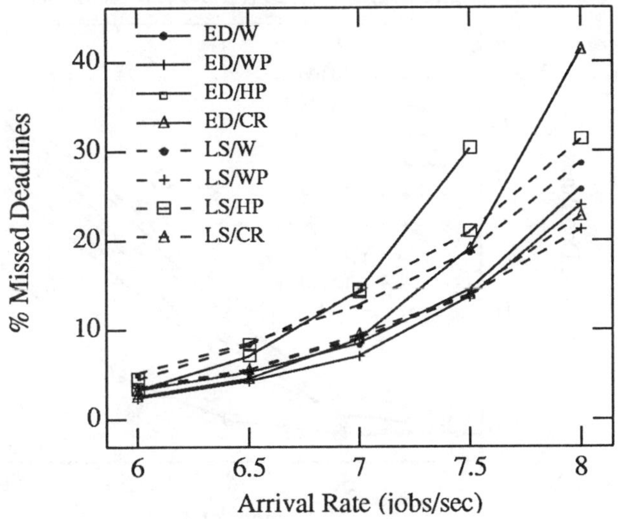

Figure 3. ED vs LS, each with W, WP, HP, CR.

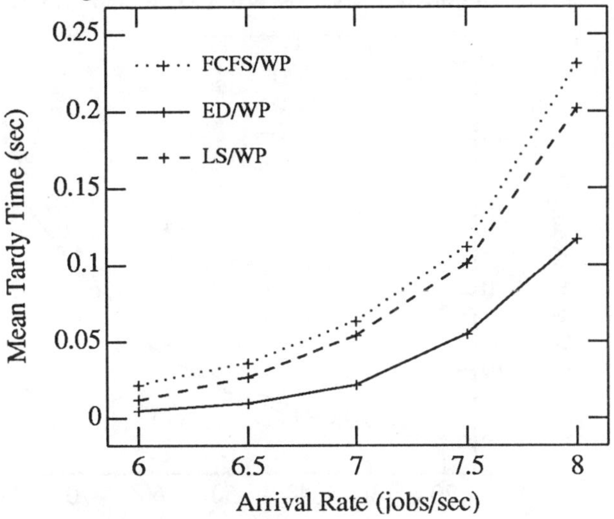

Figure 4. FCFS/WP, ED/WP, and LS/WP.

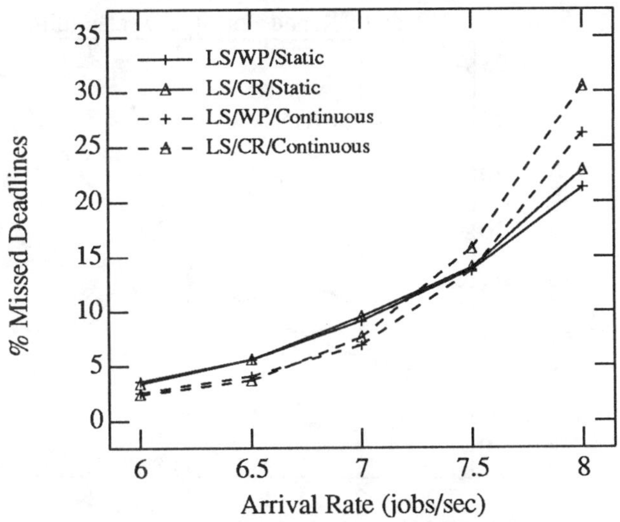

Figure 5. LS/WP and LS/CR under static and continuous evaluation.

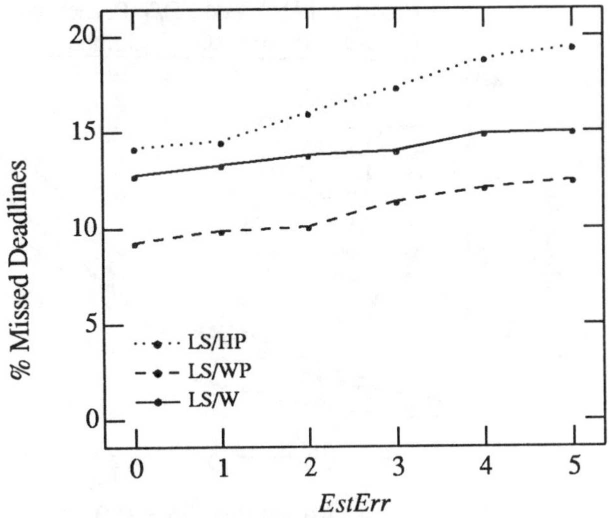

Figure 6. LS/W, LS/WP, and LS/HP under static evaluation.

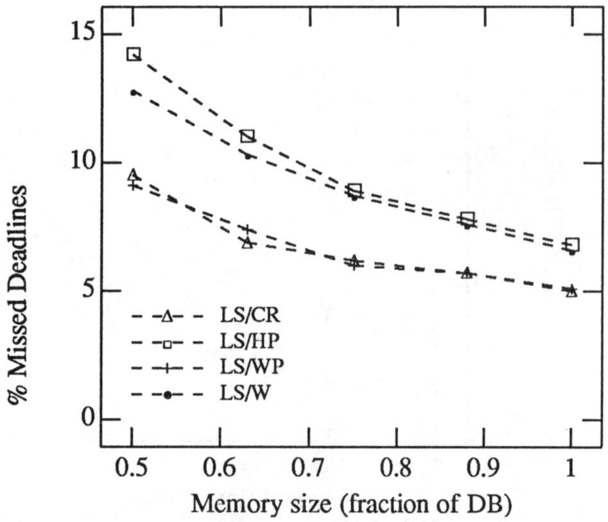

Figure 7. LS/W, LS/WP, LS/HP, and LS/CR.

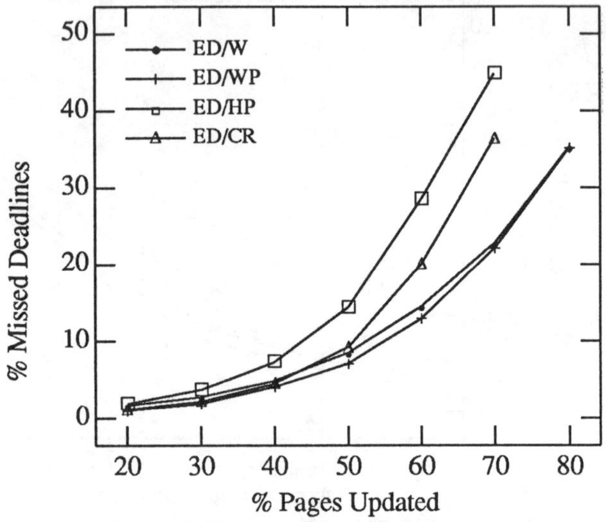

Figure 8. ED/W, ED/WP, ED/HP, and ED/CR.

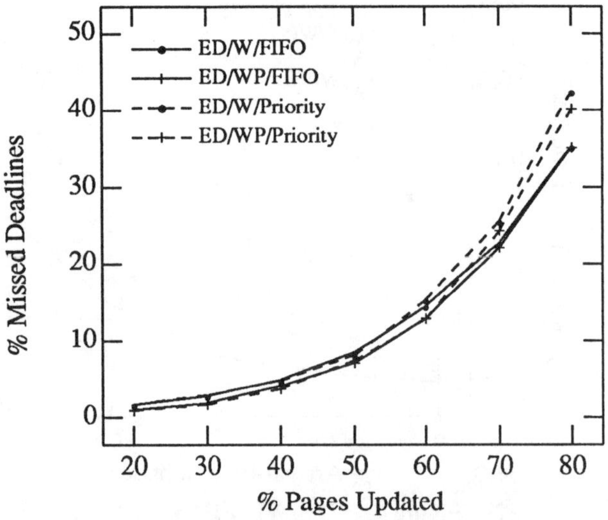

Figure 9. ED/W and ED/WP, each with FIFO and Priority IO scheduling.

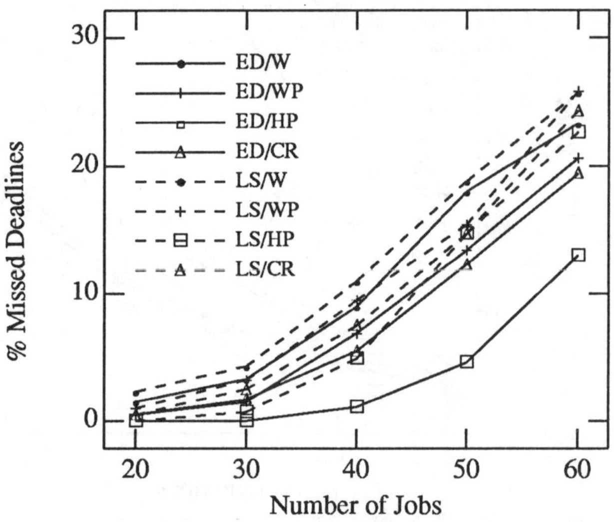

Figure 10. ED and LS, each with W, WP, HP, and CR, and Priority IO scheduling.

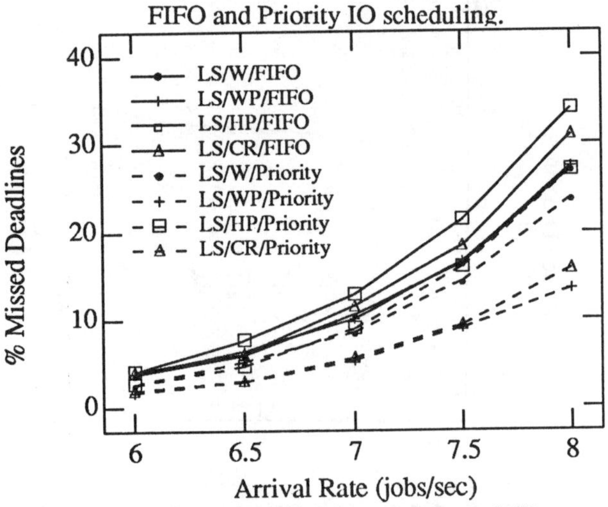

Figure 11. LS with W, WP, HP, and CR, under FIFO and Priority IO scheduling.

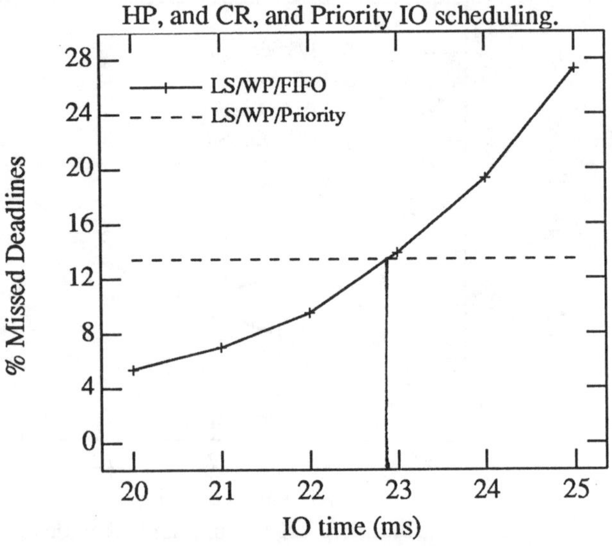

Figure 12. LS/WP.

Priority in DBMS Resource Scheduling

Michael J. Carey
Rajiv Jauhari
Miron Livny

Computer Sciences Department
University of Wisconsin
Madison, WI 53706

ABSTRACT — In this paper, we address the problem of priority scheduling in a database management system. We start by investigating the architectural consequences of adding priority to a DBMS. Specific priority-based schemes are then proposed for managing DBMS resources, including a priority-based disk scheduling algorithm and two approaches to adding priority to the buffer manager of a DBMS. We study the performance of our proposals through simulation, both individually and in combination. Our simulation results indicate that the objectives of priority scheduling cannot be met by a single priority-based scheduler. They show that, regardless of whether the system bottleneck is the CPU or the disk, priority scheduling on the critical resource must be complemented by a priority-based buffer management policy.

1. INTRODUCTION

1.1. Background

Priority scheduling of computer systems is a concept that has been studied extensively for more than two decades [Coff68, Klei76]. The priority of a task is a measure of its "importance" to the system, and the objective of priority scheduling is to provide preferential treatment to tasks with higher priority values. One use of priority is as an end in itself. In this case, task priorities are determined externally, and are used directly to guide the system in making scheduling decisions. For example, a system designer may decide to run interactive tasks at a higher level of priority than long batch jobs in order to ensure good responsiveness for interactive users. An alternative use of priority is as a means to an end. Here, task priorities are determined internally (i.e., assigned by the system itself) in order to meet some other goal. For example, priority scheduling can be used as a way of minimizing the number of missed deadlines in a soft real-time computing environment [Chan85].

This work was performed under subcontract to Xerox Advanced Information Technology, which was supported by the Defense Advanced Research Projects Agency and by the Rome Air Development Center under Contract No. F30602-87-C-0029. Additional support was provided by the National Science Foundation under grant IRI-8657323 and by the Digital Equipment Corporation through its Initiatives for Excellence program. The views and conclusions contained in this report are those of the authors and do not necessarily represent the official policies of the Defense Advanced Research Projects Agency, the Rome Air Development Center, or the U.S. Government.

Despite the attention that the concept of priority has received for CPU scheduling in operating systems, it has received very little attention in the database system area. This is somewhat surprising, as priority seems every bit as useful and applicable in a DBMS. The interactive versus batch job example is relevant; in a transaction processing system, it would seem equally desirable to ensure responsiveness for interactive DBMS users while allowing batch jobs to be run as well. In addition, the database and real-time systems communities are beginning to show interest in applying database technology to data-intensive, real-time applications such as stock trading, computer-integrated manufacturing, and command and control systems [Abbo88, SIGM88, Stan88].

Priority scheduling in a DBMS differs from the problem of priority CPU scheduling due to the heterogeneity and multiplicity of DBMS resources. Important DBMS resources include the CPU, disks, and main memory, which are physical resources, and data items, which can be viewed as logical resources. It is common for each of these resources to be managed by an independent scheduler: the underlying operating system kernel schedules the CPU; the operating system or DBMS device drivers handle disk request scheduling; the buffer manager controls the main memory resource (i.e., buffer pool page frames); and, the concurrency control manager schedules accesses to data items. Thus, in a DBMS, there are actually four schedulers into which one could consider incorporating priority.

1.2. Related Work

As mentioned above, priority CPU scheduling for centralized systems has been studied extensively. Overviews of much of this work can be found in [Coff68, Klei76]. Classical approaches to scheduling a single CPU in the presence of priority involve maintaining the CPU queue as a priority queue. Requests are served in priority order either non-preemptively, which is called "head-of-line" or "non-preemptive priority" scheduling, or preemptively, which is known as "preemptive-resume" scheduling. The objective of the latter approach is to ensure that a higher priority job is never made to wait while a lower priority job receives service. Recent work has addressed priority scheduling in the presence of preemption overheads and multiple processors (e.g., in distributed systems [Chan85, Chan87]). Static and dynamic approaches to CPU scheduling for real-time systems have also been studied; this work is reviewed briefly in [Stan88], and [Jens86] provides an excellent treatment of dynamic CPU scheduling approaches for real-time systems. Resources other than the CPU have, for the most part, been ignored with respect to dynamic priority and real-time scheduling [Coff68, Stan88].

In the database area, real-time issues are just beginning to attract attention. [Wede86] proposed a scheme for prefetching data pages for canned transactions in a real-time DBMS; detailed a priori knowledge of the workload is assumed, and static transaction pre-analysis techniques are applied to it. Others have also begun to address real-time DBMS issues [SIGM88, Daya88, Buch89], especially in the area of concurrency control algorithms [Abbo88, SIGM88, Buch89]. Most notably, [Abbo88] proposed and investigated the performance of several priority-based locking (and CPU scheduling) schemes in the context of a memory-resident, real-time DBMS. However, the question of how priority can be used in managing the *physical* resources of a DBMS has not been addressed to the best of our knowledge.

1.3. Our Work

In this paper, we explicitly address the problem of priority scheduling in a DBMS. In particular, we are interested in how priority can be employed in managing the physical resources of a DBMS. The objective of our work is to develop and evaluate policies that provide the best possible service to higher-priority transactions while minimizing the negative impact on lower-priority transactions. Essentially, we would like the DBMS to behave as much like a preemptive-resume server as possible. That is, we would like it to give transactions of a given priority the same service that they would experience in a system that does not serve transactions of lower priority. Thus, the goals of this paper are three-fold. First, we investigate the architectural consequences of adding priority to a DBMS. Second, we develop specific priority-based algorithms for managing the key DBMS resources, especially the disk(s) and the buffer pool. Third, we study the performance of our proposed algorithms, both individually and in combination with one another.

Figure 1.1 depicts the overall architecture of a DBMS from a resource perspective. From the figure, it can be seen that there are several places where a DBMS makes resource scheduling decisions into which priority could be incorporated. First, a DBMS often controls the total number of transactions allowed to be active at any one time, requiring additional transactions to wait outside the system until they can be admitted, in order to prevent thrashing due to buffer pool over-utilization [Chou85]. Second, as discussed earlier, the DBMS or the underlying operating system must make CPU scheduling decisions; similarly, disk scheduling decisions must be made. Finally, the buffer manager must make page replacement decisions when a non-resident page is requested and there are no free page frames.

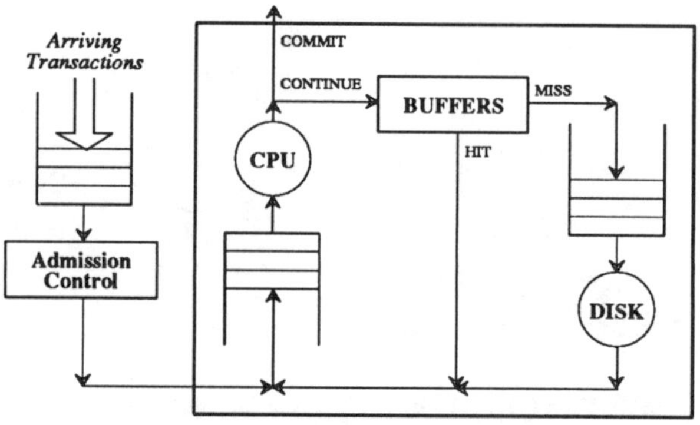

Figure 1.1: DBMS Resource Architecture.

In order to see how adding priority at some or all of these decision points might enable us to reduce the response time of high priority transactions, let us consider the composition of a transaction's response time (T_R). Basically, there are two major components of this time, the external waiting time due to admission control (T_{W_EXT}), and the time spent inside the system (T_{SYS}). T_{SYS} itself has two components, CPU time (T_{CPU}) and disk time (T_{DISK}).[1] The CPU time for a transaction consists of two subcomponents, the waiting time due to CPU contention (T_{W_CPU}) and the actual CPU service time (T_{S_CPU}). The disk time for a transaction is the product of the number of disk requests that it makes (N_{DISK}) and the average disk access time; the disk access time also consists of a waiting time (T_{W_DISK}) and a service time (T_{S_DISK}). Thus, we see that:

$$T_R = T_{W\_EXT} + T_{W\_CPU} + T_{S\_CPU} + N_{DISK}*(T_{W\_DISK} + T_{S\_DISK})$$

Incorporating priority into the admissions control decision, by ordering waiting transactions according to priority and preempting lower-priority transactions in favor of higher-priority transactions, is a way of reducing T_{W_EXT} for high-priority transactions. In terms of their CPU time, priority-based CPU scheduling can reduce T_{W_CPU} for high-priority transactions; T_{S_CPU} cannot be reduced, however. Similarly, the disk waiting time T_{W_DISK} for high-priority transactions can be reduced through the priority-based scheduling of disk requests; this may actually lead to an increase in T_{S_DISK}, though, as using a non-positional disk scheduling criterion may increase the expected seek time. Finally, buffer replacement decisions can potentially reduce N_{DISK} for high-priority transactions. Of course, none of these reductions for high-priority transactions will come for free. Rather, we must expect corresponding increases in the response time components for transactions of lower priority.

[1] Since our concern is physical resource scheduling, not concurrency control, we focus on read-only queries in this paper. Lock waiting time is therefore not a potential component of response time here.

The remainder of the paper is organized as follows: Section 2 discusses approaches to priority scheduling for the CPU, disk, and buffer resources. This section includes new, priority-based algorithms for disk scheduling and buffer management. Section 3 describes a simulation model for studying the performance consequences of our proposed algorithms. Section 4 presents a series of performance experiments that demonstrate the effectiveness and importance of the various algorithms under workloads of varying intensities for the different resources. Lastly, Section 5 summarizes the contributions of the paper and describes our plans for future work.

2. PRIORITY-BASED RESOURCE MANAGEMENT ALGORITHMS

As described in the introduction, there are three types of physical resources in a DBMS: the processors, the disks, and the main memory buffers. In this section, we describe priority scheduling algorithms for each of these resource types. We present a known algorithm for priority CPU scheduling, and we extend known algorithms for disk scheduling and buffer management to enable them to handle priorities.

2.1. CPU SCHEDULING

A number of options for priority CPU scheduling have been discussed in the operating systems literature [Coff68, Klei76, Pete86]. Preemptive-resume priority scheduling, where a high-priority job preempts a low-priority job, and the low-priority job resumes its CPU processing after all high-priority jobs have completed their processing, does not seem suitable for database systems. This is because preempting transactions while they hold short-term locks or latches may lead to the convoy problem, as described in [Blas79]. It seems more reasonable, then, to use a non-preemptive form of priority scheduling, while ensuring that high-priority requests do not suffer greatly as a result of the non-preemptability of low-priority requests. In order to reconcile priority with non-preemptability, the CPU can be scheduled according to a priority-based round-robin scheme where the length of a CPU slice is determined by the transaction and not by the scheduler. In this scheme, a transaction gives up the CPU at a "safe point" after a short burst of CPU use. Once the CPU is released, it is assigned to the transaction at the head of the CPU queue. The queue is managed in priority order with requests of the same priority being served in FCFS order.

2.2. DISK SCHEDULING

Classical disk scheduling schemes such as the *shortest seek time first* and *elevator* algorithms [Teor72, Pete86] attempt to minimize the average seek distance. Since the elevator algorithm performs especially well under high disk loads and has good fairness characteristics, it is a good candidate to serve as the basis of a priority scheduling algorithm at the disk. In the elevator algorithm, the disk head is either in an inward-seeking phase or an outward-seeking phase. While seeking inward, it services any requests it passes until there are no more requests ahead. The disk head then changes direction, seeking outward and servicing all requests in that direction as it reaches their tracks.

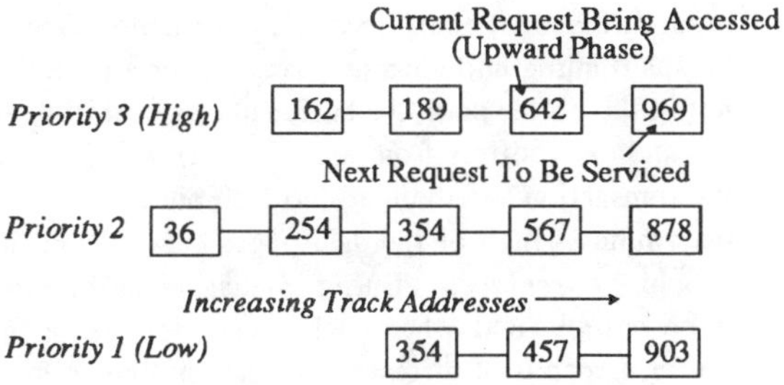

Figure 2.1: Priority-based Disk Scheduling.

In order to support priority, the elevator algorithm can be modified in the following way: disk requests are grouped on the basis of their priority, and the elevator algorithm is used within each group. There is one queue per priority level for buffering outstanding disk requests. Within each queue, requests are arranged in order of their physical (track) addresses. While seeking inward or outward, the disk services any requests that it passes in the currently-served priority queue until there are no more requests ahead. On the completion of each disk request, the scheduler checks to see whether a disk request of a higher priority is waiting for service. If such a request is found, the scheduler switches to the queue that contains the request(s) of the highest priority among those waiting and starts serving that queue. When it switches to a new queue, the request with the shortest seek distance from the head's current position is used to determine the direction in which the head will move.

Figure 2.1 shows the way that the disk scheduler organizes outstanding disk requests in the case of three priority levels. Priority increases as we move from the bottom to the top of the figure. An important side-effect of introducing priority in disk scheduling in this fashion is that the average seek time can worsen as the number of priority levels increases. As an example, consider Figure 2.1, where the disk is currently servicing the request to access track 642 in the queue of priority 3, and the disk head is moving towards higher track numbers. The next request serviced will access track 969, even though there are other requests (for track 878 of priority 2, and track 903 of priority 1) that could be handled earlier as the disk head moves in its upward phase. In the worst case, if each request had a different priority, the prioritized elevator algorithm described above would degenerate into straight priority scheduling. This suggests that in order to provide reasonable I/O performance, it will be preferable to map disk requests into a small number of priority levels at the disk scheduler (even if each transaction has a distinct priority in other parts of the system).

2.3. BUFFER MANAGEMENT

It was shown in [Effe84, Chou85, Sacc86] that the performance of a DBMS can be significantly influenced by the buffer management algorithm used. In general, buffer managers can make three types of decisions: *transaction admission*, *buffer allocation*, and *buffer replacement*. When a transaction arrives at the system, the buffer manager is asked if enough buffer space is available to allow the transaction to be *admitted* into the system.

The arriving transaction may then be blocked outside the system until sufficient memory becomes available. Once a transaction begins running and submits a request for a page that is currently not in the buffer pool, the buffer manager must determine the set of candidate buffers from among which one is to be *allocated* to the transaction. If there are no free buffers, the buffer manager determines which of the data pages currently in the buffer pool should be *replaced.* Buffer management becomes somewhat more complicated when each transaction has a priority, as the buffer demands of different transactions can no longer be treated equally. For example, when making the transaction admission decision, the buffer manager may choose to deallocate buffers from some low-priority transactions in order to allow a high-priority transaction to begin execution. Priorities may also be considered when making replacement and allocation decisions.

The Global LRU buffer management algorithm [Effe84] is simple and is the one most commonly implemented in commercial database systems. The DBMIN algorithm [Chou85], while more difficult to implement, was shown to outperform Global LRU as well as several more sophisticated algorithms (e.g., the "Hot Set" algorithm [Sacc86]). The DBMIN algorithm thus represents the sophisticated end of the buffer management algorithm spectrum, while Global LRU represents the simple, commonly used end. Together, they provide a good pair of algorithms into which the concept of priority may be introduced. In this section, we present two priority buffer management algorithms, the *Priority–LRU* policy, based on Global LRU, and the *Priority–DBMIN* policy, based on DBMIN. As this paper focuses on a query-only environment, both algorithms are described here assuming no updates.[2] *Priority–LRU* and *Priority–DBMIN* differ in their model of the information available to the buffer manager; as a consequence, the buffer pool is organized differently in the two cases. We begin by defining some terms that are needed to describe both algorithms. Then, for each priority-based algorithm, we review the basics of the original algorithm and describe our changes to the buffer pool organization and the policies for transaction admission, buffer replacement, and buffer allocation.

2.3.1. Definitions and Assumptions

At any point in time, the transactions that have accessed a particular resident page (since it was last brought into memory) and are still executing are called the *users* of the page. The *owner* of the page is the transaction with the highest priority among the *users* of that page. A page is *free* if does not have any users. The buffer manager associates a global *timestamp* with each resident page in order to keep track of the recency of usage of pages. A global counter called *bufSequence* is maintained to assign timestamps to resident pages. The *bufSequence* counter is initially set at 0 at system startup time. Each time the data in any of the buffers is accessed, *bufSequence* is incremented and its new value is inserted as the *timestamp* of the data page. Thus, the larger the value of the *timestamp* of a page, the more recently the page was accessed.

A transaction may be admitted to the system right away, or it may be blocked initially by the transaction admission policy. Once a transaction is allowed to begin execution, it continues until it commits or until it is *suspended*. A transaction is said to be *suspended* by the buffer manager if it is temporarily prohibited from making further buffer requests; the buffers owned by the transaction (except those that it has fixed) are freed. The buffer manager considers *reactivating* suspended transactions at the same decision point that it considers admitting blocked transactions, which is whenever a running transaction completes. A reactivated transaction resumes its execution at the point where it was suspended.

2.3.2. The Priority-LRU Algorithm

Global LRU [Effe84] is a buffer replacement policy based on the assumption that there is a temporal locality of data references in relational database operations. Thus, when a buffer frame is required by a transaction, and no free frame is available, the frame with the least recently accessed data (from among all the frames in the buffer pool) should be selected for replacement. The policy is global in that all the frames in the buffer pool are treated according to a single criterion (recency of usage) for allocation as well as for replacement; thus, there is no difference between the policy used for replacement and the policy used for allocation.

Priority-LRU Buffer Pool Organization

In Priority-LRU, our prioritized version of Global LRU, the buffer pool is organized dynamically into priority levels in the following way. At system startup time, all the buffer frames are free, and are arranged in a free list. When a transaction with priority P is allocated a frame from the free list, the frame is inserted in an LRU queue of frames whose owners have priority P. Thus, if there are transactions having m different priority levels at any given time, the buffer pool consists of m LRU queues (one per priority level) and a free list. Figure 2.2 shows an example of the organization of the buffer pool for Priority-LRU; there are three priority levels, and thus three LRU queues, and there are no free frames in this example. Priority increases as we move from the bottom to the top of the figure, and the least recently used page of each queue is in the rightmost frame.

Priority-LRU Transaction Admission

When a new transaction arrives at the DBMS, the buffer manager has to decide whether to allow the transaction to begin execution or not. The key idea in making the admission decision is that, at the very least, every running transaction must have sufficient buffer space to hold all of the pages that it needs to fix concurrently. Otherwise, deadlocks may occur due to contention for buffers. Thus, transactions are required to estimate the maximum number of pages that they will need to fix concurrently, and the buffer manager keeps track of the sum of these "fixing

[2] Update-related buffer management issues (such as when to flush dirty pages to the disk) are essentially orthogonal to the issue of priority-based buffer management. Details of handling updates in a conventional DBMS buffer manager are discussed in [Chen84, Effe84].

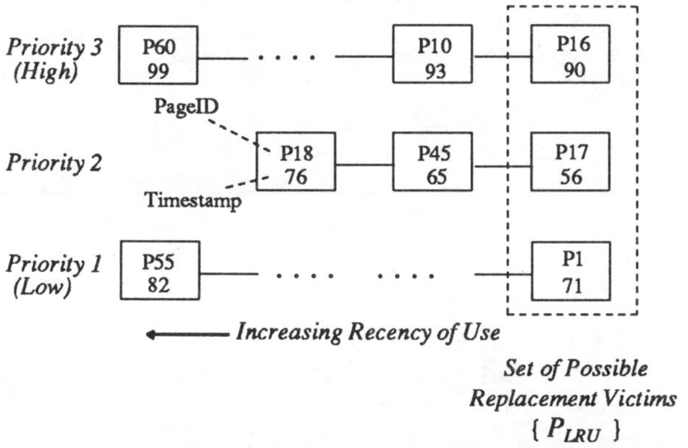

Figure 2.2: Example of Priority-LRU Buffer Pool Organization.

requirements" for all active transactions. If admitting a newly arrived transaction does not cause this sum to exceed the size of the buffer pool, then the transaction is admitted. Otherwise, if there are running transactions of lower priority than the new arrival, the one(s) with the lowest priority among them are suspended until enough buffer space is available for the transaction to be admitted.[3]

Priority-LRU Buffer Replacement and Allocation

As in the conventional Global LRU policy, the buffer allocation and replacement policies coincide in our algorithm. There are two factors that must be weighed when choosing a victim page for replacement: the likelihood that it will be accessed soon, and the priority of the transaction that will access it. These two factors are modeled by the *timestamp* field associated with resident pages and the use of the priority-based LRU queues for buffer pool partitioning, respectively. Note that if priority and recency of usage are the only factors used to choose a replacement victim, then the victim will always be the least recently used page of one of the LRU queues.[4] Let us call the set of possible victims P_{LRU}. In Figure 2.2, P_{LRU} consists of the frames containing the data pages P1, P17, and P16.

The key idea of the replacement policy is that the least recently used page of the lowest priority should be chosen as the victim, with the following caveat: the W_R (for Window of Replacement) most recently accessed data pages should not be chosen for replacement regardless of priority. W_R is a threshold parameter; by varying it, it is possible to vary the relative importance given to recency and priority when making replacement decisions. Let m be the cardinality of P_{LRU}, i.e., let there be m LRU queues in the system. In order to find a replacement victim, we must look at a maximum of m candidates. We start the search at the lowest priority queue, and check whether the candidate's timestamp falls inside the protected W_R window. If it does, we move up one priority level, and we repeat the process until we have either found a victim or else exhausted the search. If all the members of P_{LRU} fall within the window, then the default victim is simply the one with the lowest priority.

An example using Figure 2.2 illustrates the replacement policy. Let *bufSequence*, the global timestamp counter, be 100, and let W_R be 25. We start the search for a victim at P1, the LRU page of the lowest priority level. Since 100–71 > 25, P1 will be chosen as the victim. If W_R were 30, however, then P1 would fall within the window; in this case, P17 would be chosen as the victim instead.

2.3.3. The Priority-DBMIN Algorithm

As discussed in [Chou85], the primitive operations (e.g., selections, joins) of transactions in a relational DBMS can be described as a composition of a set of regular reference patterns such as sequential scans and hierarchical index lookups. These patterns are known to the query optimizer. The DBMIN buffer management policy makes use of this information using the following key ideas:

(1) Buffers should be allocated to transactions on a "per file instance" basis: i.e., since the pattern of accessing each file used by a transaction can be different, a different set of buffers (called a "locality set") should be allocated to a transaction for each file that it opens.

(2) For each file instance Fi, there is an optimum number of buffers ($OptBufs_{Fi}$) and an optimum replacement policy ($RepPol_{Fi}$). As long as the number of buffers actually allocated to file instance Fi is less than $OptBufs_{Fi}$, the admission policy (see (4)) guarantees that there will be at least one free buffer available for Fi; when the number of buffers allocated to Fi is equal to $OptBufs_{Fi}$, $RepPol_{Fi}$ is used to choose a victim from Fi's locality set when replacement is required. Thus, the replacement policy in DBMIN is local rather than global.

(3) The query optimizer can inform the buffer manager of $OptBufs_{Fi}$ and $RepPol_{Fi}$ for each file instance Fi. The buffer manager can then ensure that the maximum number of buffers allocated to a file instance is $Optbufs_{Fi}$.

(4) The buffer manager ensures that no transaction is allowed to begin running unless it can be guaranteed to get the optimum number of buffers for each of its file instances.

Priority-DBMIN Buffer Pool Organization

As in the original DBMIN algorithm, the buffer pool is organized into "locality sets," where the data pages in each set are all part of the same file instance and have the same *owner*. If a page is accessed by more than one concurrent transaction, its *owner* is the transaction with the highest priority among them. Within a locality set, pages are arranged according to the replacement policy prescribed by the optimizer. The buffer manager maintains the sum of the *OptBufs* values for the running transactions of each priority. Thus, for any priority P, it is easy to compute the

[3] Among multiple transactions of the same priority, older transactions are favored over younger ones.

[4] Also note that fixed pages cannot be considered for replacement, since this could lead to corruption of data. The admissions policy guarantees that there will always be at least one unfixed replacement candidate.

sum of the *OptBufs* values for all higher priority transactions that are running.

Priority-DBMIN Transaction Admission

As discussed earlier, DBMIN relies heavily on its Transaction Admission policy. When transactions have priorities, however, the guarantee that any transaction allowed to run will always find *OptBufs* buffers available must be made conditional as follows: Let the combined size of the locality sets of a newly arrived transaction T be $OptBufs_T$, and let its priority be P_T. Let the sum of the optimum sizes of the locality sets of running transactions with priority $\geq P_T$ be $OptBufs_{HIGHER}$, and the sum of the optimum sizes of the locality sets of transactions with priority $< P_T$ be $OptBufs_{LOWER}$. Then, if $(N - OptBufs_{HIGHER}) \geq OptBufs_T$, where N is the total number of buffers in the buffer pool, the buffer manager will admit T. The idea is that if there are sufficient buffers for all currently running transactions of priority P_T or higher and for T itself, then T should be allowed into the system. However, if $(N - OptBufs_{HIGHER} - OptBufs_{LOWER}) < OptBufs_T$, then some buffers may have to be deallocated from lower priority transactions in order to satisfy T's buffer requirements. In this case, the buffer manager will successively suspend transactions (starting with the lowest priority) until there is room enough to admit T.

Priority-DBMIN Buffer Replacement and Allocation

Buffer allocation and replacement are exactly the same as in the original DBMIN algorithm.

3. MODELING A PRIORITY-ORIENTED DBMS

In this section, we describe our performance model of a priority-oriented DBMS. The model, which we implemented using the DeNet simulation language [Livn88], consists of five components: the database itself; a *Source*, which generates the workload of the system; a *Query Manager*, which models the execution behavior of queries; a *Resource Manager*, which models the CPU, I/O, and buffer resources of the system; and a *Concurrency Control Manager*, which implements the details of a particular concurrency control algorithm. Since we will be using only read-only workloads here, we will not discuss the Concurrency Control Manager further.

3.1. MODELING THE DATABASE

The database is modeled as a collection of *relations*. In turn, each relation is modeled as a collection of pages. In addition to relations, the database model contains indices on relations. An index may be either a clustered or a nonclustered B+ Tree. Table 3.1 summarizes the key parameters of the database model. The number of relations in the database is *NumRelations*. For each relation i ($1 \leq i \leq NumRelations$), $RelSize_i$ is the relation size in pages, and $Indexed_i$ determines whether or not the relation has an index. If the relation is indexed, $IndexType_i$ indicates whether the index is clustered or non-clustered, and $Fanout_i$ indicates the fanout of the internal nodes in the index (and thus determines the number of levels of the B+ Tree).

Parameter	Meaning
NumRelations	Number of relations in database
$RelSize_i$	Number of pages in relation i
$Indexed_i$	Whether relation i has an index
$IndexType_i$	Type of index (clustered/nonclustered)
$Fanout_i$	Fanout of internal nodes of index

Table 3.1: Database Model Parameters.

3.2. THE SOURCE MODULE

The Source module is the component responsible for modeling the workload for the DBMS. Table 3.2 summarizes the key parameters of the workload model. A query may belong to any one of *NumClasses* classes, and it may have any one of *NumPriorities* priority levels. The model is that of an open system, and the arrival rate of queries of every <class, priority> combination is controlled by a matrix of arrival rates called *ArrRate*. The arrival of queries of each <class, priority> combination is a Poisson process. Among the per-class parameters is $QueryType_i$, which indicates the query type for the class. Currently, only single-relation select queries and two-relation select-join queries are supported; the model supports selections performed via sequential relation scans, selections performed using either clustered or nonclustered indices, and selections of any of these types followed by a join (which can be either a nested loops join or an index join). Queries are modeled at a logical level because one of the algorithms of interest to us is Priority-DBMIN, which makes use of the logical sequencing of the page accesses of a query. Since the looping behavior of queries can affect buffer hit ratios significantly, the particular query types supported were chosen to provide queries both with and without looping in their page access behavior.

For each query type i, a query plan is provided in the form of a set of parameters. For a join query, $JoinMethod_i$ indicates the join algorithm (nested loops or indexed join), and $Inner_i$ and $Outer_i$ indicate which relation is the inner relation and which is the outer. For each relation j accessed by query i, $AccessPath_{ij}$

Parameter	Meaning
Overall Arrival Pattern Parameters	
NumClasses	Number of query classes
NumPriorities	Number of query priority levels
$ArrRate_{ij}$	Mean exponential arrival rates of queries of class i and priority level j
Per-Class Parameters ($1 \leq i \leq NumClasses$)	
$QueryType_i$	Type of query, e.g., select or select-join
$JoinMethod_i$	Join algorithm used
$Outer_i$	Outer relation
$Inner_i$	Inner relation
$AccessPath_{ij}$	Access path used to access jth relation
$Selectivity_{ij}$	Fraction of jth relation selected
$IndexPageCPU_i$	CPU time for processing an index page
$DataPageCPU_i$	CPU time for processing a data page

Table 3.2: Workload Model Parameters.

indicates the access method used (e.g., a clustered index scan or a sequential scan). In selections, $Selectivity_{ij}$ indicates the proportion of relation j's data that satisfies the selection criterion of query i. We assume a uniform probability of access to all data pages within each relation. The per-class parameter $IndexPageCPU_i$ specifies the expected amount of CPU processing required per index page, and the parameter $DataPageCPU_i$ specifies the expected amount of CPU processing required per data page of each relation accessed for query type i. (The actual CPU processing times per page are exponentially distributed.)

Given a query plan, the Source module generates a list of page accesses that models the sequence in which pages will be accessed by the query. For example, in a selection using a clustered index, the Source uses the selectivity parameter and the size of the target relation to generate a list of page accesses that start at the root of the index, traverse the index to the leaf level, and then access a sequence of index leaf pages and corresponding data pages. In addition to generating a list of page accesses, the Source module also provides information on the number of locality sets, the optimum number of buffers for each set, and the optimal replacement policy (such as MRU or LRU) for each locality set. The maximum number of concurrently fixed pages is similarly provided. This information can be used by the buffer manager. As discussed shortly, the Source must interact with the Resource Manager to determine whether a query is allowed to enter the system right away or whether it has to wait until sufficient buffers become available.

3.3. THE QUERY MANAGER MODULE

The Query Manager is responsible for accepting queries from the Source and modeling their execution. For each page accessed by the query, the Query Manager sends a read request to the Resource Manager; the Resource Manager informs the Query Manager when the read request is completed. The Resource Manager also informs the Query Manager when a query is suspended or reactivated. When the Resource Manager decides to reactivate a suspended query, the Query Manager ensures that the reactivated query resumes execution at the point where it was suspended.

3.4. THE RESOURCE MANAGER MODULE

The Resource Manager controls the physical resources of the DBMS, including the CPU, the disk, and the buffer pool in main memory. Two versions of the Resource Manager have been implemented, supporting the Priority-LRU and Priority-DBMIN buffer management algorithms, respectively. In addition, the model allows priority to be switched on and off at each resource of interest, i.e., at the CPU, the disk, and the buffer pool. ("Switching off" priority at a resource means that all of its service requests are treated as being of equal priority.) The parameters of the Resource Manager are summarized in Table 3.3. *CPUPrio*, *DiskPrio*, and *BufferPrio* are the parameters used to switch priority on and off for the various resources.

CPU and Disk Models

The DBMS has one CPU, which is scheduled using the priority-based round-robin algorithm described in Section 2. The length of each CPU request from a query is its per-page CPU processing time, and each query voluntarily gives up the CPU after processing one page. There is one disk in the system, with requests being scheduled according to the prioritized elevator algorithm of Section 2. Each disk request requires access to one page. The track number of a disk request is chosen at random from among *NumTracks* tracks (i.e., we model the data as being uniformly distributed across all tracks). The total time required to complete a disk access is computed as the sum of its seek time, rotational latency, and transfer time components. The rotational latency and transfer time are together modeled as a single parameter called *DiskConst*. The seek time for seeking across n tracks is computed using the formula:

$$\text{Seek Time}(n) = SeekFactor * \sqrt{n}$$

SeekFactor is specified as a parameter. This square-root relationship between seek time and seek distance is based on the discussion of current disk technology in [Bitt88].

Buffer Manager Models

The buffer manager component of the resource manager encapsulates the details of the buffer management scheme employed by the DBMS. It maintains information about resident pages, and it uses the information provided by the Source module to decide when to allow queries to enter the system. The *NumBuffers* parameter specifies the number of page frames available in the buffer pool. The Priority-LRU and Priority-DBMIN algorithms are each represented by a different buffer manager model, and the Priority-LRU model has an additional parameter W_R that it uses to balance priority and recency when making replacement decisions. In order to simplify the implementation, we decided not to model fixing and unfixing explicitly. However, we do model their effects on query admission decisions in Priority-LRU (in the manner described in Section 2).

4. EXPERIMENTS AND RESULTS

In this section, we present performance results for the priority-oriented DBMS resource scheduling algorithms described earlier. Our goal is to analyze the relative importance

Parameter	Meaning
CPUPrio	Switch to turn priority on/off at CPU
DiskPrio	Switch to turn priority on/off at disk
BufferPrio	Switch to turn priority on/off at buffer pool
NumTracks	Number of tracks per disk
DiskConst	Sum of rotational and transfer delays
SeekFactor	Factor relating seek time to seek distance
NumBuffers	Number of buffer frames in buffer pool
W_R	Window of timestamps used when choosing a replacement victim in Priority-LRU

Table 3.3: Parameters of the Resource Manager.

of priority scheduling at each of the various resources and to understand the interaction between them. In order to simplify the analysis, we consider just two priority levels ("low" versus "high" priority) in the first four experiments presented here. An experiment with four priority levels is also included at the end of this section. Depending on the nature of the workload, any of the physical resources of a DBMS (e.g., the CPU, the disks, or the buffer pool) may become the bottleneck. Together, our experiments cover each of these possibilities.

4.1. Performance Metrics

As discussed earlier, we use an open queuing system to model the DBMS. Response time will thus be the primary performance metric of interest in this study. In particular, we will examine the average response time for queries at each priority level in the workload. Since our performance objective is to provide high priority queries with a preemptive-resume view of the DBMS, we will focus most of our attention on the response time for high priority queries. Two important issues regarding the response time for these queries will be the range over which the system is stable for them and the extent to which the system is able to meet our preemptive-resume performance goal.

Note that in a priority-oriented DBMS, the system can remain stable for high-priority queries long after the arrival rate has become high enough to make the system unstable for low-priority queries. This is because more and more of the system's resources are devoted to high-priority queries as the overall load on the system increases. Thus, there are two regions of operation in each of our experiments. In the first region, the system is stable for both high priority and low priority queries. In the second region, the system has become unstable for low-priority queries, but continues to be stable for high-priority queries. Consequently, we present response time results for low-priority and high-priority queries (in their stable regions) separately for each experiment.

A query's response time is computed by subtracting the time at which the query completes from the time at which it was submitted to the DBMS. As discussed in Section 1, the response time of a query can be broken down into the following components:

$$T_R = T_{W\_EXT} + T_{W\_CPU} + T_{S\_CPU} + N_{DISK}*(T_{W\_DISK} + T_{S\_DISK})$$

We measured each of these components separately in our experiments in order to aid us in analyzing the results.

In order to obtain a statistically significant sample of query response times, each experiment was run long enough for a total of 4000 high-priority queries to complete. (The number of low-priority query completions varied with the load.) Other information was also gathered in the course of each simulation, including the utilization of the CPUs and disks, the average seek time per disk access, and the average number of queries of each priority level running concurrently.

4.2. Parameter Settings

We first present the parameters that were kept constant across all workloads. These parameters are listed in Table 4.1. We then describe our representative workloads and the parameter settings that changed with each workload.

Parameter	Setting
NumRelations	40
$RelSize_i$	1000 pages, 500 pages, 6 pages, 3 pages (10 relations of each size)
$Indexed_i$	YES (1000-page & 500-page relations) NO (6-page & 3-page relations)
$IndexType_i$	Clustered (1000-page & 500-page relations)
$Fanout_i$	20 (1000-page & 500-page relations)
NumTracks	1000
DiskConst	15 ms
SeekFactor	0.6 ms
NumBuffers	50
W_R	10

Table 4.1: Workload-Independent Parameter Settings.

Workload-Independent Parameters

The database is modeled as a collection of 40 relations. We use four different relation sizes in our experiments — 1000 pages, 500 pages, 6 pages, and 3 pages — with the database containing 10 relations of each size. The 1000-page and 500-page relations each have a clustered index available, while the smaller relations are not indexed at all. The disk has 1000 tracks, and the sum of the rotational latency and the transfer time per disk access is 15 milliseconds. The factor relating seek distance to seek time is 0.6 milliseconds[5], so the expected disk access time is between 15 and 30 milliseconds. There are 50 buffer frames in the buffer pool. The W_R parameter used in the Priority-LRU algorithm is set to 10, as this value was found to work well for the range of workloads considered. (For the workloads studied here, the system turns out not to be very sensitive to W_R.)

The model's switches for determining whether or not priorities are used for scheduling at the CPU, the disk, and the buffer pool will be turned off and on as part of each experiment. In the description of the results, the following notation will be used to identify the resources where priority scheduling is turned on: the letters B, C, and D refer to the buffer manager, the CPU, and the disk respectively, while the subscripts *YES* and *NO* refer to priority being turned on and off respectively. For example, the label $B_{YES}\ C_{YES}\ D_{NO}$ refers to an experiment where priority is turned on at the buffer manager and at the CPU, but is turned off at the disk.

Workload Parameter Settings

Four different query workloads are employed in our five experiments. Each workload consists of arrival streams of a single query type, but at different levels of priority; in all of the experiments reported here, the arrival rates for each of the priority levels will be equal. The parameters used to generate the four workloads are listed in Table 4.2. The first three workloads

[5] These values for determining disk access times were chosen based on [Bitt88, Gray89].

Parameter	Type I	Type II
QueryType	Select-join	Select-join
JoinMethod	Nested Loops	Nested Loops
Outer (*RelSize*$_1$)	500-page	500-page
Inner (*RelSize*$_2$)	3-page	3-page
AccessPath$_1$	Clustered Index Scan	Clustered Index Scan
AccessPath$_2$	Sequential Scan	Sequential Scan
Selectivity$_1$	1%	1%
IndexPageCPU	15 ms	2 ms
DataPageCPU	15 ms	2 ms
Page Accesses	23	23
Locality Set Sizes (*index*, *outer*, *inner*)	1, 1, 3	1, 1, 3
Replacement Policies (*index*, *outer*, *inner*)	MRU, MRU, MRU	MRU, MRU, MRU
Fixing Requirements	3	3

Parameter	Type III	Type IV
QueryType	Select-join	Select
JoinMethod	Nested Loops	-
Outer (*RelSize*$_1$)	500-page	1000-page
Inner (*RelSize*$_2$)	6-page	-
AccessPath$_1$	Clustered Index Scan	Clustered Index Scan
AccessPath$_2$	Sequential Scan	-
Selectivity$_1$	1%	1%
IndexPageCPU	2 ms	5 ms
DataPageCPU	2 ms	5 ms
Page Accesses	38	13
Locality Set Sizes (*index*, *outer*, *inner*)	1, 1, 6	1, 1, -
Replacement Policies (*index*, *outer*, *inner*)	MRU, MRU, MRU	MRU, MRU, -
Fixing Requirements	3	2

Table 4.2: Workload Parameter Settings.

consist of select-join queries, with the result of a selection using a clustered index on a 500-page outer relation being joined to a smaller inner relation; the fourth workload consists of a clustered index selection on a 1000-page relation. For each relation accessed by a query, the actual relation accessed was chosen uniformly from among the 10 relations of that size, so there is not much data sharing in the system. To aid in understanding the nature of the workloads, Table 4.2 also includes the total number of page accesses for each query type. In addition, it lists the optimum locality set sizes and replacement policies for Priority-DBMIN, and it lists the query fixing requirements for Priority-LRU.

4.3. Experiment I (Admission Control)

In this experiment, we investigate the impact of priority scheduling on performance under the Type I workload. This workload consists of a mix of CPU-intensive, indexed-select/nested-loops-join queries. In fact, given the Type I parameter settings, the disk utilization is only about 65% when the CPU is fully utilized, so the presence or absence of priority-based disk scheduling has no effect on performance. Thus, the priority scheduling combinations studied here are those with and without CPU and buffering priority (i.e., $B_{NO} C_{NO} D_{YES}$, $B_{NO} C_{YES} D_{YES}$, $B_{YES} C_{NO} D_{YES}$, and $B_{YES} C_{YES} D_{YES}$).

Figure 4.1 shows the response time results for high priority queries using the Priority-DBMIN buffering algorithm, and Figure 4.2 shows the corresponding results for low priority queries. The arrival rate axis of these figures indicates the *combined arrival rate* of both priority levels. The arrival rates for the two priority levels are equal, so the arrival rate for each of the levels is half of this total rate. In addition to the curves showing the high priority response times for the four scheduling combinations, Figure 4.1 also includes a curve labeled HPO (for High-Priority Only) that indicates what the high priority response times would be if no low priority queries were arriving. This curve will help us evaluate how successful we are at getting the DBMS to treat high priority queries like a preemptive-resume server should; the better we do at approximating this curve, the closer we are to meeting this design goal. Figure 4.2 contains a similar curve (labeled LPO) that shows how the system would perform with low priority queries only, indicating the extent to which low priority queries suffer due to competition from high priority queries. Despite the fact that the system is CPU-bound, Figures 4.1 and 4.2 show that the use of priority in the buffer manager is extremely important. Without priority in the buffer manager, the system saturates at an arrival rate of about three queries per second for both low and high priority queries. With priority, however, the system is stable for high priority queries until the arrival rate reaches about six queries per second. These differences, as we will see, are due to the use of priority in the Priority-DBMIN admission control policy.

To understand why admission control has such an effect, let us consider how priority affects the view that queries of each priority have of a given resource. Arriving low priority queries see a resource containing other queries of both low and high priority. This view is the same whether or not priority scheduling is employed at the resource, although the quality of service that low priority queries receive is affected by the scheduling policy. This is consistent with Figure 4.2, which shows that low priority query performance is affected only slightly by the presence or absence of priority scheduling. The system always saturates at the same point from the perspective of low priority queries. High priority queries get the same view of a resource as low priority queries if it is not scheduled using priority. This explains why, when the buffer manager does not use priority to control admission in Figure 4.1 (in the B_{NO} curves), the system saturates at three queries per second for high priority queries. In this case the saturated "resource" is the DBMS itself, and if priority is not used for admission control, the external waiting time for high priority queries is limited by the response time of low priority queries. However, when the buffer manager's admission policy does favor high priority queries, the DBMS admits them preferentially and even suspends low priority queries in their favor. In this case, they see only other high priority queries in the system, so they essentially see only half of the actual system load. This

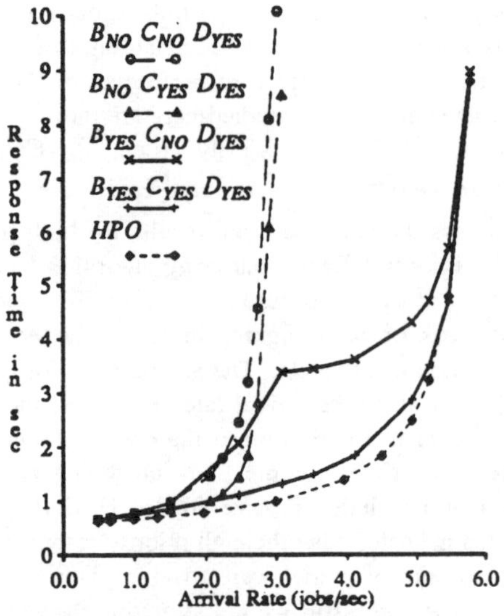

Figure 4.1: Type I: High Priority. (Priority-DBMIN)

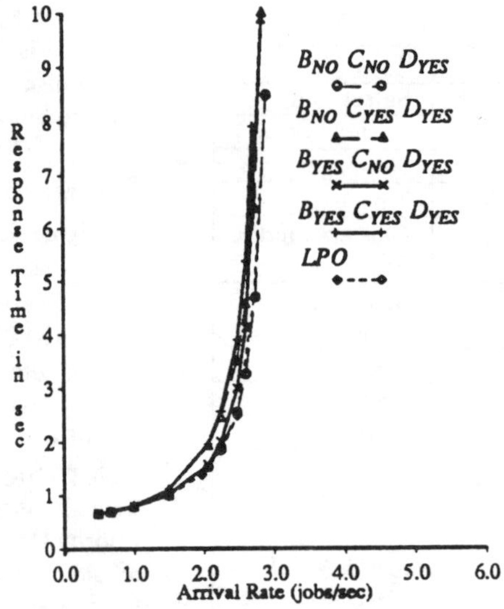

Figure 4.2: Type I: Low Priority. (Priority-DBMIN)

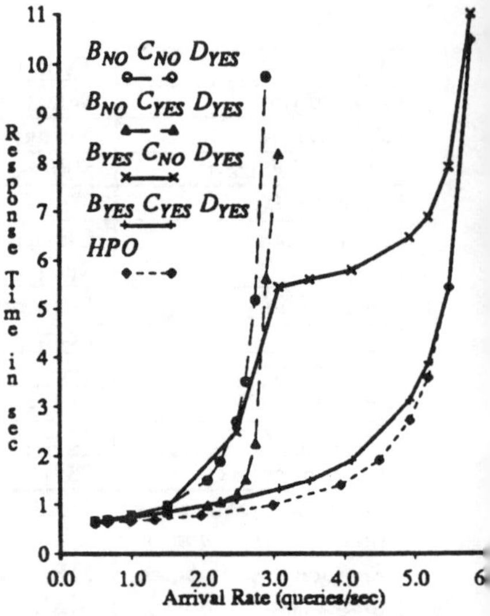

Figure 4.3: Type I: High Priority. (Priority-LRU)

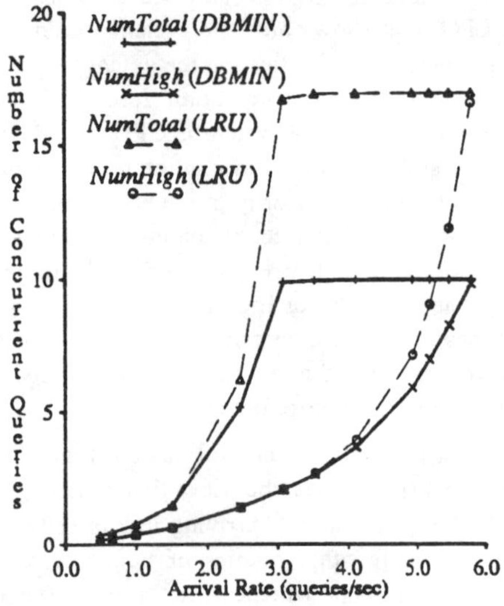

Figure 4.4: Number of Concurrent Queries.

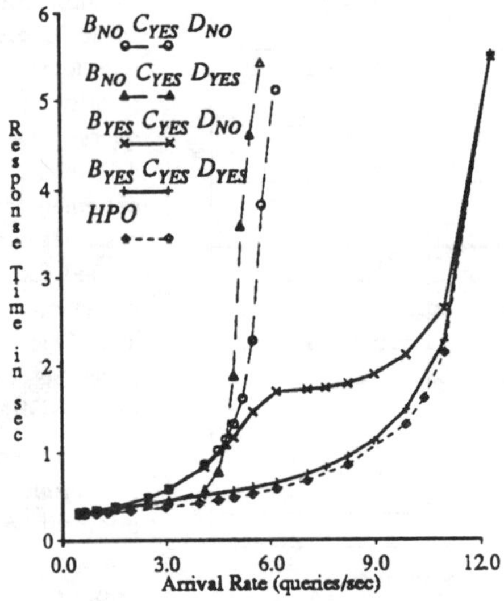

Figure 4.5: Type II: High Priority. (Priority-DBMIN)

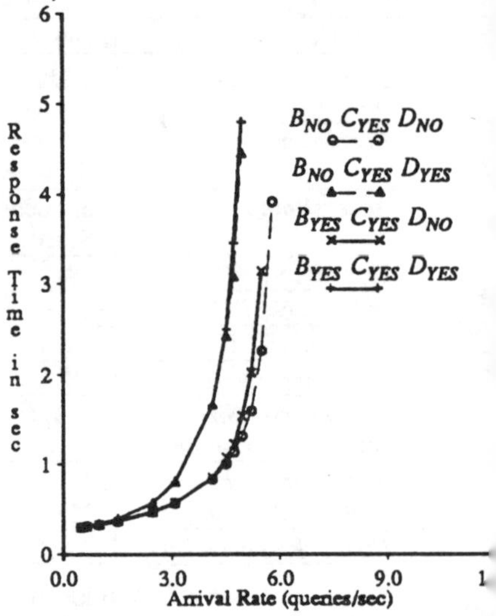

Figure 4.6: Type II: Low Priority. (Priority-DBMIN)

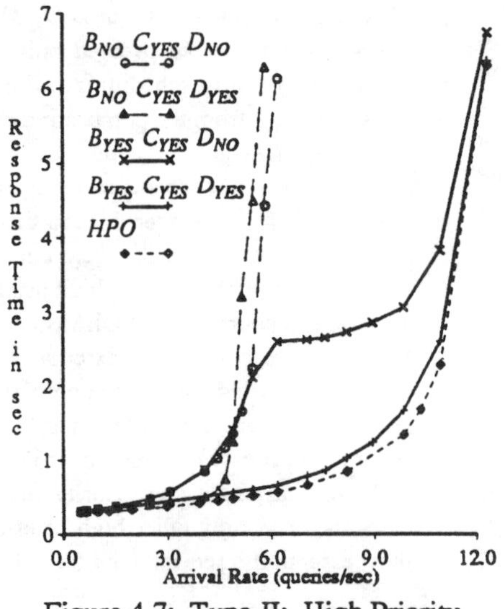

Figure 4.7: Type II: High Priority. (Priority-LRU)

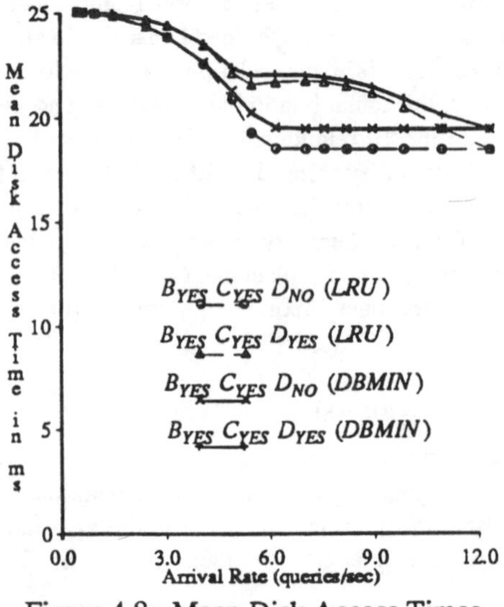

Figure 4.8: Mean Disk Access Times.

explains why the system remains stable twice as long for high priority queries under Priority-DBMIN.

Another interesting feature of Figure 4.1 is how priority scheduling at the CPU affects the response time for high priority queries under Priority-DBMIN. With priority used in the buffer manager but not at the CPU (i.e., the $B_{YES} C_{NO} D_{YES}$ curve), high priority response time is similar to low priority response time until the system saturates for low priority queries. After this point, the response time is stable for a while as more and more low priority queries are suspended due to high priority arrivals. That is, the buffer manager has already admitted as many queries as it can, so all it can do in this range is trade low priority queries for high priority ones. Since the workload is CPU-bound, the load that high priority queries see does not change in this range if CPU scheduling is not priority-based. The CPU eventually becomes saturated with high priority queries alone (at an arrival rate of six queries per second, which is three high priority queries per second). However, when CPU scheduling is also priority-based in Figure 4.1 (i.e., the $B_{YES} C_{YES} D_{YES}$ curve), we see a different trend. Here, the high priority response time behavior is very close to that of the HPO curve; recall that this curve shows the response time behavior without low priority queries. This indicates that priority scheduling at the CPU is succeeding at shielding the high priority queries from the low priority queries (i.e., giving them the impression of being the only queries in the system). Thus, a combination of Priority-DBMIN and priority CPU scheduling indeed gives high priority queries a preemptive-resume-like view of the DBMS.

Figure 4.3 shows the response time results for high priority queries using the Priority-LRU buffering algorithm; we do not show the low priority query results, as they are very similar to those for Priority-DBMIN in Figure 4.2. The Priority-LRU trends in Figure 4.3 are quite similar to those that we saw in Figure 4.1, and the explanation of the trends are the same here. The main difference between the Priority-LRU and Priority-DBMIN results is that priority CPU scheduling is more critical for Priority-LRU. The two algorithms perform almost identically when priority scheduling is used everywhere (except at the highest loads, where Priority-DBMIN is marginally better). However, when CPU scheduling is not priority-based, high priority queries suffer more under the Priority-LRU algorithm. This is due to the difference between the admission control policies of the two algorithms. Since the Priority-DBMIN policy is more conservative, it admits fewer queries than Priority-LRU does. This is evident in Figure 4.4, which shows the number of concurrent queries (both high priority and total) inside the system, as opposed to waiting outside, under the two buffer management algorithms. Priority scheduling at the CPU changes the number of competing queries seen by high priority queries from being the total number in the system to being just the high priority number, which is a bigger change in the Priority-LRU case. The general Priority-LRU conclusions are the same, though: priority is needed both in the buffer manager (for priority-based admissions control) and at the CPU in order to achieve the HPO performance objective for high priority queries.

4.4. Experiment II (Disk Priority)

In our second experiment, we focus on the impact and trade-offs involved in priority scheduling at the disk. Here we use the Type II workload, which is just like the Type I workload except that the per-page CPU time for queries is now just 2 milliseconds instead of 15 milliseconds. Thus, the workload consists of a mix of I/O-intensive, indexed-select/nested-loops-join queries. With this workload, the CPU utilization is approximately 35% when the disk becomes 100% utilized, so the presence or absence of priority-based CPU scheduling has no effect here. Thus, we examine situations with priority scheduling at the disk and the buffer manager (i.e., $B_{NO} C_{YES} D_{NO}$, $B_{NO} C_{YES} D_{YES}$, $B_{YES} C_{YES} D_{NO}$, and $B_{YES} C_{YES} D_{YES}$).

Figure 4.5 shows the response time results for high priority queries using Priority-DBMIN, and Figure 4.6 shows the corresponding results for low priority queries. Figure 4.7 presents the high priority query response time results for Priority-LRU; again, the low priority results for Priority-LRU were very much like those of Priority-DBMIN, so we omit them here. Figure 4.8 shows the mean disk access times for Priority-DBMIN and Priority-LRU with and without priority-based disk scheduling (for the two cases $B_{YES} C_{YES} D_{NO}$ and $B_{YES} C_{YES} D_{YES}$). For the most part, Figures 4.5-4.7 display the same trends that we saw in Figures 4.1-4.3, and they do so for the same reasons (albeit with a different bottleneck resource). Again, it is evident that priority must be incorporated in the buffer manager as well as the bottleneck resource (the disk) in order to provide the desired level of performance for high priority queries; and again, priority scheduling of the bottleneck resource is more important for Priority-LRU because of its less effective admission control policy. However, there are also several other interesting points to be noted from this experiment.

The first point to notice is that in Figure 4.6, unlike Figure 4.2, low priority queries clearly suffer performance-wise due to priority-based scheduling of the bottleneck resource (the disk in this case). This is due to the fact that, in the range where they suffer, priority scheduling has reduced the service capacity of the disk by increasing the mean disk access time (as shown in Figure 4.8). Combined with large waiting times, which are in excess of eight times larger than those for high priority queries, this produces earlier response time degradation for low priority queries. In contrast, high priority response time is again close to that of the HPO curve when priority scheduling is used everywhere, indicating that high priority queries are largely unaffected by the somewhat increased mean disk access time. This is because this increase is more than offset by the decrease in disk waiting times that priority-based disk scheduling produces for the high priority queries. Put another way, there is a price to be paid for doing disk scheduling based on priority, but it is the low priority queries that pay the price.

Another interesting observation can be made from Figure 4.8. At low arrival rates, where the disk load is low, there is no penalty for priority-based disk scheduling. This is because there is little or no queuing for the disk in this region. However, there is also no penalty at the highest arrival rates. The explanation is different in this case. Here, the reason that priority scheduling is able to do as well as the strict elevator algorithm is that it, too, effectively becomes the elevator algorithm; the disk is so heavily loaded due to high priority requests that it is kept busy serving their elevator queue. In the middle range of arrival rates, where there is a mean disk access time penalty, the benefits of priority scheduling far outweigh the penalty for high priority queries, as described above. Thus, priority-based disk scheduling appears to be very worthwhile for a priority-based DBMS.

4.5. Experiment III (Buffer Priority)

In our third experiment, we focus on the relative behavior of the DBMIN-based and LRU-based approaches to buffer management. Here we use the Type III workload, which is similar to the Type II workload but with an inner relation that is twice as large as before. Thus, the queries in the workload have a more significant looping behavior here. Again, we focus our attention on priority scheduling at the disk and in the buffer manager.

Figure 4.9 shows the response time results for high priority queries using Priority-DBMIN, and Figure 4.10 shows the corresponding results for low priority queries. The trends here are basically those of Experiment II, but the absolute performance is different due to the larger query size. The only apparent relative difference is that priority disk scheduling leads to somewhat less of an improvement in the high priority response time when buffering priority is used. This is because the bottleneck here is actually buffer space; the disk utilization never exceeds 90% because the admission control policy is unable to allow enough queries into the system to saturate the disk.

Figure 4.11 presents the high priority query response time results for Priority-LRU; as always, the low priority results for Priority-LRU are virtually identical to those for Priority-DBMIN, so we do not show them. For the high priority queries, we see significantly different behavior for Priority-LRU here (relative to Priority-DBMIN) than that observed in previous experiments, especially without disk priority scheduling. The difference is a consequence of the admission control policies of the two algorithms. As was shown in [Chou85], the information-based admission controller of the basic DBMIN algorithm can significantly reduce thrashing relative to the basic LRU algorithm, and the same is (of course) true of our priority-based versions of these algorithms. Figure 4.12 shows the buffer pool hit ratio for high priority queries under both the Priority-DBMIN and Priority-LRU algorithms. The hit ratio for Priority-LRU drops significantly where the high priority response times increase in Figure 4.11, indicating that this is indeed the problem. This thrashing is due to the fact that Priority-LRU admits too many high priority queries, which leads them to take buffers from one another since the number of available low priority buffers is insufficient. The result is especially drastic without priority at the disk. The low priority queries in the system have extremely low hit ratios, as their buffers are consistently chosen as replacement victims for high priority queries; this causes them to generate many more disk requests than they would under the carefully controlled Priority-DBMIN policy. Without priority at the disk, this volume of requests generates much more disk traffic for high priority queries to contend with, increasing their response times. Buffering priority alone is thus not enough to ensure stability for high priority queries in this workload under Priority-LRU.

4.6. Experiment IV (Minimal Buffer Contention)

In our final two-priority experiment, we investigate the performance of the priority-based scheduling algorithms for a workload where buffer contention is not a significant factor. Here we use the Type IV workload, which consists of indexed-selection queries with no looping behavior. The Type IV workload is I/O-bound due to its 5 millisecond page CPU time and lack of looping behavior, so we again focus on priority scheduling at the disk and in the buffer manager.

Figure 4.13 shows the response time results for high priority queries using Priority-DBMIN, and Figure 4.14 shows these results for Priority-LRU. We omit the low priority results, as they are qualitatively the same as in Experiments II and III. The main things to observe here are the impact of priority-based management of the disk and buffer resources on the performance of high priority queries. Since buffer contention is a non-issue under this workload, the Priority-DBMIN and Priority-LRU algorithms perform pretty much identically. However, they are still essential: without a priority-based buffer management algorithm (for priority-based admission control), the system saturates for high-priority queries at 4.5 queries per second; with priority in the buffer manager, the saturation point is extended to nine queries per second. Likewise, priority disk scheduling is crucial for good performance here. This is because the workload is I/O bound and buffer hits are rather rare.

4.7. Experiment V (Four Priorities)

In the last experiment of the paper, we briefly examine the performance of the priority-based scheduling algorithms for a workload consisting of four levels of priority. We return to the Type III (buffer-intensive, I/O bound) queries for this experiment. Here we look only at how queries of the different priority levels perform when priority is employed everywhere (i.e, at the CPU scheduler, the disk scheduler, and the buffer manager).

Figure 4.15 shows the query response time results for the four priority levels under Priority-DBMIN, and Figure 4.16 shows the results for the Priority-LRU algorithm. The observed trends are what we would expect based on the results of our earlier experiments. First, the system succeeds at providing preemptive-resume-like performance, as the highest priority query response times are quite close to the HPO curves. Second, the saturation points are not linearly distributed over the range of arrival rates. This is because of the system view that priority scheduling provides to the different priority levels: the highest priority queries see only themselves (i.e., one-fourth of the load), queries at the second-highest priority see themselves and the highest priority queries (i.e., one-half of the load), the second-lowest priority queries see the top three levels (i.e., three-fourths of the load), and the lowest priority queries see the entire query load as competition for system resources. Finally, as in Experiment III, Priority-DBMIN provides somewhat better performance than Priority-LRU here, as the workload is one where buffer management has a significant role to play in determining performance.

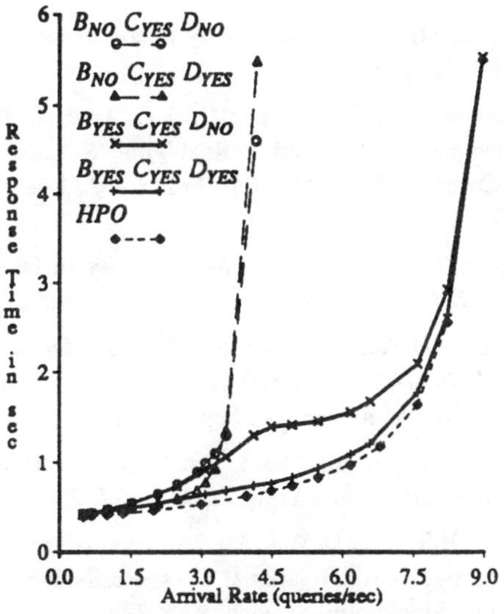

Figure 4.9: Type III: High Priority. (Priority-DBMIN)

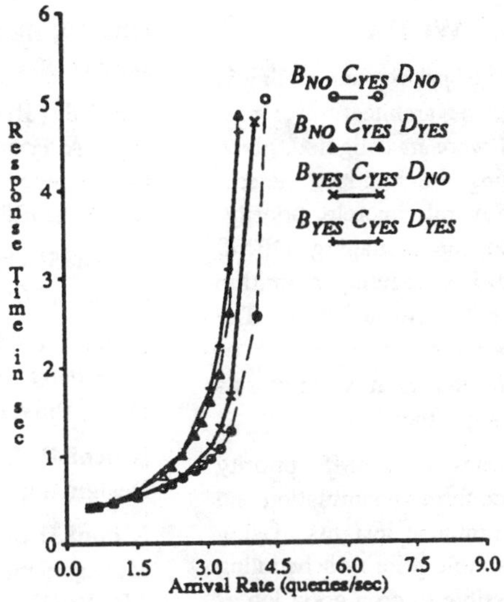

Figure 4.10: Type III: Low Priority. (Priority-DBMIN)

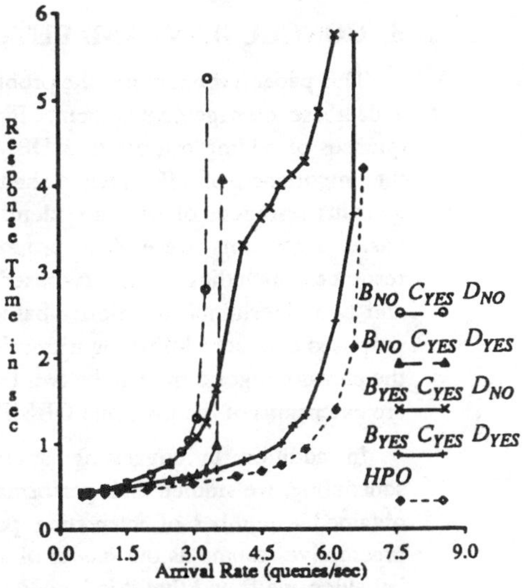

Figure 4.11: Type III: High Priority. (Priority-LRU)

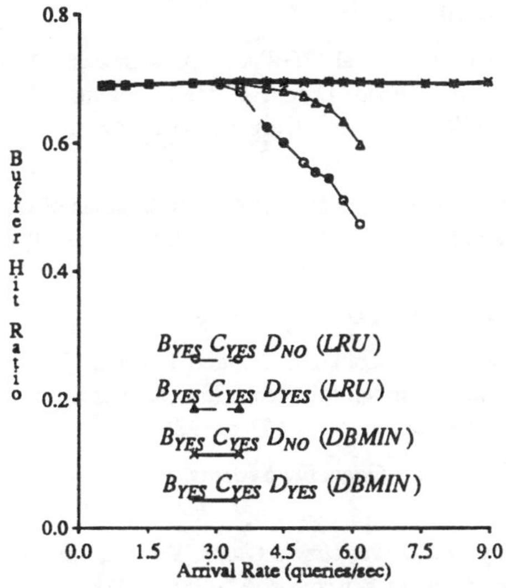

Figure 4.12: Buffer Hit Ratio. (High Priority)

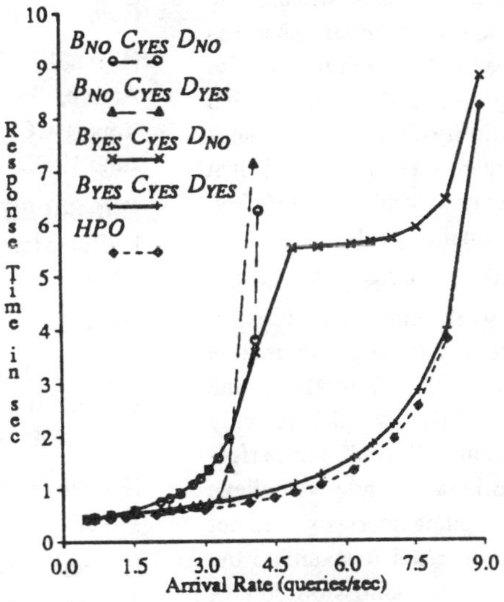

Figure 4.13: Type IV: High Priority. (Priority-DBMIN)

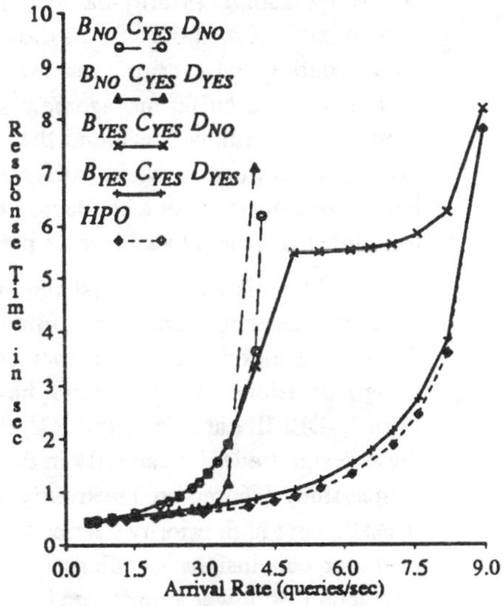

Figure 4.14: Type IV: High Priority. (Priority-LRU)

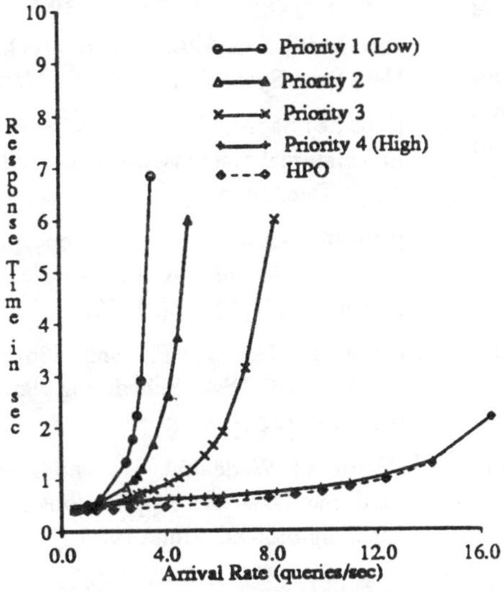

Figure 4.15: Four Priority Levels (Priority-DBMIN)

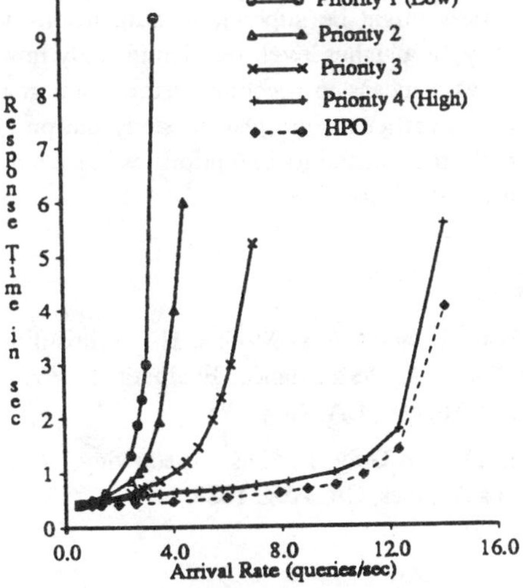

Figure 4.16: Four Priority Levels (Priority-LRU)

5. CONCLUSIONS AND FUTURE WORK

This paper has examined the problem of priority scheduling in a database management system. First, the architectural consequences of adding priority to a DBMS were investigated, given the importance of effectively scheduling the multiple, heterogeneous resources of such a system. Several concrete, priority-based algorithms were then proposed for managing DBMS resources, including a priority-based disk scheduling algorithm and two algorithms for priority-based buffer management. The proposed disk scheduling algorithm is a priority-based variant of the elevator algorithm, and the two buffer management proposals are extensions of the LRU and DBMIN algorithms.

In addition to suggesting approaches to DBMS priority scheduling, we studied their performance through simulation and obtained a number of interesting performance insights. Using preemptive-resume as our model of desirable priority scheduling behavior, we found that it is indeed possible to do a good job of priority scheduling in a DBMS context. However, our simulation results indicate that the objectives of priority scheduling cannot be met by a single priority-based scheduler. Rather, whether the system bottleneck is the CPU or the disk, it is essential that priority scheduling on the critical resource be used in conjunction with a priority-based buffer management algorithm. Between our two proposed algorithms, we found that Priority-DBMIN dominates Priority-LRU in cases where buffer contention is a factor. When buffer contention is not a key factor, either algorithm is sufficient to enable the system to achieve its performance goals.

We view this work as a first step, with a number of interesting problems and opportunities for future work remaining. First, in the area of algorithm design, there are a number of alternative design decisions that we could have made differently in the Priority-DBMIN and Priority-LRU algorithms; we plan to study these design tradeoffs carefully in the future. Second, our performance study focused on a read-only workload. While we believe that allowing high priority queries to be update queries would not alter our conclusions significantly, we do need to examine the interaction of lower priority updates and the admission control policy. One issue here is that suspending an update query and replacing its pages will carry the price of writing its updates to disk. A related issue is the impact of suspending queries that hold locks; restarts might be superior to suspensions in such situations. Finally, at a higher level, we plan to study how priority can be used as a means to meeting performance goals in a DBMS context. In particular, we plan to study the problem of mapping soft real-time constraints into priorities in such a way as to minimize missed deadlines.

REFERENCES

[Abbo88] Abbott, R., and Garcia-Molina, H., "Scheduling Real-Time Transactions: A Performance Evaluation," *Proc. 14th VLDB Conf.*, Los Angeles, CA, Aug. 1988.

[Bitt88] Bitton, D., and Gray, J., "Disk Shadowing," *Proc. 14th VLDB Conf.*, Los Angeles, CA, Aug. 1988.

[Blas79] Blasgen, M., et al, "The Convoy Phenomenon," *Operating Sys. Rev.* 13(2), April 1979.

[Buch89] Buchmann, A., et al, "Time-Critical Database Scheduling: A Framework for Integrating Real-Time Scheduling and Concurrency Control," *Proc. 5th Data Eng. Conf.*, Los Angeles, CA, Feb. 1989.

[Chan85] Chang, H.-Y., and Livny, M., "Priority in Distributed Systems," *Proc. IEEE Real-Time Sys. Symp.*, Dec. 1985.

[Chan87] Chang, H.-Y., *Dynamic Scheduling Algorithms for Distributed Soft Real-Time Systems*, Ph.D. Thesis, Comp. Sci. Dept., Univ. of Wisconsin-Madison, Sept. 1987.

[Chen84] Cheng, J., et al, "IBM Database 2 Performance: Design, Implementation, and Tuning," *IBM Sys. J.* 23(2), 1984.

[Chou85] Chou, H-T., and DeWitt, D., "An Evaluation of Buffer Management Strategies for Relational Database Systems," *Proc. 11th VLDB Conf.*, Stockholm, Sweden, Aug. 1985.

[Coff68] Coffman, E., and Kleinrock, L., "Computer Scheduling Methods and Their Countermeasures," *Proc. AFIPS Spring Joint Comp. Conf.*, April 1968.

[Daya88] Dayal, U., et al, "HiPAC: A Research Project in Active, Time-Constrained Database Management," Technical Report CCA-88-02, Computer Corporation of America, Boston, June 1988.

[Effe84] Effelsberg, W., and Haerder, T., "Principles of Database Buffer Management," *ACM Trans. on Database Sys.* 9(4), Dec. 1984.

[Gray89] Gray, J., personal commmunication.

[Jens86] Jensen, E. D., Locke, C. D., and Tokuda, H., "A Time-Driven Scheduling Model for Real-Time Operating Systems," *Proc. IEEE Real-Time Sys. Symp.*, Dec. 1986.

[Klei76] Kleinrock, L., *Queueing Systems*, John Wiley and Sons, 1976.

[Livn88] Livny, M., *DeNet User's Guide*, Version 1.0, Computer Sciences Dept., Univ. of Wisconsin, Madison, 1988.

[Pete86] Peterson, J., and Silberschatz, A., *Operating Systems Concepts*, Addison-Wesley, 1986.

[SIGM88] *SIGMOD Record* 17(1), Special Issue on Real-Time Data Base Systems, S. Son, ed., March 1988.

[Sacc86] Sacco, G.M., and Schkolnick, M., "Buffer Management in Relational Database Systems," *ACM Trans. on Database Sys.*, 11(4), Dec. 1986.

[Stan88] Stankovic, J., "Misconceptions About Real-Time Computing: A Serious Problem for Next-Generation Systems," *IEEE Computer* 21(10), Oct. 1988.

[Teor72] Teorey, T., and Pinkerton, T., "A Comparative Analysis of Disk Scheduling Policies," *Comm. ACM* 15(3), March 1972.

[Wede86] Wedekind, H., and Zoerntlein, G., "Prefetching in Realtime Database Applications," *Proc. 1986 SIGMOD Conf.*, Washington, D.C., June 1986.

The O_2 Database Programming Language

C. Lécluse and P. Richard

GIP Altaïr - Domaine de Voluceau, B. P. 105, 78153, Rocquencourt France

Abstract

In this paper, we describe the O_2 database programming language as it is currently implemented. We first show how O_2 provides the user with both objects and complex values. Then, we present the persistence management of O_2. We describe how objects are encapsulated and manipulated through methods and how values are directly accessible through operators. We also present the subtyping and inheritance relationships in O_2 together with the type-checking mechanism. Finally, we mention some interesting features which deal with exceptions and we make a comparison between O_2 and several other object-oriented database systems.

1 Introduction

The major objective of Altaïr is to prototype a complete development environment for data intensive applications. The functionalities of such a system should include those of a DBMS, those of a programming language and those of a programming environment. We decided to build an object-oriented database system, named O_2, and its programming environment. Our motivations for this choice are the following:

- We do believe that one of the main bottlenecks to the productivity of the application programmer is the impedance mismatch between the programming language and the database. This impedance mismatch cannot be solved by redefining the database box (i.e. by changing the frontier between the programming language and the database system) but by mixing database technology and programming language technology to build a complete system with the functionalities of a DBMS and of a programming language.

- We do believe that, among the available technologies produced by programming language people and among the possible approaches, the object-oriented approach is the best one to mix with database technology. This is due both to the intrinsic characteristics of the approach and to the appeal this paradigm has to programmers.

The next choice concerned the programming language of the system. Among the possible solutions (extending an existing language, designing a new language, or being language independent), we have chosen the last one mainly for marketing reasons (from a purely technical point of view the second was probably the best). The system is viewed by the user as consisting of a data definition language (DDL) by which the user can manipulate a hierarchy of classes. He/she can attach methods to classes or to objects by writing these methods in various languages. Our first target set of languages consists of C and Basic. Rather than speaking of the O_2 database programming language, one can think of the O_2 database programming languages. Programming in O_2 is done in two distinct steps. First, the programmer defines classes using O_2 commands. Then, he/she programs the code of his/her methods using one of the O_2 programming dialects. For the time being, two programming languages are specified, the CO_2 language which relies on C and the *Basic*O_2 language which relies on Basic. In this paper, we specifically report on the merge of programming language technology and database technology [AtB 87] and we shall only describe CO_2, as it is the first we have implemented. Furthermore, the approach followed for *Basic*O_2 is similar to that of CO_2. A full description of the O_2 object manager can be found in [VBD 89].

This paper is organized as follows. Section 2 contains an informal presentation of objects and values in O_2. Section 3 describes the data organization in O_2 through examples. Section 4 shows how programming is done and illustrates it using CO_2. Section 5 explains how inheritance works in O_2 and justifies its foundations through subtyping. Section 6 illustrates the features of O_2 which deal with sets and exceptions. Section 7 briefly describes how methods are type-checked in O_2 and how method safety is insured. Section 8 compares the O_2 system with several other object-oriented database systems (OODBS). Finally, we present some conclusions.

2 Objects and Values in O_2: an Informal Presentation

O_2 is object-oriented: this means that information is organized as *objects* which have an identity and encapsulate data and behaviour. Manipulation of objects is done through *methods*, which are procedures attached to the objects. Object identity is useful for supporting object sharing and updates management. The theoretical foundations for object identity as a programming language primitive can be found in [AK 89]. In classical object-oriented languages such as Smalltalk [GR 83], the value encapsulated in an object is always an atom or a tuple of other objects. In object-oriented database systems, this value is classically a tuple or a set of objects [MOP 85], [Ban et al 87], [Kup 85], [LR 88] since databases must provide flexible management of large sets of data. However, this value is always a flat value, as it can only contain identifiers of other objects, and not directly other complex values. This limitation is exactly the same as the limitation of relational systems which has motivated the introduction of nested relations and complex objects. In O_2, we provide the user with the possibility of defining, not only objects, but also *values* [1] as in standard programming languages or in the so-called complex objects[2] languages [AB 87], [Kup 85], [BK 86]. Of course, complex (nested) structures can always be modeled through the use of identifiers but we think that this solution is awkward, just as the modeling of nested relations with surrogates in relational systems. For example, we can consider the following three objects (objects identifiers are written in italics):

eiffel_tower:
tuple(`name`: "Eiffel tower",
    `address`: *eiffel_address*,
    `description`: "Paris famous monument",
    `admission_fee`: 25 FF)

eiffel_address:
tuple(`city`: *paris*,
    `street`: "Champs de Mars")

paris:
tuple(`name`: "Paris",
    `country`: "France",
    `population`: 2.6)

While both *paris* and *eiffel_address* where modeled as objects in this example, we believe that they should be treated differently: *eiffel_address* is nothing more than a pair of strings which only appears in the value of *eiffel_tower*. On the other hand, cities evolve with time (think of the population) and might be shared by other monuments, therefore we wish to model *paris* as an object. In our system, the object *eiffel_tower* should be modeled as follows, with address appearing as a *structured value* and *paris* as an *object*:

eiffel_tower:
tuple(`name`: "Eiffel tower",
    `address`: **tuple**(`city`: *paris*,
        `street`: "Champs de Mars"),
    `description`: "Paris famous monument",
    `admission_fee`: 25)

An object-oriented data base system intends to provide the application programmer with a powerful applications development support using encapsulation and inheritance. This important step forward should not be canceled by an increased complexity in structure manipulations and by losing data independence due to the navigation through objects identifiers. We claim that "pure" object-oriented database systems have severe drawbacks. The user has to define a new class every time he/she needs a complex value. This results in an undesirable growth of the class hierarchy.

Thus, the class hierarchy should only contain classes which correspond to data shared by distinct software modules. It should not be polluted by classes which are only used to describe non shared values.

Some authors already felt the need of dealing with both objects and values: in [Ban et al 87], there is a notion of exclusive relationship between an object and some of its components. When an object is exclusively owned, it can not be shared. The same notion is introduced in [CDV 88] where the programmer can specify whether he deals with a reference to a complex value or with the complex value itself, and also if an object can be shared by several objects or is exclusively owned by an object. This proposition introduces the desired distinction, but values are still implemented and manipulated as objects. In the new version of FAD [DKV 87], one can manipulate objects and values. A value in FAD is either atomic or structured. A structured value contains values. An object in FAD has an identifier and a state. A state is either an atomic value or a structure containing objects and values. Objects may be updated while values may not. Objects also allow for sharing. Exodus [CDV 88] also provides the user with a mix of objects and values similar to us (see Section 8). In O_2, we follow a similar approach. The user may choose between two kinds of organizations: *classes* whose in-

[1] The previous prototype of O_2 only dealt with objects [LRV 88]

[2] which are not objects in the object-oriented terminology but rather complex values

stances are objects and which encapsulate data and behavior and *types* whose instances are values. Values are not encapsulated, that is, their structure is known by users and they are manipulated by operators. To every class is associated a type, describing the structure of its instances. Classes are created explicitly using commands and are parts of the inheritance hierarchy. Types are not created explicitly since they only appear as components of classes and do not appear in the inheritance hierarchy. The underlying model is presented in [LR 89] and analyzed in [AK 89].

3 Types and Classes

In O_2, the user has two ways of structuring data: types and classes. Types are recursively constructed using atomic types such as integers, floats, strings, class names and the set, list and tuple constructors. Instances of types are *values*. These types are similar to classical types in programming languages. The following expression is an O_2 type:

```
tuple (name: string,
       country: string,
       population: float,
       monuments: set(Monument))
```

This type describes cities. The `monuments` attribute has a set structured value. Monument is a class name. A *value* of this type can be:

```
tuple(name: "Paris",
      country: "France",
      population: 2.6,
      monuments: set(eiffel_tower, triumph_arch))
```

Recall that we use italics to denote objects. One can see from the above example that values can be arbitrarily complex. The elements of the set value of the attribute "monuments" are objects as we shall show in the next subsection. In O_2, the user builds types using atomic types such as string, float, integer, char and boolean and three type constructors: tuple, set list. There is no restriction on the use of these constructors. We have already used the set and tuple constructors, an example of use of the list constructor is given in the next subsection.

3.1 The Schema Definition Language

In O_2, the schema is a set of *classes* related by inheritance links (see Section 5) and/or composition links. A class describes the structure *and* the behaviour of a set of objects. The structural part of a class is a type as defined above and the behavioural part is a set of methods (see Section 4). Classes are created using schema definition commands as follows:

```
add class City
    type tuple(name: string,
               country: string,
               population: integer,
               monuments: set(Monument))


add class Monument
    type tuple(name: string,
               address: tuple(street: string,
                              city: City),
               description: string,
               closing_day: list(string),
               admission_fee: integer)
```

We denote class by capitalizing the first letter. The first class has a name "City" and a type which is given after the keyword **type**. Instances of this class are objects. That is, they have a unique internal identifier and a value which is an instance of the type associated to the class. Objects are encapsulated, that is, their value is not directly accessible and they are manipulated by methods as explained in Section 4. The second class defines historical monuments. Note that classes can be mutually referencing. The "City" class references the "Monument" class which in turn references the "City" class. For every "Monument" object, the value of the "`city`" attribute is an object which may itself references the "Monument" object.

Following an approach similar to Galileo [ACO 85], the equivalence of classes is by name as opposed to type equivalence which is by structure. That is, the type of the values only depends of their structure. One the other hand, two classes are always distinct and the compatibility rule is the name equivalence rule.

3.2 Object Creation

Creation of objects is done through a system command called "new". This command "new" takes as input the name of the class corresponding to the object to be created. The object is created with a default value depending on the type associated to the class. The default values are: the empty string, the integer 0, the float 0.0, the empty list and the empty set for list and set types, and a tuple of default values for tuple types.

3.3 Naming and Persistence

In O_2, objects or values can be named. The following is an example of naming:

add object Eiffel_tower: Monument

The name Eiffel_tower will then stand for an object of class Monument. In the same way, one can name a value as follows:

add value Paris_monuments: **set**(Monument)

"Paris_monuments" is a name for a value of type (**set**(Monument)). In O_2, persistence is attached to names, that is, every named object or value is persistent. Such a name can be seen as a global variable dynamically attached to a given object or value and makes it persistent. The attached object can be changed by assignment. For instance, we can write:

Eiffel_tower = **new**(Monument)

This instruction assigns a newly created object to the name "Eiffel_tower". The initial value of the object is the tuple default value corresponding to the type. This object will always be accessible trough the name "Eiffel_tower" during the life of the system, except if the user makes another assignment.

The persistence rules are the following:

1. every named object or value is persistent,
2. every object or value which is a part of another persistent object or value is persistent.

For example, let us assume that we have made the following assignment:

Paris_monuments =
set(Eiffel_tower, *triumph_arch*)

where Eiffel_tower is a named object and *triumph_arch* denotes an object of class Monument with no name. Then, these objects are persistent. The first one is already persistent due to its name and the second is persistent because it is an element of the named value. The same holds for objects or values which appear as an attribute value in the named object "Eiffel_tower".

The *extension* of a class is the set of all objects created using the **new** command applied to that class. The system provides the user with an automatic management of class extensions. This is done using a set value which collects all the objects of a class. For instance, one can write:

```
add City with extension
    type tuple(name: string,
               country: string,
               population: integer,
               monuments: set(Monument))
```

The **with extension** clause in the class definition tells the O_2 system to create a named value of type "**set**(City)" with name "City". Moreover, every city created with the "**new**(City)" command will be automatically inserted in this set and will thus persist, as it is a component of a persistent set. Note that, according to our persistence rules, objects of a class without extension will not persist unless they are explicitly named or components of some other persistent object or value. Classes with no extension are a natural way of dealing with transient objects.

4 Objects and Values Manipulation

4.1 Methods Definition

In the object-oriented approach, objects are manipulated by *methods*. A method is a piece of code which is attached to a specific class and which can be applied to objects of this class. In O_2, method definition is done in two steps. First, the user declares the method by giving its signature, that is, its name, the type or class of the arguments and the type or class of the result if there is one. Then, he/she gives the code of the method. The following is a method declaration:

add method increase_fee (amount: integer)
in class Monument

This method increases the `admission_fee` field of a Monument object. Methods can be *private* or *public*. Private methods are only visible within their class, i.e. in the methods attached to that class. Public methods are visible by every classes and can be freely used. When declaring a method, the user can add the keywords **is public** in order to make it public. The default is private.

O_2 follows a multi-language approach. This means that method programming is done in a standard programming language such as C or Basic with manipulation of O_2 objects and values. The main idea is that most of the programming is done using the programmer's favorite language. This includes iterations, control structures and arithmetics. Access to, and manipulation of, objects and values is done using O_2 features. We give below, as an example, the code of the method "increase_fee" using the CO_2 language.

```
body increase_fee(amount: integer) in class Monument
      co2 { (*self).admission_fee += amount; }
```

The curly brackets delimit the CO_2 block of code as in pure C. The value of an object is obtained using the "dereferencing" method *, thus "self" is the object and "*self" is the associated value. This method is applied using a special syntax which follows the C "*" use. It illustrates the association between objects and values. As in standard programming languages, objects can be seen as pointers to values. In the example, the value "*self" is tuple-structured, and the access to an attribute is done using the dot operator. The assignment is done as in C and increments the integer value representing the admission fee. Notice that we stick to the C syntax for manipulating O_2 values such as dereferencing or extracting a tuple field. This way of manipulating objects is syntactically very close to what is done in C++ [Str 86]. In O_2, however this similarity is purely syntactical, as objects and values are implemented and manipulated in a special way by a persistent object manager and the CO_2 compiler generates calls to this object manager [VBD 89].

A method is applied to an object by *message passing* whose syntax is the following:

[receiver selector(arguments)]

The square brackets are used to delimit O_2 message passings. "receiver" denotes an object to which the method whose name is "selector" is applied. This eventually returns an object depending on the method code. For example, "increase_fee" is applied to a monument using the message passing:

"[Eiffel_tower increase_fee(3)]".

The keyword "self" in the above code will denote the object "Eiffel_tower" when the method is applied.

4.2 Manipulating Values

The CO_2 language allows the construction of O_2 values using the set, list and tuple constructors. We can, for instance, write a set value containing four integers as follows: **set**(1, 4, 34, -21). The following associates a value to a newly created object:

```
Eiffel_tower = new(Monument);
*Eiffel_tower =
tuple(name: "Eiffel tower",
      address: tuple(city: paris,
                     street: "Champs de Mars"),
      description: "Paris famous monument",
      closing_day: list("Christmas", "Easter");
      admission_fee: 25)
```

This assumes the * method is public for the Monument class. We have seen above that we can extract a field of a tuple value using the dot operator. All the CO_2 value manipulations are done in this way, using the classical C constructs. For instance, we shall append elements to the **closing_day** list of the Eiffel tower or modify one entry of the list as follows:

```
*Eiffel_tower.closing_day += list("June 6th");
*Eiffel_tower.closing_day[1] = "January 1st";
```

O_2 provides the user with the usual sets and lists operators (union, intersection, difference, cardinality, concatenation, ...) whose syntax follows as much as possible the C syntax.

4.3 Iterator

The iterator described here is applied on set or list structured *values*, not on objects. Indeed, objects are encapsulated and one should not know what is the structure of the encapsulated value. Of course, the values to which the iterator is applied may be a set (a list) of values or a set (a list) of objects. CO_2 provides the user with an iterator which allows for easy sets or lists manipulations.

for (x **in** S [**when** condition]) <Statement>.

This is an extension of the classical C iterator. It applies the given statement with the variable x bound to every element of the set (or list) value S satisfying the optional condition. The **when** clause adds no power to the **for** iterator, but allows some optimization when the condition is directly evaluable by the object manager. For instance, we can write:

```
co2{ o2 Monument x;
     for (x in Paris_monument
     when (*x.admission_fee <= 20.00))
     [x increase_fee(amount)];
     }
```

The above code increases the admission fee of all the monuments located in Paris, whose **admission_fee** is less or equal to 20.00FF. The expression "o2 Monument x" declares an O_2 variable which is used to denote objects of class "Monument". Recall that "Paris_monument" is a named value of type **set**(Monument) which is supposed to contain all monuments of Paris. The **for** iterator is of course less concise but more flexible than the classical join operation. The reader should notice that it is far more powerful in the context of O_2 which is a programming language and not an end-user query language.

5 Subtyping and Inheritance

Inheritance is a powerful mechanism which allows the user to define classes in an incremental way by refining already existing ones. O_2 provides the user with an inheritance mechanism based on subtyping.

5.1 Subtyping

Subtyping is a semantic relationship which connects two types. There are several ways of defining subtyping. In O_2, we defined a set inclusion semantics for subtyping. That is, a type is a subtype of another if and only if every instance of this type is also an instance of its supertype. This allows to say that a person is a human or that an employee is a person. The formal definition of the O_2 type system is given in [LR 89]. Another approach is taken by Vision [Car 87]. In this system, subtyping is expressed by means of a mapping from the objects of the subtype to objects of the supertype. We adopted a Cardelli-like approach [Car 84] for tuple subtyping. A tuple type is a subtype of another if it is more defined, that is, if it contains every attribute of its supertype plus some new ones and/or refines the type of some attributes of its supertype. The following example illustrates this.

```
tuple(name: string,
      address: tuple(street: string,
                     city: City),
      description: Text,
      closing_day: list(string),
      admission_fee: integer,
      number_rooms: integer,
      rate: integer)
```

is a subtype of:

```
tuple(name: string,
      address: tuple(street: string,
                     city: City),
      description: Text,
      closing_day: list(string),
      admission_fee: integer)
```

Another characteristic of this subtyping relationship is that a set-structured type "**set**(T)" is a subtype of "**set**(T')" if and only if T is a subtype of T'. For instance:

```
set(tuple(name: string,
          address: string))
```

is a subtype of:

```
set(tuple(name: string))
```

The same relationship holds for lists.

5.2 Inheritance

Based on this subtyping relationship, O_2 offers an inheritance mechanism. We can define the Historical_hotel class as follows:

```
add class Historical_hotel inherits Monument
        type tuple (number_rooms: integer,
                    rate: integer)
```

The effect of this declaration is the definition of an Historical_hotel class whose associated type is a subtype of the Monument type. The user only has to give the extra attributes (the other ones are taken from the definition of inherited class). The O_2 command interpreter checks whether the inheritance definition is legal, that is if there is no subtyping violation, and creates the subclass according to the subtyping rules. An object of class Historical_hotel will automatically be considered as an object of class Monument. This results in the possibility of applying any method of class Monument to Historical_hotel objects. O_2 also allows for multiple inheritance, as shown below. We first define a "Restaurant" class.

```
add class Restaurant with extension
    type tuple (name: string,
                address: tuple(city: City,
                               street: string),
                menus: set(tuple(name: string,
                                 rate: float)))
add method check_rates(float): boolean
    in class Restaurant
```

The method "check_rates" checks whether the menus rates are less than a given amount. We can now define an "Historical_restaurant" class as follows:

```
add class Historical_restaurant with extension
    inherits Monument, Restaurant
    type tuple (redefines name: string,
                redefines address:
                     tuple(city:City,
                           street: string))

add method check_rates(float): boolean
    in class Historical_restaurant
```

Class "Historical_restaurant" inherits both from Monument and from Restaurant. Here, the method "check_rates" checks whether the menus rates are less than twice the amount[3]. We shall not detail the conditions that method signatures must satisfy in order to be inherited through the subclasses, see [LRV 88].

As opposed to single inheritance, possible ambiguities may arise with multiple inheritance when an attribute or a method name is defined in two or more superclasses. There are several solutions to such ambiguities [Ban et al 87], [SCBKW 86]. We decided to follow an approach similar to that of Trellis/Owl [SCBKW 86]. That is, the user has to explicitly redefine the attribute or method name when needed. We think that, as opposed to solutions where the system solves the ambiguity by itself by ordering the superclasses, this solution is more natural and enhances the readability and maintainability of the schema. Thus, the "Historical_restaurant" class redefines the attributes "name" and "address" which are both present in the classes "Restaurant" and "Monument". The reader should note that we do not infer the subclass relationship which is user defined. The system just checks whether it is legal with respect to the subtyping rules.

5.3 Late Binding

An important feature of object-oriented systems, which is fully implemented in O_2, is late binding. The actual code of a method to be executed is not selected at compile-time but at run-time depending on the actual type of the receiver object. The main benefit is dynamicity and reuse of existing software. Indeed, existing methods do not have to be recompiled when the code of the methods they use is changed. An example of use of late binding is:

```
for(x in Restaurant) {
if(![x check_rate(120.50)])
  printf("restaurant %s is expensive", *x.name);
}
```

This iteration loop applies the "check_rate" method to every restaurant. Due to our subtyping semantics, some of them are historical restaurants. For these ones, the system automatically applies the method defined in the class "Historical_restaurant". This avoids to explicitly take into account the different status of historical restaurants. Late binding is a critical operation from the performance point of view. The O_2 choice for the implementation of late binding is described in [VBD 89].

[3]The restaurant is historical and is allowed to increase its rates!

6 Interesting Features

In this section, we describe some interesting features of O_2 which improve the expressibility of the O_2 language.

6.1 Exceptional Attributes

Due to the semantics of the subtyping relationship, a tuple value can have extra attributes. If we consider the Monument class, the "Eiffel_tower" object can have a value which also contains an attribute "height". This extra attribute will not be dealt with by the methods associated to the Monument class, however, the standard operators available on tuple values will handle it. For instance, the following is a correct CO_2 code:

```
{Eiffel_tower = new(Monument);
 *Eiffel_tower=
 tuple(name: "tour Eiffel",
       address: tuple(street: "Champs de Mars",
                      city: Paris),
       description: "Paris famous monument",
       closing_day: list("Christmas", "Easter"),
       height: 315,
       power: 15.5);
 (*Eiffel_tower). height = 320;
 return ((*Eiffel_tower).height);}
```

Assuming that de-encapsulation is allowed on Monument, this code first modifies the value of "Eiffel_tower" and adds a "height" and a "power"[4] attribute. Then, using the dot operator, the height attribute is updated and finally its new value is returned. Note that exceptional attributes are allowed for any tuple object or value, even if not named.

6.2 Exceptional Methods

One can associate specific methods to named objects. These methods are used to characterize the exceptional behaviour of an object. One can also override an existing method in the class of the object with an exceptional method. An example of this mechanism is given below:

[4]The Eiffel tower is also a radio and a TV broadcasting station

add method increase_power (amount: float)
in object Eiffel_tower

This method will be used to increase the broadcasting power of the Eiffel tower station. Note that the method is associated to the name not to a particular object, and that the actual object associated to the name "Eiffel_tower" can change at run-time. The late binding process will associate the exceptional method to the object currently bound to the name.

7 Type-Checking

O_2 emphasizes user defined classes and their associated types. This is a natural way to structure data. An other important motivation is type-checking. The goal of O_2 is to increase the productivity of business applications programmers. In this context, safety of programs is critical. Thus, O_2 offers a static type-checker which detects the illegal manipulations of O_2 objects and values when inheritance is not used. When full use of inheritance or of exceptional attributes is done, O_2 must perform some run-time type-checking. Of course, since the method code of O_2 can be written in several languages which may be loosely typed such as C, there also may be errors due to the host languages manipulations. The type checking algorithm used in O_2 is standard. It is conceptually similar to that of Trellis/Owl [SCBKW 86] in that a variable can only be assigned values (resp objects) of its declared type (resp class) or of any subtype (resp subclass). The user may modify the schema dynamically. In this case, a method which has not been recompiled may perform message passings which reference non existing methods. Of course, if the user recompiles every method which may be concerned by the schema modifications, references to non existing classes or methods are detected by the type-checker. Other run-time errors occur with exceptional attributes. At compile-time, the type-checker may not know whether an attribute, which is not present in the variable declaration, but is referred to in the code, is an exceptional attribute or not. Accepting such a manipulation implies that the method may fail at run-time because the actual value does not possess this exceptional attribute. We accept this for the sake of expressive power. The user may choose not to use exceptional attributes and have safe programs.

8 Related Works

In this section, we list the main characteristics of O_2 and see what kind of solutions others OODBS's provide. We compare O_2 to other systems on the basis of the programming language only. We shall not be concerned by query facilities, user interface or physical management.

Gemstone [MOP 85] is to our knowledge the first implementation of an OODBS. The philosophy of Gemstone was to turn Smalltalk into an database system without significant modifications of the Smalltalk programming language. Vision [Car 87] is another interesting approach. Vision models data in a way similar to Daplex [Sch 81]. All informations about an object are embodied in *functions* which map a collection of objects into another. However, function application follows a message passing mechanism using a Smalltalk-like approach. Iris [DFKLR 86] also follows a functional approach in that, to every object, is associated a set of functions which characterize its content. Orion [Ban et al 87] is another example of a functional approach since it is implemented using Lisp and has a Lisp syntax for the message passing. Vbase [AH 87] follows an approach similar to O_2 as the corresponding language (COP) is a strict superset of the C programming language. Although it is not a true OODBS, Trellis/Owl is another example of an object-oriented language with an imperative way of programming. It has a conventional programming language syntax and uses a procedure call notation to invoke operations on objects. Trellis/Owl does not have all the database functionalities but provides persistence through an object repository. A common characteristic of these approaches is that they provide compile-time type-checking. The Exodus system [CDV 88] is also an object-oriented system which allows abstract data types definitions, objects and values and a query language named Excess. Programming is also done in the E language which is a persistent C++.

We now list the main original features of O_2 and describes what is done in other OODBS's.

- O_2 *provides the user with both objects and structured values.*
  We do not follow a pure object approach as in Smalltalk or Gemstone but allow the definition of nested values built using the set, list and tuple constructors. O_2 manipulates objects using methods and values using operators. That is, full object-oriented features are available for objects (such as late binding and inheritance of methods) and values are manip-

ulated as in database systems. Most OODBS's provide object constructors similar to the set and tuple constructors. The Exodus system also gives an array constructor which is similar to the O_2 list constructor. The distinction between objects and values can also be found in Orion. In this system, however, the notion of complex value is implemented as a *dependent object* [KBCGW 87]. That is, non shared values are still objects with a constraint enforcing their privacy. The Exodus data model [CDV 88] also provides the user with this distinction. However, just as in Orion, values are second-class objects with no identity. In O_2, we enforce the distinction between objects and values in the programming language because we encapsulate objects which can only be manipulated through methods. Exodus adopts a point of view which is less object-oriented but more database oriented. In Exodus as in Orion, for the sake of query simplicity and uniformity, objects and values are manipulated in the same way. In Iris and in Vision, one only has objects. Objects are atomic items which can be printable (like the object "3") or not. If an object is not printable, its value is characterized by a set of functions which can be stored, and thus plays the role of attributes, or computed. As opposed to O_2, where the three object constructors have exactly the same rights, due to their approach, Iris and Vision manipulate complex objects which are records of functions which can however be multi-valued.

- O_2 *follows a multi-language approach.*
Classes and types are created using the O_2 schema commands, but the code of methods can be implemented using several O_2 extensions. In this paper, we concentrated on CO_2 but another extension is currently under implementation based on the Basic language. Up to now, among the existing OODBS's, O_2 is the only multi-language system.

- O_2 *has a compile-time type-checker.*
Systems such as Gemstone based on a Smalltalk like approach do not provide such a functionality nor do systems based on Lisp such as Orion. On the other hand, systems based on an imperative paradigm are statically typed. Among them, let us quote Trellis/Owl and Vbase. As in O_2, Trellis/Owl and Vbase have a strong typing. That is, every object is an instance of a type and every variable is declared of a type. A variable can only be assigned objects of its type or of a subtype of its type. In order to have statically typed languages, types and methods are not modeled as objects and manipulated by methods but are primitive constructions manipulated by schema commands.

- O_2 *provides an automatic management of persistence through named objects and values.*
Every named object or value is persistent and every component of a persistent object or value is itself persistent. The name can be seen as a handle which allows the user to access an object or value after the end of a program which has defined it. Other systems provide a somehow similar way of managing persistence. Objects in Orion also persist because they are components of persistent collections. For every user defined type, the system generates a set structured class which has at least one instance which groups the instances of the former class. In Gemstone, the management of persistence also uses reachability informations, that is, objects are persistent if they are attached to a persistence root or another persistent object. An Exodus database is a collection of named persistent objects.

- *Updates are always implicit in* O_2.
Objects are created using the "new" command. If a class is created "**with extension**" then a named set value is created which will contain every object of the class which will thus persist. If the class is not created with extension, then the created objects will only persist if attached to other persistent objects. Deletion of objects or values is obtained by removing the links which attach them to the persistence roots (the names). Classes with extension are also provided with a "delete" method which allows objects to be removed from the class extension when no other objects or values refer to it. Gemstone and Orion have a similar update policy, as they have a similar persistence policy. In Vbase, however, every object is persistent, and temporary objects have to be deleted explicitly.

- O_2 *has a set inclusion semantics for subtyping.*
Objects of a subclass are objects of the superclasses. For instance, if one performs a display on the instances of Monuments, one will also see the instances of Historical_hotel. Some systems follow this approach, such as Trellis/Owl

and Iris. On the other hand, Vision has a mapping semantics: an object of a subclass has a corresponding image object in its superclass. We find this somehow unnatural. However, this provides the same kind of functionalities at least in the context of single inheritance as provided in Vision. Iris has also a set inclusion semantics. This is even more general, since an object can have several types even if these types are not related in the specialization hierarchy. We are not aware of the way they solve ambiguities. Orion has no set inclusion semantics.

- *Multiple inheritance conflicts are solved by users.*
  Trellis/Owl proposes a similar solution. The user must solve the ambiguities which may arise. For instance, when there is an ambiguity on the inheritance of a method (operation in Trellis/Owl), the user must specify which one he/she wishes to inherit or redefine it. O_2 follows exactly the same approach as shown in Section 5. Another system which provides multiple inheritance is Orion. As opposed to O_2 or Trellis/Owl, Orion automatically solves ambiguities. Roughly speaking, the system maintains an ordering among the superclasses which desambiguates inheritance of methods. To our knowledge, other systems, such as Gemstone or Vbase do not support multiple inheritance.

- *The* O_2 *system allows exceptional methods and attributes for objects.*
  Exceptional methods can be associated to names. These methods are only accessible from the object currently attached to the name and override the methods of the class. Exceptional attributes can be added to every tuple structured objects or values. To our knowledge, no other OODBS provides such a functionality.

Another interesting approach is that of Galileo [ACO 85]. We did not put it in the collection of items above since it is not really an object-oriented data base management system, however it has some object-oriented features such as classification, abstract types and types hierarchies. As opposed to O_2, Galileo does not have the set constructor but is higher order and has a function type constructor. It has the notions of concrete and abstract types which roughly correspond to our types and classes. Galileo presents a very interesting solution to persistence which is however not yet implemented to our knowledge. Another important difference with O_2 is that Galileo does not support object identity. We did not put in this list the Damokles database system [DGL 87] which is designed for software engineering environment. As their designers say, Damokles is a "structurally object-oriented" database. That is, Damokles provides the user with object identity, complex objects based on the tuple constructor and n-ary bidirectional relationships between objects. However, Damokles does not provide encapsulation, inheritance or late binding. It is rather an " complex objects" system.

9 Conclusion

In this paper, we have described the features of the O_2 system as it is currently running. We only described the CO_2 programming language, but most of the described features are common to both CO_2 and $BasicO_2$, and the difference between the two languages is mainly syntactical.

The target applications for our language are (i) traditional applications such as business and transactional (excluding however very high performance transaction processing systems), (ii) office automation applications and (iii) spatial data management (such as geographic data management). At this stage of the game, no specific emphasis is given to CAD/CAM, CASE or knowledge base applications, but we believe that, in a later stage, the system could be enhanced to serve also these applications.

Altaïr started in September of 86. We first implemented, in December 87, a throw away prototype [Ban et al 88] whose data model is described in [LRV 88], in order to test and show the functionalities of the system.

This gave us a lot of feed back and we completely redesigned the system, its language, its data model [LR 89] and its architecture. The major differences between this version and the throw away prototype from the language point of view are: (1) complex values together with objects, (2) names for objects and values, (3) the list type constructor, (4) an automatic persistence mechanism, (5) the possibility of separating classes and method definitions from the implementation and (6) last but not least, a better merge between the O_2 syntax and the host language ones, i.e. every implementation of O_2 on a given host language follows the syntax of the host language. The current prototype runs on Sun and implements all of the functionalities listed above. The $BasicO_2$ compiler is under implementation.

10 Acknowledgments

The authors thank F. Bancilhon and P. Kanellakis for their careful reading and comments which greatly improve the quality of this paper. The O_2 compiler was implemented by the language team of the GIP Altaïr which includes D. Excoffier, L. Haux. C. Lécluse and P. Richard. The module which is in charge of storing and managing classes and methods was designed and implemented by S. Gamerman and C. Delcourt. The object manager has been implemented by the system team. Numerous suggestions and improvements on the language and its syntax have been proposed by the Altaïr team and in particular by F. Bancilhon.

References

[AB 87] S. Abiteboul and C. Beeri, "On the Power of Languages for Manipulating Complex Objects", *International Workshop on Theory and Applications of Nested Relations and Complex Objects, Darmstadt, 1987.*

[AK 89] S. Abiteboul and P. Kanellakis, " Object Identity As A Query Language Primitive", internal report, 1989.

[AN 86] H. Ait-Kaci and R. Nasr, "LOGIN: A Logic Programming Language with Built-in Inheritance", *Journal of Logic Programming, 1986.*

[ACO 85] A. Albano, L. Cardelli and R. Orsini, "Galileo: a Strongly typed, Interactive Conceptual Language", *in ACM Trans. on Database Systems, 10(2):230-260, June 1985.*

[AH 87] T. Andrews and C. Harris, "Combining language and Database Advances in an Object-Oriented Development Environment", *Proc of the OOPSLA Conference, October 1987.*

[AtB 87] M.P. Atkinson and O.P. Buneman, "Types and Persistence in Database Programming Languages", *ACM Computing Surveys, June 1987.*

[BK 86] F. Bancilhon and S. Khoshafian, "A Calculus for Complex Objects", *ACM PODS, 1986.*

[Ban et al 88] F. Bancilhon, G. Barbedette, V. Benzaken, C. Delobel, S. Gamerman, C. Lécluse, P. Pfeffer, P. Richard and F. Velez, "The design and Implementation of O_2, an Object-Oriented Database System", *in Advances in Object-Oriented Database Systems Springer-Verlag, September 1988.*

[Ban 88] F. Bancilhon, "Object-Oriented Database Systems", *Proc of PODS 88, Austin, March, 1988.*

[Ban et al 87] Banerjee J., et al., "Data Model Issues for Object-Oriented Applications", *ACM trans. on Office Information Systems, Jan 1987.*

[Ber et al 88] P. Bernstein et al., "Future directions in DBMS Research", *Workshop of the International Computer Science Institute, Feb 4–5, 1988.*

[Car 84] L. Cardelli, "A Semantics of Multiple Inheritance", Semantics of Data Types, *Lecture Notes in Computer Science, 1984.*

[CW 85] L. Cardelli and P. Wegner, "On Understanding Types, Data Abstraction, and Polymorphism", *ACM Computing Surveys, Vol 17:4, 1985.*

[CDV 88] M. Carey, D. DeWitt and S. Vandenberg, "A Data Model and Query Language for EXODUS", *Proc of the ACM-SIGMOD Conference, Chicago, 1988.*

[Car 87] M. Caruso, "The VISION Object Oriented Database Management System", *Proc of the Workshop on Database Programming Languages, Roscoff, France, Sept. 1987.*

[DKV 87] S. Danforth, S. Khoshafian and P. Valduriez, "FAD - A Database Programming Language", *MCC Technical Report, October, 1987.*

[DFKLR 86] N. P. Derrett, D. H. Fishman, W. Kent, P. Lyngbaek and T. A. Ryan, "An Object-Oriented Approach to Data Management", *Compcon 31 IEEE Computer Soc. Int. Conference, 1986.*

[DGL 87] K. Dittrich, W. Gotthard and P. Lockeman, "DAMOKLES - The Database System for the UNIBASE Software Engineering Environment", *Database Engineering, Vol 10, No 1, March 1987, pp 37-47.*

[GR 83] A. Goldberg and D. Robson, "Smalltalk80: The Language and its Implementation", *Addison Wesley, 1983.*

[KBCGW 87] W. Kim, J. Banerjee, H-T. Chou, J. F. Garza and D. Woelk, "Composite Object Support in an Object-Oriented Database System", *Proc of the OOPSLA Conference, October 1987.*

[Kup 85] G. Kuper, "The Logical Data Model: a new Approach to Database Logic", PhD thesis, Standford University, September 1985.

[LRV 88] C. Lécluse, P. Richard and F. Velez, "O_2, an Object-Oriented Data Model", *Proc of the ACM-SIGMOD Conference, Chicago, 1988.*

[LR 88] C. Lécluse, P. Richard, "Modeling Inheritance and Genericity in Object-Oriented Databases", *Proc of the ICDT 88 Conference, Brugge, Aug 31–Sep 2, 1988.*

[LR 89] C. Lécluse and P. Richard. "Modeling Complex Structures in Object-Oriented Databases", *to appear in proc of the PODS 89 Conference, Philadelphia, March 29–31, 1989.*

[MOP 85] D. Maier, A. Otis and A. Purdy, "Development of an Object-Oriented DBMS", in *a Quaterly Bulletin of the IEEE Computer Society Technical Committee on Database Engeneering, Special issue on Object-Oriented Systems, Vol 8:4, 1985.*

[Pri 84] D. Price, "Introduction to ADA", *Prentice-Hall , Inc., Englewood Cliffs, new Jersey, 1984.*

[SCBKW 86] G. Schaffert, T. Cooper, B. Bullis, M. Kilian and C. Wilpot, "An Introduction to Trellis/Owl", *in Proc of the OOPSLA Conference, Portland, 1986.*

[Sch 81] D. W. Shipman, "The Functional data Model and the Data language DAPLEX", *ACM Transactions on Database Systems 6(1), pp 140-173, March 1981.*

[Str 86] B. Stroustrup, "The C++ Programming Language", *Addison Wesley, 1986.*

[VBD 89] F. Velez, G. Bernard and V. Darnis, "The O_2 object Manager, An Overview", Altaïr Internal Report, 1989.

A MODEL OF QUERIES FOR OBJECT-ORIENTED DATABASES

Won Kim

Microelectronics and Computer Technology Corporation
3500 West Balcones Center Drive
Austin, Texas 78759

Abstract

One major source of confusion, and consequent criticisms, about object-oriented databases has been the lack of a comprehensive model of queries. Although there is a reasonable degree of agreement about an object-oriented data model, few operational systems support a query model for object-oriented databases. In this paper, we present a rather comprehensive query model which is consistent with object-oriented concepts embodied in the object-oriented data model. The model takes into account the semantics of the class hierarchy and nested objects, and as such is inherently richer than the relational or nested relational model of queries. A significant subset of the model has been cast into a query language which is supported in the ORION object-oriented database system.

1. INTRODUCTION

An object-oriented database system is a database system which directly supports an object-oriented data model. Further, an object-oriented data model includes a number of concepts found in many object-oriented programming languages and knowledge representation languages. An object-oriented database system must provide persistent storage for the objects and their descriptors (schema). The system must provide the user with an interface for schema definition and modification, and for creating and accessing objects by sending messages to the objects.

This basic system may be extended in several dimensions, just as several types of features had to be added to the relational database systems during the past decade. These features include a declarative query language; integrity features, such as transaction management and triggering; perform ance-related features, such as secondary indexing and clustering; and concurrency control and authorization for a multi-user environment.

One of the important shortcomings in most object-oriented database systems operational today is the query capabilities. In most systems, the query model captured in the query language fails to take into account some of the fundamental object-oriented concepts. Some systems, such as VBase [ANDR87], merely provide an SQL relational query interface. Some attempt to provide a new query language which is backward-compatible with a relational query language: SQL in the case of IRIS [FISH87], and QUEL [STON76] in the case of POSTGRES [STON86, ROWE87]. Other systems, such as ORION [BANE87a, KIM88a, KIM89] and GemStone [MAIE86], support a new query language which is based more on the nested-relational model [MAKI77, ABIT84, ABIT86, JAES82, IEEE88].

[BANE88] presents a query model for object-oriented databases and compares it with the relational query model; the model is subsequently extended in [KIMK89] to better account for cyclic queries. Although the model is limited in a number of important ways, to our knowledge it is the first attempt to define a query model for object-oriented databases which is *consistent* with basic object-oriented concepts. The model takes into account the two-dimensional nature of the schema of an object-oriented database: a class has a number of attributes whose types (domains) are also classes, and a class may have a number of superclasses and a number of subclasses.

The model of [BANE88, KIMK89] is not comprehensive: it does not include any consideration for operations comparable to relational joins, set operations, and views. Further, the model, even in its limited form, has not been adequately formalized. The objectives of this paper are to extend the query model to a significantly more comprehensive model, and to attempt to formalize the extended model.

2. CORE OBJECT-ORIENTED DATA MODEL

In this section we will review a core object-oriented data model, which is based on what we consider a minimal set of fundamental object-oriented concepts, and emphasize the importance of these concepts from a database perspective. A more detailed discussion of the minimal set of object-oriented concepts is given in [KIM88b] (in fact, part of this section has been taken from [KIM88b]). The reader will need to clearly understand this section in order to follow the query model we will develop.

object and object identifier

In object-oriented systems and languages any entity is uniformly modeled as an object. Further, an object is associated with a unique identifier; an identifier is not reused even when the object with which it was associated is deleted from the system. The uniform treatment of any entity as an object simplifies the user's view of the world. The object identifier is used to pin-point an object to retrieve.

attributes and methods

Every object encapsulates a state and a behavior. The state of an object is the values for the attributes of the object, and the behavior of an object is the set of methods

Proceedings of the Fifteenth International Conference on Very Large Data Bases

Amsterdam, 1989

which operate on the state of the object. The value of an attribute of an object is also an object in its own right. Further, an attribute of an object may take on a single value or a set of values.

class

All the objects which share the same set of attributes and methods are grouped into a higher level object called a class. An object must belong to only one class as an instance of that class. The relationship between an object and its class is the familiar instance-of relationship. A class is similar to an abstract data type. A class may be a primitive class, such as integer, string, and boolean; a primitive class is a class with no attributes.

The concept of a class as a set of instances provides the basis on which a query may be formulated. In relational databases, a query is issued against a relation or a collection of relations; similarly, in object-oriented databases, a query must be issued against a class or a collection of classes. Without the notion of a class, or a set of instances, to aggregate together related objects, it will be very difficult to conceptualize (and evaluate) a query.

The value of an attribute of an object, since it is necessarily an object, also belongs to some class. This class is called the domain of the attribute of the object. The domain of an attribute may be any class, including a primitive class. This represents a significant difference from the normalized relational model in which the domain of an attribute is restricted to a primitive class. The fact that the domain of an attribute may be an arbitrary class gives rise to the nested structure of the definition of a class. That is, a class consists of a set of attributes (and methods); the domains of some or all of the attributes may be classes with their own sets of attributes; and so on. Then the definition of a class results in a directed graph of classes rooted at that class; as in [KIM88b] we will call this graph a *class-composition hierarchy*. If the graph for the definition of a class is restricted to a strict hierarchy, the class becomes a nested relation.

class hierarchy and inheritance

Object-oriented systems allow the user to derive a new class from an existing class; the new class, called a subclass of the existing class, inherits all the attributes and methods of the existing class, called the superclass of the new class, and the user may specify additional attributes and methods for the subclass. A class may have any number of subclasses. Some systems allow a class to have only one superclass, while others allow a class to have any number of superclasses. In the former, a class inherits attributes and methods from only one class; this is called *single inheritance*. In the latter, a class inherits attributes and methods from more than one superclass; this is called *multiple inheritance*. In a system which supports single inheritance, the classes form a hierarchy, called a class hierarchy. If a system supports multiple inheritance, the classes form a rooted directed acyclic graph, sometimes called a class lattice.

The concept of a class hierarchy is completely orthogonal to that of a class-composition hierarchy. A class hierarchy captures the generalization relationship between one class and a set of classes specialized from it. A class-composition hierarchy has nothing to do with inheritance of attributes and methods. Figure 1 shows an

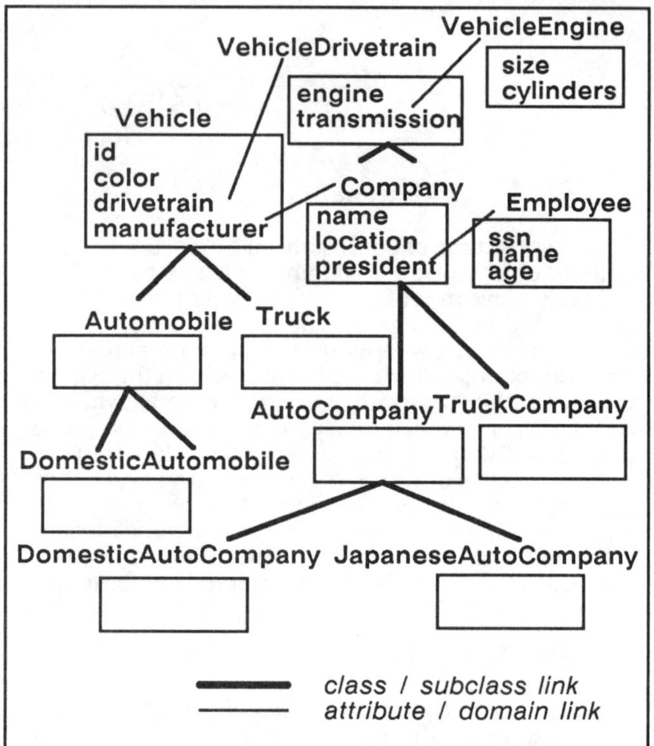

Figure 1. An Example Schema Graph

example schema. The class Vehicle is the root of a class-composition hierarchy which includes the classes VehicleDrivetrain, VehicleEngine, Company, and Employee. The class Vehicle is also the root of a class hierarchy involving the classes Automobile, DomesticAutomobile, and Truck. The class Company is in turn the root of a class hierarchy with subclasses AutoCompany, JapaneseAutoCompany, DomesticAutoCompany, and TruckCompany. It is also the root of a class-composition hierarchy involving the class Employee.

3. QUERY MODEL

The query model implemented in ORION, despite its limitations, is the first to be based on a consideration of the power and constraints of object-oriented concepts. The model restricts the target of a query to a single class or a class hierarchy rooted at that class. This is an important restriction, since this excludes operations comparable to relational joins and set operations. However, the model explicitly takes into consideration some of the important consequences of object-oriented concepts. First, it allows the user to use the directed graph model of the definition of the target class for specifying a query; predicates may be applied to any attributes of any classes on the graph. This is similar to the nested-relational extensions of the relational selection operation [JAES82, DESH88]. Second, a query may be directed against a single class or a class hierarchy rooted at the class. This is important, since a class hierarchy captures the IS-A relationship between a class and all its subclasses; and as such instances of a class may be regarded as belonging to the class and all classes on the superclass chain starting from the class. In fact, the domain of an attribute of a class is the specified class and all direct and indirect subclasses of the class.

In this section we will use the model of [BANE88] as the basis for what we hope to be a comprehensive query model for object-oriented databases. We will define our query model for a set of operations defined for relational database systems, namely, selection, projection, join, and set operations. We hasten to note that, although we will use the relational terminology, the semantics of these operations are rather different from those used in relational systems.

In Section 3.1 we first define our model for a single-operand query, that is, a query whose target is only one class or a class hierarchy rooted at one class; the relevant operations are those comparable to relational restriction and projection. In Section 3.2, we will extend the model to account for a multiple-operand query, that is, a query which captures operations comparable to relational joins and set operations. The current implementation of ORION includes single-operand acyclic queries. In the remainder of this paper, we will use, without any detailed explanation, a rather intuitive syntax for a query language for illustrative purpose. We restrict the scope of this paper to a presentation of an object-oriented query model, and defer to a future paper a full description of the query language that captures the model.

3.1 SINGLE-OPERAND QUERIES

schema graph

As we have seen, the definition of a nonprimitive class forms a two-dimensional directed graph of classes which we will call a *schema graph* for the class. Figure 1 is an example of a schema graph. The following is a formal definition of the schema graph SG for a class C.

1. SG is a rooted directed graph consisting of a set of classes N and a set of arcs E between pairs of classes. C is the root of SG, and is a nonprimitive class.
2. E has two types of arcs. One is between a pair of classes C1 and C2, such that C2 is the domain of one of the attributes of C1. Another is between a pair of classes C1 and S1, such that S1 is an immediate subclass of C1. The direction of an arc is from a class to the domain of its attribute, and from a class to its subclass.
3. An arc from an attribute of a class to the domain of the attribute may be either single-valued or set-valued. A single-valued arc means that the attribute can have only one value from its domain; while a set-valued arc means that the attribute has as its value a set of instances of its domain.
4. The set of arcs from any class Ci on SG to the direct and indirect subclasses of Ci forms a directed acyclic graph; that is, the class hierarchy rooted at Ci on SG is a directed acyclic graph.
5. The set of arcs from the root class C to the direct and indirect domains of the attributes of C forms a directed graph rooted at C; that is, the class-composition hierarchy of SG is a rooted directed graph which may be cyclic.

query graph

In relational databases, the schema graph for a relation is the relation itself. The selection operation on a relation, that is, a single-relation query on a relation with a Boolean combination of predicates on the attributes of the relation, identifies the tuples of the relation which satisfies all the predicates. In terms of the schema graph, the predicates apply on the lone node of the graph. To define the selection operation on a class in an object-oriented database, we extend this observation. The selection operation on a class C retrieves instances of the class C which satisfies a Boolean combination of predicates on a subgraph of the schema graph for C; we will call such a graph a *query graph*.

Let us consider an example query and its corresponding query graph. Using the schema for the class Vehicle shown in Figure 1, we may formulate a query to "find all blue vehicles manufactured by a company located in Detroit and whose president is under 50 years of age" as follows.

Q1: (**select** Vehicle (Color = "blue"
and Manufacturer Location = "Detroit"
and Manufacturer President Age < 50))

The query graph for this query on the class Vehicle is shown in Figure 2; it is a subgraph of the schema graph for the class Vehicle.

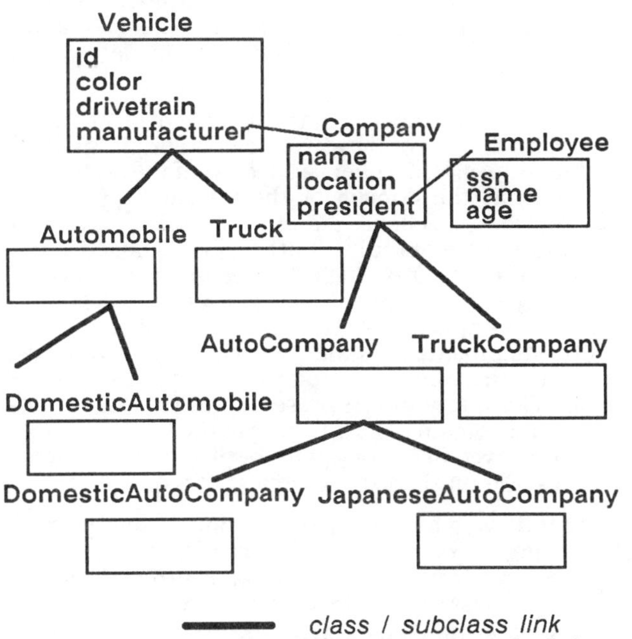

Figure 2. An Example Query Graph

The query graph includes only those classes (and the class hierarchies rooted at them) on which the predicates of the query are specified (as we will show shortly, this is not completely accurate). A predicate may be one of two types: a simple predicate and a complex predicate. Note that we define these terms differently from [BANE88]. A *simple predicate* is of the form < attribute-name operator value >. The value may be an instance of a primitive class (string, integer, etc.) or an object identifier of an instance of some class. The latter is important, since it may be used for testing the object equality [KHOS86], that is, equality of referenced objects. This is excluded by an oversight from the definition of a simple predicate given in [BANE88]. The predicate (Color = "blue") in the example query above is a simple predicate.

Another type of predicate is a *complex predicate*; this concept is also explored in [ZANI83]. The definition of a complex predicate given in [BANE88] as a predicate on a complex attribute of a class is inaccurate. A complex predicate is actually a predicate on a contiguous sequence of attributes along a branch of the class–composition hierarchy of a class. The predicate (Manufacturer Location = "Detroit") in the above example is a complex predicate. All classes (and the class hierarchies rooted at them) to which any of the attributes in the sequence of attributes specified in a complex predicate belong are included in the query graph.

So far we have assumed a predicate to simply be an expression of the form <attribute-name operator value>, without specifying what the operator is. In traditional database systems the operator is a scalar comparison operator (=, <, >, etc.) or a set comparison operator (contained-in, contains, set-equality, etc.). In object-oriented systems, the user may define methods on a class, a method may be used for any part of a predicate, that is, as the attribute-name, the operator, or the value. The use of a method in a predicate, as an attribute-name or as an operator, cause difficulties with query compilation and optimization.

Next, since any subclass of the domain of an attribute is also a valid domain of the attribute, it is certainly useful to be able to specify in a query a subclass of the domain of an attribute as the domain of the attribute. In other words, an arc in a schema graph from an attribute of a class to the domain of the attribute may be changed in the query graph to one from the attribute to any subclass of the domain. For example, it may be useful to be able to change the example query Q1 such that the domain of the attribute Manufacturer is the class DomesticAutoCompany, rather than the class Company as specified in the schema. The query model presented in [BANE88] implies that the domain of an attribute in the schema is necessarily preserved in queries, thereby placing an unnecessary restriction on the expressiveness of a query.

The following is a formal definition of a query graph QG for a single-operand query on a class C. The result of a single-operand query is a set of instances from the scope of the target class which satisfy the query predicates.

1. QG is a connected subgraph of the schema graph SG for C. C is the root of QG; that is, QG for C and SG for C have the same root. C is a nonprimitive class.
2. QG includes only those nodes of the corresponding SG on which at least one predicate of the query is specified.
3. An arc from an attribute of a class to the domain in SG may be changed in QG to one from the attribute to a subclass of the domain. Then only the class hierarchy rooted at the new domain is included in QG.
4. The set of arcs from the root class C to the direct and indirect domains of the attributes of C included in QG form a directed graph rooted at C; that is, the class–composition hierarchy of QG is a rooted directed graph, in which some branches contain cycles and others do not. (The nature of the cycles will be described further below.)
5. The leaf node of an acyclic branch has only simple predicates on it. The interior nodes of any branch (cyclic or acyclic) may have both simple and complex predicates defined on them.

cyclic branch in a class–composition hierarchy

Now we will characterize precisely the cyclic branch of the class–composition hierarchy (either in a query graph or in a schema graph). A branch of a class–composition hierarchy is cyclic, if it contains a class Ci and a class Cj on the branch, such that Cj is the (indirect) domain of an attribute of Ci, and Ci is the domain of an attribute of Cj; or Cj is the (indirect) domain of an attribute of Ci, and a superclass or a subclass of Ci is the domain of an attribute of Cj. This definition brings together four different types of cycle shown in Figure 3. Two of them, namely,

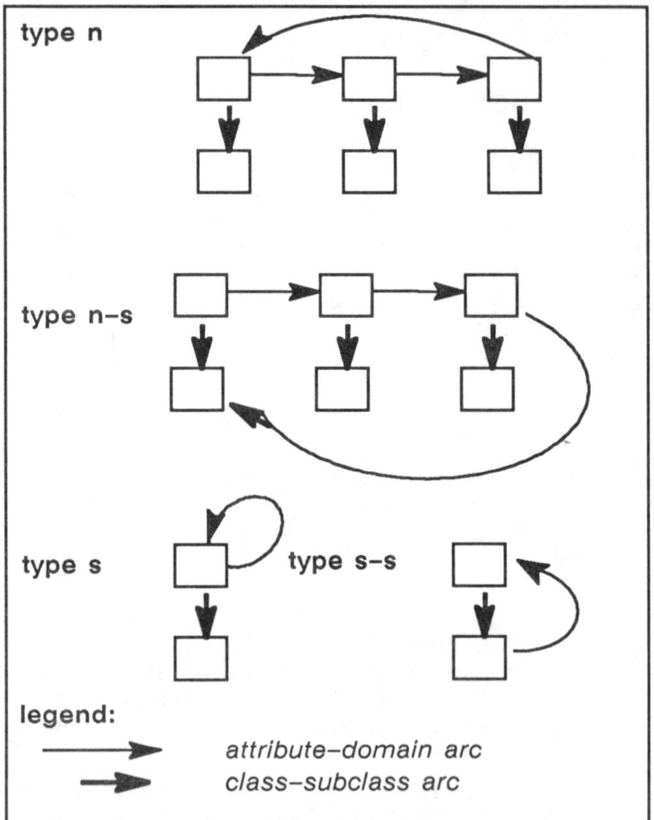

Figure 3. Cyclic Branches

type–ns and type–ss, may be regarded as quasi-cycles, since they are not cycles in conventional sense. They may be viewed as the type–n and type–s cycles, respectively, with additional conditions on the query result (we will illustrate this shortly). The following defines the four types of cycle for a branch of a class–composition hierarchy. We will assume that the branch has n nodes, such that each node may in turn be the root of a class hierarchy.

1. A type–n cycle is a cycle formed by n > 1 nodes on the branch.
2. A type–ns cycle is a quasi–cycle corresponding to a type–n cycle. It is formed by n > 0 nodes on the branch, and a superclass or a subclass of one of the n nodes.
3. A type–s cycle is a cycle formed for a single node.
4. A type–ss cycle is a quasi cycle corresponding to a type–s cycle. It is formed by a class and its superclass or a subclass.

Assuming that the class Employee has an additional attribute Drives whose domain is the class Vehicle, a meaningful query is to "find all blue vehicles driven by the president of the company that manufactured them." This is an example of a query with a type–n cycle in its query graph. The following is an intuitive syntax for expressing the query. We leave it as a simple exercise for the reader to construct the query graphs for the following example queries.

```
Q–c1:  (select (Vehicle :V)
          ( Color  =  "blue"
            and Manufacturer President Drives =  V))
```

In this query, the variable :V is used for binding each instance of the class Vehicle, as the set of instances of Vehicle is scanned one at a time. The class Vehicle is the target class.

A query to "find all blue vehicles manufactured by a company whose president drives a Japanese automobile" is an example of a query with a type–ns cycle in the query graph. This query may be expressed as follows.

```
Q–c2: (select Vehicle
         (:J is–a JapaneseAutomobile)
         ( Color  =  "blue"
           and Manufacturer President Drives  =  J))
```

The expression (:J is–a JapaneseAutomobile) adds a constraint to the value the complex attribute (Manufacturer President Drives) may take on to instances of the class JapaneseAutomobile. Note that the class Vehicle is still the target class.

For the next two examples, let us add the Manager attribute to the class Employee; the domain of Manager is the class Employee. The following is an example query with a type–s cycle. The query is to find "all managers of an employee named Johnson."

```
Q–c3:  (select Employee
          (recurse Manager)
          ( Name  =  "Johnson"))
```

The expression (recurse Manager) specifies that, once an instance of the class Employee is found which satisfies the predicate (Name = "Johnson"), the recursive values of the Manager attribute of the instance are retrieved.

For our final example, let us assume that the class Employee has a subclass FemaleEmployee. The following is an example query with a type–ss cycle, which finds "all female managers of an employee named Johnson."

```
Q–c4:  (select  Employee
          (recurse Manager :M
             (:M is–a FemaleEmployee))
          ( Name  =  "Johnson"))
```

The expression (:M is–a FemaleEmployee) restricts the result of the query to instances of the class FemaleEmployee.

query result

In relational databases, the result of a single–relation query is a subset of the tuples of the relation that satisfies the selection predicates. The concept of a class hierarchy in object–oriented databases captures the generalization abstraction, which means that a class, when used as a generalized concept, subsumes all its subclasses. This observation leads to two equally valid interpretations for the access scope of a query; the access scope of a query on a class C is either the set of instances of C, or the set of instances of the entire class hierarchy rooted at C. Therefore, the result of a single–operand query on a class C may be the set of instances of the class C or the set of instances of the entire class hierarchy rooted at C which satisfy the query predicates. For example, the access scope of a query against the class Vehicle may be restricted to the instances of Vehicle, or may be interpreted to also include the instances of all types of Vehicle, that is, all subclasses of Vehicle.

If a query involves a projection operation, only the desired attributes in the instances that satisfy the predicates will be returned, regardless of whether the access scope of the query is a single class or a class hierarchy. If the scope is a class hierarchy, we need to take into consideration the consequences of inheritance of attributes.

First, attributes may be renamed in classes which have inherited them, as observed in [BANE87b]. For the purposes of queries, a renamed attribute should be regarded as the original attribute. For example, suppose that the attribute Color, defined in the class Vehicle, is renamed as AutoColor in the class Automobile. Although the attribute has two different names, the distinction should be ignored in queries.

Second, in the case of multiple inheritance, a class may inherit an attribute from one of its superclasses [BANE87b]. If the attributes in all superclasses are *identical*, that is, they have the same name and same domain, there is no problem. However, complications arise if the attributes are only *equivalent*, that is, they were all defined in the same class somewhere higher in the class hierarchy, but they have been renamed or their domains have been changed. Suppose that the superclasses Si and Sj of a class C have equivalent attributes Ai and Aj, and that C inherits only the attribute Ai of Si. Now consider a query on Sj whose scope of evaluation is the class hierarchy rooted at Sj, and the attribute Aj is to be output. Then the values of the attribute Ai inherited into the class C (and its subclasses) should also be output, since Ai and Aj are equivalent after all.

If the access scope of a query on a class C is the class hierarchy rooted at C, the result of the query is a heterogeneous set of instances, since the instances belong to a number of different classes with different number of attributes. The application (or the user) which issues the query must obviously be prepared to deal with this heterogeneous mix of instances. One approach is for the database system to return only a list of object identifiers (along with the class identifiers) of the instances in the query result, and have the application make explicit requests for all or some of the actual instances. This approach has been implemented in ORION [KIM88a]; and it is in a sense similar to the cursor mechanism in SQL/DS [IBM81] or portals in POSTGRES [STON86]. If the query involves a projection operation, the database system may return the set of object identifiers and the values of the projected attributes of the corresponding objects.

The projection operation entails the familiar problem of eliminating duplicates from the query result. For object-oriented databases, unlike relational databases, two different criteria exist for determining the uniqueness of an instance from a collection of instances. One criterion is the equality of the object identifiers, which may be called *object equality*; while another is the equality of the contents of the instances, which may be called *value equality*, that is, the values of the *user-defined* attributes in the instances. We note that an instance may contain a number of system-defined attributes, such as the version number, update timestamp, and so on, to support various functionality of the system [CHOU88]. The values of system-defined attributes really should not be included in testing value equality of objects, since the users do not explicitly create their values. In relational databases, only value equality is used. For object-oriented databases, both may be useful; however, value equality requires some additional consideration. A class in general does not have the same set of attributes as its superclass, since a class inherits all attributes from its superclass and may have additional attributes defined on it. This means that value equality between classes with different sets of attributes need not be tested.

It is interesting to note that our type s and type s-s queries are recursive queries and that they are safe, that is, the result of such a query is a finite set of objects. This is because a primitive class is never the target of a query.

Throughout our discussion so far, we have assumed that the query result is simply returned to the application (or the user). Now we need to consider the problem of saving the result of a query. As we discussed earlier, an object belongs to a class, and a class has a position somewhere in the class hierarchy. These simple principles impose some difficult constraints on the query model for object-oriented databases. In relational databases, the result of a query is itself a relation, and it may simply be saved as a new relation. In object-oriented databases, the result of a query is a set of objects belonging to a class or a class hierarchy rooted at some class. Let us consider a query on a class C whose scope of access is the class hierarchy rooted at C. The result of the query will in general be a heterogeneous set of instances that form a separate class hierarchy each of whose class is derived from a corresponding class in the original class hierarchy; let us denote the root of the new class subhierarchy by C-new. The instances in the newly created classes must all have different object-identifiers from those in the corresponding instances of the original classes.

A difficult question is where we should place the class C-new in the class hierarchy for the entire database. Intuitively, there are two options to consider. One is to place C-new as a new subclass of the superclasses of the class C; and the other is to treat C-new as an immediate subclass of the class OBJECT, which is the root of the class hierarchy for the entire database. These options are shown in Figure 4, where the class AutoCompany-new is the root of a new class hierarchy which is created to save the result of a query on the class-hierarchy rooted at the class AutoCompany.

Let us consider option-1, where C-new is made a subclass of the superclasses of the class C. If the query involves a projection operation, this solution would result in a subclass with fewer attributes than its superclass, thereby violating the *full-inheritance invariant* in schema evolution as observed in [BANE87b]. In particular, if any attribute which C inherited from its superclasses is dropped from the query result, C-new will have fewer attributes than the superclasses of C; for example, in Figure 4, the class AutoCompany-New may have fewer attributes than the class Company. This suggests that C-new with fewer attributes than C should perhaps be made a superclass (rather than a subclass) of a superclass of C. However, this fix does not work either, if C-new contains any attribute which was defined in C (rather than inherited into C). C-new will then end up with an attribute which its subclass does not have.

There is another problem with option-1, regardless of whether the query involves a projection operation. If C-new is placed on the class hierarchy, it becomes subject to schema evolution on its superclasses. In other words, changes to the definition of any of its new superclasses propagate down to C-new, again in accordance with the full-inheritance invariant. For example, if an attribute is dropped from the class Company, the same attribute must be dropped from the class AutoCompany-New. This defeats the purpose of saving the result of a query.

Let us now consider option-2, where C-new is made an immediate subclass of the system-defined class OBJECT. One drawback of this solution is that all attributes and methods inherited into C-new must now be explicitly replicated in C-new. This requires extra storage; however, it does not appear to be a serious drawback. Option-2 is our choice.

3.2 MULTIPLE-OPERAND QUERIES

3.2.1 Join

A relational database consists of a number of relations. Join is used to correlate n different relations (n > 1) on the basis of the values in attributes common to the relations. It is important to realize that join is a crucial operation in relational databases largely because of the limitations of the relational model of data. In particular, the relational model of data restricts the attribute of a relation to a single primitive value; that is, an attribute may have only one value, and the value must be of a primitive data type (integer, float, string, boolean). These limitations have been removed in object-oriented and nested-relational models of data.

implicit joins in a class-composition hierarchy

Let us consider the three classes Vehicle, Company, and Employee in Figure 1. These classes may be defined as relations as follows.

Vehicle (ID, Color, DriveTrain,
Manufacturer)
Company (CompanyName, Location,

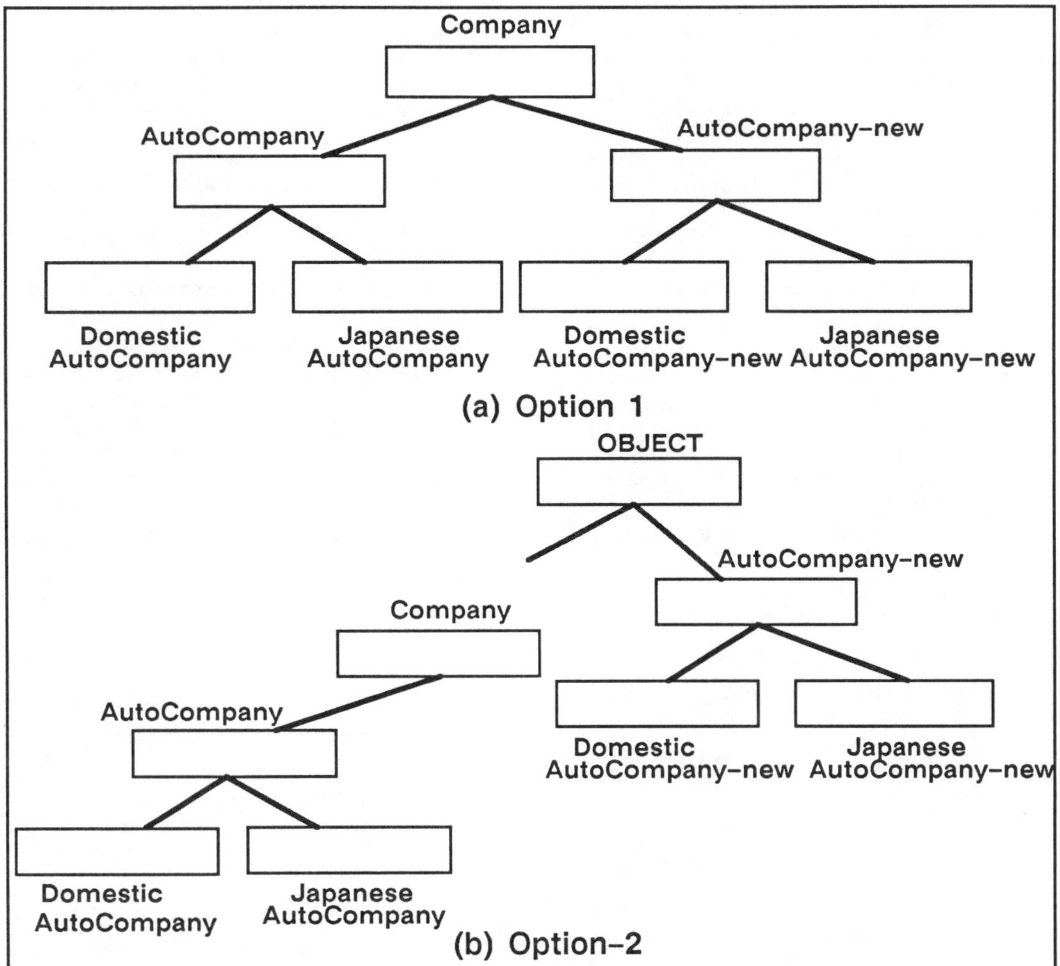

Figure 4. Disposition of the Query Result

President)
Employee (SSN, EmployeeName, Age)

The attributes Manufacturer and CompanyName have the same domain; and
the President and EmployeeName attributes share the same domain.

To find all blue vehicles manufactured by a company located in Detroit and whose president is under 50 years of age (example query Q1), the query must be formulated as a join of the above three relations, and the user must explicitly specify the join attributes. As we have seen already, however, the same query may be formulated in object-oriented databases as a retrieval from a class-composition hierarchy in which joins between the classes are defined in the schema between an attribute of a class and the domain of the attribute.

A single-operand query in object-oriented databases is then one type of join; it is an implicit join of the classes on a class-composition hierarchy rooted at the target class of the query. However, it has a few important limitations relative to the relational join. One is that the output attributes are restricted to those in the target class; that is, it is not possible to output any attribute in any class which is not the root of the query graph for a query. This may be easily remedied. The query syntax needs to be augmented with some means of specifying the output attributes; the output attributes may be specified in a single list, or they may be specified in a separate list associated with each of the classes in the query.

Another, much more serious, limitation of the implicit join in a single-operand query is that the join is restricted to the attribute-domain relationship specified in the schema; that is, the implicit join is essentially the materialization of the values of an attribute. Consider a class Ci with an attribute A whose domain is the class Cj. As observed in [BANE88], the materialization of the instances of Ci is equivalent to a join of Ci and Cj, in which the attribute A of Ci and the object-identifier attribute of Cj are the join attributes. Below, we elaborate on the limitations of the implicit join, and propose solutions to overcome these limitations.

explicit joins of classes

The limitation of a join implied in a single-operand query is essentially that it statically determines the classes to be joined, and the join ordering of the classes. Let us examine this in more detail. The implicit join statically determines the join ordering between a pair of classes Ci and Cj, where Cj is the domain of an attribute of Ci, such that Ci is always the 'outer class' and Cj the 'inner class' (analogous to the outer and inner relations in a join ordering for a pair of relations in relational databases). This means that it is not possible to formulate a query whose semantics require implicit reversal of an attribute–domain link specified in the schema. For example, the class Company is the domain of the attribute Manufacturer of the class Vehicle in Figure 1. It is possible to find all vehicles manufactured by certain companies. However, it is not possible to find all companies that manufacture certain vehicles; the class Company has no attribute, say, Manufactures, whose domain is the class Vehicle.

In relational databases, the existence of a common attribute in a pair of relations implies, correctly, the attribute–domain relationship in both directions between the relations. It is certainly useful to extend our model of single–operand queries by postulating, for any attribute of a class C whose domain is the class D, an implicit attribute of the class D whose domain is the class C; and by allowing implicit joins to be formulated along the attribute–domain links defined by the implicit attribute. For example, the attribute Manufacturer of the class Vehicle, and the class Company define an implicit attribute of the class Company whose domain is the class Vehicle. Then a single–operand query on the class Company may be formulated in which instances of the class Vehicle are materialized as values of the implicit attribute of the class Vehicle.

However, the implicit join also statically determines the classes which may be joined to those pairs of classes Ci and Cj such that Cj is the domain of an attribute of Ci. This means that it excludes joins between an arbitrary pair of classes with attributes which share a common domain. For example, suppose that, as shown in Figure 5, the class Employee has an additional attribute Address, and that the domains of Address and the attribute Location of the class Company are the same, say a class City. Then a meaningful query may be to find all employees who live in the city in which their companies are located. However, as long as there is no attribute–domain relationship between the attribute Address and the class Company, or between the Location attribute and the class Employee, it is not possible to formulate this query.

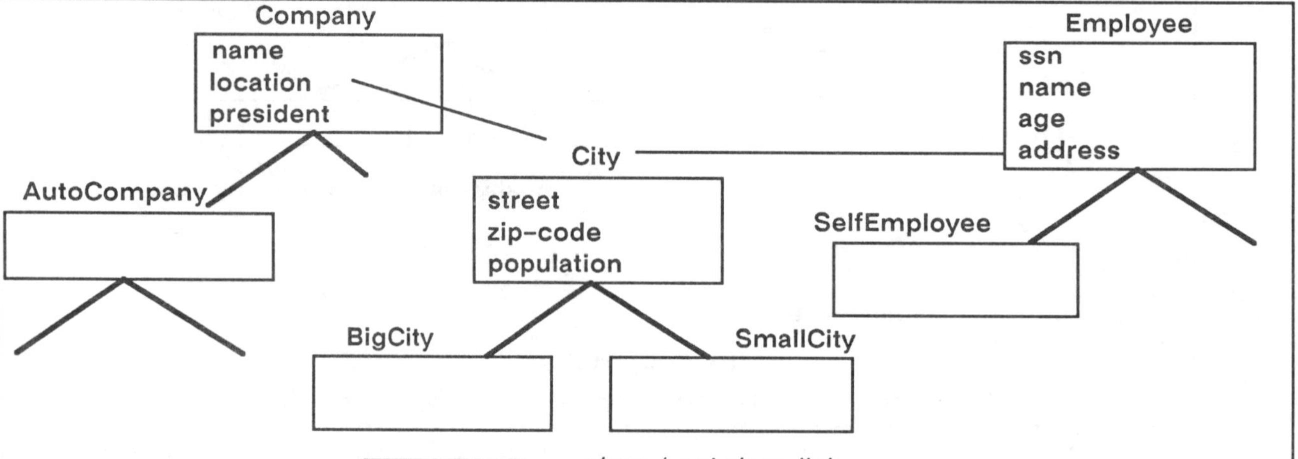

Figure 5. An Example Query Graph for a Join

We now extend our query model to admit joins of classes on user–specified arbitrary join attributes. Again, the extended model must be consistent with the object–oriented data model. To fully define the semantics of joins, we need to introduce the notion of attribute compatibility. Two attributes Ai and Aj are *compatible*, if the domain of Ai is identical to the domain of Aj, or is a superclass or a subclass of the domain of Aj. Unlike relational joins, in which the domains of the join attributes must be identical, we require the join attributes to only be compatible. For example, in Figure 5 the attributes Address and Location are compatible, since the share the same domain. Suppose now that the domain of the Location attribute is the class BigCity, which is a subclass of City. The attributes Address and Location are still compatible, since the domain of Location is a subclass of the domain of Address.

Next, we must account for the class hierarchy in the scope of query evaluation. Each class to be joined is the root of a class hierarchy. The scope of query evaluation, for each class to be joined, may be either instances of the class alone or instances of the class hierarchy rooted at the class. The query syntax must allow the user to specify the scope for each class in a join, just as was the case for a single–operand query.

The result of a join is a set of instances formed by concatenating instances from different classes. This presents a problem for systems like ORION which first return the list of object identifiers and then return specific instances on demand; this problem does not arise in systems which directly return the actual set of concatenated instances. The problem is whether to return a set of n object identifiers (one for each class containing an output attribute from the join) for each concatenated instance or to return a new object identifier for the concatenated instance. If the system simply returns a set of n object identifiers for each concatenated instance, the user must in general send the entire list back to the system to request the retrieval of the actual concatenated instance. However, if the system must return a new object identifier for each concatenated instance, it must incur the overhead of generating the identifiers, and temporarily saving all the concatenated instances along with their identifiers. If the query result is not to be saved, this is a rather high overhead; therefore we adopt the first approach.

Just as is the case with a single-operand query, if the result of a join needs to be saved as a snapshot, the instances in the result must be given unique identifiers and assigned to a class. The same considerations we had for saving the result of a single-operand query apply here, and the result of a join is to be saved in a new class which is an immediate subclass of the class OBJECT, the root of the class hierarchy for the entire database.

Figure 5 is the query graph for a query involving a join of two classes Company and Employee; the join attributes are Location and Address. The following is a formal definition of the query graph for a join of a pair of classes Ci and Cj; it is straightforward to extend the definition for a join of more than two classes, and we leave it as an exercise to the reader. Let us denote by Si the domain of the join attribute in Ci, and by Sj the domain of the join attribute in Cj; let us assume that Si is a superclass of Sj or Si = Sj. The result of the join is a set of instances formed by concatenating the instances from the scope query evaluation for Ci and Cj which satisfy the join predicate.

1. Ci and Cj each is the root of a query graph corresponding to a single-operand query on Ci and Cj, respectively. That is, each query graph is the root of a class hierarchy and a class-composition hierarchy rooted at the class to be joined.
2. The query graphs for Ci and Cj partially overlap, because of the compatibility in the join attributes. Si and Sj each is the root of a class-composition hierarchy, each of whose node is in turn the root of a class hierarchy. If Si = Sj, the entire class-composition hierarchy is shared by Ci and Cj via their respective join attributes. If Si is a superclass of Sj, the class-composition hierarchy rooted at Si is the domain of the join attribute of Ci; while the domain of the join attribute in Cj is the class-composition hierarchy rooted at Sj.

3.2.2 Set Operations

The set operations for object-oriented databases also require some extensions to the corresponding operations for relational databases. As in relational databases, these operations are useful largely for further manipulating the results of queries, that is, when the operands are queries.

The operand is a set of instances; more specifically, it may be the set of instances of a class defined in the database, or it may be a set of instances obtained as the result of a query. On the surface, the semantics of the set operations for object-oriented databases are identical to those for relational databases. Actually, however, they differ in three interesting ways.

First, for object-oriented databases, the operand, and the result of the operation, may be a heterogeneous set of instances, that is the instances of the operand may not all belong to the same class; in relational databases, the operand is a homogeneous set of tuples.

Second, as we have seen already, for object-oriented databases, two different criteria exist for determining the uniqueness of an instance from a collection of instances, namely, object equality and value equality.

Third, as usual, the scope of evaluation for an operand which is a database class may be instances of the class or the instances of a class hierarchy rooted at the class. The user may specify the scope for each class in the operation.

4. COMPARISON WITH OTHER QUERY MODELS

There are three common, and major, differences between our query model and other query models, including those for hierarchical, relational, nested relational, and object-oriented data models. First, our query model reflects the semantics of the class hierarchy. The concept of a class hierarchy is not a part of a nonobject-oriented data model, and therefore, of a query model derived from it. Unhappily, even the few query models proposed for object-oriented database systems [COPE84, BEEC87, ROWE87, ALAS88] neglect to account for the impacts of the class hierarchy on queries.

Second, the record-type hierarchy in hierarchical databases and the nested relations in nested relational databases are similar to the class-composition hierarchy in an object-oriented data model. However, the record-type hierarchy and nested relations form a directed acyclic graph, unlike our class-composition hierarchy which may include cyclic branches. The query model based on the normalized relational model does not give rise to a nested structure of relations.

Third, other query models use only value equality, that is, equality testing between entities is done on the basis of the contents of the entities, rather than object equality based on the object identifiers. This difference is significant in the definition of a predicate and the definition of the semantics of the set operations.

One additional major difference between our query model and other models proposed for object-oriented databases is the treatment of a snapshot, that is, the query result which is saved and becomes a part of the persistent database. Our model solves an important problem posed in [KIM88b].

SUMMARY

In this paper, we provided what we believe now to be a comprehensive model of queries for object-oriented databases. The work represents a significant formalization and extensiosn of the query model first proposed in [BANE88]. The model proposed in [BANE88] and elaborated somewhat in [KIMK89] is based on the view that a query model may be defined as a subschema of the database schema; the database schema is reduced to a query model by applying the selection and projection operations. To our knowledge, it is the first query model which made serious efforts to capture the semantics of object-oriented concepts. However, the model defined only limited type of a single-operand query, that is, a query whose target is a single class or a class hierarchy rooted at that class. Further, the model contained some important oversights, notably in its treatment of the projection operation, and the directionality of the arcs in the class-composition hierarchy.

In this paper, we first provided a considerably more rigorous treatment of the single-operand query, and corrected the mistakes in the model given in [BANE88]. Next, we significantly extended the model of [BANE88] to provide a formal basis for a query which involves more than one operand, namely, object-oriented equivalents of the relational join and set operations. Then, we briefly summarized the essential differences between our query model, and the query models for traditional data models and the models proposed by other researchers for object-oriented databases.

ACKNOWLEDGEMENTS

This paper is a compendium of the ideas about queries in object-oriented databases during the past three years. Past and present, both permanent and temporary, members of the ORION project have provided some of the ideas which have helped me to shape my conclusions about queries for object-oriented databases. I owe special thanks to Jay Banerjee (now with Unisys) and Fausto Rabitti (with C.N.R., Italy) for their ideas and discussions. In addition, Elisa Bertino (with C.N.R., Italy) and Fred Mellender (Eastman Kodak), and the anonymous referees, gave me a number of insightful and helpful comments on an initial draft of this paper.

REFERENCES

[ABIT84] Abiteboul, S., and N. Bidoit. "Non First Normal Form Relations to Represent Hierarchically Organized Data," *Proc. ACM Symp. on Principles of Database Systems,* Waterloo, Canada,1984.

[ABIT86] Abiteboul, S., and N. Bidoit. "Non First Normal Form Relations: An Algebra Allowing Data Restructuring," *Journal of Computer and System Sciences,* no. 33, pp 361-393, 1986.

[ALAS88] Alashqur, A., et al. "OQL - An Object-Oriented Query Language," technical report, Database Systems R/D Center, University of Florida, Gainesville, Florida, 1988.

[ANDR87] Andrews, T., and C. Harris, "Combining Language and Database Advances in an Object-Oriented Developmant Environment," *Proc. OOPSLA 87 Conference*, Orlando, Florida, 1987.

[BANE87a] Banerjee, J., et al. "Data Model Issues for Object-Oriented Applications," *ACM Trans. on Office Information Systems*, January 1987.

[BANE87b] Banerjee, J., W. Kim, H.J. Kim, and H.F. Korth. "Semantics and Implementation of Schema Evolution in Object-Oriented Databases," in *Proc. ACM SIGMOD Intl. Conf. on Management of Data,* 1987.

[BANE88] Banerjee, J., W. Kim, and K.C. Kim. "Queries in Object-Oriented Databases," in *Proc. 4th Intl. Conf. on Data Engineering,* Los Angeles, Calif. Feb. 1988.

[BEEC87] Beech, D. "OSQL: A Language for Migrating from SQL to Object Databases," working paper, Hewlett-Packard Lab., Palo Alto, Calif., 1987.

[COPE84] Copeland, G., and D. Maier. "Making Smalltalk a Database System," in *Proc. ACM SIGMOD Intl. Conf. on Management of Data*, June 1984, pp. 316-325.

[DESH88] Deshpande, A., and D. Van Gucht. "An Implementation for Nested Relational Databases," in *Proc. Intl. Conf. on Very Large Data Bases*, 1988.

[FISH87] Fishman, D., et al. "IRIS: an Object-Oriented Database Management System," *ACM Trans. on Office Information Systems*, vol. 5. no. 1, Jan. 1987, pp. 48-69.

[IBM81] SQL/Data System: Concepts and Facilities. GH24-5013-0, File No. S370-50, IBM Corporation, Jan. 1981.

[IEEE88] IEEE Computer Society, *Database Engineering,* special issue on Non-First Normal Form Relational Databases (ed. Z.M. Ozsoyoglu), Sept. 1988.

[JAES82] Jaeschke, G., and H. Schek. "Remarks on the Algebra of Non First Normal Form Relations," *Proc. ACM Symp. on Principles of Database Systems,* Los Angeles, CA,1982.

[KHOS86] Khoshafian, S., and G. Copeland. "Object Identity," in *Proc. 1st Intl. Conf. on Object-Oriented Programming Systems, Languages, and Applications,* Portland, Oregon, Oct. 1986.

[KIM88a] Kim, W., et al. "Integrating an Object-Oriented Programming System with a Database System," in *Proc. 2nd Intl. Conf. on Object-Oriented Programming Systems, Languages, and Applications,* San Diego, Calif., Sept. 1988.

[KIM88b] Kim, W. "Object-Oriented Databases: Definition and Research Directions" submitted for publication, Sept. 1988.

[KIM89] Kim, W., et al. "Features of the ORION Object-Oriented Database System," *Object-Oriented Concepts, Applications, and Databases,* (ed. W. Kim, and F. Lochovsky), Addison-Wesley, 1989.

[KIMK89] Kim, K.C., W. Kim, and A. Dale. "Cyclic Query Processing in Object-Oriented Databases," in *Proc. 5th Intl. Conf. on Data Engineering*, Los Angeles, Calif., Feb. 1989.

[MAIE86] Maier, D., et al. "Development of an Object-Oriented DBMS," in *Proc. 1st Intl. Conf. on Object-Oriented Programming Systems, Languages, and Applications,* Portland, Oregon, Oct. 1986.

[MAKI77] Makinouchi, A. "A Consideration of Normal Form of Not-necessarily Normalized Relations in the Relational Data Model," in *Proc. Intl Conf. on Very Large Data Bases*, 1977, pp. 447-453.

[ROWE87] Rowe, L., and M. Stonebraker. "The POSTGRES Data Model," in *Proc. Intl. Conf. on Very Large Data Bases*, Brighton, England, Sept. 1987, pp. 83-95.

[STEF86] Stefik, M., and D.G. Bobrow. "Object-Oriented Programming: Themes and Variations," *The AI Magazine*, January 1986, pp. 40-62.

[STON76] Stonebraker, M., E. Wong, P. Kreps, G. Held. "The Design and Implementation of INGRES," *ACM Trans. on Database Systems*, vol. 1, no. 3, Sept. 1976, pp. 189-222.

[STON86] Stonebraker, M., and L. Rowe. "The Design of POSTGRES," in *Proc. ACM SIGMOD Intl. Conf. on Management of Data,* May 1986.

[ZANI83] Zaniolo, C. "The Database Language GEM," in *Proc. ACM SIGMOD Intl. Conf. on Management of Data*, May 1983, pp. 207-218.

OQL: A Query Language for Manipulating Object-oriented Databases

A. M. Alashqur
S.Y.W. Su
H. Lam

Database Systems Research and Development Center
Electrical Engineering Department
University of Florida

Abstract

An essential property which is desirable in a query language designed for a certain data model is that queries issued in that language must produce results that are structured and modeled using the same data model. A consequence of maintaining this property in a query language is that the result of a query can be used as an operand in some other query (or queries) or can be saved as a user's view. Existing query languages that have been designed for the class of object-oriented data models do not posses this property. In this paper, we introduce the object-oriented query language (OQL), which maintains this property. An OQL query is considered as a function, which when applied to a database, returns a subdatabase whose structure consists of some selected object classes and their associations. The objects that satisfy the search conditions and participate in the patterns of object associations specified in the query constitute the extension of the resulting subdatabase. A subdatabase forms a "context" under which system-defined and/or user-defined operations can be specified and performed. Several advanced features such as branching association patterns and set operations on subdatabases are also presented.

1. Introduction

The limitations of the existing record-oriented data models have long been observed [HAM81,HUL87,SU88]. To alleviate these limitations, several object-oriented (**OO**) and semantic data models have been introduced as the potential alternatives for modeling many advanced database applications such as CAD/CAM, office automation, and multimedia databases [HAM81,KIN84,BAT85,SU86,HUL87]. **OO** and semantic data models can capture much more of the semantics of these application domains in a "natural" way. They provide a rich variety of modeling constructs, which can model most of the situations that may arise in such application domains.

The term "**OO** data model" is used to refer to a data model that is a structurally and/or behaviorally object-oriented [DIT86]. A structurally **OO** data model is one that encompasses at least the following characteristics:
1. It allows for defining aggregation hierarchies.
2. It allows for defining generalization hierarchies.
3. It supports the unique identification of objects, that is, each object is assumed to have a unique object identifier (surrogate).

The **OO** view of an application world is represented in the form of a network of classes and associations, which can be aggregation or generalization associations.

Object classes can be either **primitive** classes whose instances are of simple data types (e.g., integer, string, real) or **non-primitive** classes whose instances represent real world objects (e.g., Part, Employee). At the extensional level, instances of different classes can be related (associated) with each other forming patterns of object associations. A behaviorally object-oriented data model, on the other hand, is one in which operations that describe the behavior of the objects of a class can be defined and registered with that class.

Several query languages such as DAPLEX [SHI81], GEM [ZAN83], ARIEL [MAC85], and the object-oriented query language described in [BAN88], which are based on the **OO** view of data as described above (or variations of it), have been introduced in the literature. A query in these languages is performed by choosing one of the non-primitive classes in the schema as a central class (anchor class). The user then specifies some path expressions such that each path expression starts from the central class and ends up at a primitive class. A restriction condition can be specified on the class referenced at the end of a path expression and/or such a class can be specified in the target list (i.e., the list of attributes to be retrieved). The result of a query is defined as a set of tuples (i.e., a relation) each of which corresponds to a single instance of the central class and contains values related to that instance which are collected from the primitive classes specified in the target list.

These languages share the above characteristic even though they are based on different paradigms. For example, DAPLEX is based on the functional paradigm. An attribute of a class in DAPLEX is treated as a function, which when applied to an object of the class, returns its attribute value from the related class. Path expressions starting from a central class can be specified by composing functions. The query language of [BAN88], on the other hand, is based on the message passing paradigm. In this language, an attribute of a class is treated as a message, which when sent to an object of the class, returns its attribute value from the related class. Path expressions starting from a central class can be specified using a series of messages each of which is sent to the object returned by the previous message in the series.

A major drawback of existing data manipulation languages for **OO** and semantic models is that they do not maintain the closure property because the result of a query does not have the same structural properties as those of the original database. In these languages, the input to a query has an **OO** representation (i.e., classes and their associations) and its output is a relation. Consequently, the result of a query cannot be further operated on uniformly using the same query language operators to produce a new result. Contrary to these languages, the relational query languages such as SQL, QUEL, or the relational algebra maintain the closure property. A relational query can be a single-relation query or a multi-relation one. In either case, the result of a query is always a relation. In other words, both of the input and output of a query have a relational representation. Thus, the resulting relation can be uniformly operated on by another query in a nested fashion or can be saved as a view definition and manipulated uniformly as any of the base relations.

Motivated by this limitation of the existing **OO** query languages, we have developed a new **OO** query language called OQL, which maintains the closure property. A query in this language, can be considered as a function that, when applied to a database, returns a **subdatabase** whose structure is comprised of some selected classes of objects and their associations, i.e., it has the same structural characteristics as those of the original database. For this reason, a resulting subdatabase can be further operated on by another OQL query to produce another subdatabase or it can be saved as a view definition. The instances (objects) that satisfy the search conditions and participate in the **patterns of object associations** specified in the query (see Section 3) form the extension of the resulting subdatabase.

Another shortcoming of existing **OO** query languages is that they are oriented towards retrieval and storage manipulation operations (i.e., system-defined operations) in the sense that a query specifies data to be manipulated by such operations as Retrieve, Update, and Delete. In a behaviorally **OO** data model, user-defined operations that describe the behavior of object classes can be defined and registered with these classes. In OQL, a subdatabase represents a "context" in which objects of various classes must exist before being considered for further manipulation by different operations including system-defined operations (e.g., Update, Display) as well as user-defined operations (e.g., Hire-employee, Move). Multiple operations (messages) can be issued against the different classes of a subdatabase in the same query.

This paper introduces the OQL language and is organized as follows. In Section 2, we describe the **OO** view of a university database as represented by an **OO** model. In Section 3, the concept of semantic association pattern is introduced together with the basic operators used for defining a subdatabase. Section 4 presents some advanced features of OQL. The assignment operator of OQL and the closure property are described in Section 5. Finally, some conclusions OQL are summarized in Section 6.

2. The Object-oriented View of Databases

We shall describe first the **OO** view of a university database as modeled by the **OO** semantic association model OSAM* [SU88]. The university schema shown in Figure 2.1 is then used in the remainder of this paper to issue some example OQL queries. The concepts introduced in OQL can be applied to any **OO** data model that encompasses the three structural properties described in Section 1 and are not limited to the OSAM* model.

A database schema is represented in OSAM* as a network of associated (inter-related) object classes. Graphically, object classes are represented as nodes and associations among object classes are represented as links. The resulting diagram is called the Semantic diagram or **S-diagram**. In OSAM*, there are two types of object classes: Entity object classes (**E-class**) and Domain object classes (**D-class**) which are represented in the S-diagram as rectangular and circular nodes, respectively. The sole function of a D-class is to form a domain of possible values (e.g., integers, strings, etc.) from which descriptive attributes of objects draw their values. An E-class, on the other hand, forms a domain of objects which occur in an application's world (e.g., Faculty, Department, etc.). Each object of an E-class is represented by a unique object identifier (OID).

There are five types of links (associations) in OSAM*. Two of these association types appear in Figure 2.1, namely, Aggregation (A) and Generalization (G), which are also recognized in several other semantic and **OO** models. A class can have several types of links and more than one link of each type emanating from it. In the S-diagram, links of the same type that emanate from a class are grouped together and labeled by the letter that denotes the association type.

As an example, in Figure 2.1, the E-class Person has two types of links: Aggregation links with the D-classes SS# and Name and Generalization links with the E-classes Student and Teacher (i.e., Student and Teacher are subclasses of the superclass Person). An aggregation link represents an attribute and has the same name as the class it connects to unless specified otherwise (e.g., the link labeled Major that emanates from the class Student). Aggregation links that emanate from an E-class and connect to D-classes represent the descriptive attributes of that class (e.g., the attribute section# of the class Section). A class inherits all the aggregation associations that connect to or emanate from its superclasses. Figure 2.2 shows the actual view of the class RA in which all the associations inherited by RA are explicitly represented. A detailed description of the OSAM* model can be found in [SU88].

3. Object-oriented Query Language OQL

In this section, the general OQL syntax and semantics are described and illustrative examples are given. Throughout this paper, capital letters are used to denote E-classes (A, B,...) and small letters with an integer appended to each letter are used to denote objects (or OIDs) (e.g., a1, a2, ... and b1, b2, ... are OIDs for objects that belong to the classes A and B, respectively).

3.1 Definitions and Overview

An OQL query returns a subdatabase, which is a portion of the operand database (an operand database can be the original database or another subdatabase that has been established by another query). Figure 3.1 shows a certain subdatabase SDB of the original database of Figure 2.1. A subdatabase consists of two parts: an **intensional association pattern** and a set of **extensional association patterns**. An intensional association pattern is represented as a network of E-classes and their associations and descriptive attributes. For example, Figure 3.1a represents the intensional association pattern of a the subdatabase SDB, which, in this simple case, has a linear structure that consists of the classes Teacher, Section, and Course and their associations. By default, all the descriptive attributes of the E-classes of a subdatabase are also present in the subdatabase unless specified otherwise in a Select subclause to be described later. In Figure 3.1, however, the descriptive attributes of the classes of SDB are not shown in order to keep the figure simple.

An extensional association pattern is a network of instances and their associations that belong to the classes and association types of the intensional association pattern. The set of extensional patterns of a subdatabase can be represented in the form of an **extensional diagram**. Figure 3.1b shows a possible extensional diagram for the subdatabase SDB. (We note that Section s3 is related to more than one Course instance and Section s4 is not related to any Course instance. Naturally, there is a constraint on the database that restricts the mapping between Section and Course to N:1 and another Non-null constraint on the aggregation association of Course with Section. We assume

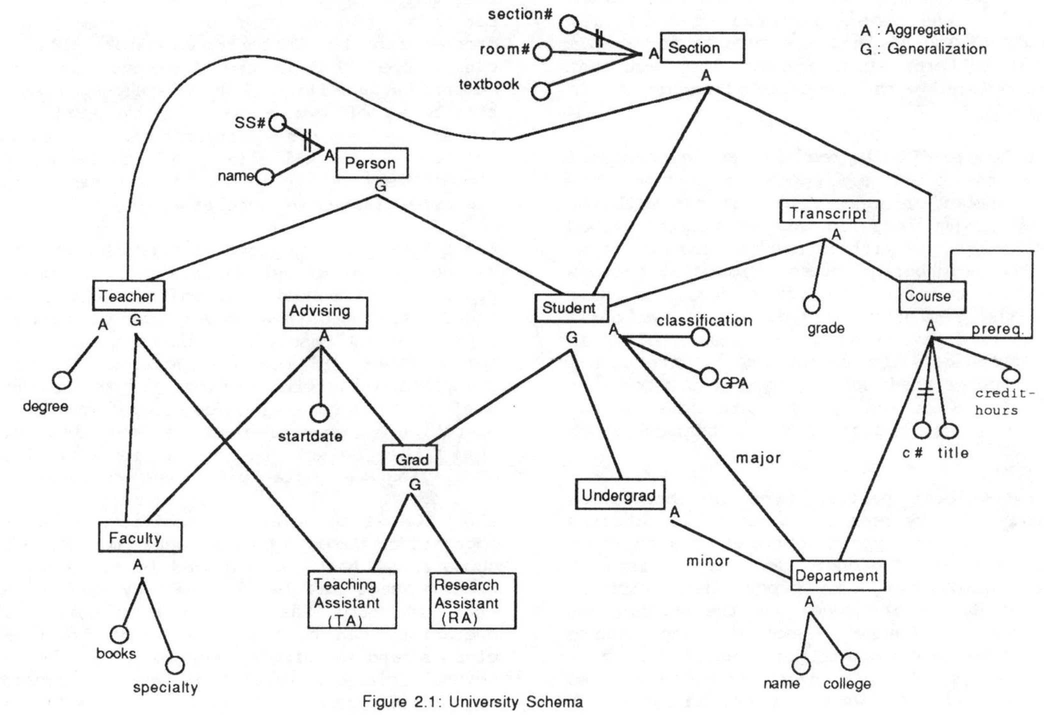

Figure 2.1: University Schema

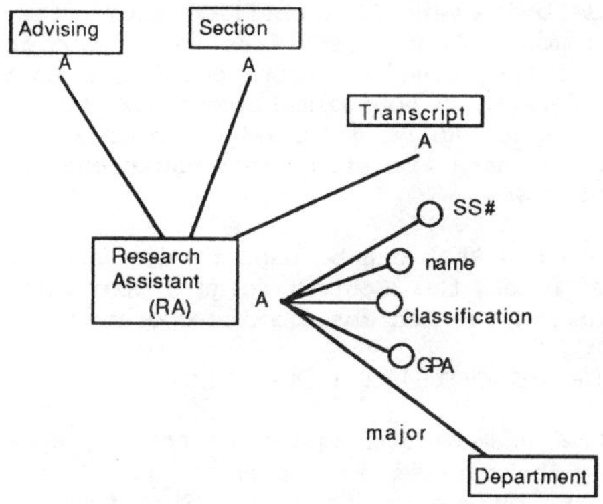

Figure 2.2: Class RA with all the Inherited Associations Explicitly Represented

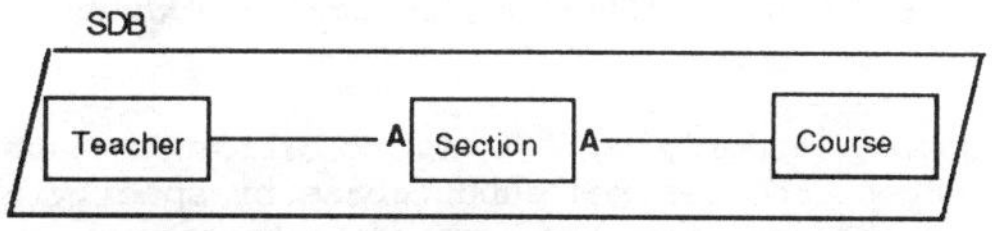

Figure 3.1a: The Intensional Pattern of a Subdatabase SDB

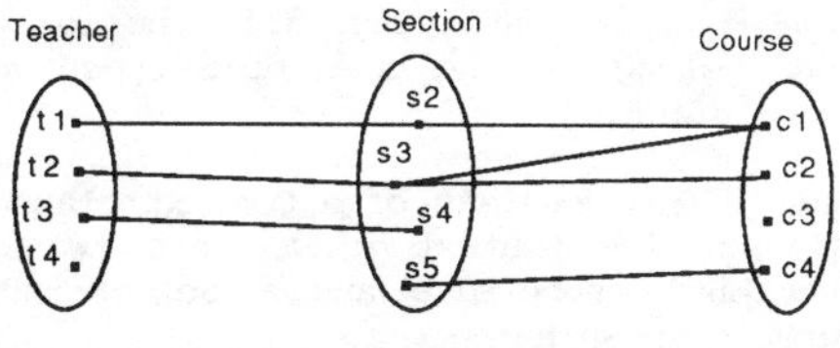

Figure 3.1b: A Possible Extensional Diagram for the Subdatabase SDB

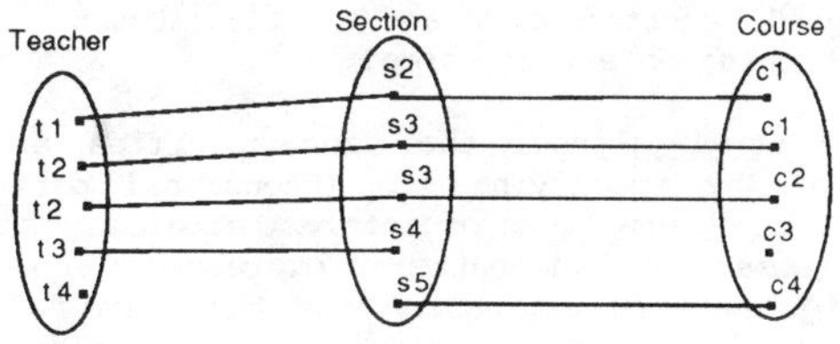

Figure 3.1c: A Normalized Extensional Diagram Corresponding to Fig. 3.1b

Figure 3.1 The Intensional and set of Extensional Patterns of a Subdatabase

that these constraints are waived here in order to be able to describe the most general case.) The interconnection of t3 and s4 in the figure is an example of an extensional pattern, which records the fact that object t3 is associated with object s4 (Teacher t3 is teaching Section s4).

A **normalized extensional diagram** is an extensional diagram in which an object may appear more than once depending on the number of associations it has with the objects of a neighboring class (in case of a multi-valued association) and a separate link is used to connect it to each object of the neighboring class. Figure 3.1c shows the normalized extensional diagram for the same subdatabase. In the remainder of this paper, we shall deal with normalized extensional diagrams only. In addition to the graphical representation, an extensional pattern may be represented as a tuple of OIDs. For example, <t1,s2,c1>, <c3> and <s5,c4> are some of the extensional patterns that appear in the subdatabase SDB (Figure 3.1c).

We define an **extensional pattern type** as the common template that is shared by several extensional patterns in a subdatabase. A pattern type is denoted by a tuple of class names. For example, <Teacher, Section, Course> is one of the extensional pattern types that exist in Figure 3.1c, which has as instances all the extensional patterns that contain Teacher, Section, and Course objects, i.e., the extensional patterns <t1,s2,c1>, <t2,s3,c1>, and <t2,s3,c2>. On the other hand, the extensional pattern <t3,s4> whose Course-component is Null (since the pattern does not contain any Course object) is of the type <Teacher,Section>. The five extensional pattern types present in the extensional diagram of Figure 3.1c are <Teacher,Section, Course>, <Teacher,Section>, <Section,Course>, <Teacher>, and <Course>.

The philosophy underlying OQL is to allow the user to specify, first, the desired subdatabase by specifying its intensional pattern and the set of extensional pattern types that are of interest and then the operation(s) to be performed on the classes of the subdatabase. The search engine of the underlying **OO** DBMS would establish the subdatabase by identifying all the extensional patterns that belong to the specified types and then perform the operation(s).

A query block in OQL consists of a Context clause and an Operation clause. The Context clause has two optional subclauses: a Where subclause and a Select subclause. This structure is shown below.

context association pattern expression
  where conditions
  select object classes and/or attributes
operation(s) object class(es)

In the context clause, the user specifies a desired subdatabase by specifying its intensional pattern and extensional pattern types of interest (both are specified in the association pattern expression). A linear association pattern expression has the form "A [intra-class conditions] **op** B [intra-class conditions] **op** C [intra-class conditions] ..." where "op" is one of the association pattern operators to be described in Sections 3.2 and 3.3. Each operator separates two E-classes that are directly associated in a schema. More complex association pattern expressions that contain branching are described in Section 4. The intra-class conditions enclosed in brackets following a class name are optional and are expressed in the form of predicates that involve the descriptive attributes of that class.

The Where subclause further causes the extensional patterns that do not satisfy some conditions to be dropped from the Context subdatabase. The conditions that can be specified in the Where subclause are inter-class comparison conditions, which are comparisons between some attributes of two classes if these attributes are type comparable, and/or comparisons between objects (equal '=' or not equal '!=') if these objects are type comparable (i.e., belonging to the same E-class or any of its superclasses or subclasses).

The Select subclause operates on the subdatabase returned by the Context clause and its optional Where subclause to produce a new subdatabase which results from "projecting" the Context subdatabase over some classes and descriptive attributes. A resulting subdatabase can be either saved for further processing (Section 5) using the same association pattern operators of OQL (since the closure property is maintained) or operated on by the operation(s) specified in an operation clause. We note that the Select operation in OQL does not imply displaying some data to the user as in SQL.

The Operation clause specifies a set of messages (operation names) to be sent to the classes of the subdatabase that is returned by the Context expression. Each message may be followed by one or more arguments that identify the recipient classes. Thus, several operations can be performed over the same or different classes and a single operation can be performed over several classes resulting from a Context clause. An operation can be either a system-defined data manipulation operation (e.g., Display, Update, Print) or a user-defined operation (e.g., Rotate, Order-part, Hire-employee).

If the Display (Print) operation is specified in the operation clause, it causes the values of the descriptive attributes that appear in the subdatabase to be displayed (printed) to the user in the form of a table. If the Display operation is not followed by a class name as an argument, the resulting table will be a first normal form table defined over all identified attributes. Otherwise, the argument class identifies a "viewpoint" based on which the resulting descriptive data is to be organized in the form of a non-normalized table (i.e., a table in which a value can be a nested table or a set). Thus, the descriptive data are structured under the objects of the argument class.

The operators that can be used in the association pattern expression of the Context clause are the **association operator** and the **non-association operator**.

3.2 The Association Operator

When the association operator (*) is applied to two directly associated E-classes A and B in a database (i.e., the expression "A * B"), it returns a subdatabase whose intensional pattern consists of the two classes A and B and their association. The resulting subdatabase contains also the set of extensional patterns drawn from the operand database such that each extensional pattern contains objects of both A and B (i.e, extensional patterns that are of the type <A,B>). B objects that are not associated with any A objects and A objects that are not associated with any B objects in the operand database are not retained in the resulting subdatabase. The following example queries illustrate the use of the association operator.

Query 3.1 Display the names of the teachers who teach some sections and the section#'s for these sections.

context Teacher * Section

```
    select name, section#
  display Teacher
```

If the Context expression in this query is applied to the subdatabase SDB of Figure 3.1, it returns a new subdatabase whose set of extensional patterns is {<t1,s2>, <t2,s3>, <t3,s4>}. The extensional pattern <t4> (or <t4, Null>) is not included in this set because its Section-component is Null (similarly the pattern <s5> is not included). Figure 3.2a shows the intensional pattern of this subdatabase where all the descriptive attributes of the classes Section and Teacher appear with them (i.e., by default). The Select subclause derives a new subdatabase from the Context subdatabase. The intensional pattern of the new subdatabase, as shown in Figure 3.2b, consists of the classes Teacher and Section and only the two attributes that are referenced in the Select subclause, namely, Name of Teacher and Section# of Section. If all the descriptive attributes of a class are to be retained in the subdatabase the following rule can be applied.

Rule #1 If all the descriptive attributes of a class are to be retained in the subdatabase derived by the Select subclause, the class name can be referenced in the subclause without specifying any of its attributes, i.e., the default is "all attributes."

A Select subclause can also select a subset of E-classes, thus producing a subdatabase that results from dropping the unreferenced E-classes from the operand subdatabase. (A class is considered "unreferenced" in a Select subclause if none of its descriptive attributes is referenced in it.) If a class to be dropped from the operand subdatabase connects two classes that are to be retained, new direct association is created between these two retained classes in the produced subdatabase. At the extensional level, direct links between the instances of the two classes are inferred. For example, the subdatabase that results from selecting the classes Teacher and Course of the subdatabase SDB of Figure 3.1 is shown in Figure 3.3.

The display operation in the above query causes the result to be displayed in a non-normalized tabular form in which each tuple consists of a teacher's name and the set of section#'s for the sections he/she teaches. This is because the argument of the Display operation indicates that the result is to be viewed from the point of view of the class Teacher. We note that the result of a Display operation does not belong to the world of subdatabases and therefore cannot be operated on using the OQL operators. However, the subdatabase that corresponds to a certain displayed result can be further operated on to produce a new subdatabase whose descriptive attributes can be also displayed.

The definition of the association operator can be easily generalized to the case when the association pattern expression contains more than two classes. For example, the expression "A * B * C" returns the extensional patterns that are of the type <A,B,C>. It is noted here that one can define a single extensional pattern type using the association operator. A mechanism for defining a richer variety of extensional pattern types in a single expression is described in Section 3.4.

Query 3.2 Display the Department names for all departments that offer 6000 level courses that have current offerings (sections). Also, display the titles of these courses and the textbooks used in each section. In addition, print the results.

```
  context Department * Course [6000 <= c# < 7000] *
        Section
    select name, title, textbook
  display
  print
```

Two operations are specified in the Operation clause of this query, namely, Display and Print. These operations are to be performed on the subdatabase returned by the Select subclause. Also, the intra-class condition on the C# attribute of Course is enclosed in brackets following the class name in the Context expression. The result of the Display or Print operation is a normalized table since neither of the two operations is followed by a viewpoint class.

As in Query 3.3 below, an association (or non-association) operator can be used between any two classes whether they are connected by a generalization or an aggregation association. An association pattern expression concerns only with whether some classes and their instances are associated with one another or not. It does not specify what types of associations relate them. This is because the types of the associations connecting them have already been explicitly defined in the schema and restating these association types in queries is unnecessary. The query processor of an **OO** DBMS can make use of the type information stored in the dictionary to properly interpret the queries and enforce the relevant semantics and constraints. For example, a link that exists between an instance of the class TA and an instance of the class Grad is an identity link. In other words, the semantics implied by the generalization association here is that the two instances are actually two different perspectives of the same real world object.

Before presenting the example query 3.3, we introduce the following two general rules that are relevant to the query.

Rule #2 An attribute that appears in the Select or Where subclause has to be qualified by its class name only if it is not unique among the attributes of the classes that are referenced in the Context clause.

Rule #3 Different aliases (range or iteration variables) of a class can be generated in OQL by appending an Underscore and an integer to the class name in an association pattern expression (e.g., Grad_1 is and alias of Grad).

Query 3.3 Print the names of graduate students who teach other graduate students in some sections. Also, print the names of those graduate students they teach. Organize the result from the point of view of the teaching graduate students.

```
  context Grad_1 * TA * Teacher * Section *
        Student * Grad_2
    select Grad_1 [name], Grad_2 [name]
  print Grad_1
```

In this query, TA inherits the status of being related to Section from both Teacher and Student with each of them having its distinctive meaning. Thus, using the expression "TA * Teacher * Section" instead of the expression "TA * Section" is to explicitly state that we are interested in TA as playing the role of Teacher rather than the role of Student of a section. Also, in this query, the Select subclause projects over the two classes Grad_1 and Grad_2 since they are the only classes referenced in it. Thus, the intensional pattern of the final subdatabase contains these two classes with a derived aggregation association between them and the attribute Name of each class.

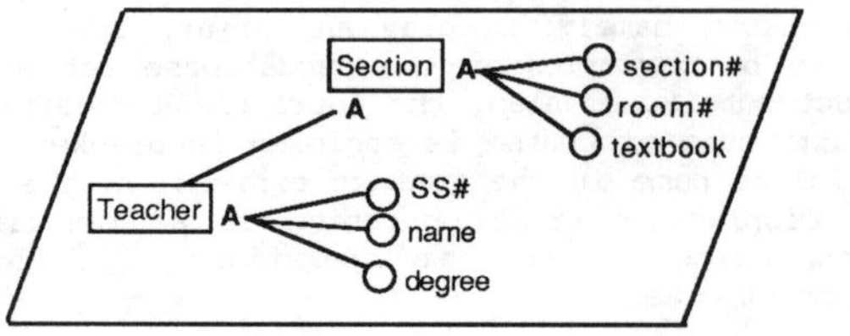

Figure 3.2a: The Result of a Context Expression of a Query

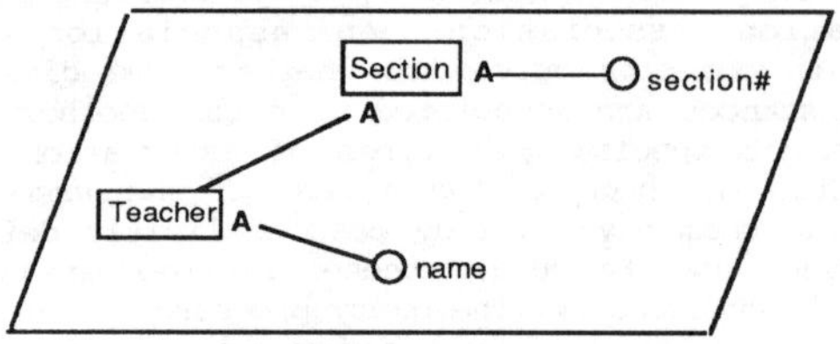

Figure 3.2b: The Projection of the Subdatabase of Fig.3.2a over some Descriptive Attributes

Figure 3.2: The Effect of a Select Subclause

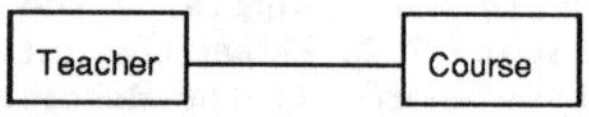

Figure 3.3a: After Projecting the Subdatabase of Figure 3.1 over Teacher and Section

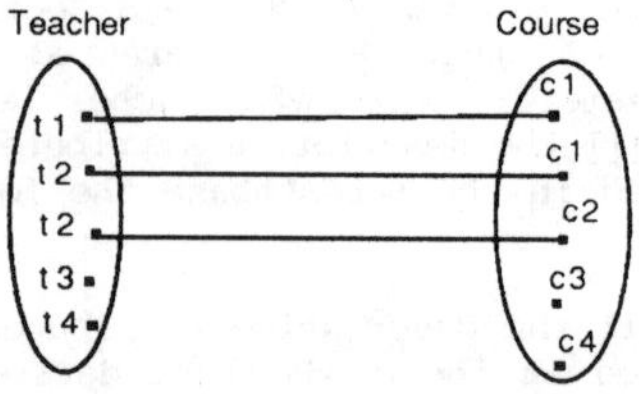

Figure 3.3b: The Extensional Diagram Corresponding to Figure 3.3a

Figure 3.3: The Result of a Projection Operation at the Intensional and Extensional Levels

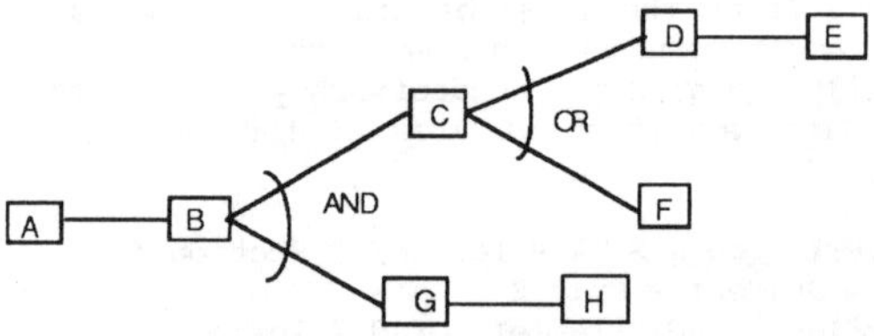

Figure 4.1: A Graphical Representation of a Branching Association Pattern

3.3 The Non-association Operator

We use the exclamation sign (!) to denote this operator. When this operator is applied to two directly associated E-classes A and B in a schema (i.e., the expression "A ! B"), it returns a subdatabase which contains only the instances of A that are <u>not</u> associated with any instances of B and the instances of B that are <u>not</u> associated with any instances of A. For example, the two instances t4 and s5 are returned in the subdatabase that results when the expression "Teacher ! Section" is applied to the subdatabase SDB of Figure 3.1 (i.e., it returns the teachers who are not assigned to any sections and the sections which are not assigned to any teacher).

The association operator has higher precedence than the non-association operator. As an example, when the association pattern expression "Teacher ! Section * Course" is applied to the classes of the subdatabase SDB of Figure 3.1, it produces a new subdatabase that contains the following set of extensional patterns: {<s5,c4>, <t3>, <t4>}. The pattern <t3> is retained in this result because of the higher precedence of the association operator over the non-association operator. When the association operator is applied first, object s4 is not retained in the result (since it is not associated with any Course object) causing object t3 to be not associated with any Section object in this result. When the non-association operator is applied next, it causes all the Teacher objects that are <u>associated</u> with any pattern of the type <Section,Course> (i.e., the objects t1 and t2) to be dropped together with these <Section,Course> patterns. Thus, the final result contains the above set of extensional patterns. The precedence of the association operator over the non-association operator can be overridden by parentheses. The following is an example query that uses the non-association operator.

<u>Query 3.4</u> Display the names of those graduate students who are TA's but not RA's.

```
context TA * Grad ! RA
   select TA [name]
display
```

3.4 Association Pattern Subexpressions

By using only the association operator in an association pattern expression, one can identify a single extensional pattern type. In some situations, extensional patterns of different types may be desired in the resulting subdatabase. This can be performed in OQL by enclosing a subexpression of the association pattern expression inside braces. This subexpression identifies a certain extensional pattern type. For example, the expression "A * {B * C} * D" returns the subdatabase whose intensional pattern consists of these four classes and whose set of extensional patterns includes all patterns that are of the types <A,B,C,D> and <B,C>. In other words, this expression means to select both the instances of A,B,C and D classes that are connected (associated) all the way through as well as those instances of B and C that are connected to each other but not necessarily connected to the instances of A and/or B. The braces around B * C capture the semantics of the Outerjoin concept [COD79].

If, in the above expression, an extensional pattern of the type <B,C> appears in the resulting subdatabase as part of a larger pattern of the type <A,B,C,D>, it will not appear independently in that resulting subdatabase. For example, if the original database contains only the two patterns <a1,b5,c5,d5> and <a3,b2,c2>, then the expression "A * {B * C} * D" returns the extensional patterns <a1,b5,c5,d5> and <b2,c2>. The extensional pattern <b5,c5> will not appear independently in the result since it already appears as a part of the extensional pattern <a1,b5,c5,d5>. In general, an extensional pattern of a certain specified type will not appear independently in the result, if it is part of a larger extensional pattern.

Subexpressions can be nested to several levels. For example, the expression "{{{A} * B} * C} * D" identifies the extensional pattern types <A>, <A,B>, <A,B,C>, and <A,B,C,D>.

<u>Query 3.5</u> Display the SS#'s of all graduate students (whether they have advisors or not) and for those graduate students who have advisors, display their advisors' names.

```
context { Grad } * Advising * Faculty
   select Grad [SS#], Faculty [name]
display
```

The attribute Name in the Select subclause of this query is qualified by its class name because it is not unique among the classes referenced in the Context clause (see Rule #2 above) and similarly the attribute SS#.

4. Advanced Features of OQL

4.1 Branching Association Patterns

An association pattern expression may contain branches expressed by an AND or an OR operator. There can be several nested levels of branching. For example, the expression "A * B * **AND** (C * **OR** (D * E, F), G * H)" is a branching association pattern expression, which corresponds to the intensional pattern shown in Figure 4.1. A class at which the branching occurs is called a **fork class** (e.g., B and C in the above expression). An AND operator means that, in the result, an instance from the fork class must be associated with instances from both of the branches, while an OR operator means that an instance from the fork class must be associated with an instance from at least one of the two branches. Figure 4.2 shows some association pattern expressions that represent networks of classes and associations together with a graphical representation of the extensional pattern types they define.

<u>Query 4.1</u> Print the name of any faculty member who is teaching any section of a course that is offered by the 'EE' department, provided that the section is taken by at least one graduate student who is an RA. Also, print the c# value(s).

```
context Faculty * Section *
        and (Course * Department [name = 'EE'], RA)
   select Faculty [name], c#
Print
```

OQL makes full use of the inheritance property of the generalization association. In this query, Faculty and RA inherit the association to Section from their superclasses. Hence, using the expression "Faculty * Section" instead of "Faculty * Teacher * Section" and the expression "Section * RA" instead of "Section * Student * Grad * RA" is legal.

4.2 Set Operators

The set operators Union, Intersection, and Difference can be applied to any two union-compatible subdatabases to produce a new subdatabase. Two subdatabases are said to be **union-compatible** if both of them have the same intensional association pattern defined in terms of E-classes only and irrespective of the D-classes that may exist in the subdatabases. Thus, the following is a legal format for a query.

```
context A * {B * C * D}
        union
        {A * B * C} * D
    select <classes and attributes>
```

The first argument to the Union operator returns a subdatabase whose extensional patterns are of the types <A,B,C,D> and <B,C,D> while the second argument returns a subdatabase whose extensional patterns are of the types <A,B,C,D> and <A,B,C>. However, both subdatabases have the same intensional pattern that consists of the four classes A, B, C, and D and their associations. The result of the Union operation is a subdatabase that contains extensional patterns of the three types <A,B,C,D>, <B,C,D>, and <A,B,C>, i.e., it contains the set of all extensional patterns that belong to either the first or second subdatabase. The Select subclause is then applied to the resulting subdatabase. The Difference operator returns a subdatabase that contains the set of all extensional patterns that belong to the first subdatabase but not to the second subdatabase. The subdatabase returned by the Intersection operator contains the set of extensional patterns that belong to both subdatabases.

Set operators can be used to create views in which some descriptive attribute values of objects do not appear with them in the view unless these objects participate in certain patterns of associations (e.g., for security reasons). This is illustrated by the following query, which produces a subdatabase that can be saved as a user's view (Section 5 describes how views can be defined and saved).

```
context Teacher * Section * Course [c# >= 5000]
   select name, degree, section#
union
context Teacher * Section * Student * Grad
   select Teacher [name, SS#], section#
```

The two arguments of the Union operator are two OQL queries that return two union-compatible subdatabases (since the Select subclause in each query projects over the two classes Teacher and Section, i.e., both subdatabases have the same intensional pattern). This means that each of the two queries returns extensional patterns of the type <Teacher, Section> but each query derives these patterns based on different conditions. The class Teacher in the first subdatabase has the descriptive attributes Name and Degree, while in the second subdatabase it has the descriptive attributes Name and SS#. The intensional patterns of these two subdatabases are shown in Figure 4.3a. In the final subdatabase, i.e., the result of the Union operation, only the Teacher instances that appeared in both operand subdatabases will have values for all the three attributes Name, Degree, and SS#. Those Teacher instances that appeared in the first subdatabase but not in the second subdatabase shall have Null values for the SS# attribute in the final subdatabase. The same goes for those Teacher instances that appeared in the second subdatabase but not in the first subdatabase with respect to Degree values. Figure 4.3b shows the intensional pattern of the final subdatabase in which the generic class Teacher has the attribute Name and each of the two subclasses Teacher_1 and Teacher_2 has one of the other two attributes. The instances of the Teacher_1 and Teacher_2 subclasses are those derived from the first and second operands of the Union operator, respectively.

Note that the non-association operator of OQL can be defined in terms of the Association and Difference operators and using braces as follows.

```
A ! B = {A} * {B} - A * B
```

Where "-" stands for the set difference operator described above. The following equation shows the equivalent expression to an expression that contains both an association operator and a non-association operator.

```
A ! B * C = {A} * {B * C} - A * B * C
```

Though the non-association operator can be defined in terms of other OQL operators, it is provided as a shorter notation for ease of use.

4.3 Queries with Multiple Expressions

In OQL, one can specify comparison conditions between attribute values or between objects of two different classes that appear in two different association pattern expressions. In this case, the two expressions shall be separated by a comma in the Context clause. Before presenting Query 4.2 which contains multiple association pattern expressions, we give the following rule that is relevant to the query.

Rule #4 If the association operator is used between two classes that are connected by more than one association (in this case the association links have to be distinctively named in the schema), then the name of the intended association needs to be specified after the association operator.

For example, there are two associations between the classes Undergrad and Department (Figure 2.1): the association labeled Minor and the inherited association labeled Major. Hence, the expression "Undergrad *Major Department" is used to refer to undergraduate students and their major departments.

Query 4.2 Display the names of all the undergraduate students minoring in the major department of the undergraduate student whose SS# = xxx. Display also the name of the department.

```
context   Undergrad_1 *Minor Department_1,
          Undergrad_2 [SS# = xxx] *Major Department_2
   where Department_1 = Department_2
   select Undergrad_1 [name], Department_1 [name]
display
```

The two association pattern expressions of the Context clause in the above query create two subdatabases. The two subdatabases are then linked by the condition stated in the Where subclause to produce a new subdatabase to which the Select subclause is applied. Also, in this query, the comparison in the Where subclause is performed between two type-comparable objects (i.e., objects that belong to the same class or to two different classes of the same generalization hierarchy).

5. Assignment Operator and Closure Property

The assignment operator ":=" can be used to save the subdatabase that is returned by a query. For example, the expression "X := **context** A * B * C" creates the subdatabase X whose intensional pattern consists of the classes A, B, and C, their associations, and their descriptive attributes and whose extensional patterns are the ones that are of the type <A,B,C>. To save a subdatabase permanently (i.e, not only for the duration of the query session), the key word **save** is used before the subdatabase name (e.g., **save** X := **Context** A * B * C).

A class that appears in a subdatabase can be referenced in the context of that subdatabase (i.e., only the instances of that class that appear in the subdatabase are considered) by qualifying its name with the subdatabase name using a colon (e.g., X:C). If not qualified by a subdatabase name the "base" class is

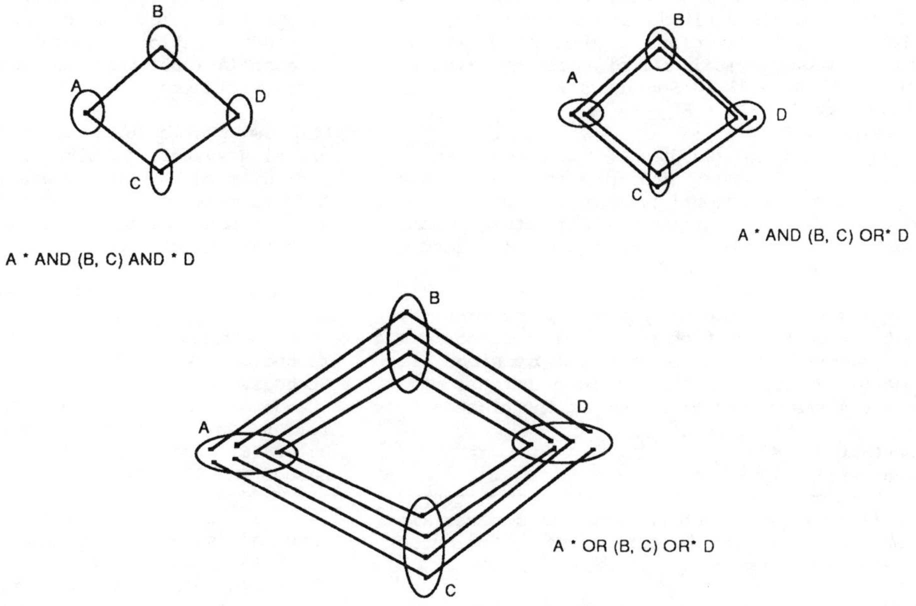

Figure 4.2: Extentional Pattern Types Specified by Context Expressions that Form Networks

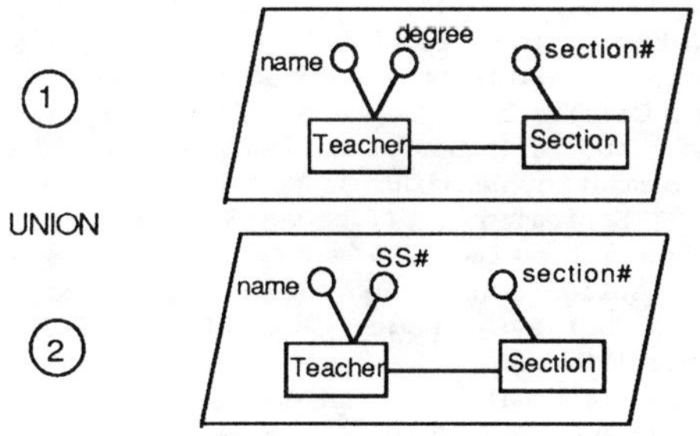

Figure 4.3a Two Union-compatible Subdatabases

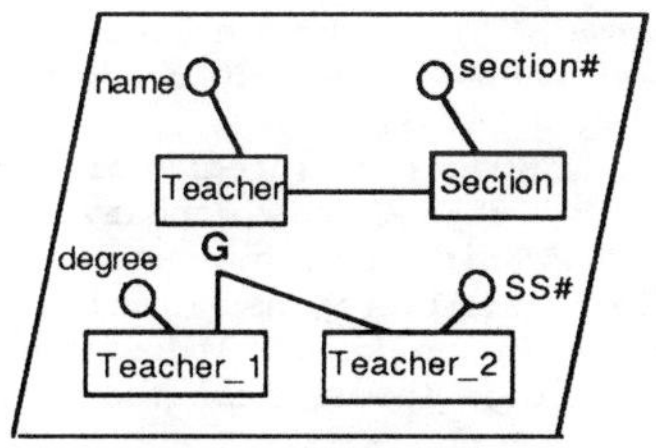

Figure 4.3b: The Result of applying the UNION Operation to the two subdatabases represented in Figure 4.3a

Figure 4.3: An example of applying a set operator to two union compatible subdatabases

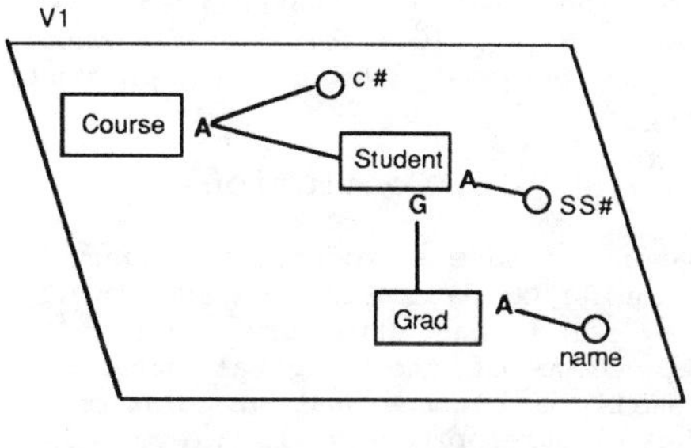

Figure 5.1: The Intensional Pattern of the View V1 Defined by an OQL Query

assumed. For example, if the class C in the original database is associated with the class D and D is associated with E, then the following query creates the subdatabase Y that consists of the class C derived from the subdatabase X and the class E derived from the original database together with all of their descriptive attributes (refer to Rule #1 in Section 3.2).

```
save Y := context X:C * D * E
          select X:C, E
```

This query operates on two databases, the subdatabase X and the original database and produces a single subdatabase (Y) whose intensional pattern consists of the classes C and E and a derived association between them. The subdatabase Y can be used by another query to produce yet another subdatabase.

Another consequence of maintaining the closure property in OQL is that nested association pattern expressions can be used. For example, the following two expressions are equivalent (where X is as defined before and assuming that B is directly associated with each of the classes A, C, J, and K).

```
Z := context J * X:B * K
Z := context J * (A * B * C):B * K
```

In the second expression, "A * B * C" is a nested association pattern expression that identifies a certain subdatabase whose B component is referenced by the outer association pattern expression.

Query 5.1 Define the user's view V1 that consists of the classes Course, Student and Grad such that a Course instance must have been taken by the Student instance who may or may not be a graduate student provided also that the Course instance belongs to the EE department. The attributes C#, SS#, and Name must appear with the classes Course, Student and Grad, respectively.

```
save V1 := context [ Department [name = 'EE'] * Course *
                     Transcript * Student ] * Grad
           select c#, Student [SS#], Grad [name]
```

The intensional pattern of the subdatabase V1 is shown in Figure 5.1, where a generalization association connects Student to Grad. In this subdatabase, only graduate students shall have values for the attribute Name, even though, in the original database, all students may have Name values.

6. Conclusion

In this paper, we have introduced the query language OQL for manipulating **OO** databases. A query in OQL returns a subdatabase, which has structural properties that are similar to those of the original database (i.e., it contains multiple classes and associations). In other words, the closure property is preserved. Thus, the result of a query can be an operand of another query.

The distinguishing features of OQL are summarized as follows.

1. A subdatabase returned by a query represents a "context" under which some operations can be specified and executed. In an OQL query, the specification of the Context subdatabase is separated from the specification of the operations on that subdatabase. This allows different operations to be performed on different object classes in the specified context.
2. Set operations can be performed on Union-compatible subdatabases. The result of a set operation is a new subdatabase that can be further manipulated in the normal way.
3. The association operators and the AND/OR operators allow very complex association patterns to be specified in a simple way. The same functionality would be specified in SQL, for example, by a complex nesting of Select-From-Where blocks.
4. Comparison operators (i.e., '=' and '!=') can be used to compare objects that belong to E-classes directly without referencing their attributes.
5. OQL makes full use of the inheritance property of the generalization association. A class inherits all the associations that emanate from or connect to its superclasses.

OQL is particularly suited for implementation on a graphics system. A query can be specified by browsing the S-diagram of object classes and pointing and traversing object classes to enter qualification conditions and association operators. An implementation of OQL on a SUN workstation is reported in a master's thesis [TY88].

Acknowledgement

This research is supported by the Navy Manufacturing Technology Program through the National Institute of Standards and Technology (formerly the National Bureau of Standards) grant number 60NANB4D0017, the Florida High Technology and Industry Council grant number UPN 85100316, and the National Science Foundation grant number DMC-8814989.

REFERENCES

BAN88 Jay Banerjee, Won Kim and Kyung-Chang Kim, "Queries in Object-Oriented Databases," Proceedings of the Fourth International Conference on Data Engineering, California, 1988, pages 31-38.

BAT85 D. Batory and W. Kim, "Modeling Concepts for VLSI CAD objects," ACM Transactions on Database Systems, vol. 10, No. 3, September 1985, pages 322-346.

COD79 E. Codd, "Extending the Database Relational Model to Capture More Meaning," ACM TODS, Vol. 4, No. 4, December 1979, Pages 397-434.

DIT86 K.R. Dittrich, "Object-oriented Database Systems: the Notion and Issues," Proceedings of the International Workshop on Object-Oriented Database Systems, California, September 1986, Pages 2-4.

HAM81 M. Hammer and D. McLeod, "Database Description with SDM: A Semantic Association Model," ACM TODS, vol. 6, No. 3, September 1981, pages 351-386.

HUL87 R. Hull and R. King, "Semantic Database Modeling: Survey, Aplications, and Research Issues," ACM Computing Surveys, pages 201-260, Vol.19, No.3, September 1987.

KIN84 Roger King, "A Database Management System Based on an Object-Oriented Model," Proceedings of FIWEDS, South Carolina, October 1984.

MAC85 R. MacGregor, "ARIEL—A Semantic Front-End to Relational DBMSs," Proceedings of VLDB 85, pages 305-315.

SHI81 D. Shipman, "The Functional Data Model and the Data Language DAPLEX," ACM TODS, Vol. 6, No. 1, March 1981, pages 140-173.

SU86 S. Su., "Modeling Integrated Manufacturing Data with SAM*," IEEE Comp., January 1986, pages 34-49.

SU88 S. Su, V. Krishnamurthy, and H. Lam, "An Object-oriented Semantic Association Model (OSAM*)," appearing in: A.I. in Industrial Engineering and Manufacturing: Theoretical Issues and Applications, S. Kumara, A. L. Soyster, and R. L. Kashyap (eds.), American Institute of Industrial Engineering, 1988.

TY88 Frederick Ty, "G-OQL: Graphics Interface to the Object-Oriented Query Language OQL," Master thesis, University of Florida, 1988.

XIA89 Daozhong Xia, "The Implementation of OSAM* KBMS Prototype in Vbase," an Internal Report, Database Systems Research and Development Center, University of Florida, 1989.

ZAN83 C. Zaniolo, "The Database Language GEM," Proceedings of the ACM SIGMOD Intl. Con. on the Management of Data, 1983.

PANEL SESSION
Database Tools and Interfaces

Panel chairman:
Dina Bitton
Stanford University and University of Illinois at Chicago

Advances in relational database technology have made available a wide range of database systems that support state of the art query languages and query optimization. However, the use and accessibility of these systems, whether they are PC based or high-end database engines, is hampered by the lack of adequate tools for database and application design. Users experience major difficulties in the process of designing a well structured relational database or writing SQL queries that are nonambiguous and perform well.

This panel will address the following questions:

1. Why is there such a wide gap between relational design theory and database design practice? The theory is sound and mature, but in practice, relational databases are often designed manually, in an ad-hoc manner. It seems that neither "higher level" data models nor diagram and form interfaces really help design well structured relational databases.

2. With the trend toward client-server architecture, will better tools and interfaces become available on client PC's and workstations? Will the entire design process be confined to the client? Will this migration help provide better user interfaces for database systems?

3. What is the functionality required for application design tools? Are there basic building blocks that all of these tools should embed, or is an application design tool for On Line Transaction Processing (OLTP) completely different from an application design tool for engineering databases or information retrieval systems? database systems more accessible?

Monitoring Database Objects

Tore Risch
Hewlett-Packard Laboratories
1501 Page Mill Rd., Palo Alto, CA 94303

Abstract

A method is described for actively interfacing an Object-Oriented Database Management System (OODBMS) to application programs. The method, called a *database monitor*, observes how values of derived or stored attributes of database objects change over time. Whenever such a value change is observed, the OODBMS invokes *tracking procedures* within running application programs. The OODBMS associates tracking procedures and the object attributes they monitor, and it invokes appropriate tracking procedures when data changes. Use is made of atomic transactions in the OODBMS.

The applicability of monitors is localized both in time and space, so that only a minimal amount of data is monitored during as short a time as possible. Such localization reduces the frequency of tracking procedure invocation, makes it easy to add and remove monitors dynamically, and permits efficient implementation.

To demonstrate these ideas, an implementation is described for the Iris OODBMS [10]. The implementation uses a technique of partial view materialization for efficient implementation.

1 Introduction

Database monitors are computer programs that observe changes in the contents of database objects, e.g. the current price of some commodity or the location of some ship. In addition to stored facts one often wants to monitor derived information, e.g. the highest paid employee in a department or the expenses of a department relative to its sales. Other monitoring applications include real time systems where a process is invoked or where a user is alerted upon pre-specified data changes. The need for similar concepts has been indicated by others [7, 18, 20, 30, 31].

The monitoring mechanism described in this paper is integrated with the Iris object-oriented database system [10]. In Iris, both derived and stored object attributes can be defined using functions expressed by an object-oriented query language, OSQL [3].

A *tracking procedure* is a procedure in a running application program to be invoked if the value of an associated object attribute has changed because of state changes to the database. We use OSQL to specify the derived or stored associated attribute. The DBMS does not transmit monitored data to tracking procedures; it merely signals that the state of derived data has changed by invoking tracking procedures. It is up to the tracking procedures to retrieve data by accessing the database. Possible tasks for tracking procedures include notifying the end user that data have changed, refreshing data browsers, or passing information to other systems. In case the application program caches data retrieved from the DBMS, tracking procedures can be used to inform the application that pieces of the cached data have been invalidated by database updates, so that caches can be refreshed and computations dependent on the old caches can be undone.

In the architecture presented here, the DBMS stores associations between tracking procedures, the attributes they monitor, identifiers of application processes, and the physical workstations in which they reside. After the DBMS has detected changes in monitored data it uses this information to call the appropriate tracking procedures. We then say that the application process has been *notified* or that the DBMS has sent a *notification* to the process.

In conventional database interfaces the DBMS behaves as a 'passive object' in that an application program always must request service from the DBMS. With monitors, the DBMS becomes an 'active object'; a request from an application process to the DBMS may result in procedure invocations in other application processes.

Tracking procedure invocations can be either *local* or *external*. An invocation is *local* if the process containing the invoked tracking procedure also requested the triggering database updates. An invocation is *external* if it is the result of a database update by some other process. External tracking procedure invocations do not occur until after the updates have been committed, while local invocations occur *before* the commit. Thus at the end of an update transaction, the DBMS first invokes local tracking procedures in the updating process and then immediately after the commit, the DBMS invokes external tracking procedures. The local tracking procedures are invoked synchronously, i.e. the system waits for them to return, while the external tracking procedures are invoked asynchronously and the system does not wait for them to return. Since a process may be performing unrelated work when the tracking procedure is invoked asynchronously, the system provides a no-interrupt option. In this mode the DBMS accumulates notifications and the application process can check for them synchronously.

2 Related work

This section discusses related techniques from several research areas. The database field has the notions of alerters, triggers, and integrity constraints, while the programming language field has the idea of active values. These ideas are somewhat related to data driven expert systems.

In [7] it is proposed that alerters be displayed when database updates cause certain boolean conditions to become true. This is a common special case of monitoring a boolean value. The RETRIEVE ALWAYS mechanism of Postgres [30] re-retrieves the results of queries into 'portals' associated with each client whenever there is a data change. Our monitoring mechanism is different from the above methods because the DBMS calls procedures in the client processes instead of sending data. In addition we employ methods, described below, that *localize* the applicability of monitors both in time and space.

Some DBMSs, e.g. System R [1], use database procedures called 'triggers' that are invoked upon updates of either user-specified base relations or by other actions [8, 16]. The ECA model [8] is a generalization of triggers in which the programmer can declare a precondition for the trigger. Triggers can be chained so that one update triggers another. This is useful for maintaining integrity constraints. However, the complicated cascading of triggers can make the database structure tangled and difficult to understand and maintain. Moreover, improper use of cascading triggers may slow down database updates considerably. In contrast, the invocation mechanism described here is non-procedural. The user specifies a database query whose value states are to be monitored, leaving the system to keep track of actions that may change a monitored state. In addition, the system creates an optimized mechanism to invoke tracking procedures when appropriate.

The technique called *access-oriented programming*, or 'active values', used in object-oriented programming languages such as Loops [28] is similar to traditional, procedural database triggers. But unlike DBMSs, active values are restricted to a single-user environment. Active values typically set properties or display some value. It is important that active values can be dynamically added to and removed from running systems [21], so that their use does not interfere with the logic of the rest of the system.

Declarative integrity constraint methods [4, 11, 18, 20, 23, 27, 29] are related to the monitor model in that they monitor boolean conditions, typically cancelling the updating transaction if the boolean conditions are unsatisfied at commit time. Tracking procedures, by contrast, are local to the application program process, are active during limited time periods, are triggered by value changes of any data type, can perform many tasks, and can be invoked both locally inside the updating transaction and asynchronously in processes outside the transaction. The tracking procedure may cancel the transaction only when invoked locally.

Database integrity constraints typically have performance problems [2]. The reason for this is that they may be very expensive to check for every update. Several techniques have been proposed to attack this problem, e.g. limit the assertions so that only those relevant to the updates of the particular file are checked [11, 19]; symbolically simplify the constraints checked per file [7, 11]; maintain intermediate aggregation data [4, 15]; exploit semantic constraint simplification [23]; simplify and check all constraints after many updates [12]; distribute constraints over several processors [24]; or automatically generate consistency maintenance code from declarative integrity constraints [18, 19, 20, 23]. These techniques are applicable to the monitor model as well. The monitor model avoids the efficiency problems of global integrity constraints, by being localized both in time and in the amount of data monitored, and by allowing asynchronous execution outside updating transactions.

Cactis [13] is an OODBMS to support software environments in which each data item is time stamped when updated. This allows the system to know what to recompile when the user retrieves an object. By contrast, our method checks if monitored data have actually changed, in which case the tracking procedures are invoked.

We may also contrast the present work with some data driven expert system tools like Syntel [9, 25, 26] and OPS5 [6]. Syntel is a tight integration of a database, an expert system tool and a user interface, specialized for financial applications. Syntel monitors a moderate amount of virtual memory data retrieved from persistent data and user input. In Syntel all shared data is passive, and there is no feedback from the persistent data to Syntel. By contrast, the method presented here monitors interactions between multiple users and their active and shared persistent large database independent of applications and user interfaces. The methods presented here may be used to extend the scope of data-driven systems like Syntel or forward chaining, rule-based systems like OPS5, making it possible to design systems where several expert systems cooperate.

3 An example

The following example of an Iris database will be used later to illustrate the monitor model. Assume a database of 'Person' and 'Department' objects with the attributes Name, Income, and Dept for objects of type Person and the attributes DeptName and Manager for objects of type Department.

The objects types can be defined by the OSQL statements:

```
create Type Person;
create Type Department;
```

The attributes of the above types are accessed using *extensional* Iris functions with OSQL definitions:

```
create Function Name(Person) -> String;
create Function Income(Person) -> Integer;
create Function Dept(Person) -> Department;
create Function DeptName(Department) -> String;
create Function Manager(Department) -> Person;
```

For example, given an object identifier (i.e. a surrogate), p, for a person, we may access its income and department name by the OSQL queries:

```
select Income(p);
select DeptName(Dept(p));
```

We may also set the value of an attribute by OSQL statements, e.g.:

```
set Income(p) = 41000;
```

These statements define the object types Person and Department. Iris also allows different *views* of objects by defining *intensional* Iris functions. For example, we may want to work with a view of the Person object with the attributes Name, Income, DeptName (as a string), MgrName (as a string), EarnsMoreThanManager (as a Boolean), and ColleaguesEarningMore (as a set of tuples of their names and incomes).

Such view of a Person object can be defined by the following Iris functions:

```
create function Name(Person) -> String;
                                /* extensional */
create function Income(Person) -> Integer;
                                /* extensional */
create function DeptName(Person p) -> String
  as select DeptName(Dept(p));
                                /* intensional */
create function MgrName(Person p) -> String
  as select Name(Manager(Dept(p)));
                                /* intensional */
create function EarnsMoreThanManager(Person p)
                                -> Boolean
  as select where
     Income(Manager(Dept(p))) < Income(p);
                                /* intensional */
create function ColleaguesEarningMore(Person p)
                      -> <Person, Integer>
  as select each Person coll, Integer inc
  where inc = Income(coll) and
        inc > Income(p) and
        Dept(coll) = Dept(p);
                                /* intensional */
```

Thus an application may have the above view of the Person object, not even knowing how attributes are physically stored.

Iris functions can also have more than one argument, e.g. to retrieve those persons earning more than a given income in a given department:

```
create function HighInc(Department d, Integer i)
                                -> Boolean
  as select each Person p
     where Income(p) > i and Dept(p) = d;
```

4 Design overview

We will now continue by discussing some critical aspects of the architecture:

- Efficiency is gained by minimizing the amount of monitored data and by allowing monitors to be active only while applications need them.
- We also gain efficiency by separately performing the expensive process of compiling monitor definitions, and storing the analyzed definitions in the database for later use.
- Application processes may receive asynchronous notifications while performing an unrelated task. Such processes will be interrupted by the DBMS to invoke the tracking procedures asynchronously. The DBMS does not wait for such asynchronous calls to return, to maximize system performance.
- Notifications of external tracking procedures must be delayed until just after transactions are committed.
- The DBMS must handle distributed notifications to client processes executing on separate machines.
- Finally, we must efficiently detect when values of monitored queries have changed.

4.1 Localizing Monitors

The monitor model handles common situations occurring in a serverized, shared, object-oriented DBMS such as Iris. Efficient execution is obtained from a relatively simple implementation.

One common situation involves monitoring values of properties of a given object. The properties may be either stored or derived from other properties. Often, monitoring is needed only during the execution of particular sections of an application process.

For example, we may wish to monitor the values of attributes `Income`, `ColleaguesEarningMore`, and `EarnsMoreThanManager` of a Person object view, p, invoking the tracking procedures `Mp1`, `Mp2`, and `Mp3`, respectively, when any of these values changes.

We provide *restricted monitors* for efficient execution of common situations, as follows:

- The monitors can be *time localized* so they are active only during a limited time, e.g. when a segment of an application program requires them to be active. We must therefore provide for efficient activation and deactivation of monitors.
- The monitors can be *client localized* by associating monitors with client processes. The system deactivates monitors when no client needs them.
- The monitors can be *object localized* by monitoring values of particular, or *focused* database objects, e.g. properties of specific persons. During the execution of an application program the monitor's focus may change from one object to another, i.e. the functions will be monitored for different arguments. Therefore the system should be reasonably efficient in refocusing from one object to another.
- The monitors can be *property localized* so that not every property of a database object is monitored. In the example, only the attributes Income, ColleaguesEarningMore, and EarnsMoreThanManager are monitored, not, e.g. Manager and Name. This restriction is important, since an object view may have many attributes.

In Iris, all the above localizations are supported.

4.2 Monitor Definition and Activation

Time localization requires efficient monitor activation and deactivation. This is achieved by separating monitor *activation* and *deactivation* from monitor *creation* and *deletion*, thereby performing as much work as early as possible at monitor creation time.

Monitor creation analyzes the monitored query and dynamically builds persistent tables and indexes to support efficient tracking. Monitor creation is therefore relatively slow. Monitors remain in the database until they are explicitly deleted. Monitors are generic in that they may be subsequently activated for different sets of arguments; we say that there are many *monitor instances* for a given monitor definition.

An application program typically focuses on different database objects during different time periods. The system thus has to support efficient changes in the set of focused objects. In Iris this amounts to an efficient method to vary arguments of monitored Iris functions. To refocus a monitor from one object to another we simply deactivate the monitor for one object and then activate it for another object.

Iris also supports overloaded functions. These currently cannot be monitored.

The following functions are provided for monitor creation and activation:

- `DefineMonitor(Fn)`, creates a new monitor; Iris data structures are created to support subsequent activation of the created monitor. Any Iris query can be monitored by first defining it as an Iris function and then defining a monitor for that function.
- `ActivateMonitor(Fn,Arglist,Proc)`, activates a monitor for a given Iris function, set of arguments, and tracking procedure within the current application process. The system will subsequently invoke the tracking procedure on changes to monitored object values.
- `DeactivateMonitor(Fn,Arglist)`, deactivates a monitor for the given Iris function and arguments in the current process.
- `DeleteMonitor(Fn)`, deletes a monitor definition permanently.

`DefineMonitor` creates a new table for each monitor definition called the *instance client table*. The instance client table is later updated by `ActivateMonitor` to store the network address and process identifier of each application process monitoring a given instance. The system also assigns a unique *tracking procedure identifier* to each tracking procedure in each process and stores it in the instance client table. Thus, by looking up the instance client table, the system will know which tracking procedures in which processes in which machines to notify when the instance is changed.

4.3 Notifying clients

After the database has been updated, one or more tracking procedures may be invoked in case their monitored function values were changed. The database server must detect value changes of monitored Iris functions, and notify client processes to invoke their tracking procedures.

The system only sends notifications after it has detected change to monitored data.

The actual time to check if updates make notifications necessary may vary. The most common case is that monitor notifications are checked only at the end of a transaction, when the database is in a consistent state. Therefore, in the current implementation, checks for notifications are made at transaction commit. However, the system provides an option for the application program to check for notifications even before a transaction is committed, by explicitly calling a procedure, `CheckMonitors`.

If `CheckMonitors` is called before a transaction is committed, and there have been updates to values monitored by that process, the system will invoke local tracking procedures of the application process. However, notifications of monitors activated by *other* processes will always be delayed until just after the transaction is committed. The system works this way because transactions hide updates from other clients until they are committed. (If 'dirty reads' were allowed we could eventually relax this delay.)

Another option is to defer checking monitors altogether. Instead, `CheckMonitors` is called explicitly after any number of transactions have been committed, or perhaps in some separately running process. This option has not been implemented.

Local tracking procedures are invoked because of local database updates. By contrast, external tracking procedures are invoked because of database updates by other processes. The external invocation may happen at any time, even when the application process is doing unrelated work. The notified client process is interrupted asynchronously when the notification arrives, in order to invoke the tracking procedure.

It is important to have the option of ignoring external monitor notifications, e.g. while processing a critical section of code. When external notifications are accepted, we say the application process is in *listening mode.* The system provides a procedure to toggle between listening mode and *not* listening mode.

When an application process is not in listening mode its monitor notifications will be collected by the system. Tracking procedure invocations will then be delayed until the procedure `CheckMonitors` is explicitly called or the process is toggled to listening mode.

4.4 Distributed Notification

Since the application process may run on a workstation other than the Iris server, we need a mechanism to handle the network communication involved in notifying a client. Such a communication architecture has been designed, by using the Network File System protocol for remote procedure calls. Figure 1 illustrates this architecture. [1]

The basic idea is that every workstation running an Iris client must also run a *monitor server* that takes care of receiving notifications and interrupting application processes as appropriate.

Given a notification, a remote procedure call, `NotifyMonitor(ProcId,MonId)`, is made to the affected monitor server, informing the monitor server that a notification has occurred to the monitor identified by `Monid` of the process identified by `ProcId` .

The monitor server saves the tracking procedure identifiers for a given process in a main memory data structure. It then interrupts the appropriate application process, signalling that some of its tracking procedures have been activated. In case the notified process has been terminated the system deletes the tracking procedure identifiers of that process from the monitor server and deactivates the monitor. When control returns back to the Iris server, the DBMS knows that the application process has been notified, but it is up to the application process to do something when the notification interrupt arrives.

If the application process is not in listening mode the interrupt signals are ignored. A later call to `CheckMonitors` first does a remote procedure call, `GetNotifications(ProcId,MonIds)`, to the monitor server to access the tracking procedure identifiers activated for the application process identifier. `GetNotifications` then deletes the tracking procedure identifiers from the global data structure of the monitor server.

Given these tracking procedure identifiers, `CheckMonitors` knows which application process procedure to invoke for the activated monitors.

If the application process is in listening mode, it catches all notification interrupts and then calls `CheckMonitors` immediately to asynchronously invoke notified tracking procedures.

If one or more notification interrupts arrive while another tracking procedure is running, the system will call `CheckMonitors` again *after* the first tracking procedure has finished executing. By this sequentialization, programmers need not worry about recursive notifications of tracking procedures.

4.5 Change detection

The purpose of the change detection algorithm is to determine if the values of a monitored Iris function have changed since the last time its monitoring clients were notified. This process is initialized by notifying the monitoring client at the end of the transaction in which the monitor was activated. Then the system determines if the value of the monitored function has changed since the previous notification.

The current algorithm is rather simple, assuming that only a few objects are monitored per function and that the values of monitored properties are not very voluminous. It also assumes that the definitions of Iris functions are simple and fast to evaluate, a reasonable assumption if the functions are used to do simple derivations such as those in the example.

The algorithm is designed to minimize the overhead of database operations when no monitor is created or active. It also minimizes the notification traffic; in particular, notification occurs only when there is actually a value change and then only at monitor checking time.

[1] The inter process communication in the current implementation is slightly more complicated than illustrated in Figure 1., because the code to detect changes is actually implemented as part of the client interface to Iris.

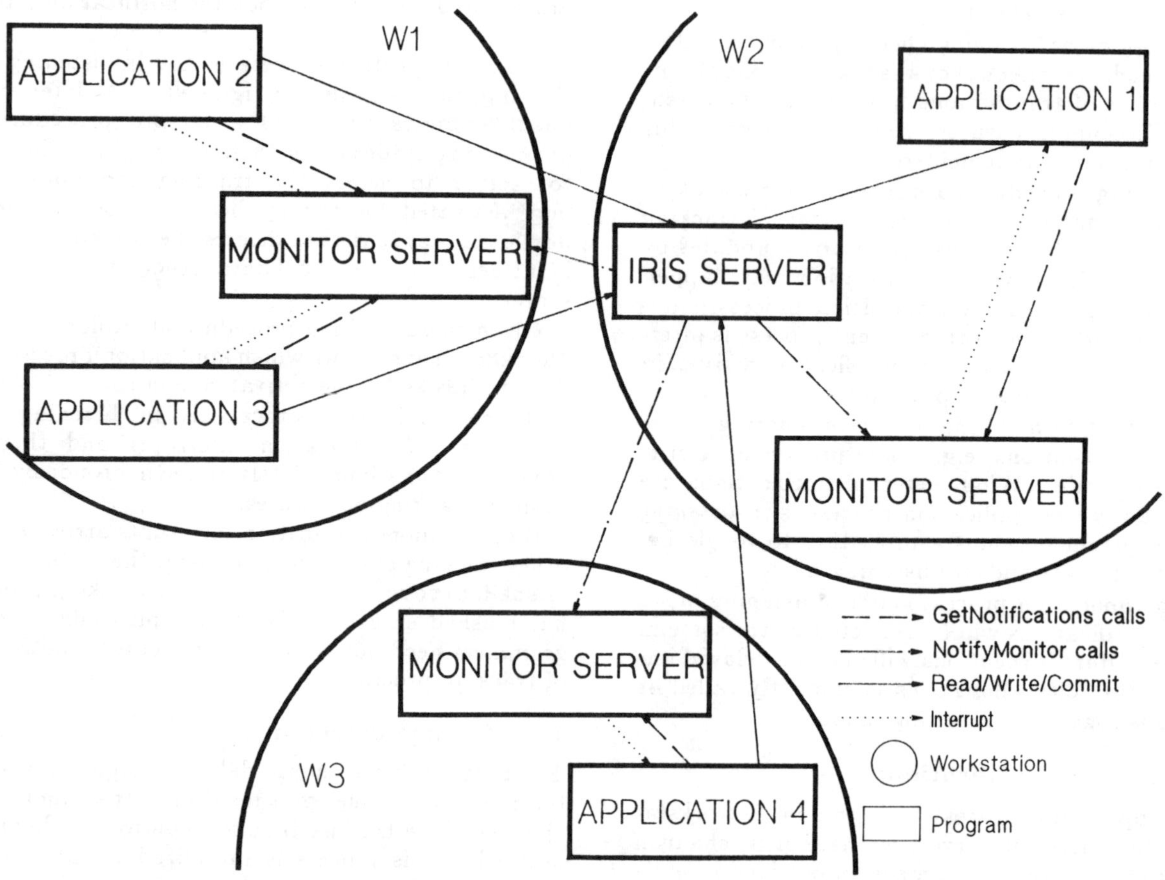

Figure 1: Inter process communication

The algorithm can be viewed as a variant of view materialization [5, 15, 17] whereby only the monitored instances of an Iris function are materialized, rather than the complete function traditionally materialized; thus we call the method *partial view materialization*. When an Iris function is defined, the system will also access-path-optimize the query to retrieve function values given that all arguments are known. If there has been an update invalidating some partial view materialization, the algorithm re-executes the access-path-optimized monitored function call for each monitored instance and compares the new values with the materialized values. If they are different, the new values are cached and the monitoring application processes are notified.

The change detection algorithm has four steps:

- A local *update detection* procedure is called for *every* Iris operation that updates the value of a stored Iris function [10], independent of whether the function is monitored or not. For every transaction, the system keeps track of Iris functions that have been updated. This procedure holds no locks and is very fast, associating with each transaction a virtual memory table of update-detected Iris functions. The size of this table is limited by the number of Iris functions updated by the transaction.
- At monitor checking time, for every update-detected Iris function, a monitor detection test is made to see if the updated functions participated in some monitor for some set of arguments. This is done by referring to a table, `SupportsMonitors`, that is updated by the system when monitors are created. This table maintains relationships between updatable Iris functions and their associated monitored Iris functions. For example these are some value sets of `SupportsMonitors`:

```
SupportsMonitors(Income)  =
      {Income, EarnsMoreThanManager,
       ColleaguesEarningMore}
SupportsMonitors(Dept)    =
      {EarnsMoreThanManager,
       ColleaguesEarningMore}
SupportsMonitors(Manager) =
      {EarnsMoreThanManager}
```

  `SupportsMonitors` is of limited size and can be kept in main memory.
- For every Iris function that was *monitor-detected*, a more expensive *instance detection* is done to test if its value changed. This detection is much slower than the previous ones, requiring re-execution of the pre-defined (and access-path-optimized) monitored Iris function in order to materialize the monitored instances and compare the new and old materialized values.

  The materialized values are stored in a system-generated table called the *instance cache table*. For example, the monitor for `ColleaguesEarningMore` will generate an instance cache table

```
_MCA_ColleaguesEarningMore(Person)
                -> <Person, Integer>
```

- For every *instance detection*, a notification is sent to the client processes that had activated that instance. The process identifier is found in the *instance client table* that maps each monitored instance to a list of its client processes and tracking procedures. Three levels of identification are needed:
  - A *client identifier* identifies the name (network address) of the workstation running each application process.
  - A *process identifier* identifies each (Unix) process per workstation.
  - A *tracking procedure identifier* identifies each tracking procedure in each process.

4.6 Fetching notified data

The notification mechanism does not transmit any data to the client. Instead the application process may decide to re-retrieve monitored instances. This can be done using the standard Iris query language. However, since the partial view materialization algorithm also caches monitored instances, we make use of the *instance cache table* when fetching notified data. This is implemented by an interface between the notification mechanism and the Iris query language, so that queries are modified to make use of the instance cache table.

For example, assume that the application program makes the OSQL query:

```
select ColleaguesEarningMore(p);
```

and that `ColleaguesEarningMore` is monitored for Person p. The system will modify the above query to access the instance cache table:

```
select _MCA_ColleaguesEarningMore(p);
```

thus avoiding any access path search. That is `ColleaguesEarningMore` is a derived function defined in terms of the stored Iris functions `Income` and `Dept`. It is not required to compute the derived value again since it is already cached. This simple query modification can be made in the client process without accessing the Iris server.

When a tracking procedure is invoked, the system passes the system-created tracking procedure identifier and a flag indicating whether the tracking procedure is invoked locally or externally. By passing the tracking procedure identifier, it is possible to write generic tracking procedures that can be used for many monitor definitions. A special primitive is provided to fetch the monitored data given such a tracking procedure identifier, using the instance cache table.

5 Summary and Discussion

We have proposed and implemented a method to interface application programs to the Iris object-oriented database system by using *monitors* that activate application program tracking procedures whenever values of derived or stored attributes of objects change. The attributes are specified declaratively, using an object-oriented query language. Monitors are localized both in space and time by limiting the amount of monitored data to be object attribute values, and by being active only while application processes need them.

The time localization strategy requires fast methods to turn monitors on and off. Therefore we separate the (slow) process of defining monitors from their activation and deactivation. Thus, monitor definition involves creation and update of database tables as well as optimizations for later monitor activations. A monitor that is created but not activated generates little overhead to database operations.

The application may run on a client workstation other than the database server. The system keeps track of which application processes on which workstations monitor which data, and automatically notifies application processes when monitored data change. Notification is done by invoking tracking procedures; external invocations may interrupt application processes and execute asynchronously, without the DBMS waiting for the tracking procedures to terminate.

Monitors may improve the concurrency of object-oriented database systems by allowing for very short transactions instead of long transactions, and letting the notification mechanism broadcast notifications to affected clients when database updates are committed.

The monitor model is also useful for the construction of systems of data-driven cooperative processes by letting the DBMS activate processes given data updates, rather than letting the cooperative processes directly call each other. In this way concurrent process invocations can be stated declaratively.

It is important that monitors execute efficiently and do not slow down the database server for applications not using them. The current implementation is efficient enough for some typical uses of database objects, where attribute definitions are specified using simple (thus fast) queries. More research is needed to extend the technique to handle complex queries.

The present change detection strategy could be combined with other strategies, such as those used for integrity constraint checking [4, 7, 11, 12, 15, 18, 19, 20, 23, 24]. Sometimes the system should not need not save values to determine that a change has occurred. For example if a monitored query specifies the sum of a set of values of which only one has changed, we may deduce that the sum has changed without doing any materialization.

The methods presented here are also applicable to relational DBMSs, by regarding Iris functions as relational views. The method proposed by Wiederhold [32] can be used to define derived objects by relational views. (Iris has many additional object features, including a persistent hierarchical type system, and persistent object identifiers.)

The monitor model could be generalized to handle integrity constraints, which can be regarded as universal monitors activated continuously for all applications. The tracking procedure for an integrity constraint would always be locally invoked whenever the constraint is violated and it would always cancel the transaction. However, more research is needed to get good performance for such integrity constraint monitors, since they would not be localized and since they could not use asynchronous tracking procedure invocations.

ACKNOWLEDGEMENTS:

I wish to thank Dan Fishman, Bill Kent, David Beech, and Waqar Hasan for helpful comments on earlier versions of this paper.

References

[1] M.Astrahan et al: System R: A Relational Approach to Database Management, *ACM Transactions on Database Systems*, 1 (2), June 1976.

[2] D.Badal, G.Popek: Cost Performance Analysis of Semantic Integrity Validation Methods, *Proc. ACM SIGMOD Conf.*, 1979, pp109-115.

[3] D.Beech: A Foundation for Evolution from Relational to Object Databases, *Advances in Database Technology - EDBT '88*, Lecture Notes in Computer Science, Springer-Verlag, 1988, pp251-270.

[4] P.Bernstein, B.Blaustein, E.Clarke: Fast Maintenance of Semantic Integrity Assertions Using Redundant Aggregate Data, *Proc. 6th VLDB Conf.*, 1980, pp126-136.

[5] J.A.Blakeley, P.A.Larson, F.W.Tompa: Efficiently Updating Materialized Views, *Proc. SIGMOD*, Washington D.C., 1986, pp61-71.

[6] L.Brownston, R.Farell, E.Kant, N.Martin: *Programming Expert Systems in OPS5*, Addison-Wesley, Reading, MA, 1985.

[7] O.P.Buneman, E.K.Clemons: Efficiently Monitoring Relational Databases, *ACM Transactions on Database Systems* 4, 3 (sept. 1979), pp368-382.

[8] U.Dayal, A.P.Buchmann, D.R.McCarthy: *Rules Are Objects Too: A Knowledge Model For An Active, Object-Oriented Database System*, Advances in Object-Oriented Database Systems, 2nd Intl. Workshop on Object-Oriented Database Systems, Sept. 1988, pp129-143.

[9] R.O.Duda, P.E.Hart, R.Reboh, J.Reiter, T.Risch: Syntel: Using a functional language for financial

risk assessment, *IEEE Expert* 2, 3 (Fall 1987), pp18-31

[10] D.H.Fishman, D.Beech, H.P.Cate, E.C.Chow, T.Connors, J.D.Davis, N.Derrett, C.G.Hoch, W.Kent, P.Lyngbaek, B.Mahbod, M.A.Neimat, T.A.Ryan, M.C.Shan: Iris: An Object-Oriented Database Management System, *ACM TOOIS*, 5, 1 (Jan 1987), pp48-69

[11] M.Hammer, D.McLeod: A Framework for Database Semantic Integrity, *Proc. 2nd Int. Conf. on Software Engineering*, San Fransisco, 1976, pp498-504.

[12] A.Hsu, T.Imielinski: Integrity Checking for Multiple Updates, *Proc. ACM SIGMOD Conf.*, Austin, 1985, pp152-168.

[13] S.E.Hudson, R.King: Object-Oriented Database Support for Software Environments, *Proc. ACM SIGMOD Conf.*, San Fransisco, 1987, pp491-502.

[14] W.Kim, D.S.Reiner, D.S.Batory: *Query Processing in Database Systems*, Springer-Verlag, New York, 1985.

[15] S.Koenig, R.Paige: A transformational framework for the automatic control of derived data, *7th VLDB Conf.*, 1981, pp306-318.

[16] A.M.Kotz, K.R.Dittrich, J.A.Mulle: Supporting Semantic Rules by a Generalized Event/Trigger Mechanism, *Advances in Database Technology - EDBT '88*, Lecture Notes in Computer Science, Springer-Verlag, 1988, pp76-91.

[17] B.Lindsay, L.Haas, C.Mohan, H.Pirahesh, P.Wilms: A Snapshot Differential Refresh Algorithm, *Proc. SIGMOD*, Washington D.C., 1986, pp53-60.

[18] M.Morgenstern: Active Databases as a Paradigm for Enhanced Computing Environments, *9th VLDB Conf.*, Florence, 1983, pp34-42.

[19] M.Morgenstern: CONSTRAINT EQUATIONS: Declarative Expression of Constraints With Automatic Enforcement, *Proc. 10th VLDB Conf.*, Singapore, 1984, pp291-300.

[20] M.Morgenstern: The Role of Constraints in Databases, Expert Systems, and Knowledge Representation, in L.Kerschberg ed.: *Expert Database Systems*, Benjamin Cummings, Menlo Park, CA, 1986.

[21] K.Osterbye: Active Objects: An Access Oriented Framework for Object-Oriented Languages, *Journal of Object-Oriented programming*, Vol. 1, No 2, (June/July 1988), pp6-10.

[22] J.Park, A.Segev: Using Common Subexpressions to Optimize Multiple Queries, *Proc. 4th Intl. Conf. on Data Engineering*, Los Angeles, 1988, pp311-318.

[23] X.Qian, G.Wiederhold: Knowledge-based Integrity Constraint Validation, *Proc. 12th VLDB Conf.*, Kyoto, Japan, 1986, pp3-12.

[24] X.Qian: Distribution Design of Integrity Constraints, *Proc. 2nd Intl. Conf. on Expert Database Systems*, Virginia, 1988, pp75-84.

[25] R.Reboh, T.Risch: Syntel: Knowledge programming using functional representations, *Proc. AAAI-86* (Philadelphia, PA, Aug), Morgan Kaufman, Los Altos, CA, 1986, pp1003-1007.

[26] T.Risch, R.Reboh, P.Hart, R.Duda: A Functional Approach to Integrating Database and Expert Systems, *Communications of the ACM* 31, 12 (Dec. 1988), pp1424-1437.

[27] A.Shepherd, L.Kerschberg: PRISM: A knowledge based system for semantic integrity specification and enforcement in database systems, *Proc. ACM SIGMOD Conf.*, Boston, MA, 1984, pp307-315.

[28] M.J.Stefik, D.G.Bobrow, K.M.Kahn: Integrating access-oriented programming into a multiparadigm environment, *IEEE Software* 3, 10 (Jan. 1986), pp10-18.

[29] M.Stonebraker: Implementation of Integrity Constraints and Views by Query Modification, *Proc. ACM SIGMOD Conf.*, San Jose CA, May 1975.

[30] M.Stonebraker: The Design of POSTGRES, *Proc. ACM SIGMOD Conf.*, Washington, D.C., 1986, pp340-355.

[31] M.Stonebraker: Future Trends in Data Base Systems, *Proc. 4th Intl. Conf. on Data Engineering*, Los Angeles, 1988, pp222-231.

[32] Wiederhold, G.: Views, objects, and databases, *IEEE Computer* 19(12) 1986, pp.37-44.

Situation Monitoring for Active Databases†

Arnon Rosenthal[1], Sharma Chakravarthy[1], Barbara Blaustein[2]
Xerox Advanced Information Technology
Email: <lastname>@xait.xerox.com

José Blakeley
Texas Instruments, Information Technologies Laboratory
MS 238, P.O. Box 655474, Dallas, TX 75265 Email: blakeley@csc.ti.com

Abstract

We present a basis for efficiently evaluating the situation (event and condition) portion of situation/action rules, either in an active database or in a standalone situation monitor. A common framework handles situations involving both database changes and nondatabase situations. We introduce ΔRelations to represent net changes to a stored or derived relation. We define an operator that computes ΔRelations for derived relations. Evaluation of expressions involving changes is optimized by defining incremental forms of relational operators and by providing a chain rule that extends incremental computation to data derived by arbitrary expressions.

1. Introduction

An *active* Database Management System monitors *situations* triggered by *events* representing database updates or occurrences external to the database. This paper is concerned with ways of specifying situations and evaluating them efficiently. The techniques described in this paper were developed as part of the HiPAC (High Performance Active) DBMS, a prototype active DBMS [DAYA88a, DAYA88b, CHAK89], parts of which have been implemented. However, the algebra and transformations described in this paper can be applied more generally. The situation evaluation mechanism can either be a component integrated tightly with a DBMS, or coupled with a heterogeneous array of applications or databases that signal it when specified events occur.

We use the relational model to specify situations and operations — all data objects will be relations, and all operators will map relations to relations. The relational model is not essential to our approach, but it simplifies the representation of database changes, allows use of the relational algebra, and avoids the need to explain the semantics of a particular object model.

1.1 Expressing Events and Situations

The problem addressed in this paper is the expression and evaluation of a single situation. A *situation* is an *event/condition* pair. An event may be a combination of more primitive pre-defined events. Informally, a situation describes a logical condition to be evaluated when one or more of a set of pre-defined events occur. Each situation is associated with a set of actions to be taken depending on the result of evaluating the condition.

An *event* is typed and associated with a relation scheme that describes certain data relevant to the event. In Example 1.1, the event of checking the status of a network link is related to a relation scheme including link identification, occurence, and severity information. An occurrence of an event is reported in a *signal* — a message that includes a *signal relation* (conforming to the event's associated relation scheme). The signal relation contains information about this particular event occurrence, and

[1]Address: 4 Cambridge Center, Cambridge MA 02142
[2]Address: 1800 Diagonal Road, Alexandria VA 22314

†This work was supported by the Defense Advanced Research Projects Agency and by Rome Air Development Center under contract No. F30602-87-C-0029. The views and conclusions contained in this paper are those of the authors and do not necessarily represent the official policies of the Defense Advanced Research Projects Agency, the Rome Air Development Center, or the U.S. Government.

may also include some information obtained from the database at the time of the event. We require that signal relations be nonempty. (If an event does not generate data, a dummy value is passed.)

The *condition* part of a situation is a relational expression whose inputs are: 1)signal relations from events in the situation's event set, and 2)zero or more database relations. The condition is evaluated when one or more events in the situation's event set have occurred. The situation *fires* (i.e., the action is invoked) if the condition result is a nonempty relation; the condition result is made available to the action.

1.2 The Evaluator

The situation evaluation component of the system, the *evaluator*, is kept simple by limiting its concerns. We depend on *event detectors* to detect events and send signals. (Any piece of software/hardware that sends event signals is an event detector.) The signals are collected and sent to the evaluator. The evaluator produces a signal that causes invocation of the appropriate action(s). The signal relation of this signal may be used by the action(s).

The major data structure used by the evaluator is a directed acyclic *signal graph* (shortened here to *graph*), that represents a situation. This is an operator graph, in which nodes represent relations: a leaf is bound either to an event (and its signal relation) or to a database relation. To aid in execution, nodes and edges on paths upward from event leaves are called *active*; the remainder are called *passive*. The *value* associated with each node is defined (though not necessarily computed) by the usual operator graph rule of bottom-up execution.

Because situation evaluation imposes overhead on every database update and other primitive event, it must be very efficient. The challenge is that the number of situations may be large, and some may require disk accesses or extensive calculations.

We show two situations that might be defined for a network-control application. We assume that a status-checking program examines the state of links, and occasionally sends a signal "Link_Status", listing all interesting test results. Note that this situation involves an event that does not necessarily do a database update, and that the condition uses an ordinary relational operator.

Example 1.1: `Severe Failures`

```
Event: Link_Status with relation scheme
         [Link#, Occurrence_Type, Severity]
```

Condition: `Select`$_{(\text{severity}>5)}$`(Link_Status)`

For the signal relation `{(link1, occ_a, 1), (link2, occ_b,6)}`, the condition evaluates to `{(link2, occ_b, 6)}`. The situation fires, and makes this result available to the action.

For the next example, assume that the database contains a relation `Link_Technology[L#, Mode]` that describes the implementation of each link. The join query below combines the incoming signal with database information.

Example 1.2: `Severe Failures on Microwave Links.`

```
Event: Link_Status with relation scheme
         [Link#, Occurrence_Type, Severity]
```

Condition:
`Join[Select`$_{(\text{Severity}>5)}$`(Link_Status),`
`Select`$_{(\text{Mode='microwave'})}$`(Link_Technology),`
`Link_Status.Link# = Link_Technology.L# ]`

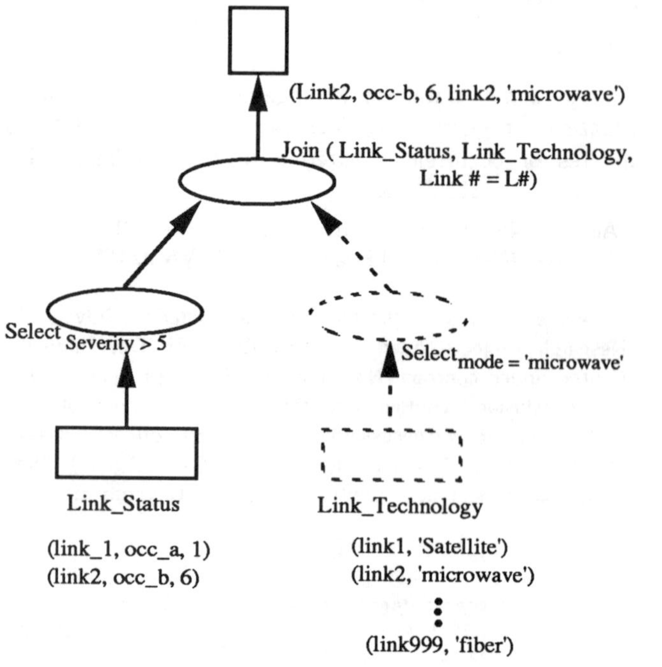

Figure 1: Signal Graph for Example 1.2

Figure 1 shows the signal graph for Example 1.2. The figure also shows the node values that would be computed during graph execution. Note that values need not be computed for all the passive nodes (shown with dashed lines). For example, if Link_Technology is a large relation, it is not advisable to explicitly evaluate $\texttt{Select}_{Mode='microwave'}$(Link_Technology). Note also that if the output of $\texttt{Select}_{(Severity>5)}$ (Link_Status) had been empty, it would be unnecessary to examine any node above it.

Section 2 describes operators that express situations involving database changes. Section 3 discusses optimization techniques for such situations, and general techniques for optimizing signal graphs. Section 4 discusses related work and presents our conclusions.

2. Database Changes

We now define a relation, called a Δ*Relation*, that can represent changes to another relation, and operators for manipulating such relations. We look at the *net* additions, deletions, and modifications to the relation instance. We do not look at individual operations. The goal is a unified treatment of changes, not separate treatments of insertions, deletions, and modifications resulting in several algorithms for generating and combining individual results. Section 2.1 describes conventions satisfied by Δrelations that represent changes, and gives some basic manipulation operators. Section 2.2 discusses an operator that examines updates to some of the objects underlying a derived relation (e.g., view), and determines the corresponding change to a derived relation. Section 2.3 describes efficient ways of implementing this operator, both for familiar relational operators and for arbitrary relational expressions.

Let **R** denote a *relation scheme* specifying a set of attributes and let R denote a *relation* (or, more precisely, a relation instance of **R**) consisting of a set of tuples whose values are taken from the domains of the set of attributes of **R**. *t.A* denotes the value of attribute A for tuple t. Each tuple has an attribute (denoted *tid*) that provides a unique immutable identifier. We use relational operators \m Select, Project (which discards tuples that are entirely null), Join, Outerjoin, Difference, and Union. The unique tid makes it easier to connect tuples that hold values of the same object before and after changes; we will assume that every updatable relation includes the tid column that provides a unique immutable identifier. For readability, pairs of parentheses may be denoted either as "(...)" or as "[...]".

2.1 Representation and Basic Manipulations

To provide a single object that captures an arbitrary change to a relation, we introduce the concept of a Δ*Relation*. For each relation scheme **R** $\equiv$ (tid, A_1, . . . , A_n), define scheme Δ**R** to be (˜tid, ˜A_1, . . . ,˜A_n, tid˜, A_1˜, . . . , A_n˜). When a ΔRelation represents changes, attribute names with a ˜-suffix (e.g., A˜) refer to new attribute values, and attribute names with a ˜-prefix (e.g., ˜A) refer to old attribute values. In any tuple t, if both t.tid˜ and t.˜tid are non-null, they must be equal; if t.tid˜ is null, so are all t.A_i˜; if t.˜tid is null, so are all t.˜A_i, $1 \le i \le n$; ˜t.tid and t˜.tid cannot both be null; no tid can appear in multiple rows (even in different columns).

ΔRelations can represent tuples (or objects) where only the tid field is nonnull. $\texttt{Join}_{tid}$ is used to indicate a join that uses predicate [˜tid=tid˜].

Example 2.1 shows a ΔRelation that expresses changes due to a transaction. A ΔRelation can be partitioned horizontally, i.e., expressed as the disjoint union of insertions, deletions, and modifications. (For insertions, the ˜-prefix attributes are null; for deletions the ˜-suffix attributes are null).

Example 2.1: Consider the Employee relation EMP and the transaction T:

EMP	tid	Name	Salary
	@123	Joe	30k
	@456	Lynn	40k
	@321	Ann	28k

Transaction T:

```
{Modify(@123, new_tuple = (@123, Joe, 33k));
 Insert  (@789,  Ed,  25k);
 Delete (@456)}
```

The following ΔRelation (here called ΔEMP) captures the changes that T makes to EMP:

˜tid	˜Name	˜Salary	tid˜	Name˜	Salary˜
@123	Joe	30k	@123	Joe	33k
-	-	-	@789	Ed	25k
@456	Lynn	40k	-	-	-

To express conditions that involve changes, we provide several operators that manipulate ΔRelations. Here we describe some basic operators; the next section uses them to define a more powerful construct.

We first define renaming functions that add or delete tildes from attribute names in a relation scheme. Let **R** = {tid, A_1, A_2, ..., A_n}. Then **pretilde(R)** = {˜tid, ˜A_1, ˜A_2, ..., ˜A_n}, and **postilde(R)** = {tid˜, A_1˜, A_2˜, ..., A_n˜}. So Δ**R** = **pretilde(R)** ∪ **postilde(R)**.

We define analogous renaming functions on relation instances. If R is a relation, pretilde(R) is the relation with the same tuples, but with all attribute names preceded by a single tilde (and similarly for postilde(R)). For relations where all attributes have pretildes, or all have postildes, untilde removes tildes from all attribute names.

The next set of operators project the old or new tuple values from a ΔRelation. Some of the operators are illustrated in Example 2.2. **Removals˜** takes a ΔRelation and returns the relation consisting of tuples removed by either deletion or modification. **Removals** returns the corresponding relation without the ˜-prefix on attribute names. **Additions˜** returns the relation consisting of tuples added by either insertion or modification, and **Additions** returns the corresponding relation without ˜-suffix. Formally,

Removals˜(ΔR) ≡ Project[ΔR, pretilde(**R**)]
Removals(ΔR) ≡ untilde[Removals˜(ΔR)]
Additions˜(ΔR) ≡ Project[ΔR, postilde(**R**)]
Additions(ΔR) ≡ untilde[Additions˜(ΔR)]

Example 2.2

Removals˜(ΔEMP)

˜tid	˜Name	˜Salary
@123	Joe	30k
@456	Lynn	40k

Additions(ΔEMP)

tid	Name	Salary
@123	Joe	33k
@789	Ed	25k

The main use of ΔRelations is to represent the net effect of a collection of updates to a relation. In such cases, if the "before value" is the relation R, the updates will be represented as ΔR, and the updated relation will be denoted R'. We now define operators that relate these three values.

Definition: The pair (R, ΔR) is *composable* if:
(1) The relation scheme for ΔR is **pretilde(R)** ∪ **postilde(R)**;
(2) **Removals(ΔR)** ⊆ R, ; and
(3) **Additions(ΔR)** ∩ R = ∅.

Composability requires that the ΔRelation contain only the net updates. The restrictions can be loosened somewhat, at the cost of some extra complications for **Compose** and other operators. If the restrictions are sufficiently loosened, a ΔRelation will resemble an undo/redo log.

Definition: Suppose the pair (R, ΔR) is composable. Then define

Compose(R, ΔR) ≡ (R - **Removals**(ΔR)) ∪ **Additions**(ΔR).

The result will often be denoted R'.

Example 2.3: Given EMP and ΔEMP as in Example 2.1:

EMP' ≡ Compose(EMP, ΔEMP)

EMP'	tid	Name	Salary
	@123	Joe	33k
	@789	Ed	25k
	@321	Ann	28k

Definition: **Δdifference** takes any pair (R_1, R_2) of relations on the same scheme, and produces a ΔRelation that describes their differences. Null values are used to pad tuples that appear in only R_1 - R_2 or R_2 - R_1. It can be expressed as the Outerjoin of tuples in R_1 - R_2 (supplying the ˜-prefix attributes) and R_2 - R_1 (supplying the ˜-suffix attributes). That is,

Δdifference(R_1,R_2) ≡
Outerjoin_{tid}[pretilde(R_1 - R_2), postilde(R_2-R_1)]

The following lemmas show some of the algebraic properties of these operators.

Lemma 1A:
ΔR= Outerjoin_{tid}[**Removals˜**(ΔR), **Additions˜**(ΔR)]

Lemma 1B: For any relations R_1 and R_2 having the same scheme, **Compose**[R_1, **Δdifference**(R_1, R_2)] = R_2

2.2 The Changes Operator

We now introduce a new high-level operator, called **Changes**, to express how a derived relation (e.g., view) changes when at least one of its input relations changes. Let E denote an algebraic expression that defines a derived relation, and let **E** denote the scheme of the relation produced by E. The output of the **Changes** operator is a ΔRelation with scheme **ΔE**. **Changes** can be useful in writing situations, and is easy for a compiler to recognize and optimize.

Notation: Let $E(R_1, ..., R_n)$ denote an expression defined on relation schemes $\mathbf{R_1}...\mathbf{R_n}$; let each R_i denote an arbitrary relation; let ΔR_i denote a relation that is composable with R_i, and let R'_i denote Compose(R_i, ΔR_i). A pair $L = \{(R_i, 1 \le i \le n), (\Delta R_i, 1 \le i \le n\}$ is called a *substitution list*.

Definition:
Changes(E; L) ≡ Δdifference($E(R_1,...,R_n)$, $E(R'_1,...,R'_n)$)

Lemma 2: Let ΔR be composable with R. Then
Changes[R; (R, ΔR)] = ΔR

The following lemma states that Changes(E; L) correctly computes the changes to the relation derived by E.

Lemma 3:
$E(R'_1,...,R'_n)$=Compose($E(R_1,...,R_n)$, Changes(E; L))

Example 2.4 High-Salaried Employees
Consider the EMP and ΔEMP relations in Example 2.1. To monitor changes to the view High_Salary_EMP ≡ $\text{Select}_{(\text{Salary}>32k)}$(EMP), define the situation:

```
Event:     Update to EMP
Condition: Changes(High_Salary_EMP; [EMP, ΔEMP])
```

Substituting the definition of High_Salary_EMP, the condition becomes Changes($\text{Select}_{(\text{Salary}>32k)}$; [EMP, ΔEMP]). The result of evaluating this condition for the updates of Example 2.1 is given by the relation

~tid	~Name	~Salary	tid~	Name~	Salary~
-	-	-	@123	Joe	33k
@456	Lynn	40k	-	-	-

If the user wants to monitor all changes to a view, there is no need to specify explicitly which base-relations' updates should be monitored — in fact, encapsulation is improved if the system rather than the user fills in that list. The explicit form allows specification of a situation that monitors changes caused only by certain kinds of events (e.g., an abstract event Hire_Employee).

3. Optimizations

Efficiency is critical to situation monitoring, but evaluation of **Changes** directly from its definition is very inefficient, requiring retrieval or materialization of R_i and R'_i, for all i. This section describes three ways in which the evaluator's performance can be improved. The first two subsections define transformations on expressions involving the **Changes** operator. Section 3.1 presents efficient "incremental" implementations of Changes(F; L) where F is Select, Project, or Join; Section 3.2 treats the case where F is an arbitrary expression. It describes a *chain rule* that obtains an incremental form of F by composing incremental forms of the operators within F. Section 3.3 discusses issues involved in determining execution strategies for signal graphs.

3.1 Incremental Operators

For many operations F, Changes(F; L) can be computed more efficiently than by evaluating the definition for Changes. These computations use ΔRelations heavily (and sometimes exclusively), instead of computing on database relations, which tend to be much larger.

We study instantiations of Changes(F; L) for particular operations F. Define the *incremental form* for F as the operator IncrF(substitution list) --> (ΔRelation) where IncrF(L) ≡ Changes(F; L)

If the optimizer has been informed of an efficient implementation for IncrF, it will use that in place of Changes(F; L). Many authors (e.g., [KOEN81, BLAK86]) have defined efficient incremental forms of Select, Project, and Join to deal with insertions or with deletions or with modifications. Here we show efficient implementations of IncrSelect, IncrProject, and IncrJoin that handle an entire ΔRelation as a unit.

An extensible active DBMS would allow users to define new incremental forms. For example, the definer of a new function could be invited to provide an incremental form, which the DBMS would then register with the optimizer.

Incremental Select: Let pred denote a predicate defined on **R**, and $\texttt{Select}_{pred}$ denote the associated selection operator. Let ˜pred and pred˜ denote the predicates obtained from pred by replacing each attribute by its ˜-prefix (respectively, ˜-suffix) form.

The following formulas give an implementation of `IncrSelect` that does not reference the base relation. Let $\Delta S \equiv \texttt{IncrSelect}_{pred}(R, \Delta R))$.

S1 = $\texttt{Select}_{\tilde{}pred}$(`Removals`˜($\Delta R$))
S2 = $\texttt{Select}_{pred\tilde{}}$(`Additions`˜($\Delta R$))
ΔS = $\texttt{Outerjoin}_{tid}$(S1, S2) /* by lemma 1A */

Note that S1 and S2 represent `Removals`˜(ΔS) and `Additions`˜(ΔS), respectively. The expression for `IncrSelect` can be implemented by one pass over ΔR, determining for each tuple of ΔR whether it induces an insertion, deletion, or modification to `Select`(R). The proof that the above formulas are indeed implementations of `Changes`($\texttt{Select}_{predicate}$ (R); [R, ΔR]) appears in the appendix.

The optimizer transforms signal graphs, replacing **Changes** by efficient incremental forms. Figure 2a shows the initial signal graph, and Figure 2b shows the final signal graph, for **Changes**(High_Salary_Emp; [Emp, ΔEMP]) from Example 2.4.

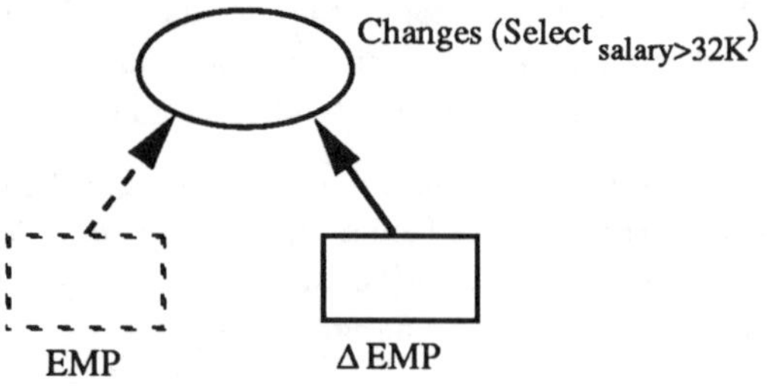

Figure 2a: Initial Signal Graph of Situation in Example 2.4

Incremental Project. Let $A = \{A_1, \ldots A_k\}$ denote a subset of the attributes in **R**, where tid $\in$ A. `IncrProj`(ΔR, A) $\equiv$ `Changes[Project`(R, A); (R, ΔR)]. Let the predicate "differ" state that for some attribute A_i in A, t.˜A_i differs from t.A˜$_i$. Then an efficient implementation of `IncrProj` is to select tuples for which at least one attribute has changed, i.e.,

`IncrProj`(ΔR,A)= $\texttt{Select}_{differ}$[`Project`(ΔR,˜A $\cup$ A˜)]

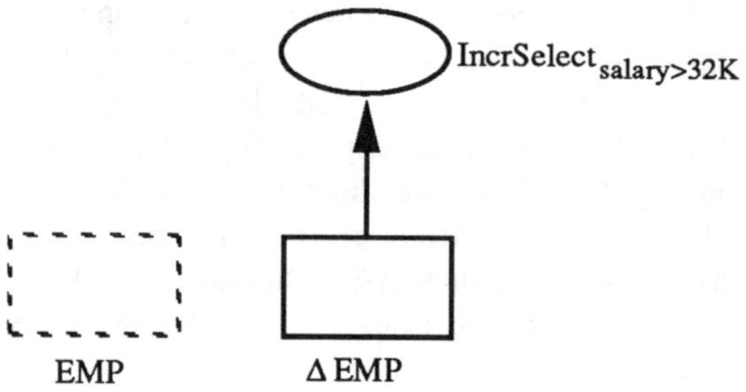

Figure 2b: Optimized Signal Graph of Situation in Example 2.4

Incremental Join. In the case of join, changes to the resulting relation can be induced by changes to either of the individual operands or to both of them. First, suppose that just R_1 changes.

Let join_pred denote a predicate on $\mathbf{R_1}$ and $\mathbf{R_2}$; and let $tid_1 \in \mathbf{R_1}$ and $tid_2 \in \mathbf{R_2}$, respectively. Let ˜join_pred (join_pred˜) denote the predicate obtained from join_pred by replacing each attribute by its ˜-prefix (˜-suffix) form. Then

`LeftIncrJoin`(ΔR_1, R_2, `join_pred`) =
 `Outerjoin(`
  `Pretilde[Join(Removals(`ΔR_1`)`, R_2, join_pred)],
  `Postilde[Join(Additions(`ΔR_1`)`, R_2, join_pred)],
  (˜tid_1 = tid_1˜) AND (˜tid_2 = tid_2˜))

`RightIncrJoin`(R_1, ΔR_2, join_pred) is defined similarly, for the case where just R_2 changes. `IncrJoin`, defined below, is used when both R_1 and R_2 change. To capture the interaction of simultaneous changes to R_1 and R_2, the notation `Both`(ΔR_1,ΔR_2,join_pred) denotes
`Outerjoin[`
 `Join(Removals`˜(ΔR_1), `Removals`˜(ΔR_2), ˜join_pred),
 `Join(Additions`˜(ΔR_1), `Additions`˜(ΔR_2), join_pred˜),
  (˜tid_1 = tid_1˜) AND (˜tid_2 = tid_2˜)]

We can now express `IncrJoin`([R_1, R_2, ΔR_1, ΔR_2], join_pred) as
 `LeftIncrJoin`(ΔR_1, R_2, join_pred) $\cup$
 `RightIncrJoin`(R_1,ΔR_2, join_pred) $\cup$
 `Both`(ΔR_1,ΔR_2, join_pred)

3.2 Transforming Expressions Inside Changes

The Chain Rule is an identity that is used to create an incremental form of an algebraic expression (denoted

Expr) from incremental forms of its constituent operators. Each application of the chain rule moves the outermost operator of Expr outside **Changes**. Where Expr is a single function F, recall that **Changes**(F, L) = **IncrF**(L). Applying the chain rule repeatedly and using this base case, the optimizer removes all appearances of the special operator **Changes**. We present the identity only for the case where the root of Expr is a unary operator (denoted F), so that Expr can be written as a composition of functions, denoted F ° G. Let [RList, ΔRList] denote the substitution list L.

Chain Rule:
Changes (*F* ° *G*; [RList, ΔRList]) =
IncrF (G(RList), **Changes**(G; [RList, ΔRList]))

The chain rule is proved in the appendix. There are three benefits to using the chain rule to open up the **Changes** expression. 1) **IncrF** now appears explicitly, and it may have an efficient implementation. 2) **IncrF** now needs to be executed only if **Changes**(G; [RList, ΔRList]) is non-empty. 3) It may be easier to derive optimizing transformations (e.g., Selections before Outerjoins) using the incremental forms that replace the **Changes** operator.

We will illustrate this with an example. Consider relation `Sensor_Data[tid, Temperature, Pressure, other_attributes]`. Let `Filter` denote a user-defined function that removes outliers and smoothes the data in a relation with schema `[tid, Temperature, Pressure]`. And let $\texttt{Proj}_{TP}$ denote the operator $\texttt{Project}_{[tid,\ Temperature,\ Pressure]}$

Example 3: `Change To Filtered Temperature/ Pressure`

`Event: Update to Sensor_Data`
`Condition:`
$\texttt{Changes}(\texttt{Filter} \circ \texttt{Proj}_{TP};$
[Sensor_Data, ΔSensor_Data])

If we apply the chain rule to the outermost operator (i.e., `Filter`), we get the Condition:
`Incr_Filter[`
$\texttt{Proj}_{TP}$(Sensor_Data), $\texttt{Changes}(\texttt{Proj}_{TP};$
[Sensor_Data, ΔSensor_Data]]

We now replace the inner projection by its incremental form, to obtain
`IncrFilter(`
$\texttt{Proj}_{TP}$(Sensor_Data), $\texttt{IncrProject}_{TP}$
(Sensor_Data, ΔSensor_Data]

$\texttt{IncrProject}_{TP}$ does not need the base relation Sensor_Data as an input. The result is the signal graph of Figure 3.

The chain rule has permitted two major improvements in the execution strategy. First, **IncrFilter** need not be evaluated if output of $\texttt{IncrProj}_{TP}$ is empty. Second, it exposed $\texttt{Changes}(\texttt{Proj}_{TP};$ [...]) so that it could be replaced by $\texttt{IncrProj}_{TP}$.

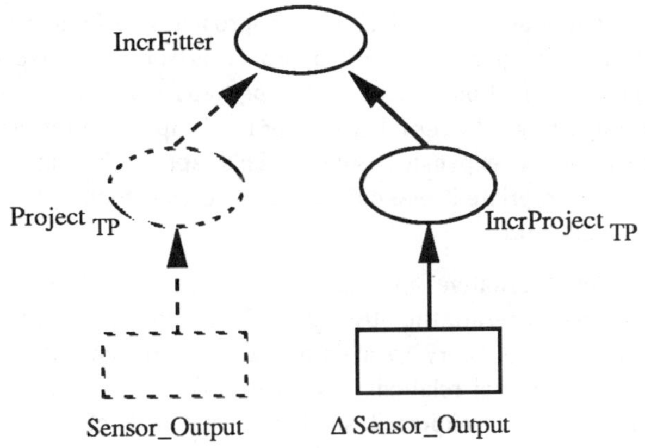

Figure 3: Optimized Signal Graph of Filtered Sensor Data

3.3 Compilation and Evaluation of Situations

Once a final signal graph has been obtained, it is necessary to determine its execution strategy. A good execution strategy will exploit the expected small size of signal relations and relations derived from them.

A leaf of the signal graph that corresponds to an event's signal relation is called *active*. Any node or edge that is on a directed path from an active node is also called active. Other nodes and edges (i.e., those defined strictly over stored relations) are called *passive*.

The passive/active distinction leads to the following simple guidelines for execution strategy:

- For many multi-input operators (e.g., **Join**, **Changes**), when the active input is ∅, the result is ∅.
- Complete evaluation of passive subgraphs is often unnecessary.
- Active nodes often have much smaller relations than passive nodes. Furthermore, they are often com-

puted directly from a signal and hence require no disk accesses. Therefore nested-loops is often a good implementation for a join operator. It requires accessing only those of the inner-relation's tuples that match a tuple from the outer relation (assuming that an index is available). For instance, in Example 1.2, the `Link_Technology` relation may contain thousands of microwave links, but only the tuple for link2 is required for the join.

There are two architectural approaches to finding execution strategies. The first approach is simply to use the heuristics mentioned above, plus perhaps some other simple ones (e.g., `Select` before `Join`, cheap selection predicates before expensive ones). This approach might be most appropriate if one did not have access to an appropriate query optimizer.

An alternative approach is to have a DBMS query optimizer generate the strategies. The signal graph can be regarded as a query to a database that consists of database plus signal relations. Given statistics on signal relations (i.e. their expected small size and lack of indexes), the optimizer will presumably select nested-loops implementations. A further benefit is that the DBMS will supply implementations of relational operators (selection, projection, etc.) and an interpreter for the strategies that are produced; otherwise we need to implement these capabilities as part of the situation evaluator.

But there are also serious drawbacks to using a conventional DBMS — software integration and efficiency. Software integration may be difficult or impossible with a non-extensible DBMS. One must find a way to add operators that manipulate ΔRelations, and to accept operator trees as input (rather than SQL). Efficiency is a problem if signals that originate in main memory need to be stored on disk to appear in DBMS queries. Also, existing optimizers may not do a very good job of exploiting a node value of $\oslash$, an occurrence that may be more common in situation evaluation than for ordinary queries.

4. Summary

4.1 Related Work

Work related to condition monitoring is found spread over several traditional problem areas of the design of database management systems. Such areas include integrity constraint enforcement, support of views, triggers and alerters, rule management, and production systems in general.

Recent contributions include [STON87, DARN87, DITT86]. [STON87] triggers condition evaluation upon database operations, and for a pre-defined set of other events (e.g., date). Conditions are expressed in QUEL, and therefore can be evaluated by the regular query processor. [DITT86] separates events from actions and triggers are defined explicitly using an event, action pair. Our work separates events, conditions, and actions explicitly and the emphasis is on optimizations and algebraic structure. [DARN87] describes a commercial implementation (Sybase). Events are limited to inserts, deletes, and modifies, described by a structure resembling a ΔRelation. Our approach generally differs in that we have provided special constructs to make it easier to express situations involving database changes, and to optimize them. Data from user-defined events may be referenced in conditions.

An algorithm to incrementally compute the changes on materialized views defined by `Select`, `Project`, and `Join` is described in [BLAK86]. However, that paper does not address the issues of representation for database changes (i.e. ΔRelations), changes to general expressions, and optimization of the expressions that compute the changes. [KOEN81] transforms programs containing database update operations so that they will maintain derived data. The transformations require recompiling the program; also, Insert, Delete, and Modify are not given a uniform treatment.

4.2 Conclusions

In this paper we have presented key concepts of a situation evaluator for HiPAC. The main contributions of this paper are:

- a unified treatment of database updates and nondatabase events. This allows the situation evaluator to be implemented in any system that provides the desired event detectors.
- ΔRelations as a compact representation for all database updates. Even transition constraints requiring values from different database states can be naturally expressed using `Changes` and Δrelations.
- an algebra for computations about database changes. The algebra includes efficient implementations of incremental forms of familiar operators and accommodates user-defined operators. Previous work generally provides a separate form for insertions, deletions, and modifications, but one transaction can include all three.

- the Chain Rule as a key transformation in optimizing changes to views. With the exception of [KOEN81], most previous work is limited to `Select/Project/Join` expressions; we make improvements even with user-defined operators.

Promising areas for future work include: materialized data for condition monitoring; rules to maintain derived data; dispatching strategies for accumulating update events before evaluation; a generalized chain rule for n-ary functions; and simultaneous evaluation of multiple conditions. Larger problems include transposing situation evaluation techniques to distributed and object-oriented systems.

5. Acknowledgements

The authors appreciate the helpful suggestions of the anonymous referees. The authors would also like to thank Dr. Alex P. Buchmann for his comments on earlier versions of this paper.

6. References

[BLAK86] J. A. Blakeley, P. Larson, and F. W. Tompa., Efficiently Updating Materialized Views, *ACM SIGMOD Conf.*, pp. 61-71, Washington, D.C. (May 1986).

[CHAK89] Chakravarthy, S., et al., HiPAC :A Research Project in Active, Time-Constrained Database Management, Final Project Report, XAIT, 1989.

[DARN87] M. Darnovsky, J. Bowman. "TRANSACT-SQL USER'S GUIDE," Document 3231-2.1, Sybase Inc., 1987.

[DAYA88a] Dayal, U., et al., The HiPAC Project: Combining Active Databases and Timing Constraints, *ACM SIGMOD Record*, March 1988.

[DAYA88b] U. Dayal, Active Database Management Systems, *Conf. of Data and Knowledge Bases*, Jerusalem, 1988.

[DITT86] K. R. Dittrich, A. M. Kotz, J. A. Mulle. "An Event/Trigger Mechanism to Enforce Complex Consistency Constraints in Design Databases," *SIGMOD Record 15*, No. 3, 1986, pp. 22-36.

[KOEN81] S. Koenig and R. Paige., A Transformational Framework for the Automatic Control of Derived Data., In *VLDB Conf.*, pp. 306-318, Cannes, (1981).

[STON87] M. Stonebraker, M. Hanson, S. Potamianos. "A Rule manager for Relational Database Systems," Technical Report, Dept. of Electrical Engineering and Computer Science, Univ. of California, Berkeley, 1987.

Appendix

Proof of IncrSelect identity: To shorten the formulas, we will use σ to denote $\texttt{Select}_{pred}$ and σR to denote $\texttt{Select}_{pred}$(R). From the definition of **Changes** we have:

Changes(σR;[R,ΔR]) $\equiv$

Δdifference[σR, σ(Compose(R, ΔR))]. We need to show that this equals the expression given for **IncrSel**. From the definition of **Compose**, this equals:

Δdifference(σR,
σ[(R -**Removals**(ΔR))$\cup$**Additions**(ΔR)])

Substituting the definition of **Δdifference**, we obtain:

$\texttt{Outerjoin}_{tid}$(
(σR - σ[(R -**Removals**(ΔR)) $\cup$ **Additions**(ΔR)]),
(σ[(R -**Removals**(ΔR)) $\cup$ **Additions**(ΔR)] - σ(R)))

Now by the distributive law of selection over difference and union, we pull out each selection, to obtain:

$\texttt{Outerjoin}_{tid}$(
σ(R - [R -**Removals**(ΔR) $\cup$ **Additions**(ΔR)]),
σ([R -**Removals**(ΔR) $\cup$ **Additions**(ΔR)] - R))

This simplifies to the formula for **IncrSel**(ΔR, pred):
Outerjoin(σ(**Removals**(ΔR)), σ(**Additions**(ΔR)))

Proof of Chain Rule: We need to show:

Changes (F $\circ$ G; [RList, ΔRList]) =
IncrF(G(RList), **Changes**(G; [RList, ΔRList])).

Let RHS denote the right hand side of the equality to be demonstrated. Substituting the definition of **IncrF**:

RHS =
Changes(F; [G(RList), **Changes**(G; [RList, ΔRList])])

Replace the "inner" **Changes** by its definition, to get:

Changes(F; [G(RList),
Δdifference(G(RList),
G(**Compose**(RList,ΔRList)))]).

Now replace the "outer" **Changes** by its definition:

RHS = Δ**difference**(F(G(RList)),
    F(**Compose**[G(RList),
        Δ**difference**(G[RList],
            G[**Compose**(RList, ΔRList)])]))

We now apply Lemma 1B with G(RList) in the role of R1 and G[**Compose**(RList, ΔRList)] in the role of R2:

RHS = Δ**difference**[F(G(RList)),
    F(G[**Compose**(RList, ΔRList)])].

This expression is just **Changes**(F ° G; [RList, ΔRList]).

PANEL SESSION

Future research directions: Evidence from this conference

Panel chairman:
Peter M.G. Apers
University of Twente, The Netherlands

Panel members:
Rakesh Agrawal
Georges Gardarin
Theo Härder
Peter Lyngbaek
Shojiro Nishio
Joachim Schmidt
Gerry Wilson

In the sixties and seventies work on relational databases created a convergence point for most database research. A theory was developed to design relational databases, database management systems were developed to show that relational databases could be implemented efficiently, relational query languages were proposed to make the relational data model acceptable for end-users, etc. Now that the relational model has become a standard for business, we find that further appicability of the database principles of persistence, sharability, concurrent use, declarative description, high-level query languages, as well as logging and backup are of sufficient value to be transferred to systems supporting engineering, manufacturing, decision-making, and planning tasks.

Research in the eighties will hence be characterized by new application domains and their database support. The strength of the relational data model and its implementation will be tested. Influences from programming languages and artificial intelligence give direction to database research. Examples of those, among many, are object-oriented databases and deductive databases. Database research is diverging in many different directions. At the end of this conference it is good to look at the many reserach directions in retrospect and see what has been achieved this last decade. For this purpose a few session chairmen are invited to report on the status of the new directions and to formulate answers to questions like:

- is the object-oriented approach the relational approach of the nineties?
- what has brought artificial intelligence to the database community?
- is the extensible database approach the right approach to support new application domain such as office automation, CAD/CAM, carthography, etc.?
- are more results to be expected from recursive query optimization?

INDEX OF AUTHORS

ORDERING INFORMATION

Morgan Kaufmann Publishers is the world wide distributor for the VLDB Conference Proceedings volumes listed below. To place your order, or receive additional information on these and other Morgan Kaufmann publications, please contact:

Morgan Kaufmann Publishers, Inc.
P.O. Box 50490
Palo Alto CA 94303-9953
USA
415-578-9911 or 415-965-4081

VLDB-89 AMSTERDAM
ISBN 1-55860-101-5
35/50 Institutions

VLDB-88 LOS ANGELES
ISBN 0-934613-75-3
35/50 Institutions

VLDB-87 BRIGHTON
ISBN 0-934613-46-X
35/50 Institutions

VLDB-KYOTO
ISBN 0-934613-18-4
35/50 Institutions

VLDB-85 STOCKHOLM
ISBN 0-934613-17-6
35/50 Institutions

VLDB-84 SINGAPORE
ISBN 0-934613-16-8
35/50 Institutions

VLDB-83 FLORENCE
ISBN 0-934613-15-X
35/50 Institutions

VLDB-82 MEXICO CITY
ISBN 0-934613-14-1
35/50 Institutions

VLDB SEVEN-YEAR SET
1982-1989
(Seven volumes)
ISBN 1-55860-102-3
210/320 Institutions